Electrical Power Systems

Electrical Power Systems

P.S.R. Murty

Butterworth-Heinemann
An imprint of Elsevier

Butterworth-Heinemann is an imprint of Elsevier
The Boulevard, Langford Lane, Kidlington, Oxford OX5 1GB, United Kingdom
50 Hampshire Street, 5th Floor, Cambridge, MA 02139, United States

Notices
Knowledge and best practice in this field are constantly changing. As new research and experience broaden our understanding, changes in research methods, professional practices, or medical treatment may become necessary.

Practitioners and researchers must always rely on their own experience and knowledge in evaluating and using any information, methods, compounds, or experiments described herein. In using such information or methods they should be mindful of their own safety and the safety of others, including parties for whom they have a professional responsibility.

To the fullest extent of the law, neither the Publisher nor the authors, contributors, or editors, assume any liability for any injury and/or damage to persons or property as a matter of products liability, negligence or otherwise, or from any use or operation of any methods, products, instructions, or ideas contained in the material herein.

British Library Cataloguing-in-Publication Data
A catalogue record for this book is available from the British Library

Library of Congress Cataloging-in-Publication Data
A catalog record for this book is available from the Library of Congress

ISBN: 978-0-08-101124-9

For Information on all Butterworth-Heinemann publications
visit our website at https://www.elsevier.com/books-and-journals

Working together
to grow libraries in
developing countries

www.elsevier.com • www.bookaid.org

Publishing Director: Joy Hayton
Senior Editorial Project Manager: Kattie Washington
Production Project Manager: Mohana Natarajan
Designer: Mark Rogers

Typeset by MPS Limited, Chennai, India

Dedicated to my late Grand Mother
Marella Ranganayakamma garu

Contents

Preface

I had been in teaching for the last 50 years—teaching undergraduate and postgraduate students mainly in Electrical Power Engineering. At the end of the day, it was felt by me that my experience in teaching Power System subjects should be incorporated in a book form, so that it is shared by teachers and students of power engineering alike.

This book is intended to serve as a textbook for undergraduate students pursuing studies in Electrical and Electronics Engineering specialization at various universities and their affiliated institutions.

Chapters 2−4 deal with basics associated with transmission lines and their electrical parameters. Chapters 5 and 6, Corona and Interference, and Performance of Transmission Lines, explain the performance of transmission lines. Various aspects of underground cables are covered in Chapter 7, Cables. Substation layout and their components and functions along with earthing practices are included in Chapter 9, Substations and Neutral Grounding. Direct current and alternating current distribution engineering principles are explained in Chapter 10, Distribution System.

Overvoltage phenomenon and protection against overvoltages are elaborated in Chapters 11 and 12, Overvoltages and Protection Against Overvoltages. For systematic analysis of Power System under faulted conditions, graph theory and network matrix formation methodology is required, and this information in provided in Chapter 13, Graph Theory and Network Matrices.

Symmetrical and unsymmetrical fault analysis is explained with appropriate phasor diagrams in Chapters 15 and 16, Balanced Fault Analysis and Unbalanced Fault Ananlysis.

Protection of Power System under fault conditions and under certain contingencies using circuit breakers and relaying practices are explained in Chapter 17, Relaying and Protection.

Stability of Power System and various aspects of steady state, transient state, and dynamic condition are discussed to create clarity to the student in Chapter 18, Power System Stability.

Load flow analysis, economic scheduling without and with losses are covered in Chapters 19 and 20, Load Flow Analysis and Economic Operation of Power Systems.

Voltage and reactive power control are explained in Chapter 21, Load Frequency Control.

A large number of problems are solved for the benefit of the student.

More problems for practice to student are given at the end of each chapter. Questions on each topic are given.

P.S.R. Murty

Acknowledgments

My sincere thanks go to Mr. Nikhil Shah, Director of BSP Books Pvt. Ltd., for his constant and encouragement to me to write and complete this book on "Electrical Power Systems."

I acknowledge gratefully the support given by the Management and Director of Sreenidhi Institute of Science and Technology without which this book could not have been written. I sincerely thank the services rendered by Mr. M.V.L. Narasimha Rao and Mrs. M. Prasanna Subha Lakshmi, office assistant in the department.

P.S.R. Murty

INTRODUCTION

Electrical energy is the most popular form of energy, since, it can be transported easily at a very high efficiency and at reasonable cost. Thomas Edison established the first power station in 1882 at Pearl street in New York City, United States. The lower Manhattan area was supplied DC power from this station. Underground cables were used for distribution. At Appleton, Wisconsin, the first water wheel generator was installed. Under Edison's patents several companies started functioning in the United States. However, these companies could supply energy to small distances due to I^2R power loss being excessive at low-voltage distribution.

In 1885 William Stanley invented the transformer which revolutionized the AC transmission. The invention of induction motor in 1888 by Nikola Tesla caused dramatic change in power consumption through AC replacing many DC motor loads.

It is an acknowledged fact that high voltage (HV) and extra light voltage (EHV) transmission alone can reduce substantially the losses and bulk power transmission is feasible at these voltages. Nevertheless, it is also well established that light voltage direct current (HVDC) is convenient and more economical from operation and control point of view under certain circumstances such as at distances more than 500 km.

1.1 STRUCTURE OF A POWER SYSTEM

The power system consists of three subsystems which have clearly demarcated functions. But, coordinated working of the three subsystems is absolutely essential as they are the parts of the same system. The subsystems are:

1. the generation subsystem
2. the transmission subsystem and
3. the distribution subsystem

The generation subsystem may be called GENCO, responsible for generating electric power as per the predicted load requirements.

The transmission subsystem may be called TRANSCO. Its essential duties are to provide facilities for transmission of the required power demand at various load centers. The grid that consists of various power transmission lines interconnected must be suitably rated and well protected with circuit breakers and relays.

The distribution subsystem may be called DISCOM which is the actual link between the system and the customers. Figure 1.1 shows a single line representation of the total system.

Electrical Power Systems. DOI: http://dx.doi.org/10.1016/B978-0-08-101124-9.00001-2

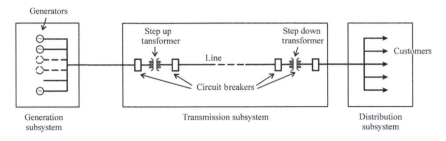

FIGURE 1.1

Structure of a power system.

Planning, design, operation, control, and protection of power system requires continuous and comprehensive analysis to evaluate the current states and remedial control, if any needed. Manual computation of power flow is extremely time consuming even for very simple networks. In 1929 AC network analyzer; an analog computer was devised. Most of the early system studies were performed on network analyzers.

Electrical power networks consist of a number of transmission lines interconnected in a fashion that is dictated by the development of load centers. This interconnected network expands continuously. A systematic procedure is needed to build a model that can be constantly updated with increasing interconnections. Network solutions can be carried out using Ohm's law and Kirchiff's laws.

Either

$$e = zi$$

or

$$i = ye$$

model can be used for steady state network solution. Accordingly either Z-bus or Y-bus models are required to be developed. To understand this process, graph theory is useful.

Power system stability is a word used in connection with alternating current power system denoting a condition, where in, the various alternators in the system remain in synchronism with each other. Owing to a variety of reasons, like sudden load changes, faults, short circuits, sudden circuit opening, etc., may create black-outs. Knowledge of behaviour of a single synchronous machine connected to infinite bus. This knowledge under the afore said condition enables to device methods to overcome the undesirable effects are to be studied by electrical engineering students.

1.2 EFFECTS OF SYSTEM VOLTAGE ON EFFICIENCY

Consider an alternating current transmission system. The power transmitted per phase, with usual notation.

$$P = VI \cos \phi \ (W) \tag{1.1}$$

where V is the phase voltage, I is the line current, and ϕ is the phase angle between V and I.

Power loss per phase

$$P = I^2R \ (W) \tag{1.2}$$

where R is the resistance of the phase conductor. The system is balanced so that there is no loss in neutral conductor.

Substituting the value of current from Eq. (1.1) into Eq. (1.2).

$$P = \frac{P^2R}{v^2 \cos^2 \phi}$$

Since resistance $R = (\rho l/A)$ where ρ is the specific resistivity, l the length of the conductor, and A the area of section, the loss can be expressed as

$$P = \frac{P^2 \rho l}{Av^2 \cos^2 \phi} \tag{1.3}$$

Volume of the conductor material $= Al$

Therefore from Eq. (1.3)

$$Al = \frac{P^2 \rho l^2}{Pv^2 \cos^2 \phi} = \left[\frac{P^2 \rho l^2}{P} \right] \frac{1}{v^2 \cos^2 \phi} \tag{1.4}$$

For transferring a given power P over a given distance l with specified power loss for a given materials, the volume of conductor material is inversely proportional to square of voltage and power factor of the load.

As power factor cannot be improved beyond unity, to minimize the volume of the conductor material, high voltage is necessary for transmission and distribution of electrical power. However, the choice of voltage is limited by the cost of insulation, transformers, switchgear, and other terminal equipments.

For a 400 kV line the average height of conductors is 15 m and phase spacing is 12 m. For 750 kV the values are 18 and 15 m, respectively. For 1200 kV we need at least 21 m height and 21 m phase-to-phase spacing. These lines are required to transmit power of 2500−12,000 MW over distances ranging from 250 to 1200 km.

THE LINE PARAMETERS

2

An electrical transmission line has four parameters that influence its operation. They are the series resistance, series inductance, shunt-connected capacitance, and shunt-connected conductance. Line conductors possess resistance, and this resistance for a specified conductor material, size, and shape that is in practice is supplied in standard tables and hand books as ohms per unit length.

The magnetic field around a conductor-carrying current is in the form of closed loops linking the conductor. The electric lines of flux originate on positive charges and are radial. They terminate on negative charges. The electric and magnetic field distributions for a single-conductor and two-conductor system are shown in Fig. 2.1.

In case of alternating currents the changing current produces change in flux linking the conductor or circuit. This induces a voltage in the conductor or circuit and is proportional to the rate of change of flux linkages. The flux linkages per ampere of changing current is called inductance of the line.

The capacitance of the line is the charge per unit potential difference between conductors or conductor and earth. Through these line charging capacitors, small leakage currents will flow to ground. To consider the effect of this leakage current in high and extra high voltage lines, the parameter conductance (G) is used.

2.1 THE LINE RESISTANCE

The effective resistance of a conductor

$$R = \frac{\text{Power loss in the conductor}}{I^2} \quad \text{(ohms)} \tag{2.1}$$

The DC resistance

$$R_{dc} = \frac{\rho l}{A} \quad (\Omega) \tag{2.2}$$

where ρ is the resistivity of conductor; l is the length of conductor; and A is the cross-sectional area of the conductor.

Due to nonuniform distribution of current density on account of skin effect, the resistance to alternating currents is slightly higher than that due to direct currents.

Furthermore the variation of resistivity of conductor material with temperature is also important. For accurate calculations these effects are to be included. In general the ac resistance

$$R_{ac} \simeq 1.2 \, R_{dc} \tag{2.3}$$

Electrical Power Systems. DOI: http://dx.doi.org/10.1016/B978-0-08-101124-9.00002-4

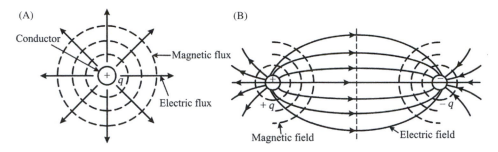

FIGURE 2.1

Electric and magnetic fields. (A) Single-conductor system and (B) two-conductor system.

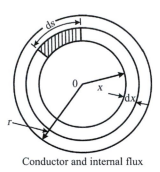

Conductor and internal flux

FIGURE 2.2

Conductor and internal flux.

2.2 THE LINE INDUCTANCE

The inductance of a transmission line conductor is composed of two parts, one due to the flux internal to the conductor and the other due to the flux external to the conductor.

2.2.1 INDUCTANCE DUE TO INTERNAL FLUX

Consider the conductor flux system shown in Fig. 2.2.

Let H be the magnetic field intensity in AT/m.

I is the total current enclosed by the conductor.

H_x is the field intensity at a radial distance x from the center of the conductor O. we have, the relation for magneto motive force

$$\text{mmf} = \oint H \cdot ds = I \ (\text{AT}) \tag{2.4}$$

Then,

$$\oint H_x \cdot ds = I_x \tag{2.5}$$

where I_x is the current enclosed up to radial distance x by the conductor.
    Assuming uniform current density

$$I_x = \frac{\pi x^2}{\pi r^2} \cdot I \ (A) \tag{2.6}$$

Substituting Eq. (2.6) into Eq. (2.5)
Therefore,

$$H_x = \frac{x}{2\pi r^2} \cdot I \ (AT/m) \tag{2.7}$$

The flux density at a radial distance x

$$B_x = \mu H_x \frac{\mu x I}{2\pi r^2} \ (wb/m^2) \tag{2.8}$$

In the tubular element of width dx the flux enclosed

$$d\phi = B_x \times (\text{area normal to flux lines})$$

$$= \frac{\mu x I}{2\pi r^2} \times (dx \times 1) \ (wb/m\text{-length})$$

The flux linkages dψ per meter length due to dϕ

$$d\psi = \frac{\pi x^2}{\pi r^2} \cdot d\phi = \frac{\mu I x^3}{2\pi r^4} \cdot dx \ (wb\text{-term}/m)$$

The total internal flux linkages

$$\psi_{Int.} = \int_0^r \frac{\mu I x^3}{2\pi r^4} dx = \frac{\mu I}{8\pi} \ (wb\text{-turns}/m) \tag{2.9}$$

If permeability of air is taken as $4\pi \times 10^{-7}$ h/m

$$\psi_{Int.} = \frac{I}{8\pi} \times 4\pi \times 10^{-7} = \frac{I}{2} \times 10^{-7} \ (wb\text{-turns}/m) \tag{2.10}$$

Inductance due to internal flux linkages

$$L_{Int.} = \frac{\psi_{int}}{I} = \frac{1}{2} \times 10^{-7} \ h/m \tag{2.11}$$

which is constant.

2.2.2 INDUCTANCE DUE TO EXTERNAL FLUX LINKAGES

Consider two points P_1 and P_2 external to a conductor-carrying current I. It distant S_1 and S_2 from the center of the conductor as shown is Fig. 2.3.

Consider the flux enclosed between the two concentric cylindrical surfaces with radii S_1 and S_2. At a radial distance x, so that $S_1 < x < S_2$, the tubular element of width dx will have a field intensity H_x given by

$$2\pi x H_x = I, \text{ by Ampere theorem} \tag{2.12}$$

The flux density correspondingly is

$$B_x = \frac{\mu I}{2\pi x} \cdot dx \times 1 \ \ (\text{wb/m}^2) \tag{2.13}$$

The differential flux $d\phi$ in the tubular element per meter length

$$d\phi = \frac{\mu I}{2\pi x} \cdot dx \times 1 \tag{2.14}$$

The differential flux $d\phi$ in the tubular element per meter length of the conductor

$$d\psi = \frac{\mu I}{2\pi x} dx \ \ (\text{wb-t/m}) \tag{2.15}$$

The total flux linkages between P_1 and P_2 linking the conductor

$$\psi_{12} = \int_{S_1}^{S_2} \frac{\mu I}{2\pi x} \cdot dx = \frac{\mu I}{2\pi} \ln \frac{S_2}{S_1} \ \ (\text{wb-t/m}) \tag{2.16}$$

Since $\mu = 1$ and $\mu_r = 4\pi \times 10^{-7}$ for free space

$$\psi_{12} = 2 \times 10^{-7} I \ \ln \frac{S_2}{S_1} \ \ (\text{wb-t/m})$$

The inductance due to flux between P_1 and P_2

$$L_{12} = \frac{2 \times 10^{-7} \cdot I}{I} \ln \frac{S_2}{S_1} = 2 \times 10^{-7} \ln \frac{S_2}{S_1} \ \ (\text{h/m}) \tag{2.17}$$

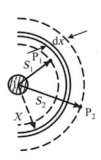

FIGURE 2.3

Flux linkages due to external flux.

2.2.3 INDUCTANCE OF A SINGLE-PHASE LINE

Consider a two-conductor system shown in Fig. 2.4.

Using Eqs. (2.11) and (2.17) the total inductance of conductor 1

$$L_1 = \left(\frac{1}{2} + 2\ln\frac{D}{r_1}\right) \times 10^{-7} \quad (h/m) \tag{2.18}$$

But $\ln e^{\frac{1}{4}} = \frac{1}{4}$. Using this relation

$$L_1 = 2 \times 10^{-7}\left[\frac{1}{4} + \ln\frac{D}{r_1}\right] \quad (h/m)$$

$$= 2 \times 10^{-7}\left[\ln e^{\frac{1}{4}} + \ln\frac{D}{r_1}\right] \quad (h/m) \tag{2.19}$$

$$= 2 \times 10^{-7}\left[\ln\frac{D}{r_1 e^{\frac{-1}{4}}}\right] \quad (h/m)$$

Let $r_1 e^{\frac{-1}{4}} = r_1^l$

Then

$$L_1 = 2 \times 10^{-7}\ln\frac{D}{r_1^l} \quad (h/m) \tag{2.20}$$

Given

$$e^{\frac{-1}{4}} = 0.7788$$

$r_1^l = 0.7788\, r_1$ is called the geometric mean radius (GMR) $\tag{2.21}$

$$L_2 = 2 \times 10^{-7}\ln\frac{D}{r_2^l} \quad (h/m)$$

The total inductance

$$L = L_1 + L_2 = 4 \times 10^{-7}\ln\frac{D}{\sqrt{r_1^l\, r_2^l}} \quad (h/m) \tag{2.22}$$

In general for practical systems $r_1 = r_2 = r_1$

Hence the inductance per loop per meter length of single-phase system (two-conductor system)

$$L = 4 \times 10^{-7}\ln\frac{D}{r^l} \quad (h/m) \tag{2.23}$$

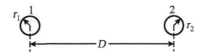

FIGURE 2.4

Two-conductor system.

2.2.4 FLUX LINKAGES IN A GROUP OF CONDUCTORS

Consider a group of n-conductors 1, 2, 3, ..., n as shown in Fig. 2.5 carrying currents I_1, I_2, I_3, ..., I_n, respectively. Consider a point P external to the group. The distances of the point P from the group of n conductors are given, respectively, by D_{1p}, D_{2p}, D_{3p}, ..., D_{np}.

The flux linkages of conductor 1 due to I_1 from Eq. (2.18) excluding all flux beyond point P.

$$\psi_{1p1} = I_1 \left(\frac{1}{2} + 2 \ln \frac{D_{1p}}{r_1} \right) \times 10^{-7} = 2 \times 10^{-7} I_1 \ln \left(\frac{D_{1p}}{r_1^1} \right) \tag{2.24}$$

where r_1 is the radius of conductor 1. In a similar way the flux-linking conductor 1 due to current I_2 in conductor 2 between point P and conductor 1

$$\psi_{1p2} = 2 \times 10^{-7} I_2 \ln \left(\frac{D_{2p}}{D_{1p}} \right) \tag{2.25}$$

Extending the argument, the total flux linkages ψ_{1p} with conductor 1 due to currents in all the n-conductors in the group, up to point P

$$\psi_{1p} = 2 \times 10^{-7} \left[I_1 \ln \frac{D_{1p}}{r_1^1} + \frac{I}{2} \ln \frac{D_{2p}}{D_{12}} + I_3 \ln \frac{D_{3p}}{D_{13}} = \cdots + I_n \ln \frac{D_{np}}{D_{1n}} \right] \tag{2.26}$$

Rearranging the terms

$$\psi_{1p} = 2 \times 10^{-7} \left[I_1 \ln \frac{1}{r_1^1} + I_2 \ln \frac{1}{D_{12}} + I_3 \ln \frac{1}{D_{13}} + \cdots + I_n \ln \frac{1}{D_{1n}} \right]$$
$$+ [I_1 \ln D_{1p} + I_2 \ln D_{2p} + I_3 \ln D_{3p} + \cdots I_n \ln D_{np}] \tag{2.27}$$

The sum of all currents $I_1 + I_2 + I_3 + \cdots + I_n = 0$
Hence

$$I_n = - [I_1 + I_2 + I_3 + \cdots + I_{n-1}) \tag{2.28}$$

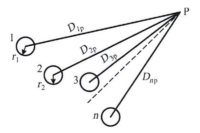

FIGURE 2.5

Conductor group near point P.

Substituting Eq. (2.28) in (2.27)

$$\psi_{1p} = 2 \times 10^{-7} \left\{ \left[I_1 \ln \frac{1}{r_1^l} + I_2 \ln \frac{1}{D_{12}} + I_3 \ln \frac{1}{D_{13}} + \cdots + I_n \ln \frac{1}{D_{1n}} \right] \right.$$
$$+ I_1 \ln D_{1p} + I_2 \ln D_{2p} + I_3 \ln D_{3p} + \cdots + I_{n-1} \ln D_{(n-1)}) \tag{2.29}$$
$$\left. - (I_1 + I_2 + I_3 + \cdots + I_{n-1}) \ln D_{np} \right\}$$

$$\psi_{1p} = 2 \times 10^{-7} \left[\left(I_1 \ln \frac{1}{r_1^l} + I_2 \ln \frac{1}{D_{12}} + I_3 \ln \frac{1}{D_{13}} + \cdots + I_n \ln \frac{1}{D_{1n}} \right) \right]$$
$$+ \left[\left(I_1 \ln \frac{D_{1p}}{D_{np}} + I_2 \ln \frac{D_{2p}}{D_{np}} + I_3 \ln \frac{D_{3p}}{D_{np}} + \cdots + I_{n-1} \ln \frac{D_{(n-1)p}}{D_{np}} \right) \right] \tag{2.30}$$

If the point P is now moved to infinity, so that conductor 1 links with all the flux generated by the n-conductors, then the ratios

$$\frac{D_{1p}}{D_{np}}, \frac{D_{2p}}{D_{np}}, \frac{D_{3p}}{D_{np}}, \dots, \frac{D_{(n-1)p}}{D_{np}}$$

tend to unity and the contribution of second term in Eq. (2.30) becomes negligible. Therefore

$$\psi_{1p} = \psi_1 = 2 \times 10^{-7} \left[I_1 \ln \frac{1}{r_1^l} + I_2 \ln \frac{1}{D_{12}} + I_3 \ln \frac{1}{D_{13}} + \cdots + I_n \ln \frac{1}{D_{1n}} \right] \text{ (wb-t/m)} \tag{2.31}$$

2.2.5 INDUCTANCE BETWEEN CONDUCTOR GROUPS: SELF AND MUTUAL GEOMETRIC MEAN DISTANCE

Consider a single-phase line consisting of two groups of conductors. Group A has n-conductors and group B has m-conductors. Conductors in each group are identical. Each conductor in group A carries a current of I/n and each conductor in group B, the return conductor, carries a current of $(-I/m)$ (Fig. 2.6).

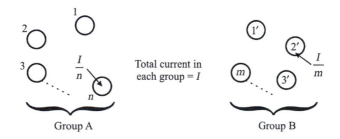

FIGURE 2.6

Two groups of conductor systems.

The distances are marked $D_{11}, D_{12}, \ldots, D_{1n}$ and $D_{11'}, D_{12'}, \ldots D_{1m}$.
Conductor 1 of group A has flux linkages

$$\psi_1 = 2 \times 10^{-7} \frac{I}{n} \left[\ln \frac{1}{r_1^l} + \ln \frac{1}{D_{12}} + \ln \frac{1}{D_{13}} + \cdots + \ln \frac{1}{D_{1n}} \right]$$
$$- 2 \times 10^{-7} \frac{I}{m} \left[\ln \frac{1}{D_{11'}} + \ln \frac{1}{D_{12'}} + \ln \frac{1}{D_{13'}} + \cdots + \ln \frac{1}{D_{1m}} \right] \text{(wb-t/m)} \quad (2.32)$$

Simplifying

$$\psi_1 = 2 \times 10^{-7} I \ln \left(\frac{m\sqrt{D_{11'} D_{12'} D_{13'} \ldots D_{1m}}}{n\sqrt{r_1^l D_{12} D_{13} \ldots D_{1n}}} \right) \text{(wb-t/m)} \quad (2.33)$$

The inductance of conductor 1 due to these flux linkages

$$L_1 = \frac{\psi_1}{I/n} = 2n \times 10^{-7} \ln \left(\frac{m\sqrt{D_{11'} D_{12'} D_{13'} \ldots D_{1m}}}{n\sqrt{r_1^l D_{12} D_{13} \ldots D_{1n}}} \right) \text{(h/m)} \quad (2.34)$$

Similarly the inductance L_2 of conductor 2 in group A.

$$L_2 = \frac{\psi_2}{I/n} = 2n \times 10^{-7} \ln \left(\frac{m\sqrt{D_{21'} D_{22'} D_{23'} \ldots D_{2m}}}{n\sqrt{r_2^l D_{21} D_{22} D_{23} \ldots D_{2n}}} \right) \text{(h/m)} \quad (2.35)$$

The average inductance of all the conductors in group A

$$L_{\text{average}} = \frac{L_1 + L_2 + L_3 + \cdots + L_n}{n} \quad (2.36)$$

All the n conductors in group A are connected in parallel and carry equal currents (I/n).
The inductance of the group or line A

$$L_A = \frac{L_{\text{average}}}{n} = \frac{L_1 + L_2 + L_3 + \cdots L_n}{n^2} \quad (2.37)$$

The inductance of line A containing n-conductors in parallel

$$L_A = 2 \times 10^{-7} \ln \frac{mn\sqrt{(D_{11'} D_{12'} \ldots D_{1m})(D_{21'} D_{22'} \ldots D_{2m}) \ldots (D_{n1'} D_{n2'} \ldots D_{nm})}}{n^2 \sqrt{(D_{11} D_{12} \ldots D_{1n})(D_{21} D_{22} \ldots D_{2n}) \ldots (D_{n1} D_{n2} \ldots D_{2n})}} \quad (2.38)$$

where $r_1^l, r_2^l, \ldots, r_n^l$ are replaced by $D_{11}, D_{22}, \ldots, D_{nn}$, respectively.

The numerator of Eq. (2.38) contains the product of $m \times n$ distances and its mn-th root. This mn-th root of $m \times n$ distances is called mutual geometric mean distance (GMD) between the groups A mean B, containing n and m conductors each in their group. It is denoted by D_{m}.

The denominator of Eq. (2.38) contains the product of $n \times n$ distances and its n^2-th, root. The n^2-th, root of n^2 distances is called self-GMD or self-geometric mean radius (GMR) and is denoted by D_{S}.

Hence,

$$L_A = 2 \times 10^{-7} \ln \frac{D_{\text{mA}}}{D_{\text{SA}}} \text{(h/m)} \quad (2.39)$$

Similarly, the inductance of the group B or line B

$$L_B = 2 \times 10^{-7} \ln \frac{D_{mB}}{D_{SB}} \quad (h/m) \tag{2.40}$$

The loop inductance

$$L = L_A + L_B \tag{2.41}$$

2.2.6 INDUCTANCE OF THREE-PHASE LINES SYMMETRICAL SPACING

Consider that the three-phase conductors a, b, and c are at the corners of an equilateral triangle of side D_m. Let the conductor radius be r (m) for each phase (Fig. 2.7).

At any instance $I_a + I_b + I_c = 0$. The flux linkages of conductor a due to current I_a, I_b, and I_c are given from Eq. (2.32) by

$$\psi_a = 2 \times 10^{-7} \left[I_a \ln \frac{1}{r^1} + I_b \ln \frac{1}{D} + I_c \ln \frac{1}{D} \right] \text{(wb-t/m)}$$

$$= 2 \times 10^{-7} \left[I_a \ln \frac{1}{r^1} + (I_b + I_c) \ln \frac{1}{D} \right] \text{(wb-t/m)}$$

Since $I_b + I_c = -I_a$

$$\psi_a = 2 \times 10^{-7} \left[I_a \ln \frac{D}{r^1} \right] \text{(wb-t/m)} \tag{2.42}$$

Inductance of conductor a

$$L_a = \frac{\psi_a}{I_a} = 2 \times 10^{-7} \ln \frac{D}{r^1} \quad (h/m) \tag{2.43}$$

Due to symmetry

$$L_a = L_b = L_c = 2 \times 10^{-7} \ln \frac{D}{r^1} \quad (h/m) \tag{2.44}$$

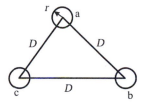

FIGURE 2.7

Three-phase line—symmetrical spacing.

2.2.7 THREE-PHASE LINE WITH UNSYMMETRICAL SPACING

Let the three-phase conductors a, b, and c be placed at the corners of an unsymmetrical triangle of sides D_{ab}, D_{bc}, and D_{ca}, respectively, as shown in Fig. 2.8.

The flux linkages ψ_a of conductor a are obtained from

$$\psi_a = 2 \times 10^{-7} \left[I_a \ln \frac{1}{r^1} + I_b \ln \frac{1}{D_{ab}} + I_c \ln \frac{1}{D_{ac}} \right]$$

Inductance of phase a

$$L_a = \frac{\psi_a}{I_a} = 2 \times 10^{-7} \left[\ln \frac{1}{r^1} + \frac{I_b}{I_a} \ln \frac{1}{D_{ab}} + \frac{I_c}{I_a} \ln \frac{1}{D_{ac}} \right]$$

Let I_a be chosen as reference phasor.

$$I_b = I_a \underline{|-120°} \text{ and } I_c = I_a \underline{|120°}$$

$$\frac{I_b}{I_a} = 1\underline{|-120°} = -\frac{1}{2} - j\frac{\sqrt{3}}{2}$$

$$\frac{I_c}{I_a} = 1\underline{|120°} = -\frac{1}{2} + j\frac{\sqrt{3}}{2}$$

Hence,

$$L_a = 2 \times 10^{-7} \left[\ln \frac{1}{r^1} + \left(-\frac{1}{2} - j\frac{\sqrt{3}}{2} \right) \ln \frac{1}{D_{ab}} + \left(-\frac{1}{2} + j\frac{\sqrt{3}}{2} \right) \ln \frac{1}{D_{ac}} \right]$$

$$= 2 \times 10^{-7} \left[\ln \frac{1}{r^1} + \ln \sqrt{D_{ab}D_{ac}} + j\sqrt{3} \ln \frac{D_{ab}}{D_{ac}} \right] (h/m) \tag{2.45}$$

In a similar way,

$$L_b = 2 \times 10^{-7} \left[\ln \frac{1}{r^1} + \ln \sqrt{D_{bc}D_{ba}} + j\sqrt{3} \ln \frac{D_{bc}}{D_{ba}} \right] \tag{2.46}$$

and

$$L_c = 2 \times 10^{-7} \left[\ln \frac{1}{r^1} + \ln \sqrt{D_{ca}D_{cb}} + j\sqrt{3} \ln \frac{D_{ca}}{D_{cb}} \right] \tag{2.47}$$

The individual phase inductances of an unsymmetrical line are complex is nature. The complex nature indicates exchange of energy between phases.

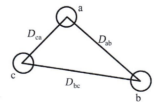

FIGURE 2.8

Three-phase line—unsymmetrical spacing.

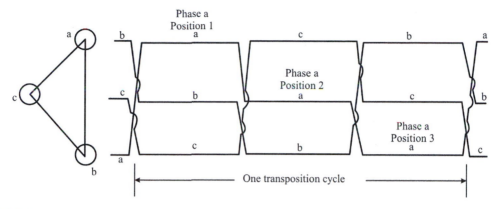

FIGURE 2.9

Transposition of three-phase conductors.

2.3 TRANSPOSITION OF TRANSMISSION LINES

Unequal flux linkages due to asymmetrical spacing causes energy exchange between phases. To eliminate this effect, in practice, transmission lines are transposed at regular intervals as shown in Fig. 2.9.

Transposition means that each phase conductor is made to occupy the original position of every other conductor over an equal distance. This is explained in Fig. 2.9. This process is repeated for the entire length of the line. This exchanging the positions of the conductors at regular intervals balances the flux linkages of each phase conductors. However, the error in inductance due to asymmetry is negligible. The average distance between conductors can be taken as

$$\sqrt[3]{(D_{ab}\, D_{bc}\, D_{ca})} = D_{equivalent} \tag{2.48}$$

and inductance may be calculated as in the symmetric case taking $D_{equivalent}$ in place of D.

2.4 INDUCTANCE OF A TRANSPOSED THREE-PHASE LINE

Let the phase conductor a occupy in one cycle of transposition all the three positions 1, 2, and 3. The flux linkages are then with conductor a in position 1.

$$\psi_{a1} = 2 \times 10^{-7} \left[I_a \ln \frac{1}{r^1} + I_b \ln \frac{1}{D_{12}} + I_c \ln \frac{1}{D_{13}} \right] \text{ (wb-t/m)}$$

Similarly when conductor a occupies position 2

$$\psi_{a2} = 2 \times 10^{-7} \left[I_a \ln \frac{1}{r^1} + I_b \ln \frac{1}{D_{23}} + I_c \ln \frac{1}{D_{21}} \right] \text{ (wb-t/m)}$$

Likewise,

$$\psi_{a3} = 2 \times 10^{-7} \left[I_a \ln \frac{1}{r^1} + I_b \ln \frac{1}{D_{31}} + I_c \ln \frac{1}{D_{32}} \right] \text{(wb-t/m)}$$

The average flux linkages are

$$\psi_a = \frac{\psi_{a1} + \psi_{a2} + \psi_{a3}}{3} = \frac{2 \times 10^{-7}}{3} \left[3I_a \ln \frac{1}{r^1} + I_b \ln \frac{1}{D_{12} D_{23} D_{31}} + I_c \ln \frac{1}{D_{12} D_{23} D_{31}} \right] \text{(wb-t/m)}$$

But $I_b + I_c = -I_a$
Therefore,

$$\psi_a = \frac{2 \times 10^{-7}}{3} \left[3I_a \ln \frac{1}{r^1} - I_a \ln \frac{1}{D_{12} D_{23} D_{31}} \right]$$

$$= 2 \times 10^{-7} I_a \ln \left(\frac{\sqrt[3]{D_{12} D_{23} D_{31}}}{r^1} \right) \text{(wb-t/m)}$$

(2.49)

The average inductance per phase a

$$L_a = 2 \times 10^{-7} \ln \left(\frac{\sqrt[3]{D_{12} D_{23} D_{31}}}{r^1} \right) \text{(h/m)}$$

$$= 2 \times 10^{-7} \ln \left(\frac{D_{eq}}{r^1} \right) \text{(h/m)}$$

(2.50)

where r^1 is the self-GMD or self-GMR of each phase conductor.

2.5 BUNDLE CONDUCTORS

In transmission lines operating at voltages greater than 220 KV, corona power loss and interference with communication lines will be excessive, if only one conductor is used per phase. The potential gradient at the conductor surface can be reduced considerably if the self-GMD or self-GMR of the conductor can be increased. This is done by using two or more conductors in close proximity per phase compared to the spacing between phases and connected in parallel. The spacing between these same phase conductors is much smaller than the spacing between inter phases. Such a line is called bundled conductor line. Two or three conductor bundles are common. The current in each conductor of the bundle will be the same, if the conductors within the bundle are also transposed. However, it is not generally required. Increase of GMR reduces reactance of the line also (Fig. 2.10).

Let r be the radius of each conductor of the bundle and d be the spacing between conductors in the bundle.

The GMR of a two-conductor bundle is

$$D_{2b} = \sqrt[4]{(0.7788)^2 \cdot d^2 \cdot r^2} = \sqrt{r^1 d}$$

(2.51)

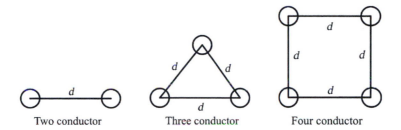

FIGURE 2.10

Different types of bundle conductors.

For a three-conductor bundle it is

$$D_{3b} = \sqrt[9]{(0.77\ \delta\delta)^3 \times d^3 \times r^3} = \sqrt[3]{r^1 d^2} \tag{2.52}$$

For a four-conductor bundle the GMR is

$$D_{4b} = \sqrt[4]{\left(r^1\ d.d \cdot \sqrt{2}d\right)^4} = 1.09\sqrt[4]{r^1 \times d^3} \tag{2.53}$$

2.6 INDUCTANCE OF A THREE-PHASE, DOUBLE-CIRCUIT LINE WITH UNSYMMETRICAL SPACING AND TRANSPOSITION

Consider a double-circuit line with conductors arranged in a vertical line for both the circuits. Let a, b, c be the conductors is one circuit and a^1, b^1 and c^1 be the conductors in another circuit. If the conductors are transposed regularly, then for one cycle of transposition the conductor positions are shown in Fig. 2.11. The distance between conductors are marked in Fig. 2.11. With reference to Fig. 2.11 the flux linkages of phase a in position 1, 2, 3 are written down.

The flux linkages of phase a in position 1

$$\psi_{a1} = 2 \times 10^{-7}\left[I_a\left(\ln\frac{1}{r^1} + \ln\frac{1}{m}\right) + I_b\left(\ln\frac{1}{h} + \ln\frac{1}{l}\right) + I_c\left(\ln\frac{1}{2h} + \ln\frac{1}{D}\right)\right]$$

Similarly in position 2 ψ_{a2} is obtained as

$$\psi_{a2} = 2 \times 10^{-7}\left[I_a\left(\ln\frac{1}{r^1} + \ln\frac{1}{D}\right) + I_b\left(\ln\frac{1}{h} + \ln\frac{1}{l}\right) + I_c\left(\ln\frac{1}{h} + \ln\frac{1}{l}\right)\right]$$

and ψ_{a3} is position 3 is obtained as

$$\psi_{a3} = 2 \times 10^{-7}\left[I_a\left(\ln\frac{1}{r^1} + \ln\frac{1}{m}\right) + I_b\left(\ln\frac{1}{2h} + \ln\frac{1}{D}\right) + I_c\left(\ln\frac{1}{h} + \ln\frac{1}{l}\right)\right]$$

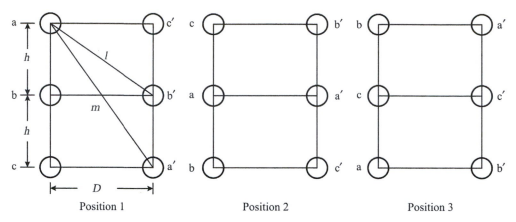

FIGURE 2.11

Double-circuit unsymmetrical transposed three-phase line.

The total average flux linkages for phase a

$$\psi_a = \frac{\psi_{a1} + \psi_{a2} + \psi_{a3}}{3}$$

$$= \frac{2 \times 10^{-7}}{3}\left[3I_a \ln \frac{1}{r^l} + I_a \ln \left(\frac{1}{D} \cdot \frac{1}{m} \cdot \frac{1}{m}\right) + I_b \ln \left(\frac{1}{2h^3}\right)\right.$$

$$\left. + I_b \ln \left(\frac{1}{l} \cdot \frac{1}{l} \cdot \frac{1}{D}\right) + I_c \ln \left(\frac{1}{2h^3}\right) + I_c \ln \left(\frac{1}{D} \cdot \frac{1}{l} \cdot \frac{1}{l}\right)\right]$$

$$= \frac{2 \times 10^{-7}}{3}\left[3I_a \ln \frac{1}{r^l} + I_a \ln \frac{1}{m^2 D} + I_b \ln \frac{1}{2h^3} + I_b \ln \frac{1}{l^2 D}\right.$$

$$\left. + I_c \ln \frac{1}{2h^3} + I_c \ln \frac{1}{Dl^2}\right]$$

$$= \frac{2 \times 10^{-7}}{3}\left[3I_a \ln \frac{1}{r^l} + I_a \ln \frac{1}{m^2 D} + (I_b + I_c) \ln \frac{1}{2h^3} + (I_b + I_c) \ln \frac{1}{l^2 D}\right]$$

Since

$$I_a + I_b + I_c = 0$$
$$(I_b + I_c) = -I_a$$

Substituting the above

$$\psi_{a1} = \frac{2 \times 10^{-7}}{3}\left[3I_a \ln \frac{1}{r^l} + I_a \ln \frac{1}{m^2 D} - I_a \ln \frac{1}{2h^3} - I_a \ln \frac{1}{l^2 D}\right]$$

$$= \frac{2 \times 10^{-7}}{3}\left[3I_a \ln \frac{1}{r^l} + I_a \ln \frac{2h^3 l^2 D}{m^2 D}\right]$$

$$= 2 \times 10^{-7}\left[\ln \left(\frac{2^{\frac{1}{3}} h l^{\frac{2}{3}}}{r^l m^{\frac{2}{3}}}\right)\right]$$

$$= 2 \times 10^{-7} I_a \ln \left[\frac{2^{\frac{1}{3}} h}{r^l}\left(\frac{l}{m}\right)^{\frac{2}{3}}\right] \quad \text{(wb-t/m)}$$

The inductance of phase a,

$$L_a = \frac{\psi_a}{I_a} = 2 \times 10^{-7} \ln \left[\frac{2^{\frac{1}{3}}h}{r^1} \left(\frac{l}{m} \right)^{\frac{2}{3}} \right] \ (h/m) \tag{2.54}$$

If $D \gg h$ then l/m tends to unity. In this case

$$L_a = 2 \times 10^{-7} \ln \left(\frac{2^{\frac{1}{3}}h}{r^1} \right) \tag{2.55}$$

Since there are two conductors per phase inductance per conductor is $L_a/2$, i.e.,

$$L = 2 \times 10^{-7} \ln \left[2^{\frac{1}{6}} \left(\frac{h}{r^1} \right)^{\frac{1}{2}} \cdot \left(\frac{l}{m} \right)^{\frac{1}{3}} \right] \ (h/m) \tag{2.56}$$

$$L \simeq 2 \times 10^{-7} \ln \left[2^{\frac{1}{6}} \cdot \left(\frac{h}{r^1} \right)^{\frac{1}{2}} \right] \ (h/m) \tag{2.57}$$

2.7 CAPACITANCE OF TRANSMISSION LINES

Capacitance always exists between conductors carrying charges in a medium of permittivity ϵ. Calculation of capacitances between line conductors and between conductor to neutral or earth is based on determining:

1. electric field strength E using Gauss's law
2. utilizing the known potential difference between conductors, and
3. by dividing the total charge with the potential difference

2.7.1 POTENTIAL DIFFERENCE BETWEEN TWO POINTS DUE TO AN ELECTRIC CHARGE

The electric flux density (D) at a point distance x (m) from a current-carrying conductor having a charge of q (c/m) is given by (Fig. 2.12)

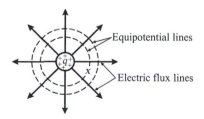

FIGURE 2.12

Electric flux lines and equipotential lines.

$$D = \frac{q}{2\pi x} \ (\text{c}/\text{m}^2) \tag{2.58}$$

The electric field intensity

$$E = \frac{q}{2\pi \varepsilon x} \ (\text{v}/\text{m}) \tag{2.59}$$

where ε is the permittivity of the medium. The potential difference between any two points P_1 and P_2 at distance D_1 and D_2 from the center of the conductor (Fig. 2.13).

$$
\begin{aligned}
v_{12} &= \int_{D_1}^{D_2} E \cdot dx = \int_{D_1}^{D_2} \frac{q}{2\pi x \varepsilon} \cdot dx \\
&= \frac{q}{2\pi \varepsilon} \ln \frac{D_2}{D_1} \ (\text{V})
\end{aligned}
\tag{2.60}
$$

2.7.2 CAPACITANCE OF A TWO-CONDUCTOR LINE

Capacitance between the two conductors of a transmission line is defined as the charge on the conductors per unit potential difference between them. Thus the capacitance per unit length of the line is (Fig. 2.14)

$$C = \frac{q}{v_{12}} \ (\text{f}/\text{m}) \tag{2.61}$$

The potential difference between the two conductors due to charges q_1 and q_2 on conductors 1 and 2 is the sum of the values due to charges q_1 and q_2.

$$V_{12} = \frac{q_1}{2\pi \varepsilon} \ln \frac{D}{r_1} + \frac{q_2}{2\pi \varepsilon} \ln \frac{r_2}{D} \tag{2.62}$$

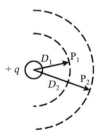

FIGURE 2.13

Potential difference.

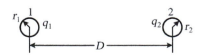

FIGURE 2.14

Capacitance of single-phase line.

where

$$\nu_{12} \text{ due to } q_1 = \frac{q_1}{2\pi\varepsilon} \ln \frac{D}{r_1}$$

and

$$\nu_{12}^1 \text{ due to } q_2 = \frac{q_2}{2\pi\varepsilon} \ln \frac{r_2}{D}$$

and

$$V_{12} = \nu_{12} + \nu_{12}^1$$

However, since $q_1 = -q_2$

$$V_{12} = \frac{q_1}{2\pi\varepsilon} \ln \left(\frac{D}{r_1} - \frac{r_2}{D} \right) \text{ (V)}$$

$$= \frac{q_1}{2\pi\varepsilon} \ln \left(\frac{D^2}{r_1 r_2} \right) \text{ (V)}$$

(2.63)

The capacitance between the two conductors

$$C_{12} = \frac{q_1}{\nu_{12}} = \frac{q_1}{\left(\dfrac{q_1}{2\pi\varepsilon} \right) \cdot \ln \left(\dfrac{D^2}{r_1 r_2} \right)} = \frac{2\pi\varepsilon}{\ln \left(\dfrac{D^2}{r_1 r_2} \right)}$$

(2.64)

In general for a two-conductor system

$$q = q_1 = -q^2 \text{ and } r = r_1 = r_2$$

So that,

$$C_{12} = \frac{2p\varepsilon}{\ln \left(\dfrac{D^2}{r^2} \right)} = \frac{\pi\varepsilon}{\ln \left(\dfrac{D}{r} \right)} \text{ (f/m)}$$

(2.65)

Capacitance to neutral or ground of each phase conductor is double this value, since the potential to neutral is half the potential difference between the two conductors.

$$C_n = \frac{2\pi\varepsilon}{\ln \left(\dfrac{D}{r} \right)} \text{ (f/m)}$$

(2.66)

2.7.3 CAPACITANCE OF A THREE-PHASE LINE WITH SYMMETRIC SPACING

Consider the three-phase system shown in Fig. 2.15. The three-phase conductors are placed at the corners of an equilateral triangle of side D (m). The radius of each conductor is r (m). The potential difference between conductors a and b due to charges q_a, q_b, and q_c is obtained as

$$V_{ab} = \frac{1}{2\pi\varepsilon} = \left[q_a \ln \frac{D}{r} + q_b \ln \frac{r}{D} + q_c \ln \frac{D}{D} \right] \text{ (V)}$$

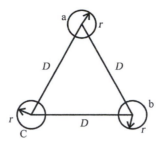

FIGURE 2.15

Symmetrically spaced three-phase line.

Similarly the potential difference between a and c is given by

$$V_{ac} = \frac{1}{2\pi\varepsilon}\left[q_a \ln\frac{D}{r} + q_b \ln\frac{D}{D} + q_c \ln\frac{r}{D}\right] \text{ (V)}$$

Then,

$$V_{ab} + V_{ac} = \frac{1}{2\pi\varepsilon}\left[2q_a \ln\frac{D}{r} + q_b \ln\frac{r}{D} + q_c \ln\frac{r}{D}\right] \text{ (V)} \qquad (2.67)$$

Assuming that there are only q_a, q_b, and q_c and no other charges are present in the vicinity.

$$q_a + q_b + q_c = 0 \text{ or } (q_b + q_c) = -q_a$$

Substituting this result is Eq. (2.67)

$$V_{ab} + V_{ac} = \frac{1}{2\pi\varepsilon}\left[2q_a \ln\frac{D}{r} - q_a \ln\frac{r}{D}\right]$$

$$= \frac{1}{2\pi\varepsilon}3q_a \ln\frac{D}{r}$$

Further, in a balanced system $V_{ab} + V_{ac} = 3V_{an}$
Hence,

$$3V_{an} = \frac{1}{2\pi\varepsilon}3q_a \ln\frac{D}{r}$$

$$V_{an} = \frac{1}{2\pi\varepsilon}q_a \ln\frac{D}{r} \text{ (V)} \qquad (2.68)$$

Capacitance to neutral or ground of each phase conductor

$$C_n = \frac{q_a}{V_{an}} = \frac{2\pi\varepsilon}{\ln\left(\dfrac{D}{r}\right)} \text{ (f/m)} \qquad (2.69)$$

It is to be noted that the radius of the conductor is not affected in capacitance calculations, while for inductance calculations the effective radius is $0.7788 \times r$ (GMR) The capacitive reactance per meter length

$$X_c = \frac{1}{2\pi f c} \text{ (}\Omega\text{)} \qquad (2.70)$$

The line charging current

$$I_c = \frac{V}{x_c} = 2\pi f_c V \ (A)$$

(2.71)

2.7.4 CAPACITANCE OF A THREE-PHASE LINE WITH UNSYMMETRICAL SPACING

Consider a three-phase line spaced at distances D_{12}, D_{23}, and D_{31} as shown in Fig. 2.16. Such lines are generally transposed so that each conductor occupies every other position over a cycle of transposition.

The potential difference V_{ab} between conductors a and b in position 1 is

$$V_{ab} = \left[q_a \ln \frac{D_{12}}{r} + q_b \ln \frac{r}{D_{12}} + q_c \ln \frac{D_{23}}{D_{31}} \right] \ (V)$$

(2.72)

when conductor a is in position 2

$$V_{ab} = \frac{1}{2\pi\varepsilon} \left[q_a \ln \frac{D_{23}}{r} + q_b \ln \frac{r}{D_{23}} + q_c \ln \frac{D_{31}}{D_{12}} \right] \ (V)$$

(2.73)

Likewise, in position 3

$$V_{ab} = \frac{1}{2\pi\varepsilon} \left[q_a \ln \frac{D_{31}}{r} + q_b \ln \frac{r}{D_{31}} + q_c \ln \frac{D_{12}}{D_{23}} \right] \ (V)$$

(2.74)

The average potential difference V_{ab} over a cycle of transposition

$$V_{ab} = \frac{1}{3} \cdot \frac{1}{2\pi\varepsilon} \left[q_a \ln \frac{D_{12} D_{23} D_{31}}{r^3} + q_b \ln \frac{r^3}{D_{12} D_{23} D_{31}} + q_c \ln \frac{D_{23} D_{31} D_{12}}{D_{31} D_{12} D_{23}} \right] \ (V)$$

$$= \frac{1}{2\pi\varepsilon} \left[q_a \ln \frac{\sqrt[3]{D_{12} D_{23} D_{31}}}{r} + q_b \ln \frac{r}{\sqrt[3]{D_{12} D_{23} D_{31}}} \right] \ (V)$$

Let

$$\sqrt[3]{D_{12} D_{23} D_{31}} = D_{eq}$$

(2.75)

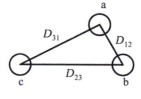

Position 1

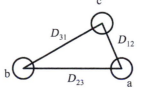

Position 2

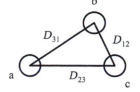

Position 3

FIGURE 2.16

Three-phase conductor over a cycle of transposition.

then

$$V_{ab} = \frac{1}{2\pi\varepsilon}\left[q_a \ln \frac{D_{eq}}{r} + q_b \ln \frac{r}{D_{eq}}\right] \text{ (V)} \tag{2.76}$$

In a similar manner the potential difference across conductors a and c over a transposition cycle can be proved to be

$$V_{ac} = \frac{1}{2\pi\varepsilon}\left[q_a \ln \frac{D_{eq}}{r} + q_c \ln \frac{r}{D_{eq}}\right] \text{ (V)} \tag{2.77}$$

$$V_{ab} + V_{ac} = 3V_{an} = \frac{1}{2\pi\varepsilon}\left[2q_a \ln \frac{D_{eq}}{r} + (q_b + q_c) \ln \frac{r}{D_{eq}}\right]$$

Since

$$q_a + q_b + q_c = 0 \tag{2.78}$$

$$3V_{an} = \frac{1}{2\pi\varepsilon}\left[3q_a \ln \frac{D_{eq}}{r}\right]$$

$$V_{an} = \frac{q}{2\pi\varepsilon} \ln \frac{D_{eq}}{r} \text{ (V)} \tag{2.79}$$

Capacitance to neutral or ground $C_n = q_a/V_{an}$
Therefore,

$$C_n = \frac{2\pi\varepsilon}{\ln\left(\dfrac{D_{eq}}{r}\right)} \text{ (f/m)} \tag{2.80}$$

2.8 CAPACITANCE OF A THREE-PHASE, DOUBLE-CIRCUIT LINE WITH SYMMETRICAL SPACING

Consider a double-circuit line spaced as shown in Fig. 2.17. The phase-to-phase distance is D (m). The potential difference V_{ab} can be written down using Fig. 2.17 as

$$\begin{aligned}
V_{ab} &= \frac{1}{2\pi\varepsilon}\left[q_a\left(\ln\frac{D}{r} + \ln\frac{\sqrt{3}D}{2D}\right) + q_b\left(\ln\frac{r}{D} + \ln\frac{2D}{\sqrt{3}D}\right) + q_c\left(\ln\frac{D}{\sqrt{3}D} + \ln\frac{\sqrt{3}D}{D}\right)\right] \\
&= \frac{1}{2\pi\varepsilon}\left[q_a \ln\frac{\sqrt{3}D}{2r} + q_b \ln\frac{2r}{\sqrt{3}D}\right]
\end{aligned} \tag{2.81}$$

In a similar manner V_{ac} can be written as

$$V_{ac} = \frac{1}{2\pi\varepsilon}\left[q_a \ln\frac{\sqrt{3}D}{2r} + q_c \ln\frac{2r}{\sqrt{3}D}\right] \tag{2.82}$$

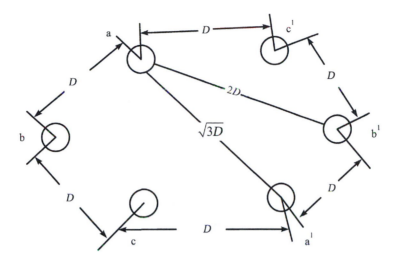

FIGURE 2.17

Capacitance of three-phase, double-circuit line-symmetrical spacing.

$$V_{ab} + V_{ac} = 3V_{an} = \frac{1}{2\pi\varepsilon}\left[2q_a \ln\frac{\sqrt{3D}}{2r} + (q_b + q_c)\ln\frac{2r}{\sqrt{3D}}\right] \tag{2.83}$$

Using the relation that $q_a + q_b + q_c = 0$

$$3V_{an} = \frac{3q_a}{2\pi\varepsilon}\ln\left(\frac{\sqrt{3D}}{2r}\right) \tag{2.84}$$

So that

$$V_{an} = \frac{q_a}{2\pi\varepsilon}\ln\frac{\sqrt{3D}}{2r} \tag{2.85}$$

Capacitance to neutral of each phase conductor

$$C_{an} = \frac{q_a}{V_{an}} = \frac{2\pi\varepsilon}{\ln\left(\dfrac{\sqrt{3D}}{2r}\right)} \quad (f/m) \tag{2.86}$$

Capacitance to neutral of each phase

$$C_a = 2\cdot C_{an} = \frac{2\cdot 2\pi\varepsilon}{\ln\left(\dfrac{\sqrt{3D}}{2r}\right)} = \frac{2\pi\varepsilon}{\ln\left(\dfrac{\sqrt{3D}}{2r}\right)^{1/2}} \quad (f/m) \tag{2.87}$$

2.9 **EFFECT OF EARTH ON THE CAPACITANCE OF TRANSMISSION LINES**

The electric lines of force and equipotential lines are orthogonal to each other for an isolated charged conductor. The presence of earth alters the electric field of a charged conductor. If earth

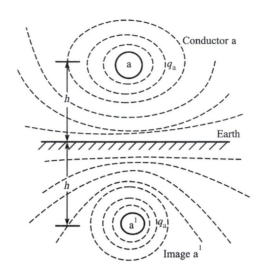

FIGURE 2.18

Conductor and its image.

is assumed as a perfect conductor in the form of a horizontal plane of infinite extent, then we notice that the electric field of the charged conductors is forced to conform to the presence of earth's equipotential surface. The potential distribution in space will remain the same if we imagine an isolated opposite charged conductor at the same depth h as the conductor is above the earth (Fig. 2.18). The capacitance between the conductor and its fictitious image

$$C_{aa'} = \frac{\pi\varepsilon}{\ln\frac{2h}{r}} \, (\text{f/m}) \qquad (2.88)$$

where r is the radius of the conductor.

Then, capacitance to earth of an isolated conductor

$$C_n = \frac{\pi\varepsilon}{\ln\left(\frac{2h}{r}\right)} \, (\text{f/m}) \qquad (2.89)$$

2.10 EFFECT OF EARTH CAPACITANCE OF A SINGLE-PHASE LINE

Consider Fig. 2.19 where a and b are the conductors carrying charges q_a and q_b. As earth is at zero potential, consider a' and b', the images of a and b carrying charges $-q_a$ and $-q_b$ at a depth of h meters. The spacing between the conductors is D (m).

The potential difference

$$V_{ab} = \frac{1}{2\pi\varepsilon}\left[\left(q_a \ln \frac{D}{r} + q_b \ln \frac{r}{D}\right) - q_a \ln \frac{\sqrt{D^2 + 4h^2}}{2h} - q_b \ln \frac{2h}{\sqrt{D^2 + 4h^2}}\right] \qquad (2.90)$$

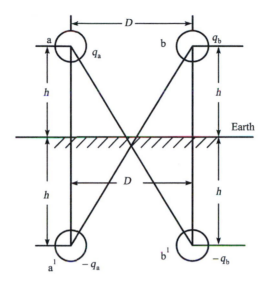

FIGURE 2.19

Effect of earth capacitance of line.

But $q_a = -q_b$, substituting this

$$
V_{ab} = \frac{1}{2\pi\varepsilon}\left[q_a \ln \frac{D}{r} + q_a \ln \frac{D}{r} - q_a \ln \frac{\sqrt{D^2 + 4h^2}}{2h} - q_a \ln \frac{\sqrt{D^2 + 4h^2}}{2h}\right]
$$

$$
= \frac{1}{2\pi\varepsilon}\left[2q_a \ln \frac{D}{r} + 2q_a \ln \frac{2h}{\sqrt{D^2 + 4h^2}}\right]
$$

$$
= \frac{q_a}{\pi\varepsilon}\left[\ln \frac{D}{r} \cdot \frac{2h}{\sqrt{D^2 + 4h^2}}\right] = \frac{q_a}{\pi\varepsilon}\left[\ln \frac{D}{r} \cdot \frac{1}{\sqrt{\frac{D^2 + 4h^2}{4h^2}}}\right] \tag{2.91}
$$

$$
= \frac{q}{\pi\varepsilon} \ln \left[\frac{D}{r} \frac{1}{\sqrt{1 + \frac{D^2}{4h^2}}}\right] V
$$

$$
C_{ab} = \frac{q_a}{V_{ab}} = \frac{\pi\varepsilon}{\ln \left(\frac{D}{r} \cdot \frac{1}{\sqrt{1 + \frac{D^2}{4h^2}}}\right)} \quad (\text{f/m}) \tag{2.92}
$$

The line capacitance is slightly increased due to the presence of earth. As h tends to infinity $D^2/4h^2$ tends to zero and

$$
C_{ab} = \left(\frac{\pi\varepsilon}{\ln\left(\frac{D}{r}\right)}\right)
$$

2.11 CAPACITANCE OF THREE-PHASE LINE INCLUDING EFFECT OF EARTH

Consider a three-phase transmission line with unsymmetrical spacing above the ground level and the images of the three-phase conductors below the earth at the same depth as the original conductors are above the ground level.

Let the conductor carry charges q_a, q_b, and q_c. Their images carry charges $-q_a$, $-q_b$, and $-q_c$, respectively. The distances are all marked as shown in Fig. 2.20. The numbers 1, 2, and 3 indicate transposition cycle positions.

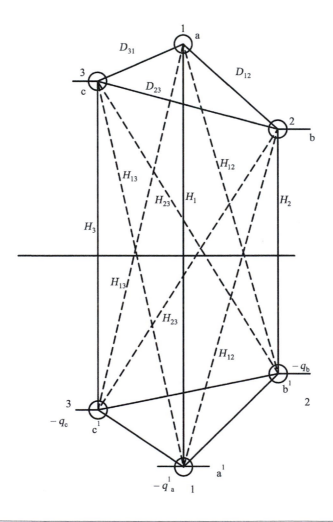

FIGURE 2.20

Capacitance of three-phase line in the phase presence of earth.

With conductor a in position 1

$$V_{ab1} = \frac{1}{2\pi\varepsilon}\left[q_a\left(\frac{D_{12}}{r} - \ln\frac{H_{12}}{H_1}\right) + q_b\left(\ln\frac{r}{D_{12}} - \ln\frac{H_2}{H_{12}}\right) + q_c\left(\ln\frac{D_{23}}{D_{31}} - \ln\frac{H_{23}}{H_{31}}\right)\right] \tag{2.93}$$

Similarly when conductor a is in position 2

$$V_{ab2} = \frac{1}{2\pi\varepsilon}\left[q_a\left(\ln\frac{D_{23}}{r} - \ln\frac{H_{23}}{H_2}\right) + q_b\left(\ln\frac{r}{D_{23}} - \ln\frac{H_3}{H_{23}}\right) + q_c\left(\ln\frac{D_{31}}{D_{12}} - \ln\frac{H_{31}}{H_{12}}\right)\right] \tag{2.94}$$

and likewise,

$$V_{ab3} = \frac{1}{2\pi\varepsilon}\left[q_a\left(\ln\frac{D_{31}}{r} - \ln\frac{H_{31}}{H_3}\right) + q_b\left(\ln\frac{r}{D_{31}} - \ln\frac{H_1}{H_{31}}\right) + q_c\left(\ln\frac{D_{12}}{D_{23}} - \ln\frac{H_{12}}{H_{23}}\right)\right] \tag{2.95}$$

$$V_{ab} = \frac{1}{3}[V_{ab1} + V_{ab2} + V_{ab3}]$$

$$= \frac{1}{3}\cdot\frac{1}{2\pi\varepsilon}q_a\left(\ln\frac{D_{23}D_{31}D_{12}}{r^3} - \ln\frac{H_{12}H_{23}H_{31}}{H_1H_2H_3}\right) + q_b\left(\ln\frac{r^3}{D_{23}D_{12}D_{31}} - \ln\frac{H_3H_1H_2}{H_{23}H_{12}H_{31}}\right)$$

$$+ q_c\left(\ln\frac{D_{31}D_{12}D_{23}}{D_{31}D_{12}D_{23}} - \ln\frac{H_{23}H_{31}H_{12}}{H_{31}H_{12}H_{23}}\right) \tag{2.96}$$

$$= \frac{1}{6\pi\varepsilon}\left[q_a\left(\ln\frac{D_{23}D_{31}D_{12}}{r^3} - \ln\frac{H_{12}H_{23}H_{31}}{H_1H_2H_3}\right) + q_b\left(\ln\frac{r^3}{D_{12}D_{23}D_{31}} - \ln\frac{H_1H_2H_3}{H_{12}H_{23}H_{31}}\right)\right]$$

In a similar manner V_{ac} can be obtained as

$$V_{ac} = \frac{1}{6\pi\varepsilon}\left[q_a\left(\ln\frac{D_{12}D_{23}D_{31}}{r^3} - \ln\frac{H_{12}H_{23}H_{31}}{H_1H_2H_3}\right) + q_c\left(\ln\frac{r^3}{D_{12}D_{23}D_{31}} - \ln\frac{H_{12}H_{23}H_{31}}{H_1H_2H_3}\right)\right.$$

$$\left. + q_c\left(\ln\frac{r^3}{D_{12}D_{23}D_{31}} - \ln\frac{H_{12}H_{23}H_{31}}{H_1H_2H_3}\right)\right] \tag{2.97}$$

We know that $V_{ab} + V_{ac} = 3V_{an}$ and $q_b + q_c = -q_a$. Substituting both,

$$3V_{an} = \frac{1}{6\pi\varepsilon}\left[2q_a\left(\ln\frac{D_{12}D_{23}D_{31}}{r^3} - \ln\frac{H_{12}H_{23}H_{31}}{H_1H_2H_3}\right) + q_a\left(\ln\frac{D_{12}D_{23}D_{31}}{r^3} - \ln\frac{H_{12}H_{23}H_{31}}{H_1H_2H_3}\right)\right]$$

$$= \frac{q_a}{2\pi\varepsilon}\left[\ln\left(\frac{D_{12}D_{23}D_{31}}{r^3}\right) - \ln\left(\frac{H_{12}H_{23}H_{31}}{H_1H_2H_3}\right)\right] \tag{2.98}$$

Capacitance of conductor to neutral with $D_{eq} = \sqrt[3]{D_{12}D_{23}D_{31}}$ (f/m)

$$C_n = \frac{V_{an}}{q_a} = \left[\frac{2\pi\varepsilon}{\ln\left(\frac{D_{eq}}{r}\right)} - \ln\frac{\sqrt[3]{H_{12}H_{23}H_{31}}}{\sqrt[3]{H_1H_2H_3}}\right] \tag{2.99}$$

2.12 SKIN EFFECT AND PROXIMITY EFFECT

Consider a conductor of cross section A. Imagine that this area A is divided into small elemental areas uniformly as shown in Fig. 2.21.

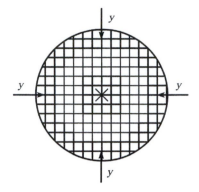

FIGURE 2.21

Conductor section divided into elemental areas.

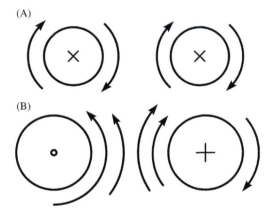

FIGURE 2.22

Electric field near two conductors carrying current (A) in opposite directions and (B) in the same direction.

Each elemental area carries a current I/n where I is the conductor current and n is the number of elemental areas. Each area current I/n produces an alternating flux of frequency determined by supply frequency. The central elemental areas link all the flux produced by inner areas (X) and also by area currents. The peripheral areas (Y) link much less flux than the inner elemental areas. Treating these elemental areas along the conductor length as filaments, the central filaments link more flux and possess more inductance and inductive reactance compared to the peripheral filaments (Y). Hence the current flowing in the central part will be less compared to current flowing in the outer or periphnal filaments, the variation in current distribution being nonuniform. Higher the frequency higher will be the reactance of central filaments, and less will be the current flowing in the central part of the conductor. This tendency to limit the flow of current at higher frequencies to only the outer most part of the conductor is called *skin effect*. The apparent resistance of the conductor is increased due to this effect. In case of alternating currents, which does not happen if direct current flows through the conductor. If two conductors are placed near each other and they carry currents in opposite directions, adjacent sides of the conductors carry more flux than the far sides. Similarly, if both conductors carry currents in opposite direction, the adjacent sides carry less flux than the far sides (Fig. 2.22).

The result in either case is that due to this effect the resistance of conductors becomes a little more than for uniform distribution of flux just as in the case of *skin effect*. Thus the effective increase in resistance due to the proximity of two alternating current-carrying conductors is called *proximity effect*.

Due to these two effects, viz., skin effect and proximity effect, the effective ac resistance is obtained as about 1.2 times the resistance for a direct current-carrying conductor.

WORKED EXAMPLES

E.2.1 Calculate the loop inductance per kilometer of a single-phase circuit comprising two parallel conductors 1 m apart and 1 cm in diameter.

Solution: $D = 1$ m

$$r^1 = \frac{0.5 \times 0.7788}{100} \text{ m}$$

$$L = 2 \times (2 \times 10^{-7}) \ln \left(\frac{1 \times 100}{0.5 \times 0.7788} \right)$$

$$= 4 \times 10^{-7} \ln 256.8 = 22.194 \times 10^{-7} \text{ h/m}$$

$$= 2.21 \text{ mh/km}$$

E.2.2 A single-phase line has two pairs of conductors. Each pair comprises two 1.25 cm diameter conductors in parallel spaced vertically and 75 cm apart. The two parallel pairs are spaced laterally by a distance of 1.5 m. Calculate the total inductance of the line per kilometer assuming the current to be equally divided.

Solution: The self-GMD or self-GMR of each pair of conductors

$$D_5 = \sqrt[4]{\frac{1.25}{2} \times 75 \times \frac{1.25}{2} \times 75 \times (0.7788)^2} = 6.042 \text{ cm} = \frac{6.042}{100} \text{ m}$$

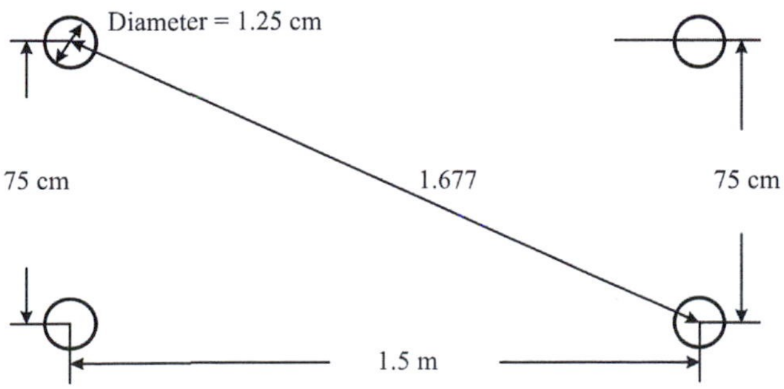

FIGURE E.2.2

The mutual GMD between the two pairs

$$D_m = \sqrt[4]{1.5 \times 1.677 \times 1.5 \times 1.677} = 1.58 \text{ m}$$

$$L = 2 \times 10^{-7} \times \log_e \left(\frac{1.56 \times 100}{6.042}\right) = 2 \times 10^{-7} \times \ln 25.819$$

$$= 2 \times 10^{\times 7} \times 3.2511 \text{ h/m}$$

Total inductance per kilometer $= 2 \times 2 \times 10^{-7} \times 3.2511 \text{ h/m}$

$$= 1.3 \text{ mh/km}$$

E.2.3 A wire of 4 mm in diameter is suspended at a constant height of 10 m above the sea level which constitutes the return conductor. Calculate the inductance of the system per kilometer.

Solution: Consider the conductor 10 m above the sea level. The image of the conductor at a depth of 10 m below the sea water level (a′).

Then the inductance of the system

$$L = 2 \times 10^{-7} \ln \left(\frac{20 \times 1000}{2 \times 0.7788}\right)$$

$$= 18.92 \times 10^{-7} \text{ h/m} = 1.892 \text{ mh/km}$$

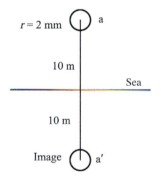

FIGURE E.2.3

E.2.4 Calculate the loop inductance per kilometer of a single-phase transmission line when the line conductors are spaced 1.2 m apart and each conductor has a diameter of 1.3 cm. Also, calculate the reactance of line, if its length is 50 km and the line is operating at 50 Hz.

Solution: Loop inductance $= 2 \times (2 \times 10^{-7}) \ln \left(\dfrac{1.2}{(1.3/2) \times \dfrac{1}{100}}\right)$

$$= 20.8 \times 10^{-7} \text{ h/m}$$

$$= 2.08 \text{ mh/km}$$

Line reactance per kilometer $= 2\pi \times 50 \times 2.08$

$$= 653.12 \times 10^{-3} \ \Omega = 0.653 \ \Omega$$

Total reactance $= 50 \times 0.653 = 32.65 \ \Omega$

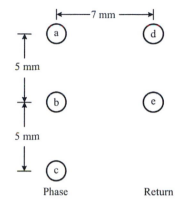

FIGURE E.2.5

E.2.5 A single-phase transmission line consists of three conductors in the phase and two conductors in the return as shown in Fig E.2.5. The radius of each phase conductor is 0.22 cm and that of return is 0.45 cm. Phase conductors are a, b, c and are in one vertical line with a spacing of 5 m and the return conductors d and e are also in a vertical line spaced 5 m apart and separated from the phase group by a horizontal spacing of 7 m. Determine the inductance of both the phase and return conductors and the inductance of the complete line in h/m.

Solution: The GMD of phase and return groups D_m is computed:

$$D_{ad} = 7 \text{ m}; D_{bd} = D_{ae} = \sqrt{5^2 + 7^2} = 8.6 \text{ m}$$

$$D_{cd} = \sqrt{50^2 + 7^2} = 12.201 \text{ m}$$

$$D_m = \sqrt[6]{D_{ad}\, D_{ae}\, D_{bd}\, D_{be}\, D_{cd}\, D_{ce}}$$

$$D_m = \sqrt[6]{7 \times 8.6 \times 7 \times 8.6 \times 8.6 \times 12.201} = \sqrt[6]{380,265.44} = 8.51 \text{ m}$$

The GMR for phase conductor group

$$D_{ph} = \sqrt[9]{D_{aa}\, D_{ab}\, D_{ac}\, D_{bb}\, D_{bc}\, D_{ba}\, D_{ca}\, D_{cb}\, D_{cc}}$$

$$= \sqrt[9]{\frac{0.22}{100} \times \frac{0.22}{100} \times \frac{0.22}{100} \times 5 \times 10 \times 5 \times 5 \times 5 \times 10 \times (0.7788)^3} = \sqrt[9]{0.0006655} = 0.408155 \text{ m}$$

$$D_{return} = \sqrt[4]{\frac{0.45}{100} \times \frac{0.45}{100} \times 5 \times 5 \times (0.7788)^2} = \sqrt[4]{0.0003071} = 0.1323794 \text{ m}$$

Self-inductance of phase conductor group

$$L_{ph} = 2 \times 10^{-7} \ln \frac{8.51}{0.408155} = 6.074 \times 10^{-7} \text{ h/m}$$

Self-inductance of return conductor group

$$L_{return} = 2 \times 10^{-7} \ln \frac{8.51}{0.132} = 8.3324 \times 10^{-7} \text{ h/m}$$

Total inductance of line per meter

$$= (6.0742 + 8.3324)10^{-7} \text{ h/m} = 14.4066 \times 10^{-4} \text{ m h/m}$$
$$= 1.44066 \text{ mh/km}$$

E.2.6 A three-phase unsymmetrical circuit has the arrangement shown in Fig E.2.6. The conductor radius is 1.5 cm. Determine the inductance per kilometer. If f 50 HZ find the line reactance.

Solution:

$$r^1 = \frac{1.5 \times 0.7788}{100} = 0.011682 \text{ m}$$

$$D_m = \sqrt[3]{6 \times 6 \times 6.63} = 6.203 \text{ m}$$

Inductance,

$$L = 2 \times 10^{-7} \ln \frac{6.203}{0.011682}$$

$$= 2 \times 10^{-7} \ln 530.988 = 2 \times 10^{-7} \times 6.27474 \text{ h/m}$$

$$= 12.549 \times 10^{-7} \text{ h/m}$$

$$= 1.2549 \text{ mh/km}$$

$$\text{Line reactance } x_L = \frac{2\pi \times 50 \times 1.2549}{1000} = 0.394 \ \Omega/\text{km}$$

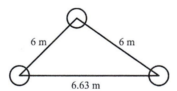

FIGURE E.2.6

E.2.7 A two-conductor, three-phase transmission line is arranged horizontally as shown in Fig E.2.7. The spacing between conductors of the bundle is 40 cm. The phase-to-phase

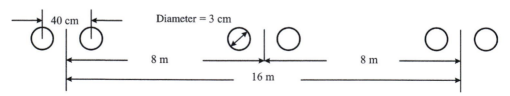

FIGURE E.2.7

spacings are 8, 8, and 16 m, respectively. Determine the inductance of the line per phase per kilometer. The conductor diameter is 3 cm.

Solution:

$$D_{2b} = \sqrt[4]{(0.7788 \times 1.5 \times 40)^2} = \sqrt[4]{2184.5} = 6.8365 \text{ cm}$$

$$D_{eq} = \sqrt[3]{8 \times 8 \times 16} = 10.08 \text{ m}$$

$$\text{Inductance per phase} = 2 \times 10^{-7} \ln\left(\frac{10.08}{0.06836}\right)$$

$$= 2 \times 10^{-7} \times 4.99 = 9.987 \times 10^{-7} \text{ h/m}$$

$$= 0.9987 \text{ mh/km}$$

The line reactance,

$$X_L = 2\pi \times 50 \times 0.9987 \times 10^{-3} \ \Omega/\text{km/phase}$$

$$= 0.3136 \ \Omega/\text{km/phase}$$

E.2.8 Determine the inductance per km of a double-circuit, three-phase line shown in Fig E.2.8. Each circuit is transposed perfectly. The diameter of the conductor is 2.0 cm.

Solution: Self-GMD of each conductor

$$= 1 \times 0.7788$$

$$= 0.7788 \times 10^{-2} \text{ m}$$

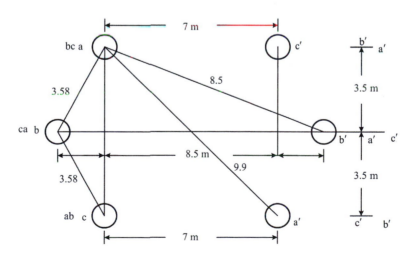

FIGURE E.2.8

$$D_{ab} = D_{bc} = D_{b'c'} = D_{b'a'}$$
$$= \sqrt{0.75^2 + 3.5^2}$$
$$= 3.58 \text{ m}$$

$$D_{ab'} = \sqrt{3.5^2 + (8.5 - 7.5)^2} = \sqrt{3.5^2 + 7.75^2} = 8.5 \text{ m}$$

$$D_{aa'} = \sqrt{7^2 + 7^2} = 9.9 \text{ m}$$

$$D_{a1} = \sqrt[4]{D_{ab} \cdot D_{ac} \cdot D_{ab'} \cdot D_{ac'}} = \sqrt[4]{3.58 \times 7 \times 7 \times 8.5} = \sqrt[4]{1491} = \sqrt{38.613} = 6.21 \text{ m}$$

$$D_{a2} = \sqrt[4]{3.58 \times 3.58 \times 8.5 \times 8.5} = \sqrt[4]{925.985} = \sqrt[2]{30.43} = 5.516 \text{ m}$$

$$D_{a3} = \sqrt[4]{7 \times 3.58 \times 7 \times 8.5} = \sqrt[4]{1491} = 6.21 \text{ m}$$

$$D_{a} = \sqrt[3]{6.21 \times 6.21 \times 5.516} = 5.96947 \text{ m} = D_{b} = D_{c}$$

$$D_{s_1} = \text{Self GMD of each phase} = \sqrt{0.007788 \times 9.9}$$
$$= \sqrt{0.0771} = 0.2777 \text{ m} = D_{s3}$$

$$D_{s2} = \sqrt{0.007788 \times 8.5} = 0.2573 \text{ m}$$

$$D_{s} = \sqrt[3]{D_{s_1} \cdot D_{s_2} \cdot D_{s_3}} = \sqrt[3]{0.2573 \times 0.2777 \times 0.2777} = 0.271 \text{ m}$$

$$\text{Inductance per phase} = 2 \times 10^{-7} \ln \frac{5.96947}{0.271}$$
$$= 2 \times 10^{-7} \times 3.0923$$
$$= 6.1846 \times 10^{-7} \text{ h/m} = 0.61846 \text{ mh/km}$$

E.2.9 The 2-cm diameter conductors of a three-phase, three-wire (line to neutral) of a three-phase system are located at the corners of a triangle, giving conductor spacings of 3.6, 5.4, and 8.1 m. The conductors are transposed at regular intervals and the load is balanced. Calculate the inductance per kilometer per phase.

Solution: The self-GMD of phase conductors

$$D_s = \frac{1 \times 0.7788}{100} \text{ m}$$

The mutual GMD of the three-phase conductors

$$D_m = \sqrt[3]{3.6 \times 5.4 \times 8.1} = 5.4 \text{ m}$$

Inductance per phase

$$L = 2 \times 10^{-7} \ln \frac{5.4 \times 100}{0.7788} = 2 \times 10^{-7} \ln 693.3749$$
$$= 13.08 \times 10^{-7} \text{ h/m}$$
$$= 1.308 \text{ mh/km}$$

E.2.10 A three-phase, three-wire system consisting of 2.5-cm diameter conductors spaced 3 m apart in a horizontal plane supplies a balanced load. Calculate the inductance per kilometer of each phase.

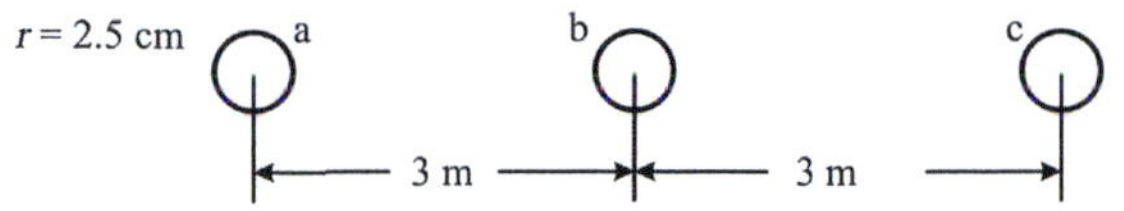

FIGURE E.2.10

Solution:

$$L_a = 2 \times 10^{-7} \left[\ln \frac{100 \times 2}{0.7788 \times 2.5} + \ln \sqrt{3 \times 6} + j \sqrt{3} \ln \frac{3}{6} \right]$$

$$= 2 \times 10^{-7} \left[4.632 + \ln 4.24264 + j \sqrt{3}(-0.3467) \right]$$

$$= 2 \times 10^{-7} [4.632 + 1.4452 - j\, 0.6005]$$

$$= 2 \times 10^{-7} [6.077 - j\, 0.6]\, h/m$$

$$= [1.2154 - j\, 0.12] mh/km$$

$$L_b = 2 \times 10^{-7} \left[\ln \frac{100 \times 2}{0.7788 \times 2.5} + \ln \sqrt{3 \times 3} \right]$$

$$= 2 \times 10^{-7} [4.632 + 1.09861] = 11.46 \times 10^{-7}\, h/m$$

$$= 1.146\, mh/km$$

$$L_c = 2 \times 10^{-7} \left[\ln \frac{\sqrt{18 \times 100 \times 2}}{2.5 \times 0.7788} + j \sqrt{3} \ln \sqrt{\frac{6}{3}} \right]$$

$$= 2 \times 10^{-7} [6.077213 + j \sqrt{3}\, 0.34642]$$

$$= 2 \times 10^{-7} [6.077 + j\, 0.6]\, h/m$$

$$= [1.2154 + j\, 0.12]\, mh/km$$

E.2.11 Calculate the capacitance of a pair of parallel conductors of 5 mm diameter spaced uniformly 20 cm apart in air. $\epsilon_o = 8.854 \times 10^{-12} = \frac{1}{36\pi} \times 10^{-9}$.

Solution:

$$D = 20\, cm \qquad \frac{D}{r} = \frac{2 \times 20 \times 10}{5} = 80$$

$$r = \frac{5}{2 \times 10}\, cm \qquad \ln \left(\frac{D}{r} \right) = \ln\, 80 = 4.382$$

$$C = \frac{2\pi \epsilon}{\ln\left(\dfrac{D}{r}\right)} = \pi \times \frac{1}{36\pi \times 10^9} \cdot \frac{1}{4.382} \ \text{f/m}$$

$$= \frac{1}{36 \times 10^9} \times \frac{1}{4.382} \times 10^6 \times 10^3 \ \mu\text{f/km}$$

$$= 0.00634 \ \mu\text{f/km}$$

E.2.12 A single–phase overhead line 32 km long consists of two parallel wires each 0.5 cm diameter, 1.5 m apart. If the line voltage is 50 kv at 50 Hz, calculate the charging current when the line is open circuited.

   Solution:

$$C = \frac{\pi \epsilon}{\ln\left(\dfrac{D}{r}\right)} = \pi \times \frac{1}{36\pi \times 10^9} \times \frac{1}{\ln\left(\dfrac{1.5 \times 100}{0.25}\right)} = \frac{1}{36 \times 10^9} \times \frac{1}{6.3969} \ \text{f/m}$$

$$X_c = \frac{1}{2\pi f c} = \frac{1}{2\pi \times 50} \times \frac{36 \times 10^9 \times 6.3969}{1} = 0.7334 \times 10^9 \ \Omega$$

Charging current

$$I_c = \frac{V}{X_C} l = \frac{50 \times 10^3 \times 32 \times 1000}{0.7334 \times 10^9}$$

$$= 2.18 \ \text{A}$$

E.2.13 Determine the capacitance per phase of a three-phase system of conductors arranged in horizontal configuration with a spacing of 3 m. The diameter of the conductors is 30 mm.

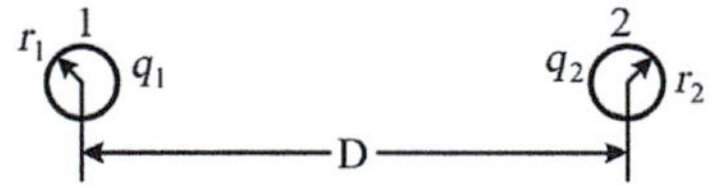

FIGURE E.2.13

   Solution:

$$D_{eq} = \sqrt[3]{3 \times 3 \times 6} = 3.78 \ \text{m}$$

$$C_m = \frac{2\pi \times 10^{-9}}{36\pi} \times \frac{1}{\ln\left(\dfrac{3.78 \times 10^3}{15}\right)} = \frac{10^{-9}}{18} \times \frac{1}{\ln \ 252} = \frac{10^{-9}}{18 \times 5.529} \ \text{f/m}$$

$$= \frac{10^{-9}}{18 \times 5.529} \times 10^3 \times 10^6 \ \mu\text{f/km} = 0.01 \ \mu\text{f/km}$$

E.2.14 A three-phase, double-circuit line is composed of 2.6-cm diameter conductors spaced vertically at a distance of 2.5 m and spaced 5.5 m apart as shown in Fig. E.2.14. Determine the capacitance of the line per kilometer.

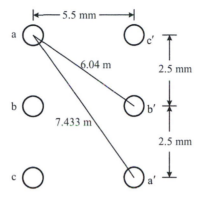

FIGURE E.2.14

Solution: GMD between phases in position 1

$$D_{ab_1} = \sqrt[4]{6.04 \times 2.5 \times 6.04 \times 2.5} = 3.8858 \text{ m}$$

$$D_{bc_1} = \sqrt[4]{2.5 \times 6.04 \times 2.5 \times 6.04} = 3.8058 \text{ m}$$

$$D_{ca_1} = \sqrt[4]{5 \times 5.5 \times 5 \times 5.5} = 5.244 \text{ m}$$

$$D_{M_1} = \sqrt[3]{5.244 \times 3.8058 \times 3.8058} = 4.294 \text{ m}$$

It can be proved that $D_{M_1} = D_{M_2} = D_{M_3} = D_M = 4.294$ m
The self-GMD

$$D_s = \sqrt[6]{\frac{1.3}{100} \times 7.433 \times \frac{1.3}{100} \times 7.433 \times \frac{1.3}{100} \times 5.5}$$

$$= \frac{1.14 \times 1.951 \times 1.3286}{10} = 0.2955$$

Therefore

$$C = \frac{2\pi}{10^9 \times 36\pi} \times \frac{1}{\ln\left(\dfrac{4.294}{0.2955}\right)} = \frac{1}{10^9 \times 18} \frac{1}{\ln 14.5373} = \frac{1}{10^9 \times 18 \times 2.6762}$$

$$= \frac{0.202076}{10^9} \text{ f/m} = 0.202076 \text{ }\mu\text{f/km}$$

E.2.15 Determine the capacitance per phase of a three-phase, double-circuit line, the conductors of which are arranged in hexagonal spacing, the distance between conductors being 2.5 m. The diameter of each conductor is 3 cm. Total length of the line is 120 km.

Solution:

$$C_a = C_b = C_c = 2C_{an} = \frac{2 \times 2\pi\,\epsilon}{\ln\left(\frac{\sqrt{3}D}{2r}\right)}\ f/m$$

$$= 2 \times 2\pi \times \frac{1}{36\pi \times 10^9} \times \frac{1}{\ln\left(\frac{\sqrt{3} \times 2.5 \times 100}{2 \times 1.5}\right)}$$

$$= \frac{1}{9 \times 10^9} \times \frac{1}{\ln 144.333} = \frac{1}{9 \times 10^9 \times 4.972}\ f/m$$

$$= \frac{10^6 \times 10^3}{9 \times 10^9 \times 4.972}\ \mu f/m$$

$$= \frac{10^6 \times 10^3 \times 120}{9 \times 10^9 \times 4.972}\ \mu f = 2.68\ f$$

E.2.16 A single-phase line constructed 13.5 m above ground has spacing between the conductors 3.9 m. The radius of the conductor is 1.78 cm. Determine the capacitance of the line per km length, considering the effect of earth and neglecting it.

Solution:

$$\sqrt{1 + \frac{D^2}{4h^2}} = \sqrt{1 + \frac{3.9^2}{4 \times 13.5^2}} = \sqrt{1 + 0.20864} = 1.09938$$

$$\ln\frac{D}{r}\sqrt{1 + \frac{D^2}{4h^2}}$$

$$= \ln\frac{3.9 \times 100}{1.78}\ \frac{1}{1.09938} = \ln 199.295 = 5.2947$$

$$C = \frac{\pi\,\epsilon}{\ln\frac{D}{r}\sqrt{1 + \frac{D^2}{4h^2}}} = \frac{\pi \times 1}{36\pi \times 10^9 \times 5.2947}\ f/m$$

$$= 0.005246\ \mu f/km$$

Without considering the effect of earth

$$C = \frac{\pi\,\epsilon}{\ln\frac{D}{r}} = \frac{\pi \times 1}{36\pi \times 10^9 \ln\frac{3.9 \times 100}{1.78}} = \frac{1}{36 \times 10^9} \times \frac{1}{5.389528}\ f/m$$

$$= \frac{10^9}{36 \times 10^9 \times 5.389528}\ \mu f/km = 0.005154\ \mu f/km$$

The effect of earth on capacitance is to increase it by
0.005246−0.005154 = 0.000092 μf/km

PROBLEMS

P.2.1 A single-phase supply is effected by conductors arranged as shown in Fig. P.2.1, the current being equally divided between both conductors forming a pair. Find the inductance per kilometer of the system.

|←——25 cm——→|←——————— 125 cm ———————→|←——25 cm——→|

FIGURE P.2.1

P.2.2 Find the inductance per kilometer per conductor (line to neutral) of a three-phase system which are placed at the corners of an equilateral triangle of side 1.49 m. The diameter of the conductor is 1.24 cm.

P.2.3 A three-phase system has its conductors placed at the corners of a triangle with spacings 3.6, 5.4, and 8.1 m. The conductors are transposed at regular intervals. Calculate the inductance per kilometer of the line, if each conductor has a diameter of 2 cm.

P.2.4 A symmetrical double-circuit line has its conductors arranged as shown in Fig. P.2.4. If the radius of each conductor is 1.25 cm, determine the effective inductance in mh/km of the line.

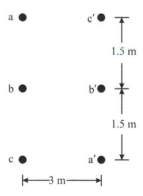

a ●   c' ● ⊤
                    ⏐
                   1.5 m
                    ⏐
b ●   b' ● ✳
                    ⏐
                   1.5 m
                    ⏐
c ●   a' ● ⊥
   |←——3 m——→|

FIGURE P.2.4

P.2.5 Calculate the capacitance per kilometer of a pair of parallel wires 5 mm is diameter spaced uniformly 25 cm apart in air.

P.2.6 A wire of 6 mm in diameter and 1 km in length is suspended at a constant height of 12 m above the sea water. Calculate the capacitance between conductor and earth.

P.2.7 A three-phase overhead line with conductors having diameter of 1.2 cm is operating at 66 kv and 50 Hz. Its conductors are arranged horizontally with a spacing of 3.2 m. Calculate the charging current, if the length of the line is 50 km.

P.2.8 Determine the capacitance per kilometer of the three-phase double-circuit line shown in Fig. P.2.8. The line is perfectly transposed. The diameter of each conductor is 2 cm. If the line operates at 200 kv, determine the charging current per kilometer.

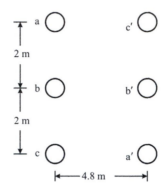

FIGURE P.2.8

QUESTIONS

Q.2.1 Derive an expression for the inductance of a single-phase line.

Q.2.2 Derive an expression for the inductance per phase of a three-phase unsymmetrical line.

Q.2.3 Explain the terms: (1) GMR and (2) GMD.

Q.2.4 What is transposition? Explain.

Q.2.5 What are bundle conductors? Why are they used?

Q.2.6 Obtain an expression for the capacitance of a single-phase line.

Q.2.7 Derive an expression for the capacitance of a three-phase line.

Q.2.8 What is the effect of earth on the capacitance of a transmission line?

Q.2.9 Explain the following: (1) Skin effect and (2) Proximity effect.

MECHANICAL DESIGN

3

The design of an overhead transmission line contains both electrical and mechanical aspects. The electrical considerations are current-carrying capacity, voltage drop, line losses, insulation coordination, tower footing resistance, etc. On the mechanical side foundation, tower structure and crossarms are important that link between the electric current—carrying conductors and the ground potential of supporting mechanical towers or poles are insulators, arranged on the crossarms. The whole mechanical structure should be properly designed to withstand all weather conditions. Furthermore the maintenance of the whole overhead transmission and supporting structure is also an important item.

The commonly used line conductor supporting structures are wooden poles, R.C.C poles, steel tubular poles, steel masts, and towers. A- and H-type poles are common. For all voltages from 66 KV and above, steel towers are preferred.

Fig. 3.1 shows some of the commonly used masts and towers. Fig. 3.1A shows single circuit-single ground wire, single circuit-double ground wire, double circuit-single ground wire, and single circuit-two conductor bundle line with double ground wire arrangements. Some of the tower structures used for EHV and UHV lines are shown in Fig. 3.1B. A HVDC guyed tower for bipolar system is shown in Fig. 3.1C.

Anchor towers with additional structural features are used wherever the transmission line changes its direction by a small angle in addition to using special insulator strings.

At the dead end of the line (starting and ending of line after or before a substation) the towers are specially fortified to withstand the unbalanced pull of the line conductors. The length of the crossarm attached to the tower or pole at the top to support the insulators is decided by the line voltage, conductor material, and configuration. The span between adjacent towers/poles, the conductor configuration, spacing between conductors and the clearances to earthed parts are all the important factors in the design of a line.

3.1 SAG AND TENSION CALCULATIONS

When a line conductor is supported between two points and allowed to hang freely, it assumes the shape of a catenary, by virtue of its own weight acting downward under gravity. This is shown in Fig. 3.2. Let us assume that the conductor is supported at both ends at the same level.

w is weight of the conductor per unit length,
T is the tension in the conductor, and
H is the horizontal component of tension at the lowest point "O."

Electrical Power Systems. DOI: http://dx.doi.org/10.1016/B978-0-08-101124-9.00003-6

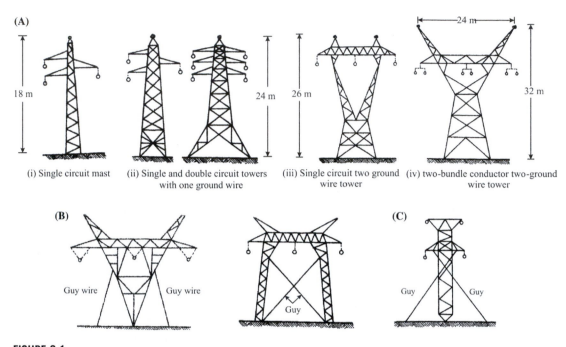

(A)

18 m

24 m   26 m

32 m

(i) Single circuit mast

(ii) Single and double circuit towers with one ground wire

(iii) Single circuit two ground wire tower

(iv) two-bundle conductor two-ground wire tower

(B)

Guy wire   Guy wire

Guy

(C)

Guy   Guy

FIGURE 3.1

Transmission towers. (A) Various single circuit, double circuit, and bundle conductor masts and towers. (B) Towers for EHV and UHV lines. (C) HVDC bipolar tower.

Let P be any point (x, y) on the conductor distant s from the lowest point "O." Let ws be the weight of the conductor of length s acting vertically downward through the center of gravity of the length OP.

T_x and T_y are the components of tension at P. Consider a small element of length ds of the conductor at P.

$$\tan \theta = \frac{T_y}{T_x}$$

For equilibrium the vertical and horizontal components of forces must balance. Hence from Figs. 3.2 and 3.3

$$\tan \theta = \frac{T_y}{T_x} = \frac{dy}{dx} = \frac{ws}{H}$$

Also,

$$\frac{ds}{dx} = \sqrt{1 + \left(\frac{dy}{dx}\right)^2} = \sqrt{1 + \frac{w^2 s^2}{H^2}} \tag{3.2}$$

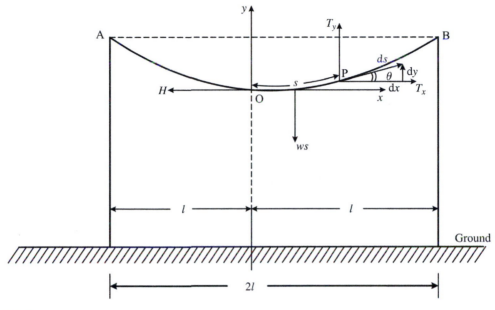

FIGURE 3.2

Sag and tension for even supports.

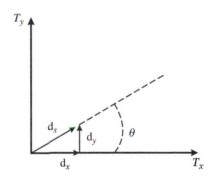

FIGURE 3.3

Tension components.

from which

$$dx = \frac{ds}{\sqrt{1 + \frac{w^2 s^2}{H^2}}}$$

(3.3)

Integrating on both sides

$$x = \frac{H}{w} \sin h^{-1} \left(\frac{ws}{H}\right) + C_1$$

At $x = 0$; $s = 0$ and therefore $C_1 = 0$

$$x = \frac{H}{w} \sin h^{-1}\left(\frac{ws}{H}\right) \tag{3.4}$$

Obtaining the value for s from Eq. (3.4)

$$s = \frac{H}{w} \sin h\left(\frac{wx}{H}\right) \tag{3.5}$$

Further at $x = 1$ (half span)

$$\text{maximum value of } s \text{ is given by } s = \frac{H}{w} \sin h\left(\frac{wl}{H}\right) \tag{3.6}$$

It is the length of half of the conductor between the supports. The total length of the conductor between supports

$$2s = \frac{2H}{w} \sin h \frac{wl}{H} \tag{3.7}$$

Again, since

$$\frac{dy}{dx} = \frac{ws}{H} = w\frac{H}{w} \sin h\frac{wx}{H}$$

$$dy = \sin h\frac{wx}{H} dx \tag{3.8}$$

Integrating both sides

$$y = \frac{H}{w} \cos h\frac{wx}{H} + c$$

At $x = 0$;

$$y = 0 = \frac{H}{w} + C_2$$

$$C_2 = -\frac{H}{w}$$

Hence,

$$y = \frac{H}{w}\left[\cos h\frac{wx}{H} - 1\right] \tag{3.9}$$

Tension at

$$P = \sqrt{T_x^2 + T_y^2} = \sqrt{H^2 + w^2 s^2}$$

$$= \sqrt{H^2 + \left[w^2\left(\frac{H}{w}\right)^2 \sin h^2 \frac{wx}{H}\right]} = \sqrt{H^2\left[1 + \sin^2\left(\frac{wx}{H}\right)\right]} \tag{3.10}$$

Therefore

$$T = H = \cos h\frac{wx}{H} \tag{3.11}$$

At $x = l$, tension for half-span length $= H \cos h \dfrac{wl}{H}$
The sag at half span

$$d = \frac{H}{w}\left[\cos h\left(\frac{wl}{H}\right) - 1\right] \tag{3.12}$$

3.2 APPROXIMATE RELATIONS FOR SAG AND TENSION

The hyperbolic functions in formulae derived in Section 3.1 can be expanded into series. Thus

$$y \;\; = \frac{H}{w}\left[1 + \frac{w^2x^2}{2!H^2} + \frac{w^4x^4}{4!H^4} + \cdots - 1\right]$$
$$\simeq \frac{H}{w}\left[\frac{w^2x^2}{2!H^2}\right] = \frac{wx^2}{2H} \tag{3.13}$$

If the sag is small $T \simeq H$

$$\text{then } y = \frac{wx^2}{2T} \tag{3.14}$$

$$\text{and } d = \frac{wl^2}{2T} \tag{3.15}$$

Similarly the length of the conductor

$$s = \frac{H}{w}\sin h\frac{wx}{H}$$

can be expanded into series

$$s = \frac{H}{w}\left[\frac{wx}{H} + \frac{w^3x^3}{3!H^3} + \cdots\right] \simeq \frac{H}{w}x\left[\frac{w}{H} + \frac{w^2x^2}{6H^2}\right]$$
$$= x\left[1 + \frac{w^2x^2}{6T^2}\right] \tag{3.16}$$

At half-span length

$$s = l + \frac{w^2l^3}{6T^2} \tag{3.17}$$

3.3 CONDUCTORS SUPPORTED AT DIFFERENT LEVELS

The conductors are supported at P_1 and P_2 as shown in Fig. 3.4. Point O is the lowest point of the conductor horizontally located at x_1 form support at P_1 and x_2 from support at P_2, where $x_1 + x_2 = 2l$ the total span. With usual notation with respect to Fig. 3.4.

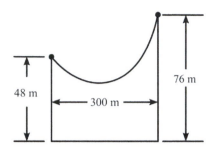

FIGURE 3.4

Sag and tension for uneven supports.

we obtain

$$d_1 = \frac{wx_1^2}{2T} \text{ and } d_2 = \frac{wx_2^2}{2T} \tag{3.18}$$

Hence

$$d_2 - d_1 = \frac{wx_2^2}{2T} - \frac{wx_1^2}{2T} = \frac{w(x_2 + x_1)(x_2 - x_1)}{2T}$$

$$\therefore h = \frac{w(x_2 - x_1) \cdot 2l}{2T} = w(x_2 - x_1)\frac{l}{T}$$

$$x_2 - x_1 = \frac{Th}{wl} \tag{3.19}$$

and

$$x_2 + x_1 = 2l \tag{3.20}$$

solving

$$x_1 = l - \frac{Th}{2wl} \tag{3.21}$$

and

$$x_2 = l + \frac{Th}{2wl} \tag{3.22}$$

The position of point O can be determined.

3.4 EFFECT OF WIND ON SAG

Wind pressure acting laterally on the conductor surface alters the sag. Let p (kgm/m^2) be the wind pressure acting horizontally on the conductor surface. The force due to wind per unit length.

$$w_w = (d \times 1) \times p \quad \text{(kg)} \tag{3.23}$$

where d is in meters and is the diameter of the conductor.

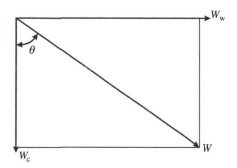

FIGURE 3.5

Effect of wind.

As is shown in Fig. 3.5, the net force due to conductor weight and wind pressure

$$W = \sqrt{w_c^2 + w_w^2} \tag{3.24}$$

where w_c is the conductor weight per unit length.

W acts at an angle θ to the vertical

$$\tan \theta = \frac{w_w}{w_c} \tag{3.25}$$

The vertical sag then is computed from

$$d = \frac{Wl^2}{2T} \cdot \cos \theta \tag{3.26}$$

3.5 EFFECT OF ICE COATING ON SAG

During winter months, in cold countries snowfall occurs and the line conductors get coated with ice and thereby the conductor weight increases (Fig. 3.6).

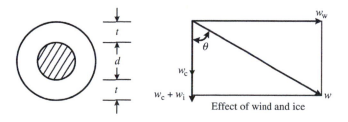

Effect of wind and ice

FIGURE 3.6

Effect of wind and ice.

It is the common practice to consider that ice deposit increases the diameter effectively by forming a coating of thickness t all around the conductor. Thus the projected surface area to horizontal wind pressure will be increased by $(d + 2t) \times 1$ m^2 where t and d, the thickness of ice and diameter of conductor are in meters. The vertical weight of the conductor is increased by:

$$w_i = \frac{\pi}{4}\left[(D^2 - d^2)\right] \times 1 \text{ m} \times 915 \text{ kg/m}^3 \qquad (3.27)$$

where the density of ice is assumed 915 kg/m^3 and

$$D = d + 2t \qquad (3.28)$$

Total vertical force $= w = w_c + w_i$

When both wind and ice effects are considered

$$\text{Vertical sag } d^1 = d \cos \theta \qquad (3.29)$$

where

$$\tan \theta = \frac{w_w}{w_c + w_i}$$

and

$$d^1 = \frac{wl^2}{2T} \cos \theta \qquad (3.30)$$

with only ice coating, but no wind pressure acting on the conductor

$$w_{c+i} = w_c + \frac{\pi}{4}\left[D^2 - d^2\right] \times 915 \text{ kg/m}^3 \qquad (3.31)$$

The

$$\text{Sag} = d_{c+i} = \frac{(w_c + w_i)}{2T} l^2 \quad \text{(m)} \qquad (3.32)$$

3.6 CONDUCTOR MATERIALS

Copper is the most suitable conductor material with high conductivity and good mechanical strength. However, since it is expensive, aluminum is used extensively as a conductor material. Aluminum is much lighter than copper (about one-third of copper), but its tensile strength is only half of that of copper and its conductivity is also about 65% of copper. Yet, since aluminum is much cheaper than copper all aluminum conductors (AAC), aluminum conductors with steel reinforcement (ACSR) and all aluminum alloy conductors (AAAC) are in wide use. ACSR have larger diameter than copper conductors for the same current-carrying capacity and are therefore less influenced by corona. Some countries used galvanized steel conductors in rural areas. In special cases and for longer spans, cadmium copper and phosphor bronze are the other materials used. AAAC contain, in addition to aluminum, magnesium silicon alloy for high conductivity. On equal diameter basis, it is claimed that AAAC are better corrosion resistant and are more stronger than ACSR. In all these, pure aluminum (i.e., electrolytically refined (99.5% pure aluminum)) is used. The minimum conductivity maintained is 61%. In case of ACSR, the steel strands may be from 6% to 40% of the total section as per the requirement of tensile strength.

3.6.1 STRANDED CONDUCTORS

Stranded conductors are more flexible, kinck resistant, break resistant, and much stronger than solid conductor. All transmission line conductors are invariably stranded. At the lowest level, a single strand of diameter d will have six strands of the same diameter and has seven strands in all. The next size will have two layers over the central strand. The outer layer contains 12 strands so that in all it has 19 strands. The overall diameter of a stranded conductor is $(2n + 1)d$ where n is the number of layers around the central strand. In case of ACSR, the central strands are of steel and the outer of aluminum. Successive layers are layered in opposite direction so as to obtain a greater cohesion and high strength for the overall conductor.

3.7 VIBRATIONS OF CONDUCTORS

It is shown that conductors are subjected to wind force and this mainly acting horizontally causes the conductors slightly swing in the horizontal plane. In addition, conductors are subject to vibrations in vertical plane. There are two types of such vibrations.

1. *Resonant Vibrations*: These are low-amplitude, high-frequency oscillations caused by light winds blowing with velocities ν ranging from 6 to 20 km/h. The amplitudes can go up to few centimeters while the frequency ranges from 5 to 50 Hz. The length of a loop or half wave length is given by $(1/2f)(\sqrt{T/w})$ where the frequency $f = v/d$, d being the conductor diameter, and ν being the wind velocity in km/h. T is the tension in the conductor and w is the weight of the conductor.

   These vibrations are generally damped by special devices such as dampers. One common form is the stock bridge damper shown in Fig. 3.7.

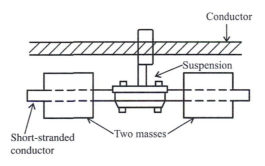

FIGURE 3.7

Stock bridge damper.

The damper consists of two masses connected at the two ends of a short-stranded conductor and suspended at the middle from the line conductor at suitable location. The energy in the vibrations will be absorbed and dissipated in friction. These vibrations are also called Aeolian vibrations.

2. *Galloping of Conductor*: These are the low-frequency, high-amplitude vibrations generally formed due to uneven coating of ice on conductors and wind pressuring the conductors from underneath. Stranding of conductors further aggravates these vibrations. When such vibrations occur, the conductors are said to be dancing. Due to the action of the wind, they may grow in amplitude becoming self-exciting. There is no way to prevent these vibrations. Outages can be limited by arranging conductors in horizontal configuration.

3.8 STRINGING CHART

While laying the transmission lines after the towers are erected and crossarms are fixed, the insulator strings are connected to the crossarm in a suitable configuration along with the conductors. This process is called stringing of conductors. The conductors are hauled up to the clamps of the suspension insulators on snatch blocks after correct sag and tension are set. The conductors are transferred from snatch blocks to insulator string clamps there after.

At higher temperatures, the conductor expands more and the stress in the conductor reduces. Thereby the tension is also lowered. At low temperatures, the sag is less, stress is more and also the tension. Further, ice and wind will increase the weight of the conductor. The sag therefore is dependent upon the temperature and other atmospheric conditions. At the time of erection, the sag to be allowed while stringing the conductors will be different from what regulations specify.

Consider the approximate formulae: At temperature $t_1°C$

$$T_1 = H_1; \quad d_1 = \frac{w_1 l^2}{2T}$$

$$y_1 = \frac{w_1 x^2}{2T_1}; \quad s_1 = l + \frac{w_1^2 l^3}{6T_1^2}$$

Let E be the Young's modulus of the conductor material and α be the coefficient of linear expansion. The stress in the conductor $f = T/a$ where a is the area of cross section of the conductor. If the temperature increases from $t_1°C$ to $t_2°C$, the conductor length is increased by

$$(t_2 - t_1)\alpha \cdot s_1 \tag{3.31}$$

In half span this will be $(t_2 - t_1)\,\alpha l$. But, increase of temperature from t_1 to t_2 reduces the stress f_1 to f_2. This causes decrease in length by

$$\frac{f_1 - f_2}{E} s_1 \simeq \frac{f_1 - f_2}{E} l \tag{3.32}$$

The new length

$$s_2 = s_1 + (t_2 - t_1)\alpha l - \frac{f_1 - f_2}{E} l \tag{3.33}$$

But

$$s_2 = l + \frac{w_2^2 l^3}{6T_2^2} = l + \frac{w_2^2 l^3}{6f_2^2 a^2} \tag{3.34}$$

Hence, from Eqs. (3.33) and (3.34)

$$l + \frac{w_2^2 l^3}{6f_2^2 a^2} = \frac{l + w_1^2 l^3}{6f_1^2 a^2} + (t_2 - t_1)\alpha l - \frac{(f_1 - f_2)}{E} \cdot l$$

$$\frac{w_2^2 l^2}{6f_2^2 a^2} = \frac{w_1^2 l^2}{6f_1^2 a^2} + (t_2 - t_1)\alpha - \frac{(f_1 - f_2)}{E} \tag{3.35}$$

$$\frac{w_1^2 l^2 E}{6f_1^2 a^2} + (t_2 - t_1)\alpha \cdot E + (f_2 - f_1) = \frac{w_2^2 l^2 E}{6f_2^2 a^2}$$

Rearranging the terms

$$f_2^2 \left[f_2 - f_1 + (t_2 - t_1)\alpha E + \frac{w_1^2 l^2 E}{6f_1^2 a^2} \right] = \frac{w_2^2 l^2 E}{6a^2} \tag{3.36}$$

The above cubic equation has to be solved to obtain f_2. Once f_2 is known, the sag can be calculated from

$$d_2 = \frac{w_2 l^2}{2T} = \frac{w_2 l^2}{2f_2 a} \tag{3.37}$$

The variation of sag and tension with temperature is called stringing chart and is shown in Fig. 3.8.
Tension $T = f \times a = $ Stress $\times$ Area of cross section
Eq. (3.36) can be rewritten, by multiplying throughout by square of area of section a^2.

$$f_2^2 a^2 \left[f_2 - f_1 + (t) \cdot \alpha E + \frac{w_1^2 l^2 E}{6f_1^2 a^2} \right] = \frac{w_2^2 l^2 E \cdot a^2}{6a^2}$$

$$T_2^2 \left[f_2 - f_1 + t \cdot \alpha E + \frac{w_1^2 l^2 E}{6T_1^2} \right] = \frac{w_2^2 l^2 E}{6} \tag{3.38}$$

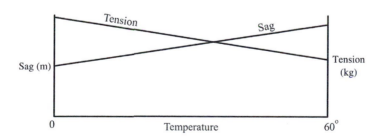

FIGURE 3.8

The stringing chart.

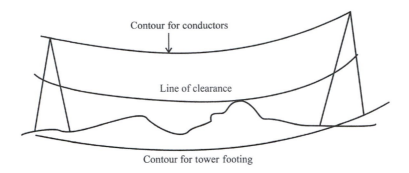

FIGURE 3.9

Sag template.

where $t_2 - t_1 = t$

Multiplying again by a

$$T_2^2\left[T_2 - T_1 + t \cdot \alpha\, E \cdot a + \frac{w_1^2 l^2 Ea}{6T_1^2}\right] - \frac{w_2^2 l^2 Ew}{6} \tag{3.39}$$

$$T_2^3 + T_2^2\left[(aE)\left(\alpha t + \frac{w_1^2 l^2}{6T_1^n}\right) - T_1\right] - \frac{w_2^2 l^2 Ea}{6} = 0 \tag{3.40}$$

Given other values, T_2 can be computed from Eq. (3.40).

3.9 SAG TEMPLATE

Sag templates are used for locating the towers correctly in the field so that the conductors follow the contour prescribed for correct tension and sag, maintaining the minimum conductor clearance (Fig. 3.9).

Such templates are prepared as drawings with all elevations marked in advance so that the stringing of the conductor does not pose any problem to the engineers while on the field.

WORKED EXAMPLES

E.3.1 An overhead transmission line has a span of 210 m. The conductor weighs 600 kg/km. Calculate the maximum sag if the ultimate strength of the conductor is 5760 kg. Assume a factor of safety of 2.

Solution: Using approximate formula

$$\text{Sag } d = \frac{wl^2}{2T} \quad w_c = \frac{600}{1000} = 0.6 \text{ kg/m}$$

$$w = w_c \qquad l = \frac{210}{2} = 105 \text{ m}$$

$$T = \frac{5760}{2}$$

$$d = \frac{0.6 \times 105^2 \times 2}{2 \times 5760} = 1.148 \text{ m}$$

E.3.2 In Example E.3.1 if wind is acting laterally at a pressure of 68 kg/m^2 and the conductor diameter is 15 mm, determine the vertical sag.

    Solution:

$$w_w = 60 \text{ kg} \times 1 \times \frac{15}{1000} = 1.02 \text{ kg/m}$$

$$w = \sqrt{w_w^2 + w_c^2} = \sqrt{1.02^2 + 0.6^2} = \sqrt{1.0404 + 0.36}$$
$$= 1.183385 \text{ kg}$$

$$d = \frac{wl^2}{2T} = \frac{1.83385 \times 105^2 \times 2}{2 \times 5760} = 2.265 \text{ m}$$

$$\tan \theta = \frac{1.02}{0.6} = 1.7$$

$$\cos \theta = 0.507$$

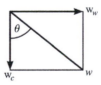

FIGURE E.3.2

$$\text{vertical sag} = d \cos \theta$$

$$\text{Vertical sag} = 2.265 \times 0.507 = 1.1485 \text{ m}$$

E.3.3 If in Example E.3.2 the conductor is covered in addition by ice of 1 cm thickness determine the vertical sag. Weight of ice is 915 kg/m^3.

Solution: Weight of ice per meter length

$$w_2 = 915 \text{ kg}/m^3 \times 1 \text{ m} \frac{\pi[(d+2t)^2 - d^2]}{4} \text{kg} = 915 \times \frac{\pi}{4}\left[\left(\frac{3.5}{100}\right)^2 - \left(\frac{1.5}{100}\right)^2\right]$$

$$= 915 \times \frac{\pi}{4} \times 0.001203$$

$$= 0.864 \text{ kg}/m$$

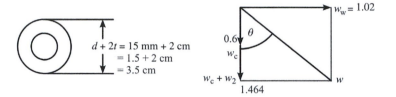

$d + 2t = 15 \text{ mm} + 2 \text{ cm}$
$\quad\quad = 1.5 + 2 \text{ cm}$
$\quad\quad = 3.5 \text{ cm}$

FIGURE E.3.3

$$w_c + w_2 = 0.6 + 0.064 = 1.464 \text{ kg}$$

$$w = \sqrt{1.464^2 + 1.02^2} = 2.14329 + 1.0404$$

$$= \sqrt{3.18369} = 1.7842 \text{ kg}$$

$$\text{Maximum sag} = \frac{wl^2}{2T} = \frac{1.7842 \times 105^2 \times 2}{2 \times 5760}$$

$$\tan\theta = \frac{1.02}{1.464} = 0.6967$$

$$\theta = 34°.88; \quad \cos\theta = 0.82$$

$$\therefore d = 3.415 \text{ m}$$

$$\text{Vertical sag} = d\cos\theta$$
$$= 3.415 \times 0.82 = 2.8 \text{ m}$$

E.3.4 An overhead line constructed across a river is supported on either side of the river by support towers of height 48 and 76 m above the water level. Find the clearance between the conductor and water at a point midway between the towers. The maximum permissible tension in the conductor is 5.800 kg. The weight of the conductor is 0.89 kg/m. The distance between the towers is 300 m. Take factor of safety 2.

Solution: $h = 76 - 48 = 28$ m

$$T = \frac{5800}{2} = 2900 \text{ kg}$$

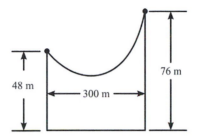

FIGURE E.3.4 (A)

$$x_1 = l - \frac{Th}{2wl} = 150 - \frac{2900 \times 28}{2 \times 0.89 \times 150} = 150 - 304.12 = -154.2$$

$$x_2 = l + \frac{Th}{2wl} = 150 + \frac{2900 \times 28}{2 \times 0.89 \times 150} = 150 + 304.12 = 454.12$$

Solving we find that both P_1 and P_2 are on the same side of O
Height of P_1 above O

$$= \frac{wx_1^2}{2T} = \frac{0.89 \times 154.2^2}{2 \times 2900} = 3.648 \text{ m}$$

The midpoint C is at a distance of $150 + 154.2 = 304.2$ m to left of O
The sag at $C = (0.89 \times 304.2^2)/(2 \times 2900) = 14.2$ m. See Fig. E.3.4B.

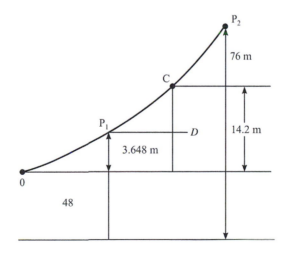

FIGURE E.3.4 (B)

$$C\,D = 14.2 - 3.648 = 10.552 \text{ m}$$

$$\text{Height of C above water level} = 48 + 10.552$$

$$= 58.552 \text{ m}$$

E.3.5 A transmission line conductor weighing 0.85 kg/m is subjected to a wind pressure of 60 kg/m^2 of projected area. The conductor has ice coating of 1 cm thick. If the maximum permissible sag is 7.5 m, determine the permissible span between two level supports. Take a factor of safety 2 and given that the density of ice is 915 kg/m^2. Ultimate strength = 8000 kg and conductor diameter 2 cm.

Solution: $w_c = 0.85$ kg/m

$$\text{Weight of ice coating per meter} = 915 \times \frac{\pi}{4}\left[\left(\frac{2+2}{100}\right)^2 - \left(\frac{2}{100}\right)^2\right] \times 1$$

$$= 718.275 \times 0.0012 = 0.86193 \text{ kg/m}$$

the wind force per meter length assumed acting horizontally

$$w_w = 60 \text{ kg/m}^2\left[1 \text{ m} \times \frac{4}{100} \text{ m}\right] = 2.4$$

$$w = \sqrt{(w_c + w_i)^2 + w_w^2}$$

$$= \sqrt{1.71193^2 + 2.4^2}$$

$$= \sqrt{2.93 + 5.7} = 8.69$$

$$= 2.948 \text{ kg/m}$$

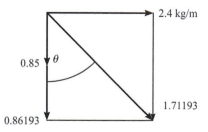

FIGURE E.3.5

The maximum sag $d = wl^2/2T$

$$7.5 = \frac{2.948 \times l^2}{2 \times (8000)/2}; \quad l^2 = \frac{7.5 \times 8000}{2.948} = 20,352.78$$

$$\text{Half span } l = \sqrt{20,352.78} = 142.663 \text{ m}$$

$$\text{Span } 2l = 285.3 \text{ m}$$

E.3.6 Find the sag and tension at erection for a transmission line at a temperature of 45°C in still air. The conductor diameter is 2 cm, and area of cross section is 314 mm². The conductor weighs 0.85 kg/m. The maximum tension is 3800 kg. The coefficient of linear expansion is 18.4×10^{-6}/°C. Modulus of elasticity $E = 9.3 \times 10^3$ kg/mm². The expected wind pressure is 60 kg/m², and ice coating could reach 1 cm thickness.

Solution: $w_c = 0.85$ kg/m

From E.3.5

$$w = \sqrt{(w_c + w_i)^2 + w_w^2}$$
$$= 2.94812 \text{ kg/m}$$

Now $w_1 = 2.948$

$$w_2 = 0.85$$

$t = 45°C - (-5°C) = 50°C$ (assuming that ice forms at $-5°C$)

$$a = 314 \text{ mm}^2 \left(\frac{\pi}{4} \times 2^2 \text{ cm} \times 100 \right)$$

$$T_2^3 + T_2^2 \left[aE \left(\alpha t + \frac{w_1^2 l^2}{6 T_1^2} \right) - T_1 \right] - \frac{w_2^2 l^2 aE}{6} = 0$$

$$T_2^3 + T_2^2 \left[314 \times 9.3 \times 10^3 \left(18.4 \times 10^{-6} \times 50 + \frac{2.948^2 \times 145^2}{6 \times 3800^2} \right) - 3800 \right]$$
$$- \frac{0.85^2 \times 145^2 \times 9.3 \times 10^3 \times 314}{6} = 0$$

$$T_2^3 + T_2^2 [(2686.584 + 6132.42) - 3800] - 7.39 \times 10^9 = 0$$

$$T_2^3 + T_2^2 [5019] - 7.39 \times 10^9 = 0$$

$$T_2^2 [T_2 + 5019] = 7.39 \times 10^9$$

$$T_2 \simeq 1100 \text{ kg}$$

$$\text{Sag at erection} = \frac{wl^2}{2T} = \frac{0.85 \times 145^2}{2 \times 1100} = 8.123 \text{ m}$$

PROBLEMS

P.3.1 An overhead line has a span of 118 m. The diameter of the conductor is 1.12 cm. The conductor is coated with ice of 0.95 cm thick. The wind pressure is 380 N/m² of projected area. The weight of the conductor is 6 N/m. Given that ice weighs 8950 N/m³ and the permissible tension is 3.6×10^4 N. Calculate the sag at midspan.

[*Note:* 1 kg force = 9.81 N]

P.3.2 An overhead line at a river crossing is supported from two towers, 22 and 16 m above the water level. The distance between the towers is 252 m. The conductor diameter is 1.038 cm.

Find the minimum clearance between conductor and water surface. Also, determine the distance of the lowest point of conductor from 16-m tower. Weight of the conductor is 595 kg/km, and its breaking load is 2725 kg. Take a factor of safety of 2.5.

P.3.3 An overhead line has a conductor diameter of 1 cm. It is suspended with a span of 150 m. When there is an ice coating of 10 mm thick at $-5°C$, and with a wind pressure of 42 kg/m^2 of projected surface area, the sag is observed to be 3.6 m. Given $E = 126 \times 10^4$ kg/cm^2, $\alpha = 16.6 \times 10^{-6}$/C, ice density 912 kg/m^3, and copper density 8900 kg/m^3. Determine the temperature at which the sag will remain the same under fair weather conditions.

P.3.4 An overhead line has a span of 215 m. The weight of the conductor is 600 kg/km. The ultimate tensile strength of the conductor is 5800 kg. Assuming a factor of safety of 2, determine the maximum sag.

QUESTIONS

Q.3.1 Why it is necessary to provide sag while stringing the transmission line conductors?

Q.3.2 Derive expressions for sag and tension using exact formulation.

Q.3.3 Derive approximate expressions for sag and tension.

Q.3.4 What is the effect of wind and ice on sag? Explain with diagrams.

Q.3.5 Derive expression for sag when wind pressure is considerable.

Q.3.6 Discuss the effect of both wind and ice on sag calculations.

Q.3.7 Explain the sag calculation when the conductors are supported at uneven levels.

Q.3.8 What are the various conductor materials use in overhead transmission lines?

Q.3.9 Explain the various factors that influence the conductor sag.

Q.3.10 What is a sag template? Explain.

Q.3.11 Discuss the importance of stringing chart.

INSULATORS

Insulators of various types are used to support overhead conductors at poles or towers. These insulators are generally made of porcelain. Porcelain is a ceramic material. It is made from a wet mixture of kaolin with other materials such as quartz (silica), feldspar, steatite, and bone ash. The mixture is molded to the required shape and heated in a kiln to a temperature of 1200°C−1400°C and then glazed. Glazing porcelain gives moisture and dust-free surface. It has good dielectric strength of about 40−280 V/mil and a relative permittivity of 5.1−5.9. Glass is another material used sometimes for voltages below 25 kV. The advantages of glass are that it is cheaper and flaws in molding, if any, can be detected easily. Some synthetic resins made of silicon compounds are also used sometimes as insulator materials.

Where greater thickness is needed, porcelain insulators are manufactured in two or more pieces of the required shape and cemented together to form a single monolithic piece. Porcelain is mechanically stronger than glass, and less affected by temperature changes. Furthermore the leakage resistance can be increased by suitable design of the pieces. Porcelain is the most widely used insulator material.

4.1 TYPES OF INSULATORS

Insulators are classified into three categories:

1. Pin type
2. Suspension type
3. Strain type

4.2 PIN TYPE INSULATORS

Pin type insulators of single piece type are used for voltage up to 25 kV. For higher voltages up to 66 kV and even beyond, two, three, and four piece constructions are used. However, as the number of pieces increase, its weight also increases and the bending moment on the pin will become more. The pin type insulator is fixed to the crossarm on the pole with a bolt. Since this type of insulator is fixed rigidly on to the crossarm, the mechanical stress must be evenly balanced. Fig. 4.1 shows two-piece and three-piece construction of pin insulators. Pin insulators perform well under pollution conditions. They provide natural cleaning by wind and rain. They are mechanically strong and possess good flash over characteristics.

Electrical Power Systems. DOI: http://dx.doi.org/10.1016/B978-0-08-101124-9.00004-8

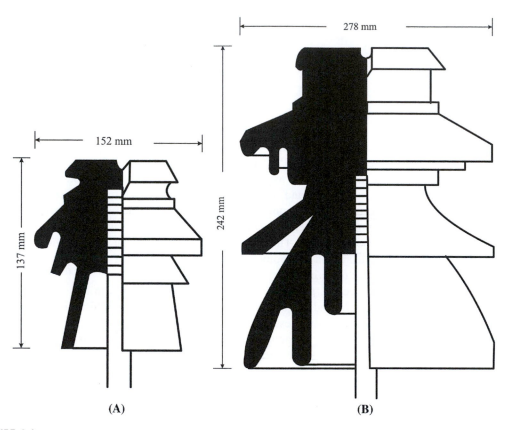

FIGURE 4.1

Pin insulators. (A) 11 kV and (B) 33 kV.

4.3 SUSPENSION TYPE OR DISC INSULATORS

There are two types of construction in disc insulators

(1) Cemented cap type and (2) Hewlett or interlinking type

With a pin insulator the conductor is connected to the insulator and is above the crossarm. In case of suspension insulators, the conductor is below the crossarm connected and suspended from the disc insulator. As the conductor is suspended freely, the stress on the insulator body is minimal. These insulators are made available as discs of about 26 cm diameter. Several discs are connected in series to form a string. Line voltage gets distributed over the number of discs connected in series, so that based on the design, suitable number of discs can be selected to form a string.

In the cemented cap type there is a metal cap at the top and a metal pin underneath. To form a string, the cap is so recessed that it can take the pin of another unit. Such a disc is shown in Fig. 4.2. The upper surface of all types of insulators are so shaped that water will drop down from the surface easily.

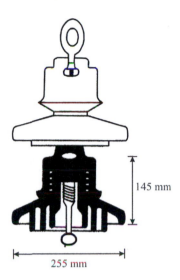

FIGURE 4.2

Disc insulator in section.

In the Hewlett type of design each disc has two curved tunnels lying in planes at right angles to each other. Steel U-shaped links covered with lead are threaded into these tunnels. They are fastened to similar links to other discs in the string. The Hewlett type insulator is more reliable than the cemented cap type, but the porcelain in this case is subjected to higher electrostatic stress and hence liable to puncture more than the cemented cap type.

4.4 STRAIN INSULATORS

Wherever transmission lines take a turning or at dead ends where the lines start-up or end, the pull of the conductors on the string becomes uneven. The tension becomes more for large spans encountered, such as at river crossings. For such cases special mechanically strong insulators are used. They are called strain insulators. For even distribution of tension, two or more strings of insulators may be used in parallel. This is shown in Fig. 4.3.

4.5 VOLTAGE DISTRIBUTION IN STRING INSULATORS

If several disc insulators are connected in series to withstand a higher voltage, ideally they should share the total potential from line to ground equally. But, in practice, since the tower is in the vicinity of the insulator and because the tower is at earth potential, each metal joint of the string has capacitance to earthed tower. These capacitances alter the potential distribution across the string

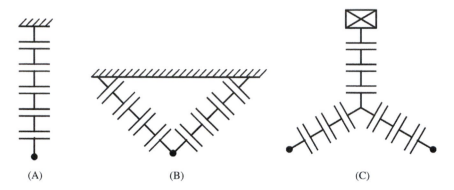

FIGURE 4.3

String insulator and arrangements.

from uniform to nonuniform. The potential across the disc nearest to the line conductor will be the maximum while the potential across the unit nearest to the crossarm end will be the least.

Let the self-capacitance of each unit be mc and the capacitance of each metal link to earth (tower) be C.

$$V = \text{Total line voltage to ground} = V_1 + V_2 + V_3 + V_4 \tag{4.1}$$

This is shown in Fig. 4.4

At junction 1

$$
\begin{aligned}
I_2 \quad &= I_1 + i_1 \\
&= V_1 mcw + V_1 cw \\
&= V_1[1 + m]cw
\end{aligned}
$$

$$V_2 mcw = V_1 cw[1 + m]$$

$$V_2 = V_1 \frac{cw(1 + m)}{mcw} = V_1 \frac{(1 + m)}{m} = V_1 \left[1 + \frac{1}{m} \right] \tag{4.2}$$

At junction 2

$$I_3 = I_2 + i_2 = (I_1 + i_1) + i_2 = V_1 mcw + V_1 cw + (V_1 + V_2)cw$$

$$V_3 mcw = V_1(1 + m)cw + V_1 cw + V_i \frac{(1 + m)}{m} cw$$

$$V_3 \quad = V_1 \frac{(1 + m)}{m} + \frac{V_1}{m} + V_1 \frac{(1 + m)}{m^2}$$

$$= V_1 \frac{[m^2 + 3m + 1]}{m^2} = V_1 \left[1 + \frac{3}{m} + \frac{1}{m^2} \right] \tag{4.3}$$

Similarly at junction 4

$$I_4 = I_3 + i_3 = V_3 mwc + (V_1 + V_2 + V_3)wc$$

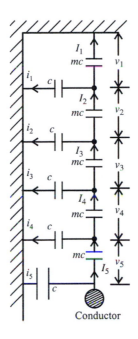

FIGURE 4.4

Voltage distribution.

$$V_4 mwc = V_3 mwc + V_1 wc + V_2 wc + V_3 wc$$
$$= V_1 wc + V_1 \frac{(m+1)}{m} wc + V_1 \frac{(m^2+3m+1)}{m^2} wc + V_3 mwc$$

$$= V_1 \left[1 + \frac{(m+1)}{m} + \frac{(m^2+3m+1)}{m^2} + \frac{(m^2+3m+1)m}{m^2} \right] wc$$

$$V_4 = \frac{1}{m} \left[\frac{m^2+m^2+m+m^2+3m+1+m^3+3m^2+m}{m^2} \right]$$

$$V_4 = V_1 \left[\frac{(m^3+m^2+5m+1)}{m^3} \right] = V_1 \left[1 + \frac{6}{m} + \frac{5}{m^2} + \frac{1}{m^3} \right] \qquad (4.4)$$

Proceeding to junction 5,

$$I_5 = V_5 mwc = I_4 + i_4 = V_4 mwc + (V_1 + V_2 + V_3 + V_4)wc$$

$$V_5 mwc = \left[V_1 + V_1 \frac{(m+1)}{m} + V_1 \frac{(m^2+3m+1)}{m^2} + \frac{V_1(m^3+6m^2+5m+1)}{m^3} \right] wc$$

$$+ V_1 \left[\frac{(m^3+6m^2+5m+1)}{m^3} \right] mwc$$

$$V_5 \quad = \frac{1}{m} V_1 \left[\frac{m^4 + 6m^3 + 15m^2 + 7m + 1}{m^3} \right]$$

$$= V_1 \left[\frac{m^4 + 6m^3 + 15m^2 + 7m + 1}{m^4} \right] \tag{4.5}$$

$$= V_1 \left[1 + \frac{6}{m} + \frac{15}{m^2} + \frac{7}{m^3} + \frac{1}{m^4} \right]$$

In this manner the potential distribution can be computed for any number of discs.

4.6 STRING EFFICIENCY

It is seen that the presence of crossarm and tower which are at earth potential in the proximity of the line conductor introduced additional capacitances providing leakage paths and this has altered the uniform potential distribution across the units into nonuniform distribution. This results in the unit nearest to the line carrying the maximum potential. The efficacy of the units is progressively reduced as the top is reached. The ratio

$$\frac{\text{Flash over voltage of a string of } n \text{ units}}{n \times \text{Flash over voltage of unit nearest to line}} \text{ is called string efficiency}$$

Various methods are there to improve the nonuniform distribution.

4.7 METHODS FOR IMPROVING STRING EFFICIENCY

From the equations derived for potential distribution across the various discs of a string, it can be concluded that potential across the unit near the line end is maximum. To make potential distribution uniform, several methods exist. They are briefly discussed in the following.

4.7.1 SELECTION OF *M*

The ratio of the self-capacitance to capacitance to earth of links m can be properly chosen. A high value of m will have the effect of equalizing the potential distribution across the units. However, a large m means a longer crossarm and large tower which proves uneconomical. In general, a value of m equal to about 10 is considered as reasonable.

4.7.2 GRADING OF UNITS

The voltage across any disc is related to capacitance C by

$$I = VwC \tag{4.6}$$

To maintain constant voltage, since a part of the current is leaking to earth, in proportion to this the value of the capacitance can be changed by selecting different types of insulator units. Since I is maximum at the line end, the capacitance of unit near the line conductor should be selected high, and progressively as the top (crossarm) is reached, the capacitances should be reduced. This enables to make the potential more uniform. However, this requires storing large quantities of insulator discs of different sizes and is not practical.

The principle of grading of insulators can be explained as follows:

Let C be the capacitance of the metal links to earth. Let mc be the capacitance of the top most unit of the string insulator nearest to the crossarm. The voltage is uniformly distributed across each disc as v.

At junction 1, with the current distribution as assumed in Fig. 4.5

$$I_2 = I_1 + I_a$$

Let $X, Y, Z, U, \ldots$ be the capacitances of the discs graded to have uniform voltage.

$$vX = vmc + vc = v[c + mc]$$

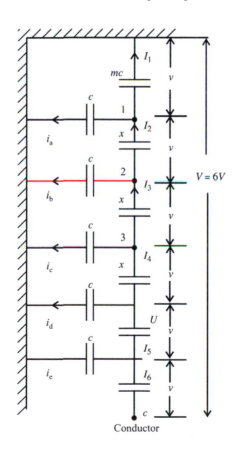

FIGURE 4.5

Insulator grading.

$$X = 1c + mc \tag{4.7}$$

At junction 2

$$I_3 = I_b + I_2 = I_b + I_a + I_1$$

$$vY = 2cv + vc + mcY$$

$$Y = (1 + 2)c + mc \tag{4.8}$$

Similarly, it can be obtained that

$$I_4 = i_c + I_3 = i_c + i_b + I_2 = i_c + i_b + i_a + I_1$$

$$vZ = 3vC + 2vcCc + vc + mcv$$

$$Z = (3 + 2 + 1)c + mc \tag{4.9}$$

and

$$U = (4 + 3 + 2 + 1)c + mc \tag{4.10}$$

Capacitance of the p-th link from the bottom is given by

$$c_p = \left[m + \frac{p(p-1)}{2} \right] c \tag{4.11}$$

where m is the ratio of mutual capacitance of the top (nearest to tower) unit to the capacitance of each link to earth.

4.7.3 STATIC SHIELDING OR GRADING RING

A large metal ring surrounding the bottom unit is connected to the line. This ring is also called guard ring. This metal piece introduces capacitances between different insulator links and the line. The effective capacitance of the bottom unit is raised. If, in addition, an arcing horn is used at the top of the string, then, in case of an over voltage, the arcing horn and the guard ring constitute a flash over path through air for the surge taking it away from the string, thus preventing damage to the insulator string.

The basic principle involved in static shielding can be explained as follows.

Consider Fig. 4.6, where C is the capacitance to earth of each metal link, $m \, c$ mutual capacitance between discs. The guard ring connected to the line is assumed to have capacitance to the links A, B, C, D, E, F .... These capacitances are so selected that the leakage currents I_a, I_b, I_c, ... are neutralized by currents flowing from shield to links I_A, I_B, I_C, I_D, etc. This is shown in Fig. 4.6.

At junction 1

$$I_a + I_1 = I_2 + i_A$$

For uniform potential distribution across the discs

$$I_1 = I_2$$

Therefore

$$I_a = I_A$$

$$EC = (n - 1)EC_A$$

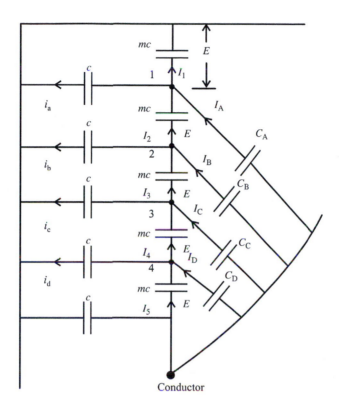

FIGURE 4.6

Static shielding.

where n is the number of discs in the string.

Then

$$C_A = \frac{C}{n-1}$$ (4.12)

At junction 2, in a similar way

$$I_b = i_B \text{ since } I_2 = I_3$$

$$C \cdot 2E = (n-2)E \cdot C_B$$

$$C_B = \frac{2C}{n-2}$$ (4.13)

Likewise,

$$C_C = \frac{3C}{n-3}$$ (4.14)

Capacitance of guard ring to p-th link

$$C_p = \frac{pc}{n-p} \tag{4.15}$$

In this way, by selecting the static shield connected to the line conductor, neutralization of the capacitance currents to earth from links can be achieved. In practice, a grading ring is used but no care is taken to achieve exact neutralization of the earth leakage currents.

4.8 TESTING OF INSULATORS

Insulators are tested as per I.S. 731 (1971) so that they withstand both electrically and mechanically field conditions or equivalent standards adopted.
Electrical Tests: They include:

1. Power frequency dry flashover test
2. Power frequency wet flashover test
3. Impulse voltage flashover test.

These tests are intended to examine the capability of the insulator to withstand both normal and storm weather conditions.

On few pieces of every batch of manufactured units break down test till puncture are carried and under specified conditions.

Mechanical Tests: Since the insulators support the line conductors which are both heavy and are subjected to large tension, it is mandatory to subject insulators for:

1. tensile strength test,
2. compression strength test,
3. torsional strength test,
4. vibration test, and
5. bending test (for pin insulators only).

Porosity test is also performed by injecting a dye under pressure to examine the penetration into the insulator body.

Pollution tests and other environmental tests are also needed depending upon the climatic and other conditions of the areas where insulators are used.

WORKED EXAMPLES

E.4.1 Calculate the maximum voltage that a string of three disc insulators can withstand, if the maximum voltage per disc unit cannot exceed 17.4 kV. Given the ratio of mutual capacitance to earth capacitance is 8.

Solution: $v_3 = 17.4$ kv

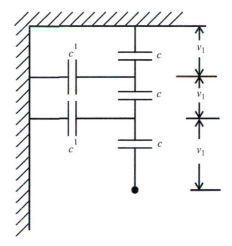

FIGURE E.4.1

$$c = 8c^1; \quad m = 8$$

$$v_2 = \frac{(m+1)}{m} v_1 \text{ and}$$

$$v_3 = v_1 = \frac{(m^2 + 3m + 1)}{m^2}$$

Therefore,

$$\frac{17.4 \times 64}{64 + 24 + 1} = v_1 = 12.512 \text{ kV}$$

$$v_2 = 12.512 \frac{(8+1)}{8} = 14.07 \text{ kV}$$

$$\text{Total voltage} \quad = 17.4 + 14.07 + 12.512$$
$$= 43.988 = 44 \text{ kV}$$

E.4.2 Each conductor of a three-phase 33-kV system is suspended by a string of three similar insulators, the mutual capacitance of which across units is 9 times the shunt capacitance between unit and earthed frame work. Calculate the voltage across each insulator. Also, calculate the string efficiency.

Solution: $m = 9; v = \frac{33}{\sqrt{3}} = 19$ kV

$$v_1 + v_2 + v_3 = 19 \text{ kV}$$

$$v_1 = v_1$$

$$v_2 = v_1 \left(\frac{m+1}{m} \right) \text{ and } v_3 = v_1 \left(\frac{m^2 + 3m + 1}{m^2} \right)$$

Hence,

$$19 = v_1 \left[1 + \left(\frac{9+1}{9} \right) + \left(\frac{81+27+1}{81} \right) \right]$$

$$= v_1 \left[1 + \frac{10}{9} + \frac{109}{81} \right] = 3.456 v_1$$

$$v_1 = 5.497 \text{ kv}$$

$$v_2 = 5.497 \left(\frac{10}{9} \right) = 6.108 \text{ kV}$$

$$v_3 = 19 - 6.108 - 5.497 = 7.395 \text{ kV}$$

$$\text{String efficiency} = \left(\frac{\text{Flashover voltage of string of } n \text{ units}}{n \times \text{Flashover voltage of one unit}} \right) \times 100$$

$$= \frac{19}{3 \times 7.395} \times 100 = \frac{19}{22.185} \times 100\% = 85.64\%$$

E.4.3 A string of six suspension insulators is to be fitted with a grading ring. If the pin to earth capacitances are equal to c, determine the line-to-pin capacitances that would give a uniform distribution of potential over the string.

Solution:

$$c_p = \frac{pc}{n-p}; \quad n = 6$$

$$c_1 = \frac{c}{6-1} = \frac{c}{5}$$

$$c_2 = \frac{2c}{6-2} = \frac{2c}{4} = \frac{1}{2}c$$

$$c_3 = \frac{3c}{6-3} = \frac{3c}{3} = c$$

$$c_4 = \frac{4c}{6-4} = \frac{4c}{2} = 2c$$

$$c_5 = \frac{4c}{6-5} = \frac{5c}{1} = 5c$$

E.4.4 A string of eight suspension insulators is to be graded to obtain uniform potential distribution across each unit. Given that the pin-to-earth capacitance c is the same for all links and the mutual capacitance of the top insulator disc is $9c$. Find the mutual capacitance of each unit in terms of c.

Solution: m = 9; *n* = 8; *p* = 7

$$c_p = \left[m + \frac{p(p-1)}{2} \right] c$$

$$c_7 = c\left[9 + \frac{7(7-1)}{2} \right] = c[9+21] = 30c$$

$$c_6 = c\left[9 + \frac{6(6-1)}{2} \right] = c[9+15] = 24c$$

$$c_5 = c\left[9 + \frac{5(5-1)}{2} \right] = c[9+10] = 19c$$

$$c_4 = c\left[9 + \frac{4(4-1)}{2} \right] = c[9+6] = 15c$$

$$c_3 = c\left[9 + \frac{3(3-1)}{2} \right] = c[9+3] = 12c$$

$$c_2 = c\left[9 + \frac{2(2-1)}{2} \right] = 10c$$

$$c_1 = c\left[9 + \frac{1(1-1)}{2} \right] = 9c$$

E.4.5 A three-disc suspension insulator string is used for a three-phase, 50-Hz system for a length of 75 km. Each disc has a self-capacitance c (F). The shunt capacitance of metal work of each insulator to earth is $0.25c$ and $0.16c$ to line. If a guard ring is fitted to increase the shunt capacitance to line of metal work for the bottom most unit to $0.36c$, determine the string efficiency.

Solution: At A

$$wcv_2 = wcv_1 + 0.25wcv_1 - 0.16wc(v_2 + v_3)$$

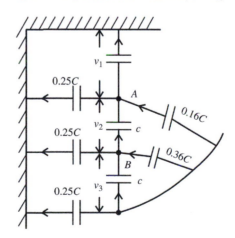

FIGURE E.4.5

$$v_2 = 1.25v_1 - 0.16v_2 - 0.16v_3$$

$$1.16v_2 = 1.25v_1 - 0.16v_3 \tag{i}$$

At B

$$wcv_3 = wcv_2 + 0.25wc(v_1 + v_2) - 0.36wcv_3$$

$$v_3 + 0.36v_3 = 1.25v_2 + 0.25v_1 \tag{ii}$$

From (i) substituting (ii) for v_3

$$1.16v_2 \quad = 1.25v_1 - \frac{0.16}{1.36}[1.25v_2 + 0.25v_1]$$

$$= 1.25v_1 - 0.147v_2 - 0.0294v_1$$

$$1.307v_2 = 1.2206v_1; \quad v_2 = 0.9339v_1$$

Again from (ii) $v_3 \quad = \dfrac{1.25v_2 + 0.25v_1}{1.36} = 0.919v_2 + 0.1838v_1$

$$= (0.919 \times 0.9339)v_1 + 0.1838v_1$$

$$= 0.857v_1 + 0.1838v_3 = 1.04v_1$$

String efficiency $\quad = \dfrac{v_1 + 0.9339v_1 + 1.04v_1}{3 \times 1.04v_1} = \dfrac{2.973}{3.12}$

$$= 95.28\%$$

PROBLEMS

P.4.1 A string of five insulator units has mutual capacitance 9 times the capacitance to earth. Determine the voltage across each unit as a ratio of the operating voltage. What is the string efficiency?

P.4.2 A suspension insulator has three discs. The capacitance of each metal part to ground is 11% of the capacitance of each unit. The voltage across each unit or disc should not exceed 11 kV. Find the operating voltage for the string.

P.4.3 A three-unit suspension insulator string is used to support the conductors for a three-phase system. If the voltage across the line unit is not to exceed 18 kv, calculate the line-to-line voltage. Given that the shunt capacitance between each unit to earth is one-ninth of the capacitance of the insulator unit. Find also the string efficiency.

QUESTIONS

Q.4.1 Explain why the voltage is not uniformly distributed across a string of disc insulators.

Q.4.2 A three-unit string insulator has mutual capacitances mc and shunt capacitance to earth of c. Determine the potential distribution across each unit.

Q.4.3 Explain how the string efficiency can be improved in practice.

Q.4.4 What do you understand by grading of insulators? Explain.

Q.4.5 Explain static shielding method of improving string efficiency.

Q.4.6 What are the various tests performed on insulators? Explain the significance of each test.

CORONA AND INTERFERENCE

5

Consider a conductor system to which voltage is applied. If the potential applied is raised gradually, beyond a limit a pale violet glow will appear. This will be accompanied by a hissing noise indicating the ionization of the surrounding air of the conductor surface with characteristic smell of ozone. This phenomenon is called corona. The high electric field at the conductor surface provides enough energy to the molecules of surrounding air that causes breakdown of the air.

In AC transmission ionized particles get attracted and repelled to conductor once during each half-cycle. Hence a charge exists near the conductor surface. In case of DC transmission the negative and positive charges move respectively to the conductor of opposite polarity. Hence the movement of charges between conductors is dependent on the electric field distribution between them. But, in case of AC it is dependent on potential gradient.

5.1 DISRUPTIVE CRITICAL VOLTAGE

The minimum potential required to start ionization of air at the conductor surface is called disruptive critical voltage, expressed in kV (rms or peak) to neutral. It is denoted by V_0. The potential gradient g_0 at which a dielectric (in this case, air) breaks down completely is called its dielectric strength expressed in kV (rms or peak per cm).

Corona occurs at AC or DC potentials. However, there will be slight difference in AC and DC corona and also for positive and negative conductors.

Corona depends on atmospheric conditions as moisture content in air influences it. For air at normal temperature and pressure of 25°C and 76 cm, Hg (NTP), the value of $g_0 = 30$ kVP/cm (21.216 kV (rms)). The relative air density factor at any other temperature and pressure is given by

$$\delta_0 = \frac{b}{76}\left(\frac{273 + 25}{273 + t}\right) = \frac{3.92\, b}{273 + t}$$

where b is the pressure of mercury and t is the temperature in °C.

5.2 VISUAL CRITICAL VOLTAGE (V_V)

Corona can be observed during twilight or in the night with naked eye. For corona to be visual a slightly higher voltage than v_0 is needed. Both time and energy are needed for visual corona after disruption occurs at the conductor surface. This higher voltage is called "visual critical voltage."

Electrical Power Systems. DOI: http://dx.doi.org/10.1016/B978-0-08-101124-9.00005-X

In case of AC, the glow is more or less uniform on both the conductors. However, for DC lines, the glow is uniform on the positive conductor but for the negative conductor it occurs as beads along the conductor.

5.3 POTENTIAL GRADIENT

Consider the two-conductor system shown in Fig. 5.1. The potential gradient at any point P distant x from conductor A and distant $(D-x)$ from conductor B.

$$g = \frac{1}{2\pi\epsilon}\left[\frac{q}{x} + \frac{q}{D-x}\right] \tag{5.1}$$

where q is the charge on the conductor per unit length.

The voltage applied to the conductors

$$v = \frac{q}{2\pi\epsilon}\left[\int_r^D \frac{dx}{x} + \int_o^{D-r} \frac{dx}{D-x}\right] = \frac{q}{\pi\epsilon}\ln\frac{D}{r} \tag{5.2}$$

Hence

$$q = \frac{v\pi\epsilon}{\log_e\left(\dfrac{D}{r}\right)} \tag{5.3}$$

From Eq. (5.1)

$$g_{max} = \frac{1}{2\pi\epsilon}\left[\frac{q}{r} + \frac{q}{D-r}\right]$$

Since, $D \gg r$

$$g_{max} \simeq \frac{q}{2\pi\epsilon r} \tag{5.4}$$

If the voltage to neutral E is considered, then $V = 2E$ and

$$g_{max} = \frac{q}{2\pi\epsilon r} = \frac{2E\pi\epsilon}{\log_e\dfrac{D}{r}}\cdot\frac{1}{2\pi\epsilon r} = \frac{E}{\log_e\dfrac{D}{r}} \tag{5.5}$$

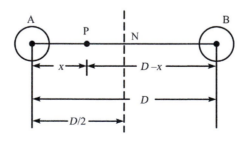

FIGURE 5.1

Potential gradient.

Even for a three-phase line, the same relationship holds true where V is the phase voltage. Hence

$$V_0 = g_{max}\, r\, \ln\frac{D}{r} \quad \text{V (to neutral)} \tag{5.6}$$

If the temperature is t ($^\circ$C) and pressure is b (cm of Hg), utilizing the relative air density factor δ

$$V_0 = \delta\, g_{max}\, r\, \ln\frac{D}{r} \quad \text{V} \tag{5.7}$$

Since, g_{max} at NTP is 30 kV (P)/cm or 21.21 kV (rms/cm)

$$V_0 = 21.21\, \delta\, r\, \ln\frac{D}{r} \quad \text{kV (rms)} \tag{5.8}$$

If the surface of the conductor is rough or the conductor is stranded, the disruptive critical voltage is lowered. This is accounted by using a surface factor m_0. The usual values for m_0 are as follows:

Smooth polished conductors: $m_0 = 1.0$
Rough or weathered conductors: $m_0 = 0.98{-}0.93$
Stranded conductors: $0.86{-}0.83$
Then, the expression for disruptive critical voltage Eq. (5.8) becomes

$$V_0 = r\, g_{max}\, \delta\, m_o\, \ln\frac{D}{r} \quad \text{kV}$$

The charged ions of air must receive some energy so that photons are released and corona can become visible. This occurs when the gradient at the conductor surface reaches a value $g_v > g_0$. According to Peek, the distance to which corona must extend into space form the conductor surface is given by

$$\left(r + 0.30\sqrt{r}\right)$$

At this gradient corona becomes visible.
The energy distance is then $r\left[1 + \frac{0.3}{\sqrt{r}}\right]$
Hence the expression for g_v at a relative air density factor of δ becomes

$$g_v = g_{max}\left[1 + \frac{0.3}{\sqrt{\delta r}}\right] \tag{5.9}$$

where r is in cm. The expression for visual critical voltage becomes

$$V_v = r\delta g_v \ln\frac{D}{r}$$

$$= r\delta g_{max}\left[1 + \frac{0.3}{\sqrt{\delta r}}\right]\ln\frac{D}{r} \quad \text{kV} \tag{5.10}$$

$$= 21.21 r\delta\left[1 + \frac{0.3}{\sqrt{\delta r}}\right]\ln\frac{D}{r} \quad \text{kV(RMS)}$$

5.4 POWER LOSS DUE TO CORONA

During ionization of air, when corona occurs energy is dissipated. It is spent as heat and light. Peek has conducted several experiments and suggested an empirical formula to determine corona loss.

Under fair weather conditions, corona loss is estimated from the formula:

$$P_c = \frac{244(f + 25)}{\delta}\sqrt{\frac{r}{D}}(V - V_0)^2 \times 10^{-5} \quad \text{(kW/phase/km)} \tag{5.11}$$

where f is the frequency of the line in Hz and V is the voltage in kV to neutral (rms). Further, it was observed that under bad weather conditions, such as rain, fog, snow, or sleet, the disruptive critical voltage will be brought down. Therefore the storm weather corona loss is estimated from

$$P_c^1 = \frac{224(f + 25)}{\delta}\sqrt{\frac{r}{D}}(V - 0.8V_0)^2 \times 10^{-5} \quad \text{(kW/phase/km)} \tag{5.12}$$

Thus in the above the disruptive critical voltage V_c is assumed to be brought down by about 20% under storm weather conditions.

Peterson has given a corona loss formula to be applicable when (V/V_0) ratio is less than 1.8. His low loss line formula is

$$P_c = \frac{21 \times 10^{-6} f V^2}{\ln\left(\frac{D}{r}\right)^2} \times F \quad \text{(kW/phase/km)} \tag{5.13}$$

The constant F varies with the ratio (v/v_0) as shown in Table 5.1.

5.5 FACTORS INFLUENCING CORONA LOSS

1. Corona loss increases with frequency. This can be seen from Eq. (5.11).
2. Corona loss increases very fast with increase in system voltage since the loss is dependent on $(V - V_0)^2$.
3. From Eq. (5.12), it is clear that the loss further increases under storm weather conditions as V_0 is lowered by 20%.
4. Old and weathered line conductors further increase corona loss.

Since formation of corona on the conductor surface is dependent upon the maximum gradient, the voltage can be raised by decreasing the gradient with increase in conductor radius. Bundled conductors with larger geometric mean radius (GMR) and hollow conductors will reduce the gradient and raise the disruptive critical voltage.

Table 5.1 Discriptive critical voltage

V/V_0	0.6	0.8	1.0	1.2	1.4	1.6	1.8
F	0.012	0.018	0.05	0.08	0.3	1.0	3.0

Corona loss, if it occurs, reduces transmission efficiency. The third harmonic current that appears in the line current due to nonsinusoidal corona current interferes with neighboring communication lines. Increase of conductor size and increasing the spacing between conductors will reduce corona effects. Corona acts as a safety valve for transient over voltages.

5.6 INTERFERENCE

Power lines and communication lines are quite often, run parallel over the same right of the way. Interference by harmonic currents and other disturbances in power lines with the neighboring communication lines occurs if no preventive action is taken. There are two types of interference that may occur.

In electromagnetic interference currents are induced in communication lines, which interfere with the communicating signal (speech, etc.).

In case of electrostatic interference the electric potential of the communication lines is raised which may be dangerous. It is therefore necessary to analyze and determine the effect of the two phenomena.

5.6.1 ELECTROMAGNETIC EFFECT

Consider the symmetrically placed power line conductors A, B, and C shown in Fig. 5.2. P and Q are the two communication line conductors running on the same conductor supports distant d_1 and d_2 from conductor A.

Consider the loop formed by A and P. The inductance of the loop due to current flowing in A only is

$$L_{AP} = 2 \times 10^{-7} \ln \frac{d_1}{r_1} \quad (\text{h/m}) \tag{5.12}$$

where $r_1 = 0.7788r$ and r is the radius of conductor A.

In a similar manner the inductance of the loop AQ due to current A is

$$L_{AQ} = 2 \times 10^{-7} \ln \frac{d_2}{r_1} \quad (\text{h/m}) \tag{5.13}$$

The mutual inductance between A and the loop PQ is

$$M_A = L_{AQ} - L_{AP} = 2 \times 10^{-7} \ln \frac{d_2}{d_1} \quad (\text{h/m}) \tag{5.14}$$

Likewise the mutual inductances of conductors B and C can be evaluated.
The net mutual inductance due to A, B, and C on loop PQ is

$$\overrightarrow{M} = \overrightarrow{M_A} + \overrightarrow{M_B} + \overrightarrow{M_C} \tag{5.15}$$

The voltage induced in the communication conductors is

$$E_m = \frac{2\pi f M I V}{m} \tag{5.16}$$

where I is the current in the power lines.

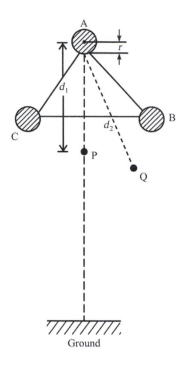

FIGURE 5.2

Electromagnetic effect.

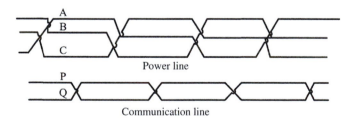

FIGURE 5.3

Transposition of power and communication lines.

It is to be noted that as the communication lines move farther and farther, away from the power line $d_2 \simeq d_1 = d$ and $\ln d_2/d_1$ tends to zero. Thus there will be no mutual induction. The induced voltage is also dependent directly on the frequency of the line. In case harmonic currents flow in power lines, the frequency of the currents being higher than power frequency, currents induced will also high and substantial distortion may be caused in communication lines. Furthermore, if the power line currents are unbalanced, the distortion could be more pronounced. Transposition of both the power and communication lines will reduce to a great extent this interference. This is shown in Fig. 5.3.

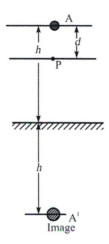

FIGURE 5.4

Electrostatic effect.

5.6.2 ELECTROSTATIC EFFECT

In Section 2.9 it is explained that a transmission line conductor A creates a potential distribution above the ground. The image of A that is, A^{1} is assumed to exist below the ground level exactly at the same depth as A is above the ground level, without altering the potential distribution (Fig. 5.4).

Potential of A with respect to earth

$$V_A = \frac{1}{2\pi \in} \int_r^h \left(\frac{q}{x} + \frac{q}{2h - x} \right) dx$$

$$= \frac{q}{r\pi \in} \ln \left(\frac{2h - r}{r} \right) \ (V)$$

(5.17)

where q is the charge as A Potential of P due to charge on A

$$V_{AP} = \frac{1}{2\pi \in} \int_d^h \left(\frac{q}{x} + \frac{q}{2h - x} \right) dx$$

$$= \frac{q}{2\pi \in} \ln \left(\frac{2h - d}{d} \right) \ (V)$$

(5.18)

$$\frac{V_{AP}}{V_A} = \frac{\left(\frac{q}{2\pi \in} \right) \ln \left(\frac{2h - d}{d} \right)}{\left(\frac{q}{2\pi \in} \right) \ln \left(\frac{2h - r}{r} \right)} = \ln \frac{\ln \frac{2h - d}{d}}{\ln \frac{2h - r}{r}}$$

(5.19)

$$V_{AP} = V_A \frac{\log \left[\frac{(2h - d)}{d} \right]}{\log \left[\frac{(2h - r)}{r} \right]} \ (V)$$

In a similar manner V_{BP} and V_{CP} can be computed.

The total potential of P due to power line conductors A, B, and C is given by

$$V_P = \overrightarrow{V_{AP}} + \overrightarrow{V_{BP}} + \overrightarrow{V_{CP}} \tag{5.20}$$

Likewise, the potential V_Q on the second communication line conductor Q is determined.

WORKED EXAMPLES

E.5.1 Determine the disruptive critical voltage for local and general corona for a three-phase 220-kV line where the conductors have a diameter of 25 mm and are spaced in 5.5 m delta formation. Given temperature of air as 27°C and pressure 73.5 cm of mercury. Surface factor $= 0.85$

Irregularity factor for local and general corona are 0.73 and 0.83, respectively.

Solution: Radius

$$r = \frac{25}{2} = 12.5 \ \text{mm} = 1.25 \ \text{cm} = \frac{1.25}{100} \ \text{m}$$

$$\delta = \frac{3.926 \times 73.5}{273 + 27} = 0.96187$$

$$m_o = 0.85; \ D = 5.5 \ \text{m}$$

$$V_0 = 21.12 \times m_0 \delta r \ \ln \frac{D}{r}$$

$$= 21.12 \times 0.85 \times 0.9618 \times 1.25 \times \ln \left[\frac{5.5 \times 100}{1.25} \right]$$

$$= 21.6763 \ \ln 440 = 21.6763 \times 6.0867747$$
$$= 131.9 \ \text{kV}$$

$$V_\vartheta(\text{local corona}) = 21.12 \times 0.73 \times 0.96187 \times 1.25$$
$$\times \left(1 + \frac{0.3}{\sqrt{(1.25 \times 0.96187)}} \right) \times \left[\frac{5.5 \times 100}{1.25} \right]$$

$$V_\vartheta = \frac{131.9 \times 0.73}{0.85} \times 1.27736 = 144.27 \ \text{kV}$$

$$V_\vartheta(\text{for general corona}) = \frac{144.27}{0.73} \times 0.83 = 164 \ \text{kV}$$

E.5.2 A three-phase 50 Hz, 110-kV line with 1-cm diameter conductors are constructed so that corona takes place if the line voltage exceeds 175 kV (rms). Determine the spacing between the conductors. Assume smooth conductors. Air density factor $= 1.0$

Solution:

$$V_0 = m_0 g_0 \delta r \ \ln \frac{D}{r}$$

$$\frac{175}{\sqrt{3}} = 1 \times 21.21 \times 1 \times 0.75 \ \ln \left(\frac{D}{0.5} \right)$$

where D is the spacing in cm.

$$101.039 = 15.9075 \ln \left(\frac{D}{8.5} \right)$$

$$\ln \frac{D}{r} = \frac{101.039}{15.9075} = 6.3516$$

$$\frac{D}{r} = e^{6.3516} = 573.41 = \frac{D}{0.5}$$

$$D = 0.5 \times 573.41 = 286.7 \text{ cm}$$
$$= 2.867 \text{ m}$$

The conductor spacing $= 2.7867$ m

E.5.3 Determine the corona characteristics of a 132-kV, three-phase line operating at 50 Hz, if the conductor diameter is 1 cm and length of the line is 150 km. The conductors are spaced in 4 m equilateral arrangement. Temperature of air is 26°C and barometric pressure is 74 cm of mercury. Surface irregularity factor $m = 0.88$. m_v for local and general corona is 0.74 and 0.84, respectively.

Solution:

$$V = \frac{132}{\sqrt{3}} = 76.21 \text{ kV}$$

The disruptive critical voltage

$$V_0 = 21.12 m \delta r \ln \frac{D}{r}$$

$$\delta = \frac{3.92b}{273 + t} = \frac{3.92 \times 74}{273 + 26} = 0.97$$

Therefore

$$V_0 = 21.12 \times 0.88 \times 0.74 \times 0.97 \times 0.5 \ln \left(\frac{4 \times 100}{0.5} \right)$$

$$= 60.255 \text{ kV/neutral (rms)}$$

$$V_v(\text{local corona}) = 21.12 m_v \delta r \left(1 + \frac{0.3}{\sqrt{\delta r}} \right) \ln \frac{D}{r}$$

$$= 21.12 \times 0.74 \times 0.97 \times 0.5 \left(1 + \frac{0.3}{\sqrt{0.5 \times 0.97}} \right) \cdot \ln \left(\frac{400}{0.5} \right)$$

$$= 7.58 \times [1 + 0.430] \times 6.6648 = 72.24 \text{ kV/neutral (rms)}$$

$$V_\vartheta(\text{for general corona}) = 72.24 \times \frac{0.84}{0.74} = 82 \text{ kV/neutral (rms)}$$

Fair weather corona loss

$$P_c = \frac{244}{\delta}(f + 25)\sqrt{\frac{r}{D}}(V - V_0)^2 \times 10^{-5} \text{ kW/phase/km}$$

$$= \frac{244(50 + 27)}{0.97}\sqrt{\frac{0.5}{4 \times 100}}(76.21 - 60.255)^2 \times 10^{-5}$$

$$= 18865.98 \times 0.03535(254.562) \times 10^{-9} = 1.69 \text{ kW/phase/km}$$

$$P_c = 150 \times 1.69 = 25.35 \text{ kW/phase}$$

Total loss,

$$P_c = 3 \times 25.35 = 76.05 \text{ kW}$$

Corona loss under storm weather conditions

$$= \frac{1.69}{254.562}(76.21 \times 0.8 \times 60.255)^2 = \frac{1.69 \times 784}{254.562}$$

$$= 5.20 \text{ kw/phase/km}$$

Total corona loss under storm condition

$$P_c^1 = 5.2 \times 3 \times 150 = 2340 \text{ kW}$$

E.5.4 A three-phase, 50-Hz, 150-km long line at 132 kW delivers a total load of 20 MW at 0.8 p.f. lagging. The conductor arrangement is shown in Fig. E.5.4. The diameter of each power line conductor is 1.2 cm. Calculate the induced voltage at fundamental frequency in the telephone circuit due to electromagnetic effect.

Solution:

$$CP = BP = \sqrt{5^2 + 2^2} = \sqrt{29}$$
$$= 5.3851 \text{ m}$$

$$CQ = BQ = \sqrt{2^2 + 6^2} = \sqrt{40}$$
$$= 6.3245 \text{ m}$$

$$M_A = 0.2 \ln\left(\frac{3.4641 + 6}{3.4641 + 5}\right)$$

$$= 0.2 \ln\left(\frac{9.4641}{8.4641}\right)$$

$$= 0.2 \times \ln 1.118146$$
$$= 0.2 \times 0.111671956 = 0.0223343913 \text{ mh/km}$$

$$M_B = M_C = 0.2 \ln\left(\frac{6.3245}{5.3851}\right)$$

$$= 0.2 \ln 1.1744443 = 0.2 \times 0.160795$$
$$= 0.032159 \text{ mh/km}$$

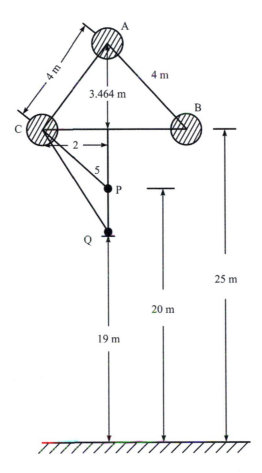

FIGURE E.5.4

$$M = 0.032159 - 0.0223343913 = 0.0098246 \text{ mh/km}$$

$$\text{Local current } I = \frac{20 \times 10^6}{\sqrt{3} \times 132 \times 10^3 \times 0.8} = 109.35 \text{ A}$$

The induced voltage due to fundamental

$$E_m = 2\pi MfI$$

$$E_m = 2 \times 3.14 \times 0.0098246 \times 109.35 \times \frac{50}{1000} \times 150$$

$$= 50.6 \text{ V}$$

E.5.5 In the previous example determine the potential of telephone conductors P and Q above the earth due to electrostatic effect only

Solution:

$$V_{AP} = V_A \frac{l\frac{2h-d}{d}}{l\frac{2h-r}{r}}$$

$$\frac{2h-d}{d} = \frac{56.9282 - 8.4641}{8.4641} = 5.7258$$

$$\frac{2h-r}{r} = \frac{56.9282 - 0.006}{0.006} = 9487$$

$$\ln 5.7258 = 1.74498$$

$$\ln 9487 = 9.15767$$

$$V_{AP} = \frac{132}{\sqrt{3}} \times \frac{1.74498}{9.15767} = 14.522 \text{ kV}$$

$$V_{BP} = V_A \frac{\ln\frac{50 - 513851}{5.3851}}{\ln\frac{50 - 0.006}{0.006}} = \frac{\ln 8.2848}{\ln 8332.33}$$

$$= \frac{2.1144}{9.0279} = 0.2342$$

$$V_{BP} = \frac{132}{\sqrt{3}} \times 0.2342 = 17.8489$$

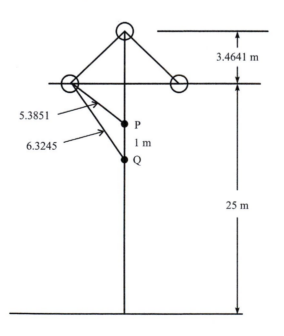

FIGURE E.5.5

$V_{AB\text{-}P}$ = Potential of p due to conductors ABC

$$= 17.8489 - 14.522 = 3.3269 \text{ kV}$$

$$= 3326.9 \text{ V}$$

$$V_{AQ} = V_A \frac{\ln\left(\dfrac{2h-d}{d}\right)}{\ln\left(\dfrac{2h-r}{r}\right)}$$

$$= V_A \frac{\ln\dfrac{56.9282 - 9.4641}{9.4641}}{\ln\dfrac{56.9282 - \dfrac{06}{100}}{\dfrac{0.6}{100}}} = V_A\left[\frac{\ln 5.015173}{\ln 9487}\right]$$

$$= 76.212 \times \left(\frac{1.6124679}{9.157677719}\right) = 76.212 \times 0.1760 = 13.4133$$

$$V_{BQ} = V_A \frac{\ln\dfrac{50 - 6.3245}{6.3245}}{\ln\dfrac{50 - \dfrac{06}{100}}{\dfrac{0.6}{100}}} = 76.212 \times \frac{\ln 6.90576}{\ln 9487}$$

$$= 76.212 \times \frac{1.932356}{9.15767} = 76.212 \times 0.211 = 16.081 \text{ kV}$$

$$V_{AB} - Q = 16.081 - 13.4133 = 2.668 \text{ kV}$$

$$= 2668 \text{ V}$$

PROBLEMS

P.5.1 Determine the disruptive critical voltage and the critical voltages for local and general corona on a three-phase overhead transmission line consisting of three copper stranded conductors spaced 2.5 m apart, at the corners of an equilateral triangle. The temperature and pressure of air are 22°C and 74 cm of mercury, respectively. Irregularity factor is 0.85; surface factors are 0.71 and 0.81. The conductor diameter is 1 cm.

P.5.2 A conductor 2.8 cm diameter is passed centrally through a bushing of relative permittivity $\varepsilon_r = 4$. The internal and external diameters of the bushing are 3 and 9 cm, respectively. The voltage between the conductor and the earthed clamp surrounding the porcelain bushing is 15 kV (rms). Determine whether or not corona will be present in the air space round the conductor.

P.5.3 A two-conductor, single-phase line of 50 Hz has its conductors placed horizontally with a spacing of 3.5 cm. A telephone line is supported on the same mast 1.8 m below the power line, again with a horizontal spacing of 0.5 m. Determine the voltage induced due to electromagnetic induction in the telephone circuit when the current in the power line is 175 A.

P.5.4 A three-phase, 6.6-kV, 50-Hz feeder has horizontal configuration with 1.15 m spacing between adjacent conductors. The height of the power line above ground level is 10 m. The vertical distance between the power lines and telephone wires is 1 m. Find the magnitude of the voltage induced in the telephone circuit due to electromagnetic induction when the power line carries a current of 285 A. The conductor arrangement is symmetrical.

QUESTIONS

Q.5.1 Explain the terms:
      (1) Disruptive critical voltage and (2) Visual critical voltage.

Q.5.2 Explain the term corona in transmission lines.

Q.5.3 What are the various factors that influence corona loss? Explain.

Q.5.4 Derive an expression for the influence of corona on communication lines due to electromagnetic effect.

Q.5.5 Derive an expression for the influence of corona on communication lines due to electrostatic effect.

PERFORMANCE OF TRANSMISSION LINES

In overhead transmission lines, conductors are suspended from the transmission tower with the support of insulators. In the preceding chapters the effect of alternating currents flowing in transmission lines has been discussed. While the alternating flux linkages result in inductance, the effect of charging electric field is modeled as capacitance. The inherent opposition to flow of electrons is included in conductor resistance. These three are the parameters for a transmission line. The performance of a transmission line depends upon the voltage and current relations. These relations is turn are influenced by line parameters, viz., resistance, inductance, and shunt-connected capacitance and its leakage conductance. Generally the sending end values are given; and it is desired to determine the receiving end values. Voltage regulation and efficiency of transmission are important in evaluating the performance of a transmission line.

6.1 CLASSIFICATION OF LINES

The line parameters are uniformly distributed along the line. However, for convenience and for practical utility, lumped parameter values are used depending upon the line length and operating voltage of the line.

6.1.1 SHORT LINES

Lines of length not exceeding 80 km (50 miles) are classified as short lines. As these lines operate at relatively lower voltages (<20 kV), the effect of capacitance and leakage conductance is negligible and therefore, can be neglected in analysis.

6.1.2 MEDIUM LINES

Lines of lengths ranging from 80 to 240 km (50−150 miles) are called medium lines. In this case lumped or localized values of inductance (L), capacitance (C), and resistance R are used.

6.1.3 LONG LINES

All lines with lengths exceeding 240 km require the use of distributed parameters for greater accuracy. In this case not only inductance, resistance, and capacitance are considered in a distributed manner but also the leakage conductance of capacitance (G) is also included in the model.

Electrical Power Systems. DOI: http://dx.doi.org/10.1016/B978-0-08-101124-9.00006-1

Analysis of transmission line performance assumes balanced operation of the three-phase system. Hence, all calculations are based on single-line or single-phase basis. The following notation is adopted.

z = series impedance per unit length per phase
r = series resistance per unit length per phase
x = series reactance per unit length per phase
y = series admittance per unit length per phase
g = shunt conductance (leakage) per unit length per phase
c = shunt capacitance per unit length per phase
l = total length of the line
Z = total series impedance per phase
  $= z \cdot l = \sqrt{r^2 + x^2} \cdot l$
Y = Total shunt admittance per phase $= yl$
X = Total reactance per phase $= xl$
R = Total resistance per phase $= rl$

6.2 THE SHORT TRANSMISSION LINE

The equivalent circuit of a short transmission line is shown in Fig. 6.1. The sending end and receiving end voltages and currents are denoted by V_s, I_s, and V_R and I_R, respectively. In case the single-phase line is represented by two conductors, then the resistance and reactance values are the loop values so that

$$R = 2 \cdot rl \ \Omega$$

and

$$X = 2 \cdot xl \ \Omega$$

The phasor diagrams for lagging, unity, and leading power factor operation are shown in Fig. 6.2.

For lagging power factor operation, from the phasor diagram (Fig. 6.3) V_s can be obtained as

$$V_S^2 = (V_R + I_R R \cos \phi_R + I_R X \sin \phi_R)^2 + (I_R X \cos \phi_R - I_R R \sin \phi_R)^2 \qquad (6.1)$$

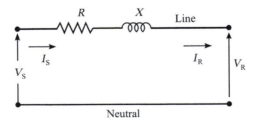

FIGURE 6.1

Short transmission line.

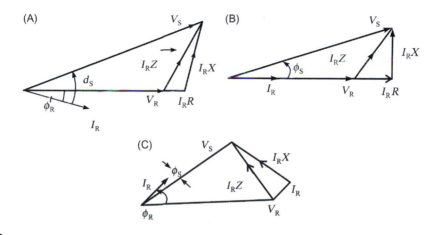

FIGURE 6.2

Phasor diagrams for short line. (A) Lagging power factor, (B) unity power factor, and (C) leading power factor.

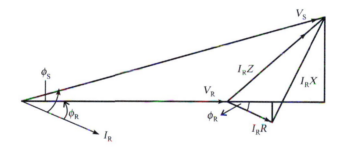

FIGURE 6.3

Short-line phasor diagram for regulation (lagging power factor)

where I_R is the load current. For clarity the voltage drops $I_R R$ and $I_R X$ are shown much large in comparison with V_R and V_S. Since the term $(I_R \times \cos \phi_R - I_R R \sin\phi_R)^2$ will be much smaller than V_S^2 and V_R^2, it can be neglected. Then, taking the positive sign

$$V_S = V_R + I_R R \cos \phi_R + I_R X \sin \phi_R \tag{6.2}$$

6.2.1 VOLTAGE REGULATION OF A TRANSMISSION LINE

It is defined as the rise in voltage at the receiving end when full load at a specified power factor is removed, expressed as a percentage of full load voltage at receiving end while the sending end voltage is kept constant.

The voltage drop in case of a short line is obtained from Eq. (6.2)

$$V_S = V_R \simeq I_R R \cos \phi_R + I_R X \sin \phi_R \tag{6.3}$$

If I_R is the full load current, then the per unit voltage regulation becomes

$$\frac{V_{R-\text{no load}} - V_{R-\text{full load}}}{V_{R \text{ full load}}} = \frac{V_S - V_R}{V_R} = \frac{IR \cos \phi_R + I_R \times \sin \phi_R}{V_R} \tag{6.4}$$

The voltage and current relations are expressed in phasor form as

$$\overline{V}_S = \overline{V}_R + \overline{I}_R \overline{Z} \cdot \overline{I}_S = \overline{I}_R \tag{6.5}$$

Expressed in matrix form

$$\begin{bmatrix} \overline{V}_S \\ \overline{I}_S \end{bmatrix} = \begin{bmatrix} 1 & Z \\ 0 & 1 \end{bmatrix} \begin{bmatrix} \overline{V}_R \\ \overline{I}_R \end{bmatrix} \tag{6.6}$$

6.2.2 EFFICIENCY OF A TRANSMISSION LINE

Since a certain amount of power is dissipated as I^2R loss in the resistance of the line conductors, the receiving end power is less than the sending end power. The efficiency of a transmission line is defined as the ratio of receiving end power to the sending end power.

$$\text{Percentage efficiency } \eta\% = 100 \times \left(\frac{3V_R I_R \cos \phi_R}{3V_S I_S \cos \phi_S} \right) = \frac{V_R I_R \cos \phi_R}{V_S I_S \cos \phi_S} \times 100\% \tag{6.7}$$

6.3 MIXED CONDITIONS

In the preceding analysis, it was assumed that all the quantities at one end are given, namely, sending end, and then the quantities at the other end can be determined. However, if the voltage at the sending end is known and load power factor is known, then it will be required to determine the receiving end voltage. Consider the phasor diagram in Fig. 6.4, where current is taken as reference.

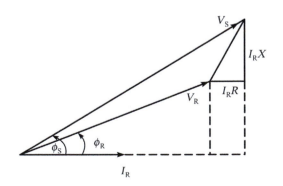

FIGURE 6.4

Mixed conditions.

From Fig. 6.4,

$$V_S^2 = (V_R \cos \phi_R + IR)^2 + (V_R \sin \phi_R + IX)^2$$
$$= V_R^2 + 2V_R I_R \cos \phi_R + 2V_R I_R \times \sin \phi_R + I^2(R^2 + X^2) \tag{6.8}$$
$$= V_R^2 + 2RI_R V_R \cos \phi_R + 2X I_R V_R \times \sin \phi_R + I^2 Z$$

We know that

$$P = V_R I_R \cos \phi_R \text{ and } Q = V_R I_R \sin \phi_R \tag{6.9}$$

Eliminating I_R from Eq. (6.8) utilizing Eq. (6.9)

$$V_S^2 = V_R^2 + 2(PR + QX) + \frac{P^2 Z^2}{V_R^2 \cos^2 \phi_R}$$

Multiplying by V_R^2 and rearranging the terms

$$V_R^4 = V_S^2 V_R^2 - 2V_R^2(PR + QX) - \frac{P^2 Z^2}{\cos^2 \phi_R}$$

or

$$V_R^4 = V_R^2 \left[V_S^2 - 2(PR + QX) \right] + \frac{P^2 Z^2}{\cos^2 \phi_R} = 0 \tag{6.10}$$

Let

$$V_S^2 - 2(PR + QX) = A$$

and

$$\frac{P^2 Z^2}{\cos^2 \phi_R} = B$$

Then solution to the quadratic in P_R^2

$$V_R^4 - AV_R^2 + B = 0 \text{ is obtained as}$$

$$V_R^2 = \frac{\left[A \pm \sqrt{A^2 - 4B} \right]}{2} \text{ and}$$

$$V_R = \sqrt{\left[\frac{A \pm \sqrt{A^2 - 4B}}{2} \right]} \tag{6.11}$$

V_R can be computed from Eq. (6.10)

6.4 MAXIMUM POWER TRANSFER

Consider Fig. 6.3 redrawn in Fig. 6.5.
From Fig. 6.5

$$V_S \cos \phi_S = V_R \cos \phi_R + I_R R$$
$$V_S \sin \phi_S = V_R \sin \phi_R + IX$$

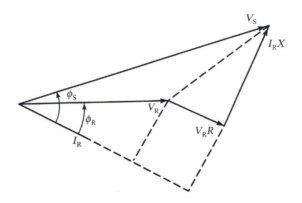

FIGURE 6.5

Phasor diagram for maximum power.

Squaring and adding

$$V_S^2 = V_R^2 + 2V_R I_R (R \cos \phi_R + X \sin \phi_R) + I^2(R^2 + X^2)$$
$$= V_R^2 + 2PR + 2Q \times + \frac{(P^2 + Q^2)(R^2 + X^2)}{V_R^2} \tag{6.12}$$

Since V_S, V_R, R, and X are constants for a given line for a given operation, keeping P and Q as variables, condition for maximum power transfer can be obtained by determining $dP/dQ = 0$.

Rearranging Eq. (6.12)

$$-V_S^2 + V_R^2 + 2PR + 2Q \times + \frac{(P^2 + Q^2)(R^2 + X^2)}{V_R^2} = 0$$

Differentiating

$$2R\frac{dP}{dQ} + 2X + \frac{(R^2 + X^2)}{V_R^2}\left(2p\frac{dP}{dQ} + 2Q\right) = 0 \tag{6.13}$$

At

$$\frac{dP}{dQ} = 0$$

$$2X + \left(\frac{R^2 + X^2}{V_R^2}\right) \cdot 2Q = 0$$
$$Q = \frac{-V_R^2 X}{R^2 + X^2} \tag{6.14}$$

Substituting this value for Q in Eq. (6.13)

$$-V_S^2 + V_R^2 + 2PR - \frac{2V_R^2 X^2}{R^2 + X^2} + \left(\frac{R^2 + X^2}{V_R^2}\right)\left[P^2 + \frac{V_R^4 X^2}{(R^2 + X^2)^2}\right] = 0 \tag{6.15}$$

This quadratic in P gives two solutions

$$P = -\frac{RV_R^2}{Z^2} \pm \sqrt{\frac{R^2 V_R^4}{Z^4} + \frac{V_S^2 V_R^2}{Z^2} - \frac{V_R^4 R^2}{Z^2}}$$

$$= -\frac{RV_R^2}{Z^2} \pm \frac{V_S V_R}{Z} = \frac{V_R^2}{Z^2}\left[-R + \frac{V_S Z}{V_R}\right] \qquad (6.16)$$

$$P_{\max} = \frac{V_R^2}{Z^2}\left[\frac{V_S^2}{V_R} - R\right]$$

If

$$V_S = V_R \text{ then } P_{\max} = \frac{V_R^2}{Z^2}(Z - R) \qquad (6.17)$$

In case R is fixed and X is allowed to vary, then the maximum power condition becomes

$$\frac{dP_{\max}}{dX} = \frac{d}{dX}\left[\frac{V_R^2}{Z^2}(Z - R)\right] = 0$$

$$= \frac{d}{dX}\left\{\frac{V_R^2}{\sqrt{R^2 + X^2}}\left[\sqrt{R^2 + X^2} - R\right]\right\} = 0 \qquad (6.18)$$

giving

$$\frac{R}{\sqrt{R^2 + X^2}} = \frac{1}{2} \text{ or } X = \sqrt{3}R \qquad (6.19)$$

6.5 MEDIUM TRANSMISSION LINES

There are three methods available for including the localized capacitance in medium transmission lines. They are:

1. Localized capacitance at the load end
2. Localized capacitance at the middle of the line or nominal T-method
3. Localizing half of the capacitance each at both ends or nominal π-method

6.5.1 LOCALIZED CAPACITANCE AT THE LOAD END

This method of representing line capacitance is shown in Fig. 6.6.
    The phasor diagram is shown in Fig. 6.7.

$$I_C = V_R Y$$

where

$$Y = j\omega C$$

$$I_S = I_R + I_C$$

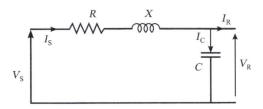

FIGURE 6.6

Localized capacitance at the load end.

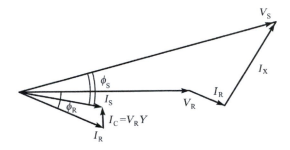

FIGURE 6.7

Phasor diagram for load end capacitance.

$$V_S = V_R + ZI_S$$
$$= V_R + Z[I_R + V_R Y] \tag{6.20}$$
$$= V_R[1 + YZ] + I_R \cdot Z$$

$$I_S = V_R Y + I_R \tag{6.21}$$

The current voltage relation can be expressed in matrix form as. . ..

$$\begin{bmatrix} V_S \\ I_S \end{bmatrix} = \begin{bmatrix} 1 + YZ & Z \\ Y & 1 \end{bmatrix} \begin{bmatrix} V_R \\ I_R \end{bmatrix} \tag{6.22}$$

Once V_S and I_S are evaluated, regulation and efficiency can be determined for all models.

6.5.2 NOMINAL T-METHOD

In the nominal T-method the entire capacitance is lumped at the middle of the line as shown in Fig. 6.8. The corresponding phasor diagram is shown in Fig. 6.9.

The series impedance Z is divided into two halves and placed on either side of the capacitance

$$V_C = V_R + I_R \frac{Z}{2} \tag{6.23}$$

$$I_C = V_C Y$$

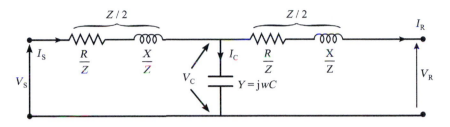

FIGURE 6.8

Medium line—nominal T-model.

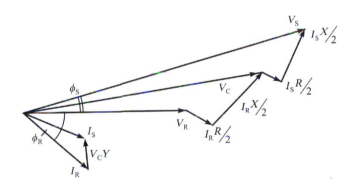

FIGURE 6.9

Phasor diagram for nominal T-line.

$$I_S = I_R + I_C = I_R + V_C Y = I_R + \left(V_R + I_R \frac{Z}{2} \right) \times Y$$

$$= V_R \cdot Y + I_R \left(1 + \frac{YZ}{2} \right) \tag{6.24}$$

$$V_S = V_C + I_S \frac{Z}{2} = \left(V_R + I_R \frac{Z}{2} \right) + \left[V_R Y + I_R \left(1 + \frac{YZ}{Z} \right) \right] \frac{Z}{2}$$

$$= V_R \left(1 + \frac{YZ}{2} \right) + I_R \left(1 + \frac{YZ}{4} \right) \tag{6.25}$$

Expressed in matrix notation Eq. (6.25) becomes

$$\begin{bmatrix} V_S \\ I_S \end{bmatrix} = \begin{bmatrix} 1 + \dfrac{YZ}{2} & Z\left(1 + \dfrac{YZ}{4}\right) \\ Y & 1 + \dfrac{YZ}{2} \end{bmatrix} \begin{bmatrix} V_R \\ I_R \end{bmatrix} \tag{6.26}$$

6.5.3 NOMINAL π-METHOD

In this representation the total line capacitance is divided into two halves and placed at sending end and receiving end, respectively, half each. This is shown in Fig. 6.10. The corresponding phasor diagram is drawn in Fig. 6.11.

From the phasor diagram and the circuit diagram, the voltage and current relations are obtained.

$$I_{C_2} = V_R \frac{Y}{2} \tag{6.27}$$

$$I = I_R + I_{C_2} = I_R + V_R \frac{Y}{2} \tag{6.28}$$

$$V_S = V_R + IZ$$

$$= V_R = \left(I_R + V_R \frac{Y}{2}\right) \cdot Z = V_R \left(1 + \frac{YZ}{2}\right) + I_R Z \tag{6.29}$$

$$I_{C_1} = V_S \frac{Y}{2} \tag{6.30}$$

$$= \left[V_R \left(1 + \frac{YZ}{2}\right) + I_R Z\right] \frac{Y}{2} \tag{6.31}$$

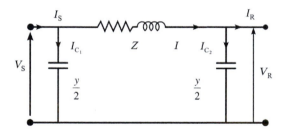

FIGURE 6.10

Nominal π-circuit.

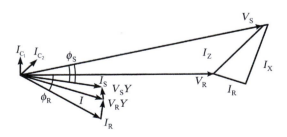

FIGURE 6.11

Phasor diagram for nominal π-model.

$$I_S = I + I_{C_1} = I_R + V_R \frac{Y}{2} + \frac{Y}{2}\left[V_R\left(1 + \frac{YZ}{2}\right) + I_R Z\right]$$

$$= V_R Y\left(1 + \frac{YZ}{4}\right) + I_R\left(1 + \frac{YZ}{2}\right)$$

(6.32)

Thus from Eqs. (6.29) and (6.32) the sending end voltage and current can be computed.

6.6 THE LONG TRANSMISSION LINE

A typical long transmission line is represented in Fig. 6.12. The distributed series impedance of the line is z ohms per unit length and $z = \sqrt{r^2 + x^2}$, r and x being the series resistance and reactance per unit length. The distributed shunt admittance $y = \sqrt{b^2 + g^2}$ where b and g are the susceptance and conductance per unit length and y is the admittance per unit length.

Consider a small length dx of the long line distant x from the load or receiving end. The voltages and currents at the beginning (sending end side) and (receiving end side) of this differential element dx are $(v + dv)$, $(I + dI)$, V and I, respectively, as shown in Fig. 6.12. The voltage drop across the line element dx.

$$dV = I \cdot z \cdot dx$$

(6.33)

from which

$$\frac{dv}{dx} = I \cdot z$$

(6.34)

The current passing through the differential shunt element is dI

$$dI = V \cdot y \cdot dx$$

(6.35)

from which

$$\frac{dI}{dx} = V \cdot y$$

(6.36)

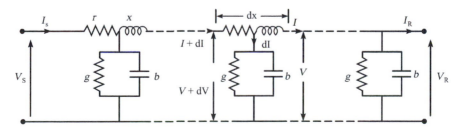

FIGURE 6.12

Long transmission line.

Differentiating (6.34) and (6.36)

$$\frac{d^2v}{dx^2} = z \cdot \frac{dI}{dx} \tag{6.37}$$

$$\frac{d^2I}{dx^2} = y \frac{dv}{dx} \tag{6.38}$$

Combining Eqs. (6.37) and (6.36)

$$\frac{d^2v}{dx^2} = yz \cdot v \tag{6.39}$$

and similarly

$$\frac{d^2I}{dx^2} = yzI \tag{6.40}$$

Solution to Eqs. (6.39) and (6.40) are of exponential type.
Let

$$v = A_1 e^{\sqrt{yz} \cdot x} + A_2 e^{-\sqrt{yz} \cdot x} \tag{6.41}$$

Then

$$\frac{dv}{dx} = \sqrt{yz} \, A_1 e^{\sqrt{yz} \, x} - \sqrt{yz} \, A_2 e^{-\sqrt{yz} \, x} \tag{6.42}$$

and

$$\frac{d^2v}{dx^2} = yz \left[A_1 e^{\sqrt{yz} \, x} + A_2 e^{-\sqrt{yz} \, x} \right]$$

$$= yz \cdot v$$

This proves that Eq. (6.41) is the correct solution to Eq. (6.39)
Substituting the values obtained in (6.33)

$$I = \frac{1}{z}\frac{dv}{dx} = \frac{1}{z}\left[\sqrt{yz} \, A_1 e^{-\sqrt{yz} \, x} - \sqrt{yz} \, A_2 e^{-\sqrt{yz} \, x} \right]$$

$$= \frac{1}{\sqrt{z/y}} A_1 e^{-\sqrt{yz} \, x} - \frac{1}{\sqrt{z/y}} A_2 e^{-\sqrt{yz} \, x} \tag{6.43}$$

From the terminal conditions, A_1 and A_2 can be evaluated.
At

$$x = 0; \quad V = V_R, \quad \text{and } I = I_R$$

From Eq. (6.41), we obtain

$$V_R = A_1 + A_2 \tag{6.44}$$

From Eq. (6.43), we obtain

$$I_R = \frac{1}{\sqrt{z/y}} A_1 - \frac{1}{\sqrt{z/y}} A_2 = \frac{1}{\sqrt{z/y}} [A_1 - A_2] \tag{6.45}$$

Solving (6.44) and (6.45), A_1 and A_2 can be obtained.

Characteristic impedance Z_C: The quantity $\sqrt{z/y} = \sqrt{(zl)/(yl)} = \sqrt{Z/Y}$ has the dimension of impedance. It is called characteristic impedance. It is also called natural impedance of the line denoted by Z_C. From Eqs. (6.44) and (6.45)

$$A_1 = \frac{V_R + I_R Z_C}{2} \text{ and } A_2 = \frac{V_R - I_R Z_C}{2}$$

Propagation constant: The quantity $\sqrt{zy} = \gamma$ is called propagation constant. As it is a complex quantity, it can be expressed as

$$r = \alpha + j\beta \tag{6.46}$$

where α is called attenuation constant and β is called phase constant.

For a lossless line, the series resistance and shunt conductance are zero.

$$\sqrt{\frac{Z}{Y}} = \sqrt{\frac{(r+jx)l}{(g-jb)l}} = \sqrt{\frac{jxl}{-jbl}} = \sqrt{\frac{j\,2\pi f L l}{j\,2\pi f C l}}$$

$$= \sqrt{\frac{L}{C}}$$

The propagation constant $\gamma = \sqrt{yz}$ reduces to $j\beta = j\omega\sqrt{LC/l}$ since the attenuation is zero for a lossless line. The quantity $\sqrt{L/C}$ has the dimensions of ohm and is called surge impedance. The characteristic impedance of a lossless line is called surge impedance of the line (surge impedance load (SIL)) denoted by Z_0. Its value for overhead lines is $400-600 \ \Omega$. For cables, it is $40-60 \ \Omega$.

For a lossless line

$$\cosh \gamma x = \cosh j\,\beta x = \cos \beta x$$

and

$$\sinh \gamma x = \sinh j\,\beta x = \sin \beta x$$

$$V(x) = \cos \beta x\, V_R + j \cdot Z_C \sin \beta x \cdot I_R$$

$$I(x) = j \cdot \frac{1}{Z_C} \sinh \beta x \cdot V_R + \cos \beta x \cdot I_R$$

Since

$$I_R = \frac{V_R}{Z_C}$$

$$V(x) = \cos \beta x\, V_R + jZ_C \frac{V_R}{Z_C} \sin \beta x = V_R\,(\cos \beta x + j \sin \beta x)$$

and

$$I(x) = j\frac{1}{Z_C} Z_C I_R \sin \beta x + I_R \cos \beta x = I_R\,(\cos \beta x + j \sin \beta x)$$

The voltage and current relations are constant at any point along the line in magnitude and hence are equal to sending end values. There is no reactive power in the line.

$$Q_S = Q_R = 0$$

The velocity of propagation $v = \frac{1}{\sqrt{LC}}$

The wavelength $\lambda = \frac{1}{f\sqrt{LC}}$

$$V = \left(\frac{V_R + I_R Z_C}{2}\right) e^{\gamma x} + \left(\frac{V_R + I_R Z_C}{2}\right) e^{-\gamma x} \tag{6.47}$$

and

$$I = \frac{\left(\dfrac{V_R}{Z_C}\right) + I_R}{2} e^{\gamma x} - \frac{\left(\dfrac{V_R}{Z_C}\right) - I_R}{2} e^{-\gamma x} \tag{6.48}$$

or

$$I = \frac{1}{Z_C}\left[\left(\frac{V_R + I_R Z_C}{2}\right) e^{\gamma x} - \left(\frac{V_R - I_R Z_C}{2}\right) e^{-\gamma x}\right]$$

Recalling that

$$\sinh \theta = \frac{e^\theta - e^{-\theta}}{2} \text{ and } \cosh \theta = \frac{e^\theta + e^{-\theta}}{2}$$

it can be expressed that

$$V = V_R \cosh \gamma x + I_R Z_C \sinh \gamma x \tag{6.49}$$

and

$$I = I_R \cosh \gamma x + \frac{V_R}{Z_C} \sinh \gamma x \tag{6.50}$$

At $x = 1$, the sending end voltage and current can be obtained from

$$V_S = V_R \cosh \gamma l + I_R Z_C \sinh \gamma l \tag{6.51}$$

and

$$I_S = I_R \cosh \gamma l + \frac{V_R}{Z_C} \sinh \gamma 1 \tag{6.52}$$

Expressed in matrix notation

$$\begin{bmatrix} V_S \\ I_S \end{bmatrix} = \begin{bmatrix} \cosh \gamma l & Z_C \sinh \gamma l \\ \dfrac{1}{Z_C} \sinh \gamma l & \cosh \gamma l \end{bmatrix} \begin{bmatrix} V_R \\ I_R \end{bmatrix} \tag{6.53}$$

6.7 GENERALIZED CIRCUIT CONSTANTS

A linear, passive, bilateral four terminal network with two input terminals and two output terminals shown in Fig. 6.13 is called a two-port network.

The network within the block is represented by four parameters A, B, C, and D. In such a network the input output relations are given by

$$V_S = AV_R + BI_R \tag{6.54}$$

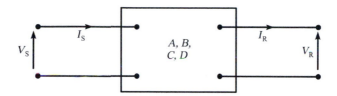

FIGURE 6.13

Two-port network.

and

$$I_S = CV_R + DI_R \tag{6.55}$$

where A and D are dimensionless, B is in ohms, and C in mhos or Siemens. It can be proved that $AD - BC = 1$. For a symmetrical network, $A = B$.

Considering the relationships derived for sending end voltages and currents from receiving conditions in previous sections, we can now compare the A, B, C, D constants with those results.

1. *Short line*

$$V_S = V_R + I_R Z; \quad A = 1, B = Z, C = 0, D = 1$$

$$I_S = R$$

2. *Medium line*
  a. Capacitance localized at load end

$$V_S = (1 + YZ)V_R + ZI_R \quad \text{and} \quad I_S = I_R + V_R Y$$
$$A = (1 + YZ); \quad B = Z; \quad C = Y; \quad D = 1$$

  b. Nominal T-method

$$V_S = V_R \left(1 + \frac{YZ}{2}\right) + I_R \left(Z + \frac{YZ^2}{4}\right)$$

$$I_S = V_R Y + I_R \left(1 + \frac{YZ}{2}\right)$$

$$A = D = 1 + \frac{YZ}{2}$$

$$B = Z \left(1 + \frac{YZ}{4}\right)$$

$$C = Y$$

  c. Nominal π-method

$$V_S = V_R \left(1 + \frac{YZ}{2}\right) + I_R Z$$

$$I_S = V_R \cdot Y \left(1 + \frac{YZ}{4}\right) + I_R \left(1 + \frac{YZ}{2}\right)$$

$$A = \left(1 + \frac{YZ}{2}\right) = D$$

$$C = Y\left(1 + \frac{YZ}{4}\right); \quad B = Z$$

3. *Long line*

$$V_S = V_R \cos\gamma l + I_R Z_C \sinh \gamma l$$

$$I_S = \frac{V_R}{Z_C} \sinh \gamma l + I_R \cosh \gamma l$$

$$A = D = \cosh \gamma l$$

$$B = Z_C \sinh \gamma l$$

$$C = \frac{V_R}{Z_C} \sinh \gamma l$$

Eqs. (6.54) and (6.55) can be solved to obtain V_R and I_R as

$$V_R = DV_S - BI_R \tag{6.56}$$

and

$$I_R = -CV_S + AI_R \tag{6.57}$$

6.8 EQUIVALENT T- AND π-CIRCUITS

Consider the equivalent T-circuit shown for a long line (Fig. 6.14).

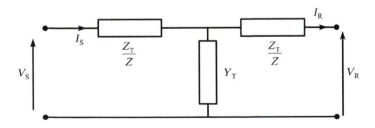

FIGURE 6.14

Equivalent T-network.

The voltage and current relations can be expressed as

$$\begin{bmatrix} V_S \\ I_S \end{bmatrix} = \begin{bmatrix} 1 + \dfrac{Y_T Z_T}{2} & Z_T\left(1 + \dfrac{Y_T Z_T}{4}\right) \\ Y_T & 1 + \dfrac{Y_T Z_T}{2} \end{bmatrix} \begin{bmatrix} V_R \\ I_R \end{bmatrix} \tag{6.58}$$

The exact solution for a long line is

$$\begin{bmatrix} V_S \\ I_S \end{bmatrix} = \begin{bmatrix} \cosh \gamma l & Z_C \sinh \gamma l \\ \dfrac{1}{Z_C} \sinh \gamma l & \cosh \gamma l \end{bmatrix} \begin{bmatrix} V_R \\ I_R \end{bmatrix} \tag{6.59}$$

Comparing the two solutions

$$Y_T = \frac{1}{Z_C} \sinh \gamma l$$

$$Z_C \sinh \gamma l = Z_T\left(1 + \frac{Y_T Z_T}{4}\right)$$

and

$$1 + \frac{Y_T Z_T}{Z} = \cosh \gamma l$$

Hence

$$Y_T = Y \frac{\sinh \gamma l}{\gamma l} \tag{6.60}$$

and

$$\frac{Z_T}{2} = Z_C \frac{\cosh \gamma l - 1}{\sinh \gamma l} = \frac{Z}{2} \frac{\tanh\left(\dfrac{\gamma l}{2}\right)}{\left(\dfrac{\gamma l}{2}\right)} \tag{6.61}$$

In a similar way if an equivalent π-network is used (Fig. 6.15), then from the network

$$\begin{bmatrix} V_S \\ I_S \end{bmatrix} = \begin{bmatrix} \left(1 + \dfrac{Z_\pi Y_\pi}{2}\right) & Z_\pi \\ Y_\pi\left(1 + \dfrac{Y_\pi Z_\pi}{4}\right) & \left(1 + \dfrac{Y_\pi Z_\pi}{2}\right) \end{bmatrix} \begin{bmatrix} V_R \\ I_R \end{bmatrix} \tag{6.62}$$

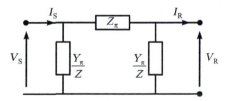

FIGURE 6.15

Equivalent π network.

From the exact solution, comparing

$$Z_\pi = Z_C \sinh \gamma l$$

$$= \sqrt{\frac{z}{y}} \sinh \gamma l \qquad (6.63)$$

$$= Z \frac{\sinh \gamma l}{\gamma l}$$

$$1 + \frac{Z_\pi Y_\pi}{2} = \cosh \gamma l \quad \text{or} \quad \frac{Y_\pi}{2} = \frac{1}{Z_C} \frac{\cosh \gamma l - 1}{\sinh \gamma l} = \frac{1}{Z_C} \tanh \frac{\gamma l}{2}$$

$$\frac{Y_\pi}{2} = \frac{Y}{2} \cdot \frac{\tanh \dfrac{\gamma l}{2}}{\dfrac{\gamma l}{2}} \qquad (6.64)$$

6.9 EVALUATION OF *A B C D* PARAMETERS

1. *Use of real angles*

$$\gamma = \alpha + j\beta$$

$$\sinh \gamma l = \sinh (\alpha + j\beta) = \sinh \alpha l \cos \beta l + j \cosh \alpha l \sin \beta l \qquad (6.65)$$

and

$$\cosh \gamma l = \cosh (\alpha + j\beta) = \cosh \alpha l \cos \beta l + j \sinh \alpha l \sin \beta l \qquad (6.66)$$

Once the hyperbolic sine and cosine are evaluated, *A, B, C, D* constants can be calculated.

2. *Use of series expansion*

The hyperbolic sine and cosine terms can be computed from Maclaurin's series

$$\sinh \gamma l = \gamma l + \frac{\gamma^3 l^3}{3!} + \frac{\gamma^5 l^5}{5!} + \cdots$$

$$= \sqrt{YZ} \left[1 + \frac{YZ}{3!} + \frac{(YZ)^2}{5!} + \cdots \right] = \sqrt{YZ} \left[1 + \frac{YZ}{6} + \frac{Y^2 Z^2}{120} + \frac{Y^3 Z^3}{5040} \right] \qquad (6.67)$$

$$\cosh \gamma l = 1 + \frac{\gamma^2 l^2}{2!} + \frac{\gamma^4 l^4}{4!} + \cdots$$

$$= 1 + \frac{YZ}{2!} + \frac{(YZ)^2}{4!} + \frac{(YZ)^3}{6!} + \cdots = 1 + \frac{YZ}{Z} + \frac{Y^2 Z^2}{24} + \frac{Y^3 Z^3}{720} + \cdots \qquad (6.68)$$

3. *Use of exponential terms*

$$\sinh \gamma l = \sinh (\alpha l + j\beta l) = \frac{e^{\alpha l} e^{\beta l} - e^{-\alpha l} e^{-j\beta l}}{2} \qquad (6.69)$$

$$= \frac{e^{\alpha l} \lfloor \beta l + e^{-\alpha l} \rfloor - \beta l}{2}$$

From the above the hyperbolic sine and cosine can be evaluated.

6.10 SURGE IMPEDANCE LOADING

It is already explained that the characteristic impedance of a lossless line is called its surge impedance. The load that can be delivered by a lossless line is called SIL. For such a line power factor is
taken as unity.

$$P_{R_{max}} = \frac{V_R^2}{Z_0} MW \tag{6.71}$$

where $Z_0 = \sqrt{L/C}$. This value lies between 400 and 600 Ω for overhead lines and 40−60 Ω for
cables as stated already. V_R is the receiving end voltage in kV. $P_{R_{max}}$ is also denoted by SIL and
natural power of the line.

6.11 FERRANTI EFFECT

A lightly loaded or unloaded line carrying negligibly small load current compared to charging current I_C of the line will be operating at leading power factor. The phasor diagram for such a line is
shown in Fig. 6.16.

   The predominantly leading current in the line flowing through line impedance produces a drop
I_X that is in opposition to V_R. The magnitude of V_S then will be less than V_R. This condition is
called Ferranti effect as Ferranti was the first person to observe it in Italy.

6.12 POWER RELATIONS IN TRANSMISSION LINES

Recalling Eqs. (6.54) and (6.55)

$$V_S = AV_R + BI_R$$

and

$$I_S = CV_R + DI_R$$

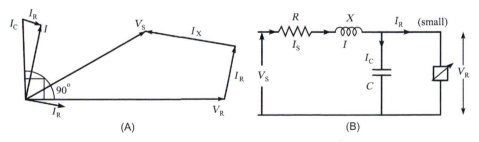

FIGURE 6.16

Ferranti effect. (A) Phasor diagram; (B) Equivalent circuit.

for the sending end and solving for receiving end, we obtain

$$V_R = DV_S - BI_S \tag{6.72}$$

$$I_R = -CV_S + AI_S$$

We can rearrange

$$
\begin{aligned}
I_R &= \frac{1}{B}V_S - \frac{A}{B}V_R \\
&= \frac{|V_S|}{|B|}\underline{|\delta - \beta} - \frac{|A||V_R|}{|B|}\underline{|\alpha - \beta}
\end{aligned}
$$

Taking the conjugate an both sides

$$I_R^* = \frac{|V_S|}{|B|}\underline{|\beta - \delta} - \frac{|A||V_R|}{|B|}\underline{|\beta - \alpha}$$

Complex power received

$$
\begin{aligned}
S_R &= P_R + jQ_R = V_R I_R^* \text{ per phase} \\
&= \frac{|V_R||V_S|}{|B|}\underline{|\beta - \delta} - \frac{|A||V_R|^2}{|B|}\underline{|\beta - \delta}
\end{aligned}
$$

Similarly

$$I_S = \frac{|A||V_S|}{|B|}\underline{|\alpha + \delta - \beta} - \frac{|V_R|}{|B|}\underline{|-\beta}$$

and

$$I_S^* = \frac{|A||V_S|}{|B|}\underline{|\beta - \alpha - \delta} - \frac{|V_R|}{|B|}\underline{|\beta}$$

The complex sending end power per phase

$$
\begin{aligned}
S_S &= P_S + jQ_S = V_S I_S^* \\
&= \frac{|A||V_S|^2}{|B|}\underline{|\beta - \alpha} - \frac{|V_R||V_S|}{|B|}\underline{|\beta - \delta}
\end{aligned}
$$

Separating out and writing the full expression for real and reactive powers at the receiving end

$$P_R = \frac{|V_S||V_R|}{|B|}\cos(\beta - \delta) - \frac{|A||V_R|^2}{|B|}\cos(\beta - \alpha) \tag{6.79}$$

and

$$Q_R = \frac{|V_S||V_R|}{|B|}\sin(\beta - \delta) - \frac{|A||V_R|^2}{|B|}\sin(\beta - \alpha) \tag{6.80}$$

At the sending end

$$P_S = \frac{|A||V_S|^2}{|B|}\cos(\beta - \alpha) - \frac{|V_S||V_R|}{|B|}\cos(\beta + \delta) \tag{6.81}$$

and

$$Q_S = \frac{|A||V_S|^2}{|B|}\sin(\beta - \alpha) - \frac{|V_S||V_R|}{|B|}\sin(\beta + \delta) \tag{6.82}$$

P_R is a maximum when $\delta = \beta$

$$P_{R_{max}} = \frac{|V_S||V_R|}{|B|} - \frac{|A|V_R|^2}{|B|}\sin(\beta - \alpha) \tag{6.83}$$

The load must draw leading VARS for maximum real power at the receiving end and this reactive power must be $((- |A||V_R|^2)/|B|)\sin(\beta - \alpha)$

For short lines where capacitive effect is neglected

$$P_R = \frac{|V_S||V_R|}{|Z|}\cos(\theta - \delta) - \frac{|V_R^2|}{|Z|}\cos\theta \tag{6.84}$$

and

$$Q_R = \frac{|V_S||V_R|}{|Z|}\sin(\theta - \delta) - \frac{|V_R^2|}{|Z|}\sin\theta \tag{6.85}$$

Correspondingly the sending end powers become

$$P_S = \frac{|V_S|^2}{|Z|}\cos\theta - \frac{|V_S||V_R|}{|Z|}\cos(\theta + \delta) \tag{6.86}$$

and

$$Q_S = \frac{|V_S|^2}{|Z|}\sin\theta - \frac{|V_S||V_R|}{|Z|}\sin(\theta + \delta) \tag{6.87}$$

Further, if resistance is neglected

$$P_R = \frac{V_S V_R}{X}\sin\delta \tag{6.88}$$

and

$$Q_R = \frac{|V_S||V_R|}{X}\cos\delta - \frac{|V_R|^2}{|X|}[\sin\theta = 90\ degree] \tag{6.89}$$

6.13 POWER CIRCLE DIAGRAMS

Consider the generalized network equations

$$V_S = AV_R + BI_R \tag{6.54}$$

and

$$I_S - CV_R + DI_R \tag{6.55}$$

where

$$A = |A|\underline{|\alpha°|}, B = |B|\underline{|\beta°|}, C = |C|\underline{|\gamma°|} \text{ and } D = |D|\underline{|\Delta°|}$$

Solving Eqs. (6.54) and (6.55), we obtain (6.72)

$$V_R = DV_S - BI_S$$

and

$$I_R = -CV_S + AI_S$$

The above relations are all voltage and current equations. By suitable operations on these equations and on their phasor diagrams, we can obtain power relations and power diagrams. These are called power circle diagrams. In the following the procedures for obtaining these diagrams are explained.

6.13.1 RECEIVING END POWER CIRCLE DIAGRAM

Eq. (6.54) is represented as phasor diagram in Fig. 6.17 for lagging power factor.

To obtain a power diagram to be of use in practical situations, the diagram needs to be converted from voltages to powers. All the phasors if multiplied by YZ/B then the phasor BI_R becomes $BI_R \cdot (V_R/B)$ giving $V_R I_R$ in voltamperes or in kVA if V_R is in kV.

Thus

$$V_S \cdot \frac{V_R}{B} = AV_R \cdot \frac{V_R}{B} + BI_R \cdot \frac{V_R}{B}$$

$$\frac{V_S V_R}{B} = \frac{AV_R^2}{B} + V_R I_R \tag{6.90}$$

Implementing this multiplication on the phasor diagram of Fig. 6.18, the magnitude of all phasors are raised by $|V_R|$ and all the phasors are turned clockwise through an angle β. This is illustrated in Fig. 6.18.

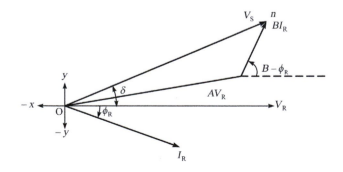

FIGURE 6.17

Phasor representation of $V_S = AV_R + BI_R$.

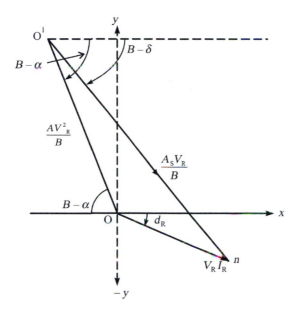

FIGURE 6.18

Phasor representation of $\dfrac{V_S V_R}{B} = \dfrac{A V_R^2}{B} + V_R I_R$.

This diagram is satisfactory. But, it is conventional to represent lagging VARS on $+y$ axis and leading VARS or $-y$ axis. For this convention the line $V_R I_R$ is rotated through an angle of $2(\pi - \phi_R)$ clockwise to obtain Fig. 6.19, where the projections of $V_R I_R$ on x- and y-axis correspond to the present lagging power factor case.

Several receiving end power circles for constant V_R, but varying V_S with O_1 as center and $V_S V_R / B$ as radii can be obtained.

If $|V_S|$ is fixed, then the receiving end power circle diagram for different V_R will have centers on the line OO_1, the distance OO_1 varying as V_R^2. The radius also changes as V_R changes.

For any torque angle δ, the power received P_R is obtained as the difference on projections on the x-axis of $(O_1 x)$ and $(O_1 O)$.

$$P_R = \frac{|V_S||V_R|}{|B|} \cos (\beta - \delta) - \frac{A|V_R^2|}{|B|} \cos (\beta - \delta) \ \text{W} \tag{6.91}$$

If $|V_S|$ and $|V_R|$ are fixed, P_R will be a maximum when $\delta = \beta$

$$P_{R_{max}} = \frac{|V_S||V_R|}{|B|} - \frac{A V_R^2}{|B|} \cos (\beta - \delta) \ \text{W} \tag{6.92}$$

6.13.2 SENDING END POWER CIRCLE DIAGRAM

Consider the equation $V_R = D V_S - B I_S$ drawn as a phasor diagram taking V_S as a reference
Multiplying $(-B I_S)$ by $\left(-\frac{V_S}{B}\right)$

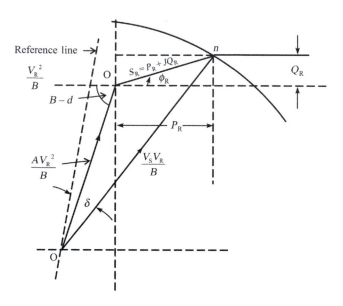

FIGURE 6.19

Receiving end power circle diagram.

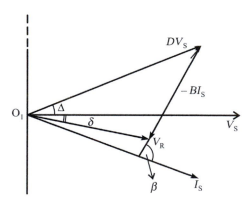

FIGURE 6.20

Construction of sending end circle diagram.

We obtain $V_S I_S$. The whole phasor diagram in Fig. 6.20 is multiplied by $(-V_S/B)$ to obtain the power diagram shown in Fig. 6.21A.

$$\frac{-V_R V_S}{B} = \frac{-V_S^2 D}{B} + V_S I_S$$

We find that Fig. 6.20 is turned clockwise to through an angle 180 degrees$-\beta$ as shown in Fig. 6.1A with O as center the projection of OX or y axis will be negative for lagging power.

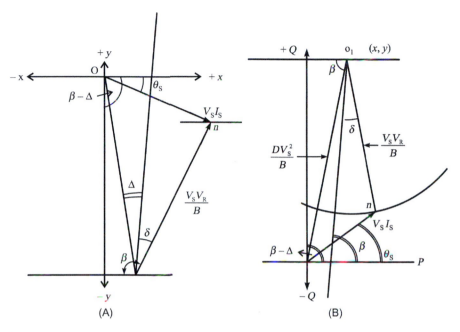

FIGURE 6.21

(A) Sending end power circle diagram and (B) ending power circle diagrams.

To adopt the convention that lagging VARS are represented on $+y$ axis, the whole diagram so turned that the reactive power is shown positive (i.e., on $+y$ axis). This is indicated in Fig. 6.21B.

For constant $|V_S|$ and $|V_R|$ the locus of the operating point n is a circle with radius $(|V_S||V_R|)/|B|$ and center O_1. The coordinates of the sending end power circle are given by

$$x = \frac{|D|}{|B|} V_S^2 \cos (\beta - \Delta) \text{ W}$$

and

$$y = \left|\frac{D}{B}\right| V_S^2 \sin (\beta - \Delta) \text{ vars}$$

For any torque angle δ the power delivered is

$$P_S = \frac{|V_S||V_R|}{|B|} \cos (\beta + \delta) + \frac{|D||V_S|^2}{|B|} \cos (\beta - \delta) \text{ W} \tag{6.93}$$

When $|V_S|$ and $|V_R|$ are fixed, P_S is a maximum when $\delta = 180 - \beta$

$$P_{S(max)} = \frac{|V_S||V_R|}{|B|} + \frac{|D||V_S|^2}{|B|} \cos (\beta - \Delta) \text{ W} \tag{6.94}$$

WORKED EXAMPLES

E.6.1 A three-phase, 50-Hz, 15-km transmission line supplying a total load of 850 kW at 0.8 power factor lagging and 11 kV has the following line constants.

$$r = 0.45 \ \Omega/\text{km}$$

$$x = 0.6 \ \Omega/\text{km}$$

Calculate the line current, receiving end voltage, voltage regulation, and efficiency of transmission.

Solution:

Receiving end current $I_R = \dfrac{850}{\sqrt{3} \times 0.8 \times 11} = 55.768$ A

The voltage drops

$$I_R R = 55.768 \times 0.45 \times 15 = 376.43 \ \text{V}$$

$$I_R X = 55.768 \times 0.6 \times 15 = 501.9 \ \text{V}$$

$$V_R \cos Q_R = \frac{11,000}{\sqrt{3} \times 0.8} = 5080.83$$

$$V_R \sin Q_R = \frac{11,000}{\sqrt{3} \times 0.6} = 3810.62$$

$$V_S = \sqrt{(V_R \cos Q_R + I_R R)^2 + (V_R \sin Q_R + I_R X)^2}$$

$$= \sqrt{(5080.83 + 376.43)^2 + (3810.62 + 501.9)^2}$$

$$= \sqrt{(5457.26)^2 + (4312.52)^2}$$

$$= 6955.5 \ \text{V}$$

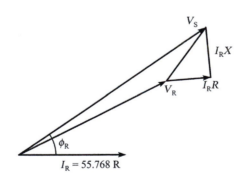

FIGURE E.6.1

Line voltage at the sending end $= \sqrt{3} \times 6.9555 = 12.047$ kV

$$\text{Percentage regulation} = \frac{I_R \times \sin \phi_R + I_R R \cos \phi_R}{V_R} \times 100$$

$$= \frac{(501.9 \times 0.6 + 376.43 \times 0.8)}{5080.83} \times 100$$

$$= \frac{602.284}{5080.83} \times 100$$

$$= 11.85\%$$

$$I^2 R \text{ loss} = (55.768)^2 \times 0.45 \times 15 \times 3 = 20992.97 \text{ W}$$

$$\text{Efficiency of transmission} = \frac{850}{850 + 20.992} \times 100$$

$$= 97.59\%$$

E.6.2 A single-phase transmission line has a resistance of 0.22 Ω and an inductive reactance of 0.36 Ω. Find the voltage at the sending end to give 500 kVA at 2000 V at the receiving end at load power factors of (a) unity and (b) at 0.707 lagging.

Solution:

$R = 0.22$ Ω

$$X = 0.36 \ \Omega$$

$$V_R = 2000 \ \text{V}$$

$$P = VI \cos \phi; \ 500 \ \text{kVA} = 2 \ \text{kV} \times I$$

Hence

$$I_R = \frac{500}{2} = 250 \ \text{A}$$

$$I_R R = 250 \times 0.22 = 55 \ \text{V}$$

$$I_R x = 250 \times 0.36 = 90 \ \text{V}$$

At unity power factor

$$V_S = \sqrt{(V_R \cos \phi_R + IR)^2 + (V_R \sin \phi_R + IX)^2}$$

$$= \sqrt{(2000 \times 1 + 55)^2 + (0 + 90)^2}$$

$$= \sqrt{4,231,125}$$

$$= 2056.97 \ \text{V}$$

At 0.707 lagging power factor

$$V_S = \sqrt{(2000 \times 0.707 + 55)^2 + (2000 \times 0.707 + 90)^2}$$

$$= \sqrt{2,157,961 + 2,262,016}$$

$$= \sqrt{4,419,977}$$

$$= 2102.37 \ \text{V}$$

E.6.3 A line having a resistance of 16 Ω and a reactance of 12 Ω supplies a load of 5 MW at voltage V_R. The supply voltage is V_S. Find the power factor of the load when $V_S = V_R$.

    Solution:

When $V_S = V_R$, voltage regulation is zero

Then, $I_R\,R\cos\phi_R = I_R \times \sin\phi_R$

$$I_R \cdot 16\cos\phi_R = I_R \cdot 12 \cdot \sin\phi_R$$

$$\tan\varphi_R = \frac{16}{12} = 1.33$$

$$\phi_R = 53°\cdot06; \quad \cos\phi_R = 0.6$$

E.6.4 Estimate the distance over which a load of 15 MW at 0.85 power factor can be delivered by a three-phase transmission line having a resistance of 0.905 Ω/km. The voltage at the receiving end is to be 132 kV and the loss in the transmission line is to the 7.5% of the load.

    Solution:

Loss $= 3I^2R = 3 \times I^2 \times 0.905 \times L$ (W) where L is the length of the line to be determined

$$7.5\% \text{ of } 15 \text{ MW}$$

$$\frac{7.5 \times 15}{100} = 1.125 \text{ MW}$$

$$P = 15 \times 10^6 \text{ W} = \sqrt{3} \times 132 \times 10^3 \times I_L \times 0.85$$

Solving,

$$I_L = \frac{15 \times 10^6}{\sqrt{3} \times 132 \times 10^3 \times 0.85} = 77.188 \text{ A}$$

$$1.125 \times 10^6 \text{ W} = 3 \times (77.188)^2 \times 0.905 \times L$$

Length of the line

$$L = \frac{1125 \times 10^3}{3 \times (77.188)^2 \times 0.905}$$

$$= 69.54 \text{ km}$$

E.6.5 A 20-km long, three-phase system delivers 6 MW at 11 kV at a power factor of 0.8 (lag). Line reactance is 5 Ω and resistance is 1.5 Ω. Calculate the sending end voltage and regulation. Also, find the power factor at which the regulation becomes zero.

    Solution:

For a short line $V_S = V_R + I_R\,R\cos\phi_R + I_R \times \sin\phi_R$

$$V_R = \frac{11,000}{\sqrt{3}} = 6351 \text{ V}$$

$$I_R = \frac{6 \times 10^6}{3 \times 6351 \times 0.8} = 393.6 \text{ A}$$

$$V_S = 6351 + 393.6 \times 1.5 \times 0.8 + 393.6 \times 5 \times 0.6$$

$$= 6351 + 42.32 + 1179 = 8002.3 \text{ V}$$

$$V_S(\text{line}) = \sqrt{3} \times 8002.3 = 13{,}059.7 = 13.06 \text{ kV}$$

$$\text{Regulation} = \frac{V_S - V_R}{V_R} = \frac{8002.3 - 6351}{6351} = 0.26 = 26\%$$

Zero regulation occurs at leading power factor
When $I_R R \cos \phi_R = I_R \times \sin \phi_R$

$$\frac{R}{X} = \tan \phi_R = \frac{1.5}{5} = 0.3$$

$$\phi_R = 16.69924 \text{ (leading)}; \quad \cos \phi_R = 0.95(\text{lead})$$

E.6.6 If zero regulation is achieved in example (E.3.6) by connecting a condenser bank at the receiving end, determine the capacitance and capacitive reactance of the condenser bank.

Solution:

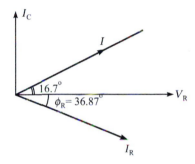

FIGURE E.6.2

At $\cos \phi_R = 0.8$; $\phi_R = 36°.87$ from phasor or diagram

$$I_R \cos 36°.87 = I \cos 16.7 \text{ degrees}$$

$$393.6 \times 0.8 = 0.95I$$

$$I = 331.45 \text{ A}$$

I_C = current in condenser bank

$$= I \sin 16°.7 + I_R \sin 36°.87$$

$$= 331.45 \times 0.2873 + 393.6 \times 0.6$$

$$= 95.245 + 236.16 = 331.4 \text{ A}$$

$$X_C = \frac{V_R}{I_C} = \frac{6351}{331.4} = 19.164 \ \Omega$$

$$X_C = \frac{1}{2\pi f c} = 19.164$$

$$C = \frac{10^6}{19.164 \times 314} = 16.182 \ \mu\text{F}$$

E.6.7 A three-phase, 33-kV line of resistance 20 Ω and reactance 10 Ω operating at 11 kV is supplying a load of 5 MW at 0.8 power factor lagging through a transformer rated at 33/11 kV. The equivalent resistance and reactance of the transformer referred to 11 kV side are 1 and 2 Ω, respectively. Find the sending end voltage and regulation.

Solution:

The resistance and reactance of the transformer referred to 33 kV side are

$$1 \times \left(\frac{33}{11}\right)^2 = 9 \ \Omega \ \text{and} \ 2 \times \left(\frac{33}{11}\right)^2 = 18 \ \Omega$$

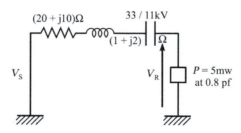

FIGURE E 6.7

Total resistance referred to 33 kV side $= 9 + 20 = 29 \ \Omega$
Total reactance referred to 33 kV side $= 18 + 10 = 28 \ \Omega$

$$I_R = \frac{5 \times 10^6}{\sqrt{3} \times 33 \times 10^3 \times 0.8} = 109.35 \ \text{A}$$

$$V_R = \frac{33,000}{\sqrt{3}} = 19.053 \ \text{kV} = 19,053 \ \text{V}$$

$$V_S = 19,053 + 109.35 \times 29 \times 0.8 + 109.35 \times 28 \times 0.6$$
$$= 19,053 + 2536.92 + 1837.08 = 23,427$$

$$V_S(\text{line}) = \sqrt{3} \times 23,427 = 40,575.56 \ \text{V}$$

$$\text{Regulation} \frac{23,427 - 19,053}{19,053} = 0.2295$$
$$= 22.95\%$$

E.6.8 A short three-phase transmission line is operating at 33 kV at the sending end and supplying a load of 8 MW at 0.85 power factor lagging. The receiving end voltage is 31.5 kV. The transmission efficiency is 0.95. Determine the resistance and reactance of the line per phase.

Solution:

$\cos \phi_R = 0.85; \ \phi_R = 31°.788; \ \sin \phi_R = 0.52678$

Receiving end voltage per phase $= \frac{31.5}{\sqrt{3}} = 18.187 \ \text{kV}$

Sending end voltage per phase $= \frac{33}{\sqrt{3}} = 19.053 \ \text{kV}$

Transmission efficiency $= \frac{\text{Output}}{\text{Input}} = \frac{\text{Output}}{\text{Output} + \text{Losses}}$

$$0.95 = \frac{8 \times 10^6}{8 \times 10^6 + P_L} \text{ solving } 7.6 \times 10^6 + 0.95P_L = 8 \times 10^6$$

$$P_L = \frac{0.4 \times 10^6}{0.95} = 0.421 \times 10^6 \text{ W} = 3I_R^2 R$$

$$I_R = \frac{8 \times 10^6}{\sqrt{3} \times 31.5 \times 0.85} = 172.51 \text{ A}$$

$$3 \times (172.51)^2 \times R = 0.421 \times 10^6 \text{ W}$$

Solving for

$$R = \frac{0.421 \times 10^6}{2 \times (172.51)^2} = 4.715 \ \Omega/\text{phase}$$

$$V_S - V_R \simeq IR \cos \phi_R + I \times \sin \phi_R$$

$$(19053 - 18187) = 172.5[4.715 \times 0.85 + X \times 0.5268]$$

$$0.86 \times 10^3 = 172.5[4 + 0.5268 \ X]$$

$$90.873 \ X = 860 - 690$$

$$X = \frac{170}{90.873} = 1.87 \ \Omega/\text{phase}$$

E.6.9 A 25-km, three-phase transmission line is supplying a load of 2000 kW at a power factor of 0.85 lagging. The line impedance is $(5 + j\,6)$ Ω/phase. If the sending end voltage is 13.2 kV, determine the receiving end voltage and the line current.

Solution:

$$\text{Line current } I_R = \frac{2000 \times 10^3}{3 \times V_R \times 0.85} = \frac{784,313.72}{V_R} A$$

$$V_S \simeq V_R + I_R R \cos \phi_R + I_R \times \sin \phi_R$$

$$V_S = \frac{13.2 \times 10^3}{\sqrt{3}} = 7621.25 \text{ V}$$

$$7621.25 = V_R + \left(\frac{784313.72}{V_R}\right) 5 \times 1.85 + \left(\frac{784313.72}{V_R}\right) \times 6 \times 0.527$$

$$7621.25 \ V_R = V_R^2 + 3,333,333.3 + 2,479,999.982$$

$$V_R^2 - 7621.25 V_R + 5,813,333.282 = 0$$

$$V_R = \frac{7621.25 \pm \sqrt{(7621.25)^2 - 4 \times 5,813,333.282}}{2}$$

$$= 3810.625 \pm \frac{\sqrt{34,830,118.44}}{2} = 3810.625 \pm 2950.85$$

$$= 6761.47 \text{ or } 859.77$$

Since the first value is possible

$$V_R = 6.76147 \text{ kV}$$

$$\text{or } V_R(\text{line}) = \sqrt{3} \times 6.76147 = 11.71 \text{ kV}$$

(This is a problem of mixed conditions)

E.6.10 A load of three impedances each $(9 + j\,21)$ Ω is supplied through a line to which a voltage of 415 is applied. The impedance of each line is $(2 + j4)$ Ω. Find the power input and output when the load is (1) star connected and when (2) delta connected.

Solution:

1. Star connected:

$$\text{Total impedance} = (2 + j4) + (9 + j21)$$
$$= (11 + j25) \, \Omega$$

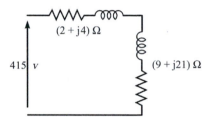

FIGURE E.6.10

$$\text{Current in the line} = \frac{415}{\sqrt{3} \, (11 + j25)} = (3.5329 - j8.025)\text{A}$$
$$= 8.768 \, A\underline{|66°.239}$$

$$\text{Power input} = \sqrt{3} \times 415 \times 8.768 \times 0.4029 = 2539.18 \text{ W}$$
$$(\cos 66.239 = 0.4029)$$

$$\text{Power loss} = 3 \times 2 \times (8.768)^2 = 461.28 \text{ W}$$

$$\text{Power output} = 2539.18 - 461.28 = 2077.9 \text{ W}$$

2. When the load is delta connected.

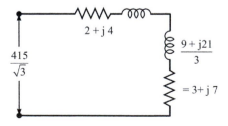

FIGURE E.6.10

Total impedance

$$= (2 + j4) + (3 + j7)$$
$$= (5 + j11)\ \Omega$$

$$\text{Current in the line}\ = \frac{415}{\sqrt{3}\ (5 + j11)} = (8.2 - j18.052)\,\text{A}$$
$$= 19.827\underline{|65°.5}$$

$$\phi_R = 65.5\ \text{degrees and}\ \cos\phi_R = 0.4135$$

$$\text{Power input} = \sqrt{3} \times 415 \times 19.827 \times 0.4135$$
$$= 5892.9\ \text{W}$$

$$\text{Losses} = 3 \times 2 \times (19.827)^2 = 2385.65\ \text{W}$$

$$\text{Power output} = 5892.9 - 2358.65$$
$$= 3534.24\ \text{W}$$

E.6.11 The sending end and receiving end voltages of a three-phase transmission line are maintained at 33 and 31.2, kV, respectively. The resistance and reactance per phase are 20 and 50 Ω, respectively. Determine the maximum power obtainable at the receiving end.

   Solution:

$$Z = \sqrt{20^2 + 50^2} = 53.85;\quad P_{max} = \frac{V_R^2}{Z^2}\left(\frac{V_S}{V_R}Z - R\right)$$

$$= \frac{(31.2)^2 \times 10^6}{(53.85)^2}\left[\frac{33}{31.2} \times 53.85 - 20\right]$$

$$= \frac{973.44 \times 10^6}{2899.82}(1.057692 \times 53.85 - 20)$$

$$= \frac{0.33568 \times 10^6 \times 36.956}{10^6}\text{MW} = 12.4\ \text{MW}$$

E.6.12 A three-phase voltage of 11 kV is applied to a line having $R = 8\ \Omega$ and $X = 11\ \Omega$ per conductor. At the end of the line, a balanced load of P kW per phase at leading power factor is connected. At what value of P is the voltage regulation zero when the power factor of the load is (1) 0.707 and (2) 0.8?

   Solution:

   From the phasor or diagram

$$V_S^2 = (V_R + I_R R \cos\phi_R - I_R \times \sin\phi_R)^2$$

$$+ (I_R \times \cos\phi_R + I_R R \sin\phi_R)^2$$

$$V_S = V_R = \frac{11}{\sqrt{3}}\ \text{kV} = 6.351\ \text{kV}$$

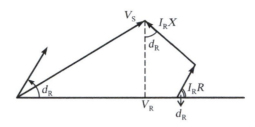

FIGURE E.6.12

1. At 0.707 leading power factor

$$(6.351 \times 10^3)^2 = [6.351 \times 10^3 + I(8 \times 0.707 - 11 \times 0.707)]^2$$
$$+ [I(11 \times 0.707 + 8 \times 0.707)]^2$$
$$(6.351 \times 10^3)^2 = (6.351 \times 10^3)^2 + [I_R(5.656 - 7.777)]^2$$
$$+ 2 \times 6.351 \times 10^3 \times 2.12\, I_R + I_R^2(7.777 + 5.656)^2$$

$$(I_R \times 2.121)^2 + 26,940.942\, I_R + I_R^2(13.433)^2 = 0$$

$$I_R^2(4.4986 + 180.445)^2 = -26,940.942\, I_R$$

$$I_R = \frac{26,940.942}{184.9436} = 145.67 \text{ A}$$

$$\text{Power per phase} = \frac{6.351 \times 10^3 \times 145.67 \times 0.707}{10^3} \text{ kW}$$

$$= 654.08 \text{ kW}$$

Total power $= 3 \times 654.08 = 1962.24$ kW

2. At 0.8 power factor leading

$$(6.351 \times 10^3)^2 = [6.351 \times 10^3 + I_R(6.4 - 6.6)]^2 + [I(6.4 + 6.6)]^2$$
$$= (6.351 \times 10^3)^2 + (0.2\, I_R)^2 + 2 \times 6.351 \times 10^3 \times 0.2 + 169\, I^2$$

$$169\, I_R^2 = 0.4 \times 6.351 \times 10^3\, I_R$$

$$I_R = \frac{0.4 \times 6.351 \times 10^3}{169} = 15.03 \text{ A}$$

Power per phase $= 6.351 \times 0.8 \times 15.03 = 76.36$ kW
Total power $= 3 \times 76.36 = 229$ kW

E.6.13 A three-phase transmission line has the following data.
   Line length 10 km
   Sending end voltage 11 kV
   Load at the receiving end 1000 KW at 0.8 (lag, power factor)
   Resistance of each conductor = 0.4 Ω/km
   Reactance of each conductor = 0.5 Ω/km
   Calculate the line current, receiving end voltage, and efficiency of transmission.

Solution:

$$R = 0.4 \times 10 = 4\ \Omega$$

$$X = 0.5 \times 10 = 5\ \Omega$$

$$V_S = \frac{11,000}{\sqrt{3}} = 6351\ \text{V}$$

$$V_S = (V_R + I_R R \cos\phi_R + I_R \times \sin\phi_R)^2 + (I_R \times \cos\phi_R - I_R R \sin\phi_R)^2$$

$$I_R = \frac{1000 \times 10^3}{\sqrt{3} \times 11 \times 10^3\ 0.8} = 65.6\ A$$

$$I_R R \cos\phi_R = 4 \times 65.6 \times 0.8 = 209.92$$

$$I_R X \sin\phi_R = 5 \times 65.6 \times 0.6 = 196.8$$

Neglecting the term $(I_R \times \cos\phi_R - I_R R \sin\phi_R)$

$$6351^2 = (V_R + 209.92 + 196.8)^2$$

$$6351 = V_R + 406.72$$

$$V_R = 6351 - 406.72 = 5944.28\ \text{V}$$

$$V_R(\text{line}) = 10,295.49\ \text{V}$$

$$\text{Losses} = 3 \times 65.6^2 \times 4 = 51,640.3\ \text{W}$$

$$\text{Efficiency of transmission}\,\frac{1,000,000}{1,000,000 + 51,640.3} \times 100$$

$$= 95.089\%$$

E.6.14 Find the sending end voltage, the percentage voltage drop, and the efficiency of a 20-km, 20-kV, 50-Hz, three-phase transmission line delivering power of 5000 kW at a power factor of 0.8 (lag). The conductors have a cross section of 95 mm^2 and an effective diameter of 12 mm. They are arranged in equilateral triangular formulation, the distance between conductor centers being 1 m.

Neglect capacitance. Temperature $= 20°C$; resistivity of copper $= \frac{1}{58}\ \Omega/\text{m}/\text{mm}^2$.

Solution:

$$\text{Resistance of the conductor} = \left(\frac{1}{58 \times 95}\right) \times 20 \times 1000$$

$$= 3.629\ \Omega$$

$$L = 2 \times 10^{-7}\ \ln\frac{D}{r^1} = 2 \times 10^{-7}\ \ln\left(\frac{1000}{0.7788 \times 6}\right)$$

$$= 2 \times 10^{-7}\ \ln 214 = 2 \times 10^{-7} \times 5.369\ \text{H/m}$$

$$\text{Total inductance} = 2 \times 10^{-7} \times 5.369 \times 20 \times 1000$$

$$= 0.0214639\ \text{H}$$

$$\text{Inductive reactance } X = 2\pi f L = 2 \times \pi \times 50 \times 0.0214639$$

$$= 6.7395\ \Omega$$

$$Z = (3.629 + j6.7395)\ \Omega$$

$$V_R = 20\ \text{kV}$$

$$I_R = \frac{5000 \times 10^3}{\sqrt{3} \times 20 \times 10^3 \times 0.8} = 180.42\ \text{A}$$

$$I_R R \cos \phi_R + I_R \times \sin \phi_R = 180.42 \times 3.629 \times 0.8 + 180.42 \times 6.7395 \times 0.6$$

$$= 180.42(2.9 + 4.0) = 6.9 \times 180.42 = 1244.76\ \text{V}$$

$$\text{Percentage voltage drop} = \frac{1264.76}{(20,000/\sqrt{3})} \times 100$$

$$= 10.78\%$$

$$V_S(\text{phase}) = \frac{20,000}{\sqrt{3}} + 244.76 = 11,547.344 + 1244.76$$

$$= 12,792.104\ \text{V}$$

$$V_S(\text{line}) = 22.156\ \text{KV}$$

$$\text{Losses} = 3 \times 3.69 \times (180.42)^2 = 360,343.736\ \text{W}$$

$$= 360.3437\ \text{kW}$$

$$\text{Input} = 5000 + 360.3437 = 5360.343\ \text{kW}$$

$$\text{Efficiency of line} = \frac{5000}{5360.343} \times 100 = 93.2776\%$$

E.6.15 A three-phase, 50-Hz overhead transmission line has the following data.

Resistance = 30 Ω
Inductive reactance = 60 Ω
Capacitive susceptance $3.5 \times 10^{-4}\ \Omega$
If the load at the receiving end is 80 MW at 0.8 power factor (lag) at 132 kV, calculate the sending end voltage, line current, sending end power factor, regulation of the line, and efficiency of transmission. Use nominal T-method.
Solution:

$$Z = (30 + j60)\ \Omega$$

$$\frac{Z}{2} = (15 + j30)\ \Omega$$

$$Y = j3.5 \times 10^{-4}\ \Omega$$

$$V_R = \frac{132 \times 10^3}{\sqrt{3}} = 76,212\ \text{V}$$

$$\text{Line current at load end } I_R = \frac{80 \times 10^6}{\sqrt{3} \times 132 \times 0.8 \times 10^3}$$

$$= 437.4\ \text{A} \angle 36°.87$$

$$= (349.92 - j262.44)$$

$$V = V_R + I_R\left(\frac{Z}{2}\right) = 76,210 + (349.92 - j262.44)(15 + j30)$$

$$= 89,332 + j6561$$

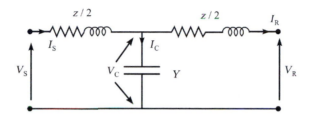

FIGURE E 6.15(A)

$$I_C = YV = j(3.5 \times 10^{-4})(89332 + j6561)$$

$$= (-2.296 + j31.2) \text{ A}$$

$$I_S = I_C + I_R = (349.92 - j262.44) - 2.296 + j31.2$$

$$= (347.624 - j231.24)A = 417.5 \ A \angle 33°.63$$

$$V_S = V + I_S\left(\frac{Z}{2}\right) = (89,332 + j6561) + (347.624 - j231.24) \times (15 + j30)$$

$$= 89332 + j6561 + 5214.36 + j10,428.72 - j3468.6 + 6937.2$$

$$= (101,483.56 + j13521)V \ \angle 7°.588$$

$$= (101.483 + j13.52) \text{ kV} \ \angle 7°.588 = 102.38 \text{ kV}$$

$$V_S \text{ (line)} = 102.38\sqrt{3} = 177.32 \text{ kV}$$

$$\text{Total angle between } V_S \text{ and } I_S = 33.63 + 7.588$$
$$= 41°.2185$$

$$\cos \phi_S = 0.7522$$

$$\text{Power at the sending end } P_S = \sqrt{3}V_SI_S \cos \phi_S$$

$$= \sqrt{3} \times 177.32 \times 417.5 \times 0.7522 \times 10^3 \text{ W}$$

$$= 96,448.487 \times 10^3 \ W = 96.448 \text{ MW}$$

$$\text{Efficiency of transmission} = \frac{80\ \text{MW}}{96.448\ \text{MW}} \times 100 = 82.94\%$$

$$\text{Percent regulation} = \frac{(177.32 - 132)}{132} \times 100$$

$$= 34.33\%$$

The phasor diagram is shown in Fig. E.6.15B

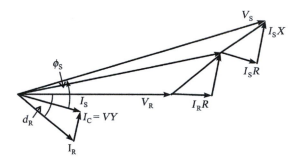

FIGURE E.6.15(B)

E.6.16 Solve Example (3.15) using nominal π-method.
   Solution:

$$Z = (30 + j60)\ \Omega$$

$$\frac{Y}{2} = \frac{j3.5 \times 10^{-4}}{2}\ \mho$$

$$= j1.75 \times 10^{-4}\ \mho$$

$$V_R = 76212$$

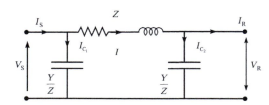

FIGURE E.6.16 (A)

$$I_R = 437.4 \angle 36°.8 = (349.93 - j262.44)\ \text{A}$$

$$I_{C_2} = V_2 \cdot \frac{Y}{2} = 76,212(j1.75 \times 10^{-4}) = j13.337\ \text{A}$$

$$\text{Line current, } I = 349.92 - j262.44 + j13.33$$
$$= (349.92 - j249.1) \text{ A}$$

Sending end voltage

$$V_S = V_R + I_Z$$
$$= 76,212 + (349.92 - j249.1)(30 + j60)$$
$$= 76,212 + 10,497.6 + j20,995.2 - j7473 + 14,946$$
$$= (101,655.6 + j13,522.2) = 102,545 \angle 7°.576$$
$$V_S = 102.545 \text{ kV} \angle 7°.576$$
$$V_S \text{ (line)} = \sqrt{3} \times 102.545 = 177.6 \text{ kV} \angle 7°.576$$
$$I_{C_1} = \frac{Y}{2}V_S = (j1.75 \times 10^{-4})(101.65 + j13.52) \times 10^3$$
$$= j177.88 \times 10^{-4} \times 10^3 - 23.66 \times 10^{-4} \times 10^3$$
$$= (-2.366 + j\,17.7) \text{ A}$$
$$I_S = I_{C_1} + I = (-2.366 + j17.7) + (349.92 - j249.1)$$
$$= (347.55 - j231.4) \text{ A} = 417.536 \angle 33°.6558$$
$$\cos\phi_S = 0.752$$

Efficiency of transmission:

$$\text{Power at the sending end} = \sqrt{3} \times 177.6 \times 417.536 \times 0.752$$
$$= 96,583 \times 10^3 \text{ kW}$$
$$= 96.583 \text{ MW}$$
$$\text{Efficiency} = \frac{80}{96.583} \times 100 = 82.83\%$$
$$\text{Percent regulation} = \frac{V_S - V_R}{V_R} \times 100 = \frac{177.6 - 132}{132} \times 100$$
$$= 34.54\%$$

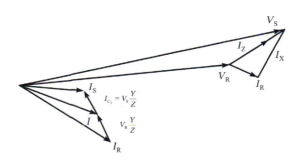

FIGURE E.6.16(B)

E.6.17 A three-phase, 50-Hz, 90-km long overhead line has the following line constants:

Resistance per kilometer = 0.15 Ω

Inductance per kilometer = 1.2 m H

Capacitance per kilometer = 0.009 μF

The line supplies 25 MW at 0.85 power factor (lag) and 170 kV. Using nominal π-representation, calculate the sending end voltage, current, power factor, regulation of the line, and efficiency of transmission.

Solution:

$$V_R = \frac{110 \times 1000}{\sqrt{3}} = 63,510 \text{ V} \angle 0 \text{ degree}$$

$$I_R = \frac{25 \times 10^6}{\sqrt{3} \times 110 \times 0.85 \times 10^3} = 154.376 \text{R} \angle 31.^\circ 788$$

$$= 154.376(0.85 - j0.5268) = (131.2 - j81.32)$$

$$R = 0.15 \times 90 = 13.5 \ \Omega$$

$$X = j2\pi \times 50 \times 1.2 \times 10^{-3} \times 90 = 33.9 \ \Omega$$

$$Y = (j2\pi \times 50 \times 0.009 \times 10^{-6}) \times 90 = 2.5434 \times 10^{-4} \angle 90^\circ \ \mho$$

$$V_S = V_R\left(1 + \frac{YZ}{2}\right) + I_R Z$$

$$YZ = j2.5434 \times 10^{-4}(13.5 + j33.9) = j0.0034 - 0.0086$$

$$\frac{YZ}{2} = -0.0043 + j0.0017$$

$$\left(1 + \frac{YZ}{2}\right) = (0.9957 + j0.0017)$$

$$V_R\left(1 + \frac{YZ}{2}\right) = 63,510(0.9957 + j0.0017)$$

$$= (63,236.9 + j107.967)$$

$$I_R Z = (131.2 - j81.32)(13.5 + j33.9)$$

$$= 1771.2 + j4447.68 - j1097.82 + 2756.748$$

$$= 4527.948 + j3349.86$$

$$V_S = 4527.948 + j3349.86 + 63,236.9 + j107.967$$

$$= (67,764.848 + j3457.827) \text{ V}$$

$$= 67,852 \angle 2^\circ.921 \text{ V}$$

$$V_S = 67.85 \text{ kV} \angle 2^\circ.921$$

$$V_S(\text{line}) = \sqrt{3} \times 67.85 = 117.52 \text{ kV} \angle 2^\circ.921$$

$$I_S = V_R Y\left(1 + \frac{YZ}{4}\right) + I_R\left(1 + \frac{YZ}{2}\right)$$

$$1 + \frac{YZ}{2} = 0.9957 + j0.0017$$

$$\frac{YZ}{4} = -0.00215 + j0.00085$$

$$1 + \frac{YZ}{4} = 1 - 0.00215 + j0.00085 = 0.99785 + j0.00085$$

$$I_R\left(1 + \frac{YZ}{2}\right) = (131.2 - j86.32)(0.9957 - j0.0017)$$

$$= (130.6358 - j0.223 - j80.944 - 0.13824)$$

$$= (130.497 - j81.082) \ A$$

$$Y\left(1 + \frac{YZ}{4}\right) = (0.99785 + j\, 0.00085) \times j\, 2.5434 \times 10^{-4}$$

$$V_R Y\left(1 + \frac{YZ}{4}\right) = 63,510 \times j2.5434 \times 10^{-4}(0.99785 + j0.00085)$$

$$= j16.15(0.99785 + j0.00085)$$

$$= (j16.11 - 0.01373)$$

$$I_S = 130.497 - j81.082 - j16.11 - 0.01373$$

$$= (130.44 - j64.97) \ A = 145.74 \angle 26°.97$$

$$\phi_S = 26.97 + 2°.92 = \angle 9°.89$$

$$\cos \phi_S = 0.8669$$

$$V_S(\text{at no load}) = \frac{V_S}{\left(1 + \dfrac{YZ}{2}\right)} = \frac{67,852}{0.9957} = 68,145$$

$$\text{Regulation} = \frac{68,145 - 63,510}{63,510} = \frac{4635}{63,510} = 0.07298$$

$$= 7.29\%$$

$$\text{Power loss} = 3 \times 13.5 \times (154.376)^2 \ W$$

$$= 965.19 \ kW$$

$$\text{Power input} = 25 \times 10^3 + 965.19 \ kW$$

$$\text{Efficiency of transmission} = \frac{25 \times 10^3}{25 \times 10^3 + 965.19}$$

$$= \frac{25}{25.965} \times 100 = 96.28\%$$

E.6.18 Determine the efficiency and regulation of a three-phase, 50-Hz transmission line which delivers a balanced load of 24 MVA at 0.8 lagging power factor. The receiving end voltage is 66 kV. Resistance, inductance, and capacitance per phase are 9.6 Ω, 0.097 H, and 0.765 μf, respectively. Neglect leakages. Use nominal T-method.

Solution:

$$Z = 9.6 + 314 \times 0.097 = 9.6 + j30.45 = 31.927 \angle 72°.474 \ \Omega$$

$$Y = j \, wC = 314 \times 0.765 \times 10^{-6} = j0.00024 \ \mho = 0.00024 \angle 90° \ \mho$$

$$YZ = 31.927 \angle 72°.474 \times 0.00024 \angle 90°$$
$$= -0.0072 + j0.0023$$

$$\frac{YZ}{2} = -0.0036 + j0.00115$$

$$\frac{YZ}{4} = -0.0018 + j0.000575$$

$$V_R = \frac{66}{\sqrt{3}} \ kV = 3816 \ V$$

$$I_R = \frac{24 \times 10^6}{\sqrt{3} \times 66 \times 10^3} = 209.95 \angle -36°.87$$

$$1 + \frac{YZ}{4} = 1 - 0.0036 + j0.00115 = 0.9964 + j0.00115$$

$$V_R \left(1 + \frac{YZ}{4}\right) = 38,106 \, [0.9964 + j0.00115] - 37.969 + j43.8$$

$$I_R Z = 209.95 \angle -36°.87 \times 31.927$$
$$= 6703.07 \angle -35°.63 = 6703.07[0.81279 + j0.5825]$$
$$= 5448.13 + j3904.53$$

$$1 + \frac{YZ}{4} = 1 - 0.0018 + j0.000575 = 0.9982 + j0.000575$$

$$I_R Z \left(1 + \frac{YZ}{4}\right) = (5448.13 + j3904.53)(0.9982 = j0.000575)$$

$$= 5438.32 + j3.1326 + j3897.5 - 2.2451$$

$$= 5436.07 + j3900$$

$$V_S = V_R \left[\left(1 + \frac{YZ}{2}\right)\right] + I_R Z \left(1 + \frac{YZ}{4}\right)$$

$$= 37,969 + j43.8 + 5436.07 + j3900$$

$$= 43,405 + j3943.8 = 43,583.8 \angle 5°.1916$$

$$\text{Regulation} = \frac{V_S - V_R}{V_R} \times 100 = \frac{43,583.8 - 38,106}{38,106} \times 100$$

$$= \frac{5477.8}{38,106} \times 100 = 14.37\%$$

$$\text{Output} = 24 \ MVA \times 0.8 = 19.2 \ MW$$

$$\text{Losses} = 3 \, I^2 R = 3 \times (209.95)^2 \times 9.6 = 1,269,475.272 \ W$$

$$\text{Input} = 19.2 \times 10^6 + 1,269,475.272$$
$$= 20,469,475 \text{ W}$$

$$\text{Efficiency} = \frac{19,200,000}{20,469,475} \times 100 = 93.79\%$$

E.6.19 A three-phase transmission line has the following constants (line to neutral):

$$R = 10 \ \Omega, \quad X = 20 \ \Omega, \quad Y = B = 4 \times 10^{-4} \ \text{℧}$$

Using nominal π-method, calculate the sending end voltage, line current, power factor, and efficiency of transmission when supplying a load of 10 MW at 66 kV, 0.8 (lag) power factor.

Solution:

$$P_R = 10 \text{ MW} = \frac{10}{3} \text{ MW/phase}$$

$$V_L = 66 \text{ kV}; \quad V_R \text{ per phase} = \frac{66 \times 10^3}{\sqrt{3}} = 38,106 \text{ V} \angle 0°$$

$$I_L = \frac{10 \times 10^6}{\sqrt{3} \times 66 \times 0.8 \times 10^3} = 109.35 \text{ A} \angle - 36°.87$$

$$Z = (10 + j20) \ \Omega = 22.36 \angle 63°.43$$

$$Y = j4 \times 10^{-4} = 4 \times 10^{-4} \angle 90° \text{℧}$$

$$YZ = 22.36 \times 4 \times 10^{-4} \angle 153°.43 = 0.00894 \angle 153°.43$$

$$= 0.00894(-0.8943 + j0.4473)$$

$$= (-0.00799 + j0.004)$$

$$\frac{YZ}{2} = -0.004 + j0.002 \text{ and } \frac{YZ}{4} = -0.002 + j0.001$$

$$1 + \frac{YZ}{2} = 1 - 0.004 + j0.002 = 0.996 + 0.002$$

$$\left(1 + \frac{YZ}{4}\right) = 1 - 0.002 + j0.001 = 0.998 + j0.001$$

$$V_R\left(1 + \frac{YZ}{4}\right) = 38,106(0.996 + j0.002) = 37,953.57 + j762$$

$$I_R Z = 109.35 \angle - 36°.87 \times 22.36 \angle 64°.43 = 2445 \angle 26°.56$$

$$= 2186.9 + j1093.1$$

$$I_R Z\left[1 + \frac{YZ}{4}\right] = (2186.9 + j1093.1)(0.998 + j0.001)$$

$$= 2182.5 + j2.1869 + j1090.9 - 1.093$$

$$= 2181.4 + j1093.0$$

$$V_S = V_R\left(1 + \frac{YZ}{4}\right) + I_R Z\left(1 + \frac{YZ}{4}\right)$$

$$= 37,953.7 + j76.2 + 2181.4 + j1093$$

$$= 40,134.97 + j1169.2 = 40,151.996 \angle 1°.668$$

$$V_S(\text{line}) = 40,151.996 \times \sqrt{3} \ \ V = 69.54 \ \text{kV}$$

$$I_S = I_R + V_C Y = I_R + YV_R + I_R\left(\frac{Z}{2}\right) \cdot Y$$

$$= 109.35(0.8 - j0.6) + j4 \times 10^{-4}(38,106) + 109.35(0.8 - j0.6)\frac{(10 + j20)}{2} \times j4 \times 10^{-4}$$

$$= 87.48 - j65.61 + j15.242 + (87.48 - j\,65.61)$$

$$(10 + j20) \times j2 \times 10^{-4}$$

$$= 87.48 - j65.61 + j15.242 + (874.8 - j656.1 + j1749.6 + 1312.2) \times j2 \times 10^{-4}$$

$$= 87.26128 - j50.457$$

$$= 100.798 \angle 30°.037746$$

Angle between V_S and $I_S = 30.03 + 1.668 = 31°.698$

$$\cos \phi_S = 0.8508$$

Efficiency of transmission $= \dfrac{V_R I_R \cos \phi_R}{V_S I_S \cos \phi_S} \times 100 \dfrac{38.106 \times 109.35 \times 0.8 \times 100}{40.152 \times 100.798 \times 0.85} = 96.67\%$

E.6.20 Find the efficiency and regulation of an 80-km, three-phase transmission line delivering 24 MVA at a power factor of 0.8 (lag) and 66 kV to a balanced load. The conductors are made of copper each having a resistance of 0.12 Ω/km and 1.5 cm outside diameter, spaced equilaterally 2.5 m between centers. Neglect leakages. Use nominal π-method.
  Solution:

Inductance of line $L = 2 \times 10^{-7} \ln \dfrac{D}{r^1}$

$$= 2 \times 10^{-7} \ln \frac{2.5 \times 2 \times 100}{0.7788 \times 1.5} = 2 \times 10^{-7} \times 6.059 \ \text{H/m}$$

$$= 2 \times 10^{-7} \times 6.059 \times 1000 \times 80 = 0.0969 \ \text{H}$$

$$X_L = 2\pi f L = 314 \times 0.0969 = 30.44 \ \Omega$$

$$R = 0.12 \times 80 = 9.6 \ \Omega$$

$$Z = R + jX_L = (9.6 + j30.44) \ \Omega$$

$$\text{Capacitance of the line } C = \frac{2\pi\epsilon}{\ln\dfrac{D}{r}} = \frac{2\pi \times 10^{-9}}{36\pi} \frac{1}{\ln\dfrac{2.5 \times 2 \times 1000}{1.5}}$$

$$= \frac{10^9}{18 \times 5.809}$$

$$Y = jwc = \frac{j314 \times 80 \times 10^{-9}}{18 \times 5.809} \times 1000 \text{ F} = j2.4024 \times 10^{-4} \ \mho$$

$$V_S = V_R\left(1 + \frac{1}{2}YZ\right) + I_R Z$$

$$V_R = 38.106 \text{ kV}$$

$$I_R = \frac{24}{3} \times \frac{10^3}{66} \times \sqrt{3} = 209.94 \text{ A}$$

$$I_R = 209.94(0.8 - j0.6) = (167.95 - j125.964) \text{ A}$$

$$I_R Z = (167.95 - j125.964)(9.6 + j30.44)$$
$$= (5441.5 + j3896.49) \text{ V}$$

$$YZ = (j2.04 \times 10^{-4})(9.6 + j30.4)$$
$$= -62 \times 10^{-4} + j19.58 \times 10^{-4}$$

$$\frac{YZ}{2} = (-31 + j9.79) \times 10^{-4}$$

$$\frac{YZ}{4} = (-15.5 + j4.895) \times 10^{-4}$$

$$\left(1 + \frac{YZ}{2}\right) = [0.9963 + j0.001152]$$

$$\left(1 + \frac{YZ}{4}\right) = [0.99845 + j0.0004895]$$

$$V_R\left(1 + \frac{1}{2}YZ\right) = 38,106(0.9963 + j0.001152)$$

$$= 3796.7 + j43.9$$

$$V_S = V_R\left(1 + \frac{1}{2}YZ\right) + I_R Z$$

$$V_S = 3796.7 + j43.9 + 5441.5 + j3896.49$$

$$= 43,408.5 + j3940.4 = 43,586.9 \text{ V} \angle 5°.186$$

$$= 43.5869 \text{ kV} \angle 5°.186$$

$$I_S = I_R Y\left(1 + \frac{1}{4}YZ\right) + I_R\left(1 + \frac{1}{2}YZ\right)$$

$$V_R Y = 38,106 \times j\,2.0402 \times 10^{-4} = j\,7.77$$

$$I_S = j7.77[0.99845 + j0.0004895] + [167.95 - j125.96][0.9963 + j0.001152]$$

$$= 167.5486 - j117.647$$

$$= 204.727 \angle 35°.075(\text{lag})$$

$$\phi_S = 5°.186 + 35°.075 = 40°.26$$

$$\cos \phi_S = \text{sending end power factor} = 0.763$$

$$\text{Efficiency of transmission} = \frac{8 \times 0.8 \times 10^3}{43.869 \times 204.727 \times 0.763}$$

$$= 93.99\%$$

$$\text{Regulation } V_{S_0} = V_R \left(1 + \frac{1}{2}YZ\right) + I_R Z(\text{at } I_R = 0)$$

$$\text{i.e., } V_{R_0} = \frac{V_S}{\left[\left(1 + \frac{1}{2}YZ\right)\right]}$$

$$V_{R_0} = \frac{V_S}{\left[1 + \frac{1}{2}YZ\right]} \times 100 = \frac{43.5869}{(0.9963 + j0.001152)} \times 100$$

$$= \frac{5.64}{38.106} \times 100 = 14.8\%$$

E.6.21 Determine the $A\ B\ C\ D$ constants for a three-phase, 50-Hz transmission line, 250-km long having the following distributed parameters:

$$l = 1.15 \times 10^{-3} \text{ H/km}$$

$$c = 7.8 \times 10^{-9} \text{ F/km}$$

$$r = 0.14 \text{ } \Omega/\text{km}$$

$$g = 0$$

Solution:

Series impedance z per kilometer

$$= (0.14 + j314 \times 1.15 \times 10^{-3}) = (0.14 + j0.3611) \text{ } \Omega/\text{km}$$

$$Z = zl = 250(0.14 \times j0.3611) = (35 + j90.275) \text{ } \Omega$$

$$= 96.82 \angle 68°.8$$

$$y = g + jwc = 0 + j314 \times 7.8 \times 10^{-9} \text{ } \Omega/\text{km}$$

$$= 2449.2 \times 10^{-9} \angle 90° \text{ } \Omega/\text{km}$$

$$Y = z \cdot l = j250 \times 2449.2 \times 10^{-9} = 0.000612 \angle 90°$$

$$\gamma l = \sqrt{zy} \cdot l = \sqrt{ZY}$$

$$= \sqrt{(96.82 \angle 68°.8 \times 0.0006123 \angle 90°)}$$

$$= \sqrt{6.0592829 \angle 158°.8}$$

$$= 0.24348 \angle 79°.4 = 0.044788 + j0.239316$$

$$YZ = 0.592825 \angle 158°.8$$

$$\frac{YZ}{2} = 0.02964 \angle 158°.8 = -0.02756 + j0.01067$$

$$\frac{Y^2Z^2}{24} = \frac{1}{12} \times \left(\frac{YZ}{2}\right) \times YZ$$

$$= \frac{1}{12} \times 0.02964 \angle 158°.8 \times 0.05928 \angle 158°.8$$

$$= 1.46.42 \times 10^{-4} \angle 317°.6 = 1.08 \times 10^{-4} - j0.9873$$

$$A = D = \cos \gamma l = 1 + \frac{YZ}{2} + \frac{Y^2Z^2}{24} + \cdots$$

$$= 1 + (-0.02756 + j0.01067) + 1.08 \times 10^{-4} - j0.9873 \times 10^{-4} + \cdots$$

$$\simeq 0.972540 - j0.01057 = 0.977258 \angle 0°.622$$

$$B = Z\left[1 + \frac{YZ}{6} + \frac{Y^2Z^2}{120} + \cdots\right]$$

$$= (35\ j90.275)$$

$$\left[1 + \frac{0.0592825 \angle 158°.8}{6} + \frac{1}{5}(108 \times 10^{-4} - j0.9873 \times 10^{-4})\right]$$

$$= [1 + (-0.009211359 + j0.0035728$$

$$+ 0.0000216 - j0.000019746][35 + j90.275]$$

$$= (35 + j90.275)(0.9908 + j0.003553)$$

$$= (34.678 + j0.124355 + j89.44 - 0.32)$$

$$= 34.358 + j89.564 = 96.04 \angle 68°9$$

$$C = y\left[1 + \frac{YZ}{6} + \frac{Y^2Z^2}{120} + \cdots\right]$$

$$= j0.0006123[0.9908 + j0.003553]$$

$$= j0.0006067 - 2.17 \times 10^{-6}$$

$$C \simeq j0.6067 \times 10^{-3}\ \mho$$

Hence

$$A = D = 0.97258 \angle 0°.622$$

$$B = 96.04 \angle 68°9 \ \Omega$$

$$C = 0.6067 \times 10^{-3} \ \mho \angle 90°$$

Alternative method:

$$\cosh \gamma_1 \ \gamma l = A = D = \cosh \alpha l \cos \beta l + j \sinh \alpha l \sin \beta l$$

$$\gamma l = 0.24348 \angle 79°.4 = 0.044788 + j0.239316$$

$$\cosh \gamma l = \cosh 0.04488 \cos 0.239316 + j \sinh 0.0444788 \sin 0.23936$$

$$= (1.001) \times (0.9715) + j(0.047898) \times (0.2370)$$

$$= 0.97247 + j0.01135 = 0.9725 \angle 0.0116 \ \text{rad}$$

$$= 0.9725 \angle 0°.6649$$

$$B = Z_C \sinh \gamma l = \sqrt{\frac{Z}{Y}} \sinh \gamma l$$

$$= \sqrt{\frac{Z}{Y}} [\sin l_1 \ \gamma l \cos \beta l + j \cosh \alpha l \sin \beta \ l]$$

$$= \sqrt{\left(\frac{35 + j90}{j0.0006123}\right)} \cdot [0.0448 \times 0.9715 + j1.001 \times 0.237]$$

$$= 397.65 \angle - 10°.51[0.0435 + j0.237237]$$

$$= 397.65 \angle - 10°.51 \times 0.241189$$

$$= 95.873 \angle 69°.139$$

$$C = \frac{1}{Z_C} \sin l_1 \ \gamma l \frac{1}{397.65} \angle 10°.51[0.241189 \angle 79°.649]$$

$$= 0.000606 \angle 90°.159$$

E.6.22 A three-phase, 50-Hz, 400-kV line is 500 km long. The line impedance is 0.93 mH/km and its capacitance is 0.012 μF/km per phase. For a lossless line, determine (1) the surge impedance Z_C, phase constant β, velocity of propagation v, and the line wavelength λ. (2) The receiving end is supplying a rated load of 750 MW at 0.8 power factor (lag) at 400 kV. Determine the sending end quantities and the voltage regulation.

Solution:

1. $\beta = \omega\sqrt{LC} = 2\pi \times 50 \times \sqrt{0.93 \times 10^{-3} \times 0.012 \times 10^{-6}}$

$$= 314 \times 0.3340 \times 10^5 = 0.00104876 \ \text{rad/km}$$

$$\text{Velocity of propagation} = \frac{1}{\sqrt{LC}}$$

$$= \frac{1}{\sqrt{\dfrac{0.93}{1000} \times \dfrac{1}{1000} \times \dfrac{0.012}{10^6} \times \dfrac{1}{1000}}} \text{ m/s}$$

$$= \frac{10^8}{\sqrt{0.1116}} = \frac{10^8}{0.334} = 2.994 \times 10^8 \text{ m/s}$$

$$\text{Wavelength } \lambda = \frac{v}{f} = \frac{2.994 \times 10^5}{50} \text{ km/s} = 5986 \text{ km}$$

2. $\beta l = 0.00104876 \times 500 \text{ km} = 0.52438 \text{ rad} = 30 \text{ degree}$

$$V_R = \frac{400\underline{|0°}}{\sqrt{3}} = 230.947\underline{|0°}\,\text{KV}$$

$$P_R = \frac{750}{0.8} \cos^{-1} 0.8 = 937.5 \angle -36°.87$$

$$I_R = \frac{937.5 \angle -36°.87 \times 10^3}{3 \times 230.947 \angle 0°} = 1353.12 \angle -36.87°$$

In a lossless line $|V_S| = |V_R|$ and $|I_S| = |I_R|$, voltage regulation is zero.

$$V_S = V_R \,(\cos \beta l + j \sin \beta l)$$

$$I_S = I_R \,(\cos \beta l + j \sin \beta l)$$

$$V_S \simeq 230.947 \,(\cos 30 + j \sin 30)$$

$$\simeq 230.947 \angle 30°$$

$$I_S = I_R \,(\cos 30 + j \sin 30)$$

$$= I_R \angle 30° = 1353.12 \angle -36°.87 + 30°$$

$$= 1353.12 \angle -6°.87$$

E.6.23 A star connected, 440-V, 50-c/s induction motor takes a line current of 40 A at a power factor of 0.8 lagging. Three mesh connected condensers are used to raise the power factor to 0.95. Find the kVA rating of the condenser bank and the capacitance of each condenser.

Solution:

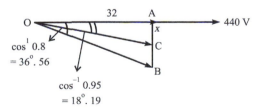

FIGURE E.6.23

$$\sin 18°19 = 0.3122498$$
$$\cos 36°861 = 0.8$$
$$\sin 36°861 = 0.6$$
$$\cos^{-1} 0.95 = 18°19$$
$$\tan 18°.19 = 0.3285898$$
$$40 \times 0.8 = 32 \ A = 0 \ A$$
$$AB = 40 \times 0.6 = 24 \ A$$
$$\frac{AC}{32} = \frac{X}{32} = 0.3285898; \quad x = 10.51488 \ A$$
$$24 - x = 13.48512 \ A$$

This is the current supplied into the line to compensate the reactive component of motor current of 24 A.

The phase current in delta-connected condenser

$$= \frac{13.48512}{\sqrt{3}} = 7.785796$$

The rating of the condenser bank

$$= \frac{3 \times 7.785796 \times 440}{1000} \text{KVA R} = 10.277 \text{ KVA R}$$

The capacitance of each condenser is c µf

$$I = V \cdot \omega = V \cdot 2\pi \times 50 \times C$$
$$C = \frac{I}{V.2\pi \times 50} = \frac{7.785796}{440 \times 2\pi \times 50} = 56.353 \ \mu f$$

E.6.24 A 200-kV, three-phase transmission line has the following constants:

$$A = D = 0.8541$$
$$B = 194$$
$$C = 0.0014 \angle 90°.5$$

The line is supplied 200 A at 0.95 power factor lagging.

Determine (1) the sending end voltage and (2) sending end current and its phase angle. Use circle diagrams.

Solution:

$$V_R = \frac{220}{\sqrt{3}} = 127.1 \text{ KV}; \quad \phi_R = \cos^{-1} 0.95 = 18°.2$$
$$(\beta - d) = 79° - 2° = 77°$$

Receiving end power circle diagram:

x-coordinate: $\frac{|A|}{|B|}|V_R|^2 \cos(\beta - \alpha) = \frac{0.854}{194} \times (127.1)^2 \times 0.207911 = 14.7851 \text{ MW}$

y-coordinate: $\dfrac{|A|}{|B|}|V_R|^2 \sin(\beta - \alpha)$ $= \dfrac{0.854}{194} \times (127.1)^2 \times 0.97437 = 69.29$ MVAR

Scale: 1 cm = 4 MW

1 cm = 4 MVAR

$x = 3.69$ cm; $y = 17.32$ cm

The receiving power circle is drawn as in Fig. E.6.24A with respect to O, the point O_1 is determined

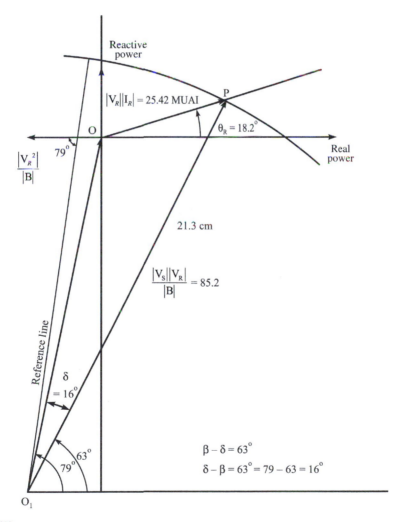

FIGURE 6.E.24(A)

$$OO_1 = \frac{|A|\,|V_R|^2}{|B|} \quad \text{and OP line is drawn at } 18.2°(\,= \theta_R)$$

$$\text{OP is set at} \frac{|V_S||V_R|}{|B|}$$

$$OP = V_R I_R = \frac{127.1 \times 200}{1000}\,\text{MW} = 25.42\,\text{MVA}$$

$$= 6.355$$

$$\text{Then} \quad \frac{V_S V_R}{B} = 21.8\,\text{cm} \times 4 = \frac{V_S \cdot 127.1}{194}$$

$$V_S = \frac{194 \times 4 \times 21.8}{127.1} = 130\,\text{kV}$$

$$V_S = (\text{Line}) = 230\,\text{kV}$$

Sending end power circle diagram

$$x\text{-coordinate:} \quad \frac{|D|}{|B|}|V_S^2|\cos(\beta - \Delta) = \frac{0.854}{194} \times 133^2 \times \cos 77°$$

$$x = 16.19$$

$$y\text{-coordinate:} \quad \frac{|D|}{|B|}|V_S^2|\sin(\beta - \Delta) = \frac{0.854}{194} \times 133^2 \times \sin 77°$$

$$y = 75.8723$$

Adopting the save scale of 1 cm = 4 MW = 4 MVAR

$$x = \frac{16.19}{4} = 4.0475\,\text{cm}$$

$$y = \frac{75.8723}{4} = 18.968\,\text{cm}$$

$$\frac{V_S V_R}{B} = \frac{133 \times 127.1}{194} = 87.1355 = 21.78\,\text{cm}$$

$$\frac{|D_1||V_S|^2}{(B)} = 77.868 = \frac{77.868}{4}\,\text{cm} = 19.467\,\text{cm}$$

$$\text{Reference line} \frac{|V_R^2|}{|B|} = \frac{(127.1)^2}{194} = 83.27 = 20.817\,\text{cm}$$

From the sending power circle diagram

$$OP = 6.725\,\text{cm} \times 4\,\text{MVA} = 26.9$$

$$OP = 26.9 = V_S I_S$$

$$I_S = \frac{26.9 \times 10^3}{133} = 202\ \text{A} \angle 20°$$

The angle of I_S is 20 degrees with respect to V_S

With respect to V_R, its total angle is

$$20 + 16 = 36°$$

Receiving end power circle diagram

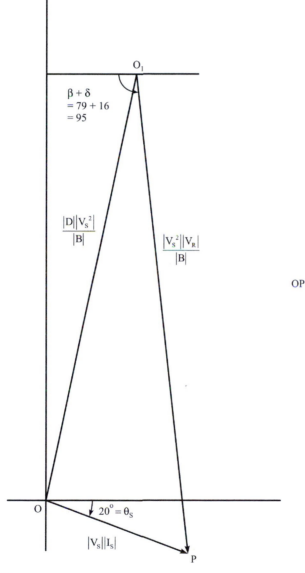

$\beta + \delta$
$= 79 + 16$
$= 95$

$\dfrac{|D||V_s^2|}{|B|}$

$\dfrac{|V_s^2||V_R|}{|B|}$

$OP = 6.725 \text{ cm} \times 4 \text{ MW}$
$26.9 = V_S I_S$

$I_S = \dfrac{26.9 \times 10^3}{133}$

$= 202 \lfloor 120°$

$20° = \theta_S$

$|V_s||I_s|$

FIGURE 6.E.24(B)

PROBLEMS

P.6.1 A three-phase transmission line of impedance $15 + j25 \, \Omega$ is fed through a 1:3 transformer whose equivalent impedance on the secondary side is $2 + j8 \, \Omega$. The load current is 100 A at power factor 0.8, while the line voltage at the midpoint of the line is 33 kV. Determine the supply voltage on the low voltage side. Determine also the equivalent resistance and reactance of each phase of the load.

P.6.2 Two short three-phase lines are operating in parallel and supply a total load of 6 MW at 33 kV. The load is balanced with a power factor of 0.8 (lag). The resistance and reactance of each line are 3 and 5 Ω, respectively. Determine the current, power, and power factor of each line.

P.6.3 A three-phase transmission line has a resistance of 10 Ω and inductive reactance of 20 Ω. Its capacitive susceptance $= 4 \times 10^{-4} \, \mho$. Calculate the sending end voltage, line current, power factor, and the efficiency of transmission when supplying a balanced load of 10 MW at 66 kV and 0.8 power factor lagging.
   1. Use nominal π-method
   2. Use nominal T-method

P.6.4 A 175 km long, three-phase, 132-kV line is delivering 60 MVA at 0.8 power factor (lag). Given $R = 0.156 \, \Omega/\text{km}$ and $x = 0.39 \, \Omega/\text{km}$. The capacitive susceptance $= 0.00874 \, \mu\text{f/km}$. Determine the A, B, C, D constants of the line operating at 50 Hz. Also, find the sending end voltage, current, power factor, efficiency of transmission, and regulation. The radius of the conductor is 1 cm.

QUESTIONS

Q.6.1 Explain the classification of transmission lines. Justify the classification.

Q.6.2 Explain the terms efficiency and regulation in relation to transmission lines.

Q.6.3 State and derive the condition for maximum power transfer on a short line.

Q.6.4 Given receiving end voltage and current, explain how the sending end values are determined using (1) nominal T-method and (2) nominal π-method.

Q.6.5 Derive the voltage and current relation for a long transmission line.

Q.6.6 Explain the following: (1) surge impedance, (2) velocity of propagation, (3) phase constant, and (4) attenuation constant.

Q.6.7 What do you understand by surge impedance loading? Explain.

Q.6.8 What is Ferranti effect? Draw the phasor diagram when this effect occurs on a transmission line.

Q.6.9 What are power circle diagrams? What is their utility? Explain.

CABLES

A cable is an insulated conductor. Cables are used for underground transmission and distribution of electrical power in densely populated areas. Since they are laid underground, external protection to withstand mechanical forces is required. Further the cable has to be protected against entry of moisture which will bring down the dielectric strength of the insulation. In addition, protection is needed in special cases against chemical reaction on insulation.

The conductor material is annealed copper in stranded form. Aluminum is also used in cables. Tin-plated copper, silver-coated copper, nickel-coated copper conductors in solid or stranded form are also used. Cables may be single conductors, multiconductor composite, and coaxial in construction. Cables are always expensive than overhead lines. Temperature rise is another restriction on the rating of the cable. At higher voltages, cables are always expensive and are used as alternatives only when it is absolutely necessary. However cables are widely used for low- and medium-voltage distribution.

7.1 TYPES OF CABLES

Low-Tension Cables: Low-tension cables up to 1 kV use paper insulation impregnated in oil; varnished cambric and vulcanized bitumen are other insulating materials used in low-voltage cables.

High-Tension Cables: Cables used up to 11 kV come into this category. In case of three-core cables, each core is insulated separately and the three cores are belted as shown in Fig. 7.1. The whole cable is covered with lead sheath through an extruding process to prevent moisture from entering into the cable. An additional steel armor is provided to withstand wear and tear if the situation demands.

H-Type Cables: These cables are designed by Hochstaedter and are used for 22 and 33 kV applications. Each insulated core is covered by a metallized paper maintained at ground potential. This eliminates the belt insulation from the influence of electrostatic field. Otherwise, in three-core cables belt insulation with filler material is subject to void formation, ionization, and breakdown under electric stress. Fig. 7.2 shows the features of H-type cable.

Polyethylene, polypropylene, polyurethane, polyester elastomer, Mylicon, cellular fluoropolymer fluoroelastomer, etc., care the modern materials used for cable insulation.

SL and HSL Cables: In SL-type cables, each core is covered separately by a lead sheath. In HSL-type cables each core has both metallized paper covering grounded to zero potential as well as a separate lead sheath for each core. Fig. 7.3 shows HSL-type cable.

EHV Cables: These cables are generally of single-core type since it is convenient to handle them and install. For such requirements three-core cables will be unwieldy to handle and install as the size of the cable becomes very large. However, in single-core cables the sheath losses are more.

Electrical Power Systems. DOI: http://dx.doi.org/10.1016/B978-0-08-101124-9.00007-3

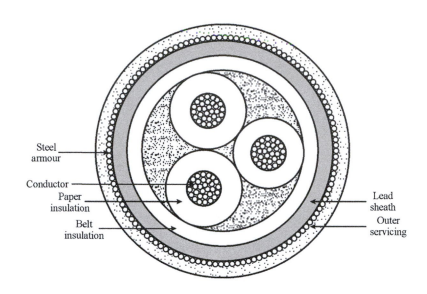

Steel armour

Conductor

Paper insulation

Belt insulation

Lead sheath

Outer servicing

FIGURE 7.1

Three-core belted cable.

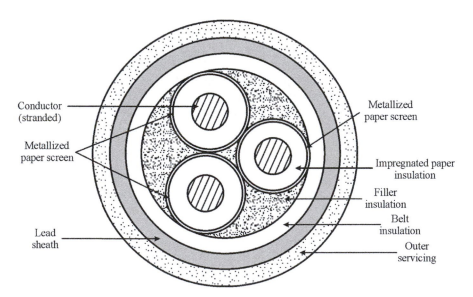

Conductor (stranded)

Metallized paper screen

Lead sheath

Metallized paper screen

Impregnated paper insulation

Filler insulation

Belt insulation

Outer servicing

FIGURE 7.2

Three-core H-type cable.

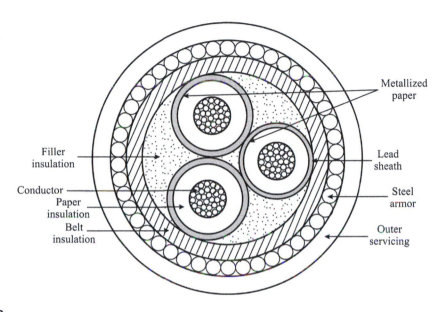

FIGURE 7.3

Three-core HSL cable.

7.2 CABLE INSULATION

Cable insulating materials must possess high dielectric strength, high insulation resistance, good mechanical properties, such as tenacity and elasticity, and high-temperature withstand capability. Furthermore the material must be nonhygroscopic and resistant to chemical action.

Vulcanized India Rubber: Natural rubber mixed mainly with sulfur is processed to become vulcanized rubber. This material has good insulating property and is moisture resistant. Vulcanized India rubber (VIR) cables are used for wiring domestic, office, and factory buildings.

Synthetic Rubber: Several synthetic materials are available which are similar to rubber in resemblance but have better moisture, fire, and oil resisting properties. Some of them are butyl rubber, silicone rubber, neoprene, and styrene rubber. They are all made of synthetic elastomeric compounds. They are used up to 60–85°C.

Polyvinyl Chloride: It is a polymer made from acetylene and depending upon the process, different grades of polyvinyl chloride (PVC) products are obtained. PVC is preferred to rubber under certain environmental conditions such as oil, alkalis, acids, and oxygen.

Cables are classified as solid cables and pressurized cables. All the cables described so far come under the category of solid cables. At higher voltages (33 kV and above), void formation is enhanced due to high electric stresses. To avoid void formation or minimize it, cables are filled with pressurized oil or gas. In case of oil it is the mineral oil used for impregnation. Nitrogen is the most common gas used to pressurize the cable. Such pressurized cables are used even up to 500 kV. Leakage of oil is the main problem in oil-filled cables. Maintenance of reservoirs of oil, connections, and couplings are other problems associated with oil-filled cables. The pressure of oil should always be maintained to be above atmospheric pressure.

In case of gas-filled cables, nitrogen gas at $1000\,\text{kV/m}^2$ pressure is used to fill the interstices between layers of insulation. Alternatively the whole cable is placed inside a steel pipe filled with nitrogen and maintained at $12-15$ atmosphere. SF_6 gas is also used as a medium for pressure.

7.3 ELECTROSTATIC STRESS IN SINGLE-CORE CABLE

Consider the single-core cable shown in Fig. 7.4. The radius of the conductor is r and inner sheath radius R. The dielectric between the conductor and the sheath has a permittivity of ε. The charge per unit length is q (c/m). The cable is rated to operate at a voltage V.

The electric field intensity or the potential gradient at a radial distance x from the conductor surface at point P in the dielectric.

$$g = \frac{q}{2\pi\varepsilon x}\ (\text{c/m}^2)$$
$$= E\ (\text{c/m}^2) \tag{7.1}$$

The potential difference V between the core and the sheath is given by

$$V = -\int_R^r E \cdot dx = -\int_r^R \frac{q}{2\pi\varepsilon x}\cdot dx$$
$$= \frac{q}{2\pi\varepsilon}\ln\frac{R}{r}\ (V) \tag{7.2}$$

But

$$g = \frac{q}{2\pi\varepsilon x} = \frac{2\pi\varepsilon}{\ln\dfrac{R}{r}}\cdot\frac{V}{2\pi\varepsilon\cdot x} = \frac{V}{x\ln\dfrac{R}{r}} \tag{7.3}$$

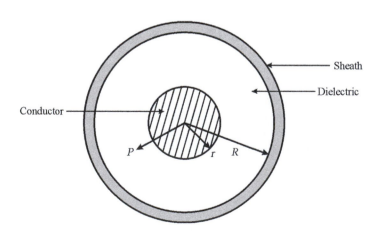

FIGURE 7.4

Single-core cable.

G is a minimum at $x = R$ and is a maximum at $x = r$

$$g_{min} = \frac{V}{R \ln \dfrac{R}{r}} \quad (V/m) \tag{7.4}$$

$$g_{max} = \frac{V}{r \ln \dfrac{R}{r}} \quad (V/m) \tag{7.5}$$

For a given operating voltage V, keeping g_{max} to a minimum, a relationship can be obtained between r and R that makes the overall size of the cable most economical. Differentiating g_{max} and equating to zero

$$\frac{d}{dr}\left[\frac{V}{r \ln\left(\dfrac{R}{r}\right)}\right] = 0$$

$$\frac{0 - V \dfrac{d}{dr} r \ln\left(\dfrac{R}{r}\right)}{\left[r \ln\left(\dfrac{R}{r}\right)\right]^2} = 0$$

gives $\ln R/r = 1$, i.e.,

$$\frac{R}{r} = e \tag{7.6}$$

$R = er = 2.71828\,r$, where e is the Naperian base.

This is the minimum value for the inner radius of the sheath for which g at the conductor radius does not exceed g_{max}. If g exceeds this value, the dielectric may fail.

7.4 GRADING OF CABLES

It is seen in Section 7.3 that the potential gradient is maximum at the conductor surface and decreases gradually as the sheath is reached. If this difference could be reduced by some means, the cable size can be reduced for the same operating voltage. In other words, with the same dimensions the cable could be used for higher voltages. There are two methods suggested for this purpose: intersheath grading and capacitance grading.

7.4.1 INTERSHEATH GRADING

Consider a single-core cable containing two metallic intersheaths between the core and the outer sheath. By external means suppose that it is possible to maintain the two intersheaths at fixed potentials V_1 and V_2 as shown in Fig. 7.5, the cable is operated at a voltage V.

The potential difference between the conductor and the first intersheath

$$V - V_1 = \int_r^{r_1} g_1 \cdot dx \tag{7.7}$$

where g_1 is the potential gradient at a distance x between the conductor and the first intersheath.

$$g_1 = \frac{q}{2\pi\varepsilon x} = \frac{c_1}{x}$$

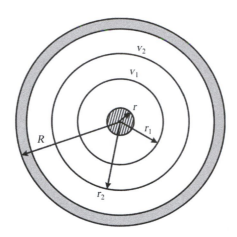

FIGURE 7.5

Intersheath grading.

where c_1 is a constant.

$$V - V_1 = \int_r^{r_1} \frac{c_1}{x} dx c_1 \ln \frac{r_1}{r}$$

$$c_1 = \frac{V - V_1}{\ln \dfrac{r_1}{r}} \tag{7.8}$$

The maximum stress at $x = r$ is given by

$$g_{1\ max} = \frac{V - V_1}{r \ln \dfrac{r_1}{r}} \tag{7.9}$$

In a similar manner it can be derived that

$$g_{2\ max} = \frac{V_1 - V_2}{r_1 \ln \dfrac{r_2}{r_1}} \quad \text{at the first intersheath surface} \tag{7.10}$$

and

$$g_{3\ max} = \frac{V_2 - 0}{r_2 \ln \dfrac{R}{r_2}} \quad \text{at the second intersheath surface} \tag{7.11}$$

By suitable choice of V_1 and V_2, it is possible to make

$$g_{1\ max} = g_{2\ max} = g_{3\ max} = g_{max} \tag{7.12}$$

Fig. 7.6 shows the variation of electric stress in the cable dielectric for (a) single core cable with no intersheaths and (b) with two intersheaths.

Consider that the thicknesses of the insulation are so selected that

$$\frac{r_1}{r} = \frac{r_2}{r_1} = \frac{R}{r_2} = k \tag{7.13}$$

If g_{max} is to be the same at various radii
then

$$\frac{V - V_1}{r \ln \dfrac{r_1}{r}} = \frac{V_1 - V_2}{r_1 \ln \dfrac{r_2}{r_1}} = \frac{V_2}{r_2 \ln \dfrac{R}{r_2}} \tag{7.14}$$

That is

$$\frac{V - V_1}{r \ln k} = \frac{V_1 - V_2}{r_1 \ln k} = \frac{V_2}{r_2 \ln k} \tag{7.15}$$

From the first two ratios

$$V_{r_1} - V_1 r_1 = V_1 r - V_2 r$$
$$V_{r_1} = V_1[r + r_2] + V_2 r \tag{7.16}$$

From the second and third ratios

$$V_1 r_2 - V_2 r_2 = V_2 r_1$$
$$V_1 r_2 = V_2\{r_1 + r_2\}$$
$$V_1 = V_2\left[\frac{r_1 + r_2}{r_2}\right] = \left(\frac{r_1}{r_2} + 1\right)V_2 \tag{7.17}$$
$$V_1 = V_2\left[\frac{1}{k} + 1\right] = \left(\frac{k + 1}{k}\right)V_2$$

From the first and third ratios

$$Vr_2 - V_1 r_2 = V_2 r$$

$$Vr_2 = V_1 r_2 + V_2 r = \left(\frac{k + 1}{k}\right)V_1 r_2 + V_2 r$$

$$V = V_2\left[\frac{k + 1}{k}r_2 + r\right]\cdot\frac{1}{r_2} = V_2\left[\frac{k + 1}{k}\frac{r_2}{r_2} + \frac{r}{r_2}\right] \tag{7.18}$$

$$= V_2\left[\frac{k + 1}{k} + \frac{r}{r_1}\cdot\frac{r_1}{r_2}\right] = V_2\left[\frac{k + 1}{k} + \frac{1}{k}\cdot\frac{1}{k}\right]$$

$$= V_2\left[\frac{k + 1}{k} + \frac{1}{k^2}\right] = V_2\left[\frac{1 + k + k^2}{k^2}\right]$$

also

$$V = \left[\frac{1 + k + k^2}{k^2}\right]\cdot V_1\left[\frac{k}{k + 1}\right] = V_1\left[\frac{1 + k + k^2}{k(k + 1)}\right] \tag{7.19}$$

Then, from Eq. (7.9),

$$g_{max} = \frac{V - V_1}{r \ln \dfrac{r_1}{r}} = V\left[\frac{1 - \dfrac{V(k + 1).k}{1 + k + k^r}}{r \ln k}\right]$$

$$= \frac{V}{r \ln k}\cdot\left[\frac{1 + k + k^2 - k^2 - k}{(1 + k + k^2)}\right] = \frac{V}{r \ln k}\cdot\frac{1}{(1 + k + k^2)}$$

$$g_{max} \text{ without intersheath} = \frac{V}{r \ln \dfrac{R}{r}} = g_{max}^0$$

Then

$$g_{max}^0 = \frac{V}{r \ln \left(\dfrac{R}{r_2} \cdot \dfrac{r_2}{r_1} \cdot \dfrac{r_1}{r} \right)} = \frac{V}{r \ln k^3} = \frac{V}{3r \ln k} \tag{7.20}$$

$$\frac{g_{max} \text{ with inter sheath}}{g_{max}^0 \text{ (without inter sheath)}} = \frac{3}{1 + k + k^2} \tag{7.21}$$

Since $k > 1$, the gradient with intersheaths is smaller than the case where no intersheaths are used.

The voltage of the cable with intersheaths is

$$V = g_{max} \left[r \ln \frac{r_1}{r} + r_1 \ln \frac{r_2}{r_1} + r_2 \ln \frac{R}{r_2} \right] \tag{7.22}$$

When intersheaths are used, it is required to keep them at the specified potential by some external means. Hence this method is not practical.

7.4.2 CAPACITANCE GRADING

In this method dielectrics with different permittivities are selected. Let there be three layers of different dielectrics of permittivities ε_1, ε_2, and ε_3; the dimensions and thicknesses are as shown in Fig. 7.6. The voltage of the cable

$$V = \int_r^{r_1} g_1 dx + \int_{r_1}^{r_2} g_2 dx + \int_{r_2}^{R} g_3 dx$$

$$= \frac{q}{2\pi} \left[\frac{1}{\varepsilon_1} \ln \frac{r_1}{r} + \frac{1}{\varepsilon_2} \ln \frac{r_2}{r_1} + \frac{1}{\varepsilon_3} \ln \frac{R}{r_2} \right] \tag{7.23}$$

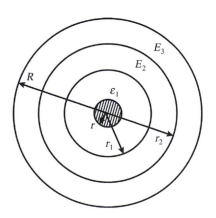

FIGURE 7.6

Capacitance grading.

Capacitance of the cable

$$C = \frac{Q}{V}$$

$$= \frac{2\pi}{\left[\frac{1}{\varepsilon_1} \ln \frac{r_1}{r} + \frac{1}{\varepsilon_2} \ln \frac{r_2}{r_1} + \frac{1}{\varepsilon_3} \ln \frac{R}{r_2}\right]} \tag{7.24}$$

$$g_{1\ max} = \frac{V}{r\left(\ln \frac{r_1}{r} + \frac{\varepsilon_1}{\varepsilon_2} \ln \frac{r_2}{r_1} + \frac{\varepsilon_1}{\varepsilon_3} \ln \frac{R}{r_2}\right)} \tag{7.25}$$

$$g_{2\ max} = \frac{V}{r_1\left(\frac{\varepsilon_2}{\varepsilon_1} \ln \frac{r_1}{r} + \ln \frac{r_2}{r_1} + \frac{\varepsilon_2}{\varepsilon_3} \ln \frac{R}{r_2}\right)} \tag{7.26}$$

and

$$g_{3\ max} = \frac{V}{r_2\left(\frac{\varepsilon_3}{\varepsilon_1} \ln \frac{r_1}{r} + \frac{\varepsilon_3}{\varepsilon_2} \ln \frac{r_2}{r_1} + \ln \frac{R}{r_2}\right)} \tag{7.27}$$

If the same maximum stress is to exist at each layer (i.e., at r, r_1 and r_2) then

$$g_{max} = \frac{q}{2\pi\varepsilon_1 r} = \frac{q}{2\pi\varepsilon_2 r_1} = \frac{q}{2\pi\varepsilon_3 R} \tag{7.28}$$

$$\varepsilon_1 r = \varepsilon_2 r_1 = \varepsilon_2 R \tag{7.29}$$

Since $r < r_1 < r_2$, we have to select

$$\varepsilon_1 > \varepsilon_2 > \varepsilon_3 \tag{7.30}$$

The permittivities of the dielectrics are to be so selected that they are in the descending order from core to sheath. Absolute grading of cables is not possible. But dielectrics of different strengths can be selected to get a better cable design.

7.5 CAPACITANCE OF THE CABLE

Consider a single-core cable carrying a charge of q (c/m). From Eq. (7.2)

$$V = \frac{q}{2\pi\varepsilon} \ln \frac{R}{r} \text{ (V)}$$

Then, capacitance per meter length of the cable

$$C = \frac{q}{V} = \frac{q}{q} \cdot \frac{2\pi\varepsilon}{\ln \frac{R}{r}} = \frac{2\pi\varepsilon}{\ln \frac{R}{r}} \text{ (f/m)} \tag{7.31}$$

Since the conductor is much nearer to the sheath which is at ground potential, the capacitance of a cable is much higher than the capacitance of an overhead line which is at a much larger distance from the ground. Furthermore the permittivity of the cable dielectric ε is much larger distance from the ground. Further, the permittivity of the cable dielectric ε is much higher than that for air.

7.6 **CAPACITANCE OF THREE-CORE CABLES**

In case of three-core cables, there is mutual capacitance between the three cores. But, in addition, there is also capacitance between each core and the sheath. This is shown in Fig. 7.7.

C_s and C_c are the core-to-sheath and core-to-core capacitances. The core-to-core mutual capacitances C_c, which are delta connected can be converted into equivalent as shown in Fig. 7.8. The two connections are related by

$$\frac{1}{\omega C_c^1} = \frac{1}{3\omega C_c}$$

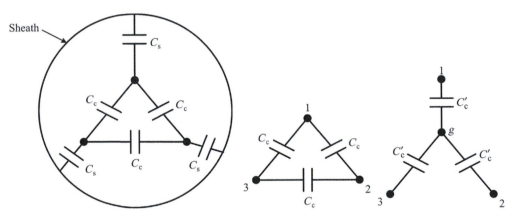

FIGURE 7.7

Capacitance in three-core cables.

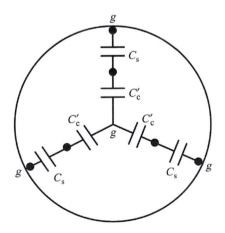

FIGURE 7.8

Three-core cable Δ-Y conversion.

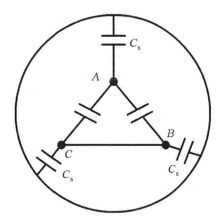

FIGURE 7.9

Three-core cable—Test I.

$$C_c^1 = 3C_c \tag{7.32}$$

The star point will be at zero potential as at ground.
The total capacitance of each conductor from Fig. 7.8 is seen as

$$C_0 = C_c^1 + C_s = 3C_c + C_s \tag{7.33}$$

Note that since the star point 0 and the sheath are both at ground potential, both C_s and C_c^1 are in parallel. These capacitances can be experimentally determined as follows:

Test 1: Conductors B and C are connected to sheath and the capacitance between the sheath and conductor is measured. This gives (Fig. 7.9)

$$C_1 = 2C_c + C_s \quad \text{(see Fig. 7.10)} \tag{7.34}$$

Test 2: The three cores are bunched together and the capacitance between this bunch and the sheath is measured. The measured capacitance

$$C_2 = 3 \, C_s \tag{7.35}$$

as all the core-to-sheath capacitance C_s are connected in parallel (see Fig. 7.10)

$$C_s = \frac{1}{3}C_2 \tag{7.36}$$

Once C_s is determined, C_c can be determined from Eq. (7.34).

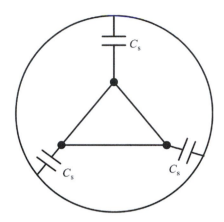

FIGURE 7.10

Three-core cable—Test II.

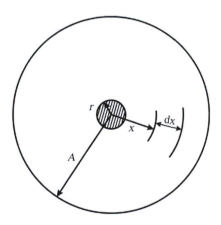

FIGURE 7.11

Insulation resistance.

7.7 INSULATION RESISTANCE OF CABLES

The resistance of a material to flow of current is given by

$$R = \frac{\rho l}{a} \tag{7.37}$$

Consider a single-core cable of conductor radius r and sheath inner radius R. Consider an annular ring of the cable at a radial distance x; $r < x < R$ as shown in Fig. 7.11 of thickness dx. The insulation resistance of this annular ring per meter length

$$dR_i = \rho \frac{1 \times dx}{2\pi x \times 1}$$

$$\int dR_i = R_i = \frac{\rho}{2\pi} \int_r^R \frac{dx}{x} \tag{7.38}$$

$$R_i = \frac{\rho}{2\pi} \ln \frac{R}{r} \quad (\Omega/m)$$

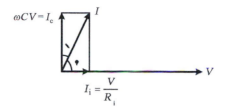

FIGURE 7.12

Cable power factor.

A cable of length L will have a leakage resistance of $(\rho/2\pi L) \ln (R/r)$ (Ω)

The leakage current flows through this insulation resistance R_i and is given by V/R_i.

This is in phase with the cable voltage V and results in thermal loss (see Fig 7.12). The capacitive current through the dielectric $I_c = \omega C V$ is leading the voltage V by 90 degrees and the resultant of both these currents I is at an angle ϕ with respect to V and is leading the voltage.

$$\cos \phi = \cos \left(\frac{\pi}{2} - \delta\right) = \sin \delta \tag{7.39}$$

when δ is small $\sin \delta \simeq \delta$ and also $\tan \delta = \delta$

$$\tan \delta = \frac{V}{R_i} \cdot \frac{1}{\omega C V} = \frac{1}{R_i \omega C} = \frac{G_c}{\omega C} \tag{7.40}$$

where G_c is the cable conductance.

7.8 BREAKDOWN OF CABLES

Cables are likely to fail in operation due to three reasons.

1. *Void Formation*: The cable is subjected to cycles of operation and therefore it undergoes alternate heating and cooling processes. This causes minute expansion and contraction of layers of insulation resulting in the formation of voids with trapped air therein. As air has low breakdown strength compared to the dielectric in the cable, ionization starts in these voids. Carbon formation starts in the oil used in paper-insulated cables. Since maximum stress occurs at the conductor surface, carbon formation starts at that place and extends slowly toward the lead sheath. This is called tracking. Majority of cable breakdowns occur due to tracking.
2. *Disruptive Breakdown*: If the dielectric used is defective, a puncture may occur and the cable may fail. Normally, such defects are detected in the laboratory tests and eliminated at the time of manufacture. Failure of cables due to breakdown (puncture) of dielectric is very rare.
3. *Failure Due to Thermal Instability*: Thermal instability may not occur in a well-laid cable system. As discussed earlier, temperature rise occurs in cables during normal operation due to core and dielectric losses. The heat produced must be properly dissipated. If the rise in temperature is not arrested, losses may cumulatively increase, resulting in thermal breakdown of the cable.

7.9 THERMAL PHENOMENON IN CABLES

Heat is generated in cables due to the following causes:

1. Core loss or copper loss in the conductor
2. Dielectric loss in the insulation
3. Sheath loss

Core Loss or Copper Loss: Core loss depends upon the resistance of the cable conductors, and hence it is necessary to estimate the resistance of the conductor under actual conditions of operation. The temperature variation of resistance is calculated from

$$R_{t_2} = R_{t_1}[1 + \alpha(t_2 - t_1)] \tag{7.41}$$

As the conductors are stranded, the effective cross section of the core may be smaller than the actual section and hence the resistance may be higher than the estimated resistance. Again, due to stranding all the strands may not have the same length as they are twisted spirally while forming the core. The central strands are smaller than the outer layer strands. For these reasons, the resistance may have to be further increased in addition to taking care of the working temperature. This is done by increasing the value obtained from Eq. (7.41) by 1.02–1.04. From the corrected resistance R, the core loss I^2R can be estimated.

Dielectric Loss: It is already explained that the cable can be represented by a pure capacitor C in parallel with a resistor R to represent the leakage loss (see Fig. 7.13).

The dielectric power loss

$$P_d = V \cdot \left(\frac{V}{R_i}\right) = V \cdot V\omega C \tan \delta$$
$$\cong V^2 \omega C \delta \ \text{(W)} \tag{7.42}$$

The power factor angle δ increases with temperature. The loss also varies as the square of the voltage.

The equivalent circuit for the coupling between the core and the sheath is shown in Fig. 7.14. It is an air-cored transformer with 1:1 turns ratio as the core is the primary and sheath is the

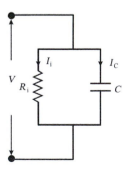

FIGURE 7.13

Cable dielectric equivalent circuit.

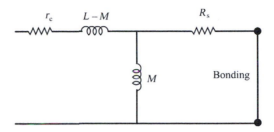

FIGURE 7.14

Cable core and sheath coupling circuit.

secondary. r_c is the resistance of the core, $L\text{-}M$ is the leakage inductance of the core, M is the mutual inductance between the core and the sheath, and R_s is the sheath resistance.

$$L = 2 \times 10^{-7} \ln \frac{D}{r} \ (\text{h/m}) \tag{7.43}$$

and

$$M = 2 \times 10^{-7} \ln \frac{D}{r_s} \tag{7.44}$$

where D is the distance between two cores, center-to-center.

$$L - M = 2 \times 10^{-7} \ln \frac{r_s}{r} \ (\text{h/m}) \tag{7.45}$$

The equivalent impedance of the circuit referred to primary

$$Z_{eq} = r_c + j\omega(L - m) + \frac{R_s j\omega M}{R_s + j\omega M}$$

$$= r_c + j\omega(L - m) + \frac{j\omega M R_s (R_s - j\omega r)}{R_s^2 + \omega^2 M^2} \tag{7.46}$$

$$= \left[r_c + \frac{\omega^2 M^2 R_s}{Rs^2 + \omega^2 M^2} \right] + j\omega \left[L - M \frac{\omega^2 M^2}{R_s^2 + \omega^2 r^2} \right]$$

$$= R_{eq} + j\omega L_{eq} \tag{7.47}$$

Bonding of cable sheath results in increasing the core resistance r_c by

$$\frac{\omega^2 M^2 R_s}{R_s^2 + \omega^2 M^2}$$

and in reducing the inductance by

$$M \frac{\omega^2 M^2}{R_s^2 + \omega^2 r^2}$$

To eliminate the sheath losses in three-core cables, the cables are transposed as in the case of overhead lines. The sheath are also cross bonded as shown in Fig. 7.15.

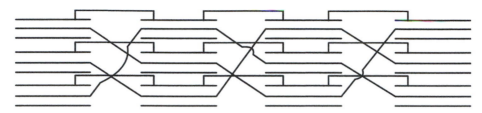

FIGURE 7.15

Cable-cross banding.

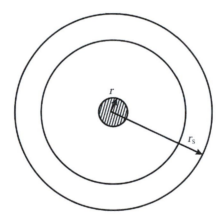

FIGURE 7.16

Sheath loss.

7.10 SHEATH LOSS

When power is transmitted or distributed using single-core cables, since the flux is alternating, it links the sheath and induces emf in it. This induced emf circulates induced currents in the sheath, which results in loss of energy. To account for the sheath loss, the AC resistance of the cable may be increased by a factor. In addition to these sheath losses, if the sheaths of other cores are all connected electrically at both ends or at different points on the way, then these currents will circulate and may result in slightly higher losses. Connecting cable sheaths in this way is called bonding (Fig. 7.16).

If M is the coefficient of mutual induction between the core and cable then, the induced emf in sheath E due to current I in the conductor

$$E = \omega M I \tag{7.48}$$

The mutual induction

$$M = 2 \times 10^{-7} \ln \frac{r}{r_s} \quad (h/m)$$

Hence

$$E = 2\omega I \times 10^{-7} \ln \frac{r}{r_s} \quad (\text{V/m}) \tag{7.49}$$

If two single-core cables are bonded at one end, the voltage at the other end between the two sheaths will be $2E$ (V/m).

When a short circuit occurs, the current I is large enough to make E substantially high so as to create a spark between two sheaths. One formula for estimating the sheath losses suggested in the literature is given by

$$P_{se} = \text{Sheath eddy current loss}$$

$$= \left[I^2 \left[\frac{3\omega^2}{R_s} \left(\frac{r_s}{2r} \right)^2 \times 10^{-18} \right] \right] \quad (\text{W/cm/phase}) \tag{7.50}$$

where I is the conductor current in amperes and R_s is the sheath resistance.

7.11 THERMAL RESISTANCE

Just as the conductor material offers resistance to flow of current and dielectric material has insulation resistance, so also for the transfer or flow of heat in materials such as dielectrics. We consider thermal resistance. Similar to specific resistivity ρ, we can define thermal resistivity ρ_{th}.

Similar to the derivation for insulation resistance in Section 7.7, we can consider an annular length of the dielectric of thickness dx a radial distance x from the conductor surface and write down expression for thermal resistance.

$$dG = \frac{\rho_{thc} dx}{2\pi x} \quad (\text{th-}\Omega/\text{m}) \tag{7.51}$$

The unit for ρ_{th} is °C-m/W. For cable insulation, it is about 5°C−6°C-m/W. The thermal resistivity per meter.

$$G_c = \int_r^R \frac{\rho_{thc} \, dx}{2\pi x} = \frac{\rho_{thc}}{2\pi} \ln \frac{R}{r} \quad (\text{th-}\Omega/\text{m}) \tag{7.52}$$

The thermal resistance of a three-core cable is given empirically by

$$G_{3c} = \frac{\rho_{thc}}{6\pi} \ln \frac{R^6 - a^6}{3R^3 a^2 r} \quad (\text{th-}\Omega/\text{m}) \tag{7.53}$$

where a is the radius of the circle through which the centers of conductors pass, R and r are the dielectric outer radius and conductor radius, respectively (Fig. 7.17).

Thermal Resistance of the Ground: It depends upon the type of soil which determines the moisture retentivity and the vegetation along the cable path. For homogeneous ground, the thermal resistance is given by

$$G_g = G_g = \frac{\rho_{thg}}{2\pi} \ln \frac{2h}{R} \quad (\text{th-}\Omega/\text{m}) \tag{7.54}$$

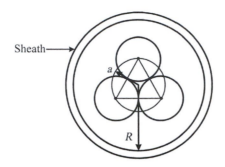

FIGURE 7.17

Thermal resistance in three-core cables.

where ρ_{thg} is the thermal resistivity of the soil and h is the depth of the cable axis below ground surface.

Thermal Resistance of the Cable Servicing: The protective covering over the sheath including the armor has a thermal resistance

$$G_s = \frac{\rho_{\text{ths}}}{2\pi} \ln \frac{R_2}{R_1} \quad (\text{th-}\Omega/\text{m}) \tag{7.55}$$

where ρ_{ths} is the thermal resistivity of the servicing which is about 5°C-m/W. R_1 is the outer radius of the sheath and R_2 is the overall radius of the cable.

7.12 CURRENT RATING OF THE CABLE

Let θ_c°C be the conductor temperature inside the cable and θ_s°C be the temperature of the sheath.

Let R be the resistance of each core of the cable (Ω/m) with n cores contained in the cable. The heat produced

$$H = nI^2R \quad (\text{W/m}) \tag{7.56}$$

where I is the cable current per core. But,

$$H = \frac{\theta_c - \theta_s}{G_c} \tag{7.57}$$

Let the sheath losses be assumed as λ times the core losses. Neglecting armor losses, total heat to be dissipated

$$= (1 + \lambda)nI^2R \quad (\text{W}) \tag{7.58}$$

This heat has to be transferred to ambience by the protective servicing and the soil.

Let θ_a°C be the ambient temperature.

$$(1 + \lambda)nI^2R = \frac{\theta_s - \theta_a}{G_g + G_s} \tag{7.59}$$

From Eqs. (7.58) and (7.59)

$$nI^2RG_{nc} = \theta_c - \theta_s \tag{7.60}$$

$$(l + \lambda)nI^2R\,(G_g + G_s) = \theta_s - \theta_a \tag{7.61}$$

Adding

$$nI^2RG_{nc} + (1 + \lambda)MI^2R(G_g + G_s) = \theta_c - \theta_s + \theta_s - \theta_a$$
$$= \theta_c - \theta_a \tag{7.62}$$

$$I^2[nRG_{nc} + (1 + \lambda)nR(G_g + G_s)] = \theta_c - \theta_a$$

$$I = \sqrt{\frac{(\theta_c - \theta_a)}{nR[G_{nc} + (1 + \lambda)(G_s + G_g)]}} \ \ (A) \tag{7.63}$$

The effect of temperature on dielectric loss, which increases with temperature can be accounted by derating further as

$$I = \sqrt{\frac{(\theta_c - \theta_a - \theta_d)}{nR[G_{nc} + (1 + \lambda)(G_s + G_g)]}} \ \ (A) \tag{7.64}$$

θ_d is the temperature rise caused at the maximum permissible temperature.

WORKED EXAMPLES

E.7.1 A single-core cable 1 km in length has a core diameter of 1.0 cm and a diameter under the sheath of 2.5 cm. The relative permittivity is 3.5. The power factor on open circuit is 0.03. Calculate (1) the capacitance of the cable, (2) its equivalent insulation resistance, (3) the charging current, (4) the dielectric loss when the cable is connected to 6600 V−50 Hz bus bars.

Solution:

1. Capacitance of the cable $= \dfrac{2\pi \in}{\ln \dfrac{R}{\mu}}$

$$= 2\pi \times \frac{10^9}{36\pi} \times \frac{3.5}{\ln \dfrac{1.25}{0.5}} = \frac{3.5}{18 \times 0.91629} = 0.2122 \ \mu f$$

2. Insulation resistance $= R_i$

$$\phi = \frac{\pi}{2} - \delta; \ \ \delta = \frac{\pi}{2} - \phi$$

$$\cos\phi = 0.03; \ \ \phi = 88°.28; \ \ \delta = 90 - 88.28 = 1.72 \text{ degrees}$$

$$\delta = \frac{1}{R_i} \cdot \frac{1}{WC} = \frac{1}{R_i} \cdot \frac{1}{314} \times \frac{10^6}{0.2122} = 0.03 \text{ rad}$$

$$\delta = \frac{1.72 \times \pi}{180} = 0.03 \text{ rad}$$

$$R_i = \frac{10^6}{0.03 \times 314 \times 0.2122} = 5 \times 10^5 \ \Omega$$

3. Charging current I_C

$$I_c = V\omega C = 6600 \times 314 \times 0.2122 \times 10^{-6} \text{ A}$$
$$= 0.439 \text{ A}$$

4. The dielectric loss P_d

$$P_d = V^2\omega C \tan \delta \cong V^2\omega C\delta$$
$$= 6600^2 \times 314 \times 0.2122 \times 10^{-6} \times 0.03$$
$$= 87 \text{ W}$$

E.7.2 A three-core cable gives on test a capacitance of 3 µf between two cores. Find the line charging current of the cable when connected to 11-kV, 50-Hz bus-bars.

Solution: The equivalent star connection is shown in Fig. E.7.2.

The series circuit between 2 and 3 has a capacitance of

$$\frac{1}{2}[C_s + 3C_c] = \text{Capacitance measure } C_m$$

$$\frac{1}{C_m} = \frac{1}{C_s + 3C_c} + \frac{1}{C_s + 3C_c}$$

$$C_m = 3\,\mu\text{F} = \frac{1}{2}(C_s + 3C_c)$$

$$\frac{1}{C_m} = \frac{2}{C_s + 3C_c}$$

$$C_m = \left[\frac{C_s + 3C_c}{2}\right]$$

$$C_s + 3C_c = \text{Capacitance of each core} = 6\,\mu\text{F}$$

$$\text{Line charging current per phase} = V\omega C = \frac{11 \times 10^3}{\sqrt{3}} \times 314 \times 6 \times 10^{-6} = 11.96 \text{ A}$$

E.7.3 In a three-phase, three-core, metal-sheathed cable the measured capacitance between any two cores is 2 µF. Calculate the Kilovoltamperes taken by the cable when it is connected to 50-Hz, 11-kV bus bars.

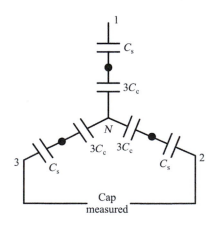

FIGURE E.7.2

Solution:

$$I_c = V\omega C = \frac{11,000}{\sqrt{3}} \times 314 \times 2 \times 2 \times 10^{-6} = 7.9769 \text{ A}$$

$$C_c + 3C_s = 2 \times 2 \text{ } \mu F$$

$$\text{KVA taken by cable} = \frac{11}{\sqrt{3}} \times 7.9769 = 151.9849$$

E.7.4 Find the diametral dimension for the *t*-core, metal-sheathed cable, giving the greater economy of insulating material for a working voltage of 85 kV, if a dielectric stress of 60 kV/cm can be allowed.

Solution:

For economy $R = er$

$$g_{max} = \frac{V}{r \ln \frac{R}{r}} = \frac{V}{r \cdot 1}; \ 60 = \frac{85}{r}$$

$$r = \frac{85}{60} = 1.4166 \text{ cm}$$

Conductor diameter $= 2 \times r = 2.833$ cm
Sheath inner diameter $= e \times d = 2.718 \times 2.833 = 7.7$ cm

E.7.5 An 85-kV, one-core, metal-sheathed cable is to be graded by means of a metallic intersheath.

  1. Find the diameter d_1 of the intersheath and the voltage at which it must be maintained in order to obtain the minimum overall cable diameter D. The insulating material can be worked at 60 kV/cm.
  2. Prove if any formula is used and compare the conductor and outside diameters (d and D) with those of an ungraded cable of the same material under the same conditions.

Solution:

$$\text{For economy } r_1 = 2.718 \text{ } r \tag{i}$$

For the part conductor and intersheath

$$g_{max} = \frac{V - V_1}{r \ln \frac{r_1}{r}} \tag{ii}$$

For the part of the cable containing intersheath to lead sheath inner diameter

$$g_{max} = \frac{V_1 - 0}{r_1 \ln \frac{R}{r_1}} \tag{iii}$$

Given that g_{max} is the same at conductor surface and intersheath surface.

$$g_{max} = \frac{V - V_1}{r \ln \frac{r_1}{r}} = \frac{V_1}{r_1 \ln \frac{R}{r_1}} \tag{iv}$$

From (iv)

$$r = \frac{V - V_1}{g_{max} \ln \frac{r_1}{r}}; \text{ but } \ln \frac{r_1}{r} = 1 \text{ for economy}$$

$$\therefore \ r_1 = \frac{(V - V_1)2.718}{g_{max} \cdot 1} \text{ and } r = \frac{r_1}{2.718}$$

Again from (iv)
Substituting the value for r_1

$$g_{max} = \frac{V_1}{r_1 \ln \frac{B}{r_1}}$$

$$g_{max} = \frac{V_1}{\frac{2.718(V - V_1)}{g_{max}} \cdot \ln \frac{R g_{max}}{2.718(V - V_1)}}$$

Cross multiplying and rearranging

$$\frac{R g_{max}}{2.718(V - V_1)} = e^{\frac{V_1}{2.718(V - V_1)}}$$

$$R = \frac{2.718}{g_{max}}(V - V_1)e^{\frac{V_1}{2.718(V - V_1)}}$$

For overall minimum cable diameter, V_1 is to vary so that

$$\frac{dR}{dV_1} = \frac{2.718}{g_{max}} \left\{ (V - V_1)e^{\frac{V_1}{2.718(V - V_1)}} \left[\frac{2.718(V - V_1) + 2.718V_1}{2.718^2(V - V_1^2)} \right] + \frac{V_1}{e^{\frac{V_1}{2.718(V - V_1)}}}[-1] \right\} = 0$$

Simplifying we obtain

$$\frac{2.718(V - V_1) + 2.718V_1}{(2.718)^2(V - V_1)^2} = 1$$

$$\frac{V}{2.718(V - V_1)} = 1 \text{ or } \frac{V}{V - V_1} = 2.718$$

$$\frac{85}{85 - V_1} = 2.718 \text{ so that } V_1 = 53.727 \text{ kV}$$

$$V_1 = 0.632 \text{ V}; \quad g_{max} = \frac{V - V_1}{r_1}$$

$$d = \frac{2(V - V_1)}{g_{max}} = \frac{2(85 - 53.727)}{60} = 1.04 \text{ cm}$$

$$d_1 = 2.718 \times 1.04 = 2.8339 \text{ cm}$$

In case of ungraded cable

$$g_{max} = \frac{V}{r \ln \dfrac{D}{r}} \quad \text{for economy} \quad \frac{D}{r} = e$$

$$\therefore \quad 60 = \frac{85}{r}; \quad r = \frac{85}{60} = 1.4166 \text{ cm}$$

Diameter $d = 2.832$ cm

Sheath inner diameter $= D = d \cdot e = 2.832 \times 2.718$

$$= 7.6973 \simeq 7.7 \text{ cm}$$

E.7.6 A single-core, lead-covered cable is to be designed for operation at 66 kV. The conductor radius is 0.5 cm and it is graded with three insulating materials A, B, and C with relative permittivities 4.5, 4, and 2.5, respectively. The respective maximum permissible stresses are 50, 40, and 30 kV/cm, respectively. Find the minimum internal diameter of the sheath.

Solution:

$$g_{1 \text{ max}} = \frac{q}{r \pi \in_0 \in_1 r}$$

$$g_{2 \text{ max}} = \frac{q}{r \pi \in_0 \in_2 r_1}$$

$$g_{3 \text{ max}} = \frac{q}{r \pi \in_0 \in_3 r_2}$$

Therefore

$$q = 2\pi \in_0 \in_1 g_{1 \text{ max}} = 2\pi \in_0 \in_2 r_1 g_{2 \text{ max}}$$

$$= 2\pi \in_0 \in_3 r_2 g_{3 \text{ max}}$$

$$\in_1 r g_{1 \text{ max}} = \in_2 r_1 g_{2 \text{ max}} = \in_3 r_2 g_{3 \text{ max}}$$

Since $r = 0.5$ cm

$$0.5 \times 4.5 \times 50 = r_1 \times 4 \times 40 = r_2 \times 2.5 \times 30$$

$$r_1 = \frac{0.5 \times 4.5 \times 50}{4 \times 4.0} = 0.7031 \text{ cm}$$

$$r_2 = \frac{0.5 \times 4.5 \times 50}{2.5 \times 30} = 1.5 \text{ cm}$$

$$V = 66 \text{ kV} = r g_{1 \text{ max}} \ln \frac{r_1}{r} + r_1 g_{max} \ln \frac{r_2}{r_1} + r_2 g_{3 \text{ max}} \ln \frac{R}{r_2}$$

$$66 \text{ kV} = 0.5 \times 50 \times \ln \frac{0.7031}{0.5} + 0.7037 \times 40 \times \ln \frac{1.5}{0.7031}$$

$$+ 1.5 \times 30 \ln \frac{R}{r_2}$$

$$66 = 0.5 \times 50 \times 0.34089 + 0.7037 \times 40 \times 0.7577 + 1.5 \times 30 \ln \frac{R}{1.5}$$

$$\ln = \frac{R}{1.5} = \frac{36.168}{45} = 0.803733$$

$$\frac{R}{1.5} = e^{0.8} = 2,233,864$$

$R = 1.5 \times 2.233864 = 3.35$ cm

Inner diameter of sheath $= 6.7$ cm

E.7.7 Find the maximum working voltage of a single-core, lead-sheathed cable joint with a conductor of 1 cm diameter and sheath of 5 cm inner diameter. Two insulating materials of dielectric strength 60 and 50 kV/cm and relative permittivities 4 and 2.5 are used for grading the cable insulation, respectively.

Solution:

$$\in_1 r_1 g_1 = \in_2 r_2 g_2$$

$$60 \times 4 \times 0.5 = 50 \times 2.5 \ r_2$$

$$r_2 = \frac{60 \times 4 \times 0.5}{50 \times 2.5} = 0.96 \text{ cm}$$

Working voltage of the cable

$$V = rg_{1 \text{ max}} \ \ln \frac{r_1}{r} + r_1 g_{2 \text{ max}} \ \ln \frac{R}{r_1}$$

$$= 0.5 \times 60 \times \ln \frac{0.96}{0.5} + 0.96 \times 50 \times \ln \frac{2.5}{0.96}$$

$$= 30 \times 0.652325 + 0.96 \times 50 \times 0.9555$$

$$= 19.5697 + 45.86 = 65.42 \ \text{kV}$$

E.7.8 The insulation resistance of a single-core cable is 515 $\mu\Omega$/km. If the core diameter is 3.6 cm and resistivity of insulation is $4.65 \times 10^{14} \ \Omega$ cm, find the thickness of insulation.

Solution: Consider 1 km length of cable

$$\text{Insulation resistance } R = \frac{\rho}{2\pi L} \log_e \frac{R}{r}$$

$$\rho = 4.65 \times 10^{14} \ \Omega \text{ cm}$$

$$= 4.65 \times 10^{12} \ \Omega \text{ m}$$

$$\frac{4.65 \times 10^{12}}{2\pi \times 1000} \log_e \frac{R}{r} = 515 \times 10^6$$

$$\therefore \quad \log_e \frac{R}{r} = \frac{515 \times 10^6 \times 2\pi \times 1000}{4.65 \times 10^{12}} = 0.6955$$

$$\frac{R}{r} = 2.0047$$

$$R = 2.0047 \ r = 2.0047 \times 1.8 = 3.60848 \ \text{cm}$$

$$R - r = \text{Thickness of insulation} = 1.808 \ \text{cm}$$

E.7.9 Find the insulation resistance per kilometer of a two-core concentric cable having an inner conductor diameter of 1.33 cm and an outer conductor of inside diameter 3.62 cm. The dielectric has a specific resistance of $8 \times 10^{12} \ \Omega$ m at the operating temperature.

Solution: Insulation resistance

$$R = \frac{\rho}{2\pi} \ln \frac{R}{r}$$

$$= \frac{8 \times 10^{12}}{2\pi} \ln \frac{3.62}{1.33} \, \Omega/m$$

$$= 1.275 \times 10^{12} \, \Omega/m = 1275 \, M\Omega/km.$$

PROBLEMS

P.7.1 A single-core cable, 1 km long has a conductor radius of 15 mm and an insulation of 5 mm. If the resistivity of the insulation is 7.2×10^{12} Ω-m, find the insulation resistance of the cable.

P.7.2 A single-core cable 21 km long has a capacitance per kilometer of 0.16 μF. If the cable operates at 11 kV and 50 Hz, determine the charging current.

P.7.3 A three-phase, three-core cable is operating at 33 kV and is 2.5 km long. The radius of each conductor is 1.25 cm and the radial thickness of insulation is 0.61 cm. The relative permittivity of the dielectric is 3.5. Determine
  1. Capacitance of the cable per phase
  2. Charging current per phase
  3. Given the power factor of the cable on no-load as 0.02, determine the dielectric loss per phase
  4. What is the total charging KVAR?

P.7.4 Determine the minimum internal diameter of the sheath of a single-core, lead-covered cable operating at 65 kV, if the conductors diameter is 1 cm. The cable is graded with three insulating materials A, B, and C of relative permittivities 4, 3, and 2.5, respectively, with peak permissible stresses of 70, 62, and 45 kV/cm.

P.7.5 A single-core cable designed to operate at 132 kV has a conductor radius of 1 cm. The inner diameter of the sheath is 12 cm. If the stress varies between the same maximum and minimum values, due to two intersheaths in the dielectric, determine the radii of the intersheaths and their voltages.

P.7.6 A three-core cable during a test for capacitance measurement gave 2.5 μF between two cores. If the cable operates at 11 kV and 50 Hz determine the line charging current.

QUESTIONS

Q.7.1 Explain the constructional features of three-core cables. Sketch a three-core cable.

Q.7.2 What are insulation materials used in cables? What specific problems are faced while operating with three-core cables?

Q.7.3 What do you understand by grading of cables? Explain.

Q.7.4 Compare intersheath grading with capacitance grading. Which is better and why?

Q.7.5 Explain the theory of intersheath grading.

Q.7.6 Explain the theory of capacitance grading.

Q.7.7 Explain thermal phenomenon in cable.

Q.7.8 Explain how the current rating of a cable is fixed.

HIGH-VOLTAGE DIRECT CURRENT TRANSMISSION

8

Power generated at conventional power stations is transferred by AC transmission lines for further utilization through distribution. However, some lines transfer power using high-voltage direct current (HVDC). For bulk transmission of power by high-voltage alternating current (HVAC) lines, there are some technical problems. Therefore it has become a common practice to consider HVDC transmission systems for transferring large power over long distances. Several large systems are interconnected through HVDC links. Undersea cable systems also prefer the use of DC transmission. High-voltage transmission has several advantages over AC transmission.

8.1 ADVANTAGES AND DISADVANTAGES OF HIGH-VOLTAGE DIRECT CURRENT TRANSMISSION

1. The most important advantage is that there is no stability problem with DC transmission systems. It is possible to transmit power between two unsynchronized AC networks. HVDC lines have the ability to operate successfully between countries that operate their power systems at different frequencies. As HVDC power lines facilitate transmission of power in either direction in a faster manner, they contribute to the stability of the connected systems. The power transfer capability of an AC line is inversely proportional to the transmission distance while it is unaffected by the length of the transmission line for DC transmission.

2. HVDC lines increase the capability of existing lines when it is not possible to add additional lines or new line construction is quite expensive.

3. Bulk transmission of power requires intermediate taps when AC is used. With HVDC no such taps are needed. Hence, they are quite suited to transmit bulk power from remote generating stations to the load centers. Nelson river DC transmission system in Canada is an example of this advantage.

4. With DC links renewable energy sources can be integrated easily with the main AC grid. Using multiple voltage source converter systems, wind energy farms, off shore power sources, or onshore generation can be provided easy access to the main transmission system.

5. Undersea transmission of power is economical with HVDC cables. Undersea cables where high capacitance causes additional losses DC is better suited. The Sweden to Germany 250-km Baltic cable and the NorNed cable system 580 km in length between Norway and Netherlands are some of the several examples that utilized this feature.

6. AC lines are subjected to line charging currents that flow through the charging capacitances. The DC cables are limited in operation only by temperature rise while charging current plays an important role in the case of AC lines. There is no restriction on the length of the line.

Electrical Power Systems. DOI: http://dx.doi.org/10.1016/B978-0-08-101124-9.00008-5

7. The AC line conductors possess skin effect which creates uneven distribution of current across the cross section of a conductor. There is no skin effect in DC transmission. Thus copper is saved when DC is used for power transmission. Conductors in DC system carry more power per conductor.

8. Due to asynchronous connection, cascaded failures or series tripping in the event of a large disturbance do not happen with HVDC transmission.

9. The towers are narrower, simpler, and cheaper in case of HVDC when compared to HVAC lines.

10. The direction of power flow can be changed faster.

11. The HVDC system requires less insulation when compared to HVAC system of the same rating.

The bipolar DC systems are found to more reliable than three-phase AC systems. Further, the corona loss is also less. It is reported from the results of a survey conducted that the scheduled unavailability of HVDC lines is about 5.39% and the forced unavailability 1.62%.

Against these advantages, there are also some disadvantages in transmitting power through HVDC lines:

1. Converter substations are expensive and complex than HVAC. The controlling and regulating mechanisms are economically feasible only for long distances (Fig. 8.1). Below a critical distance, AC lines will be cheaper. This feature is mainly due to the additional cost to be incurred for the converter−inverter and associated equipment needed.

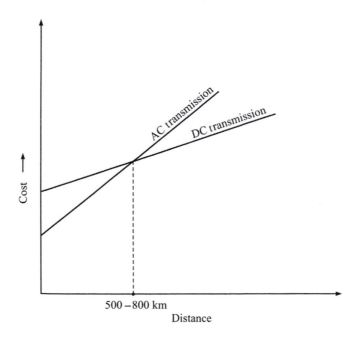

FIGURE 8.1

Comparison of HVAC and HVDC transmission cost with distance.

2. For the same reason, in practice it is also found that the availability and reliability of DC systems is lower than the AC systems.

3. In contrast to AC systems, designing and operating multiterminal HVDC is complicated. There is a problem of harmonic generation in conversion and reactive power is consumed. It is necessary to install expensive filter compensation units and VAR compensating units.

4. If short circuits occur in the AC systems near the input side of the DC terminal, voltage fall takes place.

5. The high-frequency constituents in HVDC transmissions can cause radio noise in lines that are nearer to them.

6. It is practically impossible to build HVDC systems with more than five substations.

7. Grounding HVDC lines involves complex and difficult installation. Dangerous "step voltage" has to be prevented with good grounding.

8. Reliable HVDC breaker systems are difficult to build.

In practice the merits and demerits both technological and economical are to be decided in the design of every project based on the factual and field considerations rather than on generalized considerations. There is a break even point over which distance DC transmission of power becomes cheaper than AC transmission (Fig. 8.1).

8.2 HIGH-VOLTAGE DIRECT CURRENT TRANSMISSION SYSTEM

The various components of the system are the transmission line or cable conductors, the towers, converter station equipment, rectifying valves, converter transformers, DC smoothing reactor, harmonic filters, control equipment, reactive power compensators, and grounding electrodes. A simple block schematic of DC transmission line is shown in Fig. 8.2.

HVDC transmission requires converters to convert AC into DC and again inverters to change the DC into AC for further use. Thus converter and inverter operation of thyristor valves forms an important aspect in the HVDC system. They also influence the cost structure of all HYDE systems. A converter station usually consists of

1. Converters

2. Transformers

3. Smoothing reactors

4. DC filters and

5. Tuned AC filters to suppress or reduce 11-, 13-, 23-, 25- order harmonics.

In the conversion process the converter needs reactive power which is supplied in part by the filters and in part by capacitor banks. There are two types of converters used in practice. They are current source converters and voltage source converters.

8.2.1 LINE-COMMUTATED CONVERTERS

A basic line-commutated converter can be configured with a three-phase bridge rectifier or a six-pulse bridge. The AC waves will have a phase shift of 60 degrees and hence, the harmonic content

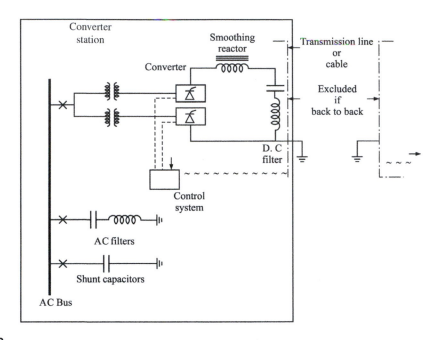

FIGURE 8.2

Block schematic of HVDC transmission. *HVDC*, high-voltage direct current.

will be higher. A six-pulse, full-wave rectifier is shown in Fig. 8.3A. A pair of SCRs connected across two lines will conduct when there is a grid signal. For every 60 degrees, one thyristor from positive limb and one thyristor from negative limb will conduct. The waveforms for six-pulse rectifier operation are illustrated in Fig. 8.3B.

A 12-pulse rectifier is shown in Fig. 8.4 reduces the harmonics considerably. The AC is split into two separate three-phase supplies before conversion. One secondary is star connected and another one is delta. This enables the two AC waveforms to attain a phase shift of 30 degrees only. The 12-pulse converter system has become standard for line-commutated converters used in HVDC transmission.

The output from the converter becomes lesser and lesser positive as the firing angle for thyristors is increased. If the firing angle $\propto$ becomes greater than 90 degrees, inversion occurs and negative DC voltage will be produced. In practice the angle is limited to about 150–160 degrees as otherwise the turnoff time becomes insufficient. Rectifier and inverter operation with change in firing angle is shown in Fig. 8.5.

Since a single thyristor cannot withstand the operating voltage levels which may be as high as 800 kV, several thyristors are connected in series to match the working voltage level of the line. Also, grading capacitors and resistors are connected in parallel with each thyristor unit. It is seen that all the thyristors share the voltage uniformly. Direct or indirect optical switching is used to turn on the thyristors. A large inductance is placed in the output circuit to smoothen and provide a constant current output. On the AC side, a line-commutated converter operates as a current source. It injects grid frequency and harmonic currents both into the AC network. A line-commutated converter is therefore called current source converter.

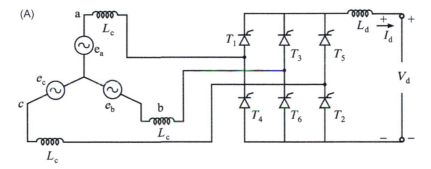

L_c : Converter transformer leakage inductance
L_d : Smoothing reactor
I_d : Constant DC
V_d : DC voltage

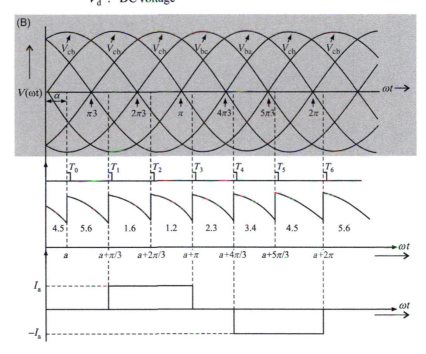

FIGURE 8.3

(A) Six-pulse bridge rectifier and (B) six-pulse rectifier and waveforms.

8.2.2 VOLTAGE SOURCE CONVERTER

The line-commutated converters have only one degree of freedom. The converter can be turned on by gate control but cannot be turned off. But, by using the insulated gate bipolar transistor (IGBT) instead of the thyristor, it is possible to control both turning on and turning off. They are said to

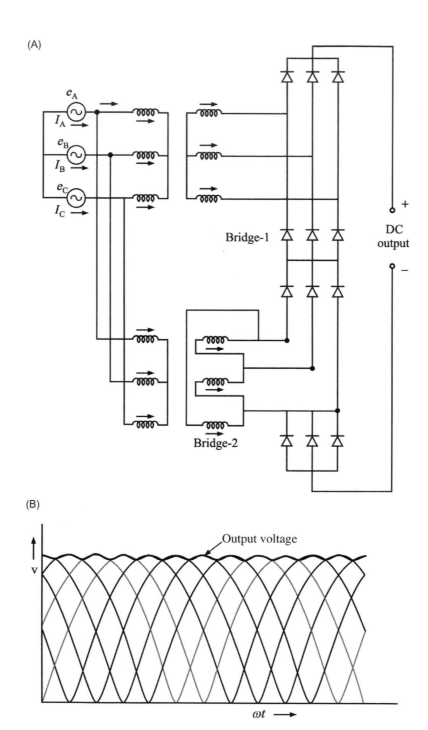

FIGURE 8.4

Twelve-pulse converter. (A and B) A 12-pulse converter circuit diagram.

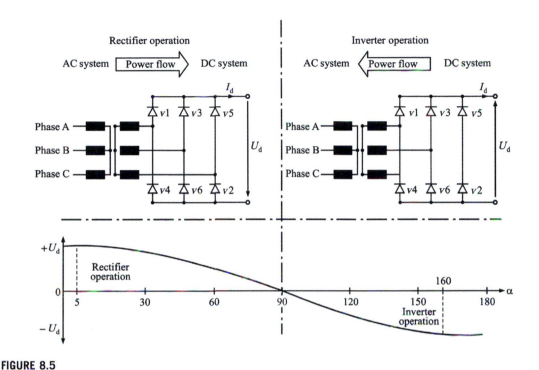

FIGURE 8.5

Control of DC voltage. *DC*, direct current.

possess two degrees of freedom. The IGBTs are used to operate as self-commutated converters in conjunction with a diode connected antiparallel as shown in Fig. 8.6A. The output voltage is smoothed by a capacitor. Due to the ability of the IGBTs to be switched on and off a number of times during a cycle, they provide a constant voltage output and better harmonic performance. Pulse width modulation (PWM) is used to reduce the harmonic content. These converters do not require a synchronous machine to provide AC supply. They are suitable to supply power to networks containing passive loads. This is not possible with line-commutated converters. As the harmonic content is less, a six-pulse connection is sufficient. Most of the six-pulse bridges that are using thyristors are now replaced by IGBTs with inverse parallel diodes and with a DC smoothing capacitor as shown in Fig. 8.6B. From 2012 onward two-level, three-level, and even multilevel converters are built with modular construction. A number of independent modules form a single valve in the modular construction. Every module is provided with its own capacitor. They operate with very little harmonic distortion.

8.2.3 CONVERTER TRANSFORMERS

Transformers are required at both ends of a HVDC transmission line since the line connects AC systems on either side. Generally, three physically separated single-phase transformers are used, which isolate the substations on either side of the line. The transformers are ungrounded. They also provide a local earth. For line-commutated converter HVDC systems, specially designed

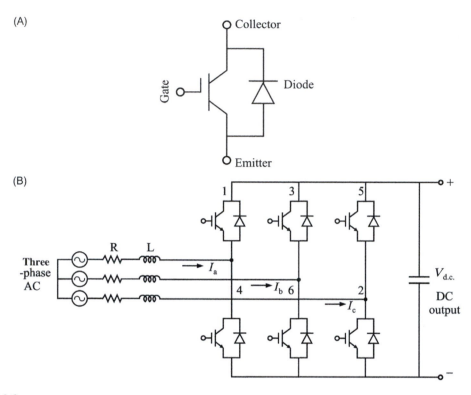

FIGURE 8.6

PWM—voltage source converter. (A) Valve and (B) converter. *PWM*, pulse width modulation.

transformers are needed. Due to the high content of harmonics, a constant DC component of voltage will be present. The secondary winding of the transformer experiences additional stress and requires more insulation. Again, 30-degree phase shift is required to be provided to effect harmonic cancelation in case of line-commutated converters. In case of voltage source converters, the transformer design is not much complicated due to lesser harmonic content.

8.2.4 CONDUCTOR SYSTEMS

Monopole System: This system consists of a single-line conductor and an earth return path (Fig. 8.7). One terminal of the converter at high potential is connected to a line conductor on the transmission line. The other terminal of the rectifier is connected to earth to constitute the return path for the current. The second terminal at the inverter end may be connected to earth. Current flows in the earth between the two stations. The current flowing through earth may cause corrosion of metallic objects buried under earth such as pipelines. In case of underwater return circuit, there may be associated effects. Further, unbalanced currents may produce a net magnetic field. This magnetic field may interact with the functioning of navigational systems of ships when employed

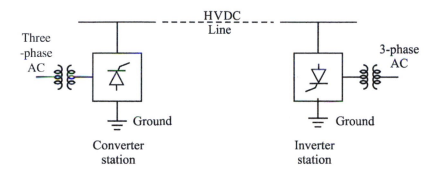

FIGURE 8.7

Monopole earth return HYDE system.

for undersea cables. These undesirable effects can be eliminated by providing a metal conductor return path between the two ends of the monopole transmission system. There is an advantage provided by such an additional conductor as it need not be insulated for the full-voltage level of the line and is thus cheaper in cost than the monoconductor on the transmission line. The selection of a return conductor is determined by environmental, technical, and economic factors. A monopolar link is operated with negative polarity. A monopolar conductor is operated with negative polarity as corona effect is considerably less with negative polarity compared with positive polarity.

Bipolar System: In this system two conductors are used one for positive polarity and the other for negative polarity (Fig. 8.8). Both the conductors are required to be insulated for full voltage. Hence, the cost is higher than monopole system using earth as the return. In this system under normal conditions the earth current is negligible as in monopole arrangement. In case of fault developing in any line, the other part of the line can still be operated as a monopole system. Half of the rated power of the line can be transmitted till the fault is cleared. When compared to the monopole system, in the bipole configuration of supply for a given power transfer, the line conductors carry only half of the current in monopole case. Thus, there is a saving in conductor cost. The bipole transmission system can also have metallic earth return conductors.

Homopolar Links: This type of the system has two or more conductors all having the same polarity. Usually they are maintained at negative polarity since corona losses are less for negative polarity. The return path is ground (Fig. 8.9).

Tripole System: In a patented tripole system two conductors operate as a bipole system and the third conductor is used as a parallel monopole system.

Multiterminal System: A HVDC link is a DC transmission line that connects two different systems. For large capacities series, parallel, or a hybrid, series - parallel combination, multiterminal converter systems are used. The converters are voltage source controlled so that reverse power flow can take place.

Back-to-Back System: Two grids operating at different frequencies can be interconnected using a back-to-back DC system (Fig. 8.10). The length of such a system is generally very small and even the voltage is limited to lower level so that the number of thyristors in the series circuit is limited. In some cases both the converter stations may be constructed even at the same location.

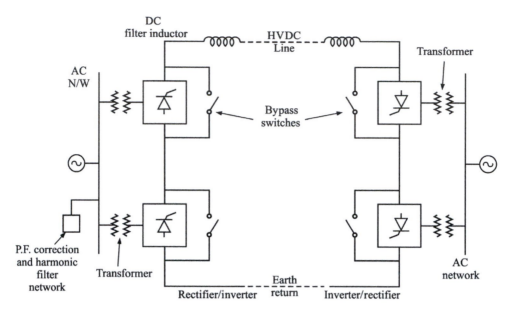

FIGURE 8.8

Bipole HVDC transmission system. *HVDC*, high-voltage direct current.

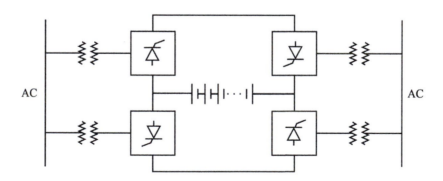

FIGURE 8.9

Homopolar system.

Smoothing Reactors: They reduce the harmonic content in both voltage and current. Generally an inductance of 1 H per pole is required in series for this purpose. These inductors also prevent commutating failures in inverters. Again when the system is operating at light load with low current, the inductance prevents discontinuous flow. It also prevents harmonic heating of capacitors.

Reactive Power: Under steady-state operation, the reactive power requirement is as high as 50% of the real power transferred. Under transient conditions, it is even more. Reactive power compensation is provided near the rectifiers.

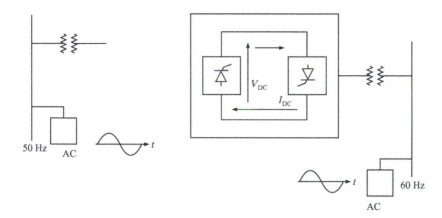

FIGURE 8.10

Back-to-back system.

Electrodes: The electrodes used for grounding have large surface area so that surface voltage gradient is limited to safe value and current density is also restricted. In monopolar and bipolar conductor systems using ground as return conductor results in economy and efficiency as the loss is less than in the case of metallic conductor return path by virtue of low resistance. But care has to be taken while installing that buried metallic objects are avoided. Also from the station, the buried electrode must be away at least by 5 km. The controllers are digitally operated and optical fibers are increasingly used.

Many of the earliest HVDC transmission system were decommissioned or are in that process, as the earlier technology became obsolete. Most of the existing lines use thyristor-based converters. For the advantages cited earlier, nowadays several of the lines in construction are using IGBT-based converter systems. HVDC lines are constructed for power transmission all over the world.

QUESTIONS

Q.8.1 Explain the various factors that make DC transmission economical over AC transmission.

Q.8.2 Why 12-pulse operation is more desired than 6-pulse converter operation?

Q.8.3 What are the advantages of voltage source converters over line-commutated converters?

SUBSTATIONS AND NEUTRAL GROUNDING

At generating stations power is generated and their power is transmitted over long distances to the load centers at high voltage. However, loads are spread over large geographic areas and from the receiving ends of transmission lines, power is distributed through feeders and distributor lines to consumers through a large number of substations where generally the voltage is stepped down further. Sometimes between main transmission and distribution, there could be subtransmission lines as per the requirement. There are several types of substation classification. The important types of classification are based on

1. Service requirements and
2. Constructional features

They are detailed in the following.

9.1 SERVICE REQUIREMENTS—BASED CLASSIFICATION

1. *Step-up substation* is a transformer substation where generated power is stepped up suitably and transmitted over an overhead line or cable at the sending end. A similar transformer substation exists at the receiving end of the line or cable from where either subtransmission or distribution of power starts.
2. *Primary grid substations* are located at suitable load centers where power is stepped down for supply to secondary substations in case of subtransmission or to primary distribution.
3. *Distribution substations* supply from primary or secondary transmission to loads through step-down transformers operating at appropriate voltages.
4. *Industrial substations* are substations specially arranged for bulk consumers such as industrial and large commercial loads.
5. *Power factor correction substations* are sustaining substation where required condenser banks are switched in or out for power factor correction. Variable control is implemented here.
6. *Converting substations:* Here AC is converted into DC or DC converted into AC.
7. *Mining substations* are special and carefully designed substations to supply underground mining operation.
8. *Mobile substations* are designed to supply power for constructional and other purposes.

Electrical Power Systems. DOI: http://dx.doi.org/10.1016/B978-0-08-101124-9.00009-7

9.2 CONSTRUCTION FEATURE-WISE CLASSIFICATION

1. Indoor substation
2. Outdoor substation
    a. Pole-mounted substation
    b. Plinth-mounted substation

Outdoor substations need more space, less capital investment, and less time for erection compared to indoor substations. Further, outdoors substations can be expanded to accommodate future load easily. A typical pole-mounted substation is shown in Fig. 9.1.

9.3 SUBSTATION EQUIPMENT

Bus Bars: Various incoming and outgoing circuits are connected to bus bars made of copper or aluminum.
    Circuit Breakers: Automatic switching for circuit interruption and reclosing.
    Isolators: Disconnecting switches on no-load only.
    Earthing Switch: Discharges voltage on the lines to earth for safety.
    Lightning Arrestor: Diverts high-surge voltages to ground or earth and protects.
    Instrument Transformers: Current transformers and potential transformers for measurement, protection, and control.
    Series Reactors: To limit short circuit current level at appropriate places.
    On-load tap changer: For voltage control
    Shunt Reactors: For extra-high-voltage lines to compensate capacitance of line.
    Shunt and Series Capacitors: For VAR compensation, power factor improvements and to increase power transfer capability.
    Coupling Capacitor: For carrier currents and line traps.
    Batteries: To supply relays and emergency lighting.
    Station Transformer: For auxiliary power supply to the station such as lighting, fans, and battery charging.
    Wave Trap: For communication purposes.
    Fire protection equipment.

9.4 FACTORS GOVERNING LAYOUT OF SUBSTATIONS

1. Number of incoming lines
2. Number of outgoing lines
3. Nature of the load to be supplied
4. Type of operation and maintenance required
5. Weather conditions
6. Purpose, function, and capacity of substations.

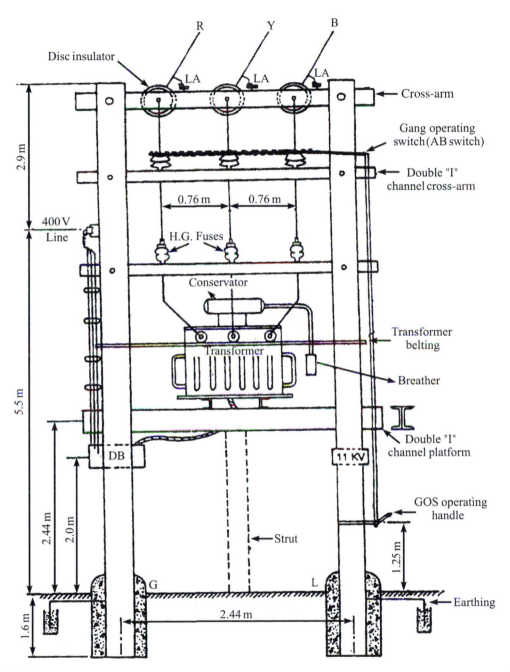

FIGURE 9.1

Pole-mounted substation (11 kV/400-V).

9.5 STATION TRANSFORMER

In both primary and secondary substations many protective devices are provided. For successful operation of these devices DC supply is needed. A station transformer is used to give supply to all the needs of the substation exclusively. It charges the station batteries all through the day. It supplies all lighting, fans, motors, refrigeration, and alarm bells in the substation.

9.6 BATTERIES

The station battery provides an independent source of supply for operation of control circuits. They are the backbone of all switchgear and protective gear. Two different types of batteries: (1) Lead—acid type and (2) Nickel-Cadmium type are used in substations. The commonly used lead—acid battery arranged in a rubber container consists of a lead peroxide anode plate and sponge lead cathode plate both immersed in an electrolyte solution of sulfuric acid. Generally the battery is maintained in good condition by trickle charge. If the specific gravity falls low or as per the instruction of the manufacturer, boost charge is done at regular intervals. The battery charger consists of a step-down transformer, rectifying device, and a rheostat. Provision is made for a voltmeter and an ammeter. The battery room should be cool, clean, dry, and well-ventilated. The flooring of the battery room should be acid resistant.

9.7 EARTHING IN SUBSTATIONS

An earthing system is provided in all substations for the following reasons:

1. To provide earth connections to all neutral points
2. To provide discharge path for lightning arrestors, gaps, etc.
3. To keep the noncurrent carrying parts such as transformer tanks, metallic structures, etc., always safe at earth potential even if insulation fails.

9.8 ELEMENTS TO BE EARTHED IN A SUBSTATION

1. All the frames, tanks, and enclosures of electrical machines, transformers, lightning arrestors, earth switches, isolators, circuit breakers, and other items of equipment.
2. The operating mechanism of switch gear.
3. Frame work of switch boards, control boards, individual panel boards, cubicals, etc.
4. The structural steel work of indoor and outdoor substation metal cable joining boxes, the metal shells of the cables, metal conduits, and similar metal work.

9.9 **POWER SYSTEM EARTHING**

System earthing is very important. All metallic components other than live equipment should be maintained at zero potential so that the operating personnel's safety is ensured. The potential rise between neutral and ground should also be limited. Again if potential gradients build up in grounds around the neutral, protective gear may not properly operate. Earthing applies to both neutral and electrical equipment and appliances.

The effect of electric current on human body depends upon the magnitude, duration, and frequency of current. Currents in the range of 1−6 mA called "let go currents" are not serious but will cause a shocking experience. But when the currents are in the range of 9−25 mA they are quite

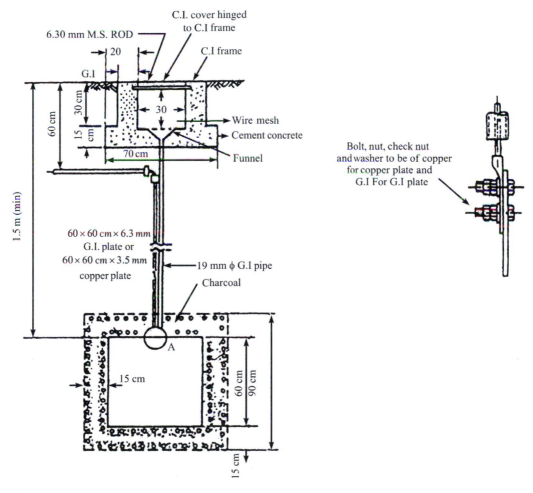

FIGURE 9.2

Plate earthing.

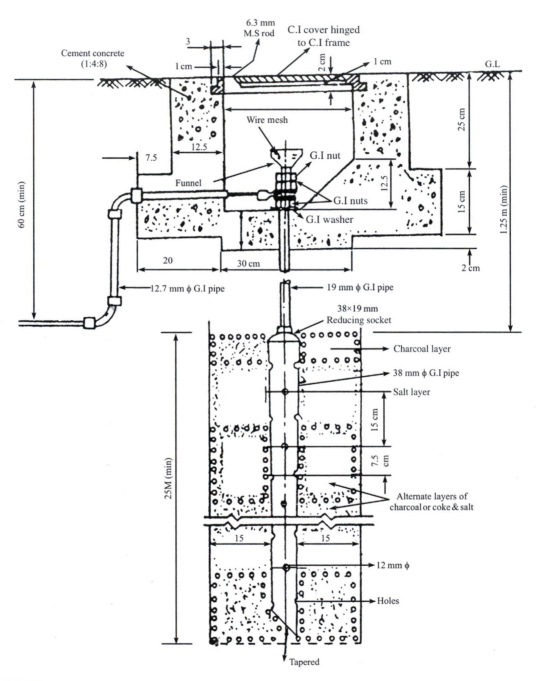

FIGURE 9.3

Pipe earthing.

dangerous. Higher currents will cause breathing problems and may affect heart and stop blood circulation; this is called ventricular fibrillation. A person of about 60 kg weight may be able to withstand a current in the body (rms) of 0.13 $\sqrt{t}$ where t is 0.03–3 s. The actual withstand capacity depends upon constitution of the body. As frequency is increased, the tolerable current also increases.

The resistivity of soil depends upon the nature of the soil, moisture content in the soil, salts in the soil, and temperature. Soil resistivity changes with depth. An earthing electrode in a distribution system may be a meter long metallic rod driven vertically into ground. Even a metallic pipe may be used. Alternately a plate may be buried as earthing electrode. Substations require an earthing grid. The electrode used for earthing dissipates the current into the soil.

If the soil is homogenous, the earthing resistance of a rod of diameter d and length

$$l \text{ is } R = \frac{\rho}{2\pi l}\left(\ln\frac{\delta l}{d} - 1\right)$$

Empirical relations are available for calculating resistance of two and more (multiple) rod electrodes.

For high-voltage and extra-high-voltage system earthing grids, with interconnected base conductors are formed and buried to a depth of 0.5–1 m. The grid is also sometimes connected to vertically driven rods in some arrangement determined from various considerations. Typical plate earthing and pipe earthing arrangements are shown in Figs. 9.2 and 9.3.

A human being while standing with his 2 ft. set aside can tolerate a potential difference of about

$$V_{\text{step}} = (R_B + 2R_f)I_B \text{ V}$$

where R_B is the body resistance (about 1000 Ω) and R_f is the earthing resistance of each foot, and I_B is body current.

Similarly the touch voltage (maximum possible horizontal reach potential difference).

$$V_{\text{touch}} = (R_B + 0.5R_f)I_B$$

9.10 **EARTHING TRANSFORMERS**

Sometimes the neutral of a system may not be available readily. This may be the case when the transformer is connected in delta or when the neutral is not brought out. An artificial neutral has to be then created. This can be done by earthing transformers. The connection diagram for such a transformer is shown in Fig. 9.4.

The three line terminals are brought out and connected to a zig-zag-connected transformer. This transformer has no secondary winding. Each limb of this transformer has two identical windings wound differentially as shown in Fig. 9.4, carrying currents in opposite directions. The total flux is negligible, and hence it carries almost negligible current. A neutral point is thus created for bus bars and delta-connected situations.

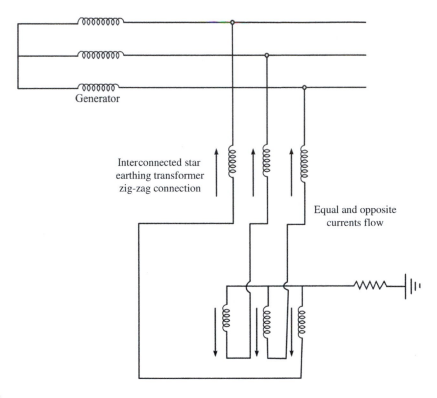

FIGURE 9.4

Earthing transformer.

9.11 BUS BAR ARRANGEMENTS

Single Bus Scheme: It is the simplest and the cheapest arrangement. This scheme with sectionalizing switches is very convenient if incoming and outgoing circuits are properly arranged as in Fig. 9.5.

Double Bus Schemes: If two identical bus bars are provided then, each load can be fed from either bus. In the scheme operational flexibility is provided. A bus tie breaker is provided for on-load change over from one bus to another bus. The arrangement does not provide for breaker maintenance without stoppage of supply. This is shown in Fig. 9.6.

Double Bus Double Breaker: For breaker or isolator maintenance without interrupting supply this scheme is the best. But it is very expensive and is seldom used. The scheme is shown in Fig. 9.7.

Main and Transfer Bus Scheme: This scheme provides facility for carrying out breaker maintenance. However, there is no provision for bus maintenance. Through bus tie breaker supply can be transferred to transfer bus and any breaker can be attended for maintenance. It should be noted that the connection to the transfer bus is through isolators (Fig. 9.8).

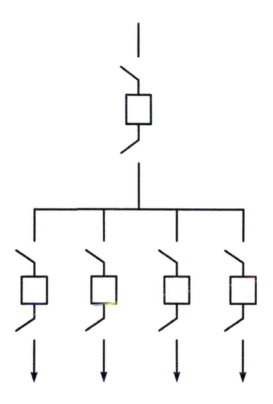

FIGURE 9.5

Single bus scheme.

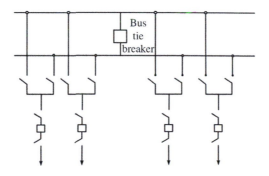

FIGURE 9.6

Double bus single breaker scheme.

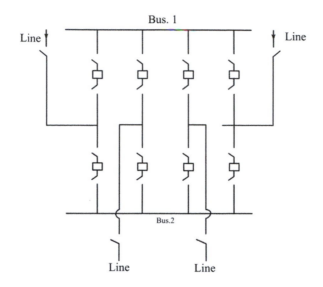

FIGURE 9.7

Double bus, double breaker scheme.

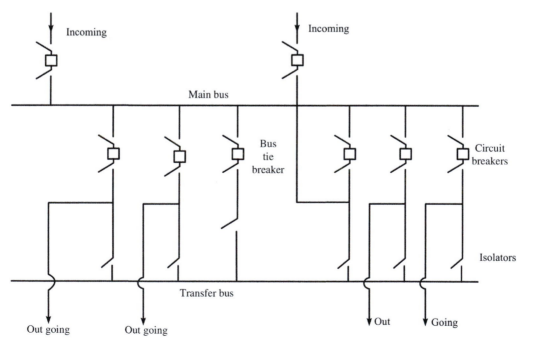

FIGURE 9.8

Main and transfer bus scheme.

There are several other bus arrangements such as breaker and a half scheme, double bus with bypass isolators.

9.12 NEUTRAL GROUNDING

Under balanced conditions of operation, neutral grounding has no effect. However, when the system is unbalanced as in the case of a line-to-ground fault, neutral grounding method assumes importance. An ungrounded neutral system will be able to clear line to ground faults without interruption. But in case of long and high voltage lines, when the line capacitance is appreciable fault currents are maintained by line-to-ground capacitances arcing grounds occur. To overcome this problem, several neutral grounding methods exist. Analysis of ungrounded and grounded neutral systems where solid, effective, resistance, and reactance grounding methods are adopted is given in the following.

9.13 UNGROUNDED SYSTEM

A three-phase ungrounded system is shown in Fig. 9.9A and the phasor diagram for a ground to line B fault is shown in Fig. 9.9B. The fault current I_F is maintained by the line-to-ground capacitances C_y and C_R and they flow. They are maintained by the closed path. These currents I_{By} and I_{BR} must be prevented from flowing through C_y and C_R so that the fault current finds no closed path. The maintenance of an arc due to fault to ground due to line capacitances gives rise to arcing grounds.

$$I_{BR} = \frac{V_{BR}}{X_C} = \frac{\sqrt{3}V_{BN}}{X_C}$$

$$I_{BY} = \frac{V_{BY}}{X_C} = \frac{\sqrt{3}V_{BN}}{X_C}$$

The fault current

$$I_F = \bar{I}_{BR} + \bar{I}_{BY} = \frac{\sqrt{3}\sqrt{3}V_{phase}}{X_C}$$

$$= \frac{3V_{phase}}{X_C}$$

where V_{ph} is the phase value of the voltage. It is seen that the magnitude of the fault current is three times the normal value of the current. The voltages on the healthy phases rises from phase values (V_{RN} and V_{YN}) to line voltages V_{YB} and V_{BR}, i.e., the magnitudes increase by $\sqrt{3}$ times. While the capacitance current increases $\sqrt{3}$ times the normal value in the two healthy phases, in the faulted line, the current increases to three times the normal value.

For the above reasons, neutrals of all three phase systems are invariably grounded through some element.

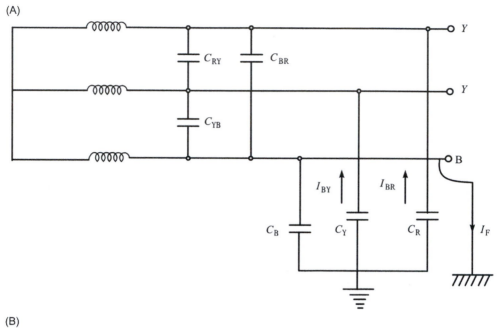

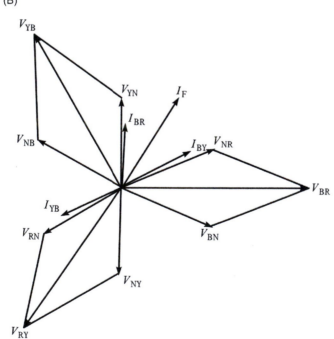

FIGURE 9.9

Arcing grounds and line-to-ground fault (A) and phasor diagram (B).

9.14 SOLID GROUNDING

This method of earthing the neutral is shown in Fig. 9.10A, and the phasor diagram is shown in Fig. 9.10B for a line-to-ground fault at phase B, when the neutral is solidly grounded. Under this condition the fault current flowing through the capacitances of the healthy lines R and Y, C_R and C_Y are I_{NR} and I_{NY}. Using symmetrical component analysis, the fault current

$$I_F = \frac{3 \cdot V_{ph}}{Z_0 + Z_1 + Z_2}$$

where Z_0, Z_1, and Z_2 are the sequence impedances of the network and the current I_F is predominantly reactive with almost a lagging angle of 90 degrees with respect to the voltage. This large lagging current will to a great extent compensate the leading nature of the currents through capacitances C_R and C_Y. This is shown in the phasor diagram Fig. 9.10B where $\bar{I}_{NR} + \bar{I}_{NY}$, the fault capacitive current is neutralized by the lagging fault current I_F to a great extent. The flow of these currents is shown in Fig. 9.10A.

9.15 RESISTANCE GROUNDING

In this method the neutral is grounded through a resistance R. This will limit the magnitude of the earth fault currents. Noninductive resistors such as liquid resistors are used. From Fig. 9.11A and B it can be seen that I_{BR} and I_{BY} lead the voltages V_{BR} and V_{BY} by 90 degrees when a line-to-ground fault occurs at phase B. The fault current I_F lags behind the faulted phase voltage V_{BN}. The capacitive fault current I_{CF} can be neutralized by $I'_F \cdot I_{CF}$ is the resultant of I_{BR} and I_{BY}. By properly adjusting the value of R, I_{CF} can be neutralized.

Peterson has suggested that the value of R may be chosen as

$$R = \frac{(1 - 2 \cdot 5) \times I}{C_R + C_Y + C_B}$$

Alternatively,

$$R = \frac{V_L}{\sqrt{3}I}$$

where V_L is the line voltage and I is the full load current of the largest machine or transformer in the system.

9.16 REACTANCE GROUNDING

In another method the neutral is grounded through a reactance coil. The coil is an iron cored reactor mounted in the neutral earthing circuit. The coil reactance can be varied by tuning it, so that it can resonate with the system capacitance and cancel its effect. The arrangement is shown in Fig. 9.12A. The corresponding phasor diagram is drawn in Fig. 9.12B. In this case I_F becomes I_{CF}.

$$I_{CF} = \sqrt{3}I_{BR} = \sqrt{3}\left(\frac{V_L}{X_C}\right) = \sqrt{3}\frac{\sqrt{3}V_{ph}}{X_C} = \frac{3V_{ph}}{X_C}$$

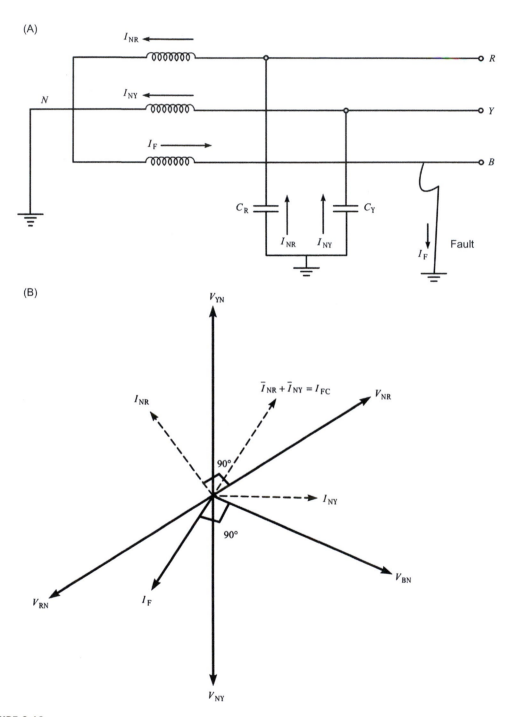

FIGURE 9.10

Solid grounding (A) and phasor diagram (B).

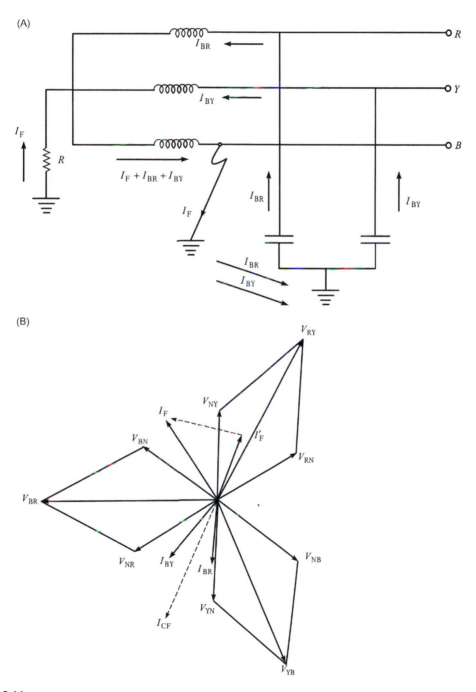

FIGURE 9.11

Resistance grounding (A) and phasor diagram (B).

(A)

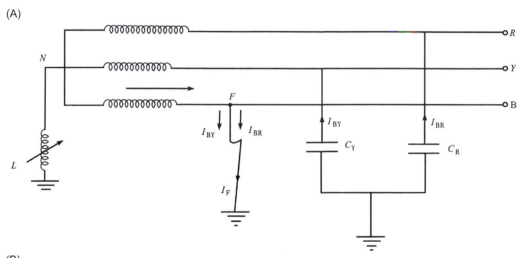

(B)

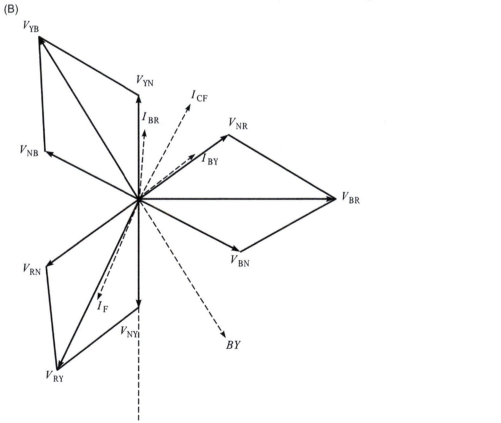

FIGURE 9.12

Arc suppression coil (A) and phasor diagram (B).

The capacitive current is three times the normal current.
It can be seen that

$$I_{CF} = I_F$$

$$\frac{3V_{ph}}{X_C} = \frac{V_{ph}}{X_L}$$

where $X_L = 2\pi flc$, the reactance of the coil in the earthing circuit
Hence

$$X_C = 3X_L$$

$$\frac{1}{\omega C} = 3\omega L$$

$$L = \frac{1}{3\omega^2 C} \text{ Henry} = \frac{1}{3(2\pi f)^2 \cdot C} \text{ Henry}$$

If the capacitive fault current I_{CF} is fully compensated by the coil inductance ωL, the earthing reactance is called Peterson coil and the grounding is called resonance or resonant grounding.

9.17 EFFECTIVE GROUNDING

The effectiveness of grounding is known from the ratio of X_0/X_1 of the power system.
A system is said to be reactance grounded if

$$\frac{X_0}{X_1} > 3.0$$

A system is said to be solidly grounded if

$$\frac{X_0}{X_1} < 3.0$$

If the X_0/X_1 is not greater than 3 and R_0/X_1 is not greater than 1, it is called effective grounding as per IEEE definition. For an effectively grounded system the voltage rise in healthy phases should not be more than 80% of line voltage. In practice, it is limited in most of the cases to 15% of line-to-line voltage.

9.18 GAS-INSULATED SUBSTATIONS

Gas-insulated substations (GISs) are widely designed and implemented over the last three decades because of their high reliability, easy maintenance, small ground space requirement, and other benefits. In a GIS, various equipments such as circuit breakers, bus bars, isolators, load-break switches, current transformers, voltage transformers, and earthing switches, are all housed in metal-enclosed modules fitted with SF_6 gas. SF_6 provides insulation to ground of all live equipment. The concept was first implemented in early 1960s. The introduction of SF_6 gas revolutionized not only the technology of circuit breakers but also the layout of substations.

The dielectric strength of SF_6 gas at atmospheric pressure is approximately three times that of air. It has arc quenching properties three to four times better than air at the same pressure. SF_6 fitted substations need only 10% of space required by a conventional substation. Hence, in big cities and where space is scarce, GIS substations are preferred.

Since the entire equipment is enclosed in pressurized SF_6 module, the installation is not subject to environmental pollution. There is an advantage in coastal and industrial areas. With GIS installation, land and construction costs are reduced substantially. Further, the substation can be located closer to the load center. In GIS cast resin support insulators are used.

WORKED EXAMPLES

E.9.1 A 220-kV, three-phase, 50 cycles, overhead line 110 km long has a capacitance to earth for each line of 0.016 μf/km. Determine the inductance and KVA rating of the arc suppression coil suitable for the system.

Solution:

$$L = \frac{1}{3\omega^2 C} = \frac{10^6}{3 \times (314)^2 \times 50 \times 0.016} = 4.22\,h$$

$$\text{Current rating of the coil} = \frac{\frac{220}{\sqrt{3}}}{X_L} \times 1000\,A$$
$$= \frac{220}{\sqrt{3}} \times \frac{1000}{2\pi \times 50 \times 4.22} = 95.86\,A$$

$$\text{Rating of the arc suppression coil} = 95.86 \times \frac{220}{\sqrt{3}}\,kVA = 12,176\,kVA$$

PROBLEMS

P.9.1 Calculate the reactance of the arc suppression coil suitable for 66-kV, three-phase transmission system which has a capacitance of 5.3 μf for each conductor to earth.

P.9.2 A 132-kV, three-phase, 50 cycle transmission line has a capacitance of 0.12 μf/phase. Determine the inductance of arc suppression coil to neutralize the effect of capacitance of (1) complete length of the line, (2) 90% of the line, and (3) 75% of the line.

QUESTIONS

Q.9.1 How are the substations classified?

Q.9.2 Explain the various equipment needed in a substation.

Q.9.3 What is a station transformer? Explain its utility.

Q.9.4 Why and where earthing is needed in substations?

Q.9.5 What is the effect of soil resistivity on earthing?

Q.9.6 Explain the necessity and operation of an earthing transformer.

Q.9.7 Discuss the various bus arrangement schemes with figures, used in substations.

Q.9.8 Why is it necessary to ground the neutral of a power system? Explain.

Q.9.9 What are arcing grounds? How are they suppressed?

Q.9.10 Explain the various methods of neutral grounding with schemes and phasor diagrams.

Q.9.11 Distinguish between solid grounding and effective grounding.

DISTRIBUTION SYSTEM

10

The distribution system consists of feeders, distributors, and mains supplied from the secondary of a distribution transformer. Feeders carry large currents to the feeding points. Generally, there are no intermediate tapping points on feeders. The size of a feeder is determined by the current-carrying capacity. Distributors, on the other hand, are those conductors from which current is tapped to supply the consumers. The service mains are small cable lengths that connect the customers to the distributors.

Distributors are designed on the basis of voltage drop along the conductor. This is because the service mains are connected to distributors and it is obligatory on the part of the utility system to supply power to the consumers at specified voltages. The feeder is designed on the basis of constant current density. A typical distribution system is shown in Fig. 10.1 showing the various components of it. A typical distribution feeder is shown in Fig. 10.2.

10.1 EFFECTS OF VOLTAGE ON THE CONDUCTOR VOLUME

Let a feeder supply a given power P to the feeding point. Now, if the voltage is raised by n times the current is reduced by $(1/n)$ times. Then, the feeder cross section need be only $(1/n)$ times the original section since current density is to be maintained the same.

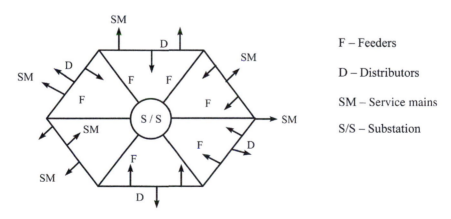

F – Feeders

D – Distributors

SM – Service mains

S/S – Substation

FIGURE 10.1

Distribution system.

Electrical Power Systems. DOI: http://dx.doi.org/10.1016/B978-0-08-101124-9.00010-3

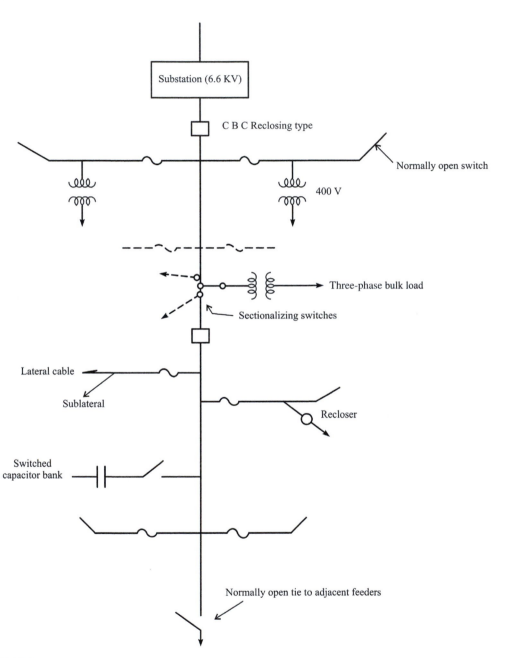

FIGURE 10.2

A typical primary distribution feeder.

In case of a distributor, if the voltage is raised by n times the original voltage V, to supply a given power P, the potential drop in any given section $V = IR$. Since voltage is increased by n times, the current is reduced by $(1/n)$ times.

$$nV = \frac{1}{n}IR^1 \text{ or } R^1 = n^2\left(\frac{V}{I}\right) = n^2R$$

where R is the original resistance of the section and R^1 is the new resistance of the sector for the same power transfer. The area of section a^1 need be $(1/n^2)$ times the original section a since $(R = \rho l/a)$ R is inversely proportional to the area of section.

10.1.1 RADIAL AND RING MAINS

A radial distribution system is shown in Fig. 10.3. In case of failure at any point on a radial system, the supply system beyond the fault gets isolated. If continuity of supply is to be ensured, to the system beyond the faulted system while it is isolated and fault rectified, an alternate path for supply of power should be provided. This is made possible in a ring main system shown in Fig. 10.3. A typical radial distribution main system is shown in Fig. 10.4.

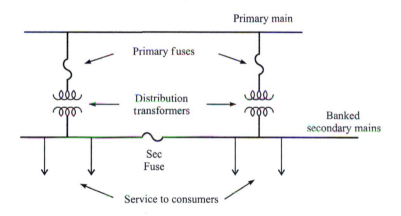

FIGURE 10.3

Radial distribution system.

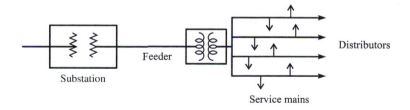

FIGURE 10.4

Typical distribution main.

There are several circuit opening devices used in distribution system.

A disconnect switch disconnects the circuit at no-load. A load break switch interrupts load current only and cannot be used under fault currents. A circuit breaker can interrupt fault currents. An automatic line sectionalizing switch is used only with backup circuit breakers or reclosers, but are not used alone.

Reclosers are automatic circuit closing devices with overcurrent protection. They trip and reclose a preset number of times to clear any fault current of transient nature or to isolate permanent faults. Fuses are ordinary overcurrent protective devices that contain a fusible member in the circuit device.

10.2 DIRECT CURRENT DISTRIBUTION SYSTEM

At present almost all the energy needs are met by AC systems only. Only in few cases like traction DC is used. However, as a basis to understand the basic principle involved in distribution, few DC distribution methods are discussed in the following.

There could be three-wire or two-wire DC distribution. The principle is illustrated in Fig. 10.5.

10.3 DISTRIBUTOR FED FROM ONE END

This is shown in Fig. 10.6 AB and A^1B^1 ... are the go and return conductors of a two-wire system. Currents in the sections AC, CD and DE are i_1, i_2, i_3,...,i_n, respectively.

The return currents are also the same in sections A^1C^1, C^1D^1 and D^1E^1..., the resistances of the section are r_1, r_2, r_3, ..., r_n. The total resistance from the feeding end AA^1 are R_1, R_2, ..., R_n. A single line representation used in practice is shown in Fig. 10.6B.

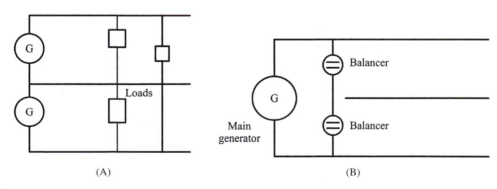

(A)   (B)

FIGURE 10.5

Three-wire distribution system. (A) Three-wire distribution using two generates and (B) three-wire distribution using balancers.

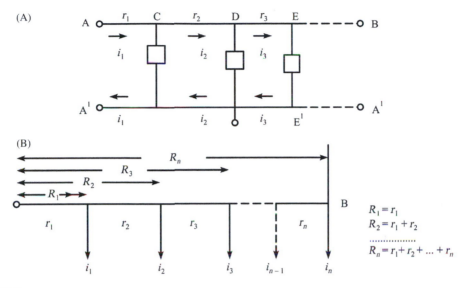

FIGURE 10.6

Distributor fed from one end. (A) Double line representation and (b) single line representation.

The voltage drop in section AC

$$= (i_1 + i_2 + \cdots + i_n) \cdot r_1$$

The voltage drop in section CD

$$= (i_2 + i_3 + \cdots + i_n)r_2$$

In a similar manner, the voltage drop in DE

$$= (i_3 + i_4 + \cdots + i_n)r_3, \text{ and so on}$$

The total voltage drop along a single conductor is

$$
\begin{aligned}
v &= r_1(i_1 + i_2 + \cdots + i_n) + r_2(i_2 + i_3 + \cdots + i_n) + r_3(i_3 + \cdots + i_n) + \cdots \\
&= i_1 r_1 + i_2(r_1 + r_2) + i_3(r_1 + r_2 + r_3) + \cdots \\
&= i_1 R_1 + i_2 R_2 + i_3 R_3 + \cdots + i_n R_n
\end{aligned}
$$

Total voltage drop in the two conductors go and return

$$= 2v = 2(i_1 R_1 + i_2 R_2 + \cdots + i_n R_n) \tag{10.1}$$

10.4 DISTRIBUTOR FED FROM BOTH ENDS AT THE SAME VOLTAGE

When loads are supplied from both ends A and B (I_A and I_B at potential V), the minimum potential will occur at one of the loads i_2 or i_3 depending upon the values of currents i_i ($i = 1, 2, 3, 4$). Let us assume that load i_3 is supplied by both A and B feeding points. The distribution of currents is shown in various sections assuming A supplies x amperes to load i_3 (Fig. 10.7).

Drop in section AC $= (i_1 + i_2 + x)\, r_1$

Drop in section CD $= (i_2 + x)$

Drop in section DE $= xr_3$

Drop in section BF $= (i_3 + i_4 - x) \cdot r_5$

Drop in section FE $= (i_3 - x) \cdot r_4$

$$V_{AE} = V_{BE}$$

Hence

$$V_A - (i_1 + i_2 + x)r_1 + (i_2 + x)r_2 + xr_3 = V_B - (i_3 + i_4 - x) - (i_3 - x)r_4 \tag{10.2}$$

Knowing all currents and resistances, x and resistances, x can be determined. The minimum potential at E is the supply voltage at A minus the drop in section AE or the supply voltage at B minus the drop in section BE. Note that $V_A = V_B = V$

10.5 DISTRIBUTOR FED FROM BOTH ENDS AT DIFFERENT VOLTAGES

This case is shown in Fig. 10.8. The only difference from the previous case is that V_A is not equal to V_B. Hence, we obtain the relation up to the point of minimum voltage along the distributor from either end.

$$V_A - [(i_1 + x)r_1 + x \cdot r_2] = V_B - [r_3(i_2 - x) + r_4(i_2 + i_3 - x)] \tag{10.3}$$

Knowing V_A, V_B, i_1, i_2, i_3, r_1, r_2, r_3, and r_4, x can be determined.

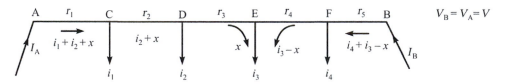

FIGURE 10.7

Distributor fed from both ends at same voltage ($V_A = V_B = V$).

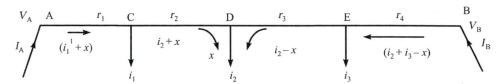

FIGURE 10.8

Distributor fed from both ends of different voltages ($V_A \neq V_B$).

10.6 UNIFORMLY LOADED DISTRIBUTOR FED FROM ONE END

In all the previous cases the loads are concentrated. If the load is uniformly distributed at i (A/m) length of the distributor and if the resistance per unit length of the conductor is r (ohm), then voltage drop can be computed by considering a differential length dx at a distance x from the feeding end as shown in Fig. 10.9.

$$v = \int_0^x i(l-x)r \cdot dx = ilr \int_0^x dx - ir \int_0^x x\,dx$$

$$v = \left[ilrx - ir\frac{x^2}{2} \right] \text{ V}$$

At $x = l$

$$v_l = il\,rl - ir\frac{l^2}{2} = \frac{irl^2}{2} = \frac{il\,rl}{2} = \frac{IR}{2} \tag{10.4}$$

where il is the total current I and rl is the total resistance R.

A uniformly distributed load on a distributor fed from both ends at the same potential will have the minimum potential at the midpoint. Each feeding end supplies half of the distributor. Hence

$$v = ir\left(\frac{l}{2}\right)^2 \frac{1}{2} = \frac{irl^2}{8} \tag{10.5}$$

Likewise, the analysis can be made for a distributor fed from both ends.

10.7 COPPER EFFICIENCIES

In the following a comparison will be made between two-wire DC system and other common systems adopted for power transport. In case of overhead lines the potential difference between line conductor to earth (crossarm and transmission tower) is important. In underground cables especially multicore cables, the electric stress between any two conductors in the intervening dielectric is important. Based on the requirement, the amount of copper (conductor material) needed for any system of distribution can be determined and compared with a DC reference system. This will give copper efficiency.

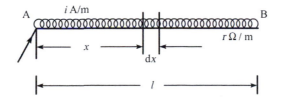

FIGURE 10.9

Uniformly loaded distributor.

10.7.1 SAME MAXIMUM VOLTAGE BETWEEN CONDUCTOR AND EARTH

1. *Two-Wire DC System*: For a two-wire DC system shown in Fig. 10.10 the output voltage $2V_{DC}$ where V_{DC} is the potential difference between either conductor and earth. I_{DC} is the line current.

Power transmitted

$$P = 2 \cdot V_{DC}I_{DC}$$

Power loss in transmission, $P = I_{DC}^2 R_{DC}$
where R_{DC} is the resistance of either of the outers.

$$P = 2R_{DC}\left(\frac{P}{2v_{DC}}\right)^2 = \frac{2P^2}{4v_{DC}^2}R_{DC}$$

Hence

$$R_{DC} = \frac{2Pv_{DC}^2}{P^2} = \frac{2Pv^2}{P^2} \tag{10.6}$$

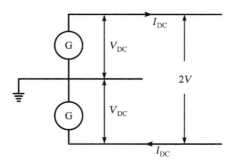

FIGURE 10.10

Two-wire DC system.

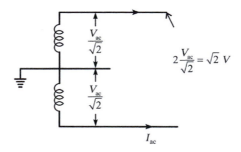

FIGURE 10.11

Single-phase, two-wire system.

2. *Single-phase, two-wire system*: The two-conductor, single-phase system shown in Fig. 10.11 transmits power

$$P = \sqrt{2}VI_{AC}$$

power loss, $P = 2I_{AC}^2 R_{AC} = 2 \dfrac{P^2}{(\sqrt{2}V \cos d)^2} R_{AC}$

$$P = \dfrac{P^2}{V^2 \cos^2 \phi} R_{AC}$$

So that

$$R_{AC} = \dfrac{PV^2 \cos^2 \phi}{P^2}$$

and

$$R_{DC} = \dfrac{2PV^2}{P^2}$$

$$\dfrac{R_{AC}}{R_{DC}} = \dfrac{\cos^2 \phi}{2}; \quad \text{But} \dfrac{R_{AC}}{R_{DC}} = \dfrac{\rho l}{a_{AC}} \cdot \dfrac{a_{DC}}{\rho l} = \dfrac{a_{DC}}{a_{AC}}$$

where a_{AC} and a_{DC} are the conductor cross sections.
Hence

$$\dfrac{a_{AC}}{a_{DC}} = \dfrac{2}{\cos^2 \phi} \quad \text{or} \quad a_{AC} = \dfrac{2}{\cos^2 \phi} a_{DC} \tag{10.7}$$

Copper efficiency is defined as the ratio of the copper required for AC supply to copper required for DC supply.
3. *Three-Phase, Three-Wire System*: Power transmitted in case of three-phase, three-wire system shown in Fig. 10.12 is

$$P = 3\dfrac{v}{\sqrt{2}}I_{AC} \cos \phi \tag{10.8}$$

Power loss in the conductors

$$P = 3I_{AC}^2 R_{AC}$$

Substituting the value of I_{AC} from Eq. (10.9)

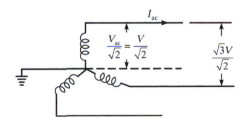

FIGURE 10.12

Three-phase, three-wire system.

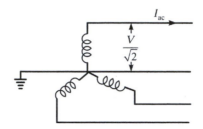

FIGURE 10.13

Three-phase, four-wire system.

$$P = 3\left(\frac{\sqrt{2}P}{3V \cos\phi}\right)^2 R_{AC}$$

Therefore

$$R_{AC} = \frac{p}{3}\frac{9v^2 \cos^2\phi}{2P^2}$$

and using (10.7)

$$\frac{R_{AC}}{R_{DC}} = \frac{3}{4}\cos^2\phi$$

Since there are only two conductors in DC system and three conductors in AC system

$$\frac{a_{AC}}{a_{DC}} = \frac{4}{3}\cos^2\phi\cdot\frac{3}{2} = \frac{2}{\cos^2\phi} \tag{10.9}$$

4. *Three-Phase, Four-Wire System*: In this system shown in Fig. 10.13 power transmitted and losses remain the same.
Hence

$$\frac{R_{AC}}{R_{DC}} = \frac{3\cos^2\phi}{4}$$

However, since there are four conductors in AC case

$$\frac{a_{AC}}{a_{DC}} = \frac{4}{3}\cos^2\phi\cdot\frac{4}{2} = \frac{8}{3\cos^2\phi} = \frac{2.67}{\cos^2\phi} \tag{10.10}$$

10.7.2 SAME MAXIMUM VOLTAGE BETWEEN TWO CONDUCTORS

1. *Two-Wire DC Supply*: In this supply system power transmitted with reference to Fig. 10.14.

$$P = V_{DC}\cdot I_{DC} = V\cdot I_{DC}$$

Power loss

$$p = 2I_{DC}^2\cdot R_{DC} = 2\cdot\frac{P^2}{V^2}\cdot R_{DC}$$

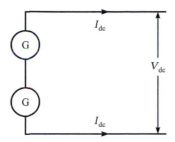

FIGURE 10.14

Two-wire DC system—voltage between conductors same.

FIGURE 10.15

Single-phase, two-wire system.

2. *Single-Phase, Two-Wire System*: From Fig. 10.15 power transmitted

$$P = \frac{V}{\sqrt{2}} I_{AC} \cos d$$

and power loss

$$P = 2I_{DC}^2 R_{AC}$$

$$P = 2\left(\frac{\sqrt{2}P}{v \cos d}\right)^2 R_{AC}$$

$$R_{AC} = \frac{P V^2 \cos^2 d}{2 \left(\sqrt{2}P\right)^2} = \frac{PV^2 \cos^2 \phi}{4P^2}$$

$$\frac{R_{AC}}{R_{DC}} = \frac{PV^2 \cos^2 \phi}{4P^2} \cdot \frac{2P^2}{PV^2} = \frac{\cos^2 \phi}{2} = \frac{a_{DC}}{a_{AC}}$$

$$\text{Copper efficiency} = \frac{a_{AC}}{a_{DC}} = \frac{2}{\cos^2 \phi} \tag{10.11}$$

3. *Three-Phase, Three-Wire System:* In this system as shown in Fig. 10.16
Power transmitted

$$P = \sqrt{3} \frac{V}{\sqrt{2}} I_{AC} \cos \phi$$

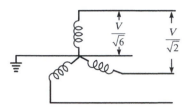

FIGURE 10.16

Three-phase, three-wire system.

Power loss

$$P = 3I_{AC}^2 R_{AC}$$

$$P = \frac{3P^2}{3V^2} \cdot \frac{2}{\cos^2 \phi} \cdot R_{AC} = \frac{2P^2}{V^2 \cos^2 \phi} \cdot R_{AC}$$

$$\frac{R_{AC}}{R_{DC}} = \cos^2 \phi \text{ and } \frac{a_{AC}}{a_{DC}} = \frac{1}{\cos^2 \phi} \cdot \left(\frac{3}{2}\right) = \frac{1.5}{\cos^2 \phi} \tag{10.12}$$

It may be remembered that the AC system contains three conductors while the DC system has only two conductors.

4. *Three-Phase, Four-Wire System:* As before (Point 4 in Section 10.7.1) since the power transmitted and losses remain the same as in Point 3 of this section

$$\frac{R_{AC}}{R_{DC}} = \cos^2 \phi$$

However, since there are four conductors in AC system

$$\frac{a_{AC}}{a_{DC}} = \frac{1}{\cos^2 \phi} \cdot \left(\frac{4}{2}\right) = \frac{2}{\cos^2 \phi} \tag{10.13}$$

From the above analysis it is seen that the DC distribution is more economical. However, for bulk transmission and distribution of electrical energy, alternating currents are suited, because of the ease with which voltage can be converted from one level to another level using transformers which operate at very high efficiency.

Example:
Compare the copper required for a two-phase, three-wire system with two-wire DC system for the same maximum voltage between conductor and earth.
Solution: The system is shown in Fig. 10.17
Power transmitted

$$P = 2\frac{V}{\sqrt{2}} I_{AC} \cos \phi = \sqrt{2} V I_{AC} \cos \phi$$

$$I_{AC} = \frac{P}{\sqrt{2} V \cos \phi}$$

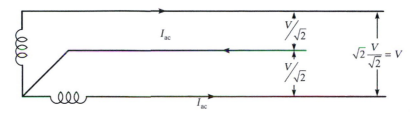

FIGURE 10.17

Two-phase, three-wire system.

Power loss

$$P = 2I_{AC}^2 R_{AC} + \left(\sqrt{2}I_{AC}\right)^2 \cdot \frac{R_{AC}}{\sqrt{2}}$$

In may be noted that the common return will carry $\sqrt{2}I_{AC}$. Hence, its cross section must be $\sqrt{2}$ times that of either outer for the same current density.

$$P = 2\frac{P^2}{2V^2 \cos^2 \phi}R_{AC} + \frac{2P^2}{2V^2 \cos^2 \phi}\frac{R_{AC}}{\sqrt{2}} - \frac{P^2}{V^2 \cos^2 \phi}\left[1 + \frac{1}{\sqrt{2}}\right] \cdot R_{AC}$$

$$\frac{R_{AC}}{R_{DC}} = \frac{\cos^2 \phi}{2\left[1 + \frac{1}{\sqrt{2}}\right]} = \frac{\cos^2 \phi}{(2 + \sqrt{2})}$$

$$\frac{a_{AC}}{a_{DC}} = \frac{(2 + \sqrt{2})}{\cos^2 \phi} \times \frac{(2 + \sqrt{2})}{2} = \frac{2 + 4 + 2 \times 2\sqrt{2}}{2} \cdot \frac{1}{\cos^2 \phi}$$

$$\text{Copper efficiency} = \frac{5.828}{\cos^2 \phi} \tag{10.14}$$

Copper required for a two-phase three-wire system is $5.828/\cos^2 \phi$ times the section of the DC system.

10.8 ALTERNATING CURRENT DISTRIBUTION

In AC distribution systems, voltage drop occurs due to both resistance and reactance. The voltage drop in a length of resistance r (Ω) and reactance x (Ω) will be $ri \cos \phi + xi \sin \phi$ where i is the current flowing and ϕ is the power factor of the load current.

It is necessary to perform all calculations on single-phase basis. All voltages and currents must have a reference and phasor notation should be used.

10.8.1 ALTERNATING CURRENT DISTRIBUTOR FED FROM ONE END

10.8.1.1 Concentrated Loads

Consider the distributor shown in Fig. 10.18.

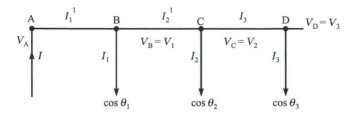

FIGURE 10.18

AC distribution. *AC*, alternating current.

V_D is the reference voltage at the receiving end

$$\bar{I}_3 = |I_3| (\cos \phi_3 - j \sin \phi_3)$$

$$\bar{I}_2 = |I_2|(\cos \phi_2 - j \sin \phi_2)$$

and

$$\bar{I}_1 = |I_1|(\cos \phi_1 - j \sin \phi_1)$$

$$\bar{I} = \bar{I}_1 + \bar{I}_2 + \bar{I}_3$$

Drop is section CD $= \nu_{CD} = I_3 (\cos \phi_3 - j \sin \phi_3)(R_3 + jX_3)$
Drop is section BC $= \nu_{BC} = I_2 (\cos \phi_2 - j \sin \phi_2) (R_2 + jX_2)$
Drop in section AB $= \nu_{AB} = I_1 (\cos \phi_1 - j \sin \phi_1) (R_1 + j X_1)$
The sending end voltage $V_A = V_D + \nu_{CD} + \nu_{BC} + \nu_{AB}$
The sending end power factor $= \cos \phi_s$

10.9 DESIGN OF FEEDER—KELVIN'S LAW

A feeder is designed on the basis of current-carrying capacity and minimum cost while a distributor is designed on the basis of voltage drop.

The cost involved in the installation of a feeder has two components:

1. The interest and depreciation on the capital cost and installation of the feeder.
2. The cost of energy loss due to resistance of the conductor material. Further, in case of cables, the losses in metallic sheath and dielectric losses in insulation used are also to be considered.

For a given length of the feeder, and its weight, the cost is proportional to the area of cross section of the feeder conductor. The annual cost of interest and depreciation is also dependent on the size of the cable and hence on the conductor cross section. If a is the area of cross section, the first component is proportional to a. If P is the constant of proportionality, the annual charges due to interest and depreciation on capital cost $= Pa$.

Again, the energy loss depends upon resistance and resistance is inversely proportional to area of section. Hence annual charges due to second component is proportional to $(1/a)$ and the cost $= Q(1/a)$ where Q is a constant of proportionality. The total annual charges due to both components

$$C = Pa + \frac{Q}{a} \tag{10.15}$$

For the total cost C to be a minimum

$$\frac{dc}{da} = P - \frac{Q}{a^2} = 0$$

Hence

$$a = \sqrt{\frac{Q}{P}} \tag{10.16}$$

Hence, each component for economy $Pa = Q/a = \sqrt{PQ}$

The most economical cross section for a feeder is that section which makes the annual charges for interest and depreciation equal to the annual cost of energy loss in the feeder that is called Kelvin's law. This is illustrated in Fig. 10.19. Kelvin's law may not be practicable for implementation. Consider a case where two-distribution system require the same current to be supplied through feeders. This means that both the feeders should have the same conductor cross section. If the interest and depreciation differ for both the systems due to different financial supports the conductor sections would be different based on Kelvin's law. This is against the engineering requirement that if both the feeders carry the same current then they cannot have different cross sections, since current density is fixed.

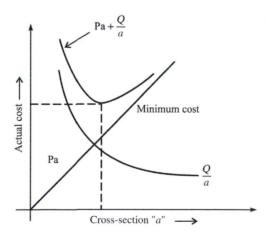

FIGURE 10.19

Kelvin's law.

WORKED EXAMPLES

E.10.1 A trolley wire supplied from one end has the following loads: 150 A at 100 m, 100 A at 300 m; 80 A at 1000 m; and 120 A at 1200 m from the feeding end. The trolley wire resistance is 200 μΩ/m and the rail return has a resistance of 150 μΩ/m. Find the voltage at the various load points if a voltage of 550 be maintained at the supply end.

Solution:

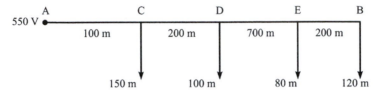

FIGURE E.10.1

The resistance per 100 m of both go and return

$$= \frac{350 \ \mu\Omega \times 100 \ m}{10^6} = 0.035 \ \Omega/100 \ m$$

$$V_A = 550 \ V$$

$$V_B = 550 - 450 \times 0.035 = 534.25 \ V$$

$$V_C = 534.25 - 2 \times 0.035 \times 300 = 513.25 \ V$$

$$V_D = 513.25 - 7 \times 0.035 \times 200 = 464.25 \ V$$

$$V_E = 464.25 - 2 \times 0.035 \times 120 = 455.85 \ V$$

E.10.2 The resistance of the two conductors of the cable loaded as shown is 0.1 Ω/1000 m for both conductors. Find the current supplied at A and B, the current in each section and the voltages at C, D, and E. Both A and B are maintained at 200 V.

Solution:

Let at load point D both ends feed. Let B feed x A. The distribution is shown in Fig. E.10.2.

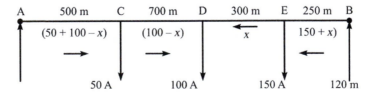

FIGURE E.10.2

Writing down the voltage drops from A to D and B to D and equating them, we obtain

$$200 - 5 \times 0.01 \times (50 + 100 - x) - 7 \times 0.01(100 - x) = 200 - 2.5 \times 0.01 \times (150 + x) - 3 \times 0.01 \times (x)$$

Simplying

$$0.175x = 10.75 \quad \text{so that } x = 61.43 \text{ A}$$
$$I_{AC} = 50 + 38.57 = 88.57 \text{ A} = I_{AC}$$
$$I_{BA} = 150 + 61.43 = 211.43 \text{ A} = I_{BA}$$
$$I_{CD} = 100 - 61.43 = 38.57 \text{ A}$$
$$I_{DE} = 61.43 \text{ A}$$
$$V_C = 200 - 5 \times 0.01 \times (50 + 100 - 61.43) = 195.57 \text{ V}$$
$$V_D = 195.57 - 7 \times 0.01 \times (100 - 61.43) = 192.87 \text{ V}$$
$$V_E = 200 - (150 + 61.43) \times 2.5 \times 0.01 = 194.71 \text{ V}$$

E.10.3 A two-conductor distributor AD is fed at A and D at 255 and 250 V, respectively and is loaded as shown. The resistances given are those of each conductor. Find the value of current in each section of the cable and the voltage at each load point.
Solution:

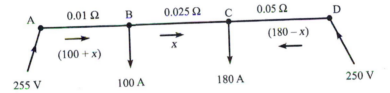

FIGURE E.10.3

Resistances are given per conductor. Hence, they need to be multiplied by two. Writing down the equation for voltage drop, assuming 180-A load point C is where both ends supply.

$$255 - 0.02(100 + x) - 0.05 \, x = 250 - 0.03(180 - x)$$

Simplifying and solving

$$0.1x = 8.4 \quad \text{or} \quad x = 84 \text{ A}$$
$$I_{AB} = 100 + 84 = 184 \text{ A}$$
$$I_{BC} = 84 \text{ A}$$
$$I_{CD} = 180 - 84 = 96 \text{ A}$$
$$V_B = 255 - (0.02 \times 184) = 251.32 \text{ V}$$
$$V_C = 250 - 0.03 \times 96 = 247.12 \text{ V}$$

E.10.4 A 250-m, 2-wire main fed from one end is loaded uniformly at the rate of 1.5 A/m, the resistance of each conductor being 0.2 Ω/km. Find the voltage necessary at the feeding end to maintain 250 V: (1) at the middle and (2) at the distant end of the cable.

Solution:

1. Let O be the midpoint

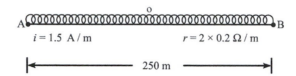

$$\frac{l}{2} = 125 \text{ m}$$

$$i = 1.5 \text{ A}; \quad I = 1.5 \times 125 \text{ A}$$

The voltage drop due to uniform load on OB is a concentrated load of 1.5×125 A flowing through AO section creating a drop of

$$(1.5 \times 125)\text{A} \times \frac{2 \times 0.2 \times 125}{1000} = 9.375 \text{ V}$$

The voltage drop along AO due to uniformly distributed load

$$v = \int_0^x i(l-x)r \, dx = \int_0^{125} 1.5(125 - x) \times \frac{0.4}{1000} \, dx$$

$$= \frac{1.5 \times 0.4}{1000}\left[125x - \frac{x^2}{2}\right]_0^{125} = \frac{1.5 \times 0.4}{1000} \times \frac{125 \times 125}{2} = 4.6875$$

or

$$v = \frac{IR}{2} = (1.5 \times 125) \times \left(\frac{2 \times 0.2 \times 125}{1000}\right) \times \frac{1}{2} = 4.6875$$

Total voltage drop along AO = 4.6875 + 9.375 = 14.06 V. The feeding end A must be maintained at 250 + 14.06 = 264.06 V to maintain 250 V at the midpoint O.

2. $$v = \int_0^{250} 1.5(250 - x) \times \frac{0.4}{1000} \, dx = \frac{1.5 \times 0.4}{1000}\left[250x\frac{-x}{2}\right]_0^{250}$$

$$= \frac{1.5 \times 0.4}{1000} \times \frac{250 \times 250}{2} = 18.75 \text{ V}$$

The feeding end A must be maintained at 250 + 18.75 = 268.75 V

$$\text{Current fed from A} = 25 + 50 + 50 \times 1$$
$$= 125 \text{ A}$$

$$\text{Current fed from B} = 200 \times 1 - 5 = 200 \text{ A}$$

Total current = 25 + 50 + 250 × 1 = 325 A

$$I_A + I_B = 125 + 200 = 325 \text{ A}$$

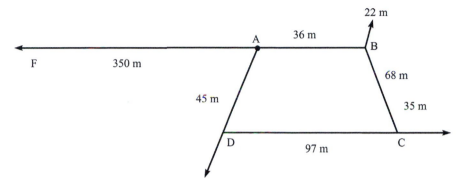

FIGURE E.10.5

E.10.5 A ring main distributor A B C D A is fed at A from a feeder F A of length 350 m. At B, C, and D loads of 22, 35, and 50 A are drawn, respectively, as shown in Fig. E.10.5.

The feeder supplies at a potential of 237 V and the minimum potential at any point in the ring main should not be less than 230 V. Specific resistivity of copper is 1.76 $\mu\Omega$/cm. Determine the ratio of feeder section to distributor section so that the volume of copper is a minimum.

Solution: Let current in section AD $= x$ A
Then current in section DC $= (x - 50)$ A
Current in section CB $= (x - 85)$ A
and Current in section BA $= (x - 107)$ A
Applying Kirchhoff's current law to the closed loop A D C B A

$$x + (x - 50) + (x - 85) + (x - 107) = 0$$

solving $x = 60.5$ A
point C supplies 35 A of which DC supplies $60.5 - 50 = 10.5$ A and BC supplies $35 - 10.5 = 24.5$ A
voltage drop from A to C is $(r \times 60.5 \times 45) + r \times 10.5 \times 97$

$$= 27,722.5r + 1018.5r = 3741r$$

where r is the resistance of ring distributor including go and return per meter length. But

$$r = \frac{\rho l}{a} = \frac{(1.76 \times 100 \times 10^{-6}) \times 2}{a_d} = \frac{1.3168}{a_d}$$

Drops in feeder

$$F A = \frac{107 \times 1.76 \times 10^{-6} \times 100 \times 2 \times 350}{a_f}$$

$$= \frac{13.1824}{a_f}$$

The voltage drop should not exceed 7 V

$$\therefore \quad 7 = \frac{13.1824}{a_f} + \frac{1.3168}{a_d}$$

solving for a_d

$$\frac{a_d}{1.317} = \frac{a_f}{7a_f - 13.18}; \quad a_d = \frac{a_f}{7a_f - 13.18}$$

Total volume of copper in ring main and feeder is proportional to ($246.\ a_d + 350\ a_f$)
For minimum copper $d(\text{volume})/d(a_f) = 0$
volume is proportional to

$$\frac{246(1.7a_f)}{7 \cdot a_f - 13.18} + 350\ a_f =$$

i.e., proportional to

$$\frac{324\ a_f}{7 \cdot a_f - 13.18} + 350\ a_f = k \left[\frac{324\ a_f + 2450\ a_f^2 - 4613\ a_f}{7 \cdot a_f - 13.18} \right]$$

$$\frac{d}{d(a_f)} k \left[\frac{2450\ a_f^2 - 4289\ a_f}{7\ a_f - 13.18} \right] = 0$$

$$\frac{[2450\ a_f^2 - 4289\ a_f][7] - (7\ a_f - 13.18)[4900\ a_f - 4289]}{(7a_f - 13.18)^2} = 0$$

$$17,150\ a_f^2 - 30,023\ a_f - 34,300\ a_f^2 - 30,023\ a_f - 64,582\ a_f + 56,529 = 0$$

$$- 17,150\ a_f^2 + 64,582a_f - 56,529 = 0$$

$$17,150\ a_f^2 + 64,582a_f + 56,529 = 0$$

Solving $a_f = 2.3818$ or 1.3838 cm^2

$$a_f = 1.3838 \text{ cm}^2 \text{ given negative value for } a_d$$

$$a_f = 2.3818 \text{ is the correct value}$$

Then

$$a_d = \frac{2.3818 \times 1.317}{7 \times 2.3818 - 13.18} = \frac{3.13683}{16.6726 - 13.18} = \frac{3.13683}{3.4926}$$

$$= 0.898136 \text{ cm}^2$$

Check: Total voltage drop

$$= \frac{13.1824}{a_f} + \frac{1.317}{a_d} = \frac{13.18}{2.3818} + \frac{1.317}{0.898136}$$

$$= 5.53 + 1.47 = 7.00 \text{ V}$$

E.10.6 The two conductors of a distributor cable 500 m long have a total resistance of 0.075 Ω. A voltage of 250 V is maintained at feeding points A and B. The cable is loaded with a uniform lighting load of 1 A/m and with additional concentrated loads as shown in

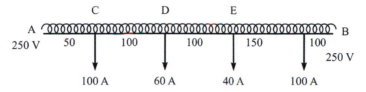

FIGURE E.10.6

Fig. E.10.6. The distances are in meters. Find: (1) the lowest voltage, (2) the point at which it occurs, and (3) the current fed into the cable at A and at B.

Solution:

It can be seen from the figure that for the distributed load of 1 A/m the point E is at the center and the voltage drop will be a minimum at E. To find the current distribution at load point E, we equate the voltage drops from both the ends, at point E. Let x be the current in amperes supplied by A into the 40 A load at E.

Then,

$$(100 + 60 + x)50 \times r + (60 + x) \times 100 \times r + 100 \times x \times r$$
$$= 100 \times (100 + 40 - x)r + 150(40 - x)r$$

Simplifying

$$8000 + 50x + 6000 + 100x + 100x = 14000 - 100x + 6000 - 150x$$

Solving for x, $500\ x = 6000$

$$x = \frac{6000}{500} = 12\ A$$
$$I_A = 250 \times 1 + 100 + 60 + 12 = 422\ A$$
$$I_B = 250 \times 1 + 100 + (40 - 12) = 378\ A$$

Minimum potential occurs at point E.

The voltage drop from A to E is given by

$$50 \times \frac{0.075}{500} \times 172 + 100 \times \frac{0.075}{500} \times 72 + 100 \times \frac{0.075}{500} \times 12 + 250 \times \frac{0.075}{500} \times \frac{250 \times 1}{2}$$
$$= 1.29 + 1.08 + 0.18 + 4.6875$$
$$= 7.2375\ V$$

$$\text{Potential at point E} = V_E = 250 - 7.2375$$
$$= 242.76\ V$$

E.10.7 A DC distributor 500 m long is fed at both ends at 220 V. The distributor is loaded as shown in Fig. E.10.7. The resistance of each conductor is 0.05 Ω/km. Find the point of minimum potential and the currents fed at both ends.

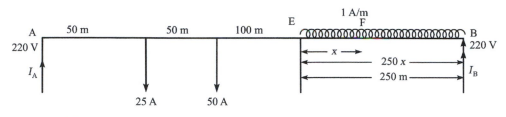

FIGURE E.10.7

Solution:

Let x be the distance from point E, where minimum potential occurs (F). Writing down the equation for potential drop from A to F.

$$v_{AF} = 50r(25 + 50 + x) + 50(50 + x)r + 150xr\frac{x^2r}{2}$$

$$v_{BF} = (250 - x)(250 - x)\frac{xr}{2}$$

$$v_{AF} = v_{BF}$$

$$3250 + 50x + 2500 + 50x + 150x + \frac{x^2}{2} = \frac{62,500}{2} + \frac{x^2}{2} - \frac{500x}{2}$$

$$\text{i.e., } 6250 + 250x = 31,250 - 250x$$

$$x = \frac{250,000}{2} = 50 \text{ m}$$

$$d_{AF} = 250 + 50 = 300 \text{ m}$$

$$d_{BF} = 250 - 50 = 200 \text{ m}$$

$$V_{BF} = 1 \times 200 \times \frac{200}{2} \times \frac{0.05}{1000} = 1 \text{ V/conductor}$$
$$\text{2 V for both conductors}$$

$$v_{AF} = \frac{0.05}{1000}\left[125 \times 50 + 50 \times 100 + 150 \times 50 + \frac{50^2}{2}\right]$$

$$= \frac{0.05}{1000}[6250 + 5000 + 7500 + 1250]$$

$$= \frac{0.05 \times 20,000}{1000} = 1/\text{conductor, 2 V for both conductors}$$

$$\text{Minimum potential} = 220 - 2 \times 1 \text{ V} = 218 \text{ V}$$

E.10.8 A three-phase, 15-km long, 33-kV overhead line supplies 5 MW at 0.85 power factor lagging for 12 h/day, 3 MW at 0.82 power factor lagging for 6 h/day, and 1.5 MW at 0.8 power factor lagging for 6 h/day. The line supplies power all through the year. The installation cost including conductors is Rs [75,850 + 1892 . a] per km length of the line for interest and depreciation where a is the cross section in mm^2. The resistance of the conductor per meter length is 0.0285 Ω/mm^2. Energy costs Re 1.00 per KWh. Determine the most economic section for the feeder line using Kelvin's law.

Solution:

$$\text{Resistance per km length} = \frac{0.0285}{a} \times 1000 \ \Omega/\text{phase}$$

where a is in mm^2

Load current for 5 MW at 0.85 power factor, 12 h

$$= \frac{5 \times 10^6}{\sqrt{3} \times 33 \times 10^3 \times 0.85} = 101.72 \text{ A}$$

Load current for 3 MW at 0.82 power factor for 6 h

$$= \frac{3 \times 10^6}{\sqrt{3} \times 33 \times 10^3 \times 0.82} = 64 \text{ A}$$

Load current for 1.5 MW at 0.8 power factor for 6 h

$$\frac{1.5 \times 10^6}{\sqrt{3} \times 33 \times 10^3 \times 0.8} = 56.018$$

Energy loss due to 101.72 A in 1 km length of line

$$= 3 \times (101.72)^2 \times \frac{0.0285}{a} \times 1000 \times \frac{12}{1000} \text{ KWh for 12 } h/\text{day}$$

$$= \frac{10,615.98}{a}$$

Energy loss due to 64 A in 1 km length of line for 6 h/day

$$= 3 \times 64^2 \times \frac{0.0285}{a} \times 1000 \times \frac{6}{1000} \text{ KWh}$$

$$= \frac{2101.248}{a} \text{ KWh}$$

Energy loss due to 56.818 A in 1 km length for 6 h/day

$$= 3 \times 56.818^2 \times \frac{0.0285}{a} \times 1000 \times \frac{6}{1000} = \frac{1656.1}{a} \text{ KWh/day}$$

Total energy loss per day per 1 km

$$= \frac{10,615.98}{a} + \frac{2101.248}{a} + \frac{1656.1}{a} = \frac{14,372.348}{a} \text{ KWh}$$

Yearly charges for energy

$$= 365 \times \frac{14,372.348}{a} \times 1 \text{ Rs}$$

$$= \frac{5,245,907.02}{a} \text{ Rs}$$

Total capital cost per km $= (75,850 + 1892 \ a)$

Total annual charges per km $= 75,850 + 1892.a + \dfrac{5,245,907.02}{a} = C$

For minimum cost C, $\dfrac{dC}{da} = 0$

i.e., $1892 = \dfrac{5,245,907.02}{a^2}$ or $a = \sqrt{\dfrac{5,245,907.02}{1892}} \text{mm}^2$

$a = 52.6565 \text{ mm}^2$

PROBLEMS

P.10.1 A direct current, 2-wire distributor 600 m long is fed at both ends A and B at 400 V. The load consists of 100 A at 100 m from A, 150 A at 150 m from A and a uniform loading of 1 A/m for the last 350 m. The resistance of each conductor is 0.05 Ω/km. At what point is the load voltage a minimum and what is its value?

P.10.2 A two-conductor street main has a total length of 500 m and is loaded as shown in Fig. P.10.2, the distances given being in meters. Both ends A and B are supplied at 245 V. If the minimum allowable voltage at the consumer's terminals is to be 240 V, find the necessary cross section of each conductor of the main. Resistivity is 1.7 μΩ cm.

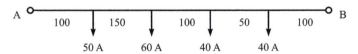

FIGURE P.10.2

P.10.3 In a direct current ring main shown in Fig. P.10.3, at A a voltage of 500 V is maintained at B a load 145 A is taken and at C a load of 210 A is taken. Find the voltages at B and C. The resistance of each conductor of the main is 0.03 Ω/km.

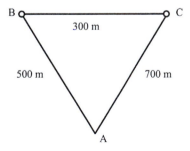

FIGURE P.10.3

P.10.4 A single-phase distributor has a resistance of 0.2 Ω and a reactance of 0.3 Ω. At the far end the voltage is $V_b = 230$ V, the current is 100 A and the power factor 0.8. At the midpoint of the distributor a, a current of 100 A is supplied at power factor 0.6 with reference to the voltage V_a at a. Find the supply voltage V_s and the phase angle between V_s and V_b.

QUESTIONS

Q.10.1 Explain the effect of change in voltage on the conductor volume in distribution.

Q.10.2 Explain the various elements in a typical distribution system with a neat sketch.

Q.10.3 What are the merits and demerits of ring mains over radial mains?

Q.10.4 Show that for a distributor fed from both ends and uniformly loaded, the maximum voltage drop is $IR/8$ where I is the total current and R is the total resistance of the distributor.

Q.10.5 Compare copper efficiencies of a two-wire DC system with single-phase AC two-wire system for overhead transmission.

Q.10.6 Determine the copper efficiency for three-phase, three-wire system with two-wire DC system for a cable system.

Q.10.7 Show that for a three-phase, four-wire system for the same maximum voltage between conductor and earth, the copper efficiency is $(2.67/\cos^2 \phi)$.

Q.10.8 State and explain Kelvin's law. What are its applications?

OVERVOLTAGES

Overvoltages are produced due to a variety of causes. They are explained in the following as externally generated overvoltages, mainly due to lightning and associated phenomena and due to internal causes such as switching and other operations.

11.1 EXTERNAL CAUSES

Dangerous overvoltages are mainly produced due to lightning. They may be due to any of the following:

1. Direct lightning
2. Electromagnetically induced currents due to lightning in the neighborhood, generally classified as side strokes
3. Electrostatically induced charges on the line conductors due to the presence of thunder clouds above the overhead lines, or
4. Electrostatic charges imparted to lines due to friction caused by particles of dust, dry snow, and flying particles due to wind action.

Electrostatic charges are also acquired by overhead line due to change in altitude.

Fig. 11.1 illustrates the above causes. Fig 11.1A shows a lighting stroke directly hitting an overhead line, Fig. 11.1B explains the effect of a side stroke, in affecting the line. The case of electrostatically induced pressure rise is shown in Fig. 11.1C.

The pressure rises could reach anywhere from 500 to 2000 kV due to side strokes. The corresponding current magnitude may vary from about 10,000 to 100,000 A. Nevertheless, these currents last only a few microseconds and find their way to ground. All the lightning discharges are more or less unidirectional. As for induced charges on the line, the potential rise could reach 10–50 kV. When the bound potential on the line is released due to a discharge elsewhere, the induced pressure rise travels along the line with the velocity of light. They constitute traveling waves.

11.2 LIGHTNING PHENOMENON

When thunder clouds are present in the velocity of a high-voltage transmission line, charges will build up as shown in Fig. 11.2. With such a charge build up, an invisible pilot streamer starts from the cloud to the line at a speed of about 0.15 m/μs. Another charge center near the discharged cloud may follow immediately along the ionized path and get it self completely discharged. This is called

Electrical Power Systems. DOI: http://dx.doi.org/10.1016/B978-0-08-101124-9.00011-5

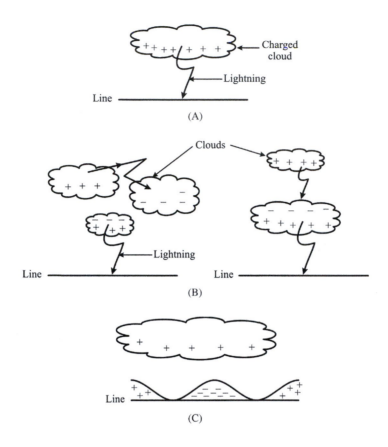

FIGURE 11.1

Overvoltages due to thunder clouds. (A) Direct stroke, (B) side stroke, and (C) induced charges on the line due to thunder clouds.

dart leader. It moves with about 3% of velocity of light carrying high current that can cause substantial damage.

Now, a return stroke starts from the line to the cloud, meeting the dart leader on its way at a velocity of about 10% of light velocity. The current in the return stroke could vary from about 1 to about 200 kA. This is shown in Fig. 11.3.

Most of the damage is caused by side strokes.

There are several theories that explain the charging process in thunder clouds.

Simpson's theory explains that raindrops while breaking up due to air currents in the space acquire positive charge and then air gets negatively charged. It is observed that large thunder clouds become negatively charged, while a positively charged layer forms on the upper side of cloud. Since moisture brings down the dielectric strength of air, humid air breaks down at about 10 kV (P)/cm instead of at 30 kV (P)/cm.

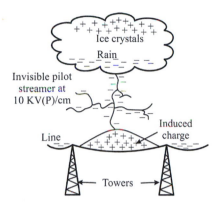

FIGURE 11.2

Buildup of pilot streamer.

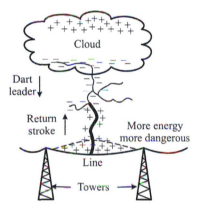

FIGURE 11.3

Return stroke.

11.3 INTERNAL CAUSES

There are several internal sources for overvoltages. These overvoltages are generally not as dangerous as those due to external causes. These are

1. switching operations on an unloaded line
2. sudden opening of a loaded line under short circuit conditions
3. resonance, and
4. arcing grounds.

Switching operations may produce voltages oscillating at 100 to several thousand cycles per second. Overvoltages will be produced between lines and line-to-ground. As the system grows in generating and transmission capacities, switching overvoltages become more and more important. Switching overvoltages are generally due to sustained earth fault on phase conductors, energization or reclosure of long lines, load, rejection at receiving end, breaker opening up on fault initiation and subsequent reclosure.

Interruption of low inductive currents produces overvoltages on high-voltage lines

Line Voltage	Overvoltage
220 kV	2.5 P.u
400 kV	1.8 P.u
750 kV	1.2 P.u

with typical values for overvoltages indicated above, the frequency may be in the range of 200–1000 Hz. All these overvoltages can be attenuated by surge arrestors. Resistors in series with circuit breakers also counter these overvoltages effectively.

Whenever transmission lines are disconnected or capacitor banks are switched off, overvoltages are produced. They are effectively handled by SF_6 and air-blast circuit breakers. Majority of the modern circuit breakers operate in about one and half cycles, after the occurrence of a fault and their contacts open in about 30 ms. Most of the switching surges, resonance, and arcing grounds do not pose any real problem.

Resonance overvoltages decide the steady-state voltage rating of devices. Switching and lightning overvoltages form the basis for selection of insulation levels for lines and equipments. Both resonance and switching overvoltages are related to system operating voltages.

Magnetizing inrush current that flows when a transformer is energized is one source for resonance overvoltage. During unsymmetrical faults, resonance at harmonic frequencies is another reason for resonance overvoltages. Lightning arrestors should not operate at these voltages. Hence it influences selection of lightning arrestor.

11.4 ATTENUATION OF TRAVELING WAVES

Consider a traveling wave of magnitude e_0 moving over an overhead line (Fig. 11.4). After traveling a distance x on the line, the power loss incurred as the wave moves over a differential length dx

$$dp = i^2 R \, dx + e^2 G \, dx$$

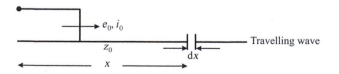

FIG. 11.4

Traveling wave.

with usual notation

Power at distance $x = ei = p = i^2 Z_0$
Differential power $dp = 2iZ_0 \, di$
Z_0 is the natural impedance of the line
Since current decreases as x increases

$$-2i \, Z_0 \, di = i^2 R \, dx + e^2 G \, dx$$

$$-2i \, Z_0 \, di = [i^2 R + (Z_0 i)^2 G] dx$$

$$\frac{di}{i} = \frac{(R + Z_0^2 G) dx}{-2Z_0} = \left(\frac{R + Z_0^2 G}{-2Z_0}\right) dx$$

Integrating both sides

$$\log i = -\left(\frac{R + Z_0^2 G}{-2Z_0}\right) x + c$$

At

$$x = 0, \quad i = i_0$$

$$\log i_0 = C \text{ or } \log i = -\left(\frac{R + Z_0^2 G}{2Z_0}\right) x + \log i_0$$

$$\therefore \quad \log \frac{i}{i_0} = -\left(\frac{R + Z_0^2 G}{2Z_0}\right) x = -Kx \tag{11.1}$$

$$i = i_0 e^{-Kx}$$

where

$$K = \left(\frac{R + Z_0^2 G}{2Z_0}\right)$$

In a similar manner it can be proved that

$$e = e_0 \, e^{-Kx} \tag{11.2}$$

11.5 SURGE IMPEDANCE AND VELOCITY OF PROPAGATION

Consider a long transmission line with distributed line parameters of inductance and capacitance L and C per unit length. Neglecting resistance (Fig. 11.5)

$$q = VCx$$

$$I = \frac{dq}{dt} = VC \frac{dx}{dt}$$

$$= VCv$$

where v is the velocity of propagation

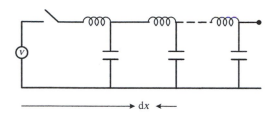

$\longrightarrow$ dx $\longleftarrow$

FIGURE 11.5

Surge impedance.

The flux linkages are

$$\Psi = ILx$$

$$V = \frac{d\Psi}{dt} = IL\frac{dx}{dt} = ILv$$

$$\frac{V}{I} = \frac{ILv}{VCv} \text{ or } \frac{V^2}{I^2} = \frac{L}{C} \tag{11.3}$$

$$\frac{V}{I} = \sqrt{\frac{L}{C}} = Z_n$$

where Z_n is called the natural impedance, characteristic impedance, or surge impedance.

Again

$$VI = VCv \cdot ILv = VILC \cdot v^2$$

$$\therefore \quad LCv^2 = 1 \text{ or } v^2 = \frac{1}{LC}$$

The velocity of propagation

$$v = \frac{1}{\sqrt{LC}} \text{ m/s} \tag{11.4}$$

where

$$L = 2 \times 10^{-7} \log_e \frac{d}{r}\frac{h}{m} \text{ and}$$

$$C = \frac{2\pi\varepsilon}{\log e\left(\frac{d}{r}\right)}$$

$$v = \frac{1}{\dfrac{2\pi\varepsilon}{\log_e\left(\dfrac{d}{r}\right)}} = \frac{1}{\sqrt{2 \times 10^{-7} \times \dfrac{2\pi \times 10^{-9}}{36\pi}}}$$

$$= \frac{1}{\sqrt{\dfrac{10^{-6}}{9}}} = 3 \times 10^8 \text{ m/s}$$

which is the velocity of light.

For overhead lines $Z_n \simeq 500 \ \Omega$ and for cables $Z_n \simeq 50 \ \Omega$.

11.6 **REFLECTION AND REFRACTION COEFFICIENTS**

Consider a point of discontinuity on a line. Let the incoming voltage and current waves be e' and i'. The reflected components at the point of discontinuity are e'' and i''. The refracted components of voltage and current are designated e_r and i_r (Fig. 11.6).

Let the surge impedance of the line be Z_1 up to the point of discontinuity and Z_2 be the surge impedance after discontinuity.

At the point of discontinuity, i.e., at the junction of lines 1 and 2 of surge impedances Z_1 and Z_2

$$e_r = e' + e'' \tag{11.5}$$

and

$$i_r = i' - i'' \tag{11.6}$$

(the reflected wave has negative sign)

Note that $i'' = -e''/Z_1$

From Eq. (11.6)

$$\frac{e_r}{Z_2} = \frac{e'}{Z_1} - \frac{e''}{Z_1}$$

and therefore, also using Eq. (11.5)

$$\frac{e''}{Z_1} = \frac{e'}{Z_1} - \frac{e_r}{Z_2} = \frac{e'}{Z_1} - \left(\frac{e' + e''}{Z_2}\right)$$

Therefore

$$e''\left[\frac{1}{Z_1} + \frac{1}{Z_2}\right] = \frac{e'}{Z_1} + \frac{e'}{Z_2} = e'\left[\frac{Z_2 - Z_1}{Z_2 + Z_1}\right]$$

Hence,

$$e'' \times \left[\frac{Z_2 - Z_1}{Z_2 Z_1}\right] = e'\left[\frac{Z_2 - Z_1}{Z_2 Z_1}\right]$$

or

$$e'' = e'\left[\frac{Z_2 - Z_1}{Z_2 + Z_1}\right] = e'\alpha \tag{11.6}$$

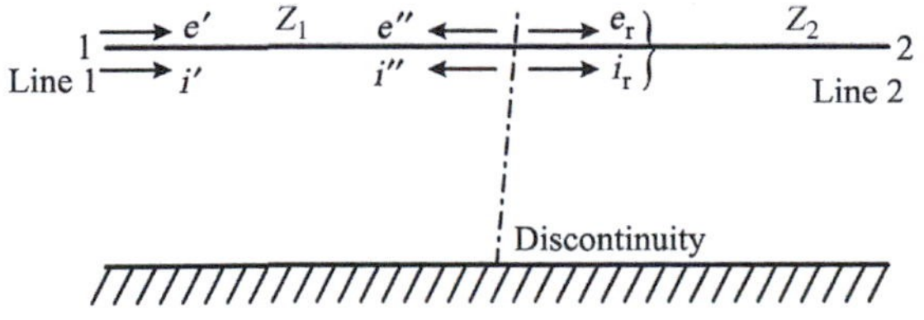

FIGURE 11.6

Reflection and refraction coefficients.

where α is called the coefficient of reflection

Again $e_r = e' + e''$

$$\frac{e_r}{Z_2} = \frac{e'}{Z_1} - \frac{e''}{Z_1} = \frac{e'}{Z_1} - \left(\frac{e' - e_r}{Z_1}\right)$$

Hence

$$\frac{e_r}{Z_2} = \frac{2e'}{Z_1} - \frac{e_r}{Z_1} \quad \text{or} \quad e_r\left[\frac{1}{Z_1} + \frac{1}{Z_2}\right] = \frac{2e'}{Z_1}$$

$$e_r = \frac{2Z_1 Z_2}{Z_1 + Z_2} \cdot \frac{e'}{Z_1} = \frac{2Z_2}{Z_1 + Z_2} e' = \beta e' \tag{11.8}$$

where β is the coefficient of refraction

Also

$$1 + \alpha = 1 + \frac{Z_2 - Z_1}{Z_2 + Z_1} = \frac{2Z_2}{Z_1 + Z_2} = \beta$$

Hence $(1 + \alpha) = \beta$

Further

$$i_r = i' - i'' = \frac{e'}{Z_1} - \frac{e''}{Z_1} = \frac{e'}{Z_1} - \left(\frac{Z_2 - Z_1}{Z_2 + Z_1}\right)\frac{e'}{Z_1}$$

$$i_r = \frac{e'}{Z_1}\left(1 - \frac{Z_2 - Z_1}{Z_2 + Z_1}\right) = \frac{e'}{Z_1}\left[\frac{Z_2 + Z_1 - Z_2 + Z_1}{Z_2 + Z_1}\right] = \frac{2e'}{Z_2 + Z_1} \tag{11.9}$$

$$i_r = \frac{2Z_1 i'}{Z_1 + Z_2}$$

Also

$$i'' = -\frac{e''}{Z_1} = -e'\left[\frac{Z_2 - Z_1}{Z_2 + Z_1}\right] = \frac{e'}{Z_1}\left[\frac{Z_1 - Z_2}{Z_1 + Z_2}\right]$$

$$i'' = -\frac{i' Z_1}{Z_1}\left[\frac{Z_1 - Z_2}{Z_1 + Z_2}\right] = i'\left[\frac{Z_1 - Z_2}{Z_1 + Z_2}\right] = -\alpha i' \tag{11.10}$$

Hence

$$i' = \frac{e'}{Z_1}$$

$$i'' = -\frac{e''}{Z_1}$$

Thus we obtain the results

$$e'' = e'\left[\frac{Z_2 - Z_1}{Z_2 + Z_1}\right] = \alpha e'$$

$$e_r = \frac{2Z_2}{Z_1 + Z_2} e' = \beta e'$$

$$i_r = \frac{2Z_1}{Z_1 + Z_2} \cdot i'$$

$$i'' = \left[\frac{Z_1 - Z_2}{Z_1 + Z_2}\right] i' = -\alpha i'$$

11.7 **LINE TERMINATED THROUGH RESISTANCE**

Consider a transmission line of natural impedance Z_0, terminated at one end through a resistance R (Fig. 11.7).

$$e_r = e' + e''$$

$$i_r = i' - i''$$

The refracted component of voltage is determined from

$$\frac{e_r}{R} = \frac{e'}{Z_0} - \frac{e''}{Z_0}$$

$$\frac{e_r}{R} = \frac{e'}{Z_0} - \left(\frac{e_r - e'}{Z_0}\right) = \frac{2e'}{Z_0} - \frac{e_r}{Z_0}$$

$$\frac{e_r}{R} + e_r\frac{1}{Z_0} = \frac{2e'}{Z_0} \quad \text{or} \quad e_r = \frac{2e'}{Z_0}\left(\frac{RZ_0}{R + Z_0}\right) \qquad (11.11)$$

$$e_r = \frac{2R}{R + Z_0}e'$$

The refracted component of current is determined from

$$i_r = \frac{e_r}{R} = \frac{2R}{R + Z_0}\frac{e'}{R} = \frac{2e'}{R + Z_0} = \frac{2Z_0 i'}{R + Z_0}$$

$$\qquad (11.12)$$

$$i_r = \frac{2Z_0 i'}{R + Z_0}$$

The reflected component of voltage e'' is given by

$$e'' = e_r - e' = \frac{2R}{R + Z_0}e' - e' = e'\left[\frac{2R}{R + Z_0} - 1\right] = \frac{e'[2R - R - Z_0]}{R + Z_0}$$

$$\qquad (11.13)$$

$$e'' = \left(\frac{R - Z_0}{R + Z_0}\right)e'$$

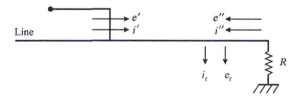

FIGURE 11.7

Line terminated through resistance.

$$i'' = -\frac{e''}{Z_0} = -\left(\frac{R-Z_0}{R+Z_0}\right)\frac{e'}{Z_0} = -\left(\frac{R-Z_0}{R+Z_0}\right)\frac{i'Z_0}{Z_0}$$

$$i'' = -\left(\frac{R-Z_0}{R+Z_0}\right)i' = -\alpha i' \tag{11.14}$$

If $R \to \infty$, the line gets open circuited and in this case

$$\underset{R \to \infty}{e_{\mathrm{r}}} = \frac{2R}{R+Z_0}e' = \frac{2}{1+\dfrac{Z_0}{R}}e' = 2e'$$

and

$$\underset{R \to \infty}{i_{\mathrm{r}}} = \frac{2Z_0 i'}{R+Z_0} = 0$$

Further

$$\underset{R \to \infty}{i''} = \frac{-(R-Z_0)}{(R+Z_0)}i' = \frac{-\left(1-\dfrac{Z_0}{R}\right)}{\left(1+\dfrac{Z_0}{R}\right)}i' = -i'$$

$$\underset{R \to \infty}{e''} = \frac{(R-Z_0)}{(R+Z_0)}e' = \frac{-\left(1-\dfrac{Z_0}{R}\right)}{\left(1+\dfrac{Z_0}{R}\right)}e' = e'$$

11.8 LINE TERMINATED BY INDUCTANCE

Consider a line of surge impedance Z_0 terminated at the other end through a pure inductor of inductance L (Fig. 11.8).

In this case

$$\alpha = \frac{Z_2 - Z_1}{Z_2 + Z_1} = \frac{sL - Z_0}{sL + Z_0}$$

and

$$\beta = \frac{2Z_2}{Z_2 + Z_1} = \frac{2sL}{sL + Z_0}$$

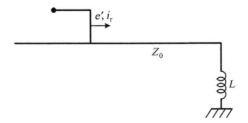

FIGURE 11.8

Line terminated through inductance.

The reflected component of voltage $e'' = ((sL - Z_0)/(sL + Z_0))(e'/s)$
where e' is the incoming wave and $Z_2 = Ls$.
Also $e'(t) = e'/s$ for step input

$$e''(s) = \frac{-\frac{Z_0}{L} + s}{\frac{Z_0}{L} + s} \cdot \frac{e'}{s} = e'\left[\frac{-\frac{Z_0}{L}}{s\left(s + \frac{Z_0}{L}\right)} + \frac{1}{s + \left(\frac{Z_0}{L}\right)}\right]$$

Let $K = Z_0/L$

$$e''(s) = e'\left[-\frac{K}{s(s+K)} + \frac{1}{s+K}\right] = e'\left[-\frac{1}{s} + \frac{1}{s+K} + \frac{1}{s+K}\right]$$

$$e''(t) = e'[-1 + e^{-Kt} + e^{-Kt}] = e'[2e^{-Kt} - 1] \tag{11.15}$$

$$= e'[2e^{-(Z_0/L)} - 1]$$

The reflected component of the current $i''(s) = -e''(s)/Z_0$
Hence

$$i''(t) = \frac{e'}{Z_0}\left[1 - 2e^{-(Z_0/L)\cdot t}\right] i'\left[1 - 2e^{-(Z_0/L)t}\right] \tag{11.16}$$

The refracted components are now computed from

$$e_r(s) = e'(t)\left(\frac{2Z_2}{Z_2 + Z_1}\right) = \frac{e'}{s}\left[\frac{2sL}{Z_0 + sL}\right] = \frac{2}{\left(\frac{Z_0}{L} + s\right)}e' \tag{11.17}$$

$$e_r(t) = 2e'e^{-(Z_0/L)t}$$

$$i_r(t) = \frac{e_r(s)}{Z_2(s)} = \frac{e'}{s} \cdot \frac{1}{Ls} \cdot \frac{2sL}{Z_0 + sL} = 2e'\left[\frac{1}{s} \cdot \frac{1}{(Z_0 + sL)}\right]$$

$$i_r(t) = \frac{2e'}{L}\left[\frac{1}{s} \cdot \frac{1}{\left(\frac{Z_0}{L} + s\right)}\right] = \frac{2e'}{L}\left[\frac{1}{(s+K)} \cdot \frac{1}{s}\right]$$

$$= \frac{2e'}{KL}\left[\frac{1}{s} - \frac{1}{s+K}\right] = \frac{2e'}{Z_0}\left[\frac{1}{s} \cdot \frac{1}{s+K}\right] = \frac{2e'}{Z_0}\left[1 - e^{-Kt}\right]$$

$$= \frac{2e'}{Z_0}\left[1 - e^{-(Z_0/L)t}\right]$$

Check:

$$e'' = e_r - e' = 2e'\left[e^{-(Z_0/L)t} - e'\right] = e'\left[2e^{-(Z_0/L)t} - 1\right]$$

$$i'' = i^r - i' = \frac{2e'}{Z_0}\left[1 - e^{-(Z_0/L)\cdot t}\right] = \frac{-e'}{Z_0} = \frac{e'}{Z_0}\left[1 - 2e^{-(Z_0/L)t}\right]$$

11.9 **LINE TERMINATED BY CAPACITANCE**

The coefficient of reflection (Fig. 11.9) is

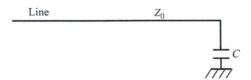

FIGURE 11.9

Line terminated through capacitance.

$$\alpha = \frac{Z_2 - Z_1}{Z_2 + Z_1}$$

The coefficient of refraction is

$$\beta = \frac{2Z_2}{Z_2 + Z_1}$$

For the line terminated by capacitance C.

$$\alpha = \frac{\frac{1}{CS} - Z_0}{\frac{1}{CS} + Z_0} = \frac{1 - Z_0\,CS}{1 + CS\,Z_0} = \frac{K - S}{K + S} \quad \text{where } K = \frac{1}{CZ_0}$$

$$\beta = \frac{2 \cdot \frac{1}{CS}}{Z_0 + \frac{1}{CS}} = \frac{2}{Z_0\,CS + 1} = \frac{\frac{2}{CZ_0}}{S + \frac{1}{CZ_0}} = \frac{2K}{S + K} \quad \text{where } K = \frac{1}{CZ_0}$$

The refracted component of voltage is

$$e_r = \frac{2Z_2}{Z_1 + Z_2} e' = \beta e' = \frac{2K}{S + K} \frac{e'}{S}$$

$$e_r(s) = 2e'\left[\frac{K}{s(s+K)}\right] = 2e'\left[\frac{1}{s} - \frac{1}{s+K}\right] \tag{11.19}$$

$$e_r(t) = 2e'(1 - e^{-Kt}) = 2e'\left(1 - e^{\frac{-t}{CZ_0}}\right)$$

The refracted component of current is

$$I_r(s) = \frac{V_r(s)}{Z_2(s)} = \frac{2Ke'}{s(s+K)} \cdot \left(\frac{1}{1/Cs}\right) = \frac{2Ke'C}{(s+K)} \tag{11.20}$$

$$I_r(t) = 2e'C\left(\frac{1}{CZ_0}\right) \cdot e^{-Kt} = \frac{2e'}{Z_0} e^{-Kt} = \frac{2e'}{Z_0} e^{\frac{-t}{CZ_0}}$$

The reflected component of voltage

$$e''(s) = \alpha \cdot e'(s) = \frac{K - s}{K + s} \cdot \frac{e'}{s}$$

$$= e'\left[\frac{K}{s(s+K)} - \frac{1}{(s+K)}\right] = e'\left[\frac{1}{s} - \frac{1}{s+K} - \frac{1}{s+K}\right] \tag{11.21}$$

$$= e''(t) = e'\left[1 - 2\,e^{-Kt}\right] = e'\left[1 - 2\,e^{-\frac{1}{CZ_0}t}\right]$$

The reflected component of current is

$$i''(t) = \frac{-e''}{Z_0} = \frac{-e'}{Z_0}\left[1 - 2e^{\frac{-t}{CZ_0}}\right]$$

$$i''(t) = \frac{e'}{Z_0}\left(2e^{\frac{-t}{CZ_0}} - 1\right)$$

(11.22)

11.10 **EFFECT OF SHUNT CAPACITANCE ON TRAVELING WAVES**

Consider that a capacitor is connected to ground at the junction of lines of surge impedances Z_1 and Z_2.

When a traveling wave reaches the junction J, the current can be shown as in Fig. 11.10 taking Z_2 as the total natural impedance to ground at the far end. The equivalent impedance at J.

$$Z_2' = \frac{1}{\frac{1}{Z_2} + Cs} = \frac{Z_2}{CsZ_2 + 1}$$

The input considered as a step wave is given by

$$e'(t) = E\,u(t)$$

The refracted component of voltage at the junction J is

$$e_r(s) = \frac{2Z_2^1}{Z_1 + Z_2^1} \cdot \frac{E}{s} = \frac{\frac{2Z_2}{(1 + CsZ_2)}}{Z_1 + \frac{Z_2}{(1 + CZ_2s)}} \frac{E}{s} = \frac{2Z_2}{Z_1 + Z_2 + CsZ_1Z_2} \cdot \frac{E}{s}$$

$$= \frac{\frac{2Z_2}{Z_1Z_2}}{\frac{2 + Z_2}{Z_1Z_2} + Cs} \frac{E}{s} = \frac{2}{Z_1} \frac{1}{\frac{Z_1 + Z_2}{ZZ_2} + Cs} \frac{E}{s}$$

where

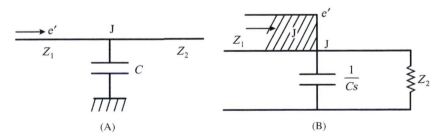

(A)　　　　　　　　　(B)

FIGURE 11.10

(A and B) Shunt capacitance on a line.

$$Z_1 = \frac{Z_1 \, Z_2}{Z_1 + Z_2}$$

$$e_r(s) = \frac{2}{Z_1} \frac{E}{C} \frac{1}{s} \cdot \left(\frac{1}{s} + \frac{1}{Z'C} \right)$$

Taking inverse Laplace transform

$$e_r(t) = e_{J_0} \left(1 - e^{\frac{-t}{Z'C}} \right) \tag{11.23}$$

where

$$e_{J_0} = 2Z$$

The transmitted wave after the junction J is attenuated with a time constant

$$\tau = Z'C = \frac{Z_2}{1 + CSZ_2} C$$

the rate of rise of the incoming wave gets reduced.

It helps in arresting the insulation failure of the apparatus connected to the line.

11.11 SHORT-CIRCUITED LINE

Consider a line short circuited at one end. In this case $Z_2 = 0$. The coefficient of reflection is

$$\alpha = \frac{Z_2 - Z_1}{Z_2 + Z_1} = -\frac{Z_1}{Z_1} = -1 \tag{11.24}$$

and the coefficient of refraction is

$$\beta = \frac{2Z_2}{Z_2 + Z_1} = 0 \tag{11.25}$$

Hence

$$e'' = \alpha e' = -e'$$

and

$$i'' = -\alpha \, i' = +i'$$

The refracted voltage is $e_r = e' + (-e') = 0$ and the current component is $i_r = i' + i'' = 2i'$.

The reflected voltage wave is the negative of the incident wave, the resultant, thus, being zero. The reflected current wave is equal to the incident wave, the resultant being twice the incident wave. The variation of voltage and current waves are shown in Fig. 11.11.

The waveforms tend always to build zero potential along the line and high short circuit current at the far end. The source voltage tends to keep the voltage at V at the incoming end.

11.12 LINE OPEN CIRCUITED AT THE OTHER END

In this as the line is open circuited, Z_2 is infinite

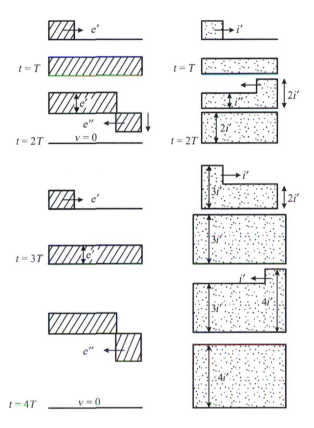

FIGURE 11.11

Voltage and current waveform when the far end is short circuited.

Hence the reflection coefficient is

$$\alpha = \frac{Z_z - Z_1}{Z_2 + Z_1} = \frac{\alpha - Z_1}{\alpha + Z_1} = 1 \qquad (11.26)$$

The refraction coefficient is

$$\beta = \frac{2 Z_z}{Z_1 + Z_2} = \frac{2}{1 + \dfrac{Z_1}{Z_2 \to \infty}} = 2 \frac{1}{1 + 0} = 2 \qquad (11.27)$$

$$e'' = \alpha \cdot e' = + e' = e'$$

$$i'' = - \alpha \cdot i' = - i'$$

The current wave is reflected with negative sign. The total potential at the open end builds up to $2e^1$. The current and voltage waves are shown at different instants of time. The length of the line is l. Time to travel length l is T.

The wave propagation always satisfies that the incoming end of the line maintained the source voltage V and the other farther end that is open cannot have any current maintained, thus making $I = 0$ (Fig. 11.12).

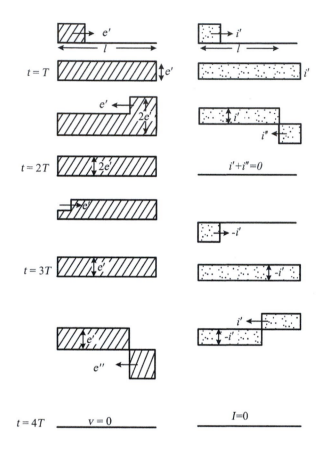

FIGURE 11.12

Variation of voltage and current in an open ended line.

11.13 LINE CONNECTED TO A CABLE

Consider an overhead line of surge impedance Z_1 connected to a cable of natural impedance Z_2 (Fig. 11.13).

In this case

$$e_r = e' \left[\frac{2Z_2}{Z_1 + Z_2} \right] \tag{11.28}$$

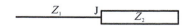

FIGURE 11.13

Line and cable junction.

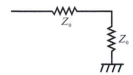

FIGURE 11.14

Line terminated with natural impedance.

$$i_r = \frac{2\,Z_1}{Z_1 + Z_2}\,i'$$ (11.29)

and

$$e'' = e'\left[\frac{Z_2 - Z_1}{Z_2 + Z_1}\right]$$ (11.30)

and

$$i'' = -\,i'\left[\frac{Z_2 - Z_1}{Z_2 + Z_1}\right]$$ (11.31)

11.14 LINE TERMINATED WITH NATURAL IMPEDANCE

In this case (Fig. 11.14)

$$e_r = \frac{2Z_0 e'}{Z_0 + Z_0} = e'$$ (11.32)

and

$$i_r = \frac{2Z_0}{Z_0 + Z_0}\,i' = i'$$ (11.33)

$$e'' = e'\,\frac{Z_0 - Z_0}{Z_0 + Z_0} = 0$$ (11.34)

and

$$i'' = -\left[\frac{Z_0 - Z_0}{Z_0 + Z_0}\right]i' = 0$$ (11.35)

The incoming waves are transmitted without any reflection. Such lines are said to be infinite length lines or matched lines. The load carried by such a line is called natural or surge impedance load.

11.15 REFLECTION AND REFRACTION AT A T-JUNCTION

Consider a line with natural impedance Z_1 connected to two different lines with natural impedances Z_2 and Z_3 as shown in Fig. 11.15.

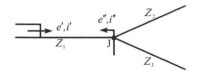

FIGURE 11.15

T-junction.

With usual notation

$$e' + e'' = e_r \tag{11.36}$$

$$i' = \frac{e'}{Z_1} \tag{11.37}$$

$$i'' = \frac{-e'}{Z_1} \tag{11.38}$$

Refracted components of current into lines 2 and 3 are given by

$$i_{r_2} = \frac{e_r}{Z_2} \tag{11.39}$$

$$i_{r_3} = \frac{e_r}{Z_3} \tag{11.40}$$

$$i' - i'' = i_{r_2} + i_{r_3} \tag{11.41}$$

$$\frac{e'}{Z_1} - \frac{e''}{Z_1} = \frac{e_r}{Z_2} + \frac{e_r}{Z_3} = e_r \left[\frac{1}{Z_2} + \frac{1}{Z_3} \right]$$

$$\frac{e'}{Z_1} - \left(\frac{e_r - e'}{Z_1} \right) = e_r \left[\frac{1}{Z_2} + \frac{1}{Z_3} \right] \tag{11.42}$$

$$\frac{e'}{Z_1} + \frac{e'}{Z_1} = e_r \left[\frac{1}{Z_1} + \frac{1}{Z_2} + \frac{1}{Z_3} \right] = \frac{2e'}{Z_1}$$

The transmitted component of voltage is

$$e_r = \frac{\left(\dfrac{2e'}{Z_1} \right)}{\left(\dfrac{1}{Z_1} + \dfrac{1}{Z_2} + \dfrac{1}{Z_3} \right)} \tag{11.43}$$

The refracted components of currents are therefore given by

$$i_{r_2} = \frac{\dfrac{2e'}{Z_1}}{\sum_{i=1}^{3} \dfrac{1}{Z_2}} \cdot \frac{1}{Z_2} = \frac{2e'}{Z_1 Z_2} \left(\dfrac{1}{\dfrac{1}{Z_1} + \dfrac{1}{Z_2} + \dfrac{1}{Z_3}} \right) \tag{11.44}$$

$$i_{r_3} = \frac{\dfrac{2e'}{Z_1}}{\sum_{i=1}^{3} \dfrac{1}{Z_2}} \cdot \frac{1}{Z_3} = \frac{2e'}{Z_1 Z_3} \left(\dfrac{1}{\dfrac{1}{Z_1} + \dfrac{1}{Z_2} + \dfrac{1}{Z_3}} \right) \tag{11.45}$$

11.16 **BEWLEY LATTICE DIAGRAM**

Consider a surge wave traveling along an overhead line or a cable. When the wave reaches the other end, where a load is located, it will get partially reflected. The reflected component reaches the source (generator or a transformer) and again it will get partially reflected. This process continues. To consider such repeated reflections, zig-zag diagram called Bewley lattice diagrams are constructed. Successive reflection of the traveling waves at the discontinuities in the system are indicated on a time−space diagram. The effect of attenuation on the incident wave can easily be understood from the Bewley lattice diagrams.

Consider a unit step voltage traveling along a line of surge impedance Z_1. Let the line is terminated into a cable of surge impedance Z_2. Given that $Z_1 = 500 \ \Omega$ and $Z_2 = 50 \ \Omega$ so that $Z_1/Z_2 = 10$

The reflection coefficient for line to cable

$$\alpha_1 = \frac{1 - 10}{1 + 10} = -0.818$$

The reflection coefficient for cable to line travel

$$\alpha_2 = \frac{10 - 1}{10 + 1} = 0.818$$

The refraction coefficient for line to cable is

$$\beta_1 = \frac{2 \times 1}{1 + 10} = \frac{2}{11} = 0.1818$$

Bewley lattice diagram is drawn assuming that the other end of the cable is open.

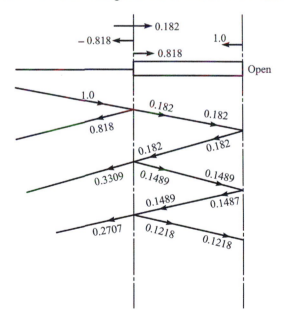

FIGURE 11.16

Bewley lattice diagram.

The reflected and refracted components are computed successively applying the formula

$$e'' = e' \left[\frac{Z_2 - Z_1}{Z_2 + Z_1} \right] = \propto e' = e' \left[\frac{1 - \dfrac{Z_1}{Z_2}}{1 + \dfrac{Z_1}{Z_2}} \right] \tag{11.46}$$

and

$$e_{\mathrm{r}} = \frac{2Z_2}{Z_2 + Z_1} \cdot e' = \beta e' = e' \left[\frac{1}{1 + \frac{Z_1}{Z_2}} \right] \tag{11.47}$$

The Bewley lattice diagram is shown in Fig. 11.16.

WORKED EXAMPLES

E.11.1 A surge of 10 kV magnitude travels along a cable toward its junction with an overhead line. The inductance and capacitance of the cable and overhead line are 0.18 mH, 0.24 μF and 0.9 mH, 0.0072 μF per kilometer, respectively. Find the voltage rise at the junction due to the surge.

Solution:

$$\text{Surge impedance of the overhead line} = \sqrt{\frac{L_{\mathrm{OHL}}}{C_{\mathrm{OHL}}}}$$

$$Z_1 = \sqrt{\frac{0.18 \times 10^{-3}}{0.24 \times 10^{-6}}} = 27.386\ \Omega$$

$$\text{Surge impedance of the cable, } Z_2 = \sqrt{\frac{L_{\mathrm{C}}}{C_{\mathrm{C}}}}$$

$$= \sqrt{\frac{0.9 \times 10^{-3}}{0.0072 \times 10^{-6}}} = 353.55\ \Omega$$

The pressure reflected at the junction due to 10-kV surge

$$e'' = e' \left[\frac{Z_2 - Z_1}{Z_2 + Z_1} \right] = \left[\frac{353.55 - 27.386}{353.55 + 27.386} \right] = 8.56\ \text{kV}$$

Hence total pressure at the junction after the first reflection $= 10 + 8.56 = 18.56$ kV.

E.11.2 A surge of 25 kV traveling in a line of surge impedance 500 Ω arrives at a junction with two lines of impedance 700 and 200 Ω, respectively. Find the surge voltages and currents transmitted into each branch line.

$$e_{\mathrm{r}} = \text{Voltage refracted as transmitted at the junction}$$

$$= \frac{2e'}{Z_1} \cdot \left[\frac{1}{Z_1} + \frac{1}{Z_2} + \frac{1}{Z_3} \right] = \frac{2 \times 25}{500} \cdot \left[\frac{1}{500} + \frac{1}{200} + \frac{1}{700} \right]$$

$$= \frac{2 \times 25}{500} \left[\frac{1}{0.0084286} \right] = 11.064\ \text{kV}$$

i_{r_1} = Current refracted or transmitted into 200-Ω line

$$= \frac{2e'}{Z_1 Z_2}\left[\frac{1}{Z_1} + \frac{1}{Z_2} + \frac{1}{Z_3}\right] = \frac{2 \times 25 \times 10^3}{500 \times 200}\left[\frac{1}{0.0084286}\right] = \frac{0.0005}{0.0084286} = 59.32 \text{ A}$$

i_{r_2} = Current transmitted into the line with surge impedance 700 Ω

$$= \frac{2e'}{Z_1 Z_3}\left[\frac{1}{\dfrac{1}{Z_1} + \dfrac{1}{Z_2} + \dfrac{1}{Z_3}}\right] = \left[\frac{2 \times 25 \times 10^3}{500 \times 700}\frac{1}{0.00842063}\right]$$

$$= 16.949 \text{ A}$$

E.11.3 A 500-kV, 2-μs, rectangular surge on a line having a surge impedance of 350 Ω approaches a station at which the concentrated earth capacitance is 3000pf. Determine the maximum value of the transmitted wave.

Solution:

The transmitted wave into the terminal capacitance charges the capacitor for a period of 2 μs.

$$e_r = 2e'\left[1 - e^{\frac{-t}{CZ_0}}\right]$$

$$e' = 500 \text{ kV} \quad \frac{t}{CZ_0} = \frac{2 \times 10^{-6}}{3000 \times 10^{-12} \times 350} = 1.90476619$$

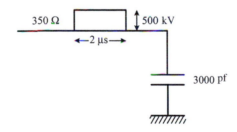

$$e^{\frac{-t}{CZ_0}} = e^{\frac{1}{t/CZ_0}} = \frac{1}{e^{1.9047619}} = \frac{1}{6.7178} = 0.148859$$

$$1 - e^{\frac{-t}{CZ_0}} = 1 - 0.148859 = 0.851$$

$$2e'(1 - e^{\frac{-t}{CZ_0}}) = 2 \times 500(0.851) = 851 \text{ kV}$$

The maximum value of the transmitted voltage = 851 kV.

E.11.4 A traveling wave of 45 kV enters an overhead line of surge impedance 350 Ω and conductor resistance of 5 Ω/km. Determine: (1) the value of the voltage wave when it has traveled through a distance of 50 km, (2) the power loss and the energy loss of the wave during the time taken to travel the distance of 50 km. Neglect leakage resistance (G). The velocity of the wave = 3×10^5 km/s.

Solution:

$$e = e_0 \, \varepsilon^{-\frac{1}{2}\left(\frac{g}{2}+GZ\right)x}$$

$$= 45\varepsilon^{\frac{-1}{2}}\left(\frac{5}{350} + 0\right).50 = 45 \, \varepsilon^{\frac{-0.714}{2}} = 45 \, \varepsilon^{-0.357}$$

$$= 45\varepsilon^{\frac{1}{0.357}} = \frac{45}{\varepsilon 1429} = 31.49 \text{ kV}$$

$$\text{Power loss} = \frac{e^r}{Z} = \frac{(31.49)^2}{350} \times 1000 \times \frac{1000}{1000} \text{ kW} = 2833.2 \text{ kW}$$

$$\text{Energy loss} = \int_0^t e.i \ dt$$

$$x = vt \ \text{ or } \ x = 3 \times 10^5 \text{ km/s} \times t(s) = 50 \text{ km}$$

$$t = \frac{50 \text{ km}}{3 \times 10^5 \text{ km}} \text{s} = \frac{50}{3 \times 10^5} \text{s}$$

$$i_0 = \frac{e_0}{Z_0} = \frac{45 \times 1000}{350} = 128.97 \text{ A}$$

E.11.5 A cable has an inner conductor of radius 0.48×10^{-2} m inside a sheath of inner radius 1.56×10^{-2} m. Find the values of inductance and capacitance per meter length, the surge impedance of the cable, and the velocity of propagation of the wave if the relative permittivity of the cable insulation $\varepsilon_r = 4$.

Solution:

$$L = 2 \ln\left[\frac{1.48 \times 10^{-12}}{0.56 \times 10^{-12}}\right] \times 10^{-7} \text{ h/m}$$

$$= 2 \ln (2.6428) \times 10^{-7} = 1.9436 \times 10^{-7} \text{ h/m}$$

$$C = \frac{4 \times 10^{-9}}{18 \ln\left(\dfrac{1.48 \times 10^{-2}}{0.56 \times 10^{-2}}\right)} \text{F/m} = \frac{4 \times 10^{-9}}{18 \times 0.9718}$$

$$= 0.2286 \times 10^{-9} \text{ F/m}$$

$$\text{Velocity of propagation } v = \frac{1}{\sqrt{LC}} = \frac{1}{\sqrt{1.9436 \times 10^{-7} \times 0.2286 \times 10^{-9}}}$$

$$= \frac{1}{0.6665} \times 10^{+8} = 1.5 \times 10^{+8} \text{ m/s}$$

$$\text{Surge impedance } Z_c = \sqrt{\frac{L}{C}} = \sqrt{\frac{1.9436 \times 10^{-7}}{0.2286 \times 10^{-9}}}$$

$$= 10 \times 2.9158 = 29.158$$

E.11.6 A surge of 110 kV travels along an overhead line toward its junction with a cable. The surge impedance of cable is 48 Ω, and the surge impedance of overhead line 420 Ω. Find the magnitude of surge wave transmitted into the cable.

Solution:

$$e_r = e' \cdot \frac{2Z_2}{Z_1 + Z_2} = 110 \times \frac{2 \times 48}{(420 + 48)} = 22.564 \text{ kV}$$

$$e'' = 110 \times \frac{48 - 420}{48 + 420} = \frac{110 \times (-372)}{468} = -87.435 \text{ kV}$$

$$i' = \frac{110 \times 10^3}{420} = 0.26 \times 10^3 \text{ A}$$

$$i_r = \frac{2e'}{Z_1 + Z_2} = \frac{2 \times 110}{420 + 48} = \frac{220}{468} = 0.47 \text{ A}$$

$$i'' = (0.26 \times 10^3) \times \frac{48 - 420}{48 + 420} = \frac{0.26 \times 10^3 \times (-372)}{468}$$

$$= 206.67 \text{ A}$$

E.11.7 A surge of 600 kV travels along a line with surge impedance $Z_1 = 450 \; \Omega$. The line is connected to a cable of 1.2 km length. The inductance of the cable is 250 μH and the capacitance of the cable is 0.150 μF. The far end of the cable is connected to a transformer of surge impedance 980 Ω. Find the surge voltage distribution 10 μs after the surge has arrived at the line-cable junction.

Solution:

Surge impedance of the cable

$$Z_c = \sqrt{\frac{L}{C}} = \sqrt{\frac{250 \times 10^{-6}}{0.150 \times 10^{-6}}} = \sqrt{1667} = 40.829 \; \Omega$$

Velocity of surge in cable $= 1/\sqrt{LC}$

$$= \frac{1}{\sqrt{250 \times 10^{-6} \times 0.15 \times 10^{-6} \times 10^{-6} \times 10^{-6} \times 10^{-6} \times 10^{-6}}}$$

$$= \frac{1}{\sqrt{250 \times 0.15 \times 10^{-6} \times 10^{-6}}} = 0.1633 \times 10^6 \text{ m/sec}$$

Distance traveled by wave in 10 μs

$$= v \times s = 10 \times 10^{-6} \times 0.1633 \times 10^6 = 1.633 \text{ km}$$
$$= 1.633 - 1.2 = 433 \text{ km}$$

$$e_{r_1} = \frac{2Z_2}{Z_1 + Z_2} = \frac{2 \times 40.829}{40.829 + 450} \times 600 \text{ kV}$$

$$= 99.82 \text{ kV}$$

$$e''_C = e_{r_1} \left[\frac{Z_3 \times Z_2}{Z_3 + Z_2} \right] = 99.82 \times \left[\frac{980 \times 40.829}{980 + 40.829} \right]$$

$$= \frac{99.82 \times 939.171}{1020.829} = 91.835 \text{ kV}$$

Voltage on CD = 99.82 + 91.835 = 191.655

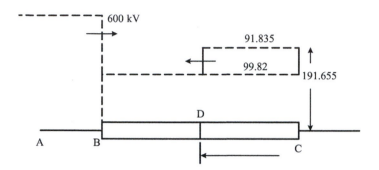

Voltage from D to B is 99.82 kV

E.11.8 An overhead line of surge impedance 400 Ω is connected in series with a cable of surge impedance 50 Ω. The cable is 4 km in length and is open at its far end. A surge of peak value 100 kV travels along the overhead line toward the junction of the line with the cable. Find the voltage distribution at the instants 1 and 40 μs after the surge reaches the junction. The velocity of propagation of the wave on the overhead line is 3×10^5 km/s and in the cable 1.5×10^5 km/s.

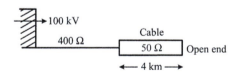

Solution:

The coefficient of reflection is

$$\alpha = \frac{Z_2 - Z_1}{Z_2 + Z_1} = \frac{50 - 400}{50 + 400} = -0.778$$

The coefficient of refraction is

$$\beta = \frac{2 Z_2}{Z_2 + Z_1} = \frac{2 \times 50}{50 + 400} = 0.222$$

$$e' = 100 \text{ kV}$$

$$e'' = -0.778 \times 100 = -77.8 \text{ kV}$$

$$e_r = 0.222 \times 100 = 22.2 \text{ kV}$$

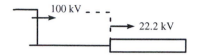

Distance traveled by the transmitted wave into the cable in 1 μs

$$= 1 \times 10^{-6} \times 1.5 \times 10^5 \text{ km} = 0.15 \text{ km} = 150 \text{ m}$$

Distance traveled by the reflected wave in 1 μs

$$= 3 \times 10^5 \times 1 \times 10^{-6} \text{ km} = 300 \text{ m}$$

The voltage distribution after 1 μs is given in the following figure

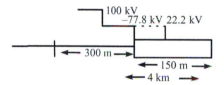

Time required by the transmitted wave to reach the open end of the cable

$$= \frac{4 \times 1000 \text{ m}}{1.5 \times 10^5 \times 10^{+3}} \times 10^6 \text{ μs} = \frac{4 \text{ km}}{1.5 \text{ km}} \times 10^{+5} \times 10^6 = \frac{40}{1.5}$$
$$= 26.67 \text{ μs}$$

The incoming voltage wave of magnitude 22.2 kV reaches the open end in 26.67 μs. Distance traveled from the open end into the cable can be found now. Time taken is 40 μs − 26.67 μs = 13.33 μs.
The voltage gets doubled, i.e., 2 × 22.2 = 44.4 kV

$$\text{Distance traveled} = 13.33 \text{ μs} \times 1.5 \times 10^5 \times 10^{-6}$$
$$= 1.9995 \text{ km from the open end toward the junction}$$

The distance traveled by the reflected wave at the junction into the overhead line

$$= 3 \times 10^5 \times 40 \text{ μs} \times 10^{-6} \text{ s} = 12 \text{ km}$$

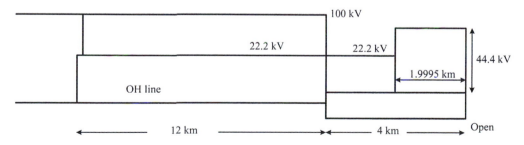

E.11.9 Two stations are connected together by an underground cable having a surge impedance of 50 Ω connected to an overhead line of surge impedance 500 Ω. If a surge wave of 100 kV amplitude travels along the cable toward the junction of the cable and line, determine the value of the reflected and refracted voltage and current waves at the junction.

Solution:

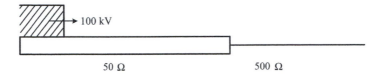

$$\text{Reflected voltage wave} = e'' = \frac{Z_2 - Z_1}{Z_2 + Z_1} e'$$

$$= 100 \cdot \frac{500 - 50}{500 + 50} = \frac{100 \times 450}{550} = 81.8181 \text{ KV}$$

$$\text{Transmitted voltage } e_r = e' + e'' = 100 + 81.8181$$

$$= 181.81 \text{ KV}$$

$$\text{Reflected current} = i'' = -K\, i' = -\frac{450}{550} \times \left(\frac{100 \times 1000}{50} \right)$$

$$= 1636.36 \text{ A}$$

$$\text{Transmitted components of current} = \frac{181.82 \times 1000 \text{ V}}{500 \; \Omega} = 363.64 \text{ A}$$

E.11.10 Two ends of two long transmission lines A and C are connected in between by a cable of length 1.2 km long. The surge impedances of A, B, and C are 450, 60, and 550 Ω, respectively. A rectangular voltage wave of 50-kV magnitude and of infinite length travels along line A and moves to line C. Determine the first and second surge voltages impressed on line C.

Solution:

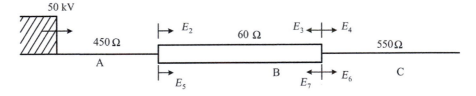

Transmitted voltage wave into the cable B

$$= 50 = \frac{50 \times 2 \times 60}{510} \left[1 + \frac{60 - 450}{60 + 450} \right]$$
$$E_2 = 11.76 \text{ kV}$$

At the junction of B and C, the transmitted voltage

$$E_4 = \frac{11.76 \times 2 \times 550}{610} = 21.20 \text{ kV}$$
$$E_4 = 21.2 \text{ kV}$$

This is the first impressed voltage at C.

The reflected voltage at B and C junction E_3

$$E_3 = \frac{11.76 \times (550 - 60)}{550 + 60} = \frac{11.76 \times 490}{610} = 9.446 \text{ kV}$$

This, on reaching the junction of B and A, is again partially reflected and this voltage is

$$E_5 = 9.446 \times \frac{450 - 60}{450 + 60} = \frac{9.446 \times 390}{550} = 6.698 \text{ kV}$$

This voltage again on reaching the junction of B and C gets partially transmitted. This is E_6

$$E_6 = 6.698 \times \frac{2 \times 550}{610} = 12.078 \text{ kV}$$

After the second impression, the voltage at C is

$$21.2 + 12.078 = 33.278 \text{ kV}$$

E.11.11 A cable with a surge impedance of 90 Ω is terminated in two parallel connected overhead lines having surge impedances 500 and 800 Ω, respectively. If a steep-fronted voltage wave of 5 kV travels along the cable, find the voltages and currents in the cable and overhead lines immediately after the traveling wave reaches the junction of cable and overhead lines. The traveling voltage wave is infinite in length.

Solution:

Combined impedance of the two parallel lines

$$Z_{\text{total}} = \frac{Z_1 Z_2}{Z_1 + Z_2} = \frac{500 \times 800}{500 + 800}$$

$$= 307.692 \text{ Ω}$$

$$e_r = 5 \text{ kV} \left[1 + \frac{307.692 - 90}{307.692 + 90} \right]$$

$$e_r = 5 \left[\frac{2 \times 307.692}{397.692} \right] = 7.7369 \text{ kV}$$

$$\frac{2^1}{r^1} = \frac{e_r}{Z_1} = \frac{7.7369}{500} \times 1000 = 15.4738 \text{ A}$$

$$\frac{2^1}{r^2} = \frac{e_r}{Z_2} = \frac{7.7369 \times 1000}{800} = 9.671 \text{ A}$$

$$\text{The reflected voltage } e^{11} = 5 \left[\frac{307.692 - 90}{307.692 + 90} \right]$$

$$= \frac{5 \times 217.692}{397.692} = 2.7369 \text{ kV}$$

$$\text{The reflected current} = - \left[\frac{2.7369 \times 1000}{90} \right] = -30.41 \text{ A}$$

E.11.12 A 5-kV surge travels along a line of surge impedance 500 Ω. (1) Find the magnitude of the incident current wave. The line is terminated by a resistance of 800 Ω. (2) Find the rate of dissipation of energy, and the values of reflected voltage and current waves. (3) What is the rate of energy reflection? (4) What should be the value of the terminating resistance so that energy in the wave is completely dissipated and also the transmitted and reflected power.

Solution:

1. The incident current wave $e^1 = \dfrac{e^1}{Z_1} = \dfrac{5 \times 10^3}{500} = 10 \text{ A}$

2. Refracted at transmitted voltage

$$e_r = e \times \frac{2Z_2}{Z_2 + Z_1} = 5 \times \frac{2 \times 800}{500 + 800} = 6.1538 \text{ A}$$

$$Z_r^i = \frac{6.1538 \times 1000}{800} = 7.692 \text{ A}$$

$$e^{11} = 5 \times \left[\frac{800 - 500}{500 + 800} \right] = \frac{50 \times 300}{1300} = 1.1538 \text{ kV}$$

$$i^{11} = - \frac{1.1538}{500} \times 1000 = 2.3076 \text{ A}$$

$$\text{Rate of energy dissipation} = e_r i_r$$
$$= 6.1538 \times 7.692 \times 1000 \text{ W} = 47335.02.96$$
$$= 47.335 \text{ kW}$$

$$\text{Rate of energy reflection} = 1.1538 \times 1000 \times 2.3076$$
$$= 2662.50,800 \text{ W} = 2.6625 \text{ kW}$$

The resistance required for the energy to be completely dissipated

$$R = Z_0 = 500 \ \Omega$$

PROBLEMS

P.11.1 A surge of 12 kV magnitude travels along a cable toward its junction with an overhead line. The inductance and capacitance of the cable and overhead line are 0.185 mH and 0.25 μF and 0.91 mH and 0.0073 μF per kilometer, respectively. Find the voltage rise at the junction due to the surge.

P.11.2 A surge of 27 kV traveling in a line of surge impedance 495 Ω arrives at a junction with two lines of impedances 720 and 230 Ω, respectively. Find the surge voltage and currents transmitted into each branch line.

P.11.3 A 47-kV, 2.2-μs rectangular surge on a line having a surge impedance of 362 Ω approaches a station at which the concentrated earth capacitance is 3200 pf. Determine the maximum value of the transmitted wave.

P.11.4 The ends of two long transmission lines A and B are connected by a cable C, 1.2 km long. The surge impedances of A, B, and C are 525,618, and 75 Ω, respectively. A rectangular voltage wave of 12 kV magnitude and of infinite length is initiated in A and travels to B. Determine the first and second voltages impressed on B and the voltage at a point on A, 0.6 km from the junction of A and C, 30 μs after the initial wave has reached the junction of C and B. The velocity of the wave in C is 10^8 m/s.

P.11.5 An inductance of 800 μH connects two sections of a transmission line each having a surge impedance of 350 Ω. A 500-kV, 2-μs rectangular surge travels along the line toward the inductance. Determine the maximum value of the transmitted wave.

QUESTIONS

Q.11.1 Explain the various external and internal causes for overvoltages on transmission lines.

Q.11.2 How are traveling waves attenuated in practice on overhead lines? Explain.

Q.11.3 Explain the following terms:
(1) Surge impedance, (2) velocity of propagation, (3) coefficient of reflection, and (4) coefficient of refraction.

Q.11.4 If a rectangular surge of magnitude e is incident on a line terminated by pure inductance, the reflected component of voltage at far end is $e_r = e'[Z \in^{-[Z_0/L]t} - 1]$, where Z_0 is the surge impedance of the line.

Q.11.5 If in question (4), if the line is terminated by a capacitor instead of inductance, show that the refracted or transmitted component of current is $I_r(t) = (2e/Z_0)e^{\frac{-t}{CZ_0}}$

Q.11.6 Explain the nature of reflected components of voltage and current at the junction of two lines of surge impedances Z_1 and Z_2.

Q.11.7 What is a Bewley lattice diagram? Explain its utility in the study of traveling waves.

PROTECTION AGAINST OVERVOLTAGES

12

Overhead lines and substation equipment can be protected against direct strokes to a considerable extent by using earth or ground wires, which shield the equipment. There are nonshielding methods, such as use of lightning arresters and spark gaps. Shielding methods do not permit an arc to form between the conductor and ground. Hence they are preventive in nature. In case of nonshielding devices, the overvoltage incident on the protective device is allowed to find a path to ground to discharge the surge and in general prevent the flow of dynamic current that follows thereafter.

12.1 GROUND WIRE

It is a conductor run parallel to the main conductors of the transmission line, placed higher than them. It is supported on the same towers and is earthed at regular intervals. The ground wire increases the capacitance between the line conductor and the ground.

In Fig. 12.1 C_1 and C_g are the capacitances between charged cloud-to-line and line-to-ground, respectively. The total voltage of the cloud above ground V is the sum of the two voltages cloud to line V_1 and line to ground V_g, respectively.

$$V_1 + V_g = V \tag{12.1}$$

If I is the charging current, then

$$IX_1 + IX_g = V$$

i.e.,

$$I\left[\frac{1}{\omega c_1} + \frac{1}{\omega c_g}\right] = V$$

$$V = \frac{I}{\omega}\left[\frac{C_g + C_1}{C_g C_1}\right] \tag{12.2}$$

But

$$C_g = \frac{2\pi\varepsilon}{\log_e \frac{D}{r}} \tag{12.3}$$

If a ground wire is used, the distance D between the cloud and the ground is reduced so that C_g is increased. The voltage across the conductor to ground $V_g = I/(wc_g)$ becomes utilizing Eq. (12.2)

Electrical Power Systems. DOI: http://dx.doi.org/10.1016/B978-0-08-101124-9.00012-7

$$V_g = \frac{VC_gC_1}{C_g + C_1} \cdot \frac{1}{C_g}$$

hence $V_g = \frac{V_1 C_1}{C_g + C_1} = V \Big/ \left(\frac{C_g}{C_1} + 1 \right)$ is reduced. Thus the ground wire located above the line conductors gives protection.

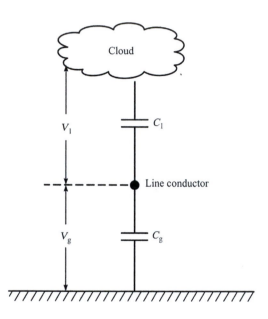

FIGURE 12.1

Effect of ground wire.

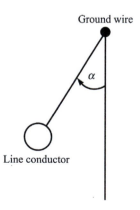

FIGURE 12.2

Angle of protection.

Being at a level higher than ground, the ground wire shields the line conductors against direct strokes. The angle of protection is defined as the angle between a vertical line through the ground wire and a slanting line connecting the ground wire and the line conductor to be protected. The angle of protection is about 30 degrees for towers of 30 m height or less. The protective angle α is

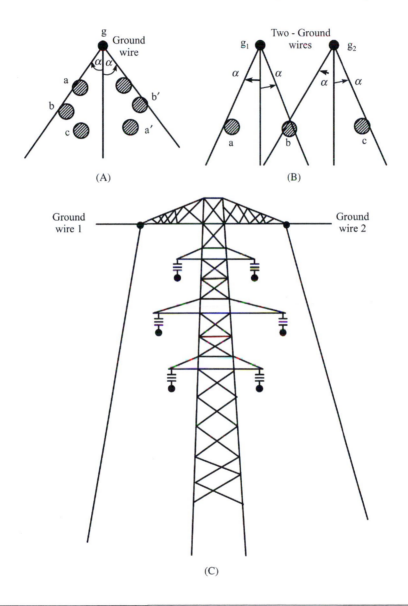

FIGURE 12.3

Ground wire protection. (A) Ground wire protection to a double-circuit line, (B) protection by two ground wires to a double-circuit line, and (C) double-ground-wire protection to a single circuit.

generally in the range of 20−45 degrees (see Fig. 12.2). Peek has proved that a single ground wire reduces the induced voltage by 50%, two ground wires reduce by 66%, three ground wires reduce by 75%, and so on, when compared to the case, where no ground wire is provided.

The ground wire acts as a short circuited secondary to the line conductors and dissipates energy in the induced wave. However, ground wires require additional cost, and further, in the event of breaking, the line conductors may get short circuited. Galvanized steel−stranded wires are used as ground wires. Ground wires function effectively, if the tower footing resistance is reduced by driven rods and counterpoises. Fig. 12.3 shows the ground wire arrangement for protection of double- and single-circuit lines.

12.2 SURGE DIVERTERS

A surge diverter is a device that will spark over at a certain predetermined voltage when a voltage surge reaches it. This is illustrated in Fig. 12.4. At point A the sparkover occurs. The surge diverter is connected between line and earth near the substation. A conducting path is thus provided to the surge from line to ground. Hence a surge diverter must possess the following features:

1. It must not allow current to pass through under normal operation.
2. Transient or surge voltages must be allowed to spark over fast and provide a path to the overvoltages to discharge to ground at a predetermined voltage.
3. The discharge current should not damage the surge diverter.
4. Voltage should not be allowed to rise above the breakdown voltage of the diverter.
5. After the surge is discharged, the power frequency (normal) current should not be permitted to flow to ground through the diverter.

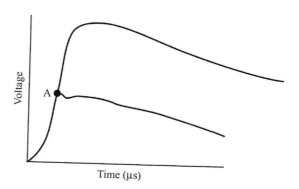

FIGURE 12.4

Protection by surge diverter.

There are three types of such protective devices
They are:

1. Protective gaps
2. Expulsion type arrester
3. Valve type lightning arresters.

12.2.1 ROD GAP

It is the most commonly used protective gap. When the magnitude of an incoming wave exceeds the gap setting of the rod gap, a sparkover occurs and the surge is diverted. However, it has several limitations. It is not capable of preventing the power frequency current that follows a discharge. It is also not effective for steep-fronted surge waves. Further, climatic conditions and polarity of surge waves will influence the operation of the gap. Heat generated during discharge damages the material of the rods. However, it is a good backup protection. The gap is shown in Fig. 12.5 mounted to protect the bushing of a transformer. It is connected between line and earth for a transformer bushing rod. Gap settings are standardized with respect to system operating voltages.

In practice the gap is set to flash over at about 30% below the withstand level of the equipment to be protected. A typical volt–time characteristic for a rod gap is shown in Fig. 12.6.

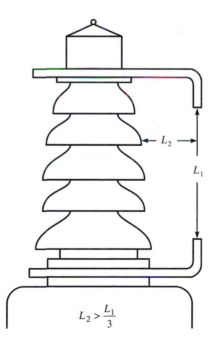

FIGURE 12.5

A rod gap on a bushing insulator.

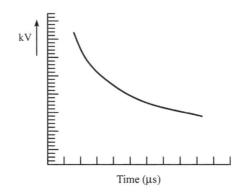

FIGURE 12.6

Characteristic of a rod gap.

12.2.2 EXPULSION TYPE ARRESTERS

Both expulsion gap arresters and protection tube arresters come under this category.

In principle an expulsion gap consists of a rod gap in series with a second gap within a fiber tube. When a surge wave reaches the arrester, both the spark gaps break down simultaneously. The dynamic current is limited by the tower footing resistance and the surge impedance of the ground wires. The internal arc in the fiber tube vaporizes a small part of the fiber material. The gases thus produced interrupt the arc driving away the ionized air and other products of the arc. At a zero of the power frequency current, the arc gets extinguished completely. In the meantime the gaps regain their dielectric strength and the circuit gets open. In this manner the flow of power frequency current is prevented. Fig. 12.7 shows the constructional features of an expulsion gap.

In case of a protector tube the hollow gap with fiber material is replaced by a nonlinear element which offers high impedance at low currents but has low impedance for high currents.

12.2.3 VALVE TYPE LIGHTNING ARRESTERS

Valve type lightning arresters are also called nonlinear arresters. They are relatively expensive but are superior in performance compared to other type of arresters.

The arrester contains a number of series gaps, coil elements, and a number of discs of a nonlinear material such as silicon carbide. The nonlinear disc has the current−voltage relationship of the form

$$I = KV^n$$

where the exponent lies between 2 and 6.

The dynamic volt−ampere characteristic is shown in Fig. 12.8. For a nonlinear arrester V_d and I_d are the discharge voltage and discharge currents for the surge. The maximum value of the surge across the diverter is V_d. V_s and I_f are the system voltage and power frequency follow on current. It is seen that the current rises rapidly with voltage. At higher currents, resistance offered is low for the nonlinear disc.

The schematic representation of a valve type lightning arrester is shown in Fig. 12.9.

The system voltage distribution across the stack of discs is made uniform by a grading ring or a high resistance connected across the discs. The nonlinear disc material is patented as Thyrite,

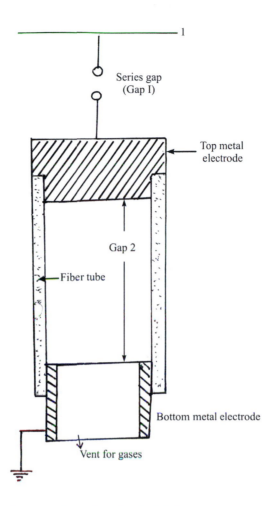

FIGURE 12.7

Expulsion type lightning arrester.

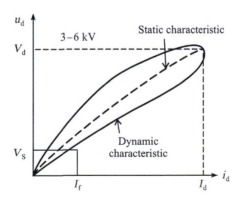

FIGURE 12.8

Dynamic volt–ampere characteristic of surge diverter.

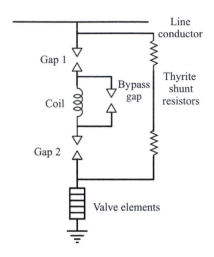

FIGURE 12.9

Valve type lightning arrester.

FIGURE 12.10

Lightning arrester.

Metrosil, etc. The dynamic and static characteristics of this type of surge diverter are shown in Fig. 12.8, and the schematic representation in Fig. 12.9. When the Gaps 1 and 2 spark over due to a surge wave, the induced emf in the coil $L(di/dt)$ is large and this causes the bypass gap to spark over and the coil is taken out of the circuit till the surge and the associated current are discharged and power frequency current follows. To the normal current the coil offers low impedance and the magnetic field it interacts and extinguishes the arcs in the gaps at a current zero.

There are several other types of arresters used in practice. For example, metal oxide arresters use zinc oxide in place of Thyrite which is relatively expensive. The metal oxide arrester does not need series gaps, is simple in construction, and has better surge-protecting features. The nonlinear element is made of zinc oxide in a ceramic body. Other patented metal oxides are also used. A typical lightning arrester in porcelain housing is shown in Fig. 12.10, and the silicon carbide block assembly inside it is shown in Fig. 12.11.

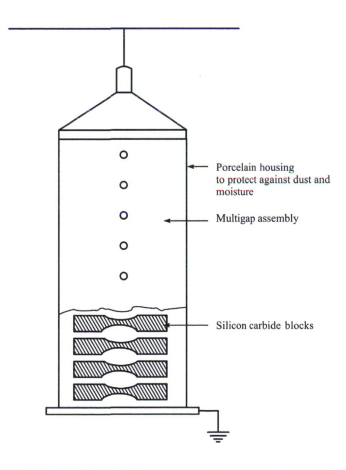

Porcelain housing to protect against dust and moisture

Multigap assembly

Silicon carbide blocks

FIGURE 12.11

Silicon carbide block assembly.

12.3 HORN GAP ARRESTER

The horn gap arrester is one of the cheap and effective surge-diverting devices used from early days. It consists of two rods shaped to form a horn gap as shown in Fig. 12.12. It is used generally in series with a resistance to limit the current.

One end of the gap is connected to the line and the other one is grounded. When a steep-fronted surge travels on the line, it sparks over at the base of the gap and is blown upward under the influence of electromagnetic forces, thereby increasing the length of the arc. The surge that is unable to maintain the increased arc length will facilitate quenching of the arc. An inductor introduced in series on the line side will reduce the steepness of the incident wave. It will also reflect the voltage surge back on to the horn gap.

However, as a protective device, it takes too long a time to be effective and is hence not used as a primary protective device.

A typical horn gap protection to an insulator is shown in Fig. 12.13.

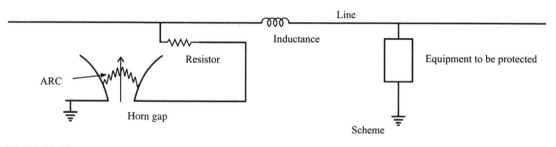

FIGURE 12.12

Protection scheme by a horn gap.

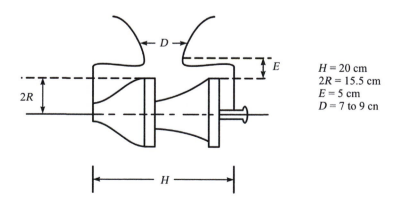

$H = 20$ cm
$2R = 15.5$ cm
$E = 5$ cm
$D = 7$ to 9 cn

FIGURE 12.13

Horn gap protection.

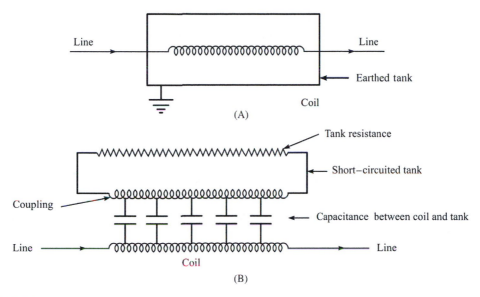

FIGURE 12.14

(A) Ferranti surge absorber and (B) equivalent circuit.

12.4 SURGE ABSORBER

A surge absorber, unlike a surge diverter, absorbs the energy in the traveling wave. Ferronti surge absorber is one such device. It consists of an inductance coil magnetically coupled to an enclosing metal tank. When a surge wave passes through the line and the coil, the closed metal tank which has high resistance acts like a short circuited secondary of an air cored transformer and dissipates the energy in the wave as heat. The capacitance between the coil and the earthed tank filters high-frequency currents in the wave. Fig. 12.14 shows Ferranti surge absorber and its equivalent circuit.

12.5 CLASSIFICATION OF LIGHTNING ARRESTERS

There are three types of lightning arresters

1. *Line Type Arresters*: These arresters are generally used for protection of distribution transformers, smaller substations, and power transformers. Their ratings are generally low (<66 kV) cheaper and are of smaller cross section. The currents may range from 60,000 to 100,000 A.
2. *Station Type Arresters*: These arresters are used to protect substations and power transformers. They are used for circuits of voltages 2.2 kV to the highest voltage. They are larger in size, dissipate higher capacities, and are meant for larger substations.

3. *Distribution Type Arresters*: These arresters are mainly designed to meet the small distribution transformers, generally pole mounted. They are used for circuits operating at voltages less than 15 kV. They can discharge currents up to 65,000 A.

12.6 RATING OF LIGHTNING ARRESTERS

The main consideration in specifying the rating of a lightning arrester is the line-to-ground dynamic voltage. However, the highest value that it may reach is the line-to-line rms value. Allowing an additional 10% margin, a 220-kV line may have a lightning arrester voltage rating of $220 \times 1.1 = 242$ kV. Further, the earthing of the system also is an important factor. Coefficient of earthing is defined as the ratio of highest rms voltage on healthy lines to earth to the nominal line-to-line rms voltage.

For effectively grounded system it is less than 0.8, whereas for noneffectively grounded system it is greater than 0.8.

Hence, the lightning arrester rating for 220 kV may be fixed at $242 \times 0.8 \simeq 194$ kV.

British standards specify four nominal discharge currents of 10, 5, 2.5, and 1.5 kA with wave shapes confirming to 8/20 μs.

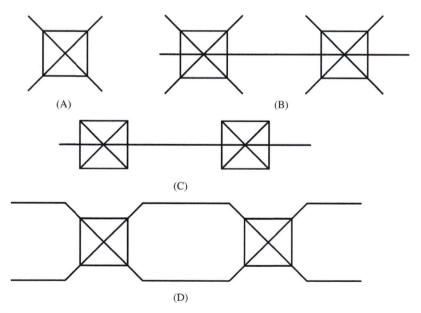

FIGURE 12.15

Counterpoise. (A) Radial counterpoise, (B) radial and continuous counterpoise, (C) single parallel continuous, (D) double parallel continuous.

12.7 **TOW FOOTING RESISTANCE**

The tower footing resistance (TFR) plays an important role in providing protection against surges. If the soil resistance is high, ground rods are used to reduce TFR. They are metallic rods of about 15 mm diameter and 2−3 m in length connected to each leg of the tower and are driven to some depth depending up on the soil. For hard soils, the depth could be even 50 m.

Further reduction of the TFR is possible by use of counterpoise. A counterpoise is a galvanized steel wire run in parallel or radial or in combination with respect to overhead lines. They are buried at a depth of about 0.5−1.0 m. Counterpoise is more effective than ground rods. They reduce TFR to about 25 Ω.

The greater the depth to which the rods are driven smaller will be the TFR.

Various arrangements of counterpoise are shown in Fig. 12.15.

12.8 **INSULATION COORDINATION**

There are several elements in a power system such as transformers, transmission lines, circuit breakers, bus-bars, switch gear, bushing, and many more that require insulation. Each element operating at a high voltage must be well protected against overvoltages. There is a need to coordinate the insulating strengths of the insulation provided to these equipments operating at the same voltage level or at different voltage levels so that the system passes over any overvoltage in a smooth manner to the protective device. To understand this, knowledge of volt−time curves of the protected equipment and the protecting device is required.

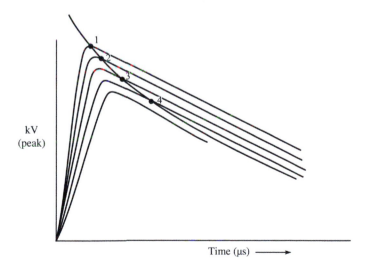

FIGURE 12.16

Determination of volt−time curve.

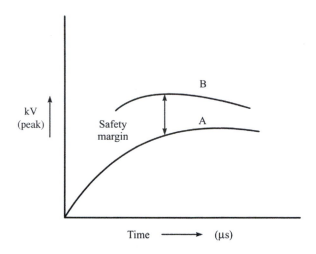

FIGURE 12.17

Insulation coordination.

Overvoltages are either of power frequency or unidirectional surges. Under impulse conditions almost instantaneous breakdown of the protecting device or the insulation under test occurs at a certain voltage. Fig. 12.16 shows an impulse wave applied to an insulated device. When the voltage reaches point 1 on its peak, let us say that the insulation on the device fails. Now, if an impulse voltage of slightly reduced value is applied to the insulation, then it may fail at a point 2 slightly to the right of point 1 as shown in the figure. If the process is repeated several times, each time reducing the voltage magnitude (peak) only slightly to a lesser value, we obtain the curve 1, 2, 3, 4. This is the volt–time curve for the insulation used for the device. A similar curve can be obtained for the protective device (gap or arrester).

It is to be noted that the breakdown voltage of an insulation or protective gap depends not only on the magnitude of the surge wave but also on the time duration of application of it.

Consider the volt–time characteristic of the protective device A and the volt–time characteristic of the equipment to be protected B. It is clear from Fig. 12.17 that if the volt–time curve A lies above the curve B, then the curve A fails to protect the equipment with characteristic B. The volt–time characteristic of the protective device must always be below the characteristic of the equipment to be protected.

12.8.1 THE STANDARD IMPULSE WAVE

The impulse waves that travel on overhead lines have a rise rate of about $0.5-10\,\mu s$ and decay to 50% of their peak value in $30-200\,\mu s$. Generally they are unidirectional in nature. The lightning overvoltages or surges can be represented by

$$v = V_0[\varepsilon^{-\alpha t} - \varepsilon^{\beta t}] \text{ different type epsilon}$$

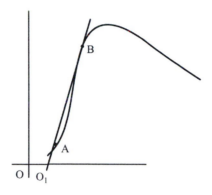

FIGURE 12.18

Standard impulse wave.

where α and β are constants. Since these waves have arbitrary shapes for testing and standardization, it has become necessary to specify a standard wave. An impulse wave has a fast rising wave front and slowly decreasing wave tail. The standard wave is shown in Fig. 12.18 and is explained in the following. The origin of the wave front or the initial portion of the wave front cannot be ascertained with certainty. It is therefore proposed to consider the time required for the wave to reach from 10% of the peak value to 90% of the peak value. The corresponding times are marked t_1 and t_2 in Fig. 12.18. A line is drawn through the two points A and B on the wave corresponding to time t_1 and t_2, which cuts the x-axis at O_1.

At point C on the wave tail, the value of the surge voltage has fallen to 50% of its peak value and the time for it is t_3 marked on the x-axis.

Wavefront and wave tail are defined as follows:

Wavefront: It is the time required for the surge wave to attain its peak value and is given by

$$1.25(t_2 - t_1)\mu s = 1.25(O_1 t_2 - O_1 t_1)\mu s$$

Wave tail: It is the value of the time that the surge wave takes to reach 50% of its peak value on the wave tail. It is the time represented by $O_1 t_3$ where t_3 corresponds to point C as the tail when the surge wave decreases to 50% of its peak value. According to India standard specification 1.2/50 μs wave chosen as the standard for lightning impulse wave.

> A   1/50 μs wave has $\alpha = 0.01477$ and $\beta = 1.933$ and
> 1.2/50 μs wave has $\alpha = 0.01460$ and $\beta = 2.467$

The corresponding values for V_0 are 1.039 and 1.046.

12.8.2 IMPULSE RATIO

It is observed while discussing about volt–time characteristic that with increasing time less voltage is needed for breakdown. While for impulse voltages, breakdown occurs almost instantaneously, for power frequency voltages where the time duration is in milliseconds, failure occurs at much lower voltages. The ratio of breakdown voltage for any insulation or gap due to an impulse voltage of specified t_1/t_2 or shape to power frequency breakdown voltage is defined as impulse ratio. For a 1.2/50 μs wave $t_1 = 1.2$ μs and $t_2 = 50$ μs.

For a sphere gap, where the electric field is uniform at the gap, the impulse ratio is unity. For all other types of gaps (rod gap, disc gap, horn gap, etc.) impulse ratio is more than unity. The flashover voltage for an insulator under dry conditions at power frequency is called dry flashover voltage.

The flashover voltage for an insulator under wet conditions at power frequency is called wet flashover voltage.

12.8.3 BASIC IMPULSE LEVELS

However, under impulse conditions (as described by a standard impulse wave), flashover will occur at a higher voltage than for power frequency.

For this reason an international standard impulse insulation that withstand voltage level is defined. This is called Basic Impulse Insulation Level (BIL). This is the withstand crest voltage of an insulation test specimen when a standard impulse wave is applied to it.

BILs are reference levels expressed in impulse crest voltage with a standard wave not larger than 1.5/40 μs (American) or 1.2/50 μs (India) standard wave.

Equipment must withstand the specified BIL. Thus a reference class kV is identified for each and every power apparatus or equipment insulation designed to operate at a specific operating voltage. Corresponding to each reference class, a standard BIL is specified in kV. The insulation chosen should withstand this BIL voltage. To understand this, some reference class voltages and the corresponding standard BIL kV are given in Table 12.1.

The breakdown voltages for the devices should be either equal to or greater than the corresponding BIL.

In a power system equipment is subject to the following:

* Highest power frequency system voltage that the equipment has to continuously withstand
* Temporary power frequency overvoltages due to load rejection, faults, etc.
* Transient overvoltages or surges due to switching phenomenon or lightning.

If a device is rated at a nominal voltage 132 kV, the highest system voltage could reach 10% of it and may be taken as $132 + 13 = 145$ kV at 50 Hz. This highest system voltage for which equipment is designed is designated as U_m (rms).

Table 12.1 Reference class and BIL voltages	
Reference Class (kV)	**Standard BIL (kV)**
23	150
46	250
92	450
161	750
230	1050

The highest power frequency overvoltage the equipment has to withstand, phase to ground $= \sqrt{2}U_m/\sqrt{3}$ kV (peak). The power frequency withstand voltage is for a duration of 1 min. The 1-min, 50-Hz withstand voltage for 132-kV nominal voltage is 300 kV.

The lightning impulse voltage withstand value (BIL) for this 132 kV is given as 450 kV (peak). It is observed that for lines and equipment operating at voltages in excess of 300 kV, the switching surges become more important than voltage surges due to lightning systems.

Switching surges in general may cause overvoltages ranging from 1.5 to 3.0 times the system line to ground voltage, even though it can go up to 3.6 times the value for high reclosing type of lines after fault clearance.

For 132-kV nominal voltage, if we take $145 \times 3 = 435$ kV rms (phase-to-phase), then the switching surge could be $435 \times (\sqrt{2}/\sqrt{3}) = 355$ kV (peak) phase to ground. It is also denoted by U_p.

Surge or overvoltage protection by a surge arrester is given in microseconds and for this the following information is needed.

1. Nominal power frequency system voltage
2. Highest power frequency voltage, the equipment is subjected
3. Lightning impulse withstand voltage level
4. Switching surge withstand voltage level
5. Protective level of the surge arrester
6. Withstand level of equipment against surge overvoltages.

QUESTIONS

Q.12.1 Explain how a ground wire can protect as a shield against direct strokes for overhead lines.

Q.12.2 Distinguish between surge diverters and surge arresters.

Q.12.3 Explain the working of the following:
(1) Rod gap and (b) horn gap.

Q.12.4 With a neat sketch explain the working principle of an expulsion type lightning arrester.

Q.12.5 Explain the working of a valve type lightning arrester.

Q.12.6 What do you understand by tower footing resistant? How can it be reduced?

Q.12.7 Describe the working of Ferranti Surge Absorber.

Q.12.8 Explain the important of insulation coordination.

Q.12.9 What are basic impulse insulation levels? Explain the importance of BIL.

Q.12.10 What do you understand by a 1/50-μs impulse wave? Explain.

GRAPH THEORY AND NETWORK MATRICES

13.1 INTRODUCTION

Graph theory has many applications in several fields such as engineering, physical, social and biological sciences, and linguistics. Any physical situation that involves discrete objects with interrelationships can be represented by a graph. In *Electrical Engineering*, graph theory is used to predict the behavior of the network in analysis. However, for smaller networks, node or mesh analysis is more convenient than the use of graph theory. It may be mentioned that Kirchhoff was the first to develop theory of trees for applications to electrical network. The advent of high-speed digital computers has made it possible to use the graph theory advantageously for larger network analysis. In this chapter a brief account of graph theory is given that is relevant to power transmission networks and their analysis.

13.2 DEFINITIONS

Element of a Graph: Each network element is replaced by a line segment or an arc while constructing a graph for a network. Each line segment or arc is called an *element*. Each potential source is replaced by a short circuit. Each current source is replaced by an open circuit.

Node or Vertex: The terminal of an element is called a *node* or a *vertex*.

Edge: An element of a graph is called an *edge*.

Degree: The number of edges connected to a vertex or node is called its *degree*.

Graph: An element is said to be incident on a node, if the node is a terminal of the element. Nodes can be incident to one or more elements. The network can thus be represented by an interconnection of elements. The actual interconnections of the elements give a graph.

Rank: The *rank of a graph* is $n - 1$ where n is the number of nodes in the graph.

Subgraph: Any subset of elements of the graph is called a *subgraph*. A subgraph is said to be proper if it consists of strictly less than all the elements and nodes of the graph.

Path: A path is defined as a subgraph of connected elements such that not more than two elements are connected to any one node. If there is a path between every pair of nodes, then the graph is said to be connected. Alternatively, a graph is said to be connected if there exists at least one path between every pair of nodes.

Planar Graph: A graph is said to be planar, if it can be drawn without cross-over of edges. Otherwise it is called nonplanar (Fig. 13.1).

Closed Path or Loop: The set of elements traversed starting from one node and returning to the same node form a closed path or loop.

Electrical Power Systems. DOI: http://dx.doi.org/10.1016/B978-0-08-101124-9.00013-9

Oriented Graph: An oriented graph is a graph with direction marked for each element. Fig. 13.2A shows the single-line diagram of a simple power network consisting of generating stations, transmission lines, and loads. Fig. 13.2B shows the positive sequence network of the system in Fig. 13.2A. The oriented connected graph is shown in Fig. 13.3 for the same system.

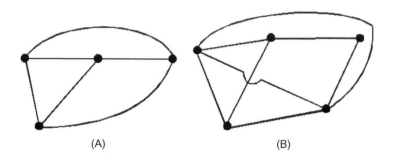

FIGURE 13.1

(A) Planar graph and (B) nonplanar graph.

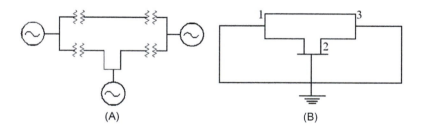

FIGURE 13.2

(A) Power system single-line diagram and (B) positive sequence network diagram.

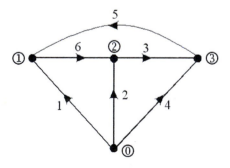

FIGURE 13.3

Oriented connected graph.

13.3 TREE AND COTREE

Tree: A tree is an oriented connected subgraph of an oriented connected graph containing all the nodes of the graph, but, containing no loops. A tree has $(n-1)$ branches where n is the number of nodes of graph G. The branches of a tree are called twigs. The remaining branches of the graph are called links or chords.

Cotree: The links form a subgraph, not necessarily connected called cotree. Cotree is the complement of tree. There is a cotree for every tree.

For a connected graph and subgraph:

1. There exists only one path between any pair of nodes on a tree,
2. Every connected graph has at least one tree,
3. Every tree has two terminal nodes, and
4. The rank of a tree is $n-1$ and is equal to the rank of the graph.

The number of nodes and the number of branches in a tree are related by

$$b = n - l \tag{13.1}$$

If e is the total number of elements, then the number of links l of a connected graph with branches b is given by

$$l = e - b \tag{13.2}$$

Hence, from Eq. (13.1), it can be written that

$$l = e - n + 1 \tag{13.3}$$

A tree and the corresponding cotree of the graph for the system shown in Fig. 13.3 are indicated in Fig. 13.4A and B.

13.4 BASIC LOOPS

A loop is obtained whenever a link is added to a tree, which is a closed path. As an example to the tree in Fig. 13.4A if the link 6 is added, a loop containing the elements 1-2-6 is obtained. Loops which contain only one link are called *independent loops* or *basic loops.*

It can be observed that the number of basic loops is equal to the number of links given by Eq. (13.2) or (13.3). Fig. 13.5 shows the basic loops for the tree in Fig. 13.4A.

13.5 CUT-SET

A cut-set is a minimal set of branches K of a connected graph G, such that the removal of all K branches divides the graph into two parts. It is also true that the removal of K branches reduces the rank of G by one, provided no proper subset of this set reduces the rank of G by one when it is removed from G.

Consider the graph in Fig. 13.6A.

(A)

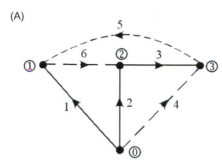

n = Number of nodes = 4

e = Number of elements = 6

$b = n - 1 = 4 - 1 = 3$

$l = e - n + 1 = 6 - 4 + 1 = 3$

(B)

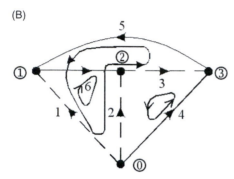

FIGURE 13.4

(A) Tree for the system in Fig. 13.3 and (B) Cotree for the system in Fig. 13.3.

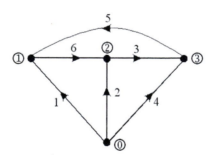

FIGURE 13.5

Basic loops for the tree in Fig. 13.4A.

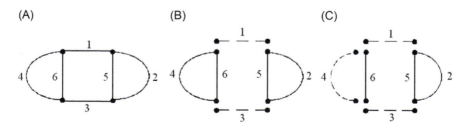

FIGURE 13.6

(A) Graph, (B) cut-sets, and (C) branches.

The rank of the graph = (no. of nodes $n - 1) = 4 - 1 = 3$. If branches 1 and 3 are removed, two subgraphs are obtained as in Fig. 13.6B. Thus 1 and 3 may be a cut-set. Also, if branches 1, 4, and 3 are removed, the graph is divided into two subgraphs as shown in Fig. 13.6C. Branches 1, 4, and 3 may also be a cut-set. In both the above cases the rank of both the subgraphs is $1 + 1 = 2$. It can be noted that (1, 3) set is a subset of (1, 4, 3) set. The cut-set is a minimal set of branches of the graph, removal of which cuts the graph into two parts. It separates nodes of the graphs into two graphs. Each group is in one of the two subgraphs.

13.6 BASIC CUT-SETS

If each cut-set contains only one branch, then these independent cut-sets are called basic cut-sets. To understand basic cut-sets select a tree. Consider a twig b_k of the tree. If the twig is removed, then the tree is separated into two parts. All the links which go from one part of this disconnected tree to the other, together with the twig b_k constitutes a cut-set called basic cut-set. The orientation of the basic cut-set is chosen as to coincide with that of the branch of the tree defining the cut-set. Each basic cut-set contains at least one branch with respect to which the tree is defined which is not contained in the other basic cut-set. For this reason, the $n - 1$ basic cut-sets of a tree are linearly independent.

Now consider the tree in Fig. 13.4A.

Consider node (1) and branch or twig 1. Cut-set A cuts branch 1 and links 5 and 6 and is oriented in the same way as branch 1. In a similar way C cut-set cuts branch 3 and links 4 and 5 and is oriented in the same direction as branch 3. Finally cut-set B cuts branch 2 and links 4, 6, and 5 and is oriented as branch 2; the cut-sets are shown in Fig. 13.7.

13.7 INCIDENCE MATRICES

There are several incidence matrices that are important in developing the various networks matrices like bus impedance matrix, branch admittance matrix, etc., using singular or nonsingular transformation.

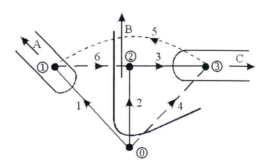

FIGURE 13.7

Cut-sets for the tree in Fig. 13.4A.

These various incidence matrices are basically derived from the connectivity or incidence of an element to a node, path, cut-set, or loop.

Incidence Matrices
The following incidence matrices are of interest in power network analysis.

1. Element-node incidence matrix
2. Bus incidence matrix
3. Branch path incidence matrix
4. Basic cut-set incidence matrix
5. Basic loop incidence matrix

Of these, the bus incidence matrices is the most important one.

13.8 ELEMENT-NODE INCIDENCE MATRIX

Element-node incidence matrix $\overline{A}$ shows the incidence of elements to nodes in the connected graph. The incidence or connectivity is indicated by the operator as follows:

$\alpha_{pq} = 1$ if the p-th element is incident to and directed away from the q node.
$\alpha_{pq} = -1$ if the p-th element is incident to and directed toward the q node.
$\alpha_{pq} = 0$ if the p-th element is not incident to the q-th node.

The element-node incidence matrix will have the dimension *exn* where "*e*" is the number of elements and n is the number of nodes in the graph. It is denoted by $\overline{A}$.

The element-node incidence matrix for the graph of Fig. 13.3 is shown in the table of Fig. 13.8 along with the graph for reference.

It is seen from the elements of the matrix that

$$\sum_{q=0}^{3} \alpha_{pq} = 0; \quad p = 1, 2, \ldots, 6 \qquad (13.4)$$

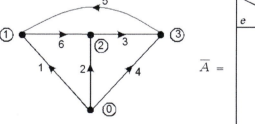

$$\overline{A} =$$

e \ n	(0)	(1)	(2)	(3)
1	1	−1		
2	1		−1	
3			1	−1
4	1			−1
5		−1		1
6		1	−1	

FIGURE 13.8

Element-node incidence matrix for the graph of Fig. 13.3.

It can be inferred that the columns of $\overline{A}$ are linearly independent. The rank of $\overline{A}$ is less than n the number of nodes in the graph.

13.9 BUS INCIDENCE MATRIX

The network in Fig. 13.2B contains a reference reflected in Fig. 13.3 as a reference node. In fact any node of the connected graph can be selected as the reference node. The matrix obtained by deleting the column corresponding to the reference node in the element-node incidence matrix $\overline{A}$ is called *bus incidence matrix A*. Thus the dimension of this matrix is $ex\ (n-1)$ and the rank will therefore be, $n-1=b$, where b is the number of branches in the graph. Deleting the column corresponding to node (0) from Fig. 13.1, the bus incidence matrix for the system in Fig. 13.2A is obtained. This is shown in Fig. 13.9.

If the rows are arranged in the order of a specific tree, the matrix A can be partitioned into two submatrices A_b of the dimension $bx\ (n-1)$ and A_1 of dimension $lx\ (n-1)$. The rows of A_b correspond to branches and the rows of A_1 correspond to links. This is shown in Fig. 13.10 for the matrix in Fig. 13.9.

$$A = \begin{array}{c|c|c|c|c}
 & e \backslash \text{(1)} & \text{(2)} & \text{(3)} \\
\text{BUS} & & & \\
\hline
1 & -1 & & \\
2 & & -1 & \\
3 & & +1 & -1 \\
4 & & & -1 \\
5 & -1 & & +1 \\
6 & 1 & -1 & \\
\end{array}$$

FIGURE 13.9

Bus incidence matrix for graph in Eq. (13.3).

$$A = \begin{array}{c|c|c|c}
e \backslash \text{BUS} & \text{(1)} & \text{(2)} & \text{(3)} \\
\hline
1 & -1 & & \\
2 & & -1 & \\
3 & & 1 & -1 \\
4 & & & -1 \\
5 & -1 & & 1 \\
6 & 1 & -1 & \\
\end{array}
\qquad
A = \begin{array}{c|c|c|c}
e \backslash \text{BUS} & \text{(1)} & \text{(2)} & \text{(3)} \\
\hline
1 & & & \\
2 & & A_b & \\
3 & & & \\
4 & & & \\
5 & & A_1 & \\
6 & & & \\
\end{array}
\begin{array}{c} \text{Branches} \\ \text{Links} \end{array}$$

FIGURE 13.10

Partitioning of matrix A.

13.10 NETWORK PERFORMANCE EQUATIONS

The power system network consists of components such as generators, transformers, transmission lines, circuit breakers, and capacitor banks, which are all connected together to perform specific function. Some are in series and some are in shunt connection.

Whatever may be their actual configuration, network analysis is performed either by nodal or by loop method. In the case of power system, generally, each node is also a bus. Thus in the bus frame of reference the performance of the power network is described by $(n - 1)$ independent nodal equations, where n is the total number of nodes. In the impedance form of the performance equation, following Ohm's law will be

$$\overline{V} = [Z_{BUS}]\, \overline{I}_{BUS} \tag{13.5}$$

where

$\overline{V}_{BUS}$ = Vector of bus voltages measured with respect to a reference bus.
$\overline{I}_{BUS}$ = Vector of impressed bus currents.
$[Z_{BUS}]$ = Bus impedance matrix.

The elements of bus impedance matrix are open-circuit driving point and transfer impedances. Consider a three-bus or three-node system. Then

$$[Z_{BUS}] = \begin{matrix} & \begin{matrix} (1) & (2) & (3) \end{matrix} \\ \begin{matrix} (1) \\ (2) \\ (3) \end{matrix} & \begin{bmatrix} Z_{11} & Z_{12} & Z_{13} \\ Z_{21} & Z_{22} & Z_{23} \\ Z_{31} & Z_{32} & Z_{33} \end{bmatrix} \end{matrix}$$

The impedance elements on the principal diagonal are called driving point impedances of the buses and the off-diagonal elements are called transfer impedances of the buses. In the admittance frame of reference

$$\overline{I}_{BUS} = [Y_{BUS}] \cdot \overline{V}_{BUS} \tag{13.6}$$

where $[Y_{BUS}]$ is the bus admittance matrix whose elements are short-circuit driving point and transfer admittances.

By definition

$$[Y_{BUS}] = [Z_{BUS}]^{-1} \tag{13.7}$$

13.11 BUS ADMITTANCE MATRIX AND BUS IMPEDANCE MATRIX

The bus admittance matrix Y_{BUS} can be obtained by determining the relation between the variables and parameters of the primitive network described in Section 2.1 to bus quantities of the network using bus incidence matrix. Consider Eq. (1.5).

$$\bar{i} + \bar{j} = [y]\bar{v}$$

Premultiplying by $[A^t]$, the transpose of the bus incidence matrix

$$[A^t]\bar{i} + [A^t]\bar{j} = A^t[y]\bar{v} \tag{13.8}$$

Matrix A shows the connections of elements to buses. $[A^t]i$ thus is a vector, wherein each element is the algebraic sum of the currents that terminate at any of the buses. Following Kirchhoff's current law, the algebraic sum of currents at any node or bus must be zero. Hence

$$[A^t]\bar{i} = 0 \tag{13.9}$$

Again $[A^t]\bar{j}$ term indicates the algebraic sum of source currents at each of the buses and must equal the vector of impressed bus currents. Hence

$$\bar{I}_{BUS} = [A^t]\bar{j} \tag{13.10}$$

Substituting Eqs. (13.14) and (13.15) into (13.13)

$$\bar{I}_{BUS} = [A^t][y]\bar{v} \tag{13.11}$$

In the bus frame power in the network is given by

$$[\bar{I}_{BUS}{}^*]^t\bar{V}_{BUS} = P_{BUS} \tag{13.12}$$

Power in the primitive network is given by

$$(\bar{j}^*)^t\bar{v} = P \tag{13.13}$$

Power must be invariant for transformation of variables to be invariant. That is to say, that the bus frame of referee corresponds to the given primitive network in performance. Power consumed in both the circuits is the same.

Therefore

$$[I_{BUS}^*]\bar{V}_{BUS} = [\bar{j}^*]\bar{v} \tag{13.14}$$

Conjugate transpose of Eq. (13.10) gives

$$[\bar{I}_{BUS}^*]^t = [j^*]^t A^* \tag{13.15}$$

However, as A is real matrix $A = A^*$

$$[\bar{I}_{BUS}^*]^t = (j^*)^t[A] \tag{13.16}$$

Substituting (13.21) into (13.19)

$$(j^*)^t[A] \cdot \bar{V}_{BUS} = (j^*)^t\bar{v} \tag{13.17}$$

i.e.,

$$[A]\bar{V}_{BUS} = \bar{v} \tag{13.18}$$

Substituting Eq. (13.17) into (13.11)

$$\bar{I}_{BUS} = [A^t][y][A]\bar{V}_{BUS} \tag{13.19}$$

From Eq. (13.6)

$$\bar{I}_{\text{BUS}} = [\bar{Y}_{\text{BUS}}]\bar{V}_{\text{BUS}} \tag{13.20}$$

Hence

$$[Y_{\text{BUS}}] = [A^t][y][A] \tag{13.21}$$

Once $[Y_{\text{BUS}}]$ is evaluated from the above transformation, (Z_{BUS}) can be determined from the relation

$$Z_{\text{BUS}} = Y_{\text{BUS}}^{-1} = \{[A^t][y][A]\}^{-1} \tag{13.22}$$

13.12 BUS ADMITTANCE MATRIX BY DIRECT INSPECTION

Bus admittance matrix can be obtained for any network, if there are no mutual impedances between elements, by direct inspection of the network. This is explained by taking an example.

Consider the three-bus power system as shown in Fig. 13.11.

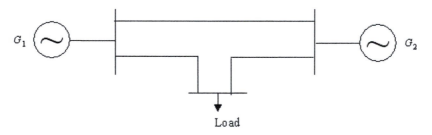

FIGURE 13.11

Three-bus system.

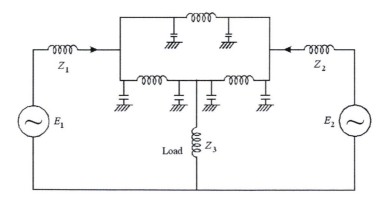

FIGURE 13.12

Equivalent circuit for the system in Fig 13.11.

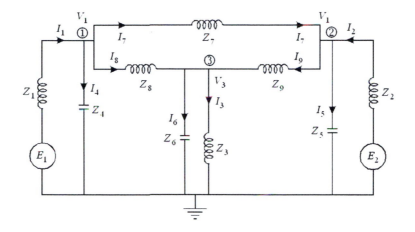

FIGURE 13.13

Reduced equivalent circuit.

The equivalent circuit is shown in Fig. 13.12. The generator is represented by a voltage source in series with the impedance. The three transmission lines are replaced by their "π equivalents".

The equivalent circuit is further simplified as in Fig. 13.13 combining the shunt admittance wherever feasible.

The three nodes are at voltages V_1, V_2, and V_3, respectively, above the ground. The Kirchhoff's nodal current equations are written as follows:

At Node 1:

$$I_1 = I_7 + I_8 + I_4$$
$$I_1 = (V_1 - V_2)Y_7 + (V_1 - V_3)Y_8 + V_1Y_4 \tag{13.23}$$

At Node 2:

$$I_2 = I_5 + I_9 - I_7$$
$$= V_2Y_5 + (V_2 - V_3)Y_9 - (V_1 - V_2)Y_7 \tag{13.24}$$

At Node 3:

$$I_3 = I_8 + I_9 - I_6$$
$$= (V_1 - V_3)Y_8 + (V_2 - V_3)Y_9 - V_3Y_6 \tag{13.25}$$

Rearranging the terms, the equations will become

$$I_1 = V_1(Y_4 + Y_7 + Y_8) - V_2Y_7 - V_3Y_8 \tag{13.26(a)}$$
$$I_2 = -V_1Y_7 + V_2(Y_5 + Y_7 + Y_9) - V_3Y_9 \tag{13.26(b)}$$
$$I_3 = V_1Y_8 + V_2Y_9 - V_3(Y_8 + Y_9 + Y_6) \tag{13.26(c)}$$

The last of the above equations may be rewritten as

$$-I_3 = -V_1Y_8 - V_2Y_9 + V_3(Y_6 + Y_8 + Y_9) \tag{13.27}$$

Thus we get the matrix relationship from the above

$$\begin{bmatrix} I_1 \\ I_2 \\ -I_3 \end{bmatrix} = \begin{bmatrix} (Y_4 + Y_7 + Y_8) & -Y_7 & -Y_8 \\ -Y_7 & (Y_5 + Y_7 + Y_9) & -Y_9 \\ -Y_8 & -Y_9 & (Y_6 + Y_8 + Y_9) \end{bmatrix} \cdot \begin{bmatrix} V_1 \\ V_2 \\ V_3 \end{bmatrix} \tag{13.28}$$

It may be recognized that the diagonal terms in the admittance matrix at each of the nodes are the sum of the admittances of the branches incident to the node. The off-diagonal terms are the negative of these admittances branchwise incident on the node. Thus the diagonal element is the negative sum of the off-diagonal elements. The matrix can be written easily by direct inspection of the network.

The diagonal elements are denoted by

$$\left. \begin{aligned} Y_{11} &= Y_4 + Y_7 + Y_8 \\ Y_{22} &= Y_5 + Y_7 + Y_9 \\ Y_{33} &= Y_6 + Y_8 + Y_9 \end{aligned} \right\} \text{ and} \tag{13.29}$$

They are called self-admittances of the nodes or driving point admittances. The off-diagonal elements are denoted by

$$\left. \begin{aligned} Y_{12} &= - Y_7 \\ Y_{13} &= - Y_8 \\ Y_{21} &= - Y_7 \\ Y_{23} &= - Y_8 \\ Y_{31} &= - Y_8 \\ Y_{32} &= - Y_9 \end{aligned} \right\} \tag{13.30}$$

using double suffix denoting the nodes across which the admittances exist. They are called mutual admittances or transfer admittances. Thus the relation in Eq. (13.28) can be rewritten as

$$\begin{bmatrix} I_1 \\ I_2 \\ -I_3 \end{bmatrix} = \begin{bmatrix} Y_{11} & Y_{12} & Y_{13} \\ Y_{21} & Y_{22} & Y_{23} \\ Y_{31} & Y_{32} & Y_{33} \end{bmatrix} \cdot \begin{bmatrix} V_1 \\ V_2 \\ V_3 \end{bmatrix} \tag{13.31}$$

$$\bar{I}_{\text{BUS}} = [Y_{\text{BUS}}] \cdot \bar{V}_{\text{BUS}} \tag{13.32}$$

In power systems each node is called a bus. Thus if there are n independent buses, the general expression for the source current toward the node i is given by

$$I_i = \sum_{j=1}^{n} Y_{ij} V_j, \quad i \neq j \tag{13.33}$$

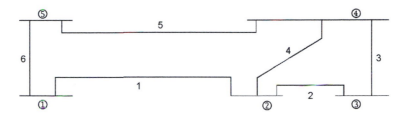

FIGURE E.13.1

WORKED EXAMPLES

E.13.1 For the network shown Fig. E.13.1, draw the graph and mark a tree. How many trees will this graph have? Mark the basic cut-sets and basic loops.

Solution:
Assume that bus (1) is the reference bus

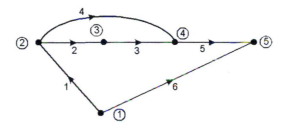

FIGURE E.13.2

Number of nodes $n = 5$
Number of elements $e = 6$
The graph can be redrawn as

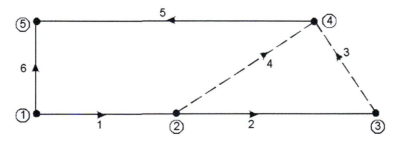

FIGURE E.13.3

Tree: A connected subgraph containing all nodes of a graph, but no closed path is called a *tree*.

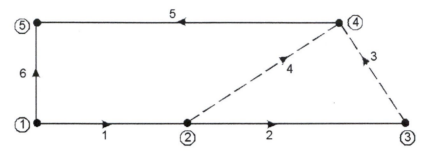

FIGURE E.13.4

Number of branches $n - 1 = 5 - 1 = 4$
Number of links $= e - b = 6 - 4 = 2$
(*Note:* Number of links = Number of cotrees).

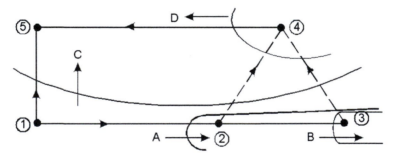

FIGURE E.13.5

The number of basic cut-sets = no. of branches = 4; the cut-sets A, B, C, D, are shown in figure.

E.13.2 Show the basic loops and basic cut-sets for the graph shown below and verify any relations that exist between them.
(Take 1-2-3-4 as tree 1).

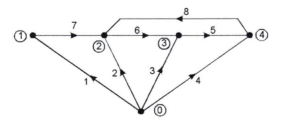

FIGURE E.13.6

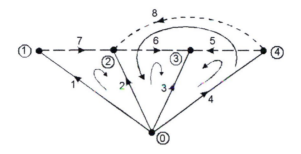

FIGURE E.13.7

Tree and cotree for the graph.

Solution:

If a link is added to the tree, a loop is formed; loops that contain only one link are called *basic loops*.

$$\text{Branches,} \qquad b = n - 1 = 5 - 1 = 4$$
$$l = e - b = 8 - 4 = 4$$

The four loops are shown in Fig. E.13.8.

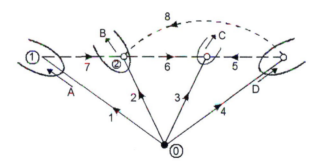

FIGURE E.13.8

Basic cut-sets A, B, C, D.

The number of basic cuts (4) = Number of branches b(4).

E.13.3 For the graph given in Figs. E.13.9 and E.13.10, draw the tree and the corresponding cotree. Choose a tree of your choice and hence write the cut-set schedule.

Solution:

The *f*-cut-set schedule (fundamental or basic)

$$\text{A:} \qquad 1,2$$

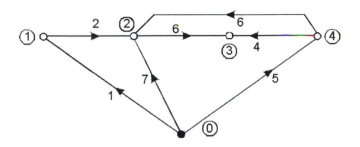

FIGURE E.13.9

Oriented connected graph.

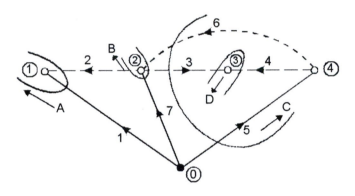

FIGURE E.13.10

Basic cut-sets A, B, C, D.

B:	2,7,3,6
C:	6,3,5
D:	3,4

E.13.4 For the network shown in figure form the bus incidence matrix, A.

Solution:

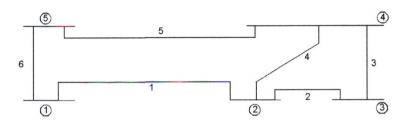

FIGURE E.13.11

Bus incidence matrix
Number of buses = Number of nodes

$$A =$$

e \ BUS	(2)	(3)	(4)	(5)
1	−1	0	0	0
2	1	−1	0	0
3	0	1	−1	0
4	1	0	−1	0
5	0	0	1	−1
6	0	0	0	−1

FIGURE E.13.12

$$A =$$

	e \ BUS	(2)	(3)	(4)	(5)
Branches	1	−1	0	0	0
	2	1	−1	0	0
	5	0	1	1	−1
	6	0	0	0	−1
Link	3	0	1	−1	1
	4	1	0	−1	0

$$=$$

e \ Bus	Buses
B branches	A_b
L links	A_1

FIGURE E.13.13

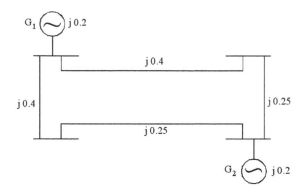

FIGURE E.13.14

E.13.5 Form the Y_{BUS} by using singular transformation for the network shown in Fig. E.13.14. including the generator buses.

Solution:

The given network is represented in admittance form

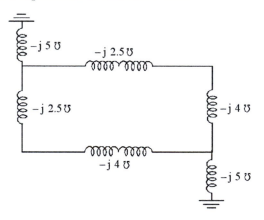

FIGURE E.13.15

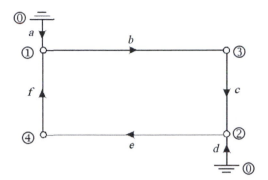

FIGURE E.13.16

The oriented graph is shown in Fig E.13.16

The above graph can be converted into the following form for convenience the element-node incidence matrix is given by

$$\hat{A} = \begin{array}{c} \text{e\textbackslash n} \\ a \\ b \\ c \\ d \\ e \\ f \end{array} \begin{array}{ccccc} 0 & 1 & 2 & 3 & 4 \\ \left[\begin{array}{ccccc} +1 & -1 & 0 & 0 & 0 \\ 0 & +1 & 0 & -1 & 0 \\ 0 & 0 & -1 & +1 & 0 \\ +1 & 0 & -1 & 0 & 0 \\ 0 & 0 & +1 & 0 & -1 \\ 0 & -1 & 0 & 0 & +1 \end{array}\right] \end{array}$$

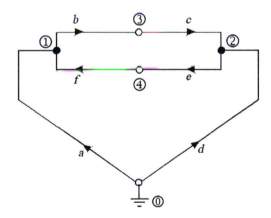

FIGURE E.13.17

Bus incidence matrix is obtained by deleting the column corresponding to the reference bus.

$$
A = \begin{array}{c} \text{e\textbackslash b} \\ a \\ b \\ c \\ d \\ e \\ f \end{array}
\begin{array}{cccc} 1 & 2 & 3 & 4 \end{array}
\begin{bmatrix} -1 & 0 & 0 & 0 \\ +1 & 0 & -1 & 0 \\ 0 & -1 & +1 & 0 \\ 0 & -1 & 0 & 0 \\ 0 & +1 & 0 & -1 \\ -1 & 0 & 0 & +1 \end{bmatrix}
$$

$$
A^t = \begin{array}{c} \text{b\textbackslash e} \\ 1 \\ 2 \\ 3 \\ 4 \end{array}
\begin{array}{cccccc} a & b & c & d & e & f \end{array}
\begin{bmatrix} -1 & 1 & 0 & 0 & 0 & -1 \\ 0 & 0 & -1 & -1 & 1 & 0 \\ 0 & -1 & 1 & 0 & 0 & 0 \\ 0 & 0 & 0 & 0 & -1 & 1 \end{bmatrix}
$$

The bus admittance matrix

$Y_{BUS} = [A]^t [y][A]$

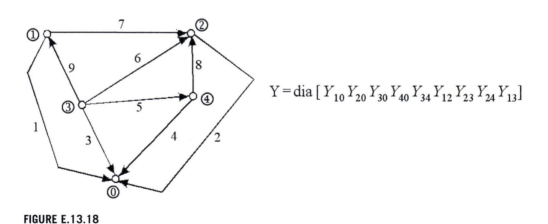

$Y = \mathrm{dia}\,[\,Y_{10}\,Y_{20}\,Y_{30}\,Y_{40}\,Y_{34}\,Y_{12}\,Y_{23}\,Y_{24}\,Y_{13}]$

FIGURE E.13.18

$$
[y]\,[A] =
\begin{array}{c}
\\
a \\ b \\ c \\ d \\ e \\ f
\end{array}
\begin{array}{cccccc}
a & b & c & d & e & f \\
\end{array}
\left[
\begin{array}{cccccc}
y_a & 0 & 0 & 0 & 0 & 0 \\
0 & y_b & 0 & 0 & 0 & 0 \\
0 & 0 & y_c & 0 & 0 & 0 \\
0 & 0 & 0 & y_d & 0 & 0 \\
0 & 0 & 0 & 0 & y_e & 0 \\
0 & 0 & 0 & 0 & 0 & y_f
\end{array}
\right]
\begin{array}{cccc}
(1) & (2) & (3) & (4) \\
\end{array}
\left[
\begin{array}{cccc}
-1 & 0 & 0 & 0 \\
1 & 0 & -1 & 0 \\
0 & -1 & 1 & 0 \\
0 & -1 & 0 & 0 \\
0 & 1 & 0 & -1 \\
-1 & 0 & 0 & 1
\end{array}
\right]
$$

$$
[y][A] =
\left[
\begin{array}{cccc}
5 & 0 & 0 & 0 \\
-2.5 & 0 & 2.5 & 0 \\
0 & 4 & -4 & 0 \\
0 & 5 & 0 & 0 \\
0 & -4 & 0 & 4 \\
2.5 & 0 & 0 & -2.5
\end{array}
\right]
$$

$$
Y_{BUS} = [A]^t\,[y]\,[A] =
\begin{array}{c}
(1) \\ (2) \\ (3) \\ (4)
\end{array}
\begin{array}{cccccc}
a & b & c & d & e & f \\
\end{array}
\left[
\begin{array}{cccccc}
-1 & 1 & 0 & 0 & 0 & -1 \\
0 & 0 & -1 & -1 & 1 & 0 \\
0 & -1 & 1 & 0 & 0 & 0 \\
0 & 0 & 0 & 0 & -1 & 1
\end{array}
\right]
\left[
\begin{array}{cccc}
2 & 0 & - & 0 \\
-2.5 & 0 & 2.5 & 0 \\
0 & 4 & -4 & 0 \\
0 & 5 & 0 & 0 \\
0 & -4 & 0 & 4 \\
2.5 & 0 & 0 & -2.5
\end{array}
\right]
$$

$$
\text{whence,} \quad Y_{BUS} =
\left[
\begin{array}{cccc}
-10 & 0 & 2.5 & 2.5 \\
0 & -13 & 4 & 4 \\
2.5 & 4 & -6.5 & 0 \\
2.5 & 4 & 0 & -6.5
\end{array}
\right]
$$

E.13.6 Find the Y_{BUS} using singular transformation for the system shown in Fig. E.13.18.

Solution:

The graph may be redrawn for convenient as follows

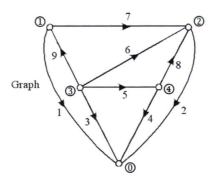

Graph

FIGURE E.13.19

A tree and a cotree are identified as shown below.

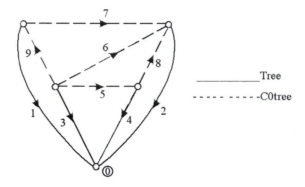

FIGURE E.13.20

The element-node incidence matrix $\hat{A}$ is given by

$$\hat{A} =$$

	(0)	(1)	(2)	(3)	(4)
1	−1	1	0	0	0
2	−1	0	1	0	0
3	−1	0	0	1	0
4	−1	0	0	0	1
5	0	0	0	1	−1
6	0	0	−1	1	0
7	0	1	−1	0	0
8	0	0	−1	0	1
9	0	−1	0	1	0

The bus incidence matrix is obtained by deleting the first column taking (0) node as reference.

$$
A = \begin{array}{c} \\ 1 \\ 2 \\ 3 \\ 4 \\ 5 \\ 6 \\ 7 \\ 8 \\ 9 \end{array}
\begin{array}{cccc}
(1) & (2) & (3) & (4) \\
\end{array}
\left[\begin{array}{cccc}
1 & 0 & 0 & 0 \\
0 & 1 & 0 & 0 \\
0 & 0 & 1 & 0 \\
0 & 0 & 0 & 1 \\
0 & 0 & 1 & -1 \\
0 & -1 & 1 & 0 \\
1 & -1 & 0 & 0 \\
0 & -1 & 0 & 1 \\
-1 & 0 & 1 & 0
\end{array}\right]
= \left[\frac{A_b}{A_1}\right] = \left[\frac{U}{A_1}\right]
$$

Given $[y] =$

$$
\left[\begin{array}{ccccccccc}
y_{10} & 0 & 0 & 0 & 0 & 0 & 0 & 0 & 0 \\
0 & y_{20} & 0 & 0 & 0 & 0 & 0 & 0 & 0 \\
0 & 0 & y_{30} & 0 & 0 & 0 & 0 & 0 & 0 \\
0 & 0 & 0 & y_{40} & 0 & 0 & 0 & 0 & 0 \\
0 & 0 & 0 & 0 & y_{34} & 0 & 0 & 0 & 0 \\
0 & 0 & 0 & 0 & 0 & y_{23} & 0 & 0 & 0 \\
0 & 0 & 0 & 0 & 0 & 0 & y_{12} & 0 & 0 \\
0 & 0 & 0 & 0 & 0 & 0 & 0 & y_{24} & 0 \\
0 & 0 & 0 & 0 & 0 & 0 & 0 & 0 & y3
\end{array}\right]
$$

$$[Y_{BUS}] = A^t[y]A$$

$$
[y][A] =
\left[\begin{array}{ccccccccc}
y_{10} & 0 & 0 & 0 & 0 & 0 & 0 & 0 & 0 \\
0 & y_{20} & 0 & 0 & 0 & 0 & 0 & 0 & 0 \\
0 & 0 & y_{30} & 0 & 0 & 0 & 0 & 0 & 0 \\
0 & 0 & 0 & y_{40} & 0 & 0 & 0 & 0 & 0 \\
0 & 0 & 0 & 0 & y_{34} & 0 & 0 & 0 & 0 \\
0 & 0 & 0 & 0 & 0 & y_{23} & 0 & 0 & 0 \\
0 & 0 & 0 & 0 & 0 & 0 & y_{12} & 0 & 0 \\
0 & 0 & 0 & 0 & 0 & 0 & 0 & y_{24} & 0 \\
0 & 0 & 0 & 0 & 0 & 0 & 0 & 0 & y3
\end{array}\right]
\left[\begin{array}{cccc}
1 & 0 & 0 & 0 \\
0 & 1 & 0 & 0 \\
0 & 0 & 1 & 0 \\
0 & 0 & 0 & 1 \\
0 & 0 & 1 & -1 \\
0 & -1 & 1 & 0 \\
1 & -1 & 0 & 0 \\
0 & -1 & 0 & 1 \\
-1 & 0 & 1 & 0
\end{array}\right]
$$

$$
= \left[\begin{array}{cccc}
y_{10} & 0 & 0 & 0 \\
0 & y_{20} & 0 & 0 \\
0 & 0 & y_{30} & 0 \\
0 & 0 & 0 & y_{40} \\
0 & 0 & y_{34} & -y_{34} \\
0 & -y_{23} & y_{23} & 0 \\
y_{12} & -y_{12} & 0 & 0 \\
0 & -y_{24} & 0 & y_{24} \\
-y_{13} & 0 & y_{13} & 0
\end{array}\right]
$$

$$[A_t]\,[y]\,[A] = \begin{matrix} & 1 & 2 & 3 & 4 & 5 & 6 & 7 & 8 & 9 \\ (1) \\ (2) \\ (3) \\ (4) \end{matrix} \begin{bmatrix} 1 & 0 & 0 & 0 & 0 & 0 & 1 & 0 & -1 \\ 0 & 1 & 0 & 0 & 0 & -1 & -1 & -1 & 0 \\ 0 & 0 & 1 & 0 & 1 & 1 & 0 & 0 & 1 \\ 0 & 0 & 0 & 1 & -1 & 0 & 0 & 1 & 0 \end{bmatrix} \begin{bmatrix} y_{10} & 0 & 0 & 0 \\ 0 & y_{20} & 0 & 0 \\ 0 & 0 & y_{30} & 0 \\ 0 & 0 & 0 & y_{40} \\ 0 & 0 & y_{34} & -y_{34} \\ 0 & -y_{23} & y_{23} & 0 \\ y_{12} & -y_{12} & 0 & 0 \\ 0 & -y_{24} & 0 & y_{24} \\ -y_{13} & 0 & y_{13} & 0 \end{bmatrix}$$

$$Y_{BUS} = \begin{bmatrix} (y_{10}+y_{12}+y_{13}) & -y_{12} & -y_{13} & 0 \\ -y_{12} & (y_{20}+y_{12}+y_{23}+y_{24}) & -y_{23} & -y_{24} \\ -y_{13} & -y_{23} & (y_{30}+y_{13}+y_{23}+y_{34}) & -y_{34} \\ 0 & -y_{24} & -y_{34} & (y_{40}+y_{34}+y_{24}) \end{bmatrix}$$

QUESTIONS

Q.13.1 Explain the following terms:
(1) Basic loops, (2) cut-set, and (3) basic cut-sets.

Q.13.2 Explain the relationship between the basic loops and links; basic cut-sets, and the number of branches.

Q.13.3 Define the following terms with suitable examples:
(1) Tree, (2) Branches, (3) links, (4) cotree, and (5) basic loop.

Q.13.4 Write down the relations between the number of nodes, number of branches, number of links, and number of elements.

Q.13.5 Define the following terms.
(1) Graph, (2) node, (3) rank of a graph, and (4) path.

Q.13.6 Derive the bus admittance matrix by singular transformation

Q.13.7 Explain how do you form Y_{BUS} by direct inspection with a suitable example.

Q.13.8 Derive the expression for bus admittance matrix Y_{BUS} in terms of primitive admittance matrix and bus incidence matrix.

SHORT-CIRCUIT ANALYSIS

Electrical networks and machines are subject to various types of faults while in operation. During the fault period, the current flowing is determined by the internal emf's of the machines in the network and by the impedances of the network and machines. However, the impedances of machines may change their values from those that exist immediately after the fault occurrence to different values during the fault till the fault is cleared. The network impedance may also change, if the fault is cleared by switching operations. It is, therefore, necessary to calculate the short-circuit current at different instants when faults occur. For such fault analysis studies and in general for power system analysis, it is very convenient to use per-unit system and percentage values. In the following the various models for analysis are explained.

14.1 PER-UNIT QUANTITIES

The per-unit value of any quantity is the ratio of the actual value in any units to the chosen base quantity of the same dimensions expressed as a decimal.

$$\text{Per-unit quantity} = \frac{\text{Actual value in any units}}{\text{Base or reference value in the same units}}$$

In power systems the basic quantities of importance are voltage, current, impedance, and power. For all per-unit calculations, a base kVA or MVA and a base kV are to be chosen. Once the base values or reference values are chosen, the other quantities can be obtained as follows:

Selecting the total or three-phase kVA as a base kVA, for a three-phase system

$$\text{Base current in Amperes} = \frac{\text{Base kVA}}{\sqrt{3}\,[\text{Base kV (line-to-line)}]}$$

$$\text{Base impedance in ohms} = \left[\frac{\text{Base kV (line-to-line)}^2 \times 1000}{\sqrt{3}[(\text{base kVA})/3]}\right]$$

$$\text{Base impedance in ohms} = \frac{(\text{Base kV (line-to-line)})^2}{\text{Base MVA}}$$

Hence,

$$\text{Base impedance in ohm} = \frac{(\text{Base kV (line-to-line)})^2 \times 1000}{\text{Base kVA}}$$

where base kVA and base MVA are the total or three-phase values.

Electrical Power Systems. DOI: http://dx.doi.org/10.1016/B978-0-08-101124-9.00014-0

If phase values are used

$$\text{Base current in Amperes} = \frac{\text{Base kVA}}{\text{Base kV}}$$

$$\text{Base impedance in ohm} = \frac{\text{Base voltage}}{\text{Base current}}$$

$$= \frac{(\text{Base kV})^2 \times 1000}{\text{Base kVA per phase}}$$

$$\text{Base impedance in ohm} = \frac{(\text{Base kV})^2}{\text{Base MVA per phase}}$$

In all the above relations the power factor is assumed unity, so that

$$\text{Base power KW} = \text{Base kVA}$$

Now

$$\text{Per unit impedance} = \frac{(\text{Actual impedance in ohm}) \times \text{kVA}}{(\text{Base kV})^2 \times 1000}$$

Sometimes, it may be required to use the relation

$$(\text{Actual impedance in ohm}) = \frac{(\text{Per unit impedance in ohms})(\text{Base kV})^2 \times 1000}{\text{Base kVA}}$$

Very often the values are in different base values. In order to convert the per-unit impedance from given base to another base, the following relation can be derived easily.

Per-unit impedance on new base

$$Z_{\text{new}} \text{ p.u} = Z_{\text{given}} \text{ p.u} \left(\frac{\text{New kVA base}}{\text{Given kVA base}} \right) \left(\frac{\text{Given kV base}}{\text{New kV base}} \right)^2$$

14.2 ADVANTAGES OF PER-UNIT SYSTEM

1. While performing calculations, referring quantities from one side of the transformer to the other side, serious errors may be committed. This can be avoided by using per-unit system.
2. Voltages, currents, and impedances expressed in per-unit do not change when they are referred from one side of transformer to the other side. This is a great advantage.
3. Per-unit impedances of electrical equipment of similar type usually lie within a narrow range, when the equipment ratings are used as base values.
4. Transformer connections do not affect the per-unit values.
5. Manufacturers usually specify the impedances of machines and transformers in per-unit or percent of name plate ratings.

14.3 THREE-PHASE SHORT CIRCUITS

In the analysis of symmetrical three-phase short circuits, the following assumptions are generally made.

1. Transformers are represented by their leakage reactances. The magnetizing current and core losses are neglected. Resistances and shunt admittances are not considered. Star-delta phase shifts are also neglected.
2. Transmission lines are represented by series reactances. Resistances and shunt admittances are neglected.
3. Synchronous machines are represented by constant voltage sources behind subtransient reactances. Armature resistances, saliency, and saturation are neglected.
4. All nonrotating impedance loads are neglected.
5. Induction motors are represented just as synchronous machines with constant voltage source behind a reactance. Smaller motor loads are generally neglected.

Per-Unit Impedances of Transformers: Consider a single-phase transformer with primary and secondary voltages and currents denoted by V_1, V_2 and I_1, I_2, respectively.

$$\text{Base impedance for primary} = \frac{V_1}{I_1}$$

$$\text{Base impedance for secondary} = \frac{V_2}{I_2}$$

$$\text{Per unit impedance referred to primary} = \frac{Z_1}{\left(\dfrac{V_1}{I_1}\right)} = \frac{I_1 Z_1}{V_1}$$

$$\text{Per unit impedance referred to secondary} = \frac{I_2 Z_2}{V_2}$$

$$\text{Again, actual impedance referred to secondary} = Z_1 \left(\frac{V_2}{V_1}\right)^2$$

$$\text{Per-unit impedance referred to secondary} = \frac{Z_1 \left(\dfrac{V_2}{V_1}\right)^2}{\left(\dfrac{V_2}{I_2}\right)} = Z_1 \cdot \frac{V_2^2}{V_1^2} \cdot \frac{I_2}{V_2} = \frac{Z_1(V_2 I_2)}{V_1^2} = Z_1 \frac{(V_1 I_1)}{V_1^2} = \frac{Z_1 I_1}{V_1}$$

$$= \text{Per-unit impedance referred to primary}$$

Thus the per-unit impedance referred remains the same for a transformer on either side.

14.4 REACTANCE DIAGRAMS

In power system analysis it is necessary to draw an equivalent circuit for the system. This is an impedance diagram. However, in several studies, including short-circuit analysis it is sufficient to consider only reactances neglecting resistances. Hence, we draw reactance diagrams. For three-phase balanced systems, it is simpler to represent the system by a single-line diagram without losing the identify of the three-phase system. Thus single-line reactance diagrams can be drawn for calculation.

This is illustrated by the system shown in Fig. 14.1A and B and by its single-line reactance diagram.

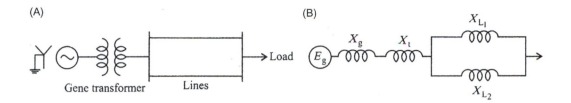

FIGURE 14.1

Power system and its equivalent circuit. (A) A power system and (B) equivalent single-line reactance diagram.

14.5 PERCENTAGE VALUES

The reactances of generators, transformers, and reactors are generally expressed in percentage values to permit quick short-circuit calculation.

Percentage reactance is defined as:

$$\%X = \frac{IX}{V} \times 100$$

where I is the full load current, V is the phase voltage, and X is the reactance in ohms per phase.

Short-circuit current I_{SC} in a circuit then can be expressed as

$$I_{SC} = \frac{V}{X} = \frac{V \cdot I}{V \cdot (\%X)} \times 100$$

$$= \frac{I \cdot 100}{\%X}$$

Percentage reactance can expressed in terms of kVA and kV as following
From equation

$$X = \frac{(\%X) \cdot V}{I \cdot 100} = \frac{(\%X)V^2}{100 \cdot V \cdot I} = \frac{(\%X)\dfrac{V}{1000} \cdot \dfrac{V}{1000} \times 1000}{100 \cdot \dfrac{V}{1000} \cdot I}$$

$$= \frac{(\%X)(kV)^2 10}{kVA}$$

Alternatively

$$(\%X) = X \cdot \frac{kVA}{10(kV)^2}$$

As has been stated already in short-circuit analysis, since the reactance X is generally greater than three times the resistance, resistances are neglected.

But, in case percentage resistance and therefore, percentage impedance values are required then, in a similar manner we can define

$$\%R = \frac{IR}{V} \times 100$$

and

$$\%Z = \frac{IZ}{V} \times 100 \text{ with usual notation.}$$

The percentage values of R and Z also do not change with the side of the transformer or either side of the transformer they remain constant. The ohmic values of R, X, and Z change from one side to the other side of the transformer.

When a fault occurs, the potential falls to a value determined by the fault impedance. Short-circuit current is expressed in term of short-circuit kVA based on the normal system voltage at the point of fault.

14.6 SHORT-CIRCUIT kVA

It is defined as the product of normal system voltage and short-circuit current at the point of fault expressed in kVA.

Let

V = normal phase voltage in Volts
I = full load current in amperes at base kVA
$\%X$ = percentage reactance of the system expressed on base kVA.

The short-circuit current is

$$I_{SC} = I \cdot \frac{100}{\%X}$$

The three-phase or total short-circuit kVA

$$= \frac{3 \cdot VI_{SC}}{1000} = \frac{3 \cdot V \cdot I \, 100}{(\%X)1000} = \frac{3VI}{1000} \cdot \frac{100}{\%X}$$

Therefore short-circuit kVA $=$ Base kVA $\times \dfrac{100}{(\%X)}$

In a power system or even in a single-power station different equipment may have different ratings. Calculations are required to be performed where different components or units are rated differently. The percentage values specified on the name plates will be with respect to their name plate ratings. Hence it is necessary to select a common base kVA or MVA and also a base kV. The following are some of the guidelines for selection of base values.

1. Rating of the largest plant or unit for base MVA or kVA.
2. The total capacity of a plant or system for base MVA or kVA.
3. Any arbitrary value.

$$(\%X)_{\text{on new base}} = \left(\frac{\text{Base kVA}}{\text{Unit kVA}}\right)(\%X \text{ at unit kVA})$$

If a transformer has 8% reactance on 50-kVA base, its value at 100-kVA base will be

$$(\%X)_{100 \text{ kVA}} = \left(\frac{100}{50}\right) \times 8 = 16$$

Similarly the reactance values change with voltage base as per the relation

$$X_2 = \left(\frac{V_2}{V_1}\right)^2 \cdot X_1$$

where X_1 is the reactance at voltage V_1 and X_2 is the reactance at voltage V_2.

For short-circuit analysis it is often convenient to draw the reactance diagrams indicating the values in per-unit.

14.7 IMPORTANCE OF SHORT-CIRCUIT CURRENTS

Knowledge of short-circuit current values is necessary for the following reasons:

1. Fault currents which are several times larger than the normal operating currents produce large electromagnetic forces and torques, which may adversely affect the stator end windings. The forces on the end windings depend on both the DC and AC components of stator currents.
2. The electrodynamic forces on the stator end windings may result in displacement of the coils against one another. This may result in loosening of the support or damage to the insulation of the windings.
3. Following a short circuit, it is always recommended that the mechanical bracing of the end windings is to be checked for any possible loosening.
4. The electrical and mechanical forces that develop due to a sudden three-phase short circuit are generally severe when the machine is operating under loaded condition.
5. As the fault is cleared within three cycles, generally the heating efforts are not considerable.

Short circuits may occur in power systems due to system overvoltages caused by lightning or switching surges or due to equipment insulation failure or even due to insulator contamination. Sometimes even mechanical causes may create short circuits. Other well-known reasons include line-to-line, line-to-ground, or line-to-line faults on overhead lines. The resultant short circuit has to the interrupted within few cycles by the circuit breaker.

It is absolutely necessary to select a circuit breaker that is capable of operating successfully when maximum fault current flows at the circuit voltage that prevails at that instant. An insight can be gained when we consider an $R-L$ circuit connected to an alternating voltage source, the circuit being switched on through a switch.

14.8 ANALYSIS OF $R-L$ CIRCUIT

Consider the circuit in Fig. 14.2.

Let $e = E_{max} \sin(\omega t + \alpha)$ when the switch S is closed at $t = 0^+$

$$e = E_{max} \sin(\omega t + \alpha) = R + L\frac{di}{dt}$$

α is determined by the magnitude of voltage when the circuit is closed.

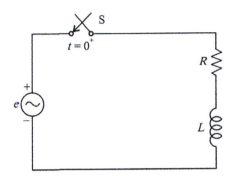

FIGURE 14.2

$R-L$ Circuit with a switch.

The general solution is

$$i = \frac{E_{max}}{|Z|}\left[\sin(\omega t + \alpha - \theta) - e^{-Rt/L}\sin(\alpha - \theta)\right]$$

where

$$|Z| = \sqrt{R^2 + \omega^2 L^2}$$

and

$$\theta = \tan^{-1}\frac{\omega L}{R}$$

The current contains two components:

$$\text{AC component} = \frac{E_{max}}{|Z|}\sin(\omega t + \alpha - \theta)$$

and

$$\text{DC component} = \frac{E_{max}}{|Z|}e^{-Rt/L}\sin(\alpha - \theta)$$

If the switch is closed when $\alpha - \theta = \pi$ or when $\alpha - \theta = 0$
    The DC component vanishes.
    The DC component is a maximum when $\alpha - \theta = \pm\pi/2$

14.9 THREE-PHASE SHORT CIRCUIT ON AN UNLOADED SYNCHRONOUS GENERATOR

If a three-phase short circuit occurs at the terminals of a salient pole synchronous generator, we obtain typical oscillograms as shown in Fig. 14.3 for the short-circuit currents the three phases. Fig. 14.4 shows the alternating component of the short-circuit current when the DC component is

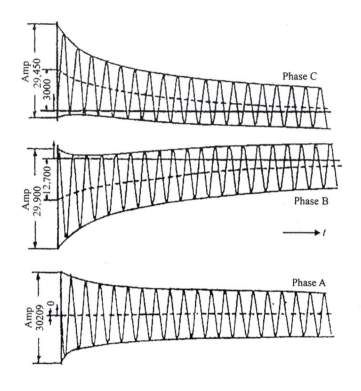

FIGURE 14.3

Oscillograms of the armature currents after a short circuit.

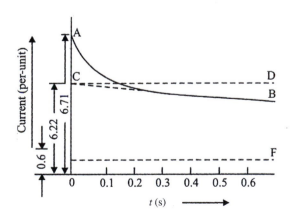

FIGURE 14.4

Alternating component of the short-circuit armature current.

eliminated. The fast changing subtransient component and the slowly changing transient components are shown at A and C. Fig. 14.5 shows the electrical torque. The changing field current is
shown in Fig. 14.6.

From the oscillogram of AC component, the quantities x''_d, x''_q, x'_d, and x'_q can be determined.

If V is the line to neutral prefault voltage, then the AC component

$$i_{AC} = \frac{V}{x''_q} = I''$$

the rms subtransient short circuit. Its duration is determined by T''_d, the subtransient direct axis time
constant. The value of i_{AC} decreases to V/x'_d when $t > T''_d$.

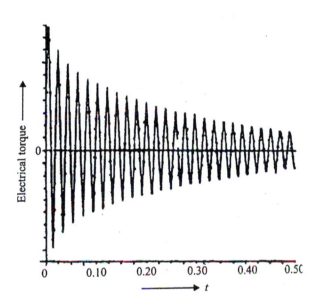

FIGURE 14.5

Electrical torque on the three-phase terminal short circuit.

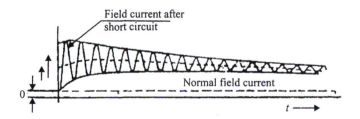

FIGURE 14.6

Oscillogram of the field current after a short circuit.

With T'_d as the direct-axis transient time constant when $t > T'_d$

$$i_{AC} = \frac{V}{x_d}$$

The maximum DC offset component that occurs in any phase at $\alpha = 0$ is

$$i_{DC_{max}}(t) = \sqrt{2}\,\frac{V}{x''_d}e^{-t/TA}$$

where T_A is the armature time constant.

14.10 EFFECT OF LOAD CURRENT OR PREFAULT CURRENT

Consider a three-phase synchronous generator supplying a balanced three-phase load. Let a three-phase fault occur at the load terminals. Before the fault occurs, a load current I_L is flowing into the load from the generator. Let the voltage at the fault be v_f and the terminal voltage of the generator be v_t. Under fault conditions, the generator reactance is x''_d.

The circuit in Fig. 14.7 indicates the simulation of fault at the load terminals by a parallel switch S.

$$E''_g = V_t + jx''_d I_L = V_f + (X_{ext} + jx''_d)I_L$$

where E''_g is the subtransient internal voltage.

For the transient state

$$E'_g = V_t + jx'_d I_L$$
$$= V_f + (Z_{ext} + jx'_d)I_L$$

E''_g or E'_g are used only when there is a prefault current I_L. Otherwise E_g, the steady-state voltage in series with the direct-axis synchronous reactance is to be used for all calculations. E_g remains the same for all I_L values and depends only on the field current. Every time, of course, a new E''_g is required to be computed.

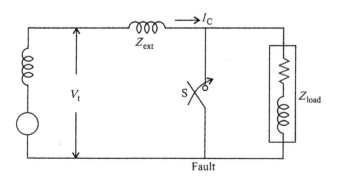

FIGURE 14.7

Fault simulation for a synchronous machine.

14.11 **REACTORS**

Whenever faults occur in power system, large currents flow. Especially, if the fault is a dead short circuit at the terminals or bus bars, enormous currents flow damaging the equipment and its components. To limit the flow of large currents under these circumstances, current-limiting reactors are used. These reactors are large coils constructed for high self-inductance.

They are also so located that the effect of the fault does not affect other parts of the system and is thus localized. From time to time, new generating units are added to an existing system to augment the capacity. When this happens, the fault current level increases and it may become necessary to change the switch gear. With proper use of reactors, addition of generating units does not necessitate changes in existing switch gear.

14.12 **CONSTRUCTION OF REACTORS**

These reactors are built with nonmagnetic core so that saturation of core with consequent reduction in inductance and increased short-circuit currents is avoided. Alternatively, it is possible to use iron core with air-gaps included in the magnetic core so that saturation is avoided.

14.13 **CLASSIFICATION OF REACTORS**

There are three types of reactors
(1) Generator reactors, (2) Feeder reactors, and (3) Bus-bar reactors
The above classification is based on the location of the reactors. Reactors may be connected in series with the generator in series with each feeder or to the bus bars.

1. *Generator Reactors*: The reactors are located in series with each of the generators as shown in Fig. 14.8 so that current flowing into a fault F from the generator is limited.
   Disadvantages:
   a. In the event of a fault occuring on a feeder, the voltage at the remaining healthy feeders also may loose synchronism requiring resynchronization later.
   b. There is a constant voltage drop in the reactors and also power loss, even during normal operation. Since modern generators are designed to withstand dead short circuit at their terminals, generator reactors are now-a-days not used except for old units in operation.
2. *Feeder Reactors:* In this method of protection each feeder is equipped with a series reactor as shown in Fig. 14.9.
   In the event of a fault on any feeder the fault current drawn is restricted by the reactor.
   Disadvantages:
   a. Voltage drop and power loss still occurs in the reactor for a feeder fault. However, the voltage drop occurs only in that particular feeder reactor.
   b. Feeder reactors do not offer any protection for bus-bar faults. Nevertheless, bus-bar faults occur very rarely.

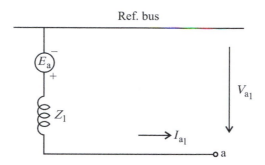

FIGURE 14.8

Generator reactors.

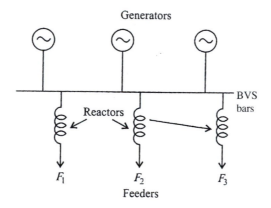

FIGURE 14.9

Feeder reactors.

As series reactors inherently create voltage drop, system voltage regulation will be impaired. Hence they are to be used only in special case such as for short feeders of large cross section.

3. *Bus-Bar Reactors:* In both the above methods the reactors carry full load current under normal operation. The consequent disadvantage of constant voltage drops and power loss can be avoided by dividing the bus bars into sections and interconnect the sections through protective reactors. There are two ways of doing this.

   a. *Ring System:* In this method each feeder is fed by one generator. Very little power flows across the reactors during normal operation. Hence, the voltage drop and power loss are negligible. If a fault occurs on any feeder, only the generator to which the feeder is connected will feed the fault and other generators are required to feed the fault through the reactor. This is shown in Fig. 14.10.

   b. *Tie-Bar System:* This is an improvement over the ring system. This is shown in Fig. 14.11. Current fed into a fault has to pass through two reactors in series between sections.

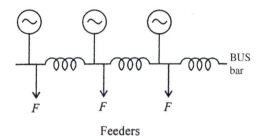

FIGURE 14.10

Ring system.

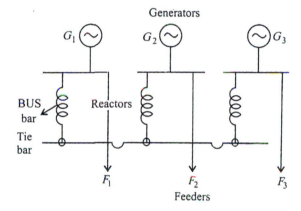

FIGURE 14.11

Tie-bar system.

Another advantage is that additional generation may be connected to the system without requiring changes in the existing reactors.

The only disadvantage is that this system requires an additional bus-bar system, the tie-bar.

WORKED EXAMPLES

E.14.1 Two generators rated at 10 MVA, 11 kV, and 15 MVA, 11 kV, respectively, are connected in parallel to a bus. The bus bars feed two motors rated 7.5 and 10 MVA, respectively. The rated voltage of the motors is 9 kV. The reactance of each generator is 12% and that of each motor is 15% on their own ratings. Assume 50-MVA, 10-kV base and draw the reactance diagram.

Solution:

The reactances of the generators and motors are calculated on 50-MVA, 10-kV base values.

Reactance of Generator 1 $= X_{G_1} = 12 \cdot \left(\dfrac{11}{10}\right)^2 \cdot \left(\dfrac{50}{10}\right) = 72.6\%$

Reactance of Generator 2 $= X_{G_2} = 12\left(\dfrac{11}{10}\right)^2 \cdot \left(\dfrac{50}{10}\right) = 48.4\%$

Reactance of Motor 1 $= X_{M_1} = 15\left(\dfrac{9}{10}\right)^2 \left(\dfrac{50}{75}\right) = 81\%$

Reactance of Motor 2 $= X_{M_2} = 15\left(\dfrac{9}{10}\right)^2 \left(\dfrac{50}{10}\right) = 60.75\%$

The reactance diagram is drawn and shown in Fig. E.14.1.

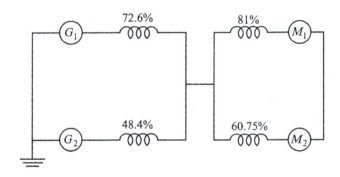

FIGURE E.14.1

E.14.2 A 100-MVA, 13.8-kV, three-phase generator has a reactance of 20%. The generator is connected to a three-phase transformer T_1 rated 100 MVA 12.5 kV/110 kV with 10% reactance. The h.v. side of the transformer is connected to a transmission line of reactance 100 Ω. The far end of the line is connected to a step-down transformer T_2, made of three single-phase transformers each rated 30 MVA, 60 kV/10 kV with 10% reactance the generator supplies two motors connected on the l.v. side T_2 as shown in Fig. E.14.2. The motors are rated at 25 and 50 MVA both at 10 kV with 15% reactance. Draw the reactance diagram showing all the values in per-unit. Take generator rating as a base.

Solution:

Base MVA $= 100$

Base kV $= 13.8$

Base kV for the line $= 13.8 \times \dfrac{110}{12.5} = 121.44$

Line-to-line voltage ratio of $T_2 = \dfrac{\sqrt{3} \times 66 \text{ kV}}{10 \text{ kV}} = \dfrac{114.31}{10}$

Base voltage for motors $= \dfrac{121.44 \times 10}{114.31} = 10.62$ kV

$\%X$ for generators $= 20 = 0.2$ p.u.

$\%X$ for transformer $T_1 = 10 \times \left(\dfrac{12.5}{13.8}\right)^2 \times \dfrac{100}{100} = 8.2$

%X for transformer T_2 on $\sqrt{3} \times 66$: 10 kV and 3×30 MVA base $= 10$

%X for T_2 on 100 MVA, and 121.44 kV: 10.62 kV is

$$\%X \text{ for } T_2 = 10 \times \left(\frac{10}{10.62}\right)^2 \times \left(\frac{100}{90}\right) = 9.85 = 0.0985 \text{ p.u.}$$

Base reactance for line $= \left(\dfrac{121.44}{100}\right)^2 = 147.47\ \Omega$

Reactance of line $= \dfrac{100}{147.47} = 0.678$ p.u.

Reactance of motor $M_1 = 10 \times \left(\dfrac{10}{10.62}\right)^2 \left(\dfrac{90}{25}\right) = 31.92\% = 0.3192$ p.u.

Reactance of motor $M_2 = 10 \times \left(\dfrac{10}{10.62}\right)^2 \left(\dfrac{90}{50}\right) = 15.96\%$

The reactance diagram is shown in Fig. E.14.2.

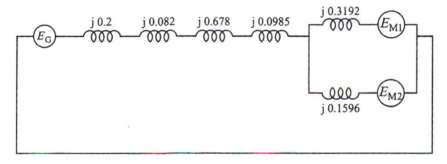

FIGURE E.14.2

E.14.3 Obtain the per-unit representation for the three-phase power system shown in Fig. E.14.3.

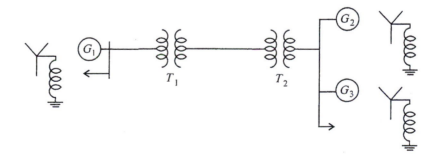

FIGURE E.14.3

Generator 1: 50 MVA, 10.5 kV, $X = 1.8\ \Omega$

Generator 2: 25 MVA, 6.6 kV, $X = 1.2\ \Omega$

Generator 3: 35 MVA, 6.6 kV, $X = 0.6\ \Omega$

Transformer T_1: 30 MVA, 11/66 kV, $X = 15\ \Omega/\text{phase}$

Transformer T_2: 25 MVA, 66/6.2 kV, as h.v. side $X = 12\ \Omega$

Transmission line: $X_L = 20\ \Omega/\text{phase}$

Solution:

Let the base MVA = 50

Base kV = 66 $(L - L)$

Base voltage on transmission as line 1 p.u. (66 kV)

Base voltage for Generator 1: 11 kV

Base voltage for Generators 2 and 3: 6.2 kV

$$\text{Reactance of transmission line} = \frac{20 \times 50}{66^2} = 0.229 \text{ p.u.}$$

$$\text{Reactance of transformer } T_1 = \frac{15 \times 50}{66^2} = 0.172 \text{ p.u.}$$

$$\text{Reactance of transformer } T_2 = \frac{12 \times 50}{66^2} = 0.1377 \text{ p.u.}$$

$$\text{Reactance of Generator 1} = \frac{1.8 \times 50}{(11)^2} = 0.7438 \text{ p.u.}$$

$$\text{Reactance of Generator 2} = \frac{1.2 \times 50}{(6.2)^2} = 1.56 \text{ p.u.}$$

$$\text{Reactance of Generator 3} = \frac{0.6 \times 50}{(6.2)^2} = 0.78 \text{ p.u.}$$

E.14.4 A single-phase, two-winding transformer is rated 20 kVA, 480/120 V at 50 HZ. The equivalent leakage impedance of the transformer referred to l.v. side is 0.0525 78.13° ohm using transformer ratings as base values, determine the per-unit leakage impedance referred to the h.v. side and l.v. side.

Solution:

Let the base kVA = 20

Base voltage on h.v. side = 480 V

Base voltage on l.v. side = 120 V

The leakage impedance on the l.v. side of the transformer

$$= Z_{l2} = \frac{V_{\text{base 2}}}{\text{VA base}} = \frac{(120)^2}{20{,}000} = 0.72\ \Omega$$

Per-unit leakage impedance referred to the l.v. of the transformer

$$= Z_{\text{p.u. 2}} = \frac{0.0525\ 78.13 \text{ degrees}}{0.72} = 0.0729\ 78.13 \text{ degrees p.u.}$$

Equivalent impedance referred to the h.v. side is

$$\left(\frac{400}{120}\right)^2 [(0.0525 \ \underline{70.13 \text{ degrees}}] = 0.84 \ \underline{78.13 \text{ degrees}}$$

The base impedance on the h.v. side of the transformer is $(480)^2/20{,}000 = 11.52 \ \Omega$
Leakage impedance referred to the h.v. side

$$= \frac{0.84 \ \underline{78.13 \text{ degrees}}}{11.52} = 0.0729 \ \underline{78.13 \text{ degrees}} \text{ p.u.}$$

E.14.5 A single-phase transformer is rated at 110/440 V, 3 kVA. Its leakage reactance measured
on 110 V side is 0.05 Ω. Determine the leakage impedance referred to 440 V side.
Solution:

Base impedance on 110 V side $= \dfrac{(0.11)^2 \times 1000}{3} = 4.033 \ \Omega$

Reactance on 110 V side $= \dfrac{0.05}{4.033} = 0.01239$ p.u.

Leakage reactance referred to 440 V side $= (0.05)\left(\dfrac{440}{110}\right)^2 = 0.8 \ \Omega$

Base impedance referred to 440 V side $= \dfrac{0.8}{64.53} = 0.01239$ p.u.

E.14.6 Consider the system shown in Fig. E.14.4. Selecting 10,000 kVA and 110 kV as base
values, find the per-unit impedance of the 200-Ω load referred to 110 and 11 kV side.

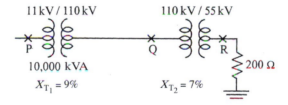

FIGURE E.14.4

Solution:

Base voltage at $\rho = 11$ kV

Base voltage at $R = \dfrac{110}{2} = 55$ kV

Base impedance at $R = \dfrac{55^2 \times 1000}{10{,}000} = 302.5 \ \Omega$

Impedance at $R = \dfrac{200 \ \Omega}{302.5 \ \Omega} = 0.661$ p.u.

Base impedance at $\phi = \dfrac{110^2 \times 1000}{10{,}000} = 1210 \ \Omega$

Load impedance referred to $\phi = 200 \times 2^2 = 800 \ \Omega$

Impedance of load referred to $\phi = \dfrac{800}{1210} = 0.661$ p.u.

Similarly base impedance at $P = \dfrac{11^2 \times 1000}{10,000} = 121.1$ Ω

Impedance of load referred to $P = 200 \times 2^2 \times 0.1^2 = 8$ Ω

Impedance of load at $P = \dfrac{8}{12.1} = 0.661$ p.u.

E.14.7 Three transformers each rated 30 MVA at 38.1/3.81 kV are connected in star-delta with a balanced load of three 0.5 ohm, star-connected resistors. Selecting a base of 900 MVA 66 kV for the h.v. side of the transformer, find the base values for the l.v. side.

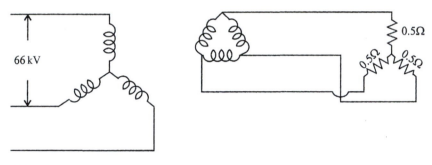

FIGURE E.14.5

Solution:

Base impedance on the l.v. side $= \dfrac{(\text{Base kV}_{L-L})^2}{\text{Base MVA}} = \dfrac{(3.81)^2}{90} = 0.1613$ Ω

Load resistance on the l.v. side $= \dfrac{0.5}{0.1613} = 3.099$ p.u.

Base impedance on the h.v. side $= \dfrac{(66)^2}{90} = 48.4$ Ω

Load resistance referred to the h.v. side $= 0.5 \times \left(\dfrac{66}{3.81}\right)^2 = 150$ Ω

Load resistance referred to the h.v. side $= \dfrac{150}{48.4} = 3.099$ p.u.

The per-unit load resistance remains the same.

E.14.8 Two generators are connected in parallel to the l.v. side of a three-phase delta-star transformer as shown in Fig. E.14.6. Generator 1 is rated 60,000 kVA, 11 kV. Generator 2 is rated 30,000 kVA, 11 kV. Each generator has a subtransient reactance of $x''_d = 25\%$. The transformer is rated 90,000 kVA at 11 kV D/66 kV g with a reactance of 10%. Before a fault occurred the voltage on the h.v. side of the transformer is 63 kV. The transformer in unloaded and there is no circulating current between the generators. Find the subtransient current in each generator when a three-phase short circuit occurs on the h.v. side of the transformer.

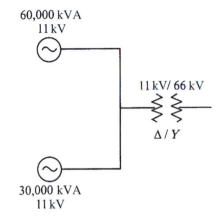

60,000 kVA
11 kV

11 kV/ 66 kV

Δ / Y

30,000 kVA
11 kV

FIGURE E.14.6

Solution:
Let the line voltage on the h.v. side be the base kV = 66 kV.
Let the base kVA = 90,000 kVA

Generator 1: $x''_d = 0.25 \times \dfrac{90,000}{60,000} = 0.375$ p.u.

For generator 2: $x''_d = \dfrac{90,000}{30,000} = 0.75$ p.u.

The internal voltage for Generator 1

$$E_{g1} = \frac{0.63}{0.66} = 0.955 \text{ p.u.}$$

The internal voltage for Generator 2

$$E_{g2} = \frac{0.63}{0.66} = 0.955 \text{ p.u.}$$

The reactance diagram is shown in Fig. E.14.7 when switch S is closed, the fault condition is simulated. As there is no circulating current between the generators, the equivalent reactance of the parallel circuit is $(0.375 \times 0.75)/(0.375 + 0.75) = 0.25$ p.u.

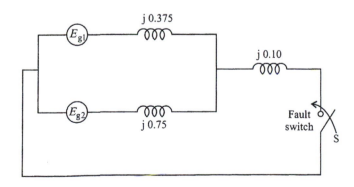

j 0.375

E_{g1}

j 0.10

E_{g2}

j 0.75

Fault
switch

S

FIGURE E.14.7

The subtransient current $I'' = \dfrac{0.955}{(j0.25 + j0.10)} = j2.7285$ p.u.

The voltage as the delta side of the transformer is $(-j2.7285)\,(j0.10) = 0.27205$ p.u.

I_1'' = the subtransient current flowing into fault from generator

$$I_1'' = \frac{0.955 - 0.2785}{j0.375} = 1.819 \text{ p.u.}$$

Similarly

$$I_2'' = \frac{0.955 - 0.27285}{j0.75} = -j1.819 \text{ p.u.}$$

The actual fault currents supplied in amperes are

$$I_1'' = \frac{1.819 \times 90,000}{\sqrt{3} \times 11} = 8592.78 \text{ A}$$

$$I_2'' = \frac{0.909 \times 90,000}{\sqrt{3} \times 11} = 4294.37 \text{ A}$$

E.14.9 R station with two generators feeds through transformers a transmission system operating at 132 kV. The far end of the transmission system consisting of 200 km long double-circuit line is connected to load from bus B. If a three-phase fault occurs at bus B, determine the total fault current and fault current supplied by each generator.

Select 75 MVA and 11 kV on the l.v. side and 132 kV on the h.v. side as base values.

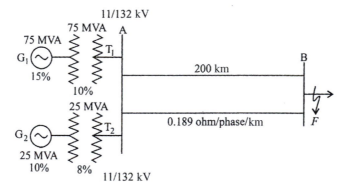

FIGURE E.14.8

Solution:

X of Generator 1 = $j0.15$ p.u.

X of Generator 2 = $j0.10\dfrac{75}{25} = j0.3$ p.u.

X of Transformer $T_1 = j0.1$ p.u.

X of Transformer $T_2 = j0.08 \times \dfrac{75}{25} = j0.24$ p.u.

$$X \text{ of each line} = \frac{j0.180 \times 200 \times 75}{132 \times 132} = j0.1549 \text{ p.u.}$$

The equivalent reactance diagram is shown in Fig. E.14.9A−C.

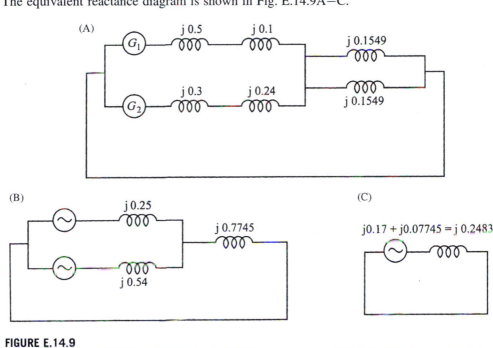

FIGURE E.14.9

Fig. E.14.9A−C can be reduced further into

$$Z_{eq} = j0.17 + j0.07745 = j0.248336$$

Total fault current $\dfrac{1 \angle 0 \text{ degree}}{j0.248336} = -j4.0268 \text{ p.u.}$

Base current for 132-kV circuit $= \dfrac{75 \times 1000}{\sqrt{3} \times 132} = 328 \text{ A}$

Hence actual fault current $= -j4.0268 \times 328 = 1321 \text{ A} \angle -90 \text{ degrees}$

Base current for the 11-kV side of the transformer $= \dfrac{75 \times 1000}{\sqrt{3} \times 11} = 3936.6 \text{ A}$

Actual fault current supplied from the 11-kV side $= 3936.6 \times 4.0248 = 15,851.9 \text{A} \angle -90$ degrees

Fault current supplied by Generator 1 $= \dfrac{1,585,139 \angle -90 \text{degrees} \times j0.54}{j0.54 + j0.25} = -j10835.476 \text{A}$

Fault current supplied by Generator 2 $= \dfrac{15,851.9 \times j0.25}{j0.79} = 5016.424 \text{ A} \angle -90 \text{ degrees}$

E.14.10 A 33-kV line has a resistance of 4 Ω and a reactance of 16 Ω, respectively. The line is connected to a generating station bus bars through a 6000-kVA step-up transformer which has a reactance of 6%. The station has two generators rated 10,000 kVA with 10%

reactance and 5000 kVA with 5% reactance. Calculate the fault current and short-circuit kVA when a three-phase fault occurs at the h.v. terminals of the transformers and at the load end of the line.

Solution:

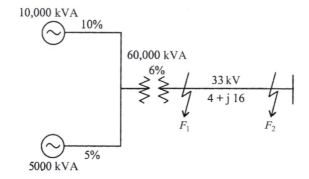

FIGURE E.14.10 (A)

Let 10,000 kVA be the base kVA

Reactance of Generator 1 $X_{G1} = 10\%$

Reactance of Generator 2 $X_{G2} = \dfrac{5 \times 10,000}{5000} = 10\%$

Reactance of transformer $X_T = \dfrac{6 \times 10,000}{6000} = 10\%$

The line impedance is converted into percentage impedance

$$\%X = \frac{\text{kVA} \cdot X}{10(\text{kV})^2}; \quad \%X_{\text{Line}} = \frac{10,000 \times 16}{10 \times (33)^2} = 14.69$$

$$\%R_{\text{Line}} = \frac{19,000 \times 4}{10(33)^2} = 3.672$$

a. For a three-phase fault at the h.v. side terminals of the transformer fault impedance

$$= \left(\frac{10 \times 10}{10 + 10}\right) + 10 = 15\%$$

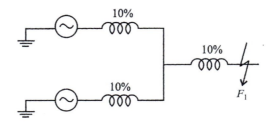

FIGURE 14.10 (B)

$$\text{Short-circuit kVA fed into the fault} = \frac{10,000 \times 100}{15} \text{ kVA}$$

$$= 66,666.67 \text{ kVA}$$

$$= 66.67 \text{ MVA}$$

For a fault at F_2 the load end of the line the total reactance to the fault

$$= 15 + 14.69$$
$$= 29.69\%$$

Total resistance to fault $= 3.672\%$

Total impedance to fault $= \sqrt{3.672^2 + 29.69^2}$

$$= 29.916\%$$

$$\text{Short-circuit kVA into fault} = \frac{100}{29.916} \times 10,000$$

$$= 33,433.63 \text{ kVA}$$

$$= 33.433 \text{ MVA}$$

E.14.11 Fig. E.14.11A shows a power system where load at bus 5 is fed by generators at bus 1 and bus 4. The generators are rated at 100 MVA; 11 kV with subtransient reactance of 25%. The transformers are rated each at 100 MVA, 11/112 kV and have a leakage reactance of 8%. The lines have an inductance of 1 mH/phase/km. Line L_1 is 100 km long while lines L_2 and L_3 are each of 50 km in length. Find the fault current and MVA for a three-phase fault at bus 5.

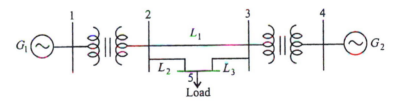

FIGURE E.14.11 (A)

Solution:
Let the base MVA $= 100$ MVA
Base voltage for the l.v. side $= 11$ kV and
Base voltage for the h.v. side $= 112$ kV
Base impedance for the h.v. side of transformer

$$= \frac{112 \times 112}{100} = 125.44 \ \Omega$$

Base impedance for the l.v. side of transformer

$$= \frac{11 \times 11}{100} = 1.21 \ \Omega$$

Reactance of line $L_1 = 2 \times p \times 50 \times 1 \times 10^{-3} \times 100 = 31.4 \ \Omega$

Reactance of line $L_1 = \dfrac{31.4}{125.44} = 0.25$ p.u.

Impedance of line $L_2 = \dfrac{2\pi \times 50 \times 1 \times 10^{-3} \times 50}{125.44} = 0.125$ p.u.

Impedance of line $L_3 = 0.125$ p.u.

The reactance diagram is shown in Fig. 14.11B.

By performing conversion of delta into star at A, B, and C, the star impedances are

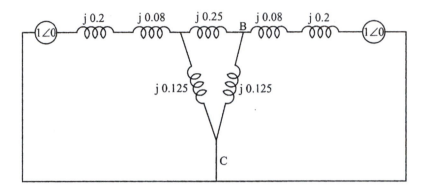

FIGURE E.14.11 (B)

$$Z_1 = \frac{j0.25 \times j0.125}{j0.25 + j0.125 + j0.125} = j0.0625$$

$$Z_2 = \frac{j0.25 \times j0.125}{j0.5} = j0.0625$$

and

$$Z_3 = \frac{j0.125 \times j0.125}{j0.5} = j0.03125$$

The following reactance diagram is obtained.

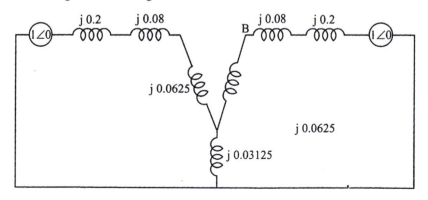

FIGURE E.14.11 (C)

This can be further reduced into Fig. E.14.11D.

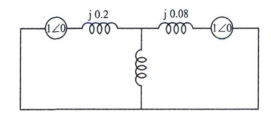

FIGURE E.14.11 (D)

Finally this can be put first into Fig. E.14.11E and later into Fig. E.14.11F.

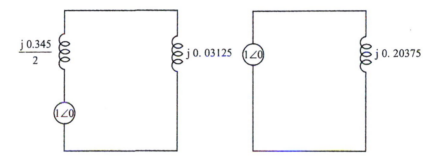

FIGURE E.14.11 (E AND F)

Fault MVA $= \dfrac{1}{0.20375} = 4.90797$ p.u.

$\qquad = 100$ MVA $\times 4.90797 = 490.797$ MVA

Fault current $= \dfrac{1}{j\,0.20375} = 4.90797$ p.u.

Base current $= \dfrac{100 \times 10^6}{\sqrt{3} \times 112 \times 10^3} = 515.5$ A

Fault current $= 4.90797 \times 515.5$

$\qquad = 2530$ A

E.14.12 Two motors having transient reactances 0.3 p.u. and subtransient reactances 0.2 p.u. based on their own ratings of 6 MVA, 6.8 kV are supplied by a transformer rated 15 MVA, 112 kV/6.6 kV and its reactance is 0.18 p.u. A three-phase short circuit occurs at the terminals of one of the motors. Calculate (1) the subtransient fault current, (2) subtransient current in circuit breaker A, (3) the momentary circuit rating of the breaker, and (4) if the circuit breaker has a breaking time of four cycles calculate the current to be interrupted by the circuit breaker A.

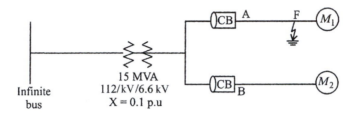

FIGURE E.14.12 (A)

Solution:

Let the base MVA = 15

Base kV for the l.v. side = 6.6 kV

Base kV for the h.v side = 112 kV

For each motor $x''_d = 0.2 \times \dfrac{15}{6} = 0.5$ p.u.

For each motor $x''_d = 0.3 \times \dfrac{15}{6} = 0.75$ p.u.

The reactance diagram is shown in Fig. E.14.12B.

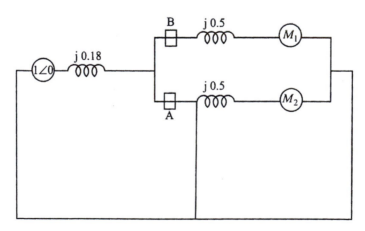

FIGURE E.14.12 (B)

Under fault condition the reactance diagram can be further simplified into Fig. E.14.12C.

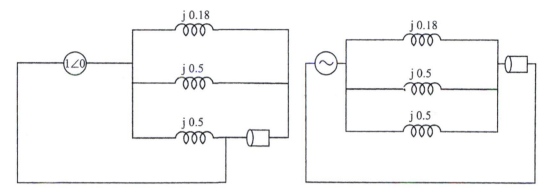

FIGURE E.14.12 (C)

$$\text{Impedance to fault} = \cfrac{1}{\cfrac{1}{j0.18} + \cfrac{1}{j0.5} + \cfrac{1}{j0.5}}$$

$$\text{Subtransient fault current} = \frac{1 \angle 0 \text{ degree}}{j0.1047} = -j9.55 \text{ p.u.}$$

$$\text{Base current} = \frac{15 \times 10^6}{\sqrt{3} \times 6.6 \times 10^3} = 1312.19 \text{ A}$$

$$\text{Subtransient fault current} = 1312.19 \times (-j9.55)$$
$$= 12,531.99 \text{ A (lagging)}$$

a. Total fault current from the infinite bus

$$\frac{-1 \angle 0 \text{ degree}}{j0.18} = -j5.55 \text{ p.u.}$$

$$\text{Fault current from each motor} = \frac{1 \angle 0 \text{ degree}}{j0.5} = -j2 \text{ p.u.}$$

Fault current into breaken A is sum of the two currents from the infinite bus and from motor 1

$$= -j5.55 + (-j2) = -j7.55 \text{ p.u.}$$

$$\text{Total fault current into breaken} = -j7.55 \times 1312.19$$
$$= 9907 \text{ A}$$

b. Manentary fault current taking into the DC offset component is approximately

$$1.6 \times 9907 = 15{,}851.25 \text{ A}$$

c. For the transient condition, i.e., after four cycles the motor reactance changes to 0.3 p.u.

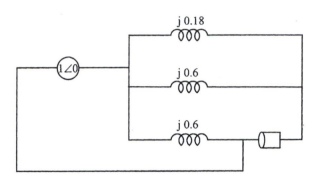

FIGURE E.14.12 (D)

The reactance diagram for the transient state is shown in Fig. E.14.12D.

The fault impedance is $\dfrac{1}{\dfrac{1}{j0.15} + \dfrac{1}{j0.6} + \dfrac{1}{j0.6}} = j0.1125$ p.u.

The fault current $= \dfrac{1 \angle 0 \text{ degree}}{j0.1125} = j8.89$ p.u.

Transient fault current $= -j8.89 \times 1312.19$

$$= 11{,}665.37 \text{ A}$$

If the DC offset current is to be considered, it may be increased by a factor of say 1.1.

So that the transient fault current $= 11{,}665.37 \times 1.1$

$$= 12{,}831.9 \text{ A}$$

E.14.13 Consider the power system shown in Fig. E.14.13A.

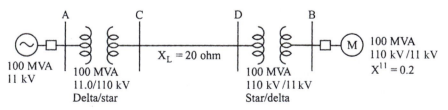

FIGURE E.14.13 (A)

The synchronous generator is operating at its rated MVA at 0.95 lagging power factor and at rated voltage. A three-phase short circuit occurs at bus A. Calculate the per-unit value of (1) subtransient fault current, (2) subtransient generator and motor currents. Neglect prefault current. Also compute (3) the subtransient generator and motor currents including the effect of prefault currents.

$$\text{Base line impedance} = \frac{(110)^2}{100} = 121 \ \Omega$$

$$\text{Line reactance} = \frac{20}{121} = 0.1652 \text{ p.u.}$$

The reactance diagram including the effect of the fault by switch S is shown in Fig. E.14.13B.

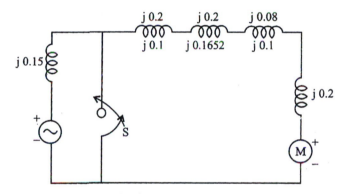

FIGURE E.14.13 (B)

Looking into the network from the fault using Thevenin's theorem

$$Z_{th} = jX_{th} = j\left(\frac{0.15 \times 0.565}{0.15 + 0.565}\right) = j0.1185$$

a. The subtransient fault current

$$I''_m = \frac{0.565}{0.565 + 0.15} \ I''_f = \frac{0.565 \times j8.4388}{0.7125} = j6.668 \text{ p.u.}$$

b. The motor subtransient current

$$I''_m = \frac{0.15}{0.715} I''_f = \frac{0.15}{0.715} \times 8.4388 = j1.770 \text{ p.u.}$$

$$\text{Generator base current} = \frac{100 \text{ MVA}}{\sqrt{3} \times 11 \text{ kV}} = 5.248 \text{ kA}$$

c. Generator prefault current $= \dfrac{100}{\sqrt{3} \times 11} \left[\cos^{-1} 0.95 \right]$

$$= 5.248 \angle -18°.19 \text{ kA}$$

$$I_{\text{load}} = \dfrac{5.248 \angle -18°.19}{5.248} = 1 \angle -18°.19$$

$$= (0.95 - j0.311) \text{ p.u.}$$

The subtransient generator and motor currents including the prefault currents are

$$I''_g = j6.668 + 0.95 - j0.311 = -j6.981 + 0.95$$
$$= (0.95 - j6.981)\text{p.u.} = 7.045 - 82.250 \text{ p.u.}$$

$$I''_m = -j1.77 - 0.95 + j0.311 = -0.95 - j1.459$$
$$= 1.74 \angle -56.93 \text{ degrees}$$

E.14.14 Consider the system shown in Fig. E.14.14A. The percentage reactance of each alternator is expressed on its own capacity determine the short-circuit current that will flow into a dead three-phase short circuit at F.

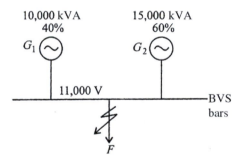

FIGURE E.14.14 (A)

Solution:

Let the base kVA $= 25,000$ and base kV $= 11$

$$\%X \text{ of Generator 1} = \dfrac{25,000}{10,000} \times 40 = 100$$

$$\%X \text{ of Generator 2} = \dfrac{25,000}{15,000} \times 60 = 100$$

$$\text{Line current at 25,000 kVA and 11 kV} = \dfrac{25,000}{\sqrt{3} \times 11} \times \dfrac{10^3}{10^3} = 1312.19 \text{ A}$$

The reactance diagram is shown in Fig. E.14.14B.

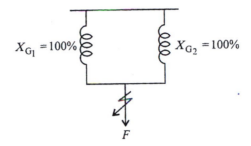

FIGURE E.14.14 (B)

The net percentage reactance upto the fault $= \dfrac{100 \times 100}{100 + 100} = 50$

Short-circuit current $= \dfrac{I \times 100}{\%X} = \dfrac{1312.19 \times 100}{50} = 2624.30$ A

E.14.15 A three-phase, 25-MVA, 11-kV alternator has internal reactance of 6%. Find the external reactance per phase to be connected in series with the alternator so that the steady-state short-circuit current does not exceed six times the full load current.

Solution:

$$\text{Full load current} = \dfrac{25 \times 10^6}{\sqrt{3} \times 11 \times 10^3} = 1312.9 \text{ A}$$

$$V_{\text{phase}} = \dfrac{11 \times 10^3}{\sqrt{3}} = 6351.039 \text{ V}$$

$$\text{Total } \%X = \dfrac{\text{Full-load current}}{\text{Short-circuit current}} \times 100 = \dfrac{1}{6} \times 100$$
$$= 16.67$$

External reactance needed $= 16.67 - 6 = 10.67\%$
Let X be the per phase external reactance required in ohms.

$$\%X = \dfrac{IX}{V} \times 100$$

$$10.67 = \dfrac{1312.19X \cdot 100}{6351.0393}$$

$$X = \dfrac{6351.0393 \times 10.67}{1312.19 \times 100} = 0.516428 \ \Omega$$

E.14.16 A three-phase line operating at 11 kV and having a resistance of 1.5 Ω and reactance of 6 Ω is connected to a generating station bus bars through a 5-MVA step-up transformer having reactance of 5%. The bus bars are supplied by a 12-MVA generator having 25% reactance. Calculate the short-circuit kVA fed into a symmetric fault (1) at the load end of the transformer and (2) at the h.v. terminals of the transformer.

Solution:

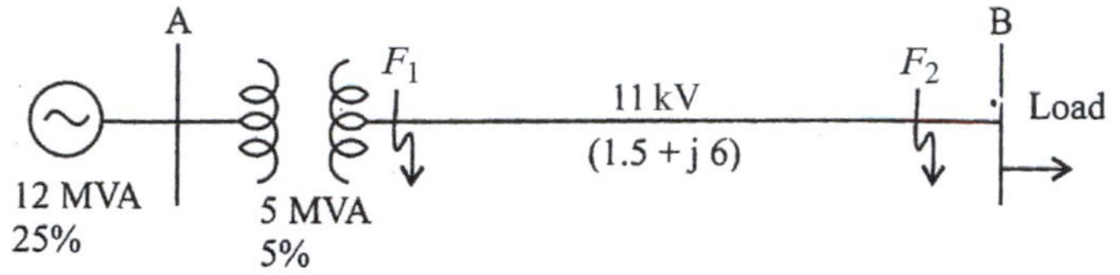

FIGURE E.14.15

Let the base kVA = 12,000 kVA
%X of alternator as base kVA = 25

$$\%X \text{ of transformer as 12,000-kVA base} = \frac{12,000}{5000} \times 5 = 12$$

$$\%X \text{ of line} \frac{12,000}{10(11)^2} \times 6 = 59.5$$

$$\%R \text{ of line} = \frac{12,000}{10(11)^2} \times 1.5 = 14.876$$

a. $\%X_{Total} = 25 + 12 + 59.5 = 96.5$

$$\%R_{Total} = 14.876$$

$$\%Z_{Total} = \sqrt{(96.5)^2 + (14.876)^2} = 97.6398$$

Short-circuit kVA at the far end or load end $F_2 = \frac{12,000 \times 100}{97.6398} = 12,290$

If the fault occurs on the h.v. side of the transformer at F_1

$$\%X \text{ upto fault} \quad \begin{aligned} F_1 &= \%X_G + \%X_T = 25 + 12 \\ &= 37 \end{aligned}$$

Short-circuit kVA fed into the fault

$$= \frac{12,000 \times 100}{37} = 32,432.43$$

E.14.17 A three-phase generating station has two 15,000-kVA generators connected in parallel each with 15% reactance and a third generator of 10,000 kVA with 20% reactance is also added later in parallel with them. Load is taken as shown from the station bus-bars through 6000-kVA, 6% reactance transformers. Determine the maximum fault MVA which the circuit breakers have to interrupt on (1) l.v. side and (2) h.v. side of the system for a symmetrical fault.

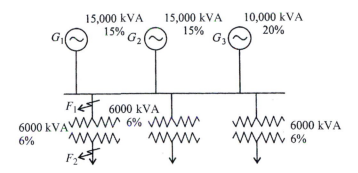

FIGURE E.14.17 (A)

Solution:

$$\%X \text{ of generator } G_1 = \frac{15 \times 15,000}{15,000} = 15$$

$$\%X \text{ of generator } G_2 = 15$$

$$\%X \text{ of generator } G_3 = \frac{20 \times 15,000}{10,000} = 30$$

$$\%X \text{ of transformer } T = \frac{6 \times 15,000}{6000} = 15$$

a. If fault occurs at F_1, the reactance is shown in Fig. E.14.17B.

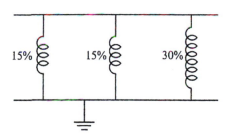

FIGURE E.14.17 (B)

The total % C upto fault $= \dfrac{1}{\dfrac{1}{15} + \dfrac{1}{15} + \dfrac{1}{30}}$

$$= 6$$

Fault MVA $= \dfrac{15,000 \times 100}{6} = 250,000 \text{ kVA}$

$$= 250$$

b. If the fault occurs at F_2, the reactance diagram will be as in Fig. E.14.17C.

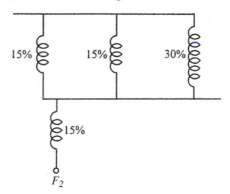

FIGURE E.14.17 (C)

The total %X upto fault 6% + 15.6 = 21

$$\text{Fault MVA} = \frac{15,000 \times 100}{21 \times 100} = 71.43$$

E.14.18 There are two generators at bus bar A each rated at 12,000 kVA, 12% reactance or another bus B, two more generators rated at 10,000 kVA with 10% reactance are connected. The two bus bars are connected through a reactor rated at 5000 kVA with 10% reactance. If a dead short circuit occurs between all the phases on bus bar B, what is the short-circuit MVA fed into the fault?

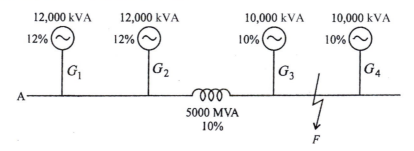

FIGURE E.14.18 (A)

Solution:
Let 12,000 kVA be the base kVA
%X of generator G_1 = 12
%X of generator G_2 = 12

$$\%X \text{ of generator } G_3 = \frac{10 \times 12,000}{10,000} = 12$$

$\%X$ of generator $G_4 = 12$

$$\%X \text{ of bus bar reactor} = \frac{10 \times 12,000}{5000} = 24$$

The reactance diagram is shown in Fig. E.14.18B.

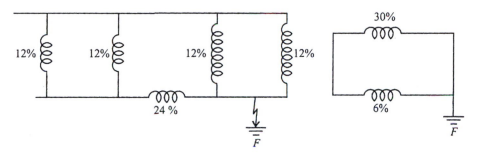

FIGURE E.14.18 (B)

$$\%X \text{ up to fault} = \frac{30 \times 6}{30 + 6} = 50$$

$$\text{Fault kVA} = \frac{12,000 \times 100}{6} = 600,000 \text{ kVA}$$

$$= 600 \text{ MVA}$$

E.14.19 A power plant has two generating units rated 3500 and 5000 kVA with percentage reactances 8% and 9%, respectively. The circuit breakers have a breaking capacity of 175 MVA. It is planned to extend the system by connecting it to the grid through a transformer rated at 7500 kVA and 7% reactance. Calculate the reactance needed for a reactor to be connected in the bus-bar section to prevent the circuit breaker from being over loaded if a short circuit occurs on any outgoing feeder connected to it. The bus-bar voltage is 3.3 kV.

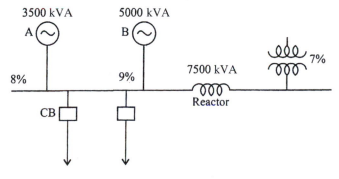

FIG. E.14.19 (A)

Solution:
Let 7500 kVA be the base kVA

$$\%X \text{ of generator A} = \frac{8 \times 7500}{3500} = 17.1428$$

$$\%X \text{ of generator B} = \frac{9 \times 7500}{5000} = 13.5$$

$\%X$ of transformer $= 7$ (as its own base)
The reactance diagram is shown in Fig. E.14.19B.

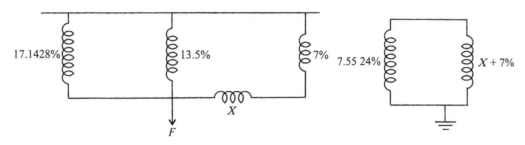

FIGURE E.14.19 (B)

$$\left[\text{Note:} \frac{1}{\left(\dfrac{1}{17.1428} + \dfrac{1}{13.5}\right)} = 7.5524 \right]$$

The short-circuit kVA should not exceed 175 MVA.

$$\text{Total reactance to fault} = \frac{1}{\left[\dfrac{1}{7.5524} + \dfrac{1}{X + 7}\right]}$$

$$= \frac{(X + 7)(7.5524)}{X + 7 + 7.5524}\% = \frac{(X + 7)(7.5524)}{X + 14.5524}\%$$

$$\text{Short-circuit kVA} = 7500 \times 100 \frac{X(X + 14.5524)}{(X + 7)(7.5524)}$$

This should not exceed 175 MVA

$$175 \times 10^3 = \frac{7500 \times 100(X + 14.5524)}{(X + 7)(7.5524)}$$

Solving

$$X = 7.02\%$$

Again

$$\%X = \frac{\text{kVA} \cdot (X)}{10(\text{kV})^2} = \frac{7500 \times (X)}{10 \times (3.3)^2}$$

$$\therefore \quad X = \frac{7.02 \times 10 \times 3.3^2}{7500} = \mathbf{0.102 \ \Omega}$$

In each share of the bus bar a reactance of 0.102 Ω is required to be inserted.

E.14.20 The short-circuit MVA at the bus bars for a power plant A is 1200 MVA and for another plant B is 1000 MVA at 33 kV. If these two are to be interconnected by a tie-line with reactance 1.2 Ω. Determine the possible short-circuit MVA at both the plants.

Solution:

Let the base MVA = 100

$$\%X \text{ of Plant 1} = \frac{\text{Base MVA}}{\text{Short-circuit MVA}} \times 100$$

$$= \frac{100}{1200} \times 100 = 8.33$$

$$\%X \text{ of Plant 2} = \frac{100}{1000} \times 100 = 10$$

$\%X$ of interconnecting tie line on base MVA

$$= \frac{100 \times 10^3}{10 \times (3.3)^2} \times 1.2 = 11.019$$

For fault at bus bars for Generator A

$$\%X = \frac{1}{\left[\dfrac{1}{8.33} + \dfrac{1}{21.019}\right]}$$

$$= 5.9657$$

$$\text{Short-circuit MVA} = \frac{\text{Base MVA} \times 100}{\%X}$$

$$= \frac{100 \times 100}{5.96576} = \mathbf{1676.23}$$

For a fault at the bus bars for Plant B

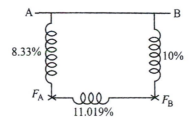

FIGURE E.14.20

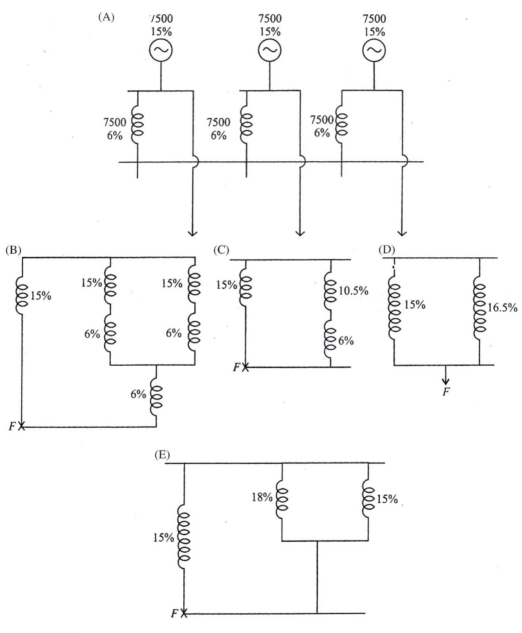

FIGURE E.14.21

$$\text{The total } \%X \text{ up to fault } F = \frac{15 \times 16.5}{15 + 16.5} = 7.857$$

$$\text{The short-circuit kVA} = \frac{7500 \times 100}{7.857} = 95,456.28 \text{ kVA} = 95.46 \text{ MVA}$$

$$\%X = \frac{1}{\left[\dfrac{1}{19.349} + \dfrac{1}{10}\right]} = 6.59$$

$$\text{Short-circuit MVA} = \frac{100 \times 100}{6.59} = 1517.45$$

E.14.21 A power plant has three generating units each rated at 7500 kVA with 15% reactance. The plant is protected by a tie-bar system. With reactances rated at 7500 MVA and 6%, determine the fault kVA when a short circuit occurs on one of the sections of bus bars. If the reactors were not present, what would be the fault kVA?

Solution:
The equivalent reactance diagram is shown in Fig. E.14.21A which reduces to Fig. E.14.21B and C.
Without reactors the reactance diagram will be as shown.

$$\text{The total } \%X \text{ up to fault } F = \frac{15 \times 7.5}{15 + 7.5} = 5$$

$$\text{Short-circuit MVA} = \frac{7500 \times 100}{5}$$

$$= 150,000 \text{ kVA}$$

$$= 150 \text{ MVA}$$

PROBLEMS

P.14.1 There are two generating stations each which an estimated short-circuit kVA of 500,000 and 600,000 kVA. Power is generated at 11 kV. If these two stations are interconnected through a reactor with a reactance of 0.4 Ω, what will be the short-circuit kVA at each station?

P.14.2 Two generators P and Q each of 6000 kVA capacity and reactance 8.5% are connected to a bus bar at A. A third generator R of capacity 12,000 kVA with 11% reactance is connected to another bus bar B. A reactor X of capacity 5000 kVA and 5% reactance is connected between A and B. Calculate the short-circuit kVA supplied by each generator when a fault occurs (1) at A and (2) at B.

P.14.3 The bus bars in a generating station are divided into three sections. Each section is connected to a tie-bar by a similar reactor. Each section is supplied by a 25,000-kVA, 11-kV, 50-Hz, three-phase generator. Each generator has a short-circuit reactance of 18%. When a short circuit occurs between the phases of one of the section bus bars, the voltage on the remaining section falls to 65% of the normal value. Determine the reactance of each reactor in ohms.

QUESTIONS

Q.14.1 Explain the importance of per-unit system.

Q.14.2 What do you understand by short-circuit kVA? Explain.

Q.14.3 Explain the construction and operation of protective reactors.

Q.14.4 How are reactors classified? Explain the merits and demerits of different types of system protection using reactors.

UNBALANCED FAULT ANALYSIS

15

Three-phase systems are accepted as the standard system for generation, transmission, and utilization of the bulk of electric power generated world over. The above holds good even when some of the transmission lines are replaced by DC links. When the three-phase system becomes unbalanced while in operation, analysis becomes difficult. Dr. C.L. Fortesque proposed in 1918 at a meeting of the American Institute of Electrical Engineers through a paper titled *Method of Symmetrical Coordinates Applied to the Solution of Polyphase Networks*, a very useful method for analyzing unbalanced three-phase networks.

Faults of various types such as line-to-ground, line-to-line, and three-phase short circuits with different fault impedances create unbalances. Breaking down of line conductors is also another source for unbalances in Power Systems Operation. The symmetrical coordinates proposed by Fortesque are known more commonly as symmetrical components or sequence components.

An unbalanced system of n phasors can be resolved into n systems of balanced phasors. These subsystems of balanced phasors are called *symmetrical components*. With reference to three-phase systems, the following balanced set of three components are identified and defined (Fig. 15.1).

1. Set of three phasors equal in magnitude, displaced from each other by 120 degrees in phase, and having the same phase sequence as the original phasors constitute *positive-sequence* components. They are denoted by the suffix 1.
2. Set of three phasors equal in magnitude, displaced from each other by 120 degrees in phase, and having a phase sequence opposite to that of the original phasors constitute *negative-sequence* components. They are denoted by the suffix 2.
3. Set of three phasors equal in magnitude and all in phase (with no mutual phase displacement) constitute *zero-sequence* components. They are denoted by the suffix 0.

Denoting the phases as R, Y, and B, V_R, V_Y, and V_B are the unbalanced phase voltages. These voltages are expressed in terms of the sequence components V_{R1}, V_{Y1}, V_{B1}, V_{R2}, V_{Y2}, V_{B2} and V_{R0}, V_{Y0}, V_{B0} as follows:-

$$V_R = V_{R1} + V_{R2} + V_{R0} \tag{15.1}$$

$$V_Y = V_{Y1} + V_{Y2} + V_{Y0} \tag{15.2}$$

$$V_B = V_{B1} + V_{B2} + V_{B0} \tag{15.3}$$

Electrical Power Systems. DOI: http://dx.doi.org/10.1016/B978-0-08-101124-9.00015-2

341

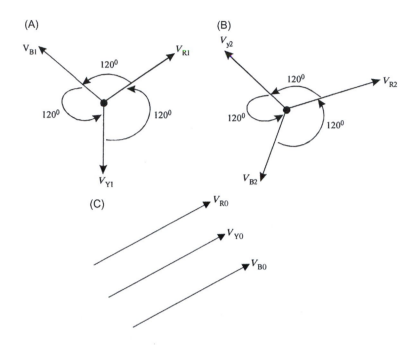

FIGURE 15.1

Sequence Components. (A) Positive-sequence components, (B) negative-sequence components, and (C) zero-sequence components.

15.1 THE OPERATOR "*a*"

In view of the phase displacement of 120 degrees, an operator "*a*" is used to indicate the phase displacement, just as j operator is used to denote 90-degree phase displacement.

$$a = 1 \angle 120 \text{ degrees} = -0.5 + j0.866$$

$$a^2 = 1 \angle 240 \text{ degrees} = -0.5 - j0.866$$

$$a^3 = 1 \angle 360 \text{ degrees} = 1 + j0$$

so that $1 + a + a^2 = 0 + j0$

The operator is represented graphically in Fig. 15.2.

Note that

$$a = 1 \angle 120 \text{ degrees} = 1 \cdot e^{j\frac{2\pi}{3}}$$

$$a^2 = 1 \angle 240 \text{ degrees} = 1 \cdot e^{j\frac{4\pi}{3}}$$

$$a^3 = 1 \angle 360 \text{ degrees} = 1 \cdot e^{j\frac{6\pi}{3}} = 1 \cdot e^{j2\pi}$$

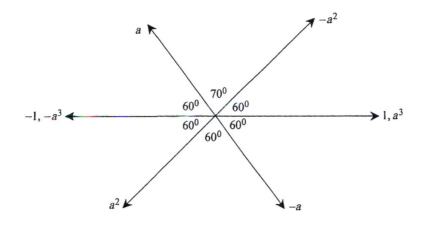

FIGURE 15.2

Operator j.

15.2 SYMMETRICAL COMPONENTS OF UNSYMMETRICAL PHASES

With the introduction of the operator "a" it is possible to redefine the relationship between unbalanced phasors of voltages and currents in terms of the symmetrical components or sequence components as they are known otherwise. We can write the sequence phasors with the operator as follows.

$$\left.\begin{array}{l} V_{R1} = V_{R1} \\ V_{R2} = V_{R2} \\ V_{R0} = V_{R0} \\ V_{Y1} = a^2 V_{R1} \\ V_{Y2} = a V_{R2} \\ V_{Y0} = V_{R0} \end{array}\right\} \left.\begin{array}{l} V_{B1} = a V_{R1} \\ V_{B1} = a^2 V_{R2} \\ V_{B0} = V_{R0} \end{array}\right\} \tag{15.4}$$

The voltage and current phasors for a three-phase unbalanced system are then represented by

$$\left.\begin{array}{l} V_R = V_{R1} + V_{R2} + V_{R0} \\ V_Y = a^2 V_{R1} + a V_{R2} + V_{R0} \\ V_B = a V_{R1} + a^2 V_{R2} + V_{R0} \end{array}\right\} \tag{15.5}$$

$$\left.\begin{array}{l} I_R = I_{R1} + I_{R2} + I_{R0} \\ I_Y = a^2 I_{R1} + a I_{R2} + I_{R0} \\ I_B = a I_{R1} + a^2 I_{R2} + I_{R0} \end{array}\right\} \tag{15.6}$$

The above equations can be put in a matrix form considering zero-sequence relation as the first for convenience.

$$\begin{bmatrix} V_R \\ V_Y \\ V_B \end{bmatrix} = \begin{bmatrix} 1 & 1 & 1 \\ 1 & a^2 & a \\ 1 & a & a^2 \end{bmatrix} \begin{bmatrix} V_{R0} \\ V_{R1} \\ V_{R2} \end{bmatrix} \tag{15.7}$$

and

$$
\begin{bmatrix} I_R \\ I_Y \\ I_B \end{bmatrix} = \begin{bmatrix} 1 & 1 & 1 \\ 1 & a^2 & a \\ 1 & a & a^2 \end{bmatrix} \begin{bmatrix} I_{R0} \\ I_{R1} \\ I_{R2} \end{bmatrix}
\tag{15.8}
$$

Eqs. (15.7) and (15.8) relate the sequence components to the phase components through the transformation matrix.

$$
C = \begin{bmatrix} 1 & 1 & 1 \\ 1 & a^2 & a \\ 1 & a & a^2 \end{bmatrix}
\tag{15.9}
$$

consider the inverse of the transformation matrix C

$$
C^{-1} = \frac{1}{3} \begin{bmatrix} 1 & 1 & 1 \\ 1 & a & a^2 \\ 1 & a^2 & a \end{bmatrix}
\tag{15.10}
$$

Then the sequence components can be obtained from the phase values as

$$
\begin{bmatrix} V_{R0} \\ V_{R1} \\ V_{R2} \end{bmatrix} = \frac{1}{3} \begin{bmatrix} 1 & 1 & 1 \\ 1 & a & a^2 \\ 1 & a^2 & a \end{bmatrix} \begin{bmatrix} V_R \\ V_Y \\ V_B \end{bmatrix}
\tag{15.11}
$$

and

$$
\begin{bmatrix} I_{R0} \\ I_{R1} \\ I_{R2} \end{bmatrix} = \frac{1}{3} \begin{bmatrix} 1 & 1 & 1 \\ 1 & a & a^2 \\ 1 & a^2 & a \end{bmatrix} \begin{bmatrix} I_R \\ I_Y \\ I_B \end{bmatrix}
\tag{15.12}
$$

15.3 POWER IN SEQUENCE COMPONENTS

The total complex power flowing into a three-phase circuit through the lines R, Y, B is

$$
S = P + jQ = \overline{V}\overline{I}^* = \overline{V}_R \overline{I}_R^* + \overline{V}_Y \overline{I}_Y^* + \overline{V}_Z \overline{I}_Z^*
\tag{15.13}
$$

Written in matrix notation

$$
S = \begin{bmatrix} V_R & V_Y & V_B \end{bmatrix} \begin{bmatrix} I_R \\ I_Y \\ I_B \end{bmatrix}^*
\tag{15.14}
$$

$$
= \begin{bmatrix} V_R \\ V_Y \\ V_B \end{bmatrix}^t \begin{bmatrix} I_R \\ I_Y \\ I_B \end{bmatrix}^*
\tag{15.15}
$$

Also

$$
\begin{bmatrix} V_R \\ V_Y \\ V_B \end{bmatrix} = C \begin{bmatrix} V_{R0} \\ V_{R1} \\ V_{R2} \end{bmatrix}
\tag{15.16}
$$

$$\begin{bmatrix} I_R \\ I_Y \\ I_B \end{bmatrix}^* = C^* \begin{bmatrix} I_{R0} \\ I_{R1} \\ I_{R2} \end{bmatrix}^* \tag{15.17}$$

$$\begin{bmatrix} V_R \\ V_Y \\ V_B \end{bmatrix}^t = \begin{bmatrix} V_{R0} \\ V_{R1} \\ V_{R2} \end{bmatrix}^t C^t \tag{15.18}$$

From Eq. (15.14)

$$S = \begin{bmatrix} V_{R0} & V_{R1} & V_{R2} \end{bmatrix} \begin{bmatrix} 1 & 1 & 1 \\ 1 & a^2 & a \\ 1 & a & a^2 \end{bmatrix} \begin{bmatrix} 1 & 1 & 1 \\ 1 & a & a^2 \\ 1 & a^2 & a \end{bmatrix} \begin{bmatrix} I_{R0} \\ I_{R1} \\ I_{R2} \end{bmatrix}^* \tag{15.19}$$

Note that $C^t \, C^* = 3 \, U$

$$S = 3 \begin{bmatrix} V_{R0} & V_{R1} & V_{R2} \end{bmatrix} \begin{bmatrix} I_{R0} \\ I_{R1} \\ I_{R2} \end{bmatrix}^* \tag{15.20}$$

Power in phase components is three times the power in sequence components.

The disadvantage with these symmetrical components is that the transformation matrix C is not power invariant or is not orthogonal or unitary.

15.4 **UNITARY TRANSFORMATION FOR POWER INVARIANCE**

It is more convenient to define "C" as a unitary matrix so that the transformation becomes power invariant.

That is power in phase components = Power in sequence components. Defining a transformation matrix T which is unitary, such that

$$T = \begin{bmatrix} \frac{1}{\sqrt{3}} \end{bmatrix} \begin{bmatrix} 1 & 1 & 1 \\ 1 & a^2 & a \\ 1 & a & a^2 \end{bmatrix} \tag{15.21}$$

$$\begin{bmatrix} V_R \\ V_Y \\ V_B \end{bmatrix} = \begin{bmatrix} \frac{1}{\sqrt{3}} \end{bmatrix} \begin{bmatrix} 1 & 1 & 1 \\ 1 & a^2 & a \\ 1 & a & a^2 \end{bmatrix} \begin{bmatrix} V_{R0} \\ V_{R1} \\ V_{R2} \end{bmatrix} \tag{15.22}$$

and

$$\begin{bmatrix} I_R \\ I_Y \\ I_B \end{bmatrix} = \begin{bmatrix} \frac{1}{\sqrt{3}} \end{bmatrix} \begin{bmatrix} 1 & 1 & 1 \\ 1 & a^2 & a \\ 1 & a & a^2 \end{bmatrix} \begin{bmatrix} I_{R0} \\ I_{R1} \\ I_{R2} \end{bmatrix} \tag{15.23}$$

so that

$$T^{-1} = \begin{bmatrix} \sqrt{3} \end{bmatrix} \begin{bmatrix} 1 & 1 & 1 \\ 1 & a & a^2 \\ 1 & a^2 & a \end{bmatrix} \tag{15.24}$$

$$\begin{bmatrix} V_{R0} \\ V_{R1} \\ V_{R2} \end{bmatrix} = \begin{bmatrix} \dfrac{\sqrt{3}}{3} \end{bmatrix} \begin{bmatrix} 1 & 1 & 1 \\ 1 & a & a^2 \\ 1 & a^2 & a \end{bmatrix} \begin{bmatrix} V_R \\ V_Y \\ V_B \end{bmatrix} \tag{15.25}$$

and

$$\begin{bmatrix} I_{R0} \\ I_{R1} \\ I_{R2} \end{bmatrix} = \begin{bmatrix} \dfrac{\sqrt{3}}{3} \end{bmatrix} \begin{bmatrix} 1 & 1 & 1 \\ 1 & a & a^2 \\ 1 & a^2 & a \end{bmatrix} \begin{bmatrix} I_R \\ I_Y \\ I_B \end{bmatrix} \tag{15.26}$$

$$S = P + jQ = VI^* \tag{15.27}$$

$$= \begin{bmatrix} V_R & V_Y & V_B \end{bmatrix} \begin{bmatrix} I_R \\ I_Y \\ I_B \end{bmatrix}^* \tag{15.28}$$

$$= \begin{bmatrix} V_R \\ V_Y \\ V_B \end{bmatrix}^t \begin{bmatrix} I_R \\ I_Y \\ I_B \end{bmatrix}^* \tag{15.29}$$

$$\begin{bmatrix} V_R \\ V_Y \\ V_B \end{bmatrix} = \begin{bmatrix} \dfrac{1}{\sqrt{3}} \end{bmatrix} \begin{bmatrix} 1 & 1 & 1 \\ 1 & a^2 & a \\ 1 & a & a^2 \end{bmatrix} \begin{bmatrix} V_{R0} \\ V_{R1} \\ V_{R2} \end{bmatrix} \tag{15.30}$$

Taking the conjugate of Eq. (15.26)

$$\begin{bmatrix} I_R \\ I_Y \\ I_B \end{bmatrix}^* = \begin{bmatrix} \dfrac{1}{\sqrt{3}} \end{bmatrix} \begin{bmatrix} 1 & 1 & 1 \\ 1 & a^2 & a \\ 1 & a & a^2 \end{bmatrix} \begin{bmatrix} I_{R0} \\ I_{R1} \\ I_{R2} \end{bmatrix} \tag{15.31}$$

Taking the transpose of Eq. (15.30)

$$\begin{bmatrix} V_R \\ V_Y \\ V_B \end{bmatrix}^t = \begin{bmatrix} V_{R0} \\ V_{R1} \\ V_{R2} \end{bmatrix} \begin{bmatrix} \dfrac{1}{\sqrt{3}} \end{bmatrix} \begin{bmatrix} 1 & 1 & 1 \\ 1 & a^2 & a \\ 1 & a & a^2 \end{bmatrix} \tag{15.32}$$

Substituting the results of (15.31) and (15.32) in (15.29)

$$S = \begin{bmatrix} V_{R0} & V_{R1} & V_{R2} \end{bmatrix} \begin{bmatrix} \dfrac{1}{\sqrt{3}} \end{bmatrix} \begin{bmatrix} 1 & 1 & 1 \\ 1 & a^2 & a \\ 1 & a & a^2 \end{bmatrix} \begin{bmatrix} \dfrac{1}{\sqrt{3}} \end{bmatrix} \begin{bmatrix} 1 & 1 & 1 \\ 1 & a & a^2 \\ 1 & a^2 & a \end{bmatrix} \begin{bmatrix} I_{R0} \\ I_{R1} \\ I_{R2} \end{bmatrix} \tag{15.33}$$

$$= \begin{bmatrix} V_{R0} & V_{R1} & V_{R2} \end{bmatrix} \begin{bmatrix} \dfrac{1}{3} \end{bmatrix} \begin{bmatrix} 1 & 1 & 1 \\ 1 & a^2 & a \\ 1 & a & a^2 \end{bmatrix} \begin{bmatrix} 1 & 1 & 1 \\ 1 & a & a^2 \\ 1 & a^2 & a \end{bmatrix} \begin{bmatrix} I_{R0} \\ I_{R1} \\ I_{R2} \end{bmatrix} \tag{15.34}$$

$$= \begin{bmatrix} V_{R0} & V_{R1} & V_{R2} \end{bmatrix} \begin{bmatrix} \dfrac{1}{3} \end{bmatrix} [3] \begin{bmatrix} I_{R0} \\ I_{R1} \\ I_{R2} \end{bmatrix} \tag{15.35}$$

Therefore

$$S = \begin{bmatrix} V_{R0} & V_{R1} & V_{R2} \end{bmatrix} \begin{bmatrix} I_{R0} \\ I_{R1} \\ I_{R2} \end{bmatrix} \tag{15.36}$$

Thus with the unitary transformation matrix

$$T = \left[\frac{1}{\sqrt{3}}\right] \begin{bmatrix} 1 & 1 & 1 \\ 1 & a^2 & a \\ 1 & a & a^2 \end{bmatrix} \tag{15.37}$$

we obtain power invariant transformation with sequence components.

15.5 SEQUENCE IMPEDANCES

Electrical equipment or components offer impedance to flow of current when potential is applied. The impedance offered to the flow of positive-sequence currents is called "positive-sequence impedance" Z_1. The impedance offered to the flow of negative-sequence currents is called negative-sequence impedance Z_2. When zero-sequence currents flow through components of power system, the impedance offered is called zero-sequence impedance Z_0.

15.6 BALANCED STAR-CONNECTED LOAD

Consider the circuit in Fig. 15.3.

A three-phase balanced load with self- and mutual-impedances Z_s and Z_m drawn currents I_a, I_b, and I_c as shown. Z_n is the impedance in the neutral circuit which is grounded draws and current in the circuit is I_n.

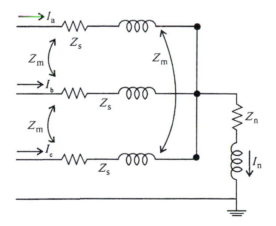

FIGURE 15.3

Three-phase balanced load with mutual impedances.

The line-to-ground voltages are given by

$$\left.\begin{array}{l} V_a = Z_s I_a + Z_m I_b + Z_m I_c + Z_n I_n \\ V_b = Z_m I_a + Z_s I_b + Z_m I_c + Z_n I_n \\ V_c = Z_m I_a + Z_m I_b + Z_s I_c + Z_n I_n \end{array}\right\} \tag{15.38}$$

Since

$$I_a + I_b + I_c = I_n$$

Eliminating I_n from Eq. (15.38)

$$\begin{bmatrix} V_a \\ V_b \\ V_c \end{bmatrix} = \begin{bmatrix} Z_s + Z_n & Z_m + Z_n & Z_m + Z_n \\ Z_m + Z_n & Z_s + Z_n & Z_m + Z_n \\ Z_m + Z_n & Z_m + Z_n & Z_s + Z_n \end{bmatrix} \begin{bmatrix} I_a \\ I_b \\ I_c \end{bmatrix} \tag{15.39}$$

Put in compact matrix notation

$$[V_{abc}] = [Z_{abc}][I_{abc}] \tag{15.40}$$

$$V_{abc} = [A] V_a^{0,1,2} \tag{15.41}$$

and

$$I_{abc} = [A] I_a^{0,1,2} \tag{15.42}$$

Premultiplying Eq. (15.40) by $[A]^{-1}$ and using Eqs. (15.41) and (15.42) we obtain

$$V_a^{0,1,2} = [A]^{-1}[Z_{abc}][A] I_a^{0,1,2} \tag{15.43}$$

Defining

$$[Z]^{0,1,2} = [A^{-1}][Z_{abc}][A] \tag{15.44}$$

$$= \frac{1}{3} \begin{bmatrix} 1 & 1 & 1 \\ 1 & a & a^2 \\ 1 & a^2 & 1 \end{bmatrix} \begin{bmatrix} Z_s + Z_n & Z_m + Z_n & Z_m + Z_n \\ Z_m + Z_n & Z_s + Z_n & Z_m + Z_n \\ Z_m + Z_n & Z_m + Z_n & Z_s + Z_n \end{bmatrix} \begin{bmatrix} 1 & 1 & 1 \\ 1 & a^2 & a \\ 1 & a & a^2 \end{bmatrix}$$

$$= \begin{bmatrix} (Z_s + 3Z_n + 2Z_m) & 0 & 0 \\ 0 & Z_s - Z_m & 0 \\ 0 & 0 & Z_s - Z_m \end{bmatrix} \tag{15.45}$$

If there is no mutual coupling

$$[Z^{0,1,2}] = \begin{bmatrix} Z_s + 3Z_n & 0 & 0 \\ 0 & Z_s & 0 \\ 0 & 0 & Z_s \end{bmatrix} \tag{15.46}$$

From the above, it can be concluded that for a balanced load the three sequences are independent, which means that currents of one sequence flowing will produce voltage drops of the same phase sequence only.

15.7 TRANSMISSION LINES

Transmission lines are static components in a power system. Phase sequence has thus no effect on the impedance. The geometry of the lines is fixed whatever may be the phase sequence. Hence, for transmission lines

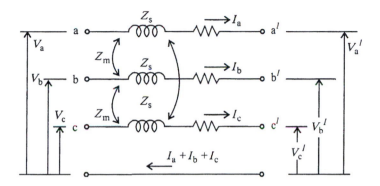

FIGURE 15.4

Three-phase transmission line with mutual impedances.

$$Z_1 = Z_2$$

we can proceed in the same way as for the balanced three-phase load for three-phase transmission lines also (Fig. 15.4).

$$\left.\begin{array}{l} V_a - = Z_s I_a + Z_m I_b + Z_m I_c \\ V_b - = Z_m I_a + Z_s I_b + Z_m I_c \\ V_c - = Z_m I_a + Z_m I_b + Z_s I_c \end{array}\right\} \tag{15.47}$$

$$\begin{bmatrix} V_a - V'_a \\ V_b - V'_b \\ V_c - V'_c \end{bmatrix} = \begin{bmatrix} Z_s & Z_m & Z_m \\ Z_m & Z_s & Z_m \\ Z_m & Z_m & Z_s \end{bmatrix} \begin{bmatrix} I_a \\ I_b \\ I_c \end{bmatrix} \tag{15.48}$$

$$[V_{abc}] = [V_{abc}] - [V_{abc'}] = [Z_{abc}][I_{abc}] \tag{15.49}$$

$$[Z^{0,1,2}] = [A^{-1}][Z_{abc}][A] \tag{15.50}$$

$$= \begin{bmatrix} Z_s + 2Z_m & 0 & 0 \\ 0 & Z_s - Z_m & 0 \\ 0 & 0 & Z_s - Z_m \end{bmatrix} \tag{15.51}$$

The zero-sequence currents are in phase and flow through the line conductors only if a return conductor is provided. The zero-sequence impedance is different from positive- and negative-sequence impedances.

15.8 SEQUENCE IMPEDANCES OF TRANSFORMER

For analysis, the magnetizing branch is neglected and the transformer is represented by an equivalent series leakage impedance.

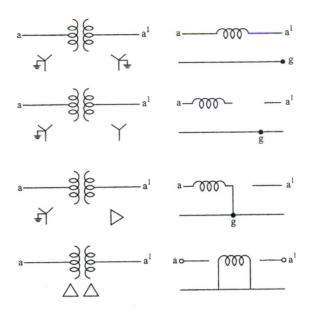

FIGURE 15.5

Zero-sequence equivalent circuits.

Since the transformer is a static device, phase sequence has no effect on the winding reactances. Hence

$$Z_1 = Z_2 = Z_l$$

where Z_l is the leakage impedance.

If zero-sequence currents flow, then

$$Z_0 = Z_1 = Z_2 = Z_l$$

In star—delta or delta—star transformers the positive-sequence line voltage on one side leads the corresponding line voltage on the other side by 30 degrees. It can be proved that the phase shift for the line voltages to be -30 degrees for negative-sequence voltages.

The zero-sequence impedance and the equivalent circuit for zero-sequence currents depends upon the neutral point and its ground connection. The circuit connection for some of the common transformer connection for zero-sequence currents are indicated in Fig. 15.5.

15.9 SEQUENCE REACTANCES OF SYNCHRONOUS MACHINE

The positive-sequence reactance of a synchronous machine may be X_d or X_d' or X_d'' depending upon the condition at which the reactance is calculated with positive-sequence voltages applied.

When negative-sequence currents are impressed on the stator winding, the net flux rotates at twice the synchronous speed relative to the rotor. The negative-sequence reactance is approximately given by

$$X_2 = X_d'' \tag{15.52}$$

The zero-sequence currents, when they flow, are identical and the spatial distribution of the mmfs is sinusoidal. The resultant air-gap flux due to zero-sequence currents is zero. Thus the zero-sequence reactance is approximately the same as the leakage flux

$$X_0 = X_1 \tag{15.53}$$

15.10 **SEQUENCE NETWORKS OF SYNCHRONOUS MACHINES**

Consider an unloaded synchronous generator shown in Fig. 15.6 with a neutral to ground connection through an impedance Z_n. Let a fault occur at its terminals which causes currents I_a, I_b, and I_c to flow through its phase a, b, and c, respectively. The generated phase voltages are E_a, E_b, and E_c. Current I_n flows through the neutral impedance Z_n.

15.10.1 **POSITIVE-SEQUENCE NETWORK**

Since the generator phase windings are identical by design and construction, the generated voltages are perfectly balanced. They are equal in magnitude with a mutual phase shift of 120 degrees. Hence the generated voltages are of positive sequence. Under these conditions a positive-sequence current flows in the generator that can be represented as in Fig. 15.7.

Z_1 is the positive-sequence impedance of the machine and I_{a1} s the positive-sequence current in phase a. The positive-sequence network can be represented for phase "a" as shown in Fig. 15.8.

$$V_{a1} = E_a - I_{a1} \cdot Z_1 \tag{15.54}$$

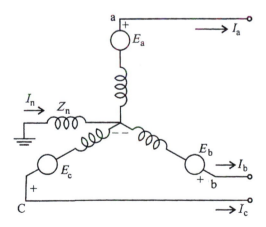

FIGURE 15.6

Unloaded synchronous machine.

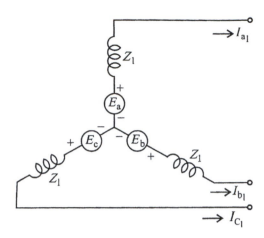

FIGURE 15.7

Positive-sequence currents in three-phase synchronous machine.

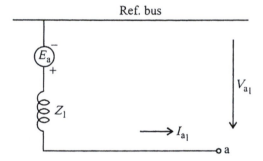

FIGURE 15.8

Positive-sequence network.

15.10.2 NEGATIVE-SEQUENCE NETWORK

Synchronous generator does not produce any negative-sequence voltages. If negative-sequence currents flow through the stator windings, then the mmf produced will rotate at a synchronous speed but in a direction opposite to the rotation of the machine rotor. This causes the negative-sequence mmf to move past the direct and quadrature axes alternately. Then, the negative-sequence mmf sets up a varying armature reaction effect. Hence the negative-sequence reactance is taken as the average of direct-axis and quadrature-axis subtransient reactances.

$$X_2 = \frac{(X_d'' + X_q'')}{2} \tag{15.55}$$

The negative-sequence current paths and the negative sequence network are shown in Fig. 15.9.

$$V_{a2} = -Z_2 I_{a2} \tag{15.56}$$

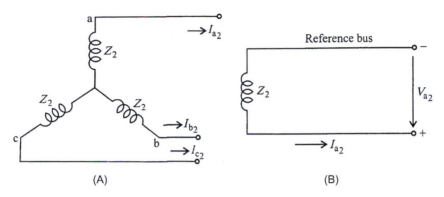

(A)   (B)

FIGURE 15.9

(A) Negative-sequence currents and (B) negative-sequence network.

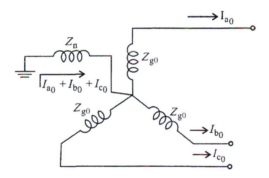

FIGURE 15.10

Zero-sequence currents.

15.10.3 ZERO-SEQUENCE NETWORK

Zero-sequence currents flowing in the stator windings produce mmfs which are in time phase. Sinusoidal space mmf produced by each of the three stator windings at any instant at a point on the axis of the stator would be zero, when the rotor is not present. However, in the actual machine leakage flux will contribute to zero-sequence impedance. Consider the circuit in Fig. 15.10.

Since $I_{a0} = I_{b0} = I_{c0}$

The current flowing through Z_n is $3I_{a0}$.

The zero-sequence voltage drop

$$V_{a0} = -3I_{a0}Z_n - I_{a0}Z_{g0} \tag{15.57}$$

where Z_{g0} is the zero-sequence impedance per phase of the generator.

Hence

$$Z_0^1 = 3Z_n + Z_{g0} \tag{15.58}$$

so that

$$V_{a0} = -I_{a0}Z_0^1 \tag{15.59}$$

The zero-sequence network is shown in Fig. 15.11.

Thus it is possible to represent the sequence networks for a power system differently as different sequence currents flow as summarized in Fig. 15.12.

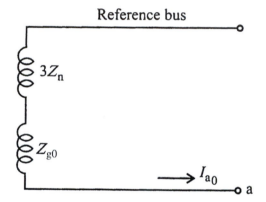

FIGURE 15.11

Zero-sequence network.

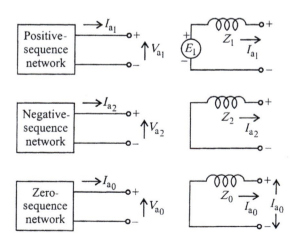

FIGURE 15.12

The three-sequence networks.

15.11 UNSYMMETRICAL FAULTS

The unsymmetrical faults generally considered are

- Line-to-ground fault
- Line-to-line fault
- Line-to-line to ground fault

Single line-to-ground fault is the most common type of fault that occurs in practice. Analysis for system voltages and calculation of fault current under the above conditions of operation is discussed now.

15.12 ASSUMPTIONS FOR SYSTEM REPRESENTATION

1. Power system operates under balanced steady-state conditions before the fault occurs. Therefore the positive-, negative-, and zero-sequence networks are uncoupled before the occurrence of the fault. When an unsymmetrical fault occurs, they get interconnected at the point of fault.
2. Prefault load current at the point of fault is generally neglected. Positive-sequence voltages of all the three phases are equal to the prefault voltage V_F. Prefault bus voltage in the positive-sequence network is V_F.
3. Transformer winding resistances and shunt admittances are neglected.
4. Transmission line series resistances and shunt admittances are neglected.
5. Synchronous machine armature resistance, saliency, and saturation are neglected.
6. All nonrotating impedance loads are neglected.
7. Induction motors are either neglected or represented as synchronous machines.

It is conceptually easier to understand faults at the terminals of an unloaded synchronous generator and obtain results. The same can be extended to a power system and results obtained for faults occurring at any point within the system.

15.13 UNSYMMETRICAL FAULTS ON AN UNLOADED GENERATOR

Single Line to Ground Fault:

Consider Fig. 15.13. Let a line-to-ground fault occur on phase a.

We can write under the fault condition the following relations.

$$V_a = 0$$

$$I_b = 0$$

and

$$I_c = 0$$

It is assumed that there is no fault impedance.

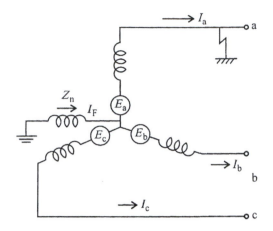

FIGURE 15.13

L-G fault on an unloaded generator.

$$I_F = I_a + I_b + I_c = I_a = 3I_{a1} \tag{15.60}$$

Now

$$\left.\begin{aligned} I_{a1} &= \frac{1}{3}(I_a + aI_b + a^nI_c) \\ I_{a2} &= \frac{1}{3}(I_a + a^nI_b + aI_c) \\ I_{a0} &= \frac{1}{3}(I_a + I_b + I_c) \end{aligned}\right\} \tag{15.61}$$

Substitute Eq. (15.61) into Eqs. (15.60)

$$I_b = I_c = 0 \tag{15.62}$$

$$I_{a1} = I_{a2} = I_{a0} = \frac{1}{3}I_a \tag{15.63}$$

Hence the three-sequence networks carry the same current and hence all can be connected in series as shown in Fig. 15.14 satisfying the relation.

$$V_a = E_a - I_{a1}Z_1 - I_{a2}Z_2 - I_{a0}Z_0 - I_FZ_n \tag{15.63a}$$

Since $V_a = 0$

$$E_a = I_{a1}Z_1 + I_{a2}Z_2 + I_{a0}Z_0 + I_FZ_n$$

$$E_a = I_{a1}Z_1 + I_{a2}Z_2 + I_{a0}Z_0 + 3I_{a1}Z_n$$

$$E_a = I_{a1}[Z_1 + Z_2 + Z_0 + 3Z_n]$$

Hence

$$I_a = \frac{3E_a}{(Z_1 + Z_2 + Z_0 + 3Z_n)} \tag{15.64}$$

The line voltages are now calculated.

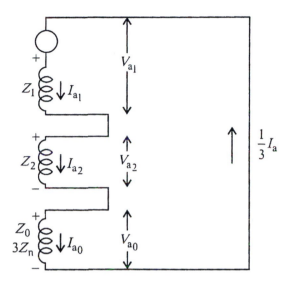

FIGURE 15.14

Sequence network connection for L-G fault.

$$V_a = 0$$

$$V_b = V_{a0} + a^2 V_{a1} + a V_{a2}$$

$$= (-I_{a0}Z_0) + a^2(E_a - I_{a1}Z_1) + a(-I_{a2}Z_2)$$

$$= a^2 E_a - I_{a1}(Z_0 + a^2 Z_1 + a Z_2)$$

Substituting the value of I_{a1}

$$V_b = a^2 E_a - \frac{E_a}{(Z_1 + Z_2 + Z_0)} \cdot (Z_0 + a_r Z_1 + a Z_2)$$

$$= E_a \left[a^2 - \frac{Z_0 + a^2 Z_1 + a Z_2}{Z_0 + Z_1 + Z_2} \right] = E_a \left[\frac{a^2 Z_0 + a^2 Z_1 + a^2 Z_2 - Z_0 - a^2 Z_1 - a Z_2}{Z_0 + Z_1 + Z_2} \right]$$

$$\therefore \quad V_b = E_a \left[\frac{(a^2 - a)Z_2 + (a^2 - 1)Z_0}{Z_0 + Z_1 + Z_2} \right] \tag{15.65}$$

$$V_c = V_{a0} + a V_{a1} + a^r V_{a2}$$

$$= (-I_{a0}Z_0) + a(E_a - I_{a1}Z_1) + a^r(-I_{a2}Z_2)$$

Since

$$I_{a1} = I_{a2} = I_{a0}$$

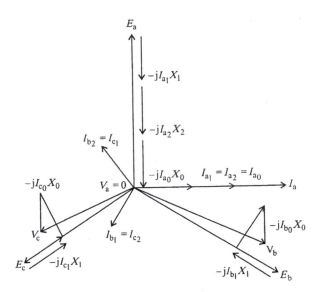

FIGURE 15.15

Phasor diagram for L-G fault.

$$V_c = aE_a - \frac{E_a}{Z_1 + Z_2 + Z_0} \cdot (Z_0 + aZ_1 + a_2 Z_2)$$

$$= E_a \left[a - \frac{(Z_0 + aZ_1 + a^2 Z_2)}{Z_1 + Z_2 + Z_0} \right]$$

$$V_c = E_a \frac{[(a-1)Z_0 + (a - a^2)Z_2]}{Z_0 + Z_1 + Z_2} \tag{15.66}$$

The phasor diagram for single line-to-ground fault is shown in Fig. 15.15.

15.14 LINE-TO-LINE FAULT

Consider a line-to-line fault across phases b and c as shown in Fig. 15.16.
From Fig. 15.16, it is clear that

$$\left. \begin{array}{c} I_a = 0 \\ I_b = -I_c \\ V_b = V_c \end{array} \right\} \tag{15.67}$$

and
Utilizing these relations

$$I_{a1} = \frac{1}{3}(I_a + aI_b + a^2 I_c) = \frac{1}{3}(a^2 - a)I_b$$

$$= j\frac{I_b}{\sqrt{3}} \tag{15.68}$$

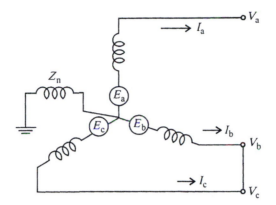

FIGURE 15.16

Line-to-line fault on an unloaded generator.

$$I_{a2} = \frac{1}{3}(I_a + a^2I_b + aI_c) = \frac{1}{3}(a^2 - a)I_b$$

$$= -j\frac{I_b}{\sqrt{3}} \tag{15.69}$$

$$I_{a0} = \frac{1}{3}(I_a + I_b + I_c) = \frac{1}{3}(0 + I_b - I_b) = 0 \tag{15.70}$$

Since $V_b = V_c$

$$a^2V_{a1} + aV_{a2} + V_{a0} = aV_{a1} + a^2V_{a2} + V_{a0}$$

$$(a^2 - a)V_{a1} = (a^2 - a)V_{a2}$$

$$\therefore \quad V_{a1} = V_{a2} \tag{15.71}$$

The sequence network connection is shown in Fig. 15.17.
From the diagram, we obtain

$$E_a - I_{a1}Z_1 = -I_{a2}Z_2$$

$$= I_{a1}Z_2$$

$$E_a = I_{a1}(Z_1 + Z_2)$$

$$I_{a1} = \frac{E_a}{Z_1 + Z_2} \tag{15.72}$$

$$I_b = (a^2 - a)I_{a1} = -j\sqrt{3}I_{a1} \tag{15.73}$$

$$I_c = (a - a^2)I_{a1} = j\sqrt{3}I_{a1} \tag{15.74}$$

Also

$$V_{a1} = \frac{1}{3}(V_a + aV_b + a^2V_c) = \frac{1}{3}\left[V_a + (a + a^2)V_b\right]$$

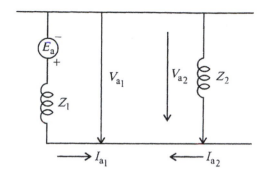

FIGURE 15.17

Sequence network connection for L-L fault.

$$V_{a2} = \frac{1}{3}(V_a + a^2 V_b + a V_c) = \frac{1}{3}\left[V_a + (a^2 + a)V_b\right]$$

Since

$$1 + a + a^2 = 0; \quad a + a^2 = -1$$

Hence

$$V_{a1} = V_{a2} = \frac{1}{3}(V_a + V_b) \tag{15.75}$$

Again

$$I_a = I_{a1} + I_{a2} = I_{a1} - I_{a1} = 0$$

$$I_b = a^2 I_{a1} + a I_{a2} = (a^2 - a)I_{a1}$$

$$= \frac{(a^2 - a)E_a}{Z_1 + Z_2}$$

$$I_c = -I_b = \frac{(a^2 - a)E_a}{Z_1 + Z_2} \tag{15.76}$$

$$V_a = V_{a1} + V_{a2} = E_a - I_{a1}Z_1 + (-I_{a2}Z_2)$$

$$= E_a - \frac{E_a}{Z_1 + Z_2}(Z1 - Z2) = E_a\left[1 - \frac{Z_1 - Z_2}{Z_1 + Z_2}\right] \tag{15.77}$$

$$= E_a\left[\frac{Z_1 + Z_2 - Z_1 + Z_2}{Z_1 + Z_2}\right] = E_a \cdot \frac{2Z_1}{Z_1 + Z_2}$$

$$V_b = a^2 V_{a1} + a V_{a2}$$

$$= a^2[E_a - I_{a1}Z_1] + a(-I_{a2}Z_2)$$

$$= a^2 E_a - I_{a1}[a^2 Z_1 - a Z_2]$$

$$= E_a\left[a^2 - \frac{(a^2 Z_1 - a Z_2)}{Z_1 + Z_2}\right] = E_a\left[\frac{a^n Z_1 + a^2 Z_2 - a^2 Z_1 + a Z_2}{Z_1 + Z_2}\right]$$

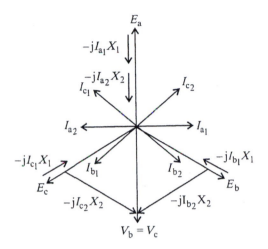

FIGURE 15.18

Phasor diagram for L-L fault.

$$= \frac{E_a Z_2 (a + a^2)}{Z_1 + Z_2} = \frac{E_a(-Z_2)}{Z_1 + Z_2} \tag{15.78}$$

$$V_c = V_b = \frac{E_a(-Z_2)}{(Z_1 + Z_2)} \tag{15.79}$$

The phasor diagram for a double-line fault is shown in Fig. 15.18.

15.15 DOUBLE LINE-TO-GROUND FAULT

Consider line-to-line fault on phases b and c also grounded as shown in Fig. 15.19.
    From Fig. 15.19.

$$\begin{aligned} I_a &= 0 \\ V_b &= V_c = 0 \\ I_b + I_c &= I_F \end{aligned} \tag{15.80}$$

$$\begin{aligned} V_{a1} &= \frac{1}{3}(V_a + aV_b + a^2V_c) \\ &= \frac{1}{3}V_a \end{aligned} \tag{15.81}$$

$$\begin{aligned} V_{a2} &= \frac{1}{3}(V_a + a^2V_b + aV_c) \\ &= \frac{1}{3}V_a \end{aligned} \tag{15.82}$$

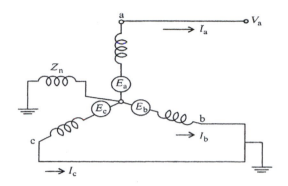

FIGURE 15.19

Double line-to-ground (L-L-G) fault on an unloaded generator.

Further

$$V_{a0} = \frac{1}{3}(V_a + V_b + V_c) = \frac{1}{3}V_a$$

Hence

$$V_{a1} = V_{a2} = V_{a0} = \frac{1}{3}V_a \tag{15.83}$$

But

$$V_{a1} = E_a - I_{a1}Z_1$$

$$V_{a2} = -I_{a2}Z_2$$

and

$$V_{a0} = -I_{a0}Z_0 - I_F Z_n$$

$$= -I_{a0}(Z_0 + 3Z_n) = -I_{a0}(Z_0^1) \tag{15.84}$$

It may be noted that

$$I_F = I_b + I_c = a^2 I_{a1} + a I_{a2} + I_{a0} + a I_{a1} + a^2 I_{a2} + a I_{a0}$$

$$= (a + a^2)I_{a1}(a + a^2)I_{a2} + 2I_{a0}$$

$$= -I_{a1} - I_{a2} + 2I_{a0} = -I_{a1} - I_{a2} - I_{a0} + 3I_{a0}$$

$$= -(I_{a1} + I_{a2} + I_{a0}) + 3I_{a0} = 0 + 3I_{a0} = 3I_{a0}$$

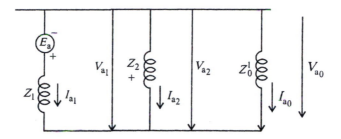

FIGURE 15.20

Sequence network connection for L-L-G fault.

The sequence network connections are shown in Fig. 15.20.

$$I_{a1} = \frac{E_a}{Z_1 + \dfrac{Z_2 Z_0^1}{Z_2 + Z_0^1}} \tag{15.85}$$

$$= \frac{E_a(Z_2 + Z_0^1)}{Z_1 Z_2 + Z_2 Z_0^1 + Z_0^1 Z_1} \tag{15.86}$$

$$V_{a2} = V_{a1}$$

$$-I_{a1}Z_2 = E_a - I_{a1}Z_1$$

$$I_{a2} = -\left(\frac{E_a - I_{a1}Z_1}{Z_2}\right)$$

$$= -\left[E_a - \frac{E_a(Z_2 + Z_0^1)\cdot Z_1}{Z_1 Z_2 + Z_2 Z_0^1 + Z_0 Z_1}\right]\cdot\frac{1}{Z_2} \tag{15.87}$$

$$= \frac{-E_a Z_0^1}{Z_1 Z_2 + Z_2 Z_0^1 + Z_0^1 Z_1}$$

Similarly

$$-I_{a0}Z_0^1 = -I_{a2}Z_2 \tag{15.88}$$

$$I_{a0} = -I_{a2}\frac{Z_2}{Z_0^1} = \frac{-E_a Z_2}{Z_1 Z_2 + Z_2 Z_0 + Z_0 Z_1} \tag{15.89}$$

$$V_a = V_{a1} + V_{a2} + V_{a0}$$

$$= E_a - I_{a1}Z_1 - I_{a2}Z_2 - I_{a0}(Z_0 + 3Z_n)$$

$$= E_a - \frac{E_a(Z_2 + Z_0)}{\Sigma Z_1 Z_2}Z_1 + \frac{E_a Z_0 Z_2}{\Sigma Z_1 Z_2} + \frac{E_a\cdot Z_2(Z_0 + 3Z_n)}{\Sigma Z_1 Z_2} \tag{15.90}$$

$$= E_a\frac{3Z_2 Z_0 + 3Z_2 Z_n}{\Sigma Z_1 Z_2} = 3E_a\left(\frac{Z_2(Z_0 + Z_n)}{Z_1 Z_2 + Z_2 Z_0 + Z_0 Z_1}\right)$$

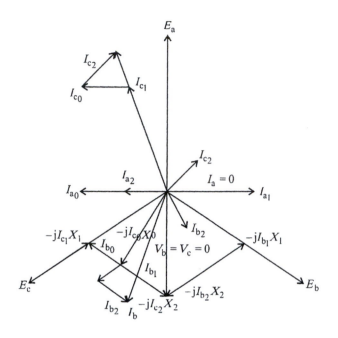

FIGURE 15.21

Phasor diagram for L-L-G fault.

$$V_b = V_{a0} + a^2 V_{a1} + a V_{a2}$$

$$= -I_{a0}(Z_0 + 3Z_n) + a^2[E_a - I_{a1}Z_1] + a[-I_{a2}Z_2]$$

$$= \frac{E_a(Z_2)(Z_0 + 3Z_n)}{\Sigma Z_1 Z_2}(Z_0 + 3Z_n) + a^2\left[E_a - \frac{(E_a Z_2 + Z_0)}{\Sigma Z_1 Z_2}\right] + a\left[\frac{E_0 Z_0 Z_2}{\Sigma Z_1 Z_2}\right]$$

$$= \frac{E_a[Z_2 Z_0 + 3Z_2 Z_n] + a^2 E_a[Z_1 Z_2 + Z_2 Z_0 + Z_0 Z_1 - Z_2 Z_1 - Z_0 Z_1]}{\Sigma Z_1 Z_2 + a E_a Z_0 Z_2}$$

$$V_b = \frac{E_a[Z_0 Z_2 + a^2 Z_0 Z_2 + a Z_0 Z_2 + 3Z_2 Z_n]}{\Sigma Z_1 Z_2}$$

$$= \frac{E_a[Z_0 Z_2(1 + a + a^2) + 3Z_2 Z_n]}{Z_1 Z_2 + Z_2 Z_0 + Z_0 Z_4} = \frac{3 \cdot E_a \cdot Z_2 Z_n}{Z_1 Z_2 + Z_2 Z_0 + Z_0 Z_1} \qquad (15.91)$$

If $Z_n = 0$; $V_b = 0$

The phasor diagram for this fault is shown in Fig. 15.21.

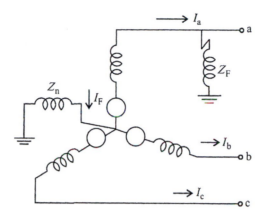

FIGURE 15.22

L-G fault with fault impedance.

15.16 SINGLE-LINE TO GROUND FAULT WITH FAULT IMPEDANCE

If in (15.13) the fault is not a dead short circuit but has an impedance Z_F, then the fault is represented in Fig. 15.22. Eq. (15.63a) will be modified into

$$V_a = E_a - J_{a1}Z_1 - J_{a2}Z_2 - I_{a0}Z_0 - I_F Z_n - I_F Z_F \tag{15.92}$$

Substituting $V_a = 0$ and solving for I_a

$$I_a = \frac{3E_a}{Z_1 + Z_2 + Z_0 + 3(Z_n + Z_F)} \tag{15.93}$$

15.17 LINE-TO-LINE FAULT WITH FAULT IMPEDENCE

Consider the circuit in Fig. (15.23) when the fault across the phases b and c has an impedance Z_F.

$$I_a = 0 \tag{15.94}$$

and

$$I_b = -I_c$$
$$V_b - V_c = Z_F I_b \tag{15.95}$$

$$(V_0 + a^2 V_1 + a V_2) - (V_0 + a V_1 + a^2 V_2) = Z_F(I_0 + a^2 I_1 + a J_2) \tag{15.96}$$

Substituting Eqs. (15.95) and (15.96) in the below equation

$$(a^2 - a)V_1 - (a^2 - a)V_2 = Z_F(a^2 - a)I_1$$

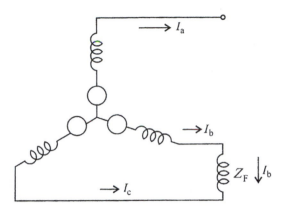

FIGURE 15.23

L-L fault with fault impedance.

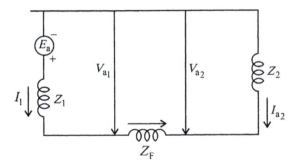

FIGURE 15.24

Sequence network connection L-L fault with impedance.

i.e.,

$$V_1 - V_2 = Z_F I_1 \tag{15.97}$$

The sequence network connection in this case will be as shown in Fig. 15.24.

15.18 DOUBLE LINE-TO-GROUND FAULT WITH FAULT IMPEDENCE

This can is illustrated in Fig. 15.25.

The representative equations are

$$\begin{aligned}
I_a &= 0 \\
V_b &= V_c \\
V_b &= (I_b + I_c)(Z_F + Z_n)
\end{aligned} \tag{15.98}$$

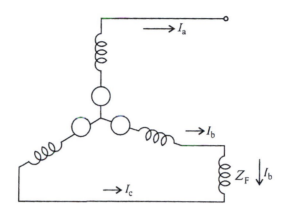

FIGURE 15.25

L-L-G fault with fault impedance.

But

$$I_0 + I_1 + I_2 = 0$$

and also

$$V_0 + aV_1 + a^2V_2 = V_a + a^2V_1 + aV_2$$

So that

$$(a^2 - a)V_1 = (a^2 - a)V_2$$

or

$$V_1 = V_2 \qquad (15.99)$$

Further,

$$(V_0 + a^2V_1 + aV_2) = (I_0 + a^2I_1 + aI^2 + I_0 + aJ_1 + a^2I_2)Z_F + Z_n)$$

Since

$$a^2 + a = -1$$

$$(V_0 - V_1) = (Z_n + Z_F)[2I_0 - I_1 - I_2)$$

But since

$$I_0 = -I_1 - I_2$$

$$V_0 - V_1 = (Z_n + Z_F)(2I_0 + I_0) = 3(Z_F + Z_n) \cdot I_0 \qquad (15.100)$$

Hence the fault conditions are given by

$$I_0 + I_1 + I_2 = 0$$

$$V_1 = V_2$$

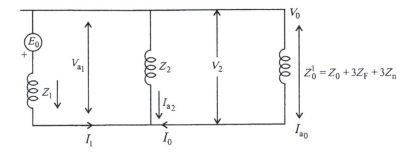

FIGURE 15.26

Sequence network connection for L-L-G fault with fault impedance.

and

$$V_0 - V_1 = 3(Z_F + Z_n) \cdot I_0$$

$$I_{a1} = \cfrac{E_a}{Z_1 + \cfrac{Z_0^1 Z_2}{Z_0^1 + Z_2}} \text{ and so on as in case (15.15)} \tag{15.101}$$

where

$$Z_0^1 = Z_0 + 3Z_F + 3Z_n$$

The sequence network connections are shown in Fig. 15.26.

WORKED EXAMPLES

E.15.1 Calculate the sequence components of the following balanced line-to-network voltages.

$$\bar{V} = \begin{bmatrix} V_{an} \\ V_{bn} \\ V_{cn} \end{bmatrix} = \begin{bmatrix} 220 & \underline{/0 \text{ degree}} \\ 220 & \underline{/-120 \text{ degrees}} \\ 220 & \underline{/+120 \text{ degrees}} \end{bmatrix} \text{kV}$$

Solution:

$$V_0 = \frac{1}{3}(V_{an} + V_{bn} + V_{cn})$$

$$= \frac{1}{3}[200 \underline{/0 \text{ degree}} + 200 \underline{/-120 \text{ degrees}} + 220 \underline{/+120 \text{ degrees}}]$$

$$= 0$$

$$V_1 = \frac{1}{3}[V_{an} + aV_{bn} + a^2V_{cn}]$$

$$= \frac{1}{3}[220\underline{/0 \text{ degree}} + 220\underline{/(-120 \text{ degrees} + 120 \text{ degrees})}$$

$$+ 220\underline{/(120 \text{ degrees} + 240 \text{ degrees})}]$$

$$= 220\underline{0 \text{ degree}} \text{ kV}$$

$$V_2 = \frac{1}{3}[V_{an} + a^2V_{bn} + a^2V_{cn}]$$

$$= \frac{1}{3}[220\underline{/0 \text{ degree}} + 220\underline{/- 120 \text{ degrees} + 240 \text{ degrees}}$$

$$+ 220\underline{/120 \text{ degrees} + 120 \text{ deg rees}}$$

$$= [220 + 220\underline{120 \text{ degrees}} + 220\underline{240 \text{ degrees}}]$$

$$= 0$$

Note: Balanced three-phase voltages do not contain negative-sequence components.

E.15.2 Prove that neutral current can flow only if zero-sequence currents are present

Solution:

$$I_a = I_{a1} + I_{a2} + I_{a0}$$
$$I_b = a^2I_{a1} + aI_{a2} + I_{a0}$$
$$I_c = aI_{a1} + a^2I_{a2} + I_{a0}$$

If zero-sequence currents are not present
then $I_{a0} = 0$
In that case

$$I_a + I_b + I_c = I_{a1} + I_{a2} + a^2I_{a1} + aI_{a2} + aI_{a1} + a^2I_{a2}$$

$$= (I_{a1} + aI_{a1} + a^2I_{a1}) + (I_{a2} + a^2I_{a2} + aI_{a2})$$

$$= 0 + 0 = 0$$

The neutral current $I_n = I_R = I_Y + I_B = 0$. Hence neutral currents will flow only in case of zero-sequence components of currents exist in the network.

E.15.3 Given the negative-sequence currents

$$\bar{I} = \begin{bmatrix} I_a \\ I_b \\ I_c \end{bmatrix} = \begin{bmatrix} 100 & 0 \text{ degree} \\ 100 & 120 \text{ degrees} \\ 100 & -120 \text{ degrees} \end{bmatrix}$$

Obtain their sequence components

Solution:

$$I_0 = \frac{1}{3}[I_a + I_b + I_c]$$

$$= \frac{1}{3}[200\underline{/0 \text{ degree}} + 100\underline{/120 \text{ degrees}} + 100\underline{/- 120 \text{ degree}} = 0 \text{ A}$$

$$I_1 = \frac{1}{3}[I_a + aI_b + a^r I_c]$$

$$= \frac{1}{3}[100\underline{/0\,\text{degree}} + 100\underline{/-120\,\text{degrees} + 120\,\text{degrees}} + 100\underline{/-120\,\text{degrees} + 240\,\text{degrees}}]$$

$$= [100\underline{/0\,\text{degree}} + 100\underline{/240\,\text{degrees}} + 100\underline{/-120\,\text{degrees}}]$$

$$= 0\,\text{A}$$

$$I_2 = \frac{1}{3}[I_a + a^2 I_b + aI_c]$$

$$= \frac{1}{3}[100\underline{/0\,\text{degree}} + 100\underline{/120\,\text{degrees} + 240\,\text{degrees}} + 100\underline{/-120\,\text{degrees} + 240\,\text{degrees}}]$$

$$= [100\underline{/0\,\text{degree}} + 100\underline{/0\,\text{degree}} + 100\underline{/0\,\text{degree}}]$$

$$= 100\,\text{A}$$

Note: Balanced currents of any sequence, positive or negative do not contain currents of the other sequences.

E.15.4 Find the symmetrical components for the given three-phase currents.

$$I_a = 10\underline{/0\,\text{degree}}$$
$$I_b = 10\underline{/-90\,\text{degrees}}$$
$$I_c = 15\underline{/135\,\text{degrees}}$$

Solution:

$$I_0 = \frac{1}{3}[10\underline{0\,\text{degree}} + 10 - 90\,\text{degrees} + 15\,135\,\text{degrees}]$$

$$= \frac{1}{3}[10(1 + j0.0) + 10(0 - j1.0) + 15(-0.707 + j0.707)$$

$$= \frac{1}{3}[10 - j10 - 10.605 + j10.605]$$

$$= \frac{1}{3}[-0.605 + j0.605] = \frac{1}{3}[0.8555]\underline{/135\,\text{degrees}}$$

$$= 0.285\underline{/135\,\text{degrees}}\,\text{A}$$

$$I_1 = \frac{1}{3}[10\underline{/0\,\text{degree}} + 10\,\underline{-/90 + 120\,\text{degree}} + 15\underline{/135\,\text{degrees} + 240\,\text{degrees}}]$$

$$= \frac{1}{3}[10(1 + j0.0) + 10\underline{/30\,\text{degrees}} + 15\underline{/15\,\text{degrees}}$$

$$= \frac{1}{3}[10 + 10(0.866 + j0.5) + 15(0.9659 + j0.2588]$$

$$= \frac{1}{3}[33.1485 + j8.849] = \frac{1}{3}[34.309298]\underline{/15\,\text{degrees}}$$

$$= 11.436\underline{/15\,\text{degrees}}\,\text{A}$$

$$I_2 = \frac{1}{3}[10\underline{/0\,\text{degree}} + 10\underline{/240\,\text{degrees} - 90\,\text{degrees}} + 15\underline{/135\,\text{degrees} + 120\,\text{degrees}}]$$

$$= \frac{1}{3}[10(1 + j0) + 10(-0.866 + j0.5)15(-0.2588 - j0.9659)]$$

$$= \frac{1}{3}[-2.542 - j9.4808]$$

$$= 3.2744\underline{/105\,\text{degrees}}\,\text{A}$$

E.15.5 In a fault study problem the following currents are measured

$$I_R = 0$$
$$I_Y = 10\,\text{A}$$
$$I_B = -10\,\text{A}$$

Find the symmetrical components
Solution:

$$I_{R1} = \frac{1}{3}\left[I_R + aI_Y + a^2I_B\right]$$

$$= \frac{1}{3}\left[0 - a(10) + a^2(-10)\right] = \frac{10}{\sqrt{3}}\text{A}$$

$$I_{R2} = \frac{1}{3}\left[I_R + a^2I_Y + aI_B\right]$$

$$= \frac{1}{3}(a^2 \cdot 10) + a(-10) = -\frac{10}{\sqrt{3}}\,\text{A}$$

$$I_{R0} = \frac{1}{3}(I_R + I_Y + I_B)$$

$$= \frac{1}{3}(10 - 10) = 0$$

E.15.6 Draw the zero-sequence network for the system shown in Fig. E.15.6.

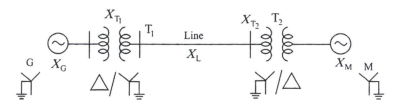

FIGURE E.15.6

Solution:
The zero-sequence network is shown in Fig. E.15.6A

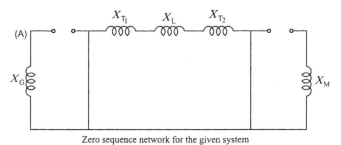

Zero sequence network for the given system

FIGURE E.15.6(A)

E.15.7 Draw the sequence networks for the system shown in Fig. E.15.7.

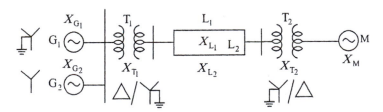

FIGURE E.15.7

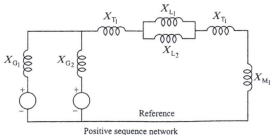

Positive sequence network

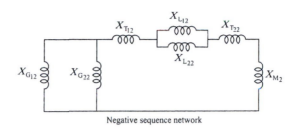

Negative sequence network

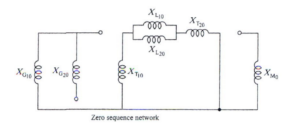

Zero sequence network

E.15.8 Consider the system shown in Fig. E.15.8. Phase b is open due to conductor break. Calculate the sequence currents and the neutral current.

$$I_a = 100 \underline{/0 \, \text{degree}} \, \text{A}$$
$$I_b = 100 \underline{/120 \, \text{degrees}} \, \text{A}$$

Solution:

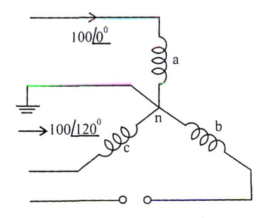

FIGURE E.15.8

$$\bar{I} = \begin{bmatrix} I_a \\ I_b \\ I_c \end{bmatrix} = \begin{bmatrix} 100 \underline{/0\,\text{degree}} \\ 0 \\ 100 \underline{/120\,\text{degrees}} \end{bmatrix} \text{A}$$

$$I_0 = \frac{1}{3}[100\underline{/0\,\text{degree}} + 0 + 100\underline{/120\,\text{degree}}]$$

$$= \frac{1}{3}[100(1 + j0) + 0 + 100(-0.5 + j0.866)$$

$$= \frac{100}{3}[0.5 + j0.866] - 33.3\underline{/60\,\text{degrees}}\,\text{A}$$

$$I_1 = \frac{1}{3}[100\underline{/0\,\text{degree}} + 0 + 100\underline{/120\,\text{degree} + 240\,\text{degree}}]$$

$$= \frac{1}{3}[100\underline{/0\,\text{degree}} + 100\underline{/0\,\text{degree}}] = \frac{200}{3} = 66.66\,\text{A}$$

$$I_2 = \frac{1}{3}[100\underline{/0\,\text{degree}} + 0 + 100\underline{/120\,\text{degrees} + 120\,\text{degrees}}]$$

$$= \frac{1}{3}[100[1 + j0 - 0.5 - j0.866]$$

$$= \frac{100}{3}[-0.5 - j0.866] - 33.33\underline{/-60\,\text{degrees}}\,\text{A}$$

Neutral current

$$\begin{aligned} I_n &= I_0 + I_1 + I_2 \\ &= 100\underline{/0\,\text{degree}} + 0 + 100\underline{/120\,\text{degree}} \\ &= 100[1 + j0 - 0.5 + j0.866] \\ &= 100\underline{/60\,\text{degree}}\,\text{A} \end{aligned}$$

Also

$$I_n = 3I_0 = 3(33.33(60\,\text{degree})) = 100\underline{/60\,\text{degree}}\,\text{A}$$

E.15.9 Calculate the subtransient fault current in each phase for a dead short circuit on one phase to ground at bus "q" for the system shown in Fig. E.15.9.

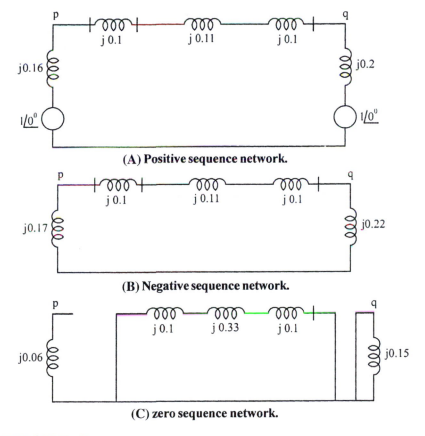

FIGURE E.15.9

All the reactances are given in p.u. on the generator base.
Solution:

(A) Positive sequence network.

(B) Negative sequence network.

(C) zero sequence network.

FIGURE E.15.9 (A–C)

The three-sequence networks are shown in Fig. E.15.9A–C. For a line-to-ground fault, a phase a, the sequence networks are connected as in Fig. E.15.9D at bus "q."

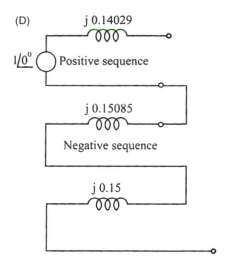

FIGURE E.15.9 (D)

The equivalent positive-sequence network reactance X_p is given form Fig. E.15.9A

$$\frac{1}{X_p} = \frac{1}{0.47} + \frac{1}{0.2}$$

$$X_p = 0.14029$$

The equivalent negative-sequence reactance X_n is given from Fig. E.15.9B

$$\frac{1}{X_n} = \frac{1}{0.48} + \frac{1}{0.22} \quad \text{or } X_n = 0.01508$$

The zero-sequence network impedence is j0.15 the connection of the three-sequence networks is shown in Fig. E.15.9D.

$$I_0 = I_1 = I_2 = \frac{1\underline{/0\,\text{degree}}}{j0.14029 + j0.150857 + j0.15}$$

$$= \frac{1\underline{/0\,\text{degree}}}{j0.44147} = -j2.2668\,\text{p.u.}$$

E.15.10 In the system given in example (E.15.9) if a line-to-line fault occurs calculate the sequence components of the fault current.

Solution:

The sequence network connection for a line-to-line fault is shown in Fig. E.15.10.

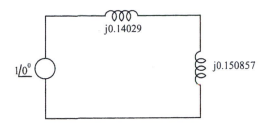

FIGURE E.15.10

From the figure

$$I_1 = I_2 = \frac{1\underline{/0}\,\text{degree}}{j0.1409 + j0.150857 + j0.15} + \frac{1\underline{/0}\,\text{degree}}{j0.291147}$$
$$= -j3.43469\,\text{p.u.}$$

E.15.11 If the line-to-line fault in example E.15.9 takes place involving ground with no fault impedance, determine the sequence components of the fault current and the neutral fault current.

Solution:

The sequence network connection is shown in Fig. E.15.11

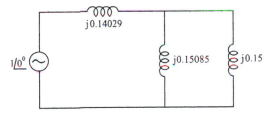

FIGURE E.15.11

$$I_1 = \frac{1\underline{/0}\,\text{degree}}{j0.14029 + \dfrac{j(0.150857)(j0.15)}{j0.150857 + j0.15}}$$
$$= \frac{1\underline{/0}\,\text{degree}}{j0.14029 + j0.0752134} = \frac{1\underline{/0}\,\text{degree}}{j0.2155034}$$
$$= -j4.64\,\text{p.u.}$$

$$I_2 = -j(4.64)\left(\frac{j0.15}{j0.300857}\right) = -j2.31339 \text{ p.u}$$

$$I_0 = -j(4.64)\left(\frac{j0.150857}{j0.300857}\right) = -j2.326608 \text{ p.u}$$

The neutral fault current $= 3j_0 = 3(-j2.326608) = -j6.91498$ p.u.

E.15.12 A dead earth fault occurs on one conductor of a three-phase cable supplied by a 5000-kVA, three-phase generator with earthed neutral. The sequence impedences of the alternator are given by

$$Z_1 = (0.4 + j4) \ \Omega; \ Z_2 = (0.3 + j0.6) \ \Omega \text{ and}$$
$$Z_0 = (0 + j0.45) \ \Omega \text{ per phase}$$

The sequence impedance of the line up to the point of fault are $(0.2 + j0.3) \ \Omega$, $(0.2 + j0.3)W$, $(0.2 + j0.3)\Omega$, and $(3 + j1) \ \Omega$. Find the fault current and the sequence components of the fault current. Also find the line-to-earth voltages on the infaulted lines. The generator line voltage is 6.6 kV.

Solution:

Total positive-sequence impedance is $Z_1 = (0.4 + j4) + (0.2 + j0.3) = (0.6 + j4.3) \ \Omega$
Total negative-sequence impedence to fault is $Z_0 = (0.3 + j0.6) + (0.2 + j0.3) = (0.5 + j0.9) \ \Omega$
Total zero-sequence impedence to fault is

$$Z_0 = (0 + j0.45) + (3 + j1.0) = (3 + j1.45) \ \Omega$$
$$Z_1 + Z_2 + Z_3 = (0.6 + j4.3) + (0.5 + j0.9) + (3.0 + j1.45)$$
$$= (4.1 + j6.65) \ \Omega$$

$$
\begin{aligned}
I_{a1} = I_{a0} = I_{a2} &= \frac{6.6 \times 1000}{\sqrt{3}} - \frac{1}{(4.1 + j6.65)} = \frac{3810.62A}{7.81233} \\
&= 487.77 - 58°.344 \text{ A} \\
&= (255.98 - j415.198)A \\
I_a &= 3 \times 487.77 \ \underline{/-\ 58°.344} \\
&= 1463.31 \text{ A} \ \underline{/-\ 58°.344}
\end{aligned}
$$

E.15.13 A 20-MVA, 6.6-kV star-connected generator has positive-, negative-, and zero-sequence reactances of 30%, 25%, and 7%, respectively. A reactor with 5% reactance based on the rating of the generator is placed in the neutral to ground connection. A line-to-line fault occurs at the terminals of the generator when it is operating at a rated voltage. Find the initial symmetrical line-to-ground rms fault current. Find also the line-to-line voltage.

Solution:

$$Z_1 = j0.3; \ Z_2 = j0.25$$
$$Z_0 = j0.07 + 3 \times j0.05 = j0.22$$

$$I_{a1} = I_{a1} = \frac{1\underline{/0\,\text{degree}}}{j(0.3) + j(0.25)} = \frac{1}{j0.55} = -j1.818\,\text{p.u.}$$

$$= -j1.818 \times \frac{20 \times 1000}{\sqrt{3} \times 6.6} = -j3180\,\text{A}$$

$$= -I_{a1}$$

$I_{a0} = 0$ as there is no ground path

$$V_a = E_a - I_{a1}Z_1 - I_{a2}Z_2$$

$$1 = -j1.818(j0.0.3 - j0.25)$$

$$= 0.9091 \times 3180 = 2890.9\ \text{V}$$

$$V_b = a^2E - (a^2I_{a1}Z_1 + aI_{a2}Z_2)$$

$$= (-0.5 - j0.866).1 + j\sqrt{3}(-j1.818)(j0.3)$$

$$= (-j0.866 - 0.5 + j0.94463)$$

$$= (-0.5 + j0.078\,6328) \times 3180$$

$$= (-1590 + j250) = 1921.63$$

$$V_c = V_b = 1921.63\ \text{V}$$

E.15.14 A balanced three-phase load with an impedance of $(6\text{-}j8)$ ohm per phase, connected in star is having in parallel a delta-connected capacitor bank with each phase reactance of $27\ \Omega$. The star point is connected to ground through an impedance of $0 + j5\ \Omega$. Calculate the sequence impedence of the load.

Solution:

The load is shown in Fig. E.15.14.

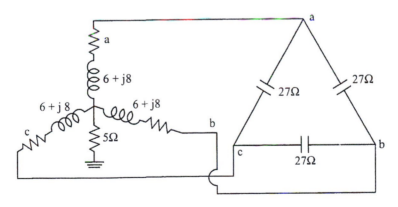

FIGURE E.15.14

Converting the delta-connected capacitor tank into star

$$C_A/\text{phase} - 27\ \Omega$$
$$C_Y/\text{phase} = \frac{1}{3}27 = a\ \Omega$$

The positive-sequence network is shown in Fig. E.15.14A

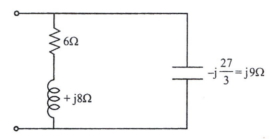

FIGURE E.15.14(A)

The negative-sequence network is also the same as the positive-sequence network.

$$Z_1 = Z_2 = Z_{\text{star}} \parallel \text{delta}$$
$$= \frac{(6 + j8)(-j9)}{6 + j8 - j9} = \frac{72 - j54}{6 - j1} = \frac{90\underline{/36.87^\circ}}{6.082\underline{/9^\circ.46}}$$
$$= 14.7977\,27^\circ 41\,\text{ohm}$$

The zero-sequence network is shown in Fig. E.15.14B

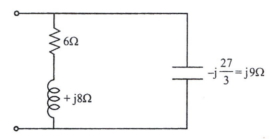

FIGURE E.15.14(B)

$$Z_0 = Z_{\text{star}} + 3Z_n = 6 + j8 + 3(j5)$$
$$= (6 + j23)\text{ohm} = 23.77\ 80^\circ.53$$

PROBLEMS

P.15.1 Determine the symmetrical components for the three-phase currents

$$I_R = 15\angle 0 \text{ degree A}, I_Y = 15\underline{/230 \text{ degrees}} \text{ A}, \text{and } I_B = 15\underline{/130 \text{ degrees}} \text{ A}$$

P.15.2 The voltages at the terminals of a balanced load consisting of three 12-Ω resistors connected in star are

$$V_{RY} = 120\angle 0 \text{ degree V}$$
$$V_{YB} = 96.96\angle - 121.44 \text{ degrees V}$$
$$V_{BR} = 108\angle 130 \text{ degrees V}$$

Assuming that there is no connection to the neutral of the load determine the symmetrical components of the line currents and also from them the line currents.

P.15.3 A 50-Hz turbo generator is rated at 500 MVA 25 kV. It is star connected and solidly grounded. It is operating at rated voltage and is on no-load. Its reactances are $x_d'' = x_1 = x_2 = 0.17$ and $x_0 = 0.06$ p.u. Find the subtransient line current for a single line to ground fault when it is disconnected from the system.

P.15.4 Find the subtransient line current for a line-to-line fault on two phases for the generator in problem (15.3)

P.15.5 A 125-MVA, 22-kV turbo generator having $x_d'' = x_1 = x_2 = 22\%$ and $x_0 = 6\%$ has a current-limiting reactor of 0.16 Ω in the neutral, while it is operating on no-load at rated voltage a double line-to-ground fault occurs on two phases. Find the initial symmetrical rms fault current to the ground.

QUESTIONS

Q.15.1 What are symmetrical components? Explain.

Q.15.2 What is the utility of symmetrical components.

Q.15.3 Derive an expression for power in a three-phase circuit in terms of symmetrical components.

Q.15.4 What are sequence impedances? Obtain an expression for sequence impedances in a balanced static three-phase circuit.

Q.15.5 What is the influence of transformer connections in single-phase transformers connected for three-phase operation?

Q.15.6 Explain the sequence networks for synchronous generator.

Q.15.7 Derive an expression for the fault current for a single line-to-ground fault on an unloaded generator.

Q.15.8 Derive an expression for the fault current for a double-line fault on an unloaded generator.

Q.15.9 Derive an expression for the fault current for a double line-to-ground fault on an unloaded generator.

Q.15.10 Draw the sequence network connections for single line-to-ground fault, double-line fault, and double line-to-ground fault conditions.

Q.15.11 Draw the phasor diagrams for
   1. Single-line-to-ground fault
   2. Double-line fault, and
   3. Double line-to-ground fault
   Conditions on unloaded generator.

Q.15.12 Explain the effect of prefault currents.

Q.15.13 What is the effect of fault impedance? Explain.

CIRCUIT BREAKERS

A circuit breaker (CB) is an automatically operated electrical switch designed to protect electrical power system circuits from damage due to either overloads or faults. It isolates the faulty part by opening its contacts which are normally closed. The opening of the breaker contacts is initiated by a relay which senses the fault or overload. Unlike a fuse which operates once and then has to be replaced, a CB can be reset either manually or automatically to resume normal operation. CBs are designed to protect individual household supply circuits to large switchgear designed to protect high-voltage circuits dealing with bulk power supply. The CB, in general contains a fixed contact and a moving contact. The contacts are enclosed in an insulating medium. When the contacts start opening by separating out, the high current flowing through them establishes an arc across the gap between the two contacts. Various techniques are used to interrupt the arc.

16.1 PRINCIPLE OF ARC EXTINCTION

The voltage drop across the arc is given empirically by

$$V_{arc} = A + \frac{B}{\sqrt{i}} \tag{16.1}$$

where A and B depend upon the length of the arc and i is the arc current. Fig. 16.1A illustrates the basic principle of CB. Fig. 16.1B shows the arc current and voltage waveforms.

Once a fault is detected, the CB contacts must open to interrupt the circuit. For this purpose some mechanically stored energy within the breaker is used, such as springs or compressed air. Small CBs are manually operated for resetting. Larger units have solenoids to trip the mechanism and electric motors to restore energy to the springs. The CB contacts should be able to carry the normal load currents without much heating. They should also withstand the heat produced by the arc while interrupting the arc. Generally copper or copper alloy contacts are used. Silver alloys and other patented materials are also used for contacts. In case of copper a nonconducting layer forms on the contact surface due to arcing and this must be wiped out. For this reason a layer of silver coating is given in many applications.

16.2 TECHNIQUES FOR ARC EXTINCTION

There are several techniques available and are used to extinguish the arc that forms between the two contacts of a CB. They are:

Electrical Power Systems. DOI: http://dx.doi.org/10.1016/B978-0-08-101124-9.00016-4

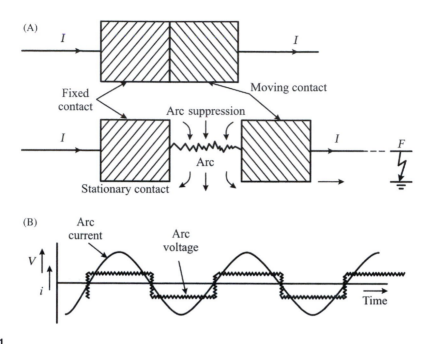

FIGURE 16.1

AC circuit breaker operation. (A) Basis of circuit breakers and (B) arc current and voltage waveforms. *AC*, alternating current.

1. Lengthening of the arc
2. Intensive cooling
3. Division into several arcs
4. Zero current quenching
5. Use of vacuum
6. Use of compressed air
7. Use of mineral oil
8. Use of special medium such as SF_6
9. Connecting capacitors in parallel with contacts in DC circuits

Application of several of these techniques will be explained later when different types of CBs are discussed.

The contacts are to be properly insulated from all metal parts including the body. For this cast epoxy resins, PVC, polystyrene, polycarbonates, and ceramics are used.

16.3 FORMATION AND MAINTENANCE OF ARC

When the arc is formed, the space between the contacts is filled with a high-temperature plasma (according to Dr. Langmuir). Plasma is ionized gas. It contains positively charged molecules of the gas and almost an equal number of electrons. In some power circuits the temperature of the plasma could reach even 50,000°K.

Ionization can occur when molecules of gas receive sufficient energy by external means so that some of the electrons leave the molecules. The molecules then become ions. This can happen due to (1) intense potential gradient applied to the molecules or (2) through supply of thermal energy or by both.

Since the contacts carry large currents, even though the arc resistance is too small, the I^2R loss is substantial and due to this thermal energy thermal emission of electrons occurs. The ions thus formed collide with neutral molecules and further ionization by collision occurs. Thus the main factors that contribute to the maintenance of the arc are:

1. the potential gradient across the opening contacts,
2. thermal energy supplied to molecules which depends upon the current in the circuit, and
3. the mean-free path for ionization by collision which depends upon the pressure applied or existing.

Since arc is maintained by field emission and thermal emission, the arc will be quenched if the rate of ionization due to these two factors is less than the rate of buildup of dielectric strength in the space between the contacts due to recombination of electrons and ions and the rate at which the products of ionization are taken out and replaced by neutral molecules.

16.4 ARC INTERRUPTION BY HIGH RESISTANCE

By increasing the resistance of the arc circuit, the arc can be interrupted. In case of alternating current, this can happen at a current zero. Arc resistance can be increased by (1) cooling (taking away the thermal energy and thereby reducing ionization by thermal emission), (2) lengthening ($R = \rho l/a$), (3) constraining (the arc to a narrow path), and (4) splitting the arc. All these modes can be utilized while designing a CB.

Whenever an arc is interrupted, the electromagnetic energy associated with it ($1/2Li^2$) will be converted into electrostatic energy ($1/2CV^2$) and thus a high voltage appears across the breaker contacts $\left(V = \sqrt{L/C}\right)$. As the separation between the contacts increases, if the breakdown strength of the gap is less than the voltage appearing across the contacts, the arc will reappear. The buildup of dielectric strength across the contacts should be faster through recombination of ions aided by cooling process and replacing ionized particles by neutral molecules with the help of blast of air, oil, SF_6, etc. This is called *recovery rate theory*. The voltage that appears across the gap is called restriking voltage. The rate of rise of restriking voltage (RRRV) should be less than the rate of buildup of dielectric strength across the contact separation for the arc to extinguish.

In a slightly modified from the extinction of arc is explained by *energy-balance theory*.

In AC circuits, at the first current zero after the contacts separate out, there is a little postzero resistance and the power is zero. After the arc is extinguished completely at a current zero, power is again zero. In between these two zeros, the power gradually increases and reaches a maximum and then reduces to zero. The energy generated in the space between contacts due to restriking voltage should be less than the rate at which heat thus produced is taken away from the space by cooling and other processes. Then the arc will be interrupted, completely.

16.5 **TRANSIENT RESTRIKING AND RECOVERY VOLTAGES**

In case of AC circuit interruption, the arc gets quenched at one of the current zeros. The voltage rises as soon as the arc is interrupted. This is the recovery voltage. At this instant the dielectric strength of the gas is improving. However, the restriking voltage is rising and if its rate of rise is more than the buildup of dielectric strength across the gap the arc may strike again. The restriking voltage is a transient voltage appearing across the gap for a very short duration of about 1/10,000 s containing high-frequency harmonics. The recovery voltage is the voltage that appears across the terminals of the pole of a CB after the fault is interrupted. It appears across that breaker pole which clears the fault first. The power frequency voltage that appears across the breaker contacts after the transient oscillation die is the recovery voltage. By recovery voltage we also mean its rms value. Fig. 16.2 illustrates the current and voltage waveform across the contact while the contacts open to interrupt the arc. In Fig. 16.3 the transient restriking and recovery voltages are clearly depicted.

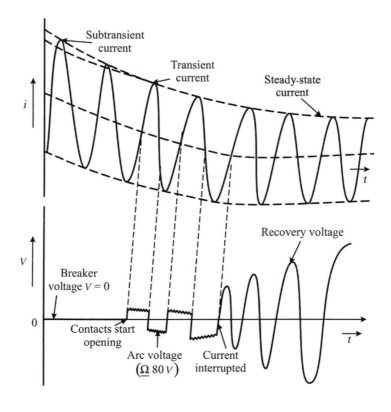

FIGURE 16.2

Arc extinction and recovery.

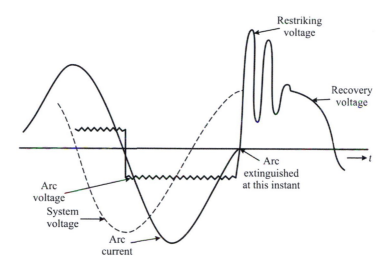

FIGURE 16.3

Restriking and recovery.

16.6 RESTRIKING VOLTAGE

Consider the system shown in Fig. 16.4A and its equivalent circuit in Fig. 16.4B.

The voltage that develops across the gap is the restriking voltage which suddenly rises and oscillates. With the fault on, when the CB is closed the capacitor is short circuited. Now, if the CB opens, the current flow is interrupted and is diverted into the capacitor. The inductance in the circuit and the capacitance constitute an oscillatory circuit and the current flowing i is given by

$$E = L\frac{di}{dt} + \frac{1}{C}\int i \cdot dt \tag{16.2}$$

where E is the system voltage.

The frequency of oscillation neglecting resistance

$$f = \frac{1}{2\pi}\sqrt{\frac{1}{LC}} \tag{16.3}$$

Since the transient is only for a very short duration, E can be considered relatively constant for the period of analysis.

$$i = \frac{dQ}{dt} = \frac{d(Cv_c)}{dt} \tag{16.4}$$

Therefore

$$\frac{di}{dt} = C\frac{d^2v_c}{dt^2} \tag{16.5}$$

where v_c is the charging voltage across the capacitor C.

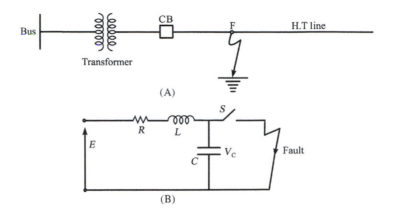

FIGURE 16.4

Fault on a system (A) and equivalent circuit for restriking voltage (B).

Also

$$\frac{1}{C}\int i\,\mathrm{d}t = \nu_C \tag{16.6}$$

Hence, Eq. (16.2) can be written using Eqs. (16.4) and (16.5)

$$LC\frac{\mathrm{d}^2\nu_C}{\mathrm{d}t^2} + \nu_C = E \tag{16.7}$$

Taking L-transform on both sides

$$LCs^2 V_C(s) + V_C(s) = \frac{E}{s} \tag{16.8}$$

Neglecting initial values

$$V_C(s)\left[LCs^2 + 1\right] = \frac{E}{s}. \quad \text{Let } \omega_n = \frac{1}{\sqrt{LC}}$$

$$V_C(s) = \frac{E}{s[LCs^2 + 1]} = \frac{E \cdot \frac{1}{LC}}{s\left[s^2 + \frac{1}{LC}\right]} = \frac{E\omega_n^2}{s[s^2 + \omega_n^2]}$$

Taking inverse L-transform

$$\mathcal{L}^{-1}V_C(s) = \mathcal{L}^{-1}\frac{E\omega_n^2}{s[s^2 + \omega_n^2]}$$

$$\nu_C(t) = E[1 - \cos\omega_n t] \tag{16.9}$$

$$= E\left[1 - \cos\frac{t}{\sqrt{LC}}\right] \tag{16.10}$$

$\nu_C(t)$ attains a maximum value at $\frac{1}{\sqrt{LC}}t = \pi$
i.e., at

$$t = \pi\sqrt{LC} \tag{16.11}$$

$\nu_C(t)$ maximum $= E\,[1 - \cos\pi] = E\,[1 + 1] = 2E$

The rate of rise of restriking voltage

$$RRRV = \frac{dE}{dt}\left[1 - \cos\frac{t}{\sqrt{LC}}\right] \qquad (16.12)$$

$$= \omega_n E \sin\omega_n t$$

The maximum value of $RRRV = \omega_n E$ and it occurs at

$$\omega_n t = \frac{\pi}{2} \quad \text{or} \quad t = \frac{\pi}{2\omega_n} = \frac{\pi}{2}\sqrt{LC} \qquad (16.13)$$

16.7 CLASSIFICATION OF CIRCUIT BREAKERS

The following is the classification of CBs based on the voltage level of operation.

1. Low voltage   –   Oil circuit breaker (OCB)

                                Miniature circuit breaker (MCB)

                                Molded case circuit breaker (MCCB)

2. Medium   –   Vacuum circuit breaker

                                Air circuit breaker

3. High voltage   – SF_6 circuit breaker

16.7.1 LOW-VOLTAGE CIRCUIT BREAKERS

They are used for voltages less than 1 kV. Generally for domestic, commercial, and industrial applications, use these CBs. They can be mounted in multitiers in L.V. switch boards or in cabinets.

Miniature Circuit Breaker: It is used for currents up to 100 A. The trip characteristics of miniature circuit breakers are not adjustable.

Molded Case Circuit Breaker (MCCB): They can be used up to 1000 A. In larger ratings the trip current could be adjusted.

Both the CBs operate either on thermal or thermal-magnetic principle. In thermal operation there is a bimetallic strip. The bimetallic strip expands due to heat and owing to different coefficients of linear expansion, the strip would bend at a certain angle which would pull the breaker lever down and severe the connection between the CB contact plate and the stationary contact plate.

In a magnetic CB an electromagnet due to increased magnetization at a high current exerts greater pressure on the CB lever and pulls it down so that the two contacts are separated.

In thermal-magnetic operation, the electromagnet pulls the lever in case of overload or fault and a bimetallic strip protects the system from prolonged overheating.

16.7.2 OIL CIRCUIT BREAKER

Oil CBs are used to interrupt currents up to 10,000 A. The trip characteristics are adjustable. Breaker contact separation is electromagnetic.

FIGURE 16.5

Air break switch.

16.7.3 AIR CIRCUIT BREAKER

They are used up to 10,000 A. Their trip characteristics are adjustable. They are electronically and through microprocessors controlled. A simple air break CB is shown in Fig. 16.5.

16.7.4 VACUUM CIRCUIT BREAKERS

They are suitable for currents up to 3000 A and extinguish the arc in a vacuum chamber. They are applied in system up to 35 kV. They have longer life than air CBs.

16.7.5 SULFUR HEXAFLUORIDE CIRCUIT BREAKER

SF_6 has excellent dielectric and arc quenching characteristics. A blast of SF_6 gas is applied to the arc.

These CBs are explained in detail in further sections.

16.8 THE PLAIN-BREAK OIL CIRCUIT BREAKER

A metal tank containing large volume of oil is used to contain the fixed and moving contacts as shown in Fig. 16.6. When a fault occurs and the relay mechanism separates out the fixed and

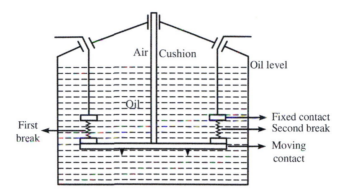

FIGURE 16.6

Principle of plain-break oil circuit breaker.

moving contacts, arc is struck between the two contacts. In Fig. 16.6 there are two contact separations and arc occurs at both the separating contacts in series. Sufficient air cushion is provided at the top of the oil level to accommodate the decomposed products of oil due to arc. The oil mainly by its ample head exerts pressure on the arc. Substantial cooling of the arc takes place and the arc will be extinguished. A vent is provided to the tank at the top, which is not shown in figure. To start with, when the contacts start opening and an arc is struck between the contacts, due to high potential gradient, oil in the vicinity of the arc gets ionized and arc temperature rises. Further, oil decomposition creates a gas bubble that restricts the arc to a narrow path and at one of the current zeros, the arc is quenched permanently as per energy-balance theory.

The plain-break oil CB is used for low-voltage DC circuits and low-voltage distribution AC circuits. It is not economical for high voltages and large current interruption as the size of the metal tank and the oil required become prohibitive.

16.9 SELF-GENERATED OIL CIRCUIT BREAKER

In further development of plain-break oil CB, consideration was given to consistent operation while the contact opening remained still small.

The arcing time is considerably reduced in the later developments in plain-break CBs. A pot made of insulating material is used to keep the fixed and moving contacts in side. The pot is placed in a metal tank.

16.10 PLAIN EXPLOSION POT

The simplest form of this type of CB is shown in Fig. 16.7. As the moving contact moves down after separating itself from the fixed contact, an arc is struck as in plain-break CB. The oil is decomposed and gas is generated. A high pressure is built up inside the pot as the moving contact

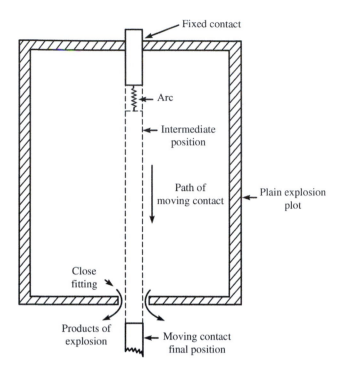

FIGURE 16.7

Principle of plain explosion pot.

has a close fitting throat at the bottom of the pot. The arc is quenched due to intense high pressure applied axially to the arc by gases and oil. The arc is surely interrupted, if it does not break while the moving contact is still in the pot at the instant when the moving contact leaves the pot.

This type of CB is good only for medium current ratings. For large currents the pot may explode and for low currents, it may take too long time to create the required pressure for arc interruption.

16.11 CROSS-JET EXPLOSION POT

In this type of CB shown in Fig. 16.8 lateral arc splitters are provided, and the arc formed between the two contacts generating high-pressure gaseous products pushes the arc across the openings between arc splitters. Thus the length of the arc is increased, increasing its resistance, and also increasing the rate of cooling. This enables faster arc interruption.

A combination of the principle of arc extinction in both the above CBs is utilized in another development called self-compensated explosion pot oil CB shown in Fig. 16.9.

The upper part of the CB is a cross-jet explosion pot, and the lower part is just like a plain explosion pot. Generally the arc is extinguished when the moving contact reaches the arc splitter as

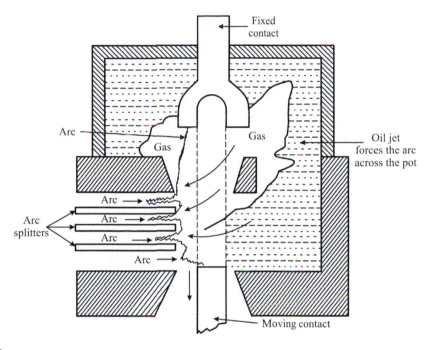

FIGURE 16.8

Cross-jet explosion pot.

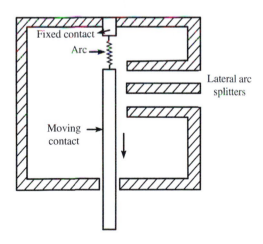

FIGURE 16.9

Self-compensated explosion pot.

in cross-jet explosion pot. In case, it is not extinguished even after moving down the arc splitters, by the time the moving contact just leaves the bottom throat, the arc is permanently extinguished as in plain explosion pot. This CB is suitable for both low- and high-current arc interruption.

16.12 MINIMUM OIL CIRCUIT BREAKER

In all the CBs described so far a large quantity of oil is used even though only a small volume of oil is needed for arc interruption. Since a metal tank is used to contain the live parts (fixed and moving contacts) large volume of oil is needed to provide insulation to ground. By choosing a ceramic container for the contacts, the size of the breaker can be substantially reduced. This type of design would be convenient for high-capacity CBs.

In a minimum oil CB this is incorporated.

In this type of CB the tank or pot is replaced by a ceramic container. The various components of this CB are illustrated in Fig. 16.10 for one pole. Three such devices will be needed in practice for three-phase operation. The ceramic container has very small volume of oil that is needed for

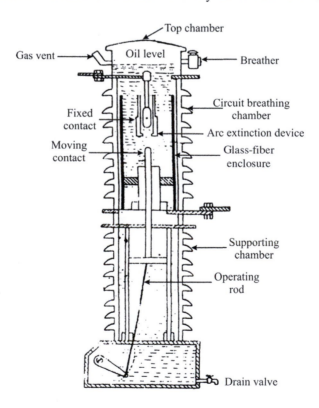

FIGURE 16.10

Minimum oil circuit breaker.

arc extinction. Arc interruption occurs inside the interrupter or circuit-breaking chamber. The container is of porcelain. A glass-fiber enclosure is provided inside the porcelain vessel. The minimum oil CB also called poor oil CB is widely in use.

16.13 ADVANTAGES AND DISADVANTAGES OF OIL CIRCUIT BREAKERS

Advantages of Oil Circuit Breakers

1. Oil acts as an insulating medium between the live parts and the metal container.
2. Oil absorbs arc energy by cooling, when it comes in contact with the arc.
3. Oil absorbs arc energy and decomposes into gases that have cooling property.
4. The oil decomposition gaseous products exert pressure on the arc and increase the rate of cooling forcing it to extinguish.

Disadvantages of Oil Circuit Breakers

1. Products of arc are inflammable. They may form explosive mixture with air.
2. Dielectric strength of oil is reduced when it absorbs moisture. Oil must be sealed against moisture.
3. When arc is struck, carbonization of oil occurs. Over long time, continuous use of oil CBs require replacement of oil due to this carbonization process. Hence, maintenance of oil is required at regular time intervals.

16.14 AIR BREAK CIRCUIT BREAKER

In these CBs, air at atmospheric pressure is used to extinguish the arc. The principle of lengthening the arc and cooling it are used. For interruption of low-voltage large currents (e.g., 460 V and 400−3500 A), oil CBs are not suitable. Even though hydrogen produced by oil in oil CBs has superior arc interrupting characteristics, air has some advantages for arc interruption. With air, there is no fire risk. Further, there is no degradation of air medium, while oil gets degraded due to usage over a long period.

Air break CB utilizes the high-resistance breaking principle. No arc control devices are used for CBs up to 1 kV. The arc is lengthened between the separating contacts to produce a high potential gradient and power loss. Arc splitters and chutes are used for this purpose as shown in Fig. 16.11. Baffles are provided at the splitters to cool the arc. For high voltages, magnetic field is used to lengthen the arc. These breakers use two pairs of contacts—one pair of main contacts made of copper and the other arcing contacts made of carbon. Initially, both the contacts are closed. When the relay operates, first the main contacts open, while the arcing contacts are still closed. Subsequently, the arcing contacts open and arc is struck across them. In the chute guided by arc runners, the arc moves up under natural draft and thereby increases in length. The splitters further lengthen the arc, and the arc gets cooled and extinguishes. In this way the main contacts are prevented from damage due to heavy arc currents.

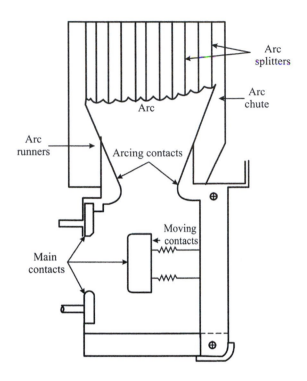

FIGURE 16.11

Air break circuit breaker.

16.15 AIR BLAST CIRCUIT BREAKERS

In air break CBs air interrupts the arc at atmospheric pressure only. But in air-blast CBs air is forced at a pressure ranging from 20 to 30 kg/cm^2 into the contact separation so that the rate of cooling increased substantially and the dielectric strength of the arc space is improved at a very fast rate. CBs operating at 132 kV and above up to 400 kV use in general air-blast CBs. The breaking capacities range from 5000 MVA at 66 kV to 60,000 MVA at 750 kV.

The following are the main reasons for using air-blast CBs

1. Air is freely available
2. Air is chemically stable and inert
3. Elimination of fire hazard
4. Possibility of high-speed operation
5. Short and consistent arcing time
6. Suitable for frequent operation
7. Facility for high-speed reclosure
8. Less maintenance

Air blast CBs have the following disadvantages:

1. An air compressor plant is required
2. Prevention of leakages in pipe fittings

3. Sensitive to restriking of arc
4. Current chopping
5. When operated, discharge of air makes noise disliked by neighborhood.

16.16 TYPES OF AIR-BLAST CIRCUIT BREAKERS

16.16.1 AXIAL BLAST CIRCUIT BREAKERS

The basic principle of axial blast CB is shown in Fig. 16.12. High-pressure air is admitted into the arcing region in the same direction as the contact separation or arc. Thus the flow of air is axial or longitudinal and parallel to the arc, compressing it into thinner arc, increasing the gradient and rate of cooling. This principle is very suitable for EHV application since at super high voltages the contacts are enclosed in porcelain chambers, single- or double-blast CBs are very useful. Fig. 16.13

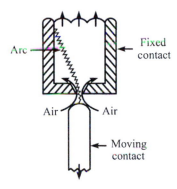

FIGURE 16.12

Axial blast CB. *CB*, circuit breaker.

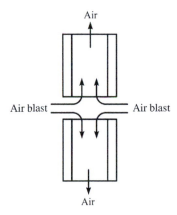

FIGURE 16.13

Radial blast.

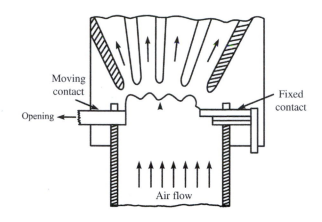

FIGURE 16.14

Cross blast.

shows the double-blast breaking principle. In this case air blast is admitted radially as shown in figure, while the flow in the arcing region is axial, along the direction of the arc formation. Increasing the number of breaks increases the efficiency of the breaker at EHV and super high voltage applications. Resistance switching is used in these CBs.

16.16.2 CROSS-BLAST CIRCUIT BREAKERS

In this case the blast of air is admitted laterally as shown in Fig. 16.14. The flow of air is perpendicular to the arc and hence the name cross-blast. As in air-break CBs, the arc is lengthened and forced into arc splitters arranged in a chute. High current interruption (up to 110 kA) at low voltage utilize cross-blast CBs since the arc resistance is increased substantially in this type of breakers, resistance switching is used (Fig. 16.15).

16.17 CURRENT CHOPPING

Consider a CB operating on no load. The currents flowing in the circuit will be only those due to no-load magnetizing currents of transformer. When these low-inductive currents are interrupted, the high interrupting force of the CB interrupts the current even before it reaches its natural zero. The energy stored in the magnetic field then appears as electrostatic energy and a high voltage appears across stray capacitance in the circuit, which restrikes the arc. This is explained in Fig. 16.16 the energy in the magnetic field $\frac{1}{2}Li^2$ where i is the instantaneous current produces an electric field of $\frac{1}{2}Cv^2$ where v is the instantaneous voltage across the capacitance. Since they are equal

$$\frac{1}{2}Li^2 = \frac{1}{2}Cv^2$$

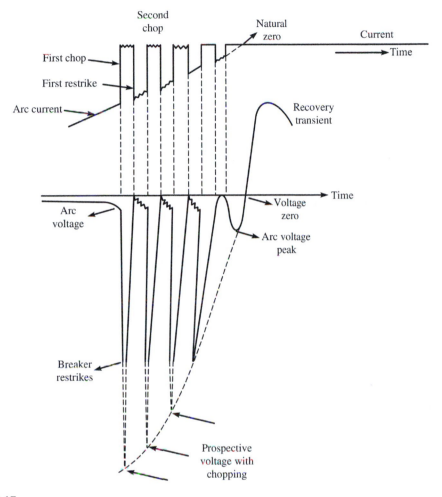

FIGURE 16.15

Current chopping.

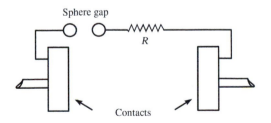

FIGURE 16.16

Resistance switching.

or

$$\nu = i\sqrt{\frac{L}{C}} \qquad (16.14)$$

Thus the current in the circuit interrupted produces an voltage across the contacts of the CB ν, and if it exceeds the breakdown strength of the gap across the contacts, the arc will reappear and a second chop of the arc take place due to CB action, this continues till the arc is finally extinguished and recovery voltage is established. The calculated value ν from equation is called prospective voltage. Current chopping occurs in air-blast CBs.

16.18 RESISTANCE SWITCHING

During current chopping that occurs while interrupting low-inductive currents, high voltage builds up across the contacts of the CB. This has been explained in Section 16.17. To reduce the high voltage that appears across the contacts of the circuit breaker, a resistance 'R" is connected across the contacts of the circuit breaker as shown in Fig. 16.16. This resistor reduces the rate of rise of the restriking voltage. It reduces also the magnitude of the transient voltage during switching out inductive or capacitive loads. A sphere gap is connected in series with the resistance so that when voltage rises to a certain level, the gap becomes conductive and the resistance gets switched in. Also, under normal conditions when the contacts are closed, the resistance is not in the circuit.

The value of the resistance R required to be connected such that the transient recovery voltage does not contain oscillations can be calculated as follows. Consider the equivalent circuit shown in Fig. 16.17 to stimulate low-inductive current breaking.

Using Kirchoff's voltage law

$$L\frac{di}{dt} + \frac{1}{C}\int i_C dt = V \qquad (16.15)$$

But

$$i = i_C + i_R \quad \text{so that} \quad \frac{di}{dt} = \frac{di_C}{dt} + \frac{di_R}{dt}$$

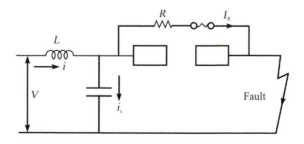

FIGURE 16.17

Low-inductive current breaking.

Further

$$i_C = \frac{dQ}{dt} = \frac{d}{dt}(Cv_C) = C\frac{dv_C}{dt}$$

and

$$\frac{di_R}{dt} = \frac{d}{dt}\left(\frac{v_C}{R}\right) = \frac{1}{R}\frac{dv_C}{dt}$$

Hence Eq. (16.15) can be rewritten as

$$L\frac{di_C}{dt} + L\frac{di_R}{dt} + \frac{1}{C}\int C\frac{dv_C}{dt} = V$$

i.e.,

$$LC\frac{d^2v_C}{dt} + \frac{L}{R}\frac{dv_C}{dt} + v_C = V \qquad (16.16)$$

Taking $\mathcal{L}$-transform on both sides of the above equation

$$LCs^2v_C(S) + \frac{L}{R}sv_C(S) + v_C(S) = \frac{V}{s}$$

$$LC\left[s^2\frac{L}{R}s + \frac{1}{LC}\right]v_C(S) = \frac{V}{s}$$

$$v_C(S) = \frac{V}{sLC\left[s^2 + \frac{1}{RC}s + \frac{1}{LC}\right]} \qquad (16.16(a))$$

Consider the equation

$$sLC\left[s + \frac{1}{RC}s + \frac{1}{LC}\right] = 0 \qquad (16.17)$$

For the transient voltage not to have oscillations, all the three roots should be real. Since one root $s = 0$ is real, the other two roots also will be real if and only if

$$\left[\left(\frac{1}{2RC}\right)^2 - \frac{1}{LC}\right] \geq 0 \qquad (16.18)$$

or

$$R \leq \frac{1}{2}\sqrt{\frac{L}{C}} \qquad (16.19)$$

For any other value of R, there will be damped oscillation with frequency

$$f = \frac{1}{2\pi}\sqrt{\frac{1}{LC} - \frac{1}{4C^2R^2}} \qquad (16.20)$$

16.19 INTERRUPTION OF CAPACITIVE CURRENTS

When an unloaded long transmission line or capacitance banks are switched off, capacitive current interruption takes place (Fig. 16.18).

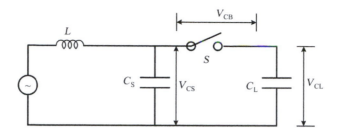

FIGURE 16.18

Circuit for capacitance current interruption.

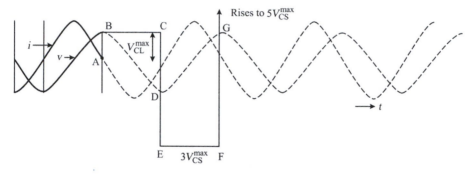

FIGURE 16.19

Interruption of capacitive current.

To understand the difficulty associated with such an operation, consider the equivalent circuit shown in Fig. 16.19. C_s is the stray capacitance of the CB and C_L is the line or capacitive bank capacitance which is much larger than C_s. The CB is represented by the switch S with voltage across the contacts shown as V_{CB}.

At the instant A, when current i is zero, the arc is extinguished. The voltage is maximum at B and at this voltage the capacitor C_L remains charged. The voltage across the gap is the difference between V_{CS} and V_{CL}. After half-a-cycle at point C, the voltage across the gap is twice the maximum value of V_{CS} (V_{CS}^{max}). If the breaker restrikes, voltage across the gap falls to zero from $2V_{CS}^{max}$. High-frequency oscillation will set in the voltage oscillates between C and E that is between $-3V_{CS}^{max}$ and V_{CS}^{max} about point D. If the arc is interrupted again at the next current zero, the capacitor C_L remains at $-3V_{CS}^{max}$. At the instant G when the voltage is positive and reaches its maximum value, the gap voltage reaches $4V_{CS}^{max}$ and capacitive current again becomes zero; the arc may be interrupted. If the arc is interrupted, then the transient restriking voltage oscillates between $-3V_{CS}^{max}$ and $+5V_{CS}^{max}$. In this way the voltage across the gap goes on building up. This is the difficulty experienced in capacitive current interruption.

16.20 VACUUM CIRCUIT BREAKERS

It is well known that an arc is produced when the two contacts of a CB separate out. The intense heat in the arc produces ionization in the medium in which the contacts are placed. At atmospheric pressure, the mean-free path for the molecules of the medium to undergo ionization is relatively small. If the medium is evacuated and a vacuum of the order of $10^{-4}-10^{-6}$ mm of mercury is maintained, then since very few molecules will be left behind, the mean-free path increases to several meters and ionization by collision becomes difficult. In such a vacuum, if the contacts are made to separate then arc will be struck initially at few points on the metallic surface of the contacts. As the probability of ionization is very much reduced, the arc will be extinguished at the subsequent current zero. There will be only electron emission due to heat (thermal emission) supplemented by field emission and little secondary emission. As current decreases to a zero since there is no further support for ionization and cooling taking place rapidly, at the zero of the current wave the arc is extinguished as shown in Fig. 16.20. In this type of CBs, the contact separation needed is only few millimeters. Extinction occurs in half-a-cycle.

Fig. 16.21 shows the physical construction of vacuum CB used in practice. Plain disc type contacts are used for currents up to 10,000 A, but for higher current ratings special geometry is used for contacts. Metal bellows shown in figure permit movement of the contacts. Metal shields prevent condensation of metal vapor on the enclosure surface. The flanges are made of nonmagnetic material. The contact tips are made of copper chromium or bismuth alloy.

Dielectric strength of space between the contacts builds up at a rate of about 20 kV/μs. These CBs are used as both indoor and outdoor interrupters with ratings of 1.2, 3.6, 7.2, 12, and 36 kV extensively. Now-a-days they are seldom used for voltages greater than 36 kV.

16.21 SF$_6$ CIRCUIT BREAKERS

Sulfur hexafluoride is an odorless, colorless, inert, nontoxic, and noninflammable gas at atmospheric conditions. It is stable up to 500°C and about five times heavier than air. It has a dielectric strength that is 2.35 times that of air and this increases further as pressure is increased. It is electronegative. It attracts free electrons and holds them. This may occur in two ways.

$$SF_6 + e \rightarrow SF_6^- \tag{16.21}$$

$$SF_6 + e \rightarrow SF_5 + F \tag{16.22}$$

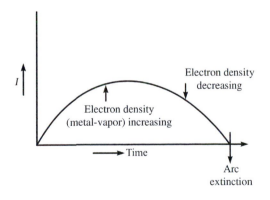

FIGURE 16.20

Vacuum circuit breaking in AC circuits. AC, alternating current.

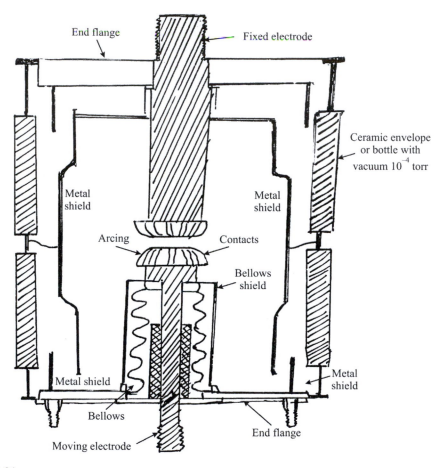

FIGURE 16.21

Vacuum circuit breaker.

The negative ions by virtue of their heaviness cannot attain energies sufficient to cause further ionization. In this way SF_6 gas molecules contribute to electron elimination. It is found that SF_6 is 100 times better than air in its efficacy to quench an arc. SF_6 gas is then very well-suited for arc extinction in CBs.

16.21.1 ADVANTAGES AND DISADVANTAGES

Advantages of SF_6 Circuit breakers

1. Arcing times are very small.
2. Consequently, contact erosion is much less.
3. Since there is no carbonization, SF_6 gas is not contaminated in any way in operation.

4. Tendency to current chopping is much less.
5. Operation is not influenced by atmospheric conditions.
6. As the dielectric strength is high, SF$_6$ CBs require smaller clearances. thereby reducing the size of the CB.

Disadvantage: The gas being heavier than air may cause suffocation to persons in case of leakage.

16.21.2 PRINCIPLE OF OPERATION

The early designs of SF$_6$ CBs followed the pattern of air-blast CBs. SF$_6$ gas was released from a high-pressure tank into the low-pressure arcing region between the breaker contacts. These models were replaced by single-pressure or impulse-type CBs. In these CBs SF$_6$ gas compressed in a cylinder moves along with the moving contact. The compressed gas is released under pressure axially into the arcing region. The arc gets extinguished at current zero. This is the Puffer principle shown in Fig. 16.22.

EHV and super high-voltage CBs have several interrupters connected in series placed on insulated supports. Such breakers are used up to 765 kV. Fig. 16.23 shows such interrupters for 470-kV SF$_6$ CB.

Using digital simulation the geometry of the arc-interrupting chamber and the connection between poles and the mechanism are optimized. Increased use of the arc energy to produce the pressure necessary to extinguish the arc and obtain current interruption was the motive behind several latest designs which included self-blast chambers in addition to puffer volume. A valve is introduced between expansion and compression volumes. The low-current interruption is achieved through puffer principle. For high-current interruption arc energy produces a high pressure in the expansion volume which leads to the closure of the valve. Thus, the two volumes, expansion and compression are separated. Action of such a breaker is shown in Fig. 16.24.

The cross section of a typical 420 kV, SF$_6$ CB is shown in Fig. 16.25.

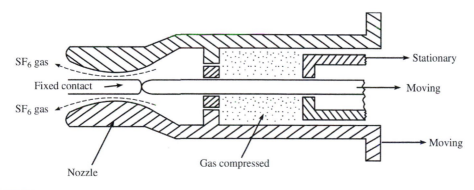

FIGURE 16.22

Puffer type SF$_6$ circuit breaker.

FIGURE 16.23

Interrupter of 420-KV SF$_6$ CB. CB, circuit breaker.

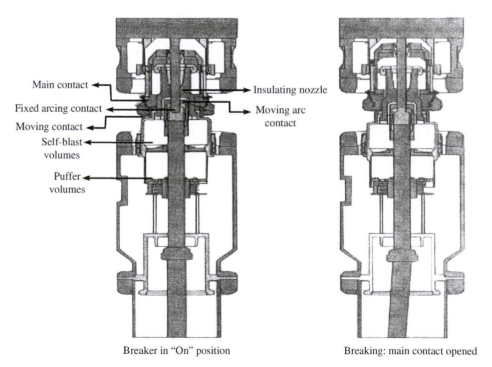

Breaker in "On" position   Breaking: main contact opened

FIGURE 16.24

Spring mechanism of 420 kV SF$_6$ circuit breaker.

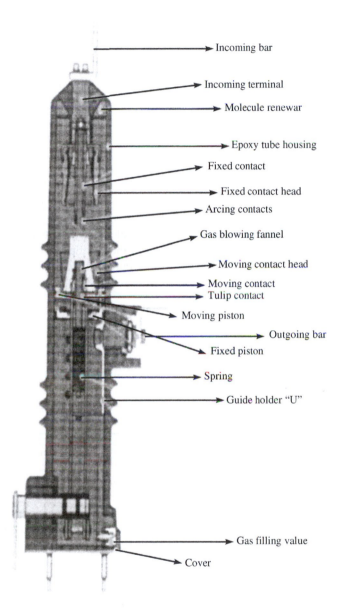

Incoming bar

Incoming terminal

Molecule renewar

Epoxy tube housing

Fixed contact

Fixed contact head

Arcing contacts

Gas blowing fannel

Moving contact head

Moving contact

Tulip contact

Moving piston

Outgoing bar

Fixed piston

Spring

Guide holder "U"

Gas filling value

Cover

FIGURE 16.25

Cross section of 420 kV SF_6 circuit breaker model.

16.22 **HIGH-VOLTAGE DIRECT CURRENT INTERRUPTION**

High-voltage direct current (HVDC) lines are used as links in AC transmission systems for several reasons such as stability, economy, and corona loss. It is possible to control the current on the

HVDC side by use of firing circuits of the thyristors used in both rectifiers and inverters. Switching operations can be performed on the AC side using AC CBs. However, if HVDC circuit breakers are available on the DC side, it gives greater scope in planning and operation. There could be parallel line operation as well as tap-off facilities on the HVDC side.

In AC circuit breakers the availability of a current zero is utilized to interrupt the arc between the contacts of the CB. If the arc is interrupted by using excessive force suddenly as in air-blast CBs, the arc energy will reappear as electrostatic energy across the contacts, restriking the arc. This was explained under current chopping. Hence, in DC circuit breakers in addition to lengthening of arc, cooling of arc, improving the dielectric strength in the arcing zone, the basic principles in arc interruption, it is necessary to design an external circuit and use it to create an artificial current zero, where the arc can be quenched.

Fig. 16.26 explains this principle, and Fig. 16.27 shows the external circuit to be designed and connected across the breaker contacts.

The parallel connected capacitor—inductor branch is closed by switch S as soon as the relay operates and the CB contacts separate out. The current in the parallel branch opposes the current in the CB circuit, i.e., the arc current creates an artificial current zero. The action of CB (say SF_6) interrupts the arc at this current zero and prevents it from restricting.

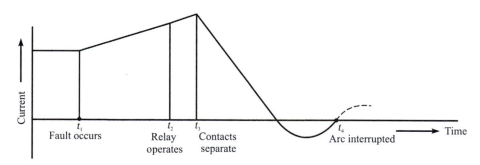

FIGURE 16.26

HVDC interruption. *HVDC*, high-voltage direct current.

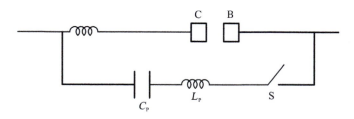

FIGURE 16.27

Creation of artificial zero.

16.23 RATING OF CIRCUIT BREAKERS

"A CB should be capable of performing the opening and closing operations as per rated operating sequence for all values of short-circuit currents up to its rated short-circuit breaking current at specified test voltage and relevant conditions of transient recovery voltage for terminal short circuits." For this reason the CB is expected to perform on short circuit the following tasks.

1. Open the breaker contacts on the occurrence of a fault
2. Close the contacts while the fault is on
3. Carry the short-circuit current for a short time and clear the fault

The abovementioned duties are incorporated in the following ratings.

1. Breaking capacity
2. Making capacity
3. Short time capacity

Breaking capacity is of two kinds

- Symmetrical breaking capacity
- Asymmetrical breaking capacity

By breaking capacity of a CB, we understand the highest current, the CB is capable of breaking at a stated recovery voltage (generally the rated voltage) and a stated reference restriking voltage under specified conditions.

If the current is symmetric, then the so-defined breaking capacity is symmetric breaking capacity. On the other hand, if the current is asymmetric the rating is asymmetric breaking capacity. Fig. 16.28 illustrates a typical short-circuit current waveform.

16.23.1 SYMMETRIC BREAKING CAPACITY

Symmetrical breaking current is the rms value of the AC component of the short-circuit current at the instant of contact separation. In Fig. 16.28 this is indicated by $AB/\sqrt{2} = x/\sqrt{2}$ under specified condition of recovery voltage. The symmetric breaking capacity is then defined as

$$\text{Symmetrical breaking capacity} = \sqrt{3} \text{ Rated voltage in KV} \times \text{Rated current in kA} \qquad (16.23)$$

16.23.2 ASYMMETRIC BREAKING CAPACITY

Asymmetric breaking current is the rms value of the total current consisting both of AC and DC components of the short-circuit current that the CB can interrupt under specified conditions of recovery voltage. The asymmetrical component or the total component of the short-circuit current at the instant of contact separation is equal to

$$\sqrt{\left(\frac{AB}{\sqrt{2}}\right)^2 + (BC)^2} = \sqrt{\left(\frac{x}{\sqrt{2}}\right)^2 + (y)^2} \text{ rms} \qquad (16.24)$$

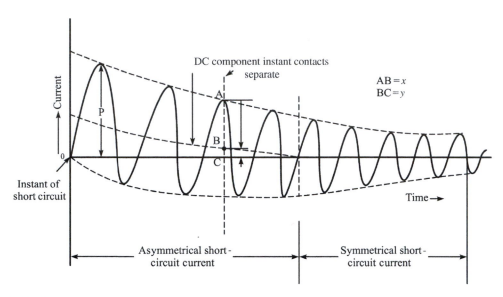

FIGURE 16.28

Short-circuit current.

Asymmetrical breaking capacity = $\sqrt{3}$ Rated voltage in kV × Rated asymmterical current in kA
Thus both the breaking capacities are expressed in MVA.

The percentage DC component of short-circuit current at the instant of contact separation is $(100 \times (y/x))$.

British practice utilizes the symmetric component of current for breaking capacity, whereas the American practice incorporates asymmetrical or total current component for specifying the breaking capacities. Thus the American rating is about 1.6 times high than the British rating.

Conditions that can be specified for CB rating are power factor, recovery voltage, and rate of rise of recovery voltage.

16.23.3 MAKING CAPACITY

This is associated with the ability of the CB to be closed on short circuit. The highest value for the current on short circuit occurs at the instant of its first peak which includes the AC component as well as the maximum DC component. The CB should be capable of being closed at this current. This value is indicated as "P" in Fig. 16.28 and occurs immediately after the short circuit and at the instant of first peak. The CB should be able to withstand the electromagnetic effects that will be maximum at this current. Since it is not possible to precisely determine the value of P, rated making current is taken as $1.8 \times \sqrt{2}$ (corresponding rated symmetric breaking current).

Thus *making capacity* $\simeq 2.55$ (*rated symmetric breaking capacity*).

16.23.4 SHORT TIME CAPACITY

CBs are required to carry short-circuit currents for a short period while predesignated CB is clearing the fault. Hence, a short time current rating based on thermal and mechanical considerations is required. The CB should be able to carry for a specified short period rated rms total (AC and DC) current. The British practice specifies this short time period as 3 s provided that

$$\frac{\text{Symmetrical breaking current}}{\text{Normal rated current}} < 40$$

and otherwise as 1 s for

$$\frac{\text{Symmetrical breaking current}}{\text{Normal rated current}} > 40$$

The American standards specify two short time ratings: The maximum current that the CB can withstand for 1 s or less and another, the maximum current that the CB can withstand for more than 1 s but less than 4 s.

Rated voltage: It is the maximum value of the voltage at which the CB is designed to operate. Generally it is slightly more than the rated nominal voltage.

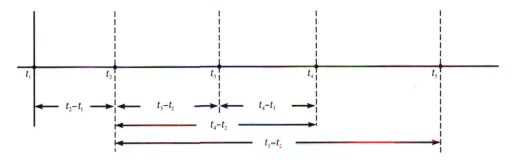

FIGURE 16.29

CB operation time sequence. *CB*, circuit breaker.

t_1	→	Fault occurs
t_2	→	Trip coil of relay energized
t_3	→	C.B. contacts open
t_4	→	Arc extinguished
t_5	→	Contacts close
$t_2 - t_1$	→	Relaying time
$t_3 - t_2$	→	Breaker opening time
$t_4 - t_2$	→	Breaker interruption time
$t_4 - t_3$	→	Arcing time
$t_5 - t_2$	→	Reclosing time

Rated current: It is the rms value of the current that the CB contacts can continuously carry without exceeding the temperature rise limit.

Rated frequency: It is the frequency at which the CB is designed to operate.

16.23.5 OPERATING DUTY

All CBs are not designed for autoreclosing.

For CBs which are not designed for autoreclosing two alternative operating duties are recommended.

$$(1)\ O - t - CO - t' - CO, \quad \text{or} \quad (2)\ O - t'' - CO$$

where "O" stands for opening, "CO" stands for closing followed by opening without time delay, and t, t' and t'' are intervals between successive operations (e.g., $t = t' = 3$ min and $t'' = 15$ s).

For CBs with autoreclosing the duty cycle is

$$O - \Delta t - CO$$

where Δt is the dead time of CBs expressed in cycles.

The time sequence for various operations is shown in Fig. 16.29.

16.24 TESTING OF CIRCUIT BREAKERS

There are a large number of tests and an elaborate procedure for each test specified in Indian Standards. There are also British and IEC specifications in addition to ASA. Testing of CBs will be very briefly discussed here.

There are two types of tests which are more common. They are type tests and routine tests.

16.24.1 TYPE TESTS

These tests are conducted on a few prototype sample breakers. Only few from each type are selected for the tests. They are conducted as per standards adopted to determine the capability of the breaker to comply with the characteristics as envisaged in its design.

16.24.2 ROUTINE TESTS

These tests are performed on each CB as per standard specification. General routine tests are mechanical, thermal, and dielectric tests.

Mechanical Tests: These are endurance tests. The CB is opened and closed a large number of times (1000). Most of these operations are performed with voltage and current applied. A few are conducted with supply on and trip circuit of the relay mechanism connected to the CB. The withstand capability of the mechanical structure is tested in this way.

Thermal Tests: These are actually temperature rise tests. Alternating current at rated value and rated frequency is applied to the breaker circuit and maintained till steady temperature is reached. The maximum temperature rise should be less than permissible limits for each component or part of the CB. Thermocouples are used to measure temperature.

DC Resistance Voltage Drop Tests: The DC resistance of each pole of CB is measured with a micro ohmmeter. A DC not more than the rated current of breaker (>100 A) is passed at ambient temperature and the resistance is measured by knowing the voltage drop.

Dielectric Tests: These tests are conducted at power frequency and also with impulse voltage applied and the withstand values are determined.

There are several routine tests performed on auxiliary and control circuits.

There are tests for breaking capacity, making capacity, and short time current.

WORKED EXAMPLES

E.16.1 In a 110-kV system, the reactance and capacitance of the system up to the location of the circuit breaker are 3.2 Ω and 0.016 μf, respectively. Calculate
   1. the frequency of oscillation,
   2. the maximum value of restricting voltage, and
   3. the maximum value of RRRV.
   Solution:

$$\text{Inductance, } L = \frac{3.2}{2\pi \times 50} = 0.01019 \text{ H}$$

1. $f_n = \dfrac{1}{2\pi\sqrt{LC}} = \dfrac{1}{2\pi}\dfrac{10^3}{\sqrt{0.01019 \times 0.016}}$
   $= 12,471.9 \text{ Hz} = 12.472 \text{ kHz}$

2. restriking voltage, $V_c = E(1 - \cos \omega_n t)$
   Maximum value of the restriking voltage

$$= 2E = 2 \times \frac{110}{\sqrt{3}} \times \sqrt{2} = 179.6 \text{ kV}$$

3. Maximum RRRV $= \omega_n E$

$$= 2\pi \times 12.472 \times 10^3 \times \frac{110 \times \sqrt{2}}{\sqrt{3}} \times 10^3$$
$$= 7033.8 \times 106 \text{ V/s}$$
$$= 7.0338 \text{ kV/}\mu\text{s}$$

E.16.2 A 50-Hz, three-phase synchronous generator has an inductance per phase of 1.86 mH and its neutral is grounded. The generator feeds through a line to a distant load. The total capacitance of the system up to the circuit breaker is 0.003 μf. If a fault occurs just after the circuit breaker on the line and the fault current is 5000 A (rms), determine the following:
   1. Natural frequency of oscillation
   2. Peak value of transient recovery voltage
   3. Time at which peak of transient recovery voltage occurs
   4. Maximum RRRV

Solution:

1. $f_n = \dfrac{1}{2\pi\sqrt{LC}} = \dfrac{1}{2\pi} \dfrac{1}{\sqrt{0.003 \times 10^{-6} \times 1.86 \times 10^3}}$

$= 67{,}430.88 \text{ c/s} = 67.43 \text{ KHz}$

2. $E = Iwl = 2\pi \times 50 \times 1.86 \times 10^{-3} \times 5000$

$= 2920.2 \text{ V} = 2.92 \text{ kV}$

$$E(\text{Peak}) = \sqrt{2} \times 2.92 = 4.129 \text{ kV}$$

Maximum value of transient recovery voltage $= 2 \times 4.129 = 8.258 \text{ kV}$

3. Time-to-peak TRV is given by

$$\frac{t_p}{\sqrt{LC}} = \pi$$

$$t_p = \pi\sqrt{LC} = \pi\sqrt{1.86 \times 10^{-3} \times 0.003 \times 10^{-6}} \text{ s}$$

$$= \pi\sqrt{1.86 \times 10^{-3} \times 0.003 \times 10^{-6}} \times 10^6 \text{ } \mu s$$

$$= 7.4173 \text{ } \mu s$$

4. Maximum RRRV $= \dfrac{E_m}{\sqrt{LC}} = \dfrac{4.129 \times 10^3 \text{ V}}{\sqrt{1.86 \times 10^{-3} \times 0.003 \times 10^{-6}}}$ V/s

$= 1747.95 \text{ } V/\mu s$

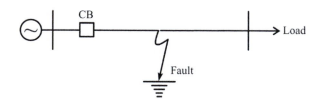

FIGURE E.16.2

E.16.3 A circuit breaker interrupts the magnetizing current of a 125-MVA transformer at 220 kV. The magnetizing current of the transformer is 6% of the full-load current. Determine the maximum voltage that may appear across the gap of the circuit breaker contacts while they are interrupting current at 60% of its peak value. The stray capacitance is estimated to 3 μf. The inductance in the circuit is 32 H.
Solution:

$$\text{Full load current } I_{FL} = \frac{125 \times 10^6}{\sqrt{3} \times 220 \times 10^3}$$

$$= 3280 \text{ A}$$

Magnetizing current $= 0.06 \times 3280 = 19.683$ A
Current value when it was chopped $= 0.6 \times 19.683 = 11.809$ A

$$\text{Instantaneous value of the current} = 11.809 \times \sqrt{2}$$
$$= 16.699 = 16.699 \text{ A}$$

$$\text{Voltage generated } V_c = i\sqrt{\frac{L}{C}}$$

$$= 16.699\sqrt{\frac{32 \times 10^6}{3}} = 54{,}538.7$$

$V_c = 54538.7 \text{ V} = 54.5387 \text{ kV}$

E.16.4 A circuit breaker is rated at 1500 A, 1800 MVA 33 kV, 3-s, three-phase oil circuit breaker. What are its normal current, breaking current, making current, and short time rating?

Solution:

Rated current = 1500 A

$$\text{Rated symmetrical breaking current} = \frac{1800 \text{ MVA}}{\sqrt{3} \times 33 \text{ kV}} = 31.492 \text{ kA}$$

$$\text{Rated making current} = 31.492 \times 2.55$$
$$= 80.3046 \text{ kA}$$

Short time rating = 31.492 kA for 3 s

PROBLEMS

P.16.1 A circuit breaker is specified as 1200 A, 1000 MVA, 33 kV, 3 s, three-phase OCB. Determine its
1. rated normal current,
2. rated symmetric breaking current,
3. rated making current, and
4. short time rating.

P.16.2 In a 110-kV transmission system, the phase-to-ground capacitance is 0.012 μF. The inductance is 6.43 H. Calculate the voltage appearing across the pole of a circuit breaker if a magnetizing current of 12.6 A is interrupted. Find the value of the resistance to be connected to eliminate the voltage transient.

P.16.3 A synchronous generator with solid neutral earthing operates at 50 Hz. It is connected to a transmission system through a circuit breaker with the following information:

emf to neutral = 8.35 kV
Reactance of generator and connected system up to circuit breaker = 4 Ω
Capacitance to neutral = 0.013 μF
Resistance is negligible

A three-phase short circuit occurs at the outgoing terminals of the circuit breaker. If B-phase clears first, determine:
1. the maximum voltage across the contacts of B-phase when the arc is interrupted at current zero
2. average RRRV up to its first peak of oscillation

3. value of the resistance required to be connected across the circuit breaker to damp the oscillation completely.

P.16.4 A 110-kV, 50-Hz, three-phase alternator has an earthed neutral. The inductance and capacitance of the system per phase are 8.5 mH and 0.015 μF. From the short-circuit test at 0.25 power factor, fault current symmetrical recovery voltage is 90% of full-line voltage. Assuming that the fault is isolated from the ground, determine the RRRV.

QUESTIONS

Q.16.1 Explain the following terms:
  1. Restriking voltage
  2. Recovery voltage
  3. Rate of rise of recovery voltage

Q.16.2 Derive an expression for restriking voltage in a circuit breaker. From the expression, prove that the maximum value of the RRRV = $W_n E$
  where E is the system voltage and W_n is the natural frequency of operation.

Q.16.3 What is resistance switching? Why is it used?

Q.16.4 Show that the value of the resistance to be connected across the CB contacts so that there are no transient oscillations is given by
  $R < \frac{1}{2}\sqrt{\frac{L}{C}}$ What is the frequency of damped oscillations?

Q.16.5 Explain the term—current chopping with neat diagrams.

Q.16.6 Explain the problems associated with capacitance breaking in circuit breakers.

Q.16.7 How are circuit breakers classified? Explain their basic principles.

Q.16.8 What are the basic principles adopted in arc interruption?

Q.16.9 Explain the following in relation to circuit breakers
  1. High-resistance interruption
  2. Current zero interruption

Q.16.10 Explain the recovery rate theory of arc interruption.

Q.16.11 What do you understand by energy-balance theory of arc interruption?

Q.16.12 What are the merits and demerits of oil circuit breakers?

Q.16.13 Explain with a neat sketch the working of a cross-jet explosion pot oil circuit breaker.

Q.16.14 What are the merits and demerits of air circuit breakers?

Q.16.15 Explain the advantages of SF_6 as a medium for arc interruption in circuit breakers.

Q.16.16 Describe the working principle of SF_6 circuit breaker with a neat sketch.

Q.16.17 How are vacuum circuit breakers superior over other type of circuit breakers? What are their demerits?

Q.16.18 How are circuit breakers rated? Explain.

Q.16.19 Explain the terms
  1. Breaking capacity
  2. Making capacity
  3. Short time capacity

Q.16.20 Write a short note on circuit breaker testing

Q.16.21 Discuss the arc interruption in HVDC circuit breakers.

RELAYING AND PROTECTION

Power systems are subject to various types of faults under normal operation. During faults, currents may flow through unwanted paths. Generally fault currents are high, since fault impedances are low. Breaking of conductor creates a series fault, while failure of insulation creates a shunt type of fault. Occurrence of a fault results in increase of current, reduction in voltage, power factor, and frequency. Even the stability of the system may be impaired. There are several causes for faults. Faults cannot be avoided in practice. To protect the various parts of the power system and the system as a whole, protective relaying is necessary.

The protective relaying should sense the abnormal condition in any part of the system and give an alarm or disconnect that part from the remaining operating system.

17.1 REQUIREMENTS OF RELAYING

Circuit breakers discussed in Chapter 15, Unbalanced Fault Analysis, are to be supplemented by protective relays. They are required to detect the existence of any abnormal condition or fault and initial action so that the circuit breaker operates. For proper functioning, the relaying scheme should possess the following features:

1. Selectivity (Discrimination)
2. Speed (Fast operation)
3. Sensitivity
4. Reliability
5. Stability
6. Adequacy
7. Simplicity
8. Adequacy

17.1.1 SELECTIVITY

The relay should be able to discriminate or distinguish between a normal state and an abnormal state and act or initiate action when abnormal condition exists in that part of the system which the relay is intended to protect, so that the affected part or component is isolated from the remaining healthy system. This property of being able to select by itself the need for operation is called selectivity. For example, the relay should distinguish between a high-fault current and transient magnetizing inrush current in a transformer.

Electrical Power Systems. DOI: http://dx.doi.org/10.1016/B978-0-08-101124-9.00017-6

17.1.2 SPEED

The relay should operate fast enough so that the high-fault currents do not cause damage to equipment by burning or cause instability and loss of synchronism. Thus the operating time for a relay is very crucial in protection. Modem relays operate in one cycle or even in half cycle.

17.1.3 SENSITIVITY

All relays are specified to operate under certain conditions. In an overcurrent relay, there is a pickup current. The overcurrent relay must definitely operate as soon as the current exceeds the preset pickup current. All relays should be sensitive to the prescribed conditions set for their operation.

17.1.4 RELIABILITY

Protection offered by relays should be absolute. That is to say that the various elements in protective relaying involved, viz., the relay, circuit breaker, the trip circuit, potential and current transformers, batteries etc., must all coordinate well and function as a single unit and act when an abnormal condition is sensed in the protected zone or part of the system. The relaying system should be reliable.

17.1.5 STABILITY

The relay should remain stable under conditions; it is not required to act even if an abnormal situation arises. For example if a fault occurs, before the concerned zone relay acts, the back-up relay should not operate. If the relay in protection zone does not operate within the time set, then only the relay in the back-up zone should operate.

17.1.6 ADEQUACY

Complete protection under all abnormal conditions that may arise may not be possible to be offered by any protective relaying scheme. However, reasonable protection must be provided for all equipment under usually occurring abnormal conditions. Here probability of occurrence of the abnormal condition, cost to be invested play a major role.

17.1.7 SIMPLICITY

The circuit, the scheme, and construction of relay should preferably be as simple as possible.

17.1.8 ECONOMY

Consistent with adequacy and reliability, the relay should be cheaper.

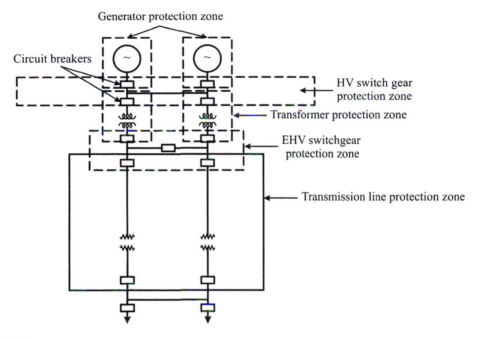

FIGURE 17.1

Zones of protection.

17.2 ZONES OF PROTECTION

Protective relaying should be a part and parcel of overall system planning, design, and operation. Circuit breakers are to be located at appropriate places. The components or equipment and parts of the system to be protected by the circuit breakers must be clearly identified.

Fig. 17.1 illustrates such an identification and demarcation of zones. Overlapping of the zones may be observed. Each zone generally covers one or two of power system elements. Location of current transformers (not shown) determines the boundaries of the zones. This is purely from protection point of view only.

17.3 PRIMARY AND BACKUP PROTECTION

Every zone identified for protection will have a suitable protection specified. If a fault occurs in that zone, it is the duty of the relays in that zone to identify and isolate the faulty element in that zone. The relay in that zone will be designated as primary relay. If for what so ever reason this primary relay fails to operate, there should be a second line of defense called back-up protection. The relay in the back-up protection is set to operate after a predetermined delay time given to primary relay so that continued existence of the fault in the system may not cause serious trouble.

It is understood from the above that when primary protection fails, the backup or secondary protection operates and saves the system. But, then a large part of the system may get isolated in such an event. This cannot be avoided.

17.4 IMPORTANT DEFINITIONS AND TERMINOLOGY

There are several terms used in relaying which need to be defined for proper understanding. Those terms that will be used frequently only will be defined here as the list of terms is large.

Relay: A relay is an automatic device that operates another electric circuit in response to a change in the same or another circuit.

Protective Relay: It is an automatic device that senses an abnormal condition in an electrical circuit and acts in such a manner that a circuit breaker isolates the faulty circuit element. The action of the circuit breaker is facilitated by closing the contacts of a trip circuit.

Primary Relay: The relay that is mainly made responsible to protect an equipment or a part of an electric circuit acting first before any other relay or device.

Measuring Relay: It is the main relay that measures the operating quantities for detecting the abnormal condition in operation.

Auxiliary Relay: It is a relay that assists the protective relay.

Back-up Relay: The relay that operates after a definite time lag after the occurrence of a fault in case the primary relay fails to initiate action.

Operating Torque: The torque that attempts to close the contacts of the relay trip circuit.

Restraining Torque: The torque that opposes the operating torque.

Pickup Level: The threshold value of the actuating quantity above which the relay operates.

Reset or drop-out level: The threshold value of the actuating quantity below which the relay gets deenergized and returns to its normal state.

Operating Time: It is the time that elapses from the instant at which the actuating quantity exceeds the pickup value to the instant when relay contacts close.

All other terms will be explained at the appropriate places.

17.5 CLASSIFICATION OF RELAYS

Based on technology, relays are classified as:

1. Electromagnetic relays
2. Static relays
3. Microprocessor-based relays

Based on their function, relays are classified as:

1. Overcurrent relays
2. Undervoltage relays
3. Impedance relays
4. Directional relays

5. Underfrequency relays
6. Thermal relays

Further the protective schemes are classified as:

1. Overcurrent protection
2. Distance protection
3. Carrier current protection
4. Differential protection

There are several principles employed in operating these relays. For example an electromagnetic relay may employ any of the following principles:

1. Attracted armature type relay
2. Balanced beam type relay
3. Induction disc relay
4. Moving coil or moving iron relay

Based on time of operation, relays are classified as:

1. Instantaneous relays
2. Definite time lag relays
3. Inverse-time relays
4. Inverse definite minimum time (IDMT) relays

A static relay many use an electronic comparator. Bucholtz relay uses gas pressure for operation.

Some important relays, their principle of operation, and application will be discussed in detail in the following sections.

17.6 BASIC PRINCIPLE OF RELAY MECHANISM

The basic principle of operation of a relay and circuit breaker is shown in Fig. 17.2. The current transformer CT senses the current in the line. Initially the switch S_1 is open. When the current in the secondary of CT exceeds a predetermined value, the relay coil gets energized and pulls the contactor down so that the switch S_1 is closed. The trip coil circuit is then energized by the battery and the normally closed switch S_2 gets open. The circuit breaker action is initiated, and the fault is eliminated. The usual ratings of current transformers are given in Table 17.1.

17.7 ELECTROMAGNETIC RELAYS

These relays generally use either plunger type (Fig. 17.2) or attracted armature type (Fig. 17.3). The actuating quantity is either current or current proportional to voltage.

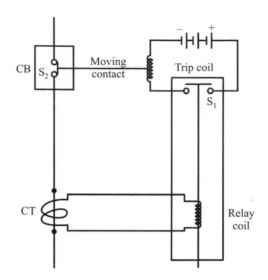

FIGURE 17.2

Basic principle of a relay operation.

Table 17.1 Standard CT ratios	
50:5	400:5
100:5	450:5
150:5	500:5
200:5	600:5
250:5	800:5
300:5	900:5
100:5	1200:5

The electromagnetic force experienced by the moving element is proportional to the square of air gap flux. The flux is proportional to the current carried by the coil, if saturation is neglected. The relay acts if the actuating force is greater than the restraining force.

i.e.,

$$K_1 I^2 > K_2$$

where K_1 is a constant depending on the coil-carrying current and K_2 is a constant of the restraining spring and friction if any.

At pickup the net force just becomes zero so that $I = \sqrt{K_2/K_1}$. The pickup current should be greater than this value.

These type of relays are faster in action, simple in construction, and are nondirectional.

Relays for action against overcurrent, overvoltage, undercurrent, and undervoltage in distribution circuits use this principle.

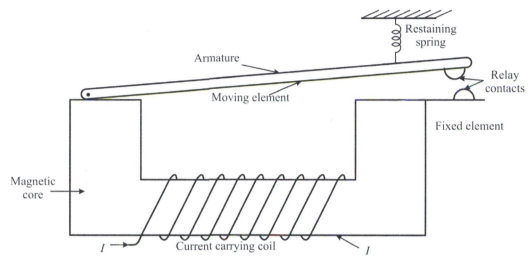

FIGURE 17.3

Attracted armature type relay.

17.8 INDUCTION RELAYS

It is well known that torque is produced whenever two fluxes interact which have displacement in both time and space. If a disc (say of aluminum) is mounted and free to rotate, and is placed in the field produced by alternating currents, then eddy currents will be induced in the disc. The interaction of the main flux and eddy current−produced flux will have both time and space displacement, and torque will be produced on the disc. Since, it is freely mounted, the disc will rotate. Rotation of the disc in one direction will cause the trip circuit to close; while rotation in the other direction is not permitted by a rest. This is the basic principle employed in induction relays.

17.8.1 TORQUE PRODUCTION IN INDUCTION DISC TYPE RELAY

Consider Fig. 17.4.

Let the flux systems be represented by

$$\phi_2 = \phi_{2max}\sin\omega t$$

$$\phi_1 = \phi_{1max}(\sin\omega t + \theta)$$

The currents associated with the flux are

$$i_{\phi1} \propto \frac{d\phi_2}{dt} \propto \phi_{2max}\cos\omega t$$

$$i_{\phi2} \propto \frac{d\phi_1}{dt} \propto \phi_{1max}\cos(\omega t + \theta)$$

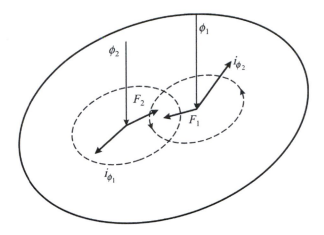

FIGURE 17.4

Torque production in an induction relay.

The forces produced by interaction of fluxes and currents are

$$F_1 \propto \phi_1 I_{\phi_2}$$

$$F_2 \propto \phi_2 I_{\phi_1}$$

The net force acting on the disc can be seen as the difference $F_1 - F_2 = F$

$$F \propto \phi_1 I_{\phi_2} - \phi_2 I_{\phi_1}$$

$$\propto \phi_{1max} \sin(\omega t + \theta)\phi_{2max} \cos \omega t - \phi_2 \sin \omega t \phi_{1max} \cos(\omega t + \theta)$$

$$\propto \phi_{1max}\phi_{2max}\sin\theta$$

The average net force or torque acting on the disc is proportional to $\phi_1\phi_2 \sin \theta$ where ϕ_1 and ϕ_2 are the rms values of fluxes.

17.8.2 INDUCTION RELAY CONSTRUCTION

Induction relays are available in three different types of construction (Fig. 17.5).

1. Shaded pole construction
2. Double winding or Watt-hour meter construction
3. Induction cup structure

1. *Shaded Pole Construction:* ϕ is the total flux. ϕ_2 is the flux through the shaded pole. ϕ_1 is the remaining flux through unshaded pole. The relevant phasor diagram is shown in Fig. 17.6. ϕ_2 lags behind ϕ_1 by α degree. V_1 and V_2 are the induced voltages lag behind their respective fluxes by 90 degrees. Eddy currents I_1 and I_2 flow in the disc due to these two fluxes. Since both the fluxes are produced by the same current

$$\text{Torque,} \quad T \propto I^2 \sin \alpha$$
$$= KI^2 \sin \alpha$$

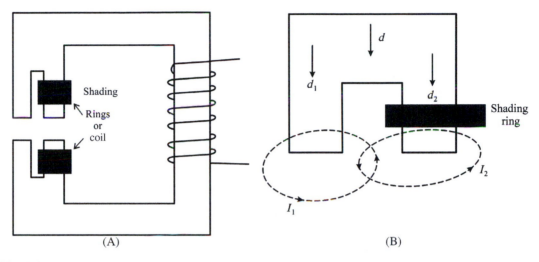

FIGURE 17.5

Shaded pole construction. (A) Structure and (B) fluxes and currents.

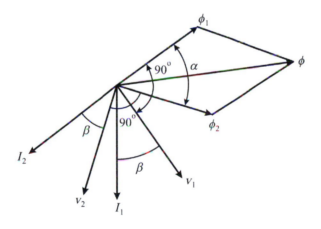

FIGURE 17.6

Phasor diagram for shaded pole type relay operation.

For a given design α is constant and therefore

$$T \alpha I^2$$

2. *Double Winding Relay Construction:* The construction of the relay is shown in Fig. 17.7. There are two electromagnets between which the disc is mounted. The upper magnet with three limb construction contains two windings on the central limb and one winding is the given supply. It carries current I_1. The second winding also on the central limb is also carried by the lower magnet which is of U-shape and is short circuited. Thus, with this arrangement, the secondary

winding carries current I_2 due to emf induced in it. The two currents have phase shift and produce fluxes through the disc which have both time and space shift. Thus torque is produced on the disc.

3. *Induction Cup Type Relay:* For high-speed induction relays, it is necessary to reduce the moment of inertia. Hence, instead of a disc element, a cup element is used with reduced inertia. A stationary iron core is placed inside a rotating cup element, so that air gap can be reduced without increasing inertia. The cup enveloped by four poles placed radially outside. The structure is shown in Fig. 17.8.

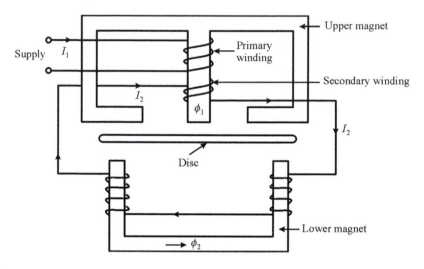

FIGURE 17.7

Double winding structure

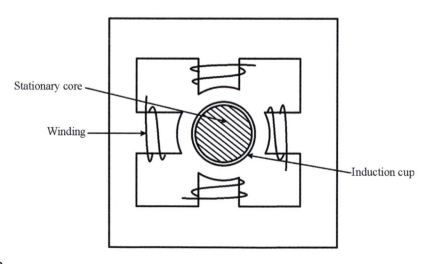

FIGURE 17.8

Induction cup type relay.

17.9 **OVERCURRENT RELAYS**

Depending on the operating characteristics, overcurrent relays are called:

1. Instantaneous overcurrent relays
2. Inverse-time current relays
3. Very inverse overcurrent relays
4. Extremely inverse overcurrent relays
5. Inverse definite minimum time current relays

Instantaneous overcurrent relays operate within 0.1 s and there is no intentional time delay introduced in the relay construction.

In an inverse-time overcurrent relay, it is so designed that the operating time decreases as the actuating quantity increases in its magnitude.

Induction relays operate basically with an inverse characteristic. The characteristic becomes flat when saturation sets in. Thus if it can be so designed that saturation occurs at a higher value of actuating quantity, the *very inverse characteristic* is obtained. If saturation occurs immediately after the pickup value, the *definite time characteristic* is obtained. The definite time, inverse, and very inverse characteristics are shown in Fig. 17.9 as (a), (b), and (c) respectively.

Likewise *an extremely inverse characteristic* may be obtained that may fit into an equation of the type

$$I^n t = \text{constant}$$

where I is the actuating current and t is the operating time curve (curve (d)). n is about $2-8$.

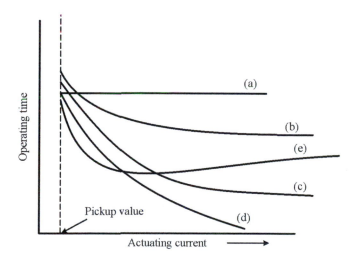

FIGURE 17.9

Time—current characteristics.

17.9.1 INVERSE DEFINITE MINIMUM TIME OVERCURRENT RELAY

In this type of relay near pickup value, the relay exhibits inverse characteristic, but at higher currents the relay operating time becomes substantially constant (curve (e)). Considering the expression

$$I^n t = k$$

An IDMT characteristic may be obtained by varying the value of n. As per British standards

$$t = \frac{K}{I^n - 1}$$

where in an IDMT curve may be obtained with $n = 0.02$ and $K = 0.14$, very inverse curve with $n = 1$ and $K = 13.5$ and an extremely inverse curve with $n = 2$ and $K = 80$.

17.10 NONDIRECTIONAL OVERCURRENT RELAY

An induction type nondirectional overcurrent relay is shown in Fig. 17.10. At the lower part of the figure the disc type induction relay elements discussed in Section 17.8.1 can be seen. The primary winding connected to the upper electromagnet has tappings on the central limb and supply is given to the two terminals of this winding. The secondary winding as both the upper and lower electromagnets is short circuited. The flux due to main current and the flux produced by induced currents create torque as discussed earlier. The moving contact attached to the spindle of the disc is free to rotate, and the angle of travel before the trip contacts (fixed contacts) are closed is adjustable. In this way time to close the trip circuit of the circuit breaker is controlled.

The number of turns in the primary winding is varied through the tappings which are connected to a plug setting bridge as shown. Using the plug setting provision, the overcurrent specified for operation is selected. This is usually in steps of say 10%, 20% or 25% or 50% to 200%. A current transformer is used as mentioned earlier with 5-A secondary rating. A 5-A secondary current with 75% setting will operate at 3.75 A. In this manner by selecting the plug setting in the plug setting bridge, the same relay can be operated for different current ratings. The plug setting multiplier is defined as

$$PSM = \frac{\text{Primary current}}{\text{Primary setting current}}$$

PSM is also called current setting multiplier.

$$PSM = \frac{\text{Primary current}}{\text{Relay current setting} \times \text{CT ratio}}$$

It has been said that by adjusting the disc position, the angle of travel or time required for closing the trip circuit contacts is varied. A time setting dial calibrated from 0 to 1 attached to it indicates the multiplier to be used to convert the time from the relay specifications to obtain actual operating time.

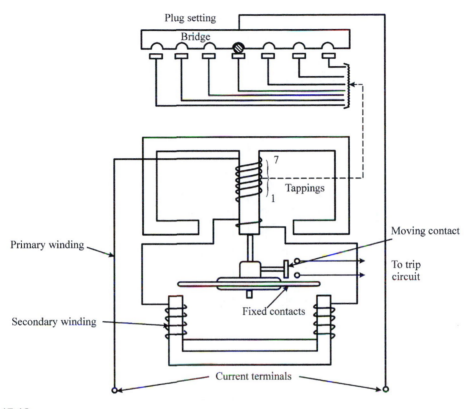

FIGURE 17.10

Nondirectional overcurrent relay (induction type).

Fig. 17.11 shows such time–current characteristics on logarithmic scale.

The relay operating time from PSM and TMS can be determined as explained in the following.

Let the relay be rated at 5 A. Let relay current setting = 150% and TMS = 0.3. Let CT ratio 400/5 = 80. If the fault current is say, 7500 A then

Relay current setting = $5 \times 1.5 = 7.5$ A

$$PSM = \frac{7500}{7.5 \times 80} = 12.5$$

The operating time for TMS = 1.0, for PSM = 12.5 is about 2.8 s. Then the actual operating time for TMS = 0.3 is $2.8 \times 0.3 = 0.84$ s.

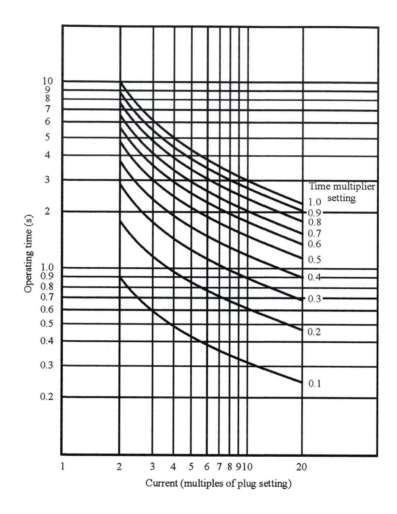

FIGURE 17.11

Time–current characteristics.

17.11 DIRECTIONAL RELAY

A directional relay is also called reverse power relay. Fig. 17.12 shows an electromagnetic directional relay. The relay is energized by two fluxes, one produced by current and the other produced by voltage applied to the two coils as shown in figure. The interaction of the two fluxes ϕ_1 and ϕ_2 produce torque T.

$$T \; \alpha \; VI \cos \phi \; \alpha \; \phi_1 \, \phi_2 \sin \theta$$

where ϕ is the power factor angle (angle between voltage and current) and

$$\theta = (90 - \phi).$$

The relay can be set to operate when the direction of flow of current is reversed.

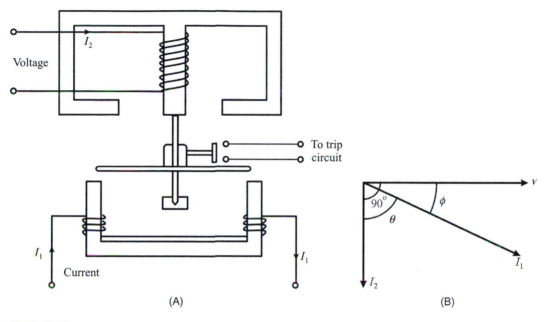

FIGURE 17.12

Directional relay. (A) Directional relay connections and (B) phasor diagram.

17.12 DIRECTIONAL OVERCURRENT RELAY

A directional overcurrent relay operates when current exceeds a preset value in a given direction. The schematic diagram for such a relay is shown in Fig. 17.13. The upper two electromagnets and the windings perform the same function as in a directional relay. The lower two electromagnets are related to overcurrent relay operation. The contacts of the directional relay element are connected in series with the lower magnet of the nondirectional overcurrent unit at the lower part. In the event of a fault the fault current in the current coil of the relay produces flux in the directional unit while the voltage coil in the directional unit produces another flux. If the interaction of these fluxes produces torque that closes the trip contacts, then the relay coil in the overcurrent relay gets energized. This flux induces voltage in the winding and the interaction of the two fluxes in the lower unit closes the circuit breaker trip circuit.

17.13 FEEDER PROTECTION

Feeder protection, in fact, encompasses all types of line protection including transmission, subtransmission, primary and secondary distribution level, as well as feeders. Overcurrent protection is conveniently applied for feeder protection since it is simple and economical. Overcurrent protection is applied for both phase faults and earth faults in station service. A fault which involves earth is an earth fault. A line-to-ground, double line-to-ground fault are typical earth faults. A line-to-line fault

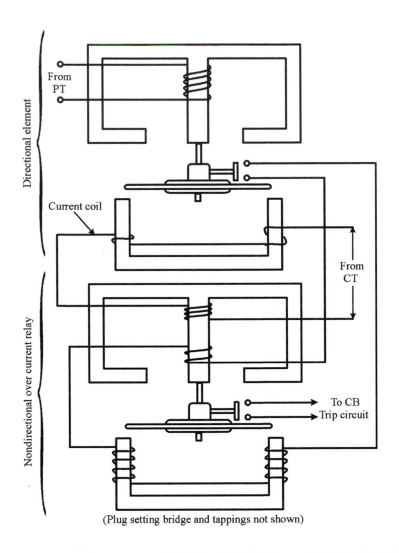

FIGURE 17.13

Directional overcurrent relay.

is a phase fault. These protective schemes will be elaborated later. Discrimination in feeder protection using overcurrent relays is obtained either by time-graded system or by current-graded system or a combination of both.

17.13.1 TIME-GRADED SYSTEM

In this method operation of relays is selected by using different time settings for relay operation. Consider the radial feeder shown in Fig. 17.14 with four sections. Each section is protected by

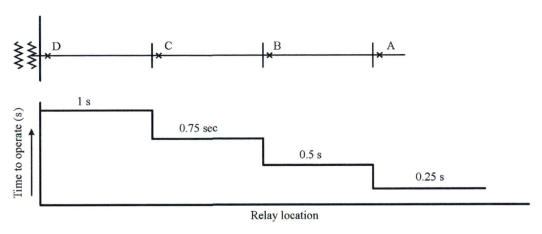

FIGURE 17.14

Time-graded feeder protection.

overcurrent relays A, B, C, and D, respectively, as shown. The relay at A is set to operate for a fault in section beyond A in 0.25 s. The relay at B will operate if the relay A does not operate 0.25 s later (i.e.,) after 0.5 s of occurring of fault and trip the circuit breaker at B.

If the relay at B also fails to operate, then the relay at C operates and the circuit breaker at C opens and clears the fault in 0.75 s.

The drawback of this simple scheme is that if a fault occurs near the source, i.e., in section DC, then the relay at D has to wait for 1 s to clear the fault, since its setting is for 1 s in a time-graded system. Heavy short-circuit current flowing for a longer time could cause more destruction and needs to be cleared as soon as possible. Further the number of sections for time grading are to be limited for the reason explained above.

17.13.2 CURRENT-GRADED SYSTEM

It is generally acknowledged that the fault current is maximum near the source and gradually reduced as we move away from the source. The pickup currents are set to increase progressively as the source is approached from the farther side of the feeder. If high-speed instantaneous overcurrent relays are used, the disadvantage of time-graded system can be avoided. However, in practice, it is difficult to determine the magnitude of the fault current, hence selectivity of relay suffers. In principle all relays are to be set to operate at the same time setting, but the current setting follows the fault current level expected for the feeder section to be protected. A typical current-graded protection is shown in Fig. 17.15.

17.13.3 COMBINED TIME–CURRENT GRADING

This type of protection is applied generally to distribution lines with IDMT relay. If they are found to be slow at low level of overloads, extremely inverse and very inverse characteristic relays are

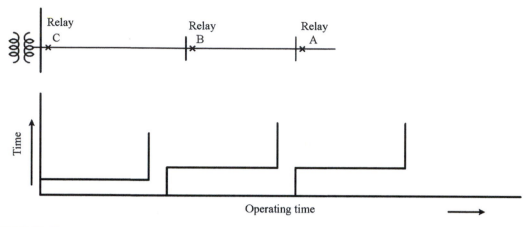

FIGURE 17.15

Current-graded system.

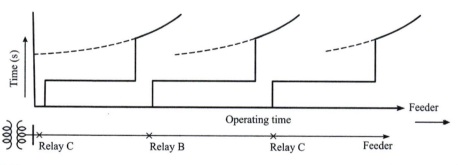

FIGURE 17.16

Combined time–current-graded protection.

also used. The current setting is made as per the fault current level of the section to be protected. Further the relays are set to pickup in a progressive manner higher current levels as the source is reached. The characteristics for a combined protection of this nature are shown in Fig. 17.16.

17.14 EARTH FAULT PROTECTION USING OVERCURRENT RELAYS

Fig. 17.17 shows the connection of relay and CTs operating on residual current principle. When there is no earth fault, the sum of the secondary currents in R, Y, B phases I_{RS}, I_{YS}, and I_{BS} is given by

$$I_{RS} + I_{YS} + I_{BS} = 0$$

During an earth fault, the sum which is called the residual current is not zero and hence, the relay carries current and trips the circuit breaker.

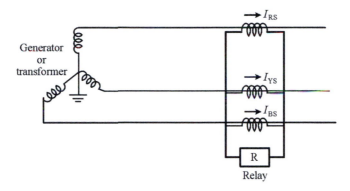

FIGURE 17.17

Earth fault protection with overcurrent relays.

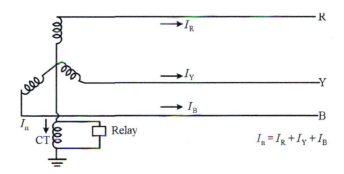

FIGURE 17.18

Earth fault protection.

An alternate method of earth fault protection is to place the relay in the neutral-to-earth circuit as shown in Fig. 17.18.

17.15 COMBINED EARTH FAULT AND PHASE FAULT PROTECTION

In principle such a protection can be offered using the relays for three phases and one relay for earth fault. However, since when phase faults occur current in any two phases must increase, only two relay units are sufficient for phase faults. This commonly employed scheme of protection is shown in Fig. 17.19.

17.16 PHASE FAULT PROTECTION

The simple phase fault protection scheme is illustrated in Fig. 17.20.

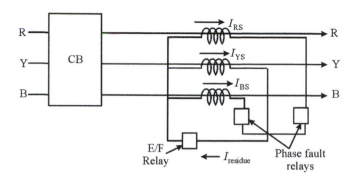

FIGURE 17.19

Combined earth fault and phase fault relay protection scheme.

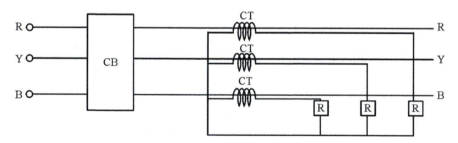

FIGURE 17.20

Phase fault protection.

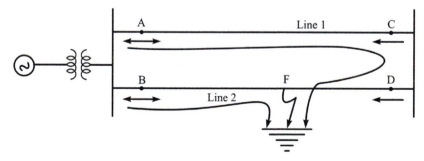

FIGURE 17.21

Parallel feeder protection by overcurrent relays.

17.17 PROTECTION OF PARALLEL FEEDERS

Overcurrent relays can be used for parallel feeder protection. This is illustrated in Fig. 17.21. Two feeders are connected in parallel. Both the feeders are supplied from one end.

A, B, C, and D are the relays located at both ends of the two feeders. It is desired that in the event of a fault on any feeder, that feeder should be isolated, and supply should continue through the other healthy feeder. If a fault occurs on feeder 2 at F, the fault current trips the breaker at B first. Later the current in feeder 1 finding its way into feeder 2 will be detected by the directional relay D which by tripping will completely isolate the faulty feeder 2. The supply continues overfeeder 1. A and B are nondirectional overcurrent relays while C and D are directional relays.

17.18 PROTECTION OF RING MAINS

Protection of ring mains is generally a complex process and is also expensive compared to radial feeder protection. Consider the ring main system shown in Fig. 17.22. One infeed or source and the three substations are interconnected. Overcurrent relay protection with a time grading of 0.2 s is used with a definite minimum time of 0.1 s with a maximum time of 0.7 s, only four sections forming the main can be protected. Since the source end lines carry the full current, the maximum delay in operation cannot be set too high.

The time grading of relays placed from A to B in clockwise direction is shown in Fig. 17.22B, and the time grading of relays from B to A counter clockwise is indicated in Fig. 17.22C. It can be easily seen that the number of section that can be protected is very much limited since the time setting at the source side cannot be increased beyond a tolerable value from safety point of view.

17.19 UNIVERSAL TORQUE EQUATION

In all electromagnetic relays torque is produced by the interaction of two flux systems. They can be produced by current flowing through a coil and current proportional to voltage applied to another coil.

Torque produced by current winding is proportional to I^2 where I is the current in the coil i.e.,

$$\text{Torque} \propto K_1 I^2$$

Torque produced by voltage is proportional to V^2 where V is the voltage applied. This torque $= K_2 V^2$. Generally the torque produced by current is positive and is called operating torque. The torque produced by voltage is so arranged as to produce a negative torque and is called restraining torque.

Torque produced by the interaction of both current I and current proportional to voltage V is proportional to $VI \cos (\theta - \alpha)$ where θ is the angle between V and Z and α is a design constant of the relay and is also called torque angle. Hence the total torque equation for a relay can be written as

$$T = K_1 I^2 - K_2 V^2 + K_3 VI \cos (\theta - \alpha) - K_4$$

where K_1, K_2, K_3, and K_4 are constants.

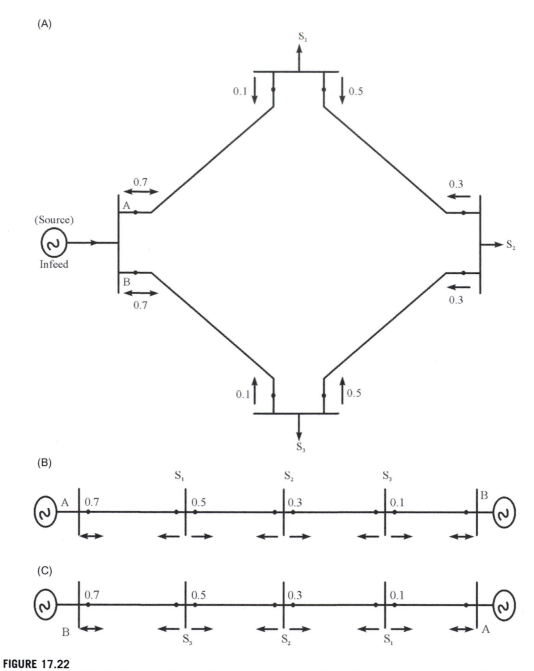

FIGURE 17.22

Ring main protection. (A) Ring main protection, (B) clockwise, and (C) counter clockwise.

The relay thus has a threshold condition given by

$$K_1 I^2 - K_2 V^2 + K_3 VI \cos(\theta - \alpha) - K_4 = 0$$

Operating torque that makes the above quantity just positive enables the relay to act.

17.19.1 OVERCURRENT RELAYS

In overcurrent relays there is no voltage coil. Hence

$$T = K_1 I^2 - K_4 = 0$$

$$I^2 = \frac{K_4}{K_1} \text{ or } I = \sqrt{\frac{K_4}{K_1}}$$

For current greater than this setting, the relay acts.

17.19.2 DIRECTIONAL RELAYS

In directional relays the torque is interactive of V and I.
    Hence

$$T = K_3 \, VI \cos(\theta - \alpha) - K_4 = 0.$$

The relay will pick up if $VI = \cos(\theta - \alpha) > K_4/K_3$

17.20 DISTANCE PROTECTION

Distance protection of transmission lines is universal. In this protection torque is produced by both current and voltage in the circuit. It is a high-speed protection and is simple to apply.

Consider a transmission line AB of length L. Its impedance is Z when measured for the entire length L (Fig. 17.23).

It gradually decreases along the line linearly, till the end B is reached where it is zero. The line carrying current I, produces a voltage drop of $I \cdot Z$ at B and this drop is zero at A and rises gradually to $v = I \cdot Z$ at B. If a fault occurs at F on the line, and the fault impedance is zero, the voltage at F falls to zero. Impedance of the line is a measure of the distance of the line from the relay location. If the impedance seen by the relay falls below a certain value, the relay will operate. If the line current and line voltage are stepped down by suitable CT and PT and are applied to the balanced beam relay shown in Fig. 17.24, then V/I is an impedance set for relay operation. If V/I is less than Z_S, the relay operates, otherwise remains open where Z_S is the relay setting.

Consider the universal torque equation simplified for this case,

$$T = K_1 I^2 - K_2 V^2 - K_4$$

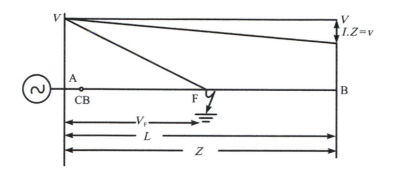

FIGURE 17.23

Distance protection.

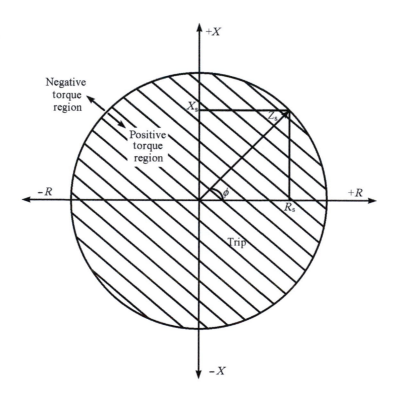

FIGURE 17.24

Balanced beam type impedance relay.

For operation, neglecting K_4,

$$K_1 I^2 - K_2 V^2 = 0$$

$$\frac{V^2}{I_2^2} = \frac{K_1}{K_2} \text{ or } \frac{V}{I} = \sqrt{\frac{K_1}{K_2}} = Z_S$$

The relay operates if the impedance falls below Z_S. This is illustrated in Fig. 17.25.

The locus of impedance $Z_S = \sqrt{R_S^2 + X_S^2}$ is a circle with Z_S as radius. Any value of Z less than Z_S will operate the relay. For $Z > Z_S$, the relay will restrain.

Z is plotted on $R - X$ diagram

$$R_S = Z_S \cos \phi \text{ and } X_S = Z_S \sin \phi$$

For values of impedance outside the circle with radius Z_S, the relay will not operate. The relay characteristic is a circle in $R-X$ plane.

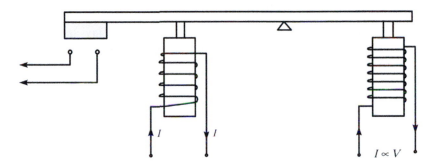

FIGURE 17.25

Principal of impedance relay operation.

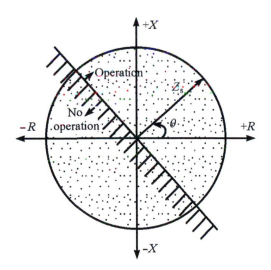

FIGURE 17.26

Impedance relay characteristic with directional unit.

17.20.1 IMPEDANCE RELAY WITH DIRECTIONAL UNIT

Impedance relays use either balanced beam structure or induction cup structure as they have no directional property. However, by incorporating a directional element, they can be made sensitive to direction of power flow. In such a case the operating characteristic of an impedance relay will have the directional feature incorporated as shown in Fig. 17.26.

17.21 THREE-ZONE PROTECTION WITH IMPEDANCE RELAYS

For three-zone protection of a line, three units of an impedance relay are required. The first unit protects 75%−90% of the line. This is shown in Fig. 17.27, where there are three line sections AB, BC, and CD. The relay at A offers primary protection to section AB, but only 75% to at most 80% of the line. Its operation should be instantaneous (1 or 2 cycles). Hundred percent protection cannot be ensured as the relay measurement of impedance is not accurate toward the end of the line and tripping should be avoided due to overreach. (This will be explained later.)

The second zone of protection given by the second unit of the relay takes care of underreach due to faults at the line end (arcing resistance), intermediate current sources, as well as errors in CT and PT measurements.

The second unit protects the remaining length of first section (i.e.) about 25%−20% of AB and up to 50% of the next shortest adjoining section. It operates with a time delay of about 0.2−0.4 s.

The third unit gives protection as a backup relay. It extends its zone of protection to the entire next section of the line BC and up to about 25% of the third section of the line CD so that underreach is completely taken care.

The starting relay initiates action and time is counted. Z_1 closes instantaneously and the trip circuit is closed. If Z_1 does not operate Z_2 closes after t_2 s delay and the trip coil is activated. Even if it does not act the backup protection provided by unit 3 closes T_3 set by Z_3. A directional unit in series for operation is an additional feature. For fault in zone Z_1, Z_2 and Z_3 will start, but Z_2 operates in time t_2. For faults outside Z_2, Z_3 only will start and removes the fault in time t_3.

17.22 IMPEDANCE TIME RELAY

An induction type impedance time relay is shown in Fig. 17.28. The lower electromagnets with current coil and a short-circuited secondary winding produce torque on the disc proportional to current I, the operating quantity. A restraining torque is applied on the spindle 2 shown above the lower part. When the operating torque exceeds the restraining torque due to voltage coil, the pull exerted by restraining magnet and coil must be exceeded by the rotation of spindle 1 coupled to spindle 2

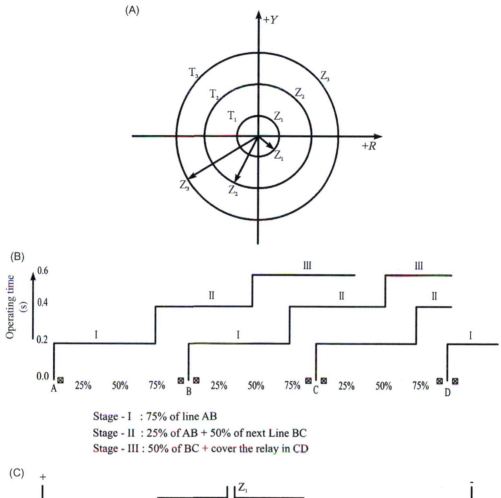

(A)

(B)

Stage - I   : 75% of line AB
Stage - II : 25% of AB + 50% of next Line BC
Stage - III : 50% of BC + cover the relay in CD

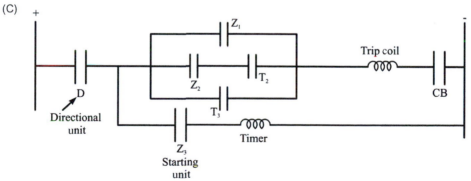

(C)

FIGURE 17.27

Three-zone protection with impedance relays. (A) Impedance relay characteristics for three zones, (B) distance covered in the three zones, and (C) the relay connection diagram.

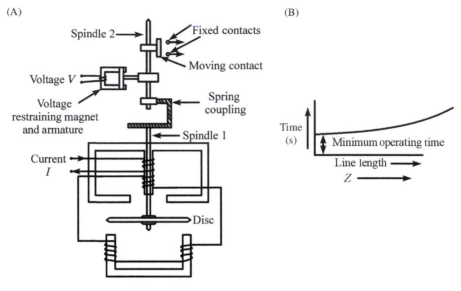

FIGURE 17.28

(A) Impedance time relay and (B) relay characteristic.

through a spring coupling. Greater the restraining force, greater would be the travel of disc to close the trip circuit.

Time to close the trip circuit t is proportional to V/I

$t \propto (V/I)$ (i.e.) $t \propto Z$ or distance of the line.

The characteristic of such a relay is shown in Fig. 17.28B.

17.23 REACTANCE RELAY

A reactance relay has a reactance measuring unit. This reactance measuring unit contains an overcurrent sensing unit and a directional unit. A reactance relay is an overcurrent relay with directional unit. Consider the universal torque equation applied to reactance relay:

$$T = K_1 I^2 - (+K_2)VI \cos (90 - \theta) - K_4$$

Neglecting the term K_4 the threshold condition for operation is

$$K_1 I^2 - (+K_2)VI \sin \theta = 0$$

The operating torque is produced by the interaction of fluxes on Poles 1, 2, and 3 (see Fig. 17.29). Current through windings on Poles 1 and 3 produce polarizing flux. The current through operating coil on Pole 2 which is in phase quadrature with the flux produced by the polarizing windings on 1 and 3, due to the phase shifting winding added, will produce an operating torque proportional to I^2 or equal to $K_1 \cdot I^2$. The flux produced by the interaction of the currents in winding on Pole 4 and the polarizing windings on 1 and 3 produce a restraining torque. The supply to restraining coil is derived from voltage and is also connected to a phase adjustment circuit. Thus the restraining torque is made proportional to $VI \sin \theta$ or equal to $K_2 VI \sin \theta$, since it opposes the operating torque. The relay operates if

$$K_1 I^2 > + K_2 VI \sin \theta$$

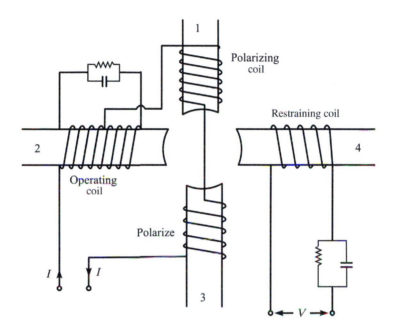

FIGURE 17.29

Reactance relay.

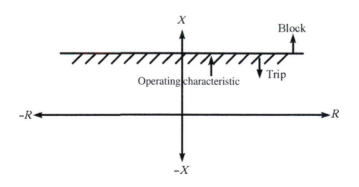

FIGURE 17.30

Operating characteristic of a reactance relay.

or

$$\left|\frac{V}{I}\right|\sin\theta < \frac{K_1}{+K_2}$$

(i.e.,) $Z\sin\theta < k$ where $|V/I| = z$ and $K_1/K_2 = K$

The relay operates, if the reactance $X < k$. The characteristic of the relay is shown in Fig. 17.30. The reactance relay is a nondirectional relay. To make it directional, a starting relay which is directional has to be used. A mho relay which has a circular characteristic is used for this purpose.

17.24 THREE-ZONE PROTECTION WITH REACTANCE RELAY

Protection in z-plane is shown in Fig. 17.31 and the relay connections in Fig. 17.32.

If a fault occurs on the line in zone 1 protection, all the units S, x_1, and x_2 start. As x_1 is set at the lowest, contacts in x_1 unit close first, which causes the CB to operate. Otherwise, at the next time setting x_2 operates and closes the trip circuit. In zone 3 when the timer in starting unit reaches its set value T_3 closes and trip the circuit breaker circuit and operates it.

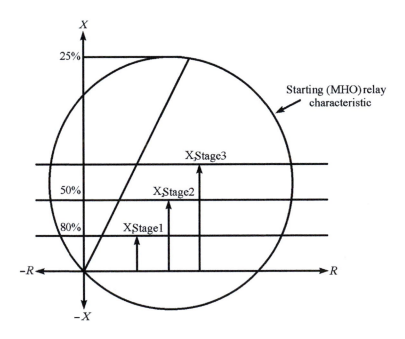

FIGURE 17.31

Three-zone line protection using reactance relay.

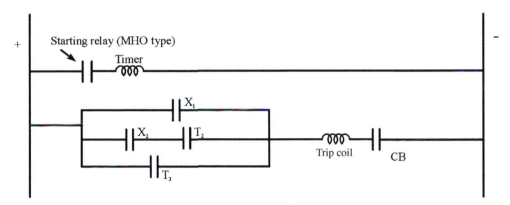

FIGURE 17.32

Reactance relay connection diagram. X_1—High-speed unit-Protects. 80% of line; X_2—Protects up to 50% of next line; X_3—Backup to protect the adjacent line completely.

17.25 MHO RELAY

In this relay (shown in Fig. 17.33) the polarizing flux is produced by current proportional to voltage flowing through Poles 1 and 3. Pole 4 carrying current produces operating torque. Proportional $VI \cos(\theta - \alpha)$ by interacting with the polarizing flux. The operating torque can be written as

$$T_{OP} = K_1 VI \cos(\theta - \alpha)$$

α can be adjusted by varying the resistance in the phase shifting circuit. The restraining torque produced by current on Pole 2 interacting with polarizing flux is equal to $K_2 V^2$. The torque equation.

$$T = K_1 VI \cos(\theta - \alpha) - K_2 V^2 + K_3$$

Neglecting K_3 at balance,
when $K_1 VI \cos(\theta - d) > K_2 V^2$

$$\frac{I}{V} \cos(\theta - \alpha) > \frac{k_2}{k_1}$$

i.e., for relay to operate

$$y \cos(\theta - \alpha) > \frac{k_2}{k_1} \quad \text{or} \quad \frac{1}{y\cos(\theta - \alpha)} < K$$

where

$$\frac{k_2}{k_1} = K$$

Also, when $z/(\cos(\theta - \alpha)) < K$ the relay operates

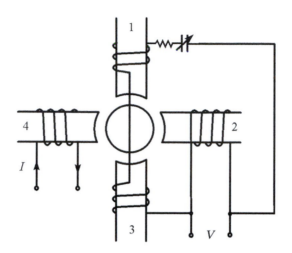

FIGURE 17.33

Induction cup type MHO relay.

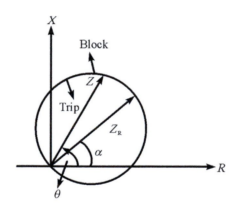

FIGURE 17.34

Characteristic of MHO relay.

Let $K = Z_R$ the relay setting value

$$z < z_R \cos(\theta - \alpha)$$

The relay operates as follows

$$|z - z_R| < |z_R| \text{—Trips}$$
$$|z - z_R| > |z_R| \text{—Blocks}$$

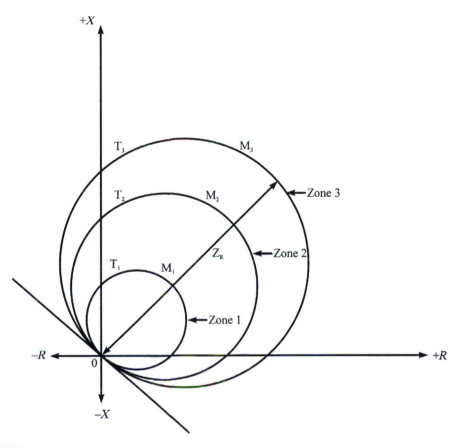

FIGURE 17.35

Live protection with MHO relay.

The characteristic is shown in Fig. 17.34. Three-zone protection using MHO relay is shown in Fig. 17.35. The MHO relay having a circular characteristic is inherently directional. It does not need a starting unit. The relay connections are similar to reactance relay.

17.26 OVERREACH AND UNDERREACH IN DISTANCE RELAYS

It has been explained that a distance relay operates when the impedance or a component of the impedance as seen by the relay falls below a preset value. This preset impedance or component of impedance is called reach of the relay. However, it is the general practice to call the corresponding distance on the protected line by the relay as reach of the relay.

Under some conditions, the relay may operate even when a fault location is not within the reach of the relay, but lies beyond it. This is called overreach. On the other side, the relay may fail to operate sometimes, even though the fault is located within the reach of relay. This is called underreach.

17.27 EFFECT OF ARC RESISTANCE ON DISTANCE RELAY PERFORMANCE

The arc resistance is given empirically by

$$R_{arc} = \frac{29 \times 10^3}{I^{1.4}} \cdot l \ \Omega$$

where l is the length of the arc and I is the fault current. The arc resistance which is variable and is also dependent upon cross-winds is added to the line impedance. A distance relay sees impedance of the line and the fault resistance both of which the latter component is variable. For ground faults, earth resistance and tower footing resistance also matter in addition to arc resistance. Fig. 17.36A explains the effect of arc resistance on the operation of impedance relays.

If a fault occurs at any point F on a line protected by distance relay and if Z_F is the imped-ance up to F, then with an arc resistance R the relay sees a total impedance of $Z_F + R$. If the arc resistance is more than R, the value of impedance will be outside the circle and the relay fails to operate. The relay can protect the line for a length corresponding to Z_F only when arc resistance is R. For any arc resistance greater than R, the relay fails to operate. With R, the relay is said to underreach, as it cannot protect the line beyond the point F up to its end. Z_L is the impedance of the total length of the line and $Z_L - Z_F$ is the (impedance of) the line length left unprotected.

With MHO type relay (Fig. 17.36B), the relay protects a line of impedance Z_F only, while the relay sees an impedance of $Z_F + R$. The relay operates so long $Z_F + R$ is less than the diameter of the circle, even if $Z_F + R$ is greater than Z_L the total line impedance.

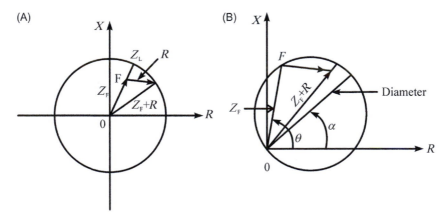

FIGURE 17.36

Effect of arc resistance on distance relay operation.

17.28 **STATIC RELAYS**

The first-generation static relays contained discrete components fitted on printed circuit boards. The second-generation static relays used integrated circuits. Compact relays with ICs and PCBs are now common. Analog circuit relays are being replaced by digital circuits. Thus digital or numerical relays have come into use.

With the development of microprocessor-based multifunctional systems, microprocessor-controlled relays came into existence. They perform measurement, data transmission, protection, as well as control functions. Programmable relays are being developed instead of fixed wired relays.

In a static relay comparison of electrical quantities is performed by a static circuit. The circuit then gives an output signal that trips the circuit breaker. A large number of electronic components and devices are used in static relays. They include the following:

1. Diodes and transistors of silicon-type Zener diodes are used as voltage-regulating diodes.
2. Junction transistors (NPN and PNP) type are used in amplifiers, level detectors, and switching circuits.
3. PNPN devices are used in switching or tripping circuits. There are several thyristors such as reverse blocking thyristors, bidirectional thyristor, turn-off thyristor, and p-gate thyristor.
4. Triac which is a further development of thyristor is used as an output element.
5. Thermistors are used for temperature measurement and temperature compensation.
6. A pair of transistors may be used as Darlington circuit or Schmitt trigger circuit.
7. An unijunction transistor may be used for firing a thyristor.

The principles, operation, characteristics, and other information related to these components and elements will not be described here as the student is expected to have knowledge about them elsewhere.

Induction cup type relays widely used for distance and directional relays are replaced mostly by rectifier bridge type static relays. For overcurrent relaying induction, disc relays are still more commonly used. However, static relays are slowly gaining ground and replacing electromagnetic relays.

17.28.1 **COMPARATORS**

Upon the occurrence of a fault, the magnitudes of voltage and current, and the phase angle between them change from their values under normal condition. Static relaying is designed to detect these changes and recognize a faulted condition from a healthy condition. For this purpose in one method the magnitudes of voltage or current are compared. In another method the phase angles between voltage and current are compared so that a faulty situation is recognized and action is initiated to trip the circuit breaker circuit. In an amplitude comparator the magnitudes of two actuating quantities are compared, while in a phase comparator, the phase angles are compared.

17.28.2 **AMPLITUDE COMPARATORS**

An amplitude comparator compares the magnitudes of two input quantities. The two input quantities are operating or actuating quantity and the other is restraining quantity. If the operating quantity (A) is greater than the restraining quantity (B), then the relay sends a signal to trip the circuit breaker. In an amplitude comparator the phase angle between A and B is of no consequence.

The comparator operates when $|A| > |B|$ (see Fig. 17.37). It is interesting to note the duality between amplitude and phase comparators. Consider an amplitude comparator with inputs $A + B$ and $A - B$ as shown in Fig. 17.37. The relay in this case operates if $|A + B| > |A - B|$ only when $\phi < 90$ degrees. This is illustrated in Fig. 17.38A−C.

Thus with an amplitude comparator it is possible to obtain phase discrimination.

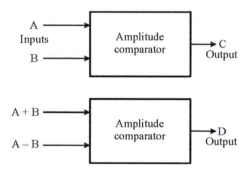

FIGURE 17.37

Amplitude comparator.

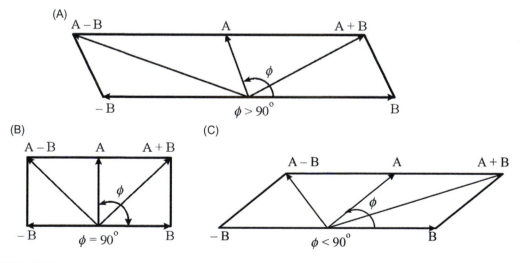

FIGURE 17.38

Amplitude comparator for phase comparison. (A) Operates when $|A + B| < |A-B|$, (B) $|A + B| = |A - B|$, and (C) $|A + B| > |A - B|$.

17.28.3 PHASE COMPARATOR

The block schematic of a phase comparator is shown in Fig. 17.39.

Let the phase comparator compare the phases of two input signals X and Y. The comparator operates if the phase angle between X and Y is less than 90 degrees.

If the input signals are changed to X + Y and X − Y, with these changed input signals, it can be verified from Fig. 17.40A−C that the relay operates only when the phase angle between X + Y and X − Y is less than 90 degrees, in which case |X| > |Y|.

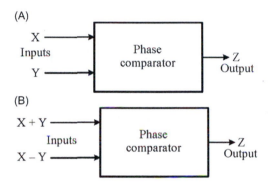

FIGURE 17.39

Phase comparator. (A) Phase comparator with inputs X and Y and (B) phase comparator with inputs X + Y and X − Y.

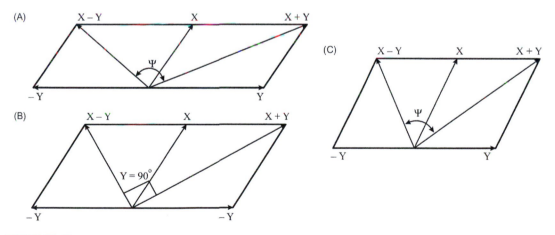

FIGURE 17.40

Phase comparator for amplitude comparison. (A) |X| > |Y|; Ψ > 90 degrees, (B) |X| = |Y|, and (C) |X| > |Y|; Ψ < 90 degrees.

17.28.4 RECTIFIER BRIDGE TYPE AMPLITUDE COMPARATORS

A typical rectifier bridge type amplitude comparator is shown in Fig. 17.41A. They are used in the construction of static overcurrent relays and distance relays. The operating and restraining inputs

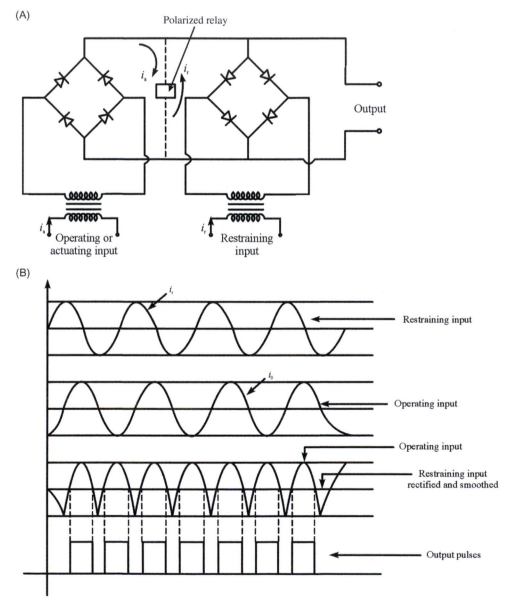

FIGURE 17.41

(A) Rectifier bridge amplitude comparator and (B) rectifier bridge type amplitude comparator waveforms.

are both rectified and as they are connected in opposition, it enables the polarized (slave) relay to operate when the operating signal exceeds the restraining signal. The waveforms are shown in Fig. 17.41B.

17.28.5 PHASE SPLITTING IN AMPLITUDE COMPARATORS

In phase splitting methods the input signals are split into components with phase difference. In one method the input is split into six components with 60 degrees apart. This enables the output to be smoothed within 5%. Both the operating and restraining signals are smoothed before they go to level detector. A continuous output signal is made available. The phase shifting circuit is shown in Fig. 17.42.

There are several types of phase comparators; coincidence type phase comparators and vector product type phase comparators are commonly used.

17.28.6 COINCIDENCE TYPE PHASE COMPARATORS

Consider two sine wave signals A and B with a phase shift of ϕ degrees (see Fig. 17.43).

The coincidence of the two waveforms for positive polarity is shown in Fig. 17.43A. The period of coincidence is $180 - \phi = \psi$ degrees. It can be seen clearly that if $\phi < 90$ degrees then $\psi > 90$ degrees. The relay is so designed to trip for $\phi < 90$ degrees, or $\psi > 90$ degrees.

The period of coincidence (ψ degree) is measured by various techniques. In phase splitting type phase comparator the input is split into two components with ± 45 degrees from the input wave.

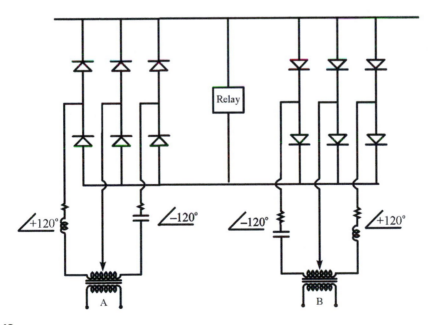

FIGURE 17.42

Phase splitting type amplitude comparator.

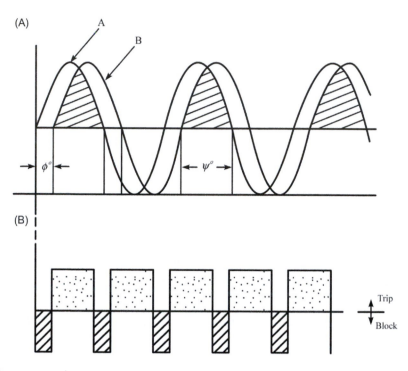

FIGURE 17.43

Coincidence type phase comparator. (A) Coincidence of two sine waves and (B) application to relay.

The four components of the two quantities, operating signal A and restraining signal B are fed into an AND gate. Tripping takes place when all the four components are positive as shown in Fig. 17.44.

There are several other phase comparators such as integrating type phase comparators where the coincidence time is measured for each cycle by integrating the output from the coincidence detector.

In vector product phase comparators magneto resistivity and Hall effect are used. In certain semiconductor materials if two voltages V_1 and V_2 are applied, the current flowing in the material will be proportional to $V_1 V_2 \cos \phi$ where ϕ is the phase angle between V_1 and V_2. A disc of such a material is used for phase detection.

17.29 STATIC OVERCURRENT RELAY

The functional block diagram for a static instantaneous overcurrent relay is shown in Fig. 17.45.

The input circuit can be adjusted for different current settings. The rectified and smoothed output includes a Zener diode to limit the input current to safe value. The level detector has a preset pickup value. The output element may be any relay or thyristor circuit.

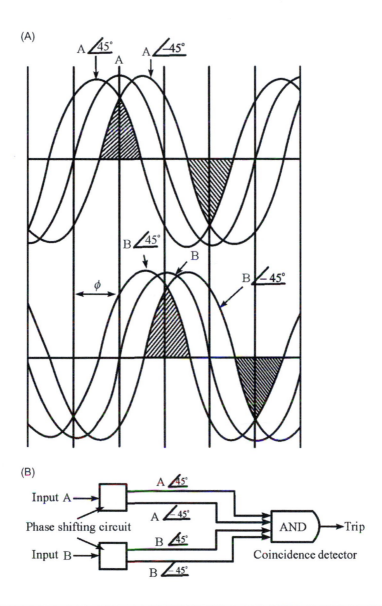

FIGURE 17.44

Phase splitting type phase comparator.

17.30 **DEFINITE-TIME OVERCURRENT RELAY**

To obtain the time—current characteristic of desired relationship, a timing circuit can be incorporated as shown in Fig. 17.46.

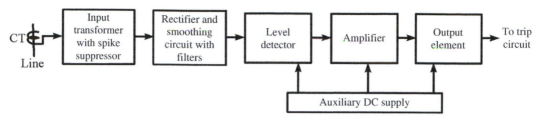

FIGURE 17.45

Static overcurrent relay block diagram.

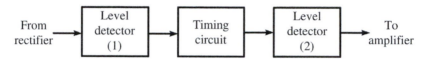

FIGURE 17.46

Definite-time overcurrent relay.

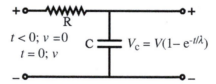

FIGURE 17.47

Timing circuit.

When the input from the rectifier to level detector 1 exceeds the preset pickup value for the level detector 1, it gives an output voltage, which charges the capacitor C of the timing circuit (shown in Fig. 17.47). When the voltage across the capacitor exceeds the preset pickup value for the level detector 2, an output signal goes to the amplifier and from there to the output element that actuates the trip signal. In this way a definite-time delay is incorporated in the operation of an instantaneous relay.

17.31 STATIC DIRECTIONAL OVERCURRENT RELAY

The directional relay senses the direction of power flow and operates when it is flowing in the wrong direction such as in the case of a parallel feeder protection. The principle depends upon the phase angle between the voltage and currents. It is given two inputs, one proportional to the current and the other proportional to the voltage. The block schematic is shown in Fig. 17.48.

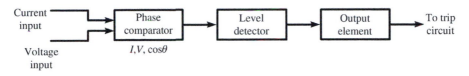

FIGURE 17.48

Block schematic of directional overcurrent relay.

17.32 STATIC IMPEDANCE RELAY

A static impedance type relay can be realized using an amplitude comparator. Consider Fig. 17.41A. If the two inputs to the relay are derived from current and voltage quantities, the relay will act as an impedance relay. It is also possible to realize as an impedance relay using a phase comparator.

17.33 STATIC REACTANCE RELAY

Using again a rectifier bridge type comparator comparing the amplitudes of

$$\left| I - \frac{V}{2X_r} \right| \quad \text{and} \quad \left| \frac{V}{2X_r} \right|$$

a reactance relay operation is achieved where X_r is the reactance of the line to be protected. To obtain the condition that $\left| I - \frac{V}{2X_r} \right| > \frac{V}{2X_r}$ for the relay to operate, the rectifier bridge comparator is arranged as shown in Fig. 17.49.

17.34 ADVANTAGES OF STATIC RELAYS OVER ELECTROMAGNETIC RELAYS

1. Static relays consume much less power (about 1/10th) than electromagnetic relays. Hence CT burden is less.
2. Static relays require much less space. The panels will be smaller and miniaturization is possible. In other words they are compact in size.
3. Operation is faster with static relays.
4. Static relays have larger life.
5. They need less maintenance due to absence of moving parts and bearings.
6. Frequent operations cause no deterioration.
7. The static relays have greater sensitivity.
8. Static relays are amenable for quick resetting, there is also no overshoot in static relays.In time−current characteristic they possess greater accuracy.
9. Logic circuits can be used for complex protective schemes.
10. Complex relaying characteristics can be obtained.
11. They possess high resistance to shock and vibration.

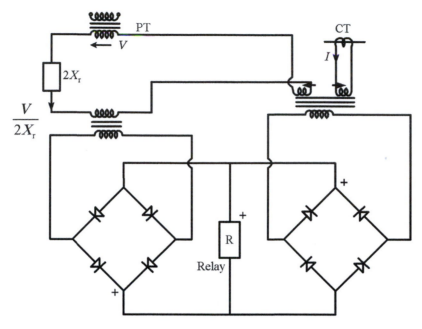

FIGURE 17.49

Principle of static reactance relay.

17.35 DISADVANTAGES OF STATIC RELAYS

1. Static relays are temperature sensitive.
2. Static relays are sensitive to voltage transients.
3. Static relays use auxiliary power supply.

However, some remedial measures are available. Thermistors can be used for temperature compensation. Digital techniques can be used for measurements. Filters and shielding can be used to protect against voltage spikes. Batteries or stabilized power supplies can be used. Due to large number of components used in relay construction, reliability will be a premium.

17.36 DIFFERENTIAL PROTECTION

In differential protection the vector difference of two or more similar electrical quantities are compared with a predetermined value, and the relay operates if the difference exceeds the set value. There are two different ways of comparing the two electrical quantities. Fig. 17.50 shows the two principles. Fig. 17.50A illustrates the basic principle of current differential protection.

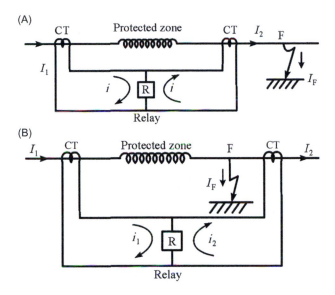

FIGURE 17.50

(A) Current differential protection. Fault outside CT's $I_1 = I_2$. Relay carries no current and (B) fault between the CT's $I_1 \neq I_2$; $i_1 \neq i_2$. Relay carries current.

Fig. 17.50A shows the case where the incoming and outgoing currents remain the same, when a fault occurs outside the zone of protection. The relay R does not carry any current and is inoperative. However, if a fault occurs within the zone of protection as in Fig. 17.50B, $I_1 \neq I_2$, and hence $i_1 - i_2$ acts on the relay, and if this value is greater than the relay setting, the relay operates.

17.37 VOLTAGE DIFFERENTIAL PROTECTION

In the voltage differential protection shown in Fig. 17.51 so long there is no fault to ground in the protected zone, the emf induced across the secondaries of the two CTs will be the same and the relay R will have net voltage applied $e_1 - e_2 = 0$. In case of any fault to ground within the protected zone, $I_1 \neq I_2$ and $e_1 > e_2$ so that $e_1 - e_2$ acts upon the relay and the relay operates for a set value of voltage difference. Faults outside the zone of protection are called external or through faults.

Under fault conditions the short-circuit current may be offset, containing both AC and DC components. If the two CTs used differ even slightly in their magnetic characteristics, under transient condition this difference may be magnified, one of the CTs may saturate earlier and due to unequal CT burden, the secondary currents may differ to that extent that the relay operates even for a through fault if this difference equals the pickup value. To overcome this problem, percentage differential protective scheme is used.

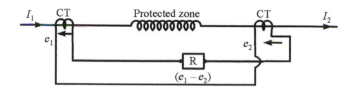

FIGURE 17.51

Voltage differential protection.

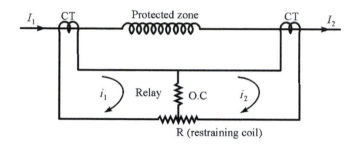

FIGURE 17.52

Percentage differential protection.

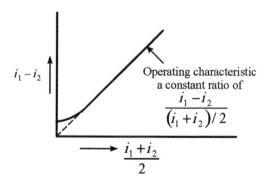

FIGURE 17.53

Operating characteristic of the relay.

17.38 PERCENTAGE DIFFERENTIAL PROTECTION

In percentage differential protection a restraining coil is introduced as shown in Fig. 17.52. The operating coil of the relay R is connected to the midpoint of the restraining coil.

In this case, the operating coil (O) of the relay carries a current $i_1 - i_2$, while the restraining coil carries on the whole a current of $(i_1/2) + (i_2/2) = ((i_1 + i_2)/2)$. The operating characteristic of the relay will be as shown in Fig. 17.53.

The relay operates when the operating current $(i_1 - i_2)$ exceeds the fixed ratio with respect to $(i_1 + i_2)/2$ the restraining current (except at low currents). Hence, this protection is called percentage differential protection. It may be noticed that the pickup value also increases with the increase in through current. The relay is biased against maloperation. Hence, it is called biased differential protection.

Differential protection is used for the protection of alternator stator windings and transformer windings (λ this type of protection is also called Merz-price Protection).

17.39 PILOT WIRE PROTECTION

In this method two wires are run parallel to the power line and they carry information signals from one end to the other end of the line to be protected. The principle of differential protection is used. The currents in the secondaries of two CTs are compared in current differential protection. In the voltage differential scheme balanced voltage principle is employed.

In general amplitude comparison and current differential method are cheaper and convenient to employ. The basic principle of current differential is illustrated in Fig. 17.54A

17.40 TRANSLAY SCHEME OF PROTECTION

In this scheme which employs voltage balance principle, the two relays used at both ends of the line to be protected are induction disc type relays. In the later versions the induction disc relays are replaced by solid-state relays.

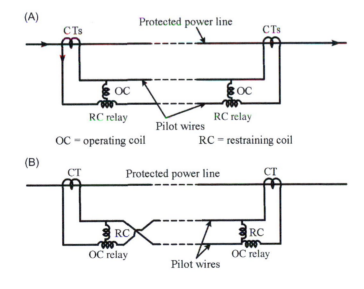

FIGURE 17.54

(A) Circulating current type pilot wire protection and (B) balanced voltage principle applied to pilot wire protection.

17.41 CARRIER CURRENT PROTECTION

EHV and UHV power transmission lines are widely protected by carrier current protection. In this method a high-frequency carrier channel directly coupled to the power line is employed.

The carrier signal is 20–700 kHz. The power line carrier protection is much faster than distance protection. For high-voltage lines employing autoreclosing, simultaneous tripping of the circuit breaker at both ends is desired. In carrier current protection the carrier signal is used to initiate or prevent the operation of the relay.

The advantage of carrier current protection is that the same carrier signal can be used for communication, supervisory control, and telemetering. In this method either phase comparison or directional comparison is used. In the former method the phase angle of current entering the line and leaving it are compared to initiate tripping action. In case of fault there will be a phase difference of about 180 degrees for internal faults. In directional comparison the direction of power flow at either end of the line is used for initiating relay action. This method is economical than pilot wire protection as there is no need for pilot wire to carry the signals separately. However, the initial cost for the terminal equipment could be on the higher side.

17.42 TRANSFORMER PROTECTION

Transformer is the most important component in a power system. Small distribution transformers are protected by fuses for earth faults and phase faults. Overcurrent protection is not provided. However, for important distribution transformers of 500 kVA and above rating overcurrent protection and earth fault protection are provided. For transformers of 1 MVA and above several protective schemes are applied including differential, restricted earth fault, and overcurrent protection. Faults that occur can be broadly divided into two types—external or through faults and internal faults. Main protection is given against internal faults.

17.43 DIFFERENTIAL PROTECTION

Differential protection is applied to transformers to detect internal faults. This is illustrated in Fig. 17.55A when the transformer is connected star–star (grounded) and the current transformers are connected in delta–delta. It can easily be verified that when an external fault occurs on one of the lines outside the zone of protection, the relays do not operate. The current distribution is shown for this condition. However, if the current transformers are connected in star–star, all the relays carry currents and operate even for a fault outside the transformer (see Fig. 17.55B). For successful differential protection, the current transformers must be connected as follows for different transformer connections.

Transformer	CT connections
Y – Y	Δ – Δ
Δ – Δ	Y – Y
Δ – Y	Y – Δ
Y – Δ	Δ – Y

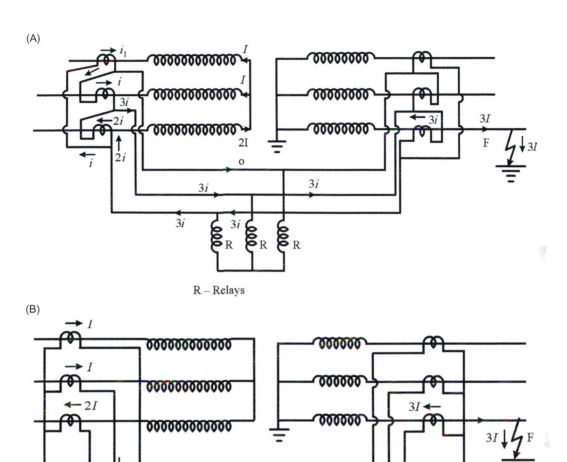

FIGURE 17.55

(A) Transformer protection—transformer Y/Y; CTs Δ/Δ—External fault-no-operation and (B) transformer protection—transformer Y/Y; CTs Y—Y connected. Relay operates for external fault.

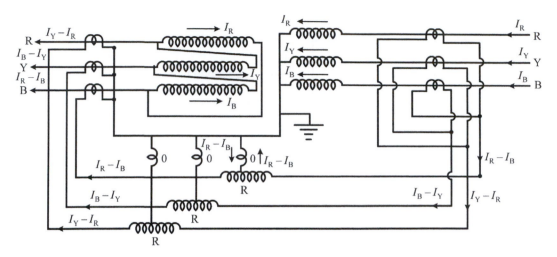

FIGURE 17.56

Percentage differential protection to Y-Δ transformer.

17.44 PERCENTAGE DIFFERENTIAL PROTECTION

Application of percentage differential protection is shown in Fig. 17.56 for a Δ-Y connected transformer. The current distribution is indicated in the figure.

The currents carried by operating coils (O) and restraining coil (R) of the three relays are shown. The incoming currents are I_R, I_Y, and I_B. The outgoing currents are $I_Y - I_R$, $I_B - I_Y$, and $I_R - I_B$, respectively. Currents flow in the CT's secondaries in opposite direction to that in the primaries. On the delta-connected side, the CTs are connected in star, and on the star-connected side, the CTs are connected in delta.

17.45 MAGNETIZING INRUSH CURRENTS IN TRANSFORMERS AND HARMONIC RESTRAINT

Whenever a transformer is switched on, as there is no induced emf initially, a large magnetizing "inrush" current, which could be several times the full-load current, flows. The inrush current depends upon the size of the transformer, the instant of switching on the voltage wave, the circuit conditions, residual flux in the transformer core, etc.

This inrush current (6—8 times the full-load current) decays in a few seconds. This current contains harmonics of which the second and third are considerable in magnitude. The second harmonic

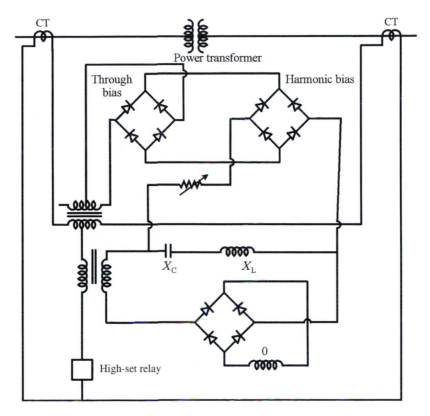

FIGURE 17.57

Harmonic restraint relay.

varies from 30% to 70%, and the third harmonic varies from 10% to 30%. In addition there may be a DC component varying from 40% to 60%.

In case of a fault current, which could also be of high value (several times full-load current), there is no second harmonic current as high as in case of inrush current. Thus it is possible to distinguish between a fault current and inrush current and prevent tripping in the later case.

The principle of harmonic restraint is applied by filtering out the harmonics from the differential unit (operating coil) and then by adding the filtered component (DC and second harmonic mainly) to the restraining unit. The relay is so adjusted that it does not operate if the second harmonic exceeds about 15% of the fundamental. A bridge type relay utilizing this principle is shown in Fig. 17.57. The tuning elements X_C and X_L allow only fundamental component of current to flow into operating coil "O."

17.46 BUCHHOLZ RELAY

Buchholz relay is a gas actuated relay. This relay is used to detect incipient faults and protects the transformer. If minor faults are not detected in time, they may develop into a major fault in course of time. Buchholz relay is a very simple relay for protection of transformer against internal faults.

Whenever a fault occurs, sparking takes place and the heat generated thereby, decomposes the transformer oil into hydrogen and hydrocarbon gases. These gases being very light push the oil downward and rise upward and they get collected at the top. The gas formation is used to protect the transformer.

The relay consists of two hinged chambers as shown in Fig. 17.58 containing mercury switches. The float at the top when pushed by the upward rising gases closes the mercury switch inside it and an alarm signal is sent to the operator.

There is another hinged float at the bottom, also containing a mercury switch. When a severe fault occurs, the decomposition of the oil and generation of gases is fast and the gases pushing the lower float close the mercury switch. The trip circuit is closed and the transformer gets disconnected.

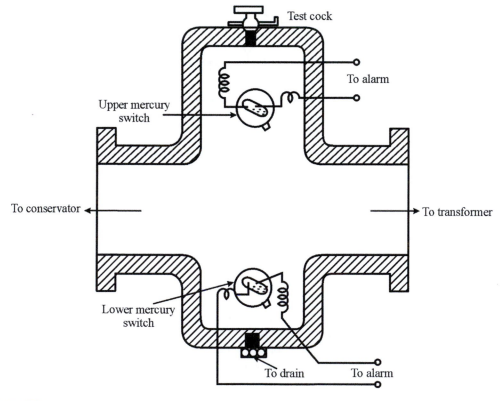

FIGURE 17.58

Buchholz relay.

17.47 **ALTERNATOR PROTECTION**

Stator Winding Protection: Percentage differential protection can be applied to stator windings for phase-to-ground and phase-to-phase faults. This is illustrated in Fig. 17.59A and B for star-connected and delta-connected alternator windings. The principle of operation has been explained

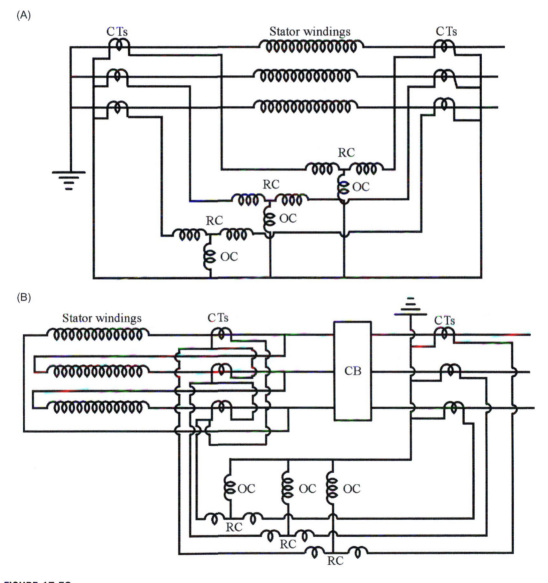

FIGURE 17.59

Merz-price protection to alternator stator windings. (A) Alternator stator Y connected and (B) alternator stator Δ-connected.

already. This scheme does not give protection against external faults and overloading. Interterm faults also cannot be detected. While differential protection gives complete protection for phase-to-phase faults, for phase-to-earth faults, the protection is 80%−85% of the winding only. This is because the magnitude of the earth fault current depends upon the method of neutral grounding.

If the neutral point of a star-connected alternator is earthed through a coil of high resistance, the earth fault current will be limited and the sensitivity of the differential relay to operate is impaired (Fig. 17.60). The relay must be set to operate for very low currents. This is not possible. Hence, an earth fault relay is needed to be incorporated as shown in Fig. 17.61 so that three relays operate for phase-to-phase fault and one relay operates for earth fault.

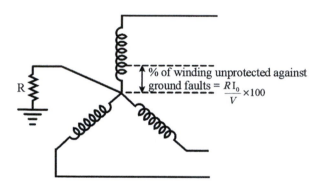

FIGURE 17.60

Percentage of unprotected winding.

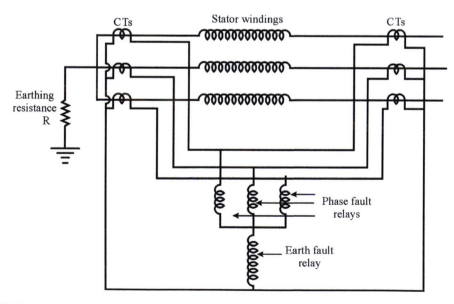

FIGURE 17.61

Alternator protection with phase fault and earth fault relays.

17.48 **RESTRICTED EARTH FAULT PROTECTION**

In smaller generators the neutral point of the star-connected stator windings is generally not available for connecting the CTs. In such cases differential protection can be applied only for earth faults. This is shown in Fig. 17.62.

For faults within the zone of protection such as at F, the neutral CT and line CTs will carry unequal currents and the relay operates. For fault at F_2, outside the protection zone the CT currents are equal and the relay does not operate (Fig. 17.62).

17.49 **ROTOR EARTH FAULT PROTECTION**

Consider the rotor circuit of an alternator. The field winding is supplied by a DC source. Consider an earth fault in the rotor circuit, F as shown in Fig. 17.63. The operating coil of an instantaneous relay is connected on one side to the field circuit through a current-limiting resistance R and on the other side a small DC power source is connected with polarity as indicated. Any earth fault occurring in the field circuit will flow through the relay coil and operates the relay, if the current is greater than the pickup value.

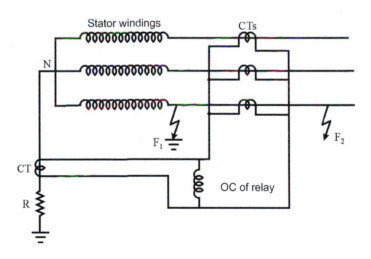

FIGURE 17.62

Restricted earth fault relay.

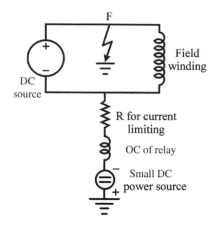

FIGURE 17.63

Rotor earth fault protection.

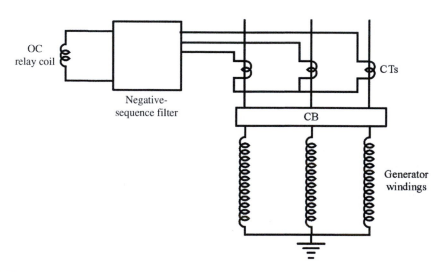

FIGURE 17.64

Protection against negative-sequence currents.

17.50 NEGATIVE-SEQUENCE PROTECTION

Unbalanced operation of an alternator causes double-frequency currents to be induced in rotor. Rotor will get overheated. Further, these unbalanced currents may cause vibrations and heating of stator. A negative-sequence filter is used for protection. The scheme is shown in Fig. 17.64.

WORKED EXAMPLES

E.17.1 The current rating of a relay is 5 A. The current setting is 200%. Time multiplier setting is 0.5. The current transformer ratio is 500/5 A. Determine the operating time for an overcurrent relay when the fault current is 8000 A.

Solution:

$$\text{CT ration} = \frac{500}{5} = 100$$

$$\text{Relay current setting} = 5A \times \frac{200}{100} = 10 \text{ A}$$

$$\text{PSM} = \frac{\text{Secondary current}}{\text{Relay current setting}} = \frac{\text{Primary current (fault current)}}{(\text{Relay current setting} \times \text{CT ratio})}$$

$$= \frac{8000 \text{ A}}{100 \times 10} = 8$$

Utilizing the data from the IDMT curve and extrapolating for PSM of 8, the operating time is 3.2 s (Table E.17.1).

Table E.17.1 Data From IDMT Curve

P.S.M	2	3.5	5	10	15	20
Time in seconds for time multiplier of 1	10	6	4	3	2	2

For a time multiplier setting of 0.5, the operating time is $0.5 \times 3.2 = 1.6$ s.

E.17.2 An 11-kV, 5-MVA star-connected synchronous generator has a reactance of 1.3 Ω/phase. The stator windings are protected by Merz-price method. The relays operate if the out-of-balance currents exceed 20% of the full-load current. Currents exceed 20% of the full-load current. The neutral of the generator is grounded through a coil of 7 Ω resistance. What percentage of winding of the stator is protected against earth fault. The stator winding resistance is negligible.

Solution: The phase voltage

$$V = \frac{11}{\sqrt{3}} = 6.351 \text{ kV}$$

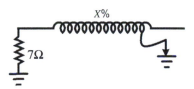

Let x be the % of winding unprotected.
Voltage of the unprotected winding $= 6351 \times \dfrac{x}{100}$
The fault current $= 6351 \times \dfrac{x}{100} \times \dfrac{1}{7}$ A
Full-load current $= \dfrac{5000}{\sqrt{3} \times 11} = 262.4396$ A
Out-of-balance current required for the relay to operate

$$= 262.4396 \times 0.20 = 52.4879 \text{ A}$$

Then

$$6351 \cdot \frac{x}{100} \times \frac{1}{7} = 52.4879$$

$$x = \frac{100 \times 7 \times 52.4879}{6351} = 5.785\%$$

(*Note*: The effect of stator winding reactance is ignored.)

E.17.3 A 6.6-kV synchronous generator supplying 3 MW at 0.85 power factor has a reactance of 12% per phase of stator winding. If differential protection is implemented and the relay operates for a fault current of 250 A, determine the value of neutral earthing resistance so that 90% of the winding is protected always.
Solution:
Generator full-load current

$$= \frac{3000 \times 1000}{0.05 \times \sqrt{3} \times 6600} = 308.75 \text{ A}$$

Let x be the reactance of the generator stator winding per phase

$$\frac{12}{100} = \frac{\sqrt{3} \times x \times 3.8.75}{6600}$$

$$x = \frac{12 \times 6600}{100 \times 308.75 \times \sqrt{3}} = 1.481 \ \Omega$$

Reactance of unprotected winding

$$= \frac{(100 - 90)}{100} = 1.481 \ \Omega$$
$$= 0.1481 \ \Omega$$

Voltage induced in the winding ND

$$= \frac{6600}{\sqrt{3} \times 10} = 381 \text{ V}$$

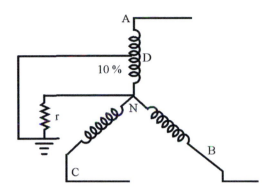

For operation at 250 A

$$250 = \frac{381}{\sqrt{r^2 + (0.1481)^2}}$$

where r is the resistance in the grounding circuit

$$r^2 + (0.1481)^2 = \left(\frac{381}{250}\right)^2 = 2.3225$$
$$r^2 = 2.3225 - 0.0219 = 2.3$$
$$r = 1.5175 \ \Omega$$

E.17.4 A star connected, three-phase synchronous generator has its stator windings protected by current differential protection. The relay pickup current is 0.18 A. The differential current has a slope of 10% for operation. A fault occurs near the neutral while the generator is operating a current difference of only 50 A occurred between the incoming and outgoing current in a phase winding due to the high-resistance fault. The actual currents are 450and 400 A. The CTs have a ratio of 500/5 A. Determine for the fault explained whether the relay will operate or not?
Solution:

$$i_1 = 450 \times \frac{5}{500} = 4.5 \ \text{A}$$

$$i_2 = 400 \times \frac{5}{500} = 4 \ \text{A}$$

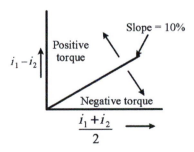

$$i_1 - i_2 = 4.5 - 4 = 0.5 \text{ A}$$

$$\frac{i_1 + i_2}{2} = \frac{4.5 + 4}{2} = 4.25 \text{ A}$$

$$\frac{i_1 + i_2}{\left(\frac{i_1 + i_2}{2}\right)} = \frac{0.5}{4.25} = 0.1176 = 11.76\%$$

since the value is greater than 10%, the relay operates.

E.17.5 A three-phase transformer has a voltage ratio of 66 kV/6.6 kV. It is star–delta connected. The protective CTs on the 6.6 kV side have a current ratio of 50. What should be the ratio of CTs on the 66-kV side.

Solution:

The transformer is $Y - \Delta$ connected

CTs must be $\Delta - Y$ connected

Let the line current be 50 A

$$\sqrt{3} \times 6.6 \times 50 = \sqrt{3} \times 66 \times I_1$$

$$I_1 = \frac{50 \times \sqrt{3} \times 6.6}{66 \times \sqrt{3}} = 5 \text{ A}$$

The pilot wires also carry 1 A, since the star-connected secondary of CTs carry 1 A (50/1 ratio).

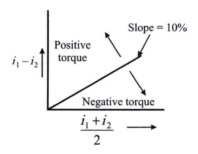

1 A on the star side is $\frac{1}{\sqrt{3}}$ A on the delta side

the CTs as the 66-kV should have a ratio of $\frac{20}{1/\sqrt{3}}:1 = 20\sqrt{3}:1 = 34.64:1$ or $35:1$.

E.17.6 A 25-MVA, 11-kV/66-kV star–delta connected transformer is given differential protection. The h.v side lags behind the l.v side by 30 degrees in phase. The restraining coils of the relay should not carry current exceeding 5 A. On the 11 kV side, the CT ratio is 2000/5. Determine the CT ratio on the 66 kV side.

Solution:

Full-load current $I_{\text{FL}_1} = \dfrac{25,000}{\sqrt{3} \times 11} = 1312.2 \text{ A}$

CT ratio on the 11-kV side $= \dfrac{2000}{5} = 400$

Secondary current $= \dfrac{1312.2}{400} = 3.28 \text{ A}$

As the 11-kV side is star connected, CT secondaries will be delta connected.

Current fed into pilot wires from the 11-kV side
CT secondaries $= \sqrt{3} \times 3.28 = 5.68$ A
On the 66-kV side primary current

$$I_{FL_2} = \frac{25,000}{\sqrt{3} \times 66} = 218.7 \text{ A}$$

Since the 66-kV side CTs are connected in star, the current in pilot wires $= 5.68$ A
Hence, CT ratio $= \dfrac{218.7}{5.68} = 38.5$
Selecting a CT ratio of 40
Secondary current of 5 A gives $5 \times 40 = 200$ A on the primary side.
Hence CT ratio on the 66-kV side $= 200{:}5$

E.17.7 A generator is provided with restricted earth fault protection. The generator is rated 11 kV, 6000 KVR. The percentage of stator winding to be protected against phase-to-ground fault is 85%. The relays are set to trip for an out-of-balance current of 25%. Calculate the resistance to be inserted in the neutral connection to ground.

Solution:
Phase voltage $= \dfrac{11 \ kV}{\sqrt{3}} = 6.35$ kV
The load current $I_{FL} = \dfrac{6000}{\sqrt{3} \times 11} = 314.9$ A
25% of full-load current $= \dfrac{25 \times 314.9}{100} = 78.73$ A
Percentage of winding left unprotected $= 15$
Voltage drop in it $= \dfrac{R \times 78.73}{6350} \times 100 = 15$

$$R = \frac{6350 \times 15}{78.73 \times 100} = 12.09 \ \Omega$$

where R is the resistance to be inserted in the neutral connection to ground.

PROBLEMS

P.17.1 A 15-MVA, 6.6-kV, three-phase alternator is protected by Merz—Price protection system. The CTs are rated 1500/5. The star point of the alternator is earthed through an 8-Ω resistance. Calculate the percentage of winding unprotected in each phase against earth faults. The pickup value of the relay current is 0.5 A.

P.17.2 A three-phase, 33 kV/6.6 kV transformer is connected star—delta and is protected by Merz-Price scheme. The current transformers on the low voltage side have a ratio of 400/5. Determine the ratio of current transformers to be used on the high-voltage side.

P.17.3 An overcurrent relay rated 5 A and with 200% setting is connected to CTs of ratio 500/5. Calculate the current in the lines for which the relay picks up.

POWER SYSTEM STABILITY

18.1 ELEMENTARY CONCEPTS

Maintaining synchronism between the various elements of a power system has become an important task in power system operation as systems expanded with increasing interconnection of generating stations and load centers. The electromechanical dynamic behavior of the prime mover—generator—excitation systems, various types of motors and other types of loads with widely varying dynamic characteristics can be analyzed through somewhat oversimplified methods for understanding the processes involved. There are three modes of behavior generally identified for the power system under dynamic condition. They are:

1. Steady-state stability
2. Transient stability
3. Dynamic stability

Stability is the ability of a dynamic system to remain in the same operating state even after a disturbance that occurs in the system.

Stability when used with reference to a power system is that attribute of the system or part of the system, which enables it to develop restoring forces between the elements thereof, equal to or greater than the disturbing force so as to restore a state of equilibrium between the elements.

A power system is said to be steady-state stable for a specific steady-state operating condition, if it returns to the same steady-state operating condition following a disturbance. Such disturbances are generally small in nature.

A stability limit is the maximum power flow possible through some particular point in the system, when the entire system or part of the system to which the stability limit refers is operating with stability.

Larger disturbances may change the operating state significantly, but still into an acceptable steady state. Such a state is called a transient state.

The third aspect of stability, viz., dynamic stability is generally associated with excitation system response and supplementary control signals involving excitation system. This will be dealt with later.

Instability refers to a condition involving loss of "synchronism" which is also the same as "falling out of the step" with respect to the rest of the system.

18.2 ILLUSTRATION OF STEADY-STATE STABILITY CONCEPT

Consider the synchronous generator—motor system shown in Fig. 18.1. The generator and motor have reactances X_g and X_m, respectively. They are connected through a line of reactance X_e. The various voltages are indicated.

Electrical Power Systems. DOI: http://dx.doi.org/10.1016/B978-0-08-101124-9.00018-8

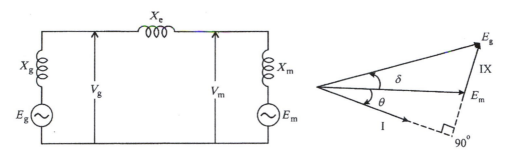

FIGURE 18.1

Synchronous generator–motor system connected through reactance along with its phasor diagram.

From Fig. 18.1

$$E_g = E_m + j \times I$$

$$I = \frac{E_g - E_m}{jX}$$

where

$$X = X_g + X_e + X_m$$

Power delivered to motor by the generator is

$$
\begin{aligned}
P \quad &= \mathrm{Re}[EI^*] \\
&= \mathrm{Re}[E_g \underline{/\delta}] \frac{[E_g \underline{/-\delta} - E_m \underline{/0\,\text{degree}}]}{X \underline{/-90\,\text{degrees}}} \\
&= \frac{E_g^2}{X} \cos 90\,\text{degree} - \frac{E_g E_m}{X} \cos(90 + \delta)
\end{aligned}
$$

$$P = \frac{E_g E_m}{X} \sin \delta \qquad (18.1)$$

P is maximum when $\delta = 90$ degrees

$$P_{\max} = \frac{E_g E_m}{X} \qquad (18.2)$$

The graph of P versus δ is called power angle curve and is shown in Fig. 18.2. The system will be stable so long $dP/d\delta$ is positive. Theoretically, if the load power is increased in very small increments from $\delta = 0$ to $\delta = \pi/2$, the system will be stable. At $\delta = \pi/2$, the steady-state stability limit will be reached. $P_{\max}$ is dependent on E_g, E_m, and X. Thus we obtain the following possibilities for increasing the value of $P_{\max}$ indicated in the next section.

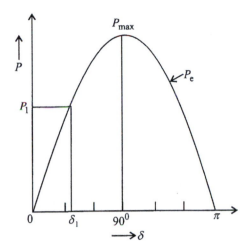

FIGURE 18.2

Power angle characteristic.

18.3 **METHODS FOR IMPROCESSING STEADY-STATE STABILITY LIMIT**

1. Use of higher excitation voltages, thereby increasing the value of E_g.
2. Reducing the reactance between the generator and the motor. The reactance $X = X_g + X_m + X_e$ is called the transfer reactance between the two machines and this has to be brought down to the possible extent.

18.4 **SYNCHRONIZING POWER COEFFICIENT**

We have

$$P = \frac{E_g E_m}{X} \sin \delta$$

The quantity

$$\frac{dP}{d\delta} = \frac{E_g E_m}{X} \cos \delta \qquad (18.3)$$

is called synchronizing power coefficient or stiffness.

For stable operation $dP/d\delta$, the synchronizing coefficient must be positive.

18.5 TRANSIENT STABILITY

Steady-state stability studies often involve a single machine or the equivalent to a few machines connected to an infinite bus undergoing small disturbances. The study includes the behavior of the machine under small incremental changes in operating conditions about an operating point on small variation in parameters.

When the disturbances are relatively larger or faults occur on the system, the system enters transient state. Transient stability of the system involves nonlinear models. Transient internal voltage E_i' and transient reactances X_d' are used in calculations.

The first swing of the machine (or machines) that occur in a shorter time generally does not include the effect of excitation system and load—frequency control system. The first swing transient stability is a simple study involving a time space not exceeding 1 s. If the machine remains stable in the first swing, it is presumed that it is transient stable for that disturbance. However, where disturbances are larger and require study over a longer period beyond 1 s, multiswing studies are performed taking into effect the excitation and turbine-generator controls. The inclusion of any control system or supplementary control depends upon the nature of the disturbances and the objective of the study.

18.6 STABILITY OF A SINGLE MACHINE CONNECTED TO AN INFINITE BUS

Consider a synchronous motor connected to an infinite bus. Initially the motor is supplying a mechanical load P_{m0} while operating at a power angle δ_0. The speed is the synchronous speed ω_s. Neglecting losses power input is equal to the mechanical load supplied. If the load on the motor

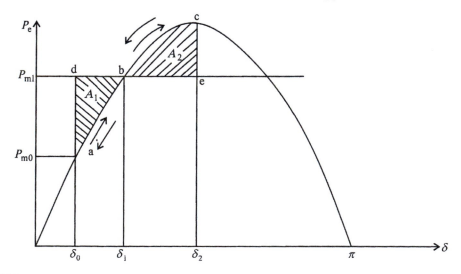

FIGURE 18.3

Stability of synchronous motor connected to an infinite bus.

is suddenly increased to P_{m1}, this sudden load demand will be met by the motor by giving up its stored kinetic energy and the motor, therefore, slows down. The torque angle δ increases from δ_0 to δ_1 when the electrical power supplied equals the mechanical power demand at b as shown in Fig. 18.3. Since the motor is decelerating, the speed, however, is less than N_s at b. Hence the torque "angle δ_1" increases further to δ_2 where the electrical power P_e is greater than P_{m1}, but $N = N_s$ at point c. At this point c further increase of δ is arrested as $P_e > P_{m1}$ and $N = N_s$. The torque angle starts decreasing till δ_1 is reached at b but due to the fact that till point b is reached P_e is still greater than P_{m1}, speed is more than N_s. Hence, δ decreases further till point "a" is reacted where $N = N_s$ but $P_{m1} > P_e$. The cycle of oscillation continues. But, due to the damping in the system that includes friction and losses, the rotor is brought to the new operating point b with speed $N = N_s$.

In Fig. 18.3 area "abd" represents deceleration and area "bce" acceleration. The motor will reach the stable operating point b only if the accelerating energy A_1 represented by "bce" equals the decelerating energy A_2 represented by area "abd."

18.7 THE SWING EQUATION

The interconnection between electrical and mechanical side of the synchronous machine is provided by the dynamic equation for the acceleration or deceleration of the combined-prime mover (turbine)—synchronous machine rotor. This is usually called swing equation.

The net torque acting on the rotor of a synchronous machine

$$T = \frac{WR^2}{g}\alpha \tag{18.4}$$

where T is the algebraic sum of all torques in kg-m; a is the mechanical angular acceleration; and WR^2 is the moment of inertia in kg-m^2

$$\text{Electrical angle } \vartheta_e = \vartheta_m \frac{P}{2} \tag{18.5}$$

where ϑ_m is mechanical angle and P is the number of poles.

$$\text{The frequency } f = PN/120 \tag{18.6}$$

where N is the rpm.

$$f = \frac{P}{2}\left(\frac{\text{rpm}}{60}\right)$$

$$\frac{60f}{\text{rpm}} = \frac{P}{2}$$

$$\vartheta_e = \left(\frac{60f}{\text{rpm}}\right)\vartheta_m \tag{18.7}$$

The electrical angular position δ_e in radians of the rotor with respect to a synchronously rotating reference axis is

$$\delta = \vartheta_e - \omega_0 t \tag{18.8}$$

where w_0 is the rated synchronous speed in rad./s and t is the time in seconds (*Note:* $\delta + \omega_0 t = \vartheta_e$)

The angular acceleration taking the second derivative of Eq. (18.8) is given by

$$\frac{d^2 \delta}{dt^2} = \frac{d^2 \vartheta_e}{dt^2}$$

From Eq. (18.7) differentiating twice

$$\frac{d^2 \vartheta_e}{dt^2} = \left(\frac{60f}{\text{rpm}}\right) \frac{d^2 \vartheta_m}{dt^2}$$

$$\therefore \quad \frac{d^2 \vartheta_m}{dt^2} = \alpha = \left(\frac{\text{rpm}}{60f}\right) \frac{d^2 \vartheta_e}{dt^2}$$

From Eq. (18.4)

$$T = \frac{WR^2}{g} \left(\frac{\text{rpm}}{60f}\right) \frac{d^2 \vartheta_e}{dt^2} = \frac{WR^2}{g} \left(\frac{\text{rpm}}{60f}\right) \frac{d^2 \delta}{dt^2} \tag{18.9}$$

Let the base torque be defined as

$$T_{\text{Base}} = \frac{\text{Base KVA}}{2\pi \left(\dfrac{\text{rpm}}{60}\right)} \tag{18.10}$$

Torque in per unit

$$T \text{ p.u.} = \frac{T}{T_{\text{Base}}} = \frac{WR^2}{g} \left(\frac{\text{rpm}}{60f}\right) \frac{d^2 \delta}{dt^2} \cdot \frac{2\pi \left(\dfrac{\text{rpm}}{60}\right)}{\text{Base KVA}}$$

$$= \frac{WR^2}{g} \left(\frac{\text{rpm}}{60}\right)^2 \frac{2\pi}{f} \cdot \frac{1}{\text{Base KVA}} \frac{d^2 \delta}{dt^2} \tag{18.11}$$

$$\text{Kinetic energy KE} = \frac{1}{2} \frac{WR^2}{g} \omega_0^2 \tag{18.12}$$

where

$$\omega_0 = 2\pi \frac{\text{rpm}}{60}$$

Defining

$$H = \frac{\text{Kinetic energy at rated speed}}{\text{base KVA}}$$

$$= \underbrace{\frac{1}{2} \frac{WR^2}{g} \left(2\pi \frac{\text{rpm}}{60}\right)^2}_{\text{KE at rated speed}} \frac{1}{\text{Base KVA}}$$

Per unit torque,

$$T = \frac{H}{\pi f} \cdot \frac{d^2 \delta}{dt^2} \tag{18.13}$$

The torque acting on the rotor of a generator includes the mechanical input torque from the prime mover, torque due to rotational losses (i.e., friction, windage, and core loss), electrical output torque and damping torques due to prime mover, generator and power system.

The electrical and mechanical torques acting on the rotor of a motor are of opposite sign and are a result of the electrical input and mechanical load. We may neglect the damping and rotational losses, so that the accelerating torque.

$$T_a = T_m - T_e$$

where T_e is the air-gap electrical torque and T_m the mechanical shaft torque.

$$\frac{H}{\pi f} \frac{d^2\delta}{dt^2} = T_m - T_e \tag{18.14}$$

i.e.,

$$\frac{d^2\delta}{dt^2} = \frac{\pi f}{H}(T_m - T_e) \tag{18.15}$$

Torque in per unit is equal to power in per unit if speed deviations are neglected. Then

$$\frac{d^2\delta}{dt^2} = \frac{\pi f}{H}(P_m - P_e) \tag{18.16}$$

Eqs. (18.15) and (18.16) are called swing equations.

It may be noted, that, since $\delta = \vartheta - \omega_0 t$

$$\frac{d\delta}{dt} = \frac{d\vartheta}{dt} - \omega_0$$

Since the rated synchronous speed in rad/s is $2\pi f$

$$\frac{d\vartheta}{dt} = \frac{d\delta}{dt} + \omega_0$$

we may put the equation in another way.

$$\text{Kinetic energy} = \frac{1}{2}I\omega^2 \text{ J}$$

The moment of inertia I may be expressed in Js^2/(rad)^2 since ω is in rad/s. The stored energy of an electrical machine is more usually expressed in MJ and angles in degrees. Angular momentum M is thus described by MJs /electrical degree

$$M = I\omega$$

where ω is the synchronous speed of the machine and M is called inertia constant. In practice ω is not synchronous speed while the machine swings and hence M is not strictly a constant.

The quantity H defined earlier as inertia constant has the units MJ.

$$H = \frac{\text{Stored energy in MJ}}{\text{Machine rating in MVA } (G)} \tag{18.17}$$

$$\text{But stored energy} = \frac{1}{2}I\omega^2 = \frac{1}{2}M\omega$$

In electrical degrees

$$\omega = 360f(= 2\pi f) \tag{18.18}$$

$$GH = \frac{1}{2}M(360f) = \frac{1}{2}M2\pi f = M\pi f$$
$$M = \frac{GH}{\pi f} \text{ MJ-s/elecdegree} \tag{18.19}$$

In the per-unit systems

$$M = \frac{H}{\pi f} \tag{18.20}$$

So that

$$\frac{d^2\delta}{dt^2} = \frac{\pi f}{H}(P_m - P_e) \tag{18.21}$$

which may be written also as

$$M\frac{d^2\delta}{dt^2} = P_m - P_e \tag{18.22}$$

This is another form of swing equation.
Further

$$P_e = \frac{EV}{X}\sin\delta$$

So that

$$M\frac{d^2\delta}{dt^2} = P_m - \frac{EV}{X}\sin\delta \tag{18.23}$$

with usual notation.

18.8 EQUAL AREA CRITERION AND SWING EQUATION

Equal area criterion is applicable to single machine connected to infinite bus. It is not directly applicable to multimachine system. However, the criterion helps in understanding the factors that influence transient stability.

The swing equation connected to an infinite bus is given by

$$\frac{Hd^2\delta}{\pi fdt^2}P_m - P_e = P_a \tag{18.24}$$

or

$$\frac{2Hd^2\delta}{w_sdt^2} = P_m - P_e = P_a \tag{18.25}$$

Also

$$\frac{d^2\delta}{dt^2}M = P_a \tag{18.26}$$

Now as t increases to a maximum value δ_{max} where $d\delta/dt = 0$, multiplying Eq. (18.11) on both sides by $2(d\delta/dt)$ we obtain

$$2\frac{d^2\delta}{dt^2}\frac{d\delta}{dt} = \frac{P_a}{M}2\frac{d\delta}{dt}$$

Integrating both sides (Fig. 18.4)

$$\left(\frac{d\delta}{dt}\right)^2 = \frac{2}{M}\int P_a \, d\delta$$

$$\frac{d\delta}{dt} = \sqrt{\frac{2}{M}\int_{\delta_0}^{\delta} P_a \, d\delta}$$

δ_0 is the initial rotor angle from where the rotor starts swinging due to the disturbance. For stability

$$\frac{d\delta}{dt} = 0$$

Hence

$$\sqrt{\frac{2}{M}\int_{\delta_0}^{\delta} P_a \, d\delta} = 0$$

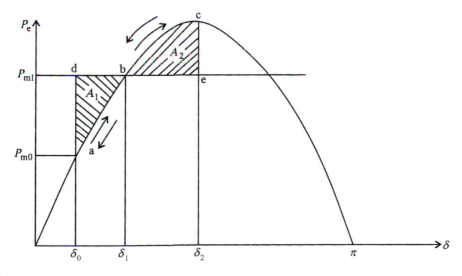

FIGURE 18.4

Equal area criterion.

i.e.,

$$\int_{\delta_0}^{\delta} P_a \, d\delta \int_{\delta_0}^{\delta} (P_m - P_e) \, d\delta = 0$$

The system is stable, if we could locate a point c on the power angle curve such that areas A_1 and A_2 are equal. Equal area criterion states that whenever, a disturbance occurs, the accelerating and decelerating energies involved in swinging of the rotor of the synchronous machine must equal so that a stable operating point (such as b) could be located.

$$A_1 - A_2 = 0 \text{ means that,}$$
$$\int_{\delta_0}^{\delta_1} (P_{m1} - P_e) d\delta - \int_{\delta_1}^{\delta_2} (P_e - P_m) d\delta = 0$$

But

$$P_e = P_{max} \sin \delta$$

$$\int_{\delta_0}^{\delta_1} (P_{m1} - P_{max}\sin \delta) d\delta - \int_{\delta_1}^{\delta_2} (P_{max} \sin \delta - P_{m1}) \, d\delta = 0$$

$$P_{m1}(\delta_1 - \delta_0) + P_{max}(\cos \delta_1 - \cos \delta_0)$$

$$P_{max}(\cos \delta_1 - \cos \delta_2) + P_{m1}(\delta_2 - \delta_1) = 0$$

i.e.,

$$P_{m1}[\delta_2 - \delta_0] = P_{max}[\cos \delta_0 - \cos \delta_2)$$

$$\cos \delta_0 - \cos \delta_2 = \frac{P_{m1}}{P_{max}}[\delta_2 - \delta_0]$$

But

$$\frac{P_{m1}}{P_{max}} = \frac{P_{max} \sin \delta_1}{P_{max}} = \sin \delta_1$$

Hence

$$(\cos \delta_0 - \cos \delta_2) = \sin \delta_1[\delta_2 - \delta_0] \tag{18.27}$$

The above is a transcendental equation and hence cannot be solved using normal algebraic methods.

18.9 TRANSIENT STABILITY LIMIT

Now consider that the change in P_m is larger than the change shown in Fig. 18.5. This is illustrated in Fig. 18.5.

In the case $A_1 > A_2$, i.e., we fail to locate an area A_2 that is equal to area A_1. Then, as stated the machine will loose its stability since the speed cannot be restored to N_s.

Between these two cases of stable and unstable operating cases, there must be a limiting case where A_2 is just equal to A_1 as shown in Fig. 18.6. Any further increase in P_{m1} will cause A_2 to be less than A_1. $P_{m1} - P_{m0}$ in Fig. 18.6 is the maximum load change that the machine can sustain synchronism and is thus the transient stability limit.

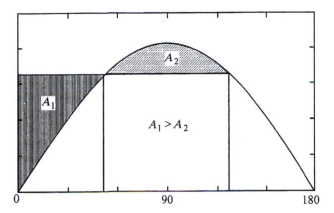

FIGURE 18.5

Unstable system ($A_1 > A_2$).

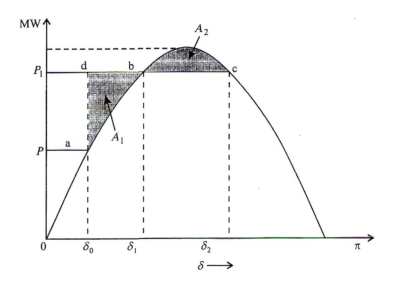

FIGURE 18.6

Transient stability limit.

18.10 FREQUENCY OF OSCILLATIONS

Consider a small change in the operating angle δ_0 due to a transient disturbance by $\Delta\delta$. Corresponding to this, we can write

$$\delta = \delta_0 + \Delta\delta$$

and

$$P_e = P_e^0 + \Delta P_e$$

where ΔP_e is the change in power and P_e^0 is the initial power at δ_0

$$(P_e + \Delta P_e) = P_{max}\sin \delta_0 + (P_{max}\cos \delta_0)\Delta\delta$$

Also

$$P_m = P_e^0 = P_{max} \sin \delta_0$$

Hence

$$(P_m - P_e^0 + \Delta P_e) = P_{max}\sin \delta_0 - [P_{max}\sin \delta_0 - (P_{max}\cos \delta_0)\Delta\delta]$$
$$= (P_{max}\cos \delta^0)\Delta\delta$$

$dP_e/d\delta$ is the synchronizing coefficient S.

The swing equation is

$$\frac{2H}{\omega}\frac{d^2\delta_0}{dt^2} = P_a = P_m - P_e^0$$

Again

$$\frac{2H}{\omega}\frac{d^2(\delta_0 + \Delta\delta)}{dt^2} = P_m - (P_e^0 + \Delta P_e)$$

Hence

$$\frac{2H}{\omega}\frac{d^2(\Delta\delta)}{dt^2} = -P_{max}(\cos \delta_0)\cdot\Delta\delta = -S^0\cdot\Delta\delta$$

where S^0 is the synchronizing coefficient at P_e^0.

Therefore

$$\frac{d^2(\Delta\delta)}{dt^2} + \left(\frac{\omega S^0}{2H}\right)\Delta\delta = 0$$

which is a linear second-order differential equation. The solution depends upon the sign of δ_0. If δ_0 is positive, the equation represents simple harmonic motion.

The frequency of the undamped oscillation is

$$\omega_n = \sqrt{\frac{\omega\delta_0}{2H}} \tag{18.28}$$

The frequency f is given by

$$f = \frac{1}{2\pi}\sqrt{\frac{\omega\delta_0}{2H}} \tag{18.29}$$

Transient Stability and Fault Clearance Time: Consider the electrical power system shown in Fig. 18.7. If a three-phase fault occurs near the generator bus on the radial line connected to it, power transmitted over the line to the infinite bus will become zero instantaneously. The mechanical input power P_m remains constant. Let the fault be cleared at $\delta = \delta_1$. All the mechanical input energy represented by area abcd $= A_1$, will be utilized in accelerating the rotor from δ_0 to δ_1. Fault

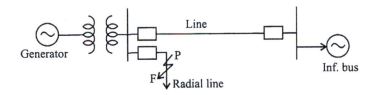

FIGURE 18.7

Generator connected to infinite bus with fault at generator end.

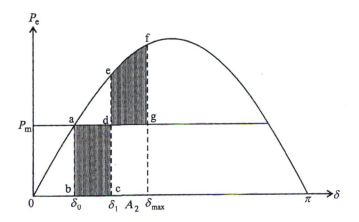

FIGURE 18.8

Equal area criterion when fault is cleared at $\delta < \delta_c$.

clearance at δ_1 angle or point c will shift the operating point from c to e instantaneously on the $P-\delta$ curve. At point f, an area $A_2 = d\,e\,f\,g$ is obtained which is equal to A_1 (Fig. 18.8). The rotor comes back from f and finally settles down at "a" where $P_m = P_e$. δ_1 is called the clearing angle and the corresponding time t_1 is called the critical clearing time t_c for the fault from the inception of it at δ_0.

18.11 **CRITICAL CLEARING TIME AND CRITICAL CLEARING ANGLE**

If, in the previous case, the clearing time is increased from t_1 to t_c such that δ_1 is δ_c as shown in Fig. 18.9, where A_1 is just equal to A_2. Then, any further increase in the fault cleaning time t_1 beyond t_c, would not be able to enclose an area A_2 equal to A_1. This is shown in Fig. 18.10. Beyond δ_c, A_2 starts decreasing. Fault clearance cannot be delayed beyond t_c. This limiting fault clearance angle δ_c is called critical clearing angle and the corresponding time to clear the fault is called critical clearing time t_c.

From Fig. 18.9

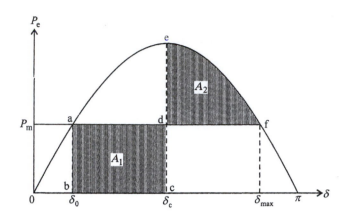

FIGURE 18.9

Equal area criterion when fault is cleared at $\delta = \delta_c$.

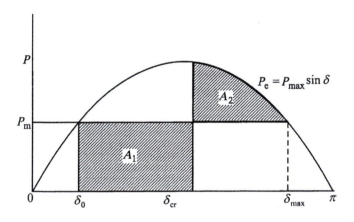

FIGURE 18.10

Equal area criterion when fault is cleared at $\delta > \delta_c$.

$$\delta_{max} = \pi - \delta_0$$

$$P_m = P_{max} \sin \delta_0$$

$$A_1 = \int_{\delta_0}^{\delta_c} (P_m - 0)d\delta = P_m[\delta_c - \delta_0]$$

$$A_2 = \int_{\delta_c}^{\delta_{max}} (P_{max}\sin \delta - P_m)\, d\delta$$

$$= P_{max}(\cos \delta_c - \cos \delta_{max}) - P_m(\delta_{max} - \delta_c)$$

$A_1 = A_2$ gives

$$\cos \delta_c - \cos \delta_m = \frac{P_m}{P_{max}} [\delta_{max} - \delta_0]$$

$$\cos \delta_c = \frac{P_m}{P_{max}} [(\pi - \delta_0) - \delta_0] + \cos(\pi - \delta_0)$$

$$= \frac{P_m}{P_{max}} [(\pi - 2\delta_0)] - [\cos \delta_0]$$

$$\delta_c = \cos^{-1} \left[\frac{P_m}{P_{max}} (\pi - 2\delta_0) - (\cos \delta_0) \right] \tag{18.30}$$

During the period of fault, the swing equation is given by

$$\frac{d^2\delta}{dt^2} = \frac{\pi f}{H} (P_m - P_e)$$

But since $P_e = 0$
During the fault period

$$\frac{d^2\delta}{dt^2} = \frac{\pi f}{H} P_m$$

Integrating both sides

$$\int_0^t \frac{d^2\delta \, dt}{dt^2} = \int_0^t \frac{\pi f}{H} P_m \, dt$$

$$\frac{d\delta}{dt} = \frac{\pi f}{H} P_m t$$

and integrating once again

$$\delta_c = \frac{\pi f}{2H} P_m t^2 + K$$

At $t = 0$; $\delta = \delta_0$, hence $K = \delta_0$
Hence

$$\delta_c = \frac{\pi f}{2H} P_m t^2 + \delta_0 \tag{18.31}$$

Hence the critical cleaning time

$$t_c = \sqrt{\frac{2H(\delta_0 - \delta_c)}{P_m \pi f}} \text{ s} \tag{18.32}$$

18.12 FAULT ON A DOUBLE-CIRCUIT LINE

Consider a single generator or generating station supplying power to a load or an infinite bus through a double-circuit line as shown in Fig. 18.11.

The electrical power transmitted is given by $P_{e12} = (EV/x'_d + x_{12}) \sin \delta$ where $(1/x_{12}) = (1/x_1) + (1/x_2)$ and x'_d is the transient reactance of the generator. Now, if a fault occurs on line 2, e.g., then the two circuit breakers on either side will open and disconnect the line 2. Since, $x_1 > x_{12}$ (two lines in parallel), the $P-\delta$ curve for one line in operation is given by

$$P_{e1} = \frac{EV}{x'_d + x_1} \sin \delta$$

will be below the $P-\delta$ curve P_{e12} as shown in Fig. 18.12. The operating point shifts from a to b on $P-\delta$ curve P_{e1} and the rotor accelerates to point c where $\delta = \delta_1$. Since the rotor speed is not synchronous, the rotor decelerates till point d is reached at $\delta = \delta_2$ so that area A_1 (=area abc) is equal to area A_2 (=area cde). The rotor will finally settle down at point c due to damping. At point c

$$P_m = P_{e1}$$

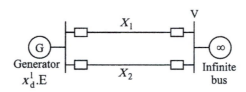

FIGURE 18.11

Double-circuit line and fault.

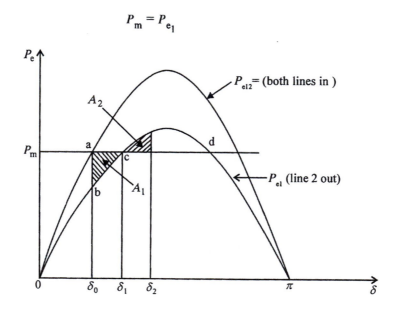

FIGURE 18.12

Three-phase generator with L-G fault and impedance in N—G circuit.

18.13 TRANSIENT STABILITY WHEN POWER IS

18.13.1 TRANSMITTED DURING THE FAULT

Consider the case where during the fault period some load power is supplied to the load or to the infinite bus. The $P-\delta$ curve during the fault is represented by Curve 3 in Fig. 18.13.

Upon the occurrence of fault, the operating point moves from a to b and on to c during the fault curve 3, when the fault is cleared at $\delta = \delta_1$, the operating point moves from b to c along the curve P_{e3} and then shifts to point e. If area defgd could equal area abcd ($A_2 = A_1$), then the system will be stable.

If the fault clearance is delayed till $\delta_1 = \delta_c$ as shown in Fig. 18.14 such that area abcd (A_1) is just equal to and e d f (A_2) then

$$\int_{\delta_0}^{\delta_c} (P_{max}\sin\delta - P_m)\, d\delta = \int_{\delta_c}^{\delta_{max}} (P_{max}\sin\delta - P_m)\, d\delta$$

It is clear from Fig. 18.14 that $\delta_{max} = \pi - \delta_0 = \pi - \sin^{-1} P_m/P_{max\,2}$
Integrating

$$(P_m \cdot \delta + P_{max}\cos\delta)\Big|_{\delta_0}^{\delta_c} + (P_{max2}\cos\delta - P_m \cdot \delta)\Big|_{\delta_c}^{\delta_{max}} = 0$$

$$P_m(\delta_c - \delta_0) + P_{max3}(\cos\delta_c - \cos\delta_0) + P_m(\delta_{max} - \delta_c) + P_{max2}(\cos\delta_{max} - \cos\delta_c) = 0$$

$$\cos\delta_c = \frac{P_m(\delta_{max} - \delta_0) - P_{max3}\cos\delta_0 + P_{max2}\cos\delta_{max}}{P_{max2} - P_{max3}} \tag{18.33}$$

The angles are all in radians.
The critical clearing angle is obtained from

$$\delta_c = \cos^{-1}\left[\frac{P_m(\delta_{max} - \delta_0) - P_{max3}\cos\delta_0 + P_{max2}\cos\delta_{max}}{P_{max2} - P_{max3}}\right] \tag{18.33a}$$

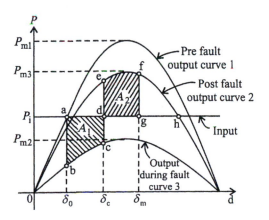

FIGURE 18.13

Equal area criterion applied to double-line system—power transmitted during fault.

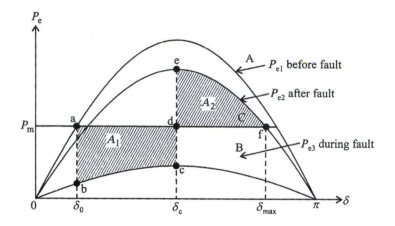

FIGURE 18.14

Critical clearing angle—power transmitted during fault.

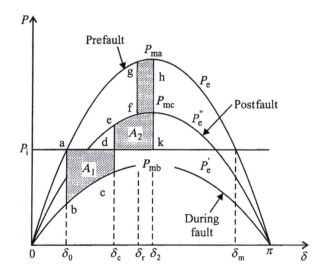

FIGURE 18.15

Fault clearance and reclosing.

18.14 FAULT CLEARANCE AND RECLOSURE IN DOUBLE-CIRCUIT SYSTEM

Consider a double-circuit system as in Section 18.12. If a fault occurs on one of the lines while supplying a power of P_{m0}; as in the previous case then an area $A_2 = A_1$ will be located and the operating characteristic changes from prefault to during the fault. If the faulted line is removed, then power transfer will be again be shifted to postfault characteristic where line 1 only is in operation. Subsequently, if the fault is cleared and line 2 is reclosed, the operation once again shifts back to prefault characteristic and normalcy will be restored. For stable operation area A_1 (=area abcd) should be equal to area A_2 (=area defghk). The maximum angle the rotor swings is δ_3. For stability δ_2 should be less than δ_m. The illustration in Fig. 18.15 assumes fault clearance and instantaneous reclosure.

18.15 SOLUTION TO SWING EQUATION STEP-BY-STEP METHOD

Solution to swing equation gives the change in δ with time. Uninhibited increase in the value of δ will cause instability. Hence it is desired to solve the swing equation to see that the value of δ starts decreasing after an initial period of increase, so that at some later point in time, the machine reaches the stable state. Generally 8, 5, 3, or 2 cycles are the times suggested for circuit breaker interruption after the fault occurs. A variety of numerical step-by-step methods are available for solution to swing equation. The plot of δ versus t in seconds is called the swing curve. The step-by-step method suggested here is suitable for hand calculation for a single machine connected to a system or an infinite bus.

The change in the angular position of the rotor $\Delta\delta$ during a short interval of time Δt is calculated under the following assumptions:

1. The accelerating power P_a computed at the beginning of any interval, Δt remains constant from the middle of the preceding interval to the middle of the interval under consideration.
2. The angular velocity w remains constant for the entire interval which is computed at the middle of the interval.

Since δ is changing continuously, both the assumptions are not true. When Δt is made very small, the calculated values become more accurate.

Let the time intervals be Δt.

Consider $(n-2)$, $(n-1)$, and nth intervals. The accelerating power P_a is computed at the end of these intervals and plotted at circles in Fig. 18.16A.

Note that these are the beginnings for the next intervals, viz., $(n-1)$, n, and $(n+1)$. P_a is kept constant between the midpoints of the intervals.

Likewise, w_r, the difference between w and w_s is kept constant throughout the interval at the value calculated at the midpoint. The angular speed therefore is assumed to change between $(n-3/2)$ and $(n-1/2)$ ordinateswe know that

$$\Delta w = \frac{dw}{dt} \cdot \Delta t$$

Hence

$$w_{r(n-\frac{1}{2})} - w_{r(n-3/2)} = \frac{d^2\delta}{dt^2} \cdot \Delta t = \frac{180f}{H} P_{a(n-1)} \cdot \Delta t \tag{18.34}$$

Again change in δ

$$\Delta\delta = \frac{d\delta}{dt} \cdot \Delta t$$

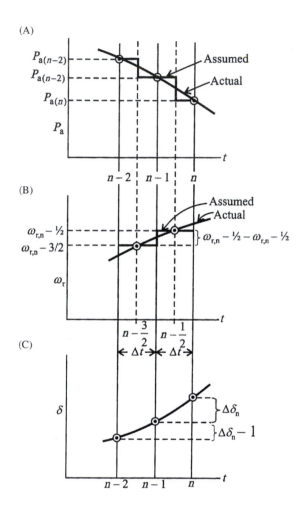

FIGURE 18.16

Plotting swing curve.

i.e.,

$$\Delta\delta_{n-1} = \delta_{n-1} - \delta_{n-2} = \omega_{r(n-3/2)} \cdot \Delta t \text{ for } (n-1)^{\text{th}} \text{ interval} \tag{18.35}$$

and

$$\Delta\delta_n = \delta_n - \delta_{n-1} = \omega_{r(n-1/2)} \cdot \Delta t \tag{18.36}$$

From Eqs. (18.35) and (18.36) we obtain

$$\Delta\delta_n = \delta_{n-1} + \left(\frac{180f}{H}\right)\Delta t^2 \cdot P_{a(n-1)} \tag{18.37}$$

Thus the plot of δ with time increasing after a transient disturbance has occurred or fault has taken place can be plotted as shown in Fig. 18.16C.

18.16 **FACTORS AFFECTING TRANSIENT STABILITY**

Transient stability is very much affected by the type of the fault. A three-phase dead short circuit is the most severe fault; the fault severity decreasing with two-phase fault and single line-to-ground fault in that order.

If the fault is farther from the generator, the severity will be less than in the case of a fault occurring at the terminals of the generator.

Power transferred during fault also plays a major role. When part of the power generated is transferred to the load, the accelerating power is reduced to that extent. This can easily be understood from the curves of Fig. 18.16.

Theoretically an increase in the value of inertia constant M reduces the angle through which the rotor swings farther during a fault. However, this is not a practical proposition because increasing M means increasing the dimensions of the machine, which is uneconomical. The dimensions of the machine are determined by the output desired from the machine and stability cannot be the criterion. Also, increasing M may interfere with speed governing system. Thus looking at the swing equations

$$M\frac{\mathrm{d}^2\delta}{\mathrm{d}t^2} = P_\mathrm{a} = P_\mathrm{m} - P_\mathrm{e} = P_\mathrm{m} - \frac{EV}{X_{12}}\sin\delta \tag{18.38}$$

the possible methods that may improve the transient stability are:

1. Increase of system voltages, and use of automatic voltage regulators.
2. Use of quick response excitation systems.
3. Compensation for transfer reactance X_{12} so that P_e increases and $P_\mathrm{m} - P_\mathrm{e} = P_\mathrm{a}$ reduces.
4. Use of high-speed circuit breakers which reduce the fault duration time and hence the accelerating power.

When faults occur, the system voltage drops. Support to the system voltages by automatic voltage controllers and fast-acting excitation systems will improve the power transfer during the fault and reduce the rotor swing.

Reduction in transfer reactance is possible only when parallel lines are used in place of single line or by use of bundle conductors. Other theoretical methods such as reducing the spacing between the conductors and increasing the size of the conductors are not practicable and are uneconomical.

Quick opening of circuit breakers and single-pole reclosing is helpful. Since majority of the faults are line-to-ground faults, selective single-pole opening and reclosing will ensure the transfer of power during the fault and improve stability.

18.17 **DYNAMIC STABILITY**

Consider a synchronous machine with terminal voltage V_t. The voltage due to excitation acting along the quadrature axis is E_q and is the voltage along this axis. The direct-axis rotor angle with respect to a synchronously revolving axis is δ. If a load change occurs and the field current I_f is not changed, then the various quantities mentioned change with the real power delivered P as shown in Fig. 18.17.

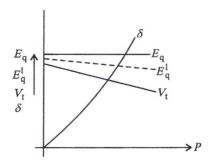

FIGURE 18.17

Effect of excitation with variation in load-excitation constant.

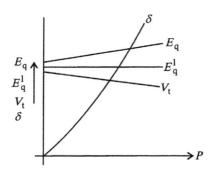

FIGURE 18.18

Effect of excitation varied to maintain field-flux linkages (E_q) constant.

In case the field current I_f is changed such that the transient flux linkages along the q-axis E_q^1 proportional to the field-flux linkages is maintained constant, the power transfer could be increased by 30%–60% greater than Fig. 18.17 and the quantities for this case are plotted in Fig. 18.18.

If the field current I_f is changed along with P simultaneously so that V_t is maintained constant, then it is possible to increase power delivery by 50%–80% more than Fig. 18.17. This is shown in Fig. 18.19.

It can be concluded from the above, that excitation control has a great role to play in power system stability and the speed with which this control is achieved is very important in this context.

Note that

$$P_{max} = \frac{EV}{X}$$

and increase of E matters in increasing P_{max}.

In Russia and other countries control signals utilizing the derivatives of output current and terminal voltage deviation have been used for controlling the voltage in addition to propositional control signals. Such a situation is termed "forced excitation" or "forced field control." Not only the

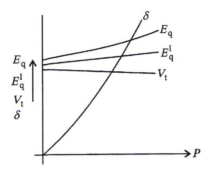

FIGURE 18.19

Effect of excitation varied to maintain terminal voltage (V_t) constant.

first derivatives of ΔI and ΔV are used, but also higher derivatives have been used for voltage control on load changes.

These controllers have not much control on the first swing stability but have effect on the operation subsequent swings.

This way of system control for satisfactory operation under changing load conditions using excitation control comes under the purview of dynamic stability.

18.17.1 POWER SYSTEM STABILIZER

A voltage regulator in the forward path of the exciter–generator system will introduce a damping torque and under heavy load conditions this damping torque may become negative. This is a situation where dynamic instability may occur and cause concern. It is also observed that the several time constants in the forward path of excitation control loop introduce large-phase lag at low frequencies just below the natural frequency of the excitation system.

To overcome these effects and to improve the damping, compensating networks are introduced to produce torque in phase with the speed.

Such a network is called "Power System Stabilizer."

18.18 NODE ELIMINATION METHODS

In all stability studies buses which are excited by internal voltages of the machines only are considered. Hence load buses are eliminated. As an example consider the system shown in Fig. 18.20.

The transfer reactance between the two buses (1) and (3) is given by

$$X_{13} = jx_d^1 + x_t + x_{l_1 l_2}$$

where

$$\frac{1}{x_{l_1 l_2}} = \frac{1}{x_{l_1}} + \frac{1}{x_{l_2}}$$

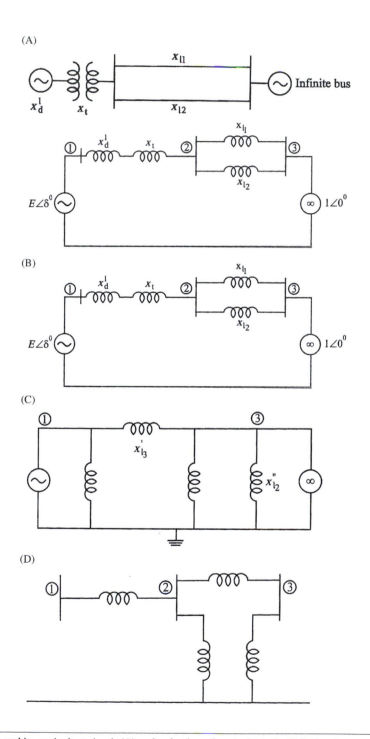

FIGURE 18.20

Double-line system and its equivalent circuit (A) and reduction of equivalent circuit further in stages (B), (C), and (D).

If a fault occurs on one of the two parallel lines, say, line 2, then the reactance diagram will become as shown in Fig. 18.20B.

Since, no source is connected to bus (2), it can be eliminated. The three reactances between buses (1), (2), (3) and (g) become a star network, which can be converted into a delta network using the standard formulas. The network will be modified into Fig. 18.20C.

X'_{13} is the transfer reactance between buses (1) and (3).

Consider the same example with delta network reproduced as in Fig. 18.20D.

For a three-bus system, the nodal equations are

$$\begin{bmatrix} I_1 \\ I_2 \\ I_3 \end{bmatrix} = \begin{bmatrix} Y_{11} & Y_{12} & Y_{13} \\ Y_{21} & Y_{22} & Y_{23} \\ Y_{31} & Y_{32} & Y_{33} \end{bmatrix} \begin{bmatrix} V_1 \\ V_2 \\ V_3 \end{bmatrix} \tag{18.39}$$

Since no source is connected to bus (2), it can be eliminated.
i.e., I_2 has to be made equal to zero.

$$Y_{21} V_1 + Y_{22} V_2 + Y_{23} V_3 = 0$$

Hence

$$V_2 = -\frac{Y_{21}}{Y_{22}} V_1 - \frac{Y_{23}}{Y_{22}} V_3$$

This value of V_2 can be substituted in the other two equations of (18.39) so that V_2 is eliminated

$$I_1 = Y_{11} V_1 + Y_{12} V_2 + Y_{13} V_3$$

$$= Y_{11} V_1 + Y_{12} \left[\frac{-Y_{21}}{Y_{22}} V_1 \right] + Y_{13} \left[\frac{Y_{23}}{Y_{22}} V_3 \right] + Y_{13} V_3$$

$$I_3 = Y_{31} V_1 + Y_{32} V_2 \left[\frac{-Y_{21}}{Y_{22}} V_1 - \frac{Y_{23}}{Y_{22}} V_3 \right] + Y_{33} V_3$$

$$= Y_{31} V_1 + Y_{32} + Y_{33} V_3$$

Thus Y_{BUS} changes to $\begin{bmatrix} Y^1_{11} & Y^1_{12} \\ Y^1_{31} & Y^1_{33} \end{bmatrix}$

where

$$Y^1_{11} = Y_{11} - Y_{12} \frac{Y_{21}}{Y_{22}}$$

and

$$Y^1_{13} = Y^1_{31} = Y_{13} - \frac{Y_{23} Y_{12}}{Y_{22}}$$

$$Y^1_{33} = Y_{33} - \frac{Y_{32} Y_{23}}{Y_{22}}$$

18.19 **VOLTAGE STABILITY**

Reactive power flow depends mainly on the difference between voltage magnitudes and it flows from higher voltage to lower voltage. Reactive power, therefore, cannot be transmitted over long distances as the line regulation cannot exceed the prescribed limit. Further, if the load is suddenly disconnected at the receiving end, there will be a temporary overvoltage in the line. Induction motors, excitation system for generators, switched capacitors and reactors, static Var compensators are some of the elements that cause fast dynamics in voltage stability along with protective relaying. On the other hand, slow changes in generation and loads, boiler dynamics are some of the factors involved in slower dynamics in voltage stability.

Voltage stability is an integral part of the power system response and is an important aspect of system stability and security. Voltage instability has been detected well before the onset of angle instability in many cases. If the problem is not corrected, it can lead to voltage collapse and system wide disturbance.

The loss of synchronism of generators is called angle instability. Voltage stability also called load stability is a subset of overall stability of a power system and is a dynamic problem that occurs due to monotonically changing voltages.

In industrialized areas increase of load demand is met without a corresponding increase in transmission capacity leading to severe problem including voltage stability.

Voltage stability is the ability to maintain the voltage so that when load is increased, load power will increase and so both power and voltage are controllable.

A power system at a given operating state and subjected to a given disturbance is voltage stable if voltage near loads approaches postdisturbance equilibrium values. Following voltage stability, a system undergoes voltage collapse if the postdisturbance equilibrium voltages near loads are below acceptance limits. Both voltage stability and collapse may occur in a time period of a fraction of a second to a few minutes.

Consider the plot of voltage V at a load bus as a function of load power P shown in Fig. 18.21 as ABCDE.

It can be observed that there are two voltages V_1 and V_2 for a constant power load. The high-voltage solution is stable while the other is unstable. The maximum loadability is determined by point C and the part CDE is uncontrollable.

Mathematically the voltage V_R can be expressed by (see Fig. 18.22).

$$V_R = \left[\frac{-2QX + V_S^2}{2} \pm \frac{1}{2} \sqrt{(2QX - V_S^2)^2 - 4X^2(P^2 + Q^2)} \right] \tag{18.40}$$

from which it is clear that V_R is a double-valued function.

This can be proved as follows:

$$V_S = V_R + jIX \quad \text{(neglecting the resistance)}$$

$$= V_R + j\frac{(P + jQ)}{V_R}X$$

$$V_S = V_R^2 + \frac{(-Q + jP)X}{V_R}$$

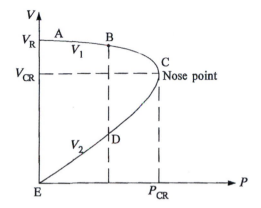

FIGURE 18.21

PV diagram at the load bus.

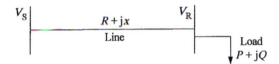

FIGURE 18.22

Transmission line with load.

$$V_S V_R = (V_R^2 - QX) + j(PX)$$

$$V_S V_R^2 = V_R^4 + Q^2 X^2 - 2V_R^2 QX + P^2 X^2$$

$$V_R^4 - V_R^2(V_S^{\tau} - 2QX) + X^2(P^2 + Q^2) = 0$$

$$V_R^2 = \frac{(V_S^2 - 2QX) \pm \sqrt{(-V_S^2 + 2QX)^2 - 4X^2(P^2 + Q^2)}}{2}$$

$$V_R = \left[\frac{-2QX + V_S^2}{2} \pm \frac{1}{2}\sqrt{(2QX - V_S^2)^2 - 4(P^2 + Q^2)X^2} \right]^{1/2}$$

Consider a line supplying only reactive power

$$Q_R = V_R I_R$$

$$= V_R \left(\frac{V_S - V_R}{X} \right)$$

$$Q_R = \frac{V_R V_S}{X} - \frac{VR^2}{X}$$

$$\frac{dQ_r}{dV_R} = 0 \text{ when } \frac{1}{X}(V_S - 2V_R) = 0$$

The critical receiving end voltage $V_{R,crit}$

$$V_{R,crit} = \frac{V_S}{2}$$

At this critical voltage

$$Q_{R,crit} = \frac{V_S}{X}\frac{V_S}{2} - \frac{V_S^2}{4X} = \frac{V_S^2}{4X}$$

The sending-end reactive power, when no real power is transmitted

$$Q_S = Q_R + I^2 X$$

$$= Q_R + \left(\frac{Q_S}{V_S}\right)^2 X$$

Simplifying

$$Q_S^2 - Q_S\frac{V_S^2}{X} + Q_R\frac{V_S^2}{X} = 0$$

$$Q_S = \frac{1}{2}\left[\frac{V_S^2}{X} \pm V_S\sqrt{\left(\frac{V_S^2}{X^2} - \frac{4Q_R}{X}\right)}\right]$$

$$\frac{dQ_S}{dQ_R} = \left(\frac{1}{4X}\right)\frac{-4V_S}{\left(\frac{V_S^2}{X^2} - \frac{4Q_R}{X}\right)^{1/2}}$$

$$= \frac{1}{\left(\frac{X^2V_S^2}{V_S^2X^2} - \frac{4Q_RX^2}{V_S^2X}\right)^{1/2}} = \frac{1}{\left(1 - \frac{4Q_RX}{V_S^2}\right)^{1/2}}$$

But

$$Q_{R,crit} = \frac{V_S^2}{4X}$$

Hence

$$\frac{dQ_S}{dQ_R} = \frac{1}{\left(1 - \frac{Q_R}{Q_{R,crit}}\right)^{1/2}}$$

dQ_S/dQ_R is called voltage collapse proximity indicators. At no load it is unity, when $Q_R = Q_{R,crit}$, the voltage collapse indicator tends to infinity. Voltage may collapse in such a situation.

18.19.1 VOLTAGE STABILITY LIMIT

Consider the power relation

$$P_R = \frac{V_S V_R}{X} \sin \delta$$

$$\frac{dP_R}{dV_S} = \frac{V_R \sin \delta}{x} + \frac{V_S V_R}{X} \cos \delta \frac{d\delta}{dV_S}$$

$$\text{If } \frac{dP_R}{dV_S} = 0, \quad \text{then } \frac{d\delta}{dV_s} = -\frac{1}{V_R} \tan \delta$$

At $P_{R_{max}}$, $\delta = 90$ degrees $\frac{d\delta}{dV_S}$ so that tends to infinity.
Again, since

$$Q_R = \frac{V_S V_R}{X} \cos \delta - \frac{VR^2}{X}$$

$$\frac{dQ_R}{dV_R} = \frac{V_S}{X} \cos \delta - \frac{2V_R}{X} - \frac{V_S V_R}{X} \sin \delta \frac{d\delta}{dV_R}$$

$$\frac{dQ_R}{dV_R} = \frac{1}{X} \left[V_S \cos \delta + V_S \frac{\sin^2 \delta}{\cos \delta} - 2V_R \right]$$

$$= \frac{1}{X} \left[\frac{V_S}{\cos \delta} - 2V_R \right]$$

If $\delta = 90$ degrees for steady-state power angle stability, then as $\cos \delta$ tends to zero, dQ_R/dV_R tends to infinity, i.e., dV_R/dQ_R tends to zero.

Thus, on no load, voltage instability occurs when power angle instability takes place.

For different power factors, the *PV* characteristic can have the shape as shown in Fig. 18.23.

Static analysis is sufficient to assess voltage stability, but for accurate prediction dynamic analysis must be carried out.

The probability of voltage instability increases as the system is operated close to its maximum loadability limit. Environmental and economic expansion of transmission network and for obvious reasons distant location of the generators from the load centers all result in overloading of the existing networks. The present trend is to optimally utilize the inherent margins available with flexible AC transmission system controllers.

Reactive power compensation close to the load centers as well as at the critical buses in the network is essential to overcome voltage instability.

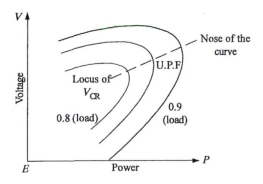

FIGURE 18.23

PV curves for different power factors.

The location size and speed of control of FACTS controllers determine the maximum benefit that can be obtained without getting into voltage instability.

18.20 METHODS FOR PREVENTION OF VOLTAGE COLLAPSE

1. *Load Shedding:* Underfrequency and undervoltage load shedding provides protection for unusual load disturbances. This should be resorted only under extreme situations.
2. *VAR Devices:* Use of reactive power compensating devices will generally reduce the voltage stability problem.
3. *Transformer Taps:* By controlling the transformer taps using on-load tap changers, it is possible to some extent to overcome the voltage stability problem with proper planning, control, and coordination of protective equipment. The operator at the console should not find it difficult to overcome most of the voltage stability problems.

WORKED EXAMPLES

E.18.1 A four-pole, 50-Hz, 11-kV turbo generator is rated 75 MW and 0.86 power factor lagging. The machine rotor has a moment of inertia of 9000 kg-m^2. Find the inertia constant in MJ/MVA and M constant or momentum in MJs/electrical degree

Solution:

$$\omega = 2\pi f = 100\pi \text{ rad/s}$$

$$\text{Kinetic energy} = \frac{1}{2}I\omega^2 = \frac{1}{2} \times 9000 + (100\pi)^2$$

$$= 443.682 \times 10^6 \text{ J}$$

$$= 443.682 \text{ MJ}$$

$$\text{MVA rating of the machine} = \frac{75}{0.86} = 87.2093$$

$$H = \frac{\text{MJ}}{\text{MVA}} = \frac{443.682}{87.2093} = 8.08755$$

$$M = \frac{GH}{180f} = \frac{87.2093 \times 5.08755}{180 \times 50}$$

$$= 0.0492979 \text{ MJs/electrical degree}$$

E.18.2 Two generators rated at 4-pole, 50 Hz, 50 MW 0.85 p.f (lag) with moment of inertia 28,000 kg-m^2 and 2-pole, 50 Hz, 75 MW 0.82 p.f (lag) with moment of inertia 15,000 kg-m^2 are connected by a transmission line. Find the inertia constant of each machine and the inertia constant of single equivalent machine connected to an infinite bus. Take 100 MVA base.

Solution:

For machine I

$$KE = \frac{1}{2} \times 28,000 \times (100\pi)^2 = 1380.344 \times 10^6 \text{ J}$$

$$MVA = \frac{50}{0.85} = 58.8235$$

$$H_1 = \frac{1380.344}{58.8235} = 23.46586 \text{ MJ/MVA}$$

$$M_1 = \frac{58.8235 \times 23.46586}{180 \times 50} = \frac{1380.344}{180 \times 50}$$
$$= 0.15337 \text{ MJs/electrical degree}$$

For the second machine

$$KE = \frac{1}{2} \times 15,000 \times (100\,\pi)^2 = 739,470,000 \text{ J}$$
$$= 739.470 \text{ MJ}$$

$$MVA = \frac{75}{0.82} = 91.4634$$

$$H_2 = \frac{739.470}{91.4634} = 8.0848$$

$$M_2 = \frac{91.4634 \times 8.0848}{180 \times 50} = 0.082163 \text{ MJs/electrical degree}$$

$$\frac{1}{M} = \frac{1}{M_1} = \frac{1}{M_2}$$

$$\therefore \qquad M = \frac{M_1 M_2}{M_1 + M_2} = \frac{0.082163 \times 0.15337}{0.082163 + 0.15337}$$
$$= \frac{0.0126}{0.235533} = 0.0535 \text{ MJs/electrical degree}$$

$$GH = 180 \times 50 \times M = 180 \times 50 \times 0.0535$$
$$= 481.5 \text{ MJ}$$

on 100-MVA base, inertia constant.

$$H = \frac{481.5}{100} = 4.815 \text{ MJ/MVA}$$

E.18.3 A four-pole synchronous generator rated at 110 MVA, 12.5 kV, 50 Hz has an inertia constant of 5.5 MJ/MVA

1 Determine the stored energy in the rotor at a synchronous speed.

2 When the generator is supplying a load of 75 MW, the input is increased by 10 MW. Determine the rotor acceleration, neglecting losses.

3 If the rotor acceleration in (**2**) is maintained for 8 cycles, find the change in the torque angle and the rotor speed in rpm at the end of 8 cycles.

Solution:

1 Stored energy $= GH = 110 \times 5.5 = 605$ MJ where $G =$ Machine rating

2 $P_a =$ The accelerating power $= 10$ MW

$$10 \text{ MW} = M\frac{d^2\delta}{dt^2} = \frac{GH}{180f}\frac{d^2\delta}{dt^2}$$

$$= \frac{605}{180 \times 50}\frac{d^2\delta}{dt^2} = 10$$

$$0.0672\frac{d^2\delta}{dt^2} = 10 \text{ or } \frac{d^2\delta}{dt^2} = \frac{10}{0.0672} = 148.81$$

$$\alpha = 148.81 \text{ electrical degrees}/s^2$$

3 8 cycles $= 0.16$ s

Change in $\delta = \frac{1}{2} \times 148.81 \times (0.16)^2$

Rotor speed at the end of 8 cycles

$$= \frac{120f}{P} \cdot (\delta) \times t = \frac{120 \times 50}{4} \times 1.90476 \, \delta \times 0.16$$
$$= 457.144 \text{ rpm}$$

E.18.4 Power is supplied by a generator to a motor over a transmission line as shown in Fig. E.18.4A. To the motor bus a capacitor of 0.8 p.u. reactance per phase is connected through a switch. Determine the steady-state power limit with and without the capacitor in the circuit.

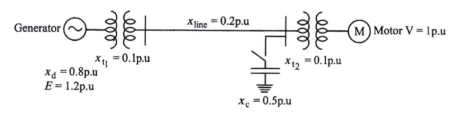

FIGURE E.18.4A

Steady-state power limit without the capacitor

$$P_{max1} = \frac{1.2 \times 1}{0.8 + 0.1 + 0.2 + 0.8 + 0.1} = \frac{1.2}{2.0} = 0.6 \text{ p.u.}$$

With the capacitor in the circuit, the following circuit is obtained.

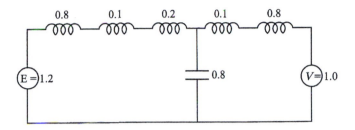

FIGURE E.18.4B

Simplifying, we obtain the circuit in Fig. E.18.4C.

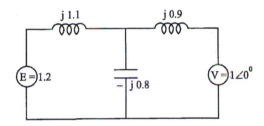

FIGURE E.18.4C

Converting the star to delta network, the transfer reactance between the two nodes X_{12} shown in Fig. E.18.4D is given by

$$\text{Steady-state power limit} = \frac{1.2 \times 1}{0.7625} = 1.5738 \text{ p.u.}$$

$$X_{12} = \frac{(j1.1)(j0.9) + (j0.9)(-j0.8) + (-j0.8 \times j1.1)}{-j0.8}$$

$$= \frac{-0.99 + 0.72 + 0.88}{-j0.8} = \frac{-0.99 + 1.6}{-j0.8} = \frac{j0.61}{0.8}$$

$$= j0.7625 \text{ p.u.}$$

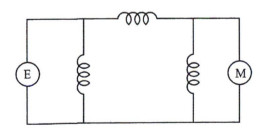

FIGURE E.18.4D

E.18.5 A generator rated 75 MVA is delivering 0.8 pu power to a motor through a transmission line of reactance j0.2 p.u. The terminal voltage of the generator is 1.0 p.u. and that of the motor is also 1.0 p.u. Determine the generator emf behind transient reactance. Find also the maximum power that can be transferred.

Solution:

When the power transferred is 0.8 p.u.

$$0.8 = \frac{1.0 \times 1.0 \sin\theta}{(0.1 + 0.2)} = \frac{1}{0.3}\sin\theta$$
$$\sin\theta = 0.8 \times 0.3 = 0.24$$
$$\theta = 13.°8865$$

Current supplied to motor

$$I = \frac{1\angle 13.°8865 - 1\angle 0\,\text{degree}}{j0.3} = \frac{(0.9708 + j0.24) - 1}{j0.3}$$

$$= \frac{-0.0292 + j0.24}{j0.3} = j0.0973 + 0.8 = 0.8571\underline{/\tan^{-1}0.1216}$$

$$I = 0.8571\underline{/6.°934}$$

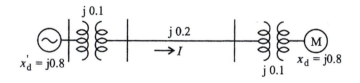

FIGURE E.18.5

Voltage behind transient reactance

$$= 1\angle 0\,\text{degree} + j1.2(0.8 + j0.0973)$$
$$= 1 + j0.96 - 0.11676$$
$$= 0.88324 + j0.96$$
$$= 1.0496\ 47°.8$$

$$P_{\max} = \frac{EV}{X} = \frac{1.0496 \times 1}{1.2} = 0.8747\ \text{p.u.}$$

E.18.6 Determine the power angle characteristic for the system shown in Fig. E.18.6A. The generator is operating at a terminal voltage of 1.05 p.u. and the infinite bus is at 1.0 p.u. voltage. The generator is supplying 0.8 p.u. power to the infinite bus.

Solution:

The reactance diagram is drawn in Fig. E.18.6B.

The transfer reactance between V_t and V is $= j\,0.1 + (j0.4/2) = j0.3$ p.u.

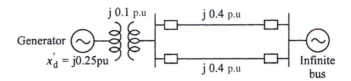

FIGURE E.18.6A

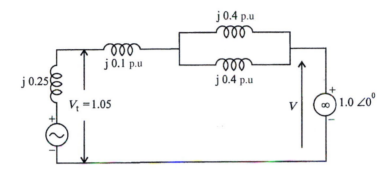

FIGURE E.18.6B

we have

$$\frac{V_t V}{X} \sin \delta = \frac{(1.05)(1.0)}{0.3} \sin \delta = 0.8$$

Solving for δ, $\sin \delta = 0.22857$ and $\delta = 13°.21$

$$\text{The terminal voltage is} \quad 1.05 \underline{/13°.21}$$
$$= 1.022216 + j0.24$$

The current supplied by the generator to the infinite bus

$$I = \frac{1.022216 + j0.24 - (1 + j0)}{j0.3}$$
$$= \frac{(0.022216 + j0.24)}{j0.3} = 0.8 - j0.074$$
$$= 1.08977 \underline{/5.°28482} \text{ p.u.}$$

The transient internal voltage in the generator

$$E^1 = (0.8 - j0.074) j0.25 + 1.22216 + j0.24$$
$$= j0.2 + 0.0185 + 1.02216 + j0.24$$
$$= 1.040 + j0.44$$
$$= 1.1299 \underline{/22°.932}$$

The total transfer reactance between E^1 and V

$$= j0.25 + j0.1 + = j0.55 \text{ p.u.}$$

The power angle characteristic is given by

$$P_e = \frac{E^1 V}{X} \sin \delta = \frac{(1.1299) \cdot (1.0)}{j0.55} \sin \delta$$

$$P_e = 2.05436 \sin \delta$$

P_e can be computed and plotted for different values of δ from zero to 90 degrees.

E.18.7 Consider the system in E.18.1 shown in Fig. E.18.7. A three-phase fault occurs at point P as shown at the midpoint on line 2. Determine the power angle characteristic for the system with the fault persisting.

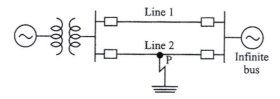

FIGURE E.18.7

Solution:
The reactance diagram is shown in Fig. E.18.7A.
The admittance diagram is shown in Fig. E.18.7B.
The buses are numbered and the bus admittance matrix is obtained.

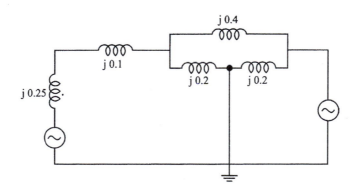

FIGURE E.18.7A

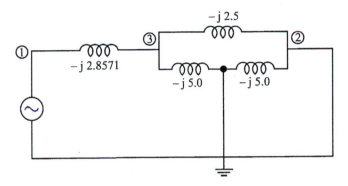

FIGURE E.18.7B

	①	②	③
①	−j1.85271	0.0	j2.85271
②	0.0	−j7.5	j2.5
③	j2.8271	j2.5	−j10.3571

Node 3 or bus 3 has no connection to any source directly, it can be eliminated.

$$Y_{11(\text{modified})} = Y_{11(\text{old})} - \frac{Y_{13}Y_{31}}{Y_{33}}$$

$$= -j2.8571 - \frac{(2.8527)(2.85271)}{(-10.3571)} = -2.07137$$

$$Y_{12(\text{modified})} = 0 - \frac{(2.85271)(2.5)}{(-10.3571)} = 0.6896$$

$$Y_{22(\text{modified})} = Y_{22(\text{old})} - \frac{Y_{32}Y_{23}}{Y_{33}}$$

$$= -7.5 - \frac{(2.5)(2.5)}{(-10.3571)} = -6.896549$$

The modified bus admittance matrix between the two sources is

	①	②
①	−2.07137	0.06896
②	0.6896	−6.89655

The transfer admittance between the two sources is 0.6896 and the transfer reactance = 1.45

$$P_2 = \frac{1.05 \times 1}{1.45} \sin \delta \text{ p.u.}$$

or

$$P_e = 0.7241 \sin \delta \text{ p.u.}$$

E.18.8 For the system considered in E.18.6 if the H constant is given by 6 MJ/MVA obtain the swing equation

Solution:

The swing equation is $\dfrac{H}{\pi f}\dfrac{d^2\delta}{dt^2} = P_m - P_e = P_a$, the accelerating power

If δ is in electrical radians

$$\frac{d^2\delta}{dt^2} = \frac{180 \times f}{H}P_a = \frac{180 \times 50}{6}P_a = 1500\,P_a$$

E.18.9 In E18.7 if the three-phase fault is cleared on line 2 by operating the circuit breakers on both sides of the line, determine the postfault power angle characteristic.

Solution:

The net transfer reactance between E^1 and V_a with only line 1 operating is

$$j0.25 + j0.1 + j0.4 = j0.75 \text{ p.u.}$$

$$P_e = \frac{(1.05)(1.0)}{j0.75}\sin \delta = 1.4 \sin \delta$$

E.18.10 Determine the swing equation for the condition in E.18.9 when 0.8 p.u. power is delivered.

Given

$$M = \frac{1}{1500}$$

Solution:

$$\frac{180f}{H} = \frac{180 \times 50}{6} = 1500$$

$$\frac{1}{1500}\frac{d^2\delta}{dt^2} = 0.8 - 1.4 \sin \delta \text{ is the swing equation}$$

where δ in electrical degrees.

E.18.11 Consider example E.18.6 with the swing equation

$$P_e = 2.05 \sin \delta$$

If the machine is operating at 28 degrees and is subjected to a small transient disturbance, determine the frequency of oscillation and also its period.

Given

$$H = 5.5 \text{ MJ/MVA}$$

$$P_e = 2.05 \sin 28 \text{ degrees} = 0.9624167$$

Solution:

$$\frac{dP_e}{d\delta} = 2.05 \cos 28 \text{ degrees} = 1.7659$$

The angular frequency of oscillation $= \omega_n$

$$\omega_n = \sqrt{\frac{\omega S^0}{2H}} = \sqrt{\frac{2\pi \times 50 \times 1.7659}{2 \times 5.5}}$$

$$= 7.099888 = 8 \text{ electrical radians/s}$$

$$f_n = \frac{1}{2\pi} \times 8 = \frac{4}{\pi} = 1.2739 \text{ Hz}$$

$$\text{Period of oscillation} = T = \frac{1}{f_n} = \frac{1}{1.2739} = 0.785 \text{ s}$$

E.18.12 The power angle characteristic for a synchronous generator supplying infinite bus is given by

$$P_e = 1.25 \sin \delta$$

The H constant is 5 s and initially it is delivering a load of 0.5 p.u. Determine the critical angle.

Solution:

$$\cos \delta_c = \frac{P_{m0}}{P_{max}}[(\pi - 2\delta_0) + \cos(\pi - \delta_0)]$$

$$\frac{P_{m0}}{P_{max}} = \frac{0.5}{1.25} = 0.4 = \sin d \, \delta_0; \, \delta_0 = 23°.578$$

$$\cos \delta_0 = 0.9165$$

$$\delta_0 \text{ in radians} = 0.4113$$

$$2\delta_0 = 0.8226$$

$$\pi - 2\delta_0 = 2.7287$$

$$\frac{P_{m0}}{P_{max}}(\pi - 2\delta_0) = 1.09148$$

$$\cos \delta_c = 1.09148 - 0.9165 = 0.17498$$

$$\delta_c = 79°.9215$$

E.18.13 Consider the system shown in Fig. E.18.13.

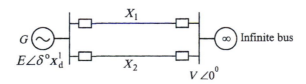

FIGURE E.18.13

$$x_d^1 = 0.25 \text{ p.u.}$$

$$|E| = 1.25 \text{ p.u. and } |V| = 1.0 \text{ p.u.; } X_1 = X_2 = 0.4 \text{ p.u.}$$

Initially the system is operating stable while delivering a load of 1.25 p.u. Determine the stability of the system when one of the lines is switched off due to a fault.

Solution:

When both the lines are working

$$P_{e_{max}} = \frac{1.25 \times 1}{0.25 + 0.2} = \frac{1.25}{0.45} = 2.778 \text{ p.u.}$$

When one line is switched off

$$P_{e_{max}}^1 = \frac{1.25 \times 1}{0.25 + 0.4} = \frac{1.25}{0.65} = 1.923 \text{ p.u.}$$

$$P_{e_0} = 2.778 \sin \delta_0 = 1.25 \text{ p.u.}$$

$$\sin \delta_0 = 0.45$$

$$\delta_0 = 26°.7437 = 0.4665 \text{ radians}$$

At point C

$$P_e^1 = 1.923 \quad \sin \delta_1 = 1.25$$

$$\sin \delta_1 = 0.65$$

$$\delta_1 = 40°.5416$$

$$= 0.7072 \text{ radians}$$

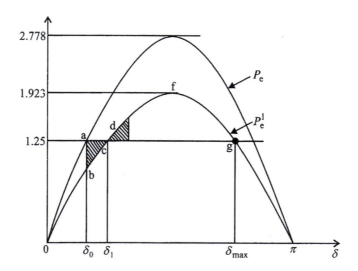

$$A_1 = \text{area abc} = \int_{\delta_0}^{\delta_1} (P_2 - P_e^1) d\delta = \int_{0.4665}^{0.7072} (1.25 - 1.923 \sin \delta) d\delta$$

$$= 1.25 \Big|_{0.4665}^{0.7072} + 1.923 \cos \delta \Big|_{26°.7437}^{40°.5416}$$

$$= 0.300625 + (-0.255759) = 0.0450$$

Maximum area available = area cdfgc

$$A_{2\,max} \int_{\delta_1}^{\delta_{max}} (P_e^1 - P_i)d\delta = \int_{0.7072}^{\pi-0.7072} (1.923 \sin \delta - 1.25) \, d\delta$$

$$A_{2\,max} = \int_{\delta_1}^{\delta_{max}} (P_e^1 - P_i)d\delta = \int_{0.7072}^{\pi-0.7072} (1.923 \sin \delta - 1.25)d\delta$$

$$= -1.923 \cos \delta \Big|_{40°.5416}^{139°.46} - 1.25(2.4328 - 0.7072)$$

$$= 0.7599 - 1.25 \times 1.7256$$

$$= 0.7599 - 2.157 = -1.3971 \gg A_1$$

The system is stable

[*Note:* Area A_1 is below $P_2 = 1.25$ line and area A_2 is above $P_2 = 1.25$ line; hence the negative sign]

E.18.14 Determine the maximum value of the rotor swing in example E18.13.

Solution:

Maximum value of the rotor swing is given by condition

$$A_1 = A_2$$

$$A_1 = 0.044866$$

$$A_2 = \int_{\delta_1}^{\delta_2} (-1.25 + 1.923 \sin \delta) \, d\delta$$

$$= (-1.25\delta_2 + 1.25 \times 0.7072) - 1.923(\cos \delta_2 - 0.76)$$

i.e.,

$$= +1.923 \cos \delta_2 + 1.25 \, \delta_2 = 2.34548 - 0.0450$$

i.e.,

$$= 1.923 \cos \delta_2 + 1.25 \, \delta_2 = 2.30048$$

By trial and error $\delta_2 = 55°.5$

E.18.15 The M constant for a power system is 3×10^{-4} s^2/electrical degree

The prefault, during the fault, and postfault power angle characteristics are given by

$$P_{e1} = 2.45 \sin \delta$$

$$P_{e2} = 0.8 \sin \delta$$

and

$$P_{e3} = 2.00 \sin \delta, \text{ respectively}$$

choosing a time interval of 0.05 s obtain the swing curve for a sustained fault on the system. The prefault power transfer is 0.9 p.u.

Solution:

$$P_{e_1} = 0.9 = 2.45 \sin \delta_0$$

The initial power angle $\delta_0 = \sin^{-1}\left(\dfrac{0.9}{2.45}\right)$

$$= 21.55 \text{ degrees}$$

At $t = 0_-$ just before the occurrence of fault.

$$P_{max} = 2.45$$

$$\sin \delta_0 = \sin 21°.55 = 0.3673$$

$$P_e = P_{max} \sin \delta_0 = 0.3673 \times 2.45 = 0.9$$

$$P_a = 0$$

At $t = 0_+$, just after the occurrence of fault

$$P_{max} = 0.8; \ \sin \delta_0 = 0.6373 \text{ and hence}$$

$$P_e = 0.3673 \times 0.8 = 0.2938$$

P_a, the accelerating power $= 0.9 - P_e$

$$= 0.9 - 0.2938 = 0.606$$

Hence the average accelerating power at $t = 0_{ave}$

$$= \frac{0 + 0.606}{2} = 0.303$$

$$\frac{(\Delta t)^2}{M} P_a = \frac{(0.05 \times 0.05)}{3 \times 10^{-4}} = 8.33 P_a = 8.33 \times 0.303 = 2°.524$$

$$\Delta \delta = 2°.524 \text{ and } \delta_0 = 21°.55$$

The calculations are tabulated up to $t = 0.4$ s (Table E.18.15).

Table E.18.15

S.No	t (s)	P_{max} (p.u.)	$\sin \delta$	$P_e = P_{max} \sin \delta$	$P_a = 0.9 - P_e$	$\frac{(\Delta t)^2}{M} P_a = 8.33 \times P_a$	$\Delta \delta$	δ
1.	$0_-2.45$	0.3673	0.9	0	—	—	21.55 degrees	
	$0_+0.8$	0.3673	0.2938	0.606	—	—	21.55 degrees	
	0_{ave}		0.3673	—	0.303	2.524	$2°.524$	$24°.075$
2.	0.05	0.8	0.4079	0.3263	0.5737	4.7786	$7°.3$	$24°.075$
3.	0.10	0.8	0.5207	0.4166	0.4834	4.027	$11°.327°$	31.3766
4.	0.15	0.8	0.6782	0.5426	0.3574	2.977	$14°.304$	$42°.7036$
5.	0.20	0.8	0.8357	0.6709	0.2290	1.9081	$16°.212$	$57°.00$
6.	0.25	0.8	0.9574	0.7659	0.1341	1.1170	$17°.329$	$73°.2121$
7.	0.30	0.8	0.9999	0.7999	0.1000	0.8330	$18°.1623$	90.5411
8.	0.35	0.8	0.9472	0.7578	0.1422	1.1847	$19°.347$	108.70
9.	0.40	0.8	0.7875	0.6300	0.2700	2.2500	$21°.596$	128.047
							$149°.097$	

Table of results for E18.15.
From the table it can be seen that the angle δ increases continuously indicating instability.

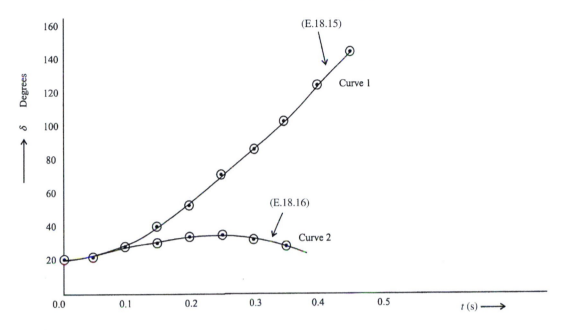

E.18.16 If the fault in the previous example E.18.14 is cleared at the end of 2.5 cycles, determine the swing curve and examine the stability of the system.

Solution:
As before

$$\frac{(\Delta t^2)}{M} P_a = 8.33 P_a$$

$$\text{Time to clear the fault} = \frac{2.5 \text{ cycles second}}{50 \text{ cycles}}$$

$$= 0.05 \text{ s}$$

In this the calculations performed in the previous example E.18.14 hold good for O_{ave}. However, since the fault is cleared at 0.05 s, there will be two values for P_{a1} one for $P_{e2} = 0.8 \sin \delta$ and another for $P_{e3} = 2.00 \sin \delta$.

At $t = 0.5 -$ (just before the fault is cleared)

$$P_{max} = 0.5; \ \sin \delta = 0.4079, \text{ and}$$

$$P_e = P_{max} \sin d \ \delta = 0.3263, \text{ so that } P_a = 0.9 - P_e = 0.57367$$

giving as before $\delta = 24°.075$

But, at $t = 0.5 +$ (just after the fault is cleared) P_{max} becomes 2.0 p.u. at the same δ and $P_e = P_{max} \sin \delta = 0.8158$. This gives a value for $P_a = 0.9 - 0.815 \ \delta = 0.0842$. Then for $t = 0.05$ are the average accelerating power at the instant of fault clearance becomes

$$P_{a_{ave}} = \frac{0.57367 + 0.0842}{2} = 0.8289$$

$$\frac{(\Delta t)^2}{M} \cdot P_a = 8.33 \times 0.3289 = 2°.74$$

and

$$\Delta \delta = 5.264$$

$$\delta = 5.264 + 24.075 = 29°.339$$

These calculated results and further calculated results are tabulated in Table E.18.16.

Table E.18.16

S. No	t	P_{max}	$\sin \delta$	$P_e = P_{max} \sin \delta$	$P_a = 0.9 - P_e$	$\frac{(\Delta t)^2}{M}$ $P_a = 8.33 \times P_a$	$\Delta \delta$	δ
1.	$0_-2.45$	0.3673	0.9	0	—	—	21.55 degrees	
	$0_+0.8$	0.3673	0.2938	0.606	—	—	21.55 degrees	
	0_{ave}		0.3673	—	0.303	2.524	2.524	24.075
2.	0.05_-	0.8	0.4079	0.3263	0.5737	—	—	—
	0.05_+	2.0	0.4079	0.858	0.0842	—	—	—
	0.05_{ave}		0.4079	—	0.3289	2.740	5.264	29.339
3.	0.10	2.0	0.49	0.98	−0.08	−0.6664	4.5976	33.9367
4.	0.15	2.0	0.558	1.1165	−0.2165	−1.8038	2.7937	36.730
5.	0.20	2.0	0.598	1.1196	−0.296	−2.4664	0.3273	37.05
6.	0.25	2.0	0.6028	1.2056	−0.3056	−2.545	−2.2182	34°.83
7.	0.30	2.0	0.5711	1.1423	−0.2423	−2.018	−4.2366	30°.5933

Table of results for E18.15.

The fact that the increase of angle δ, started decreasing indicates stability of the system.

E.18.17 A synchronous generator represented by a voltage source of 1.1 p.u. in series with a transient reactance of j0.15 p.u. and an inertia constant $H = 4$ s is connected to an infinite bus through a transmission line. The line has a series reactance of j0.40 p.u. while the infinite bus is represented by a voltage source of 1.0 p.u.

The generator is transmitting an active power of 1.0 p.u. when a three-phase fault occurs at its terminals. Determine the critical clearing time and critical clearing angle. Plot the swing curve for a sustained fault.

Solution:

$$P = \frac{EV}{X}\sin \delta_0; \, 1.0 \frac{1.1 \times 1.0}{(0.45 + 0.15)}\sin \delta_0; \, \delta_0 = 30°$$

$$\delta_c = \cos^{-1}[(\pi - 2\delta_0)\sin \delta_0 - \cos \delta_0]$$
$$= \cos^{-1}[(180° - 2 \times 30°)\sin 30° - \cos 30°]$$
$$= \cos^{-1}\left[\frac{\pi}{3} - 0.866\right] = \cos^{-1}[1.807]$$
$$= 79°.59$$

Critical clearing angle $= 79°.59$

$$\text{Critical clearing time} = \sqrt{\frac{2H}{P_m}\frac{(\delta_c - \delta_0)}{\pi f}}$$

$$\delta_c - \delta_0 = 79°.59 - 30° = 49.59° = \frac{49.59 \times 3.14}{180}\text{rad}$$

$$= 0.86507 \text{ rad}$$

$$t_c = \sqrt{\frac{2 \times 4 \times 0.86507}{1 \times 3.14 \times 50}} = 0.2099 \text{ s}$$

Calculation for the swing curve

$$\Delta \delta_n = \delta_{n-1} + \left(\frac{180f}{H}\right)\Delta t^2 \, P_{a(n-1)}$$

Let

$$\Delta t = 0.05 \, s$$

$$\delta_{n-1} = 30°$$

$$\frac{180f}{H} = \frac{180 \times 50}{4} = 2250$$

$$M = \frac{H}{180f} = \frac{1}{2250} = 4.44 \times 10^{-4}$$

$$\frac{(\Delta t)^2}{M}P_a = \frac{(0.05 \times 0.05)}{(4.44 \times 10^{-4})}P_a = 5.63P_a$$

Accelerating power before the occurrence of the fault $= P_{a-} = 2 \sin \delta_0 - 1.0 = 0$
Accelerating power immediately after the occurrence of the fault

$$P_{a+} = 2 \sin \delta_0 - 0 = 1 \text{ p.u.}$$

Average accelerating power $= \frac{0+1}{2} = 0.5$ p.u.

Change in the angle during 0.05 s after fault occurrence.

$$\Delta\delta_1 = 5.63 \times 0.5 = 2°.81$$

$$\delta_1 = 30° + 2°.81 = 32°.81$$

The results are plotted in Fig. E.18.17.

(A)

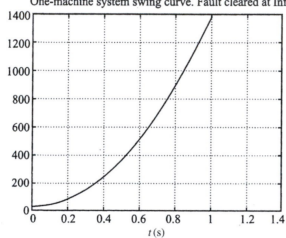

(B)

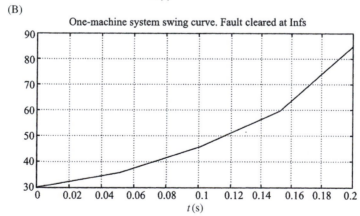

FIGURE E.18.17

The system is unstable.

E.18.18 In example E.18.17, if the fault is cleared in 100 ms, obtain the swing curve.

Solution:

The swing curve is obtained using MATLAB and plotted in Fig. E.18.18.

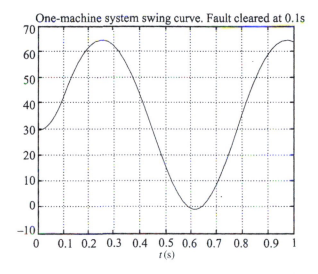

FIGURE E.18.18

The system is stable.

PROBLEMS

P.18.1 A two-pole, 50-Hz, 11-kV synchronous generator with a rating of 120 MW and 0.87 lagging power factor has a moment of inertia of 12,000 kg-m². Calculate the constants H and M.

P.18.2 A four-pole synchronous generator supplies over a short line a load of 60 MW to a load bus. If the maximum steady-state capacity of the transmission line is 110 MW, determine the maximum sudden increase in the load that can be tolerated by the system without losing stability.

P.18.3 The prefault power angle characteristic for a generator infinite bus system is given by

$$P_{e1} = 1.62 \sin \delta$$

and the initial load supplied is 1 p.u. During the fault power angle characteristic is given by

$$P_{e2} = 0.9 \sin \delta$$

Determine the critical clearing angle and the clearing time.

P.18.4 Consider the system operating at 50 Hz.

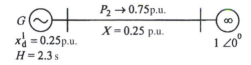

$$P_2 \rightarrow 0.75\text{p.u.}$$

G $X = 0.25$ p.u. ∞

$x_d^1 = 0.25$p.u. $1 \angle 0^0$
$H = 2.3$ s

If a three-phase fault occurs across the generator terminals plot the swing curve.
Plot also the swing curve, if the fault is cleared in 0.05 s.

QUESTIONS

Q.18.1 Explain the terms
    (1) Steady-state stability, (2) transient stability, (3) dynamic stability
Q.18.2 Discuss the various methods of improving steady-state stability.
Q.18.3 Discuss the various methods of improving transient stability.
Q.18.4 Explain the terms: (1) critical clearing angle and (2) critical clearing time.
Q.18.5 Derive an expression for the critical clearing angle for a power system consisting of a single machine supplying to an infinite bus, for a sudden load increment.
Q.18.6 A double-circuit line feeds an infinite bus from a power station. If a fault occurs on one of the lines and the line is switched off, derive an expression for the critical clearing angle.
Q.18.7 Explain the equal area criterion.
Q.18.8 What are the various applications of equal area criterion? Explain.
Q.18.9 State and derive the swing equations.
Q.18.10 Discuss the method of solution for swing equation.

LOAD FLOW ANALYSIS

Load flow or power flow is the solution for the power system under static conditions of operation. Load flow studies are undertaken to determine:

1. The line flows
2. The bus voltages and system voltage profile
3. The effect of changes on circuit configuration, and incorporating new circuits on system loading
4. The effect of temporary loss of transmission capacity and (or) generation on system loading and accompanied effects
5. The effect of in-phase and quadrature boost voltages on system loading
6. Economic system operation
7. System transmission loss minimization
8. Transformer tap settings for economic operation
9. Possible improvements to an existing system by change of conductor sizes and system voltages.

For the purpose of load flow studies, a single-phase representation of the power network is used since the system is generally balanced. When systems had not grown to the present size, networks were simulated on network analyzers for power flow studies. These analyzers are of analog type, scaled-down miniature models of power systems with resistances, reactances, capacitances, autotransformers, transformers, loads, and generators. The generators are just supply sources operating at a much higher frequency than 50 Hz to limit the size of the components. The loads are represented by constant impedances. Meters are provided on the panel board for measuring voltages, currents, and powers. The load flow solution is obtained directly from measurements for any system simulated on the analyzer.

With the advent of the modern digital computer possessing large storage and high speed, the mode of load flow studies has changed from analog to digital simulation. A large number of algorithms are developed for digital power flow solutions. Some of the generally used methods are described in this chapter. The methods basically distinguish between themselves in the rate of convergence, storage requirement, and time of computation. The loads are generally represented by constant power.

In the network at each bus or node, there are four variables, namely

1. Voltage magnitude
2. Voltage phase angle
3. Real power
4. Reactive power.

Out of these four quantities, two of them are specified at each bus and the remaining two are determined from the load flow solution.

Electrical Power Systems. DOI: http://dx.doi.org/10.1016/B978-0-08-101124-9.00019-X

19.1 BUS CLASSIFICATION

Generator Bus: A generator bus has generators connected to it. In general, at all the generators terminal voltage is maintained constant with the help of excitation control. For this reason a generator bus is also called a voltage-controlled bus. The generator prime mover is controlled in such a way that the desired real power is generated at the bus. Hence this bus is also referred to as *PV*-bus. The real power to be generated P_G and the voltage magnitude to be maintained constant $|V|$ are specified at a *PV*-bus. The reactive power required to be supplied by the generator to maintain the voltage constant is not known in advance and therefore must be computed. The bus voltage angle δ is also not known and needs to be calculated. After computing the reactive generation at any generator bus i, Q_{Gi}, it must be verified that it is within the capability range of reactive generation source at that bus i.

Load Bus: Every bus where there is load and no generation exists is called a load bus. Here the MW demand of the load connected and also the reactive power needed for the load (MVAr) are specified. These are known either from records or from measurements. Otherwise, they are to be predicted in advance before a load flow study is undertaken. In general, the real power demand is known and then assuming a suitable power factor, the reactive power demand can be estimated. Hence at a load bus the power demands P_{Li} and Q_{Li} are known at any load bus i. Then it is required to determine the voltage magnitude $|V_i|$ and its phase angle δ_i, at every load bus i.

Swing Bus: All voltage phasors require a reference axis with respect to which the phase angles of the voltage magnitude can be measured. The reference angle itself is not important as it is the difference of angles between two voltage phasors that matters. Further, the real and reactive power losses that occur in the network lines cannot be known in advance. They will be known only at the end of the load flow solution. The line flows can be computed only after all the bus voltages are estimated or known. From them the losses in the network are estimated.

For the above two reasons, a generator bus is selected to supply all the line losses. This bus is called *swing bus* or *slack bus*. Here the voltage phase angle is assumed as zero, so that all other bus voltage phase angles are measured with reference to this bus. The voltage magnitude is also fixed at this bus. This bus is also called as *reference bus*. The classification is given in Table 19.1.

The three types of buses are shown in Fig. 19.1, which illustrates a typical three-bus system.

Table 19.1 Bus Classification and Variables		
Bus	**Specified Variables**	**Computed Variables**
Slack bus	Voltage magnitude and its phase angle	Real and reactive powers
Generator bus (*PV* bus or voltage controlled bus)	Magnitudes of bus voltages and real powers (limit on reactive powers)	Voltage phase angle and reactive power
Load bus	Real and reactive powers	Magnitude and phase angle of bus voltages

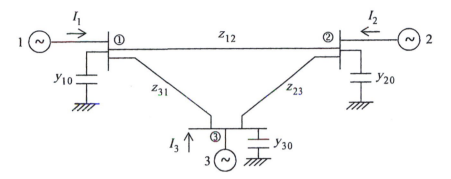

FIGURE 19.1

Three-bus transmission system.

19.2 MODELING FOR LOAD FLOW STUDIES

The bus self- and mutual-admittances constitute the bus admittance matrix Y_{Bus}. The driving point and transfer impedances compose the bus impedance matrix Z_{Bus}. For solving the power flow problem, both Y_{Bus} and Z_{Bus} can be used. However, it is more convenient to use Y_{Bus} model that will become clear as we understand the application of both the models to power (load) flow solution. Initially Y_{Bus} model is considered.

Bus admittance formation: Consider the transmission system shown in Fig. 19.1.

The line impedances joining Buses 1, 2, and 3 are denoted by z_{12}, z_{23}, and z_{31}, respectively. The corresponding line admittances are y_{12}, y_{23}, and y_{31}.

The total capacitive susceptances at the buses are represented by y_{10}, y_{20}, and y_{30}.

Applying Kirchoff's current law at each bus

$$I_1 = V_1\,y_{10} + (V_1 - V_2)y_{12} + (V_1 - V_3)y_{13}$$
$$I_2 = V_2\,y_{20} + (V_2 - V_1)y_{21} + (V_2 - V_3)y_{23}$$
$$I_3 = V_3\,y_{30} + (V_3 - V_1)y_{31} + (V_3 - V_2)y_{32}$$

In matrix form

$$
\begin{bmatrix} I_1 \\ I_2 \\ I_3 \end{bmatrix}
=
\begin{bmatrix}
y_{10} + y_{12} + y_{13} & -y_{12} & -y_{13} \\
-y_{12} & y_{20} + y_{12} + y_{23} & -y_{23} \\
-y_{13} & -y_{23} & y_{30} + y_{13} + y_{23}
\end{bmatrix}
\times
$$

$$
\begin{bmatrix} V_1 \\ V_2 \\ V_3 \end{bmatrix}
=
\begin{bmatrix}
y_{11} & y_{12} & y_{13} \\
y_{21} & y_{22} & y_{23} \\
y_{31} & y_{32} & y_{33}
\end{bmatrix}
,
\begin{bmatrix} V_1 \\ V_2 \\ V_3 \end{bmatrix}
$$

where

$$Y_{11} = y_{10} + y_{12} + y_{13}$$
$$Y_{22} = y_{20} + y_{12} + y_{23}$$
$$Y_{33} = y_{30} + y_{13} + y_{23}$$

are the self-admittances forming the diagonal terms and

$$Y_{12} = Y_{21} = -y_{12}$$
$$Y_{13} = Y_{31} = -y_{13}$$
$$Y_{23} = Y_{32} = -y_{23}$$

are the mutual-admittances forming the off-diagonal elements of the bus admittance matrix. For a n-bus system, the elements of the bus admittance matrix can be written down merely by inspection of the network as

Diagonal terms

$$Y_{ii} = y_{i0} + \sum_{\substack{k=1 \\ k \neq i}}^{n} y_{ik}$$

Off-diagonal terms

$$Y_{ik} = -y_{ik}$$

If the network elements have mutual-admittance (impedance), the above formulae will not apply. For a systematic formation of the y-bus, linear graph theory with singular transformations may be used.

19.2.1 SYSTEM MODEL FOR LOAD FLOW STUDIES

The variables and parameters associated with bus i and a neighboring bus k are presented in the usual notation as follows:

The voltage at a typical bus i of the system is given by

$$V_i = |V_i| \exp j \, \delta_i = |V_i| (\cos \delta_i + j \sin \delta_i) \tag{19.1}$$

The bus admittance is

$$Y_{ik} = |Y_{ik}| \exp j \, \theta_{ik} = |Y_{ik}| (\cos \theta_{ik} + j \sin \theta_{ik})$$
$$= G_{ik} + jB_{ik}V \tag{19.2}$$

The complex power

$$S_i = P_i + jQ_i = V_i I_i^* \tag{19.3}$$

The current injected into the bus i, I_i is given by

$$I_i = Y_{i1}V_1 + Y_{i2}V_2 + \cdots + Y_{in}V_n$$
$$= \sum_{k=1}^{n} Y_{ik}V_k \tag{19.4}$$

The bus current is given by

$$I_{\text{Bus}} = Y_{\text{Bus}} V_{\text{Bus}} \tag{19.5}$$

Using the indices G and L for generation and load

$$P_i = P_{Gi} - P_{Li} = R_e[V_i I_i^*]$$
$$Q_i = Q_{Gi} - Q_{Li} = I_m[V_i I_i^*]$$

(19.6)

Hence, from Eqs. (19.3) and (19.4) for an n-bus system

$$I_i^* = \frac{P_i - jQ_i}{V_i^*} = Y_{ii} V_i + \sum_{\substack{k=1 \\ k \neq i}}^{n} Y_{ik} V_k$$

(19.7)

Solving Eq. (19.7) for V_i

$$V_i = \frac{1}{Y_{ii}} \left[\frac{P_i - jQ_i}{V_i^*} - \sum_{\substack{k=1 \\ k \neq i}}^{n} Y_{ik} V_k \right]$$

(19.8)

Further, from Eq. (19.3) substituting for

$$P_i + jQ_i = V_i \sum_{k=1}^{n} Y_{ik}^* V_k^*$$

(19.9)

In the polar form

$$P_i + jQ_i = \sum_{k=1}^{n} |V_i \quad V_k \quad Y_{ik}| \exp j (\delta_i \quad -\delta_k \quad -\theta_{ik})$$

(19.10)

so that

$$P_i = \sum_{k=1}^{n} |V_i \quad V_k \quad Y_{ik}| \cos (\delta_i \quad -\delta_k \quad -\theta_{ik})$$

(19.11)

and

$$Q_i = \sum_{k=1}^{n} |V_i \quad V_k \quad Y_{ik}| \sin (\delta_i \quad -\delta_k \quad -\theta_{ik})$$

(19.12)

$$i = 1, 2, \ldots n; \ i \neq \text{slack bus}$$

The power flow equations (19.11 and 19.12) are nonlinear, and it is required to solve $2(n - 1)$ such equations involving $|V_i|$, δ_i, P_i, and Q_i at each bus i for the load flow solution. Finally the powers at the slack bus may be computed from which the losses and all other line flows can be ascertained. Y-matrix interactive methods are based on solution to power flow relations using their current mismatch at a bus given by

$$\Delta I_i = I_i - \sum_{k=1}^{n} Y_{ik} V_k$$

(19.13)

or using the voltage from

$$\Delta V_i = \frac{\Delta I_i}{Y_{ii}}$$

(19.14)

The convergence of the iterative methods depends on the diagonal dominance of the bus admittance matrix. The self-admittances of the buses are usually large, relative to the mutual-admittances and thus, usually convergence is obtained. Junctions of very high and low series impedances and large capacitances obtained in cable circuits long, EHV lines, series and shunt compensation are detrimental to convergence as these tend to weaken the diagonal dominance in the Y-matrix. The choice of slack bus can affect convergence considerably. In difficult cases, it is possible to obtain convergence by removing the least diagonally dominant row and column of Y. The salient features of the Y-matrix iterative methods are that the elements in the summation terms in Eq. (19.7) or (19.8) are on the average only three even for well-developed power systems. The sparsity of the Y-matrix and its symmetry reduces both the storage requirement and the computation time for iteration (Section 19.4). For a large, well-conditioned system of n-buses, the number of iterations required are of the order of n and total computing time varies approximately as n^2.

Instead of using Eq. (19.6), one can select the impedance matrix and rewrite the equation as

$$V = Y^{-1}I = Z \cdot I \tag{19.15}$$

The Z-matrix method is not usually very sensitive to the choice of the slack bus. It can easily be verified that the Z-matrix is not sparse. For problems that can be solved by both Z-matrix and Y-matrix methods, the former are rarely competitive with the Y-matrix methods.

19.3 GAUSS—SEIDEL ITERATIVE METHOD

In this method, voltages at all buses except at the slack bus are assumed. The voltage at the slack bus is specified and remains fixed at that value. The $(n-1)$ bus voltage relations from Eq. (19.8) are as follows:

$$V_i = \frac{1}{Y_{ii}} \left[\frac{P_i - jQ_i}{V_i^*} - \sum_{\substack{k=1 \\ k \neq 1}}^{n} Y_{ik} V_k \right] \qquad i = 1, 2, \ldots, n; \; i \neq \text{slack bus} \tag{19.16}$$

are solved simultaneously for an improved solution.

Consider a five-bus power system shown in Fig. 19.2.

Applying Eq. (19.8) for the five-bus system, we obtain during $(k+1)$th iteration.

$$V_1^{k+1} = \frac{1}{Y_{11}} \left[\frac{P_1 - iQ_1}{(V_1^k)^*} - Y_{12}V_2^k - Y_{13}V_3^k - Y_{15}V_5^k \right] \tag{19.16a}$$

$$V_2^{k+1} = \frac{1}{Y_{22}} \left[\frac{P_2 - jQ_2}{V_2^*} - Y_{21}V_1^{k+1} - V_{23}V_3^k \right] \tag{19.16b}$$

$$V_3^{k+1} = \frac{1}{Y_{33}} \left[\frac{P_3 - jQ_3}{(V_3^k)^*} - Y_{31}V_1^{k+1} - Y_{32}V_2^{k+1} - Y_{34}V_4^k \right] \tag{19.16c}$$

and

$$V_4^{k+1} = \frac{1}{Y_{44}} \left[\frac{P_4 - jQ_4}{(V_4^k)^*} - Y_{43}V_3^{k+1} - Y_{45}V_5^k \right] \tag{19.16d}$$

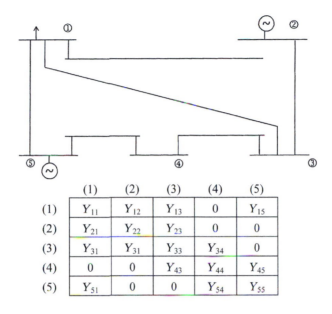

FIGURE 19.2

Five-bus system.

Bus 5 is selected as slack bus or swing bus.

To start with all voltages at load buses are assumed $(1 + j0)$ p.u

i.e.,

$$|V_1| = |V_3| = |V_4| = 1 \text{ p.u.}$$

and

$$\delta_1 = \delta_3 = \delta_4 = 0 \text{ degree}$$

$|V_5|$ = voltage magnitude at swing bus is fixed at say, 1.05 p.u. with $\delta_5 = 0$ degree as reference

Eqs. (19.16a), (19.16c), and (19.16d) are solved iteratively till convergence is reached. But, Eq. (19.16b) is solved for Q_2 and its limits are checked at each iteration. If Q_2 reaches its limit (upper or lower), then the reactive generation at Bus 3 is fixed at that Q_2, and Bus 2 is also considered as a load bus in treatment and is solved for $|V_3|$. The equation for Q_2 is

$$Q_2 = I_m \left(V_i \sum_{k=1}^{4} Y_{ik}^* V_k^* \right) \tag{19.16e}$$

Substituting this value for Q_2, V_2^{k+1} is updated

To accelerate the convergence, all newly computed values of bus voltages are substituted in Eq. (19.16). The bus voltage equation for the $(m + 1)$th iteration may then be written as

$$V_i^{(m+1)} = \frac{1}{Y_{ii}} \left[\frac{P_i - jQ_i}{V_i^{(m)*}} - \sum_{\substack{k=1 \\ k \neq 1}}^{i-1} Y_{ik} V_k^{(m+1)} - \sum_{k=i\div1}^{n} Y_{ik} V_k^{(m)} \right] \tag{19.17}$$

The method converges slowly because of the loose mathematical coupling between the buses. The rate of convergence of the process can be increased by using acceleration factors to the solution obtained after each iteration.

Acceleration Factor: In Gauss–Seidel method of solution to the load flow problem, the iterative process is continued until the amount of correction to voltage at every bus is less than some prespecified precision index. However the number of iterations required can be considerably reduced, if the correction to voltage at each bus is multiplied by some constant. This increases the amount of correction thus making the process of convergence faster. The multiplier that produces the improved convergence is called *acceleration factor*.

Let at any bus i the voltage at the end of $(k-1)$th iteration be V_i^{k-1} and at the end of the kth iteration be V_i^h. Then, if α is the acceleration factor

$$V_i^{(k)}(\text{accelerated}) = + V_i^{(k-1)} + \alpha(V_i^{(k)} - V_i^{(k-1)})$$

The accelerated value of $V_i^{(k)}$ is used for $(k+1)$th iteration. Generally the value of α is chosen as $1 < \alpha < 2$.

A fixed acceleration factor is sometimes used, using the relation

$$\Delta V_i = \alpha \frac{\Delta S_i^*}{V_i^* Y_{ii}} \tag{19.18}$$

The use of acceleration factor amounts to a linear extrapolation of V_i. For a given system it is quite often found that a near optimal choice of α exists. Some researchers have suggested even a complex value for α. But it is found convenient to use real values for α (usually 1.5 or 1.6).

Alternatively, different acceleration factors may be used for real and imaginary parts of the voltage.

Treatment of a PV bus

The method of handling a PV bus requires rectangular coordinate representation for the voltages. Lettering

$$V_i = v_i' + jv''_i \tag{19.19}$$

where v_i' and v''_i are the real and imaginary components of V_i. The relationship

$$v_i'2 + v''^2_i = |V_i|^2_{\text{scheduled}} \tag{19.20}$$

must be satisfied so that the reactive bus power required to establish the scheduled bus voltage can be computed. The estimates of voltage components, $v_i'(m)$ and $v''^{(m)}_i$ after m iterations must be adjusted to satisfy Eq. (19.20). The phase angle of the estimated bus voltage is

$$\delta_i^{(m)} = \tan^{-1}\left[\frac{v''^{(m)}_i}{v_i'(m)}\right] \tag{19.21}$$

Assuming that the phase angles of the estimated and scheduled voltages are equal, then the adjusted estimates of $V'^{(m)}$ and $V''^{(m)}_i$ are as follows:

$$v_{i(\text{new})}^{i(m)} = |V_i|_{\text{scheduled}} \cos \delta_i^{(m)} \tag{19.22}$$

and

$$v''^{(m)}_{i(\text{new})} = |V_i|_{\text{scheduled}} \sin \delta_i^{(m)} \tag{19.23}$$

These values are used to calculate the reactive power $Q_i^{(m)}$. Using these reactive powers $Q_i^{(m)}$ and voltages $V_{i(\text{new})}^{(m)}$, a new estimate $V_i^{(m+1)}$ is calculated. The flowchart for computing the solution of load flow using Gauss–Seidel method is given in Fig. 19.3.

While computing the reactive powers, the limits on the reactive source must be taken into consideration. If the calculated value of the reactive power is beyond limits, then its value is fixed at the limit that is violated and it is no longer possible to hold the desired magnitude of the bus voltage, the bus is treated as a *PQ* bus or load bus.

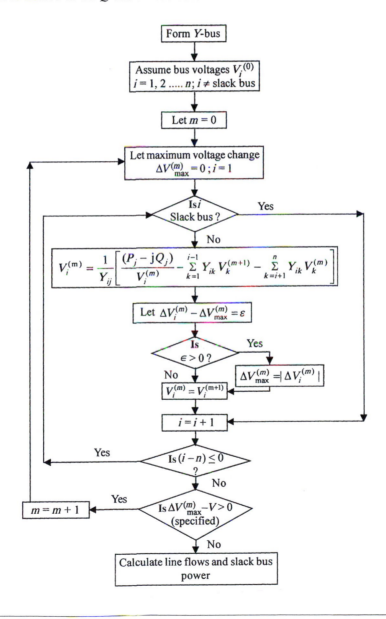

FIGURE 19.3

Flowchart for Gauss–Seidel iterative method for load flow solution using Y_{Bus}.

19.4 NEWTON–RAPHSON METHOD

The generalized Newton–Raphson method is an interactive algorithm for solving a set of simultaneous nonlinear equations in an equal number of unknowns. Consider the set of nonlinear equations.

$$f_i(x_1, x_2, \ldots, x_n) = y_i, \quad i = 1, 2, \ldots, n \tag{19.24}$$

with initial estimates for x_i

$$x_1^{(0)}, x_2^{(0)}, \ldots, x_n^{(0)}$$

which are not far from the actual solution. Then using Taylor's series and neglecting the higher order terms, the corrected set of equations is as follows:

$$f_i(x_1^{(0)} + \Delta x_1, x_2^{(0)} + \Delta x_2, \ldots, x_n^{(0)} + \Delta x_n) = y_i \tag{19.25}$$

where Δx_i are the corrections to x_i ($i = 1, 2, \ldots, n$).

A set of linear equations, which define a tangent hyperplane to the function $f_i(x)$ at the given iteration point $(x_i^{(0)})$ are obtained as

$$\Delta Y = J \Delta X \tag{19.26}$$

where ΔY is a column vector determined by

$$y_i - f_i(x_1^{(0)}, \ldots, x_n^{(0)})$$

ΔX is the column vector of correction terms Δx_i, and J is the Jacobian matrix for the function f given by the first-order partial derivatives evaluated at $x_i^{(0)}$. The corrected solution is obtained as

$$x_i^{(1)} = x_i^{(0)} + \Delta x_i \tag{19.27}$$

The square Jacobian matrix J is defined by

$$J_{ik} = \frac{\partial f_i}{\partial x_k}$$

The above method of obtaining a converging solution for a set of nonlinear equations can be used for solving the load flow problem. It may be mentioned that since the final voltage solutions are not much different from the nominal values, Newton–Raphson method is particularly suited to the load flow problem. The matrix J is highly sparse and is particularly suited to the load flow application and sparsity—programmed ordered triangulation and back substitution methods result in quick and efficient convergence to the load flow solution. This method possesses quadratic convergence and thus converges very rapidly when the solution point is close.

There are two methods of solution for the load flow using Newton–Raphson method. The first method uses rectangular coordinates for the variables, whereas the second method uses the polar coordinate formulation.

19.4.1 RECTANGULAR COORDINATES METHOD

The power entering the bus i is given by $S_i = P_i + jQ_i$

$$V_i I_i^* = V_i \sum_{k=1}^{n} Y_{ik}^* V_k^*, \quad i = 1, 2, \ldots, n \tag{19.28}$$

where

$$V_i = v'_i + jv''_i$$

$$Y_{ik} = G_{ik} + jB_{ik}$$

$$(P_i + jQ_i) = (v'_i + jv''_i) \sum_{k=1}^{n} (G_{ik} - jB_{ik})(v'_k - jv''_k) \tag{19.29}$$

Expanding the right side of the above equation and separating out the real and imaginary parts

$$P_i = \sum_{k=1}^{n} [v'_i(G_{ik}v'_k - B_{ik}v''_k) + v''_i(G_{ik}v''_k + B_{ik}v'_k)] \tag{19.30}$$

$$Q_i = \sum_{k=1}^{n} [v''_i(G_{ik}v'_k - B_{ik}v''_k) - v'_i(G_{ik}v''_k + B_{ik}v'_k)] \tag{19.31}$$

These are the two power relations at each bus and the linearized equations of the form (19.27) are written as (19.33).

$$
\begin{bmatrix} \Delta P_1 \\ \vdots \\ \Delta P_{n-1} \\ \Delta Q_1 \\ \vdots \\ \Delta Q_{n-1} \end{bmatrix}
=
\begin{bmatrix}
\dfrac{\partial P_1}{\partial v'_1} & \cdots & \dfrac{\partial P_1}{\partial v'_{n-1}} & \dfrac{\partial P_1}{\partial v''_1} & \cdots & \dfrac{\partial P_1}{\partial v''_{n-1}} \\
\vdots & \cdots & \vdots & \vdots & \cdots & \vdots \\
\dfrac{\partial P_{n-1}}{\partial v'_1} & \cdots & \dfrac{\partial P_{n-1}}{\partial v'_{n-1}} & \dfrac{\partial P_{n-1}}{\partial v''_n} & \cdots & \dfrac{\partial P_{n-1}}{\partial v''_{n-1}} \\
\dfrac{\partial Q_1}{\partial v'_1} & \cdots & \dfrac{\partial Q_1}{\partial v'_{n-1}} & \dfrac{\partial Q_1}{\partial v''_1} & \cdots & \dfrac{\partial Q_1}{\partial v''_{n-1}} \\
\vdots & \cdots & \vdots & \vdots & \cdots & \vdots \\
\dfrac{\partial Q_{n-1}}{\partial v'_1} & \cdots & \dfrac{\partial Q_{n-1}}{\partial v'_{n-1}} & \dfrac{\partial Q_{n-1}}{\partial v''_1} & \cdots & \dfrac{\partial Q_{n-1}}{\partial v''_{n-1}}
\end{bmatrix}
\begin{bmatrix} \Delta v'_1 \\ \vdots \\ \Delta v'_{n-1} \\ \Delta v''_1 \\ \vdots \\ \Delta v''_{n-1} \end{bmatrix}
\tag{19.32}
$$

Matrix equation (19.32) can be solved for the unknowns $\Delta v'_i$ and $\Delta v''_i$ ($i = 1,2,\ldots, n-1$), leaving the slack bus at the nth bus where the voltage is specified. Eq.(19.32) may be written compactly as

$$\begin{bmatrix} \Delta P \\ \Delta Q \end{bmatrix} = \begin{bmatrix} H & N \\ M & L \end{bmatrix} \begin{bmatrix} \Delta v' \\ \Delta v'' \end{bmatrix} \tag{19.33}$$

where H, N, M, and L are the submatrices of the Jacobian. The elements of the Jacobian are obtained by differentiating Eqs. (19.30) and (19.31). The off-diagonal and diagonal elements of the H-matrix are given by

$$\frac{\partial P_i}{\partial v'_k} = G_{ik}v'_i + B_{ik}v''_i, \quad i \neq k \tag{19.34}$$

$$\frac{\partial P}{\partial v'_i} = 2G_{ii}\, v'_i + \sum_{\substack{k=1 \\ k \neq i}}^{n} (G_{ik}v'_k - B_{ik}v''_k), \quad i = 1, 2, \ldots, n, \quad k = 1, 2, \ldots, n, \quad k \neq i \tag{19.35}$$

The off-diagonal and diagonal elements of N are as follows:

$$\frac{\partial P_i}{\partial v''_k} = G_{ik}v''_i - B_{ik}v'_i, \quad k \neq i \tag{19.36}$$

$$\frac{\partial P_i}{\partial v''_i} = 2v''_i G_{ii} + \sum_{\substack{k=1 \\ k \neq i}}^{n} (G_{ik}v''_k + B_{ik}v'_k), \quad i = 1, 2, \ldots, n, \quad k = 1, 2, \ldots, n, \quad k \neq i \tag{19.37}$$

The off-diagonal and diagonal elements of submatrix M are obtained as:

$$\frac{\partial Q_i}{\partial v'_k} = G_{ik}v''_i - B_{ik}v'_i, \quad k \neq i \tag{19.38}$$

$$\frac{\partial Q_i}{\partial v'_i} = -2B_{ii}v'_i - \sum_{\substack{k=1 \\ k \neq i}}^{n} (v''_k G_{ik} + v'_k B_{ik}), \quad i = 1, 2, \ldots, n, \quad k = 1, 2, \ldots, n, \quad i \neq k \tag{19.39}$$

Finally, the off-diagonal and diagonal elements of L are given by

$$\frac{\partial Q_i}{\partial v''_k} = -v''_i B_{ik} - v'_i G_{ik} \tag{19.40}$$

$$= -2B_{ii}v''_i = \sum_{\substack{k=1 \\ k \neq i}}^{n} (G_{ik}v'_k - B_{ik}v''_k) \tag{19.41}$$

$$i = 1, 2, \ldots, n, \quad k = 1, 2, \ldots, n, \quad i \neq k$$

It can be noticed that

$$L_{ik} = -H_{ik}$$

and

$$N_{ik} = M_{ik}$$

This property of symmetry of the elements reduces computer time and storage.

19.4.1.1 Treatment of generator buses

At all generator buses other than the swing bus, the voltage magnitudes are specified in addition to the real powers. At the ith generator bus

$$|V_i|^2 = v'^2_i + v''^2_i \tag{19.42}$$

Then, at all the generator nodes, the variable ΔQ_i will have to be replaced by

$$\Delta |V_i|^2$$

But

$$|\Delta V_i|^2 = \frac{\partial(|\Delta_i|^2)}{\partial v'_i}\Delta V'_i + \frac{\partial(|\Delta_i|^2)}{\partial v''_i}\Delta V''_i$$
$$= 2 v''_i \Delta v'_i + 2v''_i \Delta v''_i \tag{19.43}$$

This is the only modification required to be introduced in Eq. (19.41).

19.4.2 THE POLAR COORDINATES METHOD

The equation for the complex power at node i in the polar form is given in Eq. (19.10) and the real and reactive powers at bus i are indicated in Eqs. (19.11) and (19.12). Reproducing them here once again for convenience.

$$P_i = \sum_{k=1}^{n} \left| V_i \quad V_k \quad Y_{ik} \right| \cos(\delta_i - \delta_k - \theta_{ik})$$

and

$$Q_i = \sum_{k=1}^{n} \left| V_i \quad V_k \quad Y_{ik} \right| \sin(\delta_i - \delta_k - \theta_{ik})$$

The Jacobian is then formulated in terms of $|V|$ and δ instead of V'_i and V''_i in this case. Eq. (19.33) then takes the form

$$\begin{bmatrix} \Delta P \\ \Delta Q \end{bmatrix} = \begin{bmatrix} H & N \\ M & L \end{bmatrix} \begin{bmatrix} \Delta \delta \\ \Delta |V| \end{bmatrix} \tag{19.44}$$

The off-diagonal and diagonal elements of the submatrices H, N, M, and L are determined by differentiating Eqs. (19.11) and (19.12) with respect to δ and $|V|$ as before. The off-diagonal and diagonal elements of the H-matrix are as follows:

$$\frac{\partial P_i}{\partial \delta_k} = \left| V_i \quad V_k \quad Y_{ik} \right| \sin(\delta_i - \delta_k - \theta_{ik}), \quad i \neq k \tag{19.45}$$

$$\frac{\partial P_i}{\partial \delta_i} = -\sum_{\substack{k=1 \\ k \neq 1}}^{n} \left| V_i \quad V_k \quad V_{ik} \right| \sin(\delta_i - \delta_k - \theta_{ik}) \tag{19.46}$$

The off-diagonal and diagonal elements of the N-matrix are as follows:

$$\frac{\partial P_i}{\partial |V_k|} = V_i \, Y_{ik} \cos(\delta_i - \delta_k - \theta_{ik})(\delta_i - \delta_k - \theta_{ik}) \tag{19.47}$$

$$\frac{\partial P_i}{\partial |V_i|} = 2\left| V_i \quad Y_{ii} \right| \cos \theta_{ii} + \sum_{\substack{k=1 \\ k \neq i}}^{n} \left| V_k \quad Y_{ik} \right| \cos(\delta_i - \delta_k - \theta_{ik}) \tag{19.48}$$

The off-diagonal and diagonal elements of the M-matrix are as follows:

$$\frac{\partial Q_i}{\partial \delta_k} = -\left| V_i \quad V_k \quad Y_{ik} \right| \cos(\delta_i - \delta_k - \theta_{ik}) \tag{19.49}$$

$$\frac{\partial Q_i}{\partial \delta_i} = \sum_{\substack{k=1 \\ k \neq i}}^{n} \left| V_i \quad V_k \quad Y_{ik} \right| \cos(\delta_i - \delta_k - \theta_{ik}) \tag{19.50}$$

Finally the off-diagonal and diagonal elements of the L-matrix are as follows:

$$\frac{\partial Q_i}{\partial |V_k|} = \left| V_i \quad Y_{ik} \right| \sin(\delta_i - \delta_k - \theta_{ik}) \tag{19.51}$$

$$\frac{\partial Q_i}{\partial |V_i|} = -2\left| V_i \quad Y_{ii} \right| \sin \theta_{ii} + \sum_{\substack{k=1 \\ k \neq i}}^{n} \left| V_k \quad Y_{ik} \right| \sin(\delta_i - \delta_k - \theta_{ik}) \tag{19.52}$$

It is seen from the elements of the Jacobian in this case that the symmetry that existed in the rectangular coordinates case is no longer present now. By selecting the variable as $\Delta\delta$ and $\Delta|V|/|V|$ instead Eq. (19.44) will be in the form

$$\begin{bmatrix} \Delta P \\ \Delta Q \end{bmatrix} = \begin{bmatrix} H & N \\ M & L \end{bmatrix} \begin{bmatrix} \Delta\delta \\ \frac{\Delta|V|}{|V|} \end{bmatrix} \tag{19.53}$$

The terms of $H_{ik} = \partial P_i/\partial\delta_k = |V_i \quad V_k \quad Y_{ik}| \sin(\delta_i - \delta_k - \theta_{ik})$, $i \neq k$ $i = k$ (Eq. 19.46) remain unchanged with the modification $|\Delta V/V|$ introduced. However, the terms of L_{ik},

$$\frac{\partial Q_i}{\partial|V_k|} = |V_i \quad Y_{ik}| \sin(\delta_i - \delta_k - \theta_{ik})$$

will get changed as

$$\frac{\partial Q}{\partial|V_k|} \text{modified} = \frac{\frac{\partial Q_i}{\partial|V_k|} \cdot |V_k|}{|V_k|} |V_i \quad Y_{ik}| \sin(\delta_i - \delta_k - \theta_{ik})$$

$$= H_{ik} \tag{19.54a}$$

Hence

$$L_{ik} = H_{ik}$$

In a similar manner

$$M_{ik} = \frac{\partial Q_i}{\partial\delta_k} = -|V_i \quad V_k \quad Y_{ik}| \cos(\delta_i - \delta_k - \theta_{ik})$$

remains the same even with $\Delta V/|V|$ variable but N_{ik} changes from

$$N_{ik} = |V_i \quad Y_{ik}| \cos(\delta_i - \delta_k - \theta_{ik})$$

into

$$N_{ik}(\text{modified}) = |V_k \quad V_i \quad Y_{ik}| \cos(\delta_i - \delta_k - \theta_{ik}), \quad i \neq k$$

$$= -M_{ik} \tag{19.54b}$$

Hence

$$N_{ik} = -M_{ik}$$

or, in other words, the symmetry is restored. The number of elements to be calculated for an n-dimensional Jacobian matrix are only $n + n^2/2$ instead of n^2, thus again saving computer time and storage. The flowchart for computer solution is given in Fig. 19.4.

19.4.2.1 Treatment of generator nodes

For a *PV*-bus, the reactive power equations are replaced at the *i*th generator bus by

$$|V_i|^2 = v_i'^2 + v_i''^2$$

The elements of M are given by

$$M_{ik} = \frac{\partial(|V|)^2}{\partial\delta_k} = 0, \quad i \neq k$$

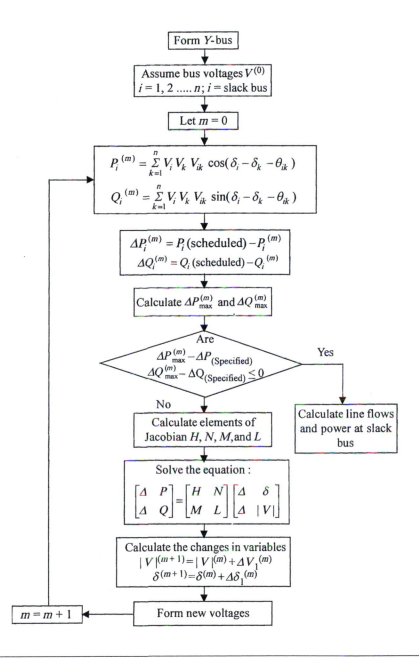

FIGURE 19.4

Flowchart for Newton–Raphson method (polar coordinates) for load flow solution.

and

$$M_{ii} = \frac{\partial |V_i|^2}{\partial \delta_i} = 0$$

Then elements of L are given by

$$L_{ik} = \frac{\partial(|V_i|^2)}{\partial |V_k|} |V_k| = 0, \quad i \neq k$$

and

$$L_{ii} = \frac{\partial(|V_i|)^2}{\partial |V_i|} |V_i| = 2|V_i|^2$$

Newton's method converges in two to five iterations from a flat start ($\{V\} = 1.0$ p.u. and $\delta = 0$) independent of system size. Previously stored solution can be used as starting values for rapid convergence. Iteration time can be saved by using the same triangulated Jacobian matrix for two or more iterations. For typical, large size systems, the computing time for one Newton–Raphson iteration is roughly equivalent to seven Gauss–Seidel iterations.

The rectangular formulation is marginally faster than the polar version because there are no time-consuming trigonometric functions. However, it is observed that the rectangular coordinates method is less reliable than the polar version.

19.5 SPARSITY OF NETWORK ADMITTANCE MATRICES

For many power networks, the admittance matrix is relatively sparse, whereas the impedance matrix is full. In general, both matrices are nonsingular and symmetric. In the admittance matrix each nonzero off-diagonal element corresponds to a network branch connecting the pair of buses indicated by the row and column of the element. Most transmission networks exhibit irregularity in their connection arrangements, and their admittance matrices are relatively sparse. Such sparse systems possess the following advantages:

1. Their storage requirements are small, so that larger systems can be solved.
2. Direct solutions using triangularization techniques can be obtained much faster unless the independent vector is extremely sparse.
3. Round-off errors are very much reduced.

The exploitation of network sparsity requires sophisticated programming techniques.

19.6 TRIANGULAR DECOMPOSITION

Matrix inversion is a very inefficient process for computing direct solutions, especially for large, sparse systems. Triangular decomposition of the matrix for solution by Gaussian elimination is more suited for load flow solutions. Generally the decomposition is accomplished by elements below the main diagonal in successive columns. Elimination by successive rows is more advantageous from computer programming point of view.

Consider the system of equations

$$Ax = b \tag{19.55}$$

where A is a nonsingular matrix, b is a known vector containing at least one nonzero element, and x is a column vector of unknowns.

To solve Eq. (19.55) by the triangular decomposition method, matrix A is augmented by b as shown

$$\begin{bmatrix} a_{11} & a_{12} & \cdots & a_{1n} & b_1 \\ a_{21} & a_{22} & \cdots & a_{2n} & b_2 \\ \vdots & \vdots & \vdots & \vdots & \vdots \\ a_{n1} & a_{n2} & \cdots & a_{nn} & b_n \end{bmatrix}$$

The elements of the first row in the augmented matrix are divided by a_{11} as indicated by the following step with superscripts denoting the stage of the computation.

$$a_{1j}^{(1)} = \left(\frac{1}{a_{11}}\right) a_{1j}, \quad j = 2, \ldots, n \tag{19.56}$$

$$b_1^{(1)} = \left(\frac{1}{a_{11}}\right) b_1 \tag{19.57}$$

In the next stage a_{21} is eliminated from the second row using the relations

$$a_{2j}^{(1)} = a_{2j} - a_{21}\, a_{1j}^{(1)}, \quad j = 2, \ldots, n \tag{19.58}$$

$$b_2^{(1)} = b_2 - a_{21}\, b_1^{(1)} \tag{19.59}$$

$$a_{2j}^{(2)} = \left(\frac{1}{a_{22}^{(2)}}\right) a_{2j}^{(1)}, \quad j = 3, \ldots, n$$

$$b_2^{(2)} = \left(\frac{1}{a_{22}^{(1)}}\right) b_2^{(1)} \tag{19.60}$$

The resulting matrix then becomes

$$\begin{bmatrix} 1 & 1_{12}^{(1)} & a_{13}^{(1)} & \cdots & a_{1n}^{(1)} & b_1^{(1)} \\ 0 & 1 & a_{23}^{(2)} & \cdots & a_{2n}^{(2)} & b_2^{(2)} \\ \vdots & \vdots & \vdots & \vdots & \vdots & \vdots \\ a_{n1} & a_{n2} & a_{n3} & \cdots & a_{nn} & b_n \end{bmatrix}$$

using the relations

$$b_{3j}^{(1)} = a_{3j} - a_{31} a_{1j}^{(1)}, \quad j = 2, \ldots, n \tag{19.61}$$

$$b_{(3)}^{1} = b_3 - a_{31} b_1^{(1)} \tag{19.62}$$

$$a_{3j}^{(2)} = a_{3j}^{(1)} - a_{32}^{(1)} a_{2j}^{(2)}, \quad j = 3, \ldots, n \tag{19.63}$$

$$b_3^{(2)} = b_3^{(1)} - a_{32}^{(1)} b_3^{(2)}, \quad j = 4, \ldots, n \tag{19.64}$$

$$a_3^{(3)} = \left(\frac{1}{a_{33}^{(2)}}\right) a_{3j}^{(1)}, \quad j = 4,\ldots,n \tag{19.65}$$

$$b_3^{(3)} = \left(\frac{1}{a_{33}^{(2)}}\right) b_3^{(2)} \tag{19.66}$$

The elements to the left of the diagonal in the third row are eliminated and further the diagonal element in the third row is made unity.

After n steps of computation for the nth order system of Eq. (19.55), the augmented matrix will be obtained as

$$\begin{bmatrix} 1 & a_{12}^{(1)} & a_{1n}^{(1)} & (b)_1^{(1)} \\ 0 & 1 & a_{2n}^{(2)} & (b)_2^{(2)} \\ \vdots & \vdots & \vdots & \vdots \\ 0 & 0 & 1 & b_n^{(n)} \end{bmatrix}$$

By back substitution, the solution is obtained as

$$x_n = b_n^{(n)} \tag{19.67}$$

$$x_{n-1} = b_{n-1}^{(n-1)} - a_{n-1}^{(n-1)}, n\ldots x_n \tag{19.68}$$

$$x_i = b_i^{(1)} - \sum_{j=i+1}^{n} a_{ij}^{(i)} x_j \tag{19.69}$$

For matrix inversion of a nth-order matrix, the number of arithmetical operations required is n^3 while for the triangular decomposition it is approximately $(n^3/3)$.

19.7 OPTIMAL ORDERING

When the A-matrix in Eq. (19.55) is sparse, it is necessary to see that the accumulation of nonzero elements in the upper triangle is minimized. This can be achieved by suitably ordering the equations, which is referred to as optimal ordering.

Consider the network system having five nodes as shown in Fig. 19.5

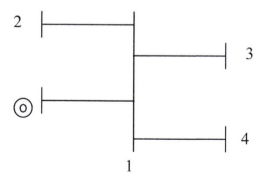

FIGURE 19.5

A five-bus system.

The *y*-bus matrix of the network will have entries as follows:

$$
\begin{array}{c}
 & \begin{array}{cccc} 1 & 2 & 3 & 4 \end{array} \\
\begin{array}{c} 1 \\ 2 \\ 3 \\ 4 \end{array} &
\left[\begin{array}{cccc}
\times & \times & \times & \times \\
\times & \times & 0 & 0 \\
\times & 0 & \times & 0 \\
\times & 0 & 0 & \times
\end{array} \right] = Y
\end{array}
\tag{19.70}
$$

After triangular decomposition the matrix will be reduced to the form

$$
\begin{array}{c}
 & \begin{array}{cccc} 1 & 2 & 3 & 4 \end{array} \\
\begin{array}{c} 1 \\ 2 \\ 3 \\ 4 \end{array} &
\left[\begin{array}{cccc}
1 & \times & \times & \times \\
0 & 1 & \times & \times \\
0 & 0 & 1 & \times \\
0 & 0 & 0 & 1
\end{array} \right] = Y
\end{array}
\tag{19.71}
$$

By ordering the nodes as in Fig. 19.6, the bus admittance matrix will be of the form:

$$
\begin{array}{c}
 & \begin{array}{cccc} 1 & 2 & 3 & 4 \end{array} \\
\begin{array}{c} 1 \\ 2 \\ 3 \\ 4 \end{array} &
\left[\begin{array}{cccc}
\times & 0 & 0 & \times \\
0 & \times & 0 & \times \\
0 & 0 & \times & \times \\
\times & \times & \times & \times
\end{array} \right] = Y
\end{array}
\tag{19.72}
$$

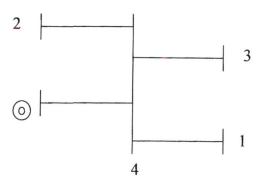

FIGURE 19.6

Renumbered five-bus system.

As a result of triangular decomposition, the Y-matrix will be reduced to

$$
\begin{array}{c c}
 & \begin{array}{cccc} 1 & 2 & 3 & 4 \end{array} \\
\begin{array}{c} 1 \\ 2 \\ 3 \\ 4 \end{array} &
\left[\begin{array}{cccc}
1 & 0 & 0 & \times \\
0 & 1 & 0 & \times \\
0 & 0 & 1 & \times \\
0 & 0 & 0 & 1
\end{array}\right] = Y
\end{array}
\tag{19.73}
$$

Thus comparing the matrices in Eqs. (19.71) and (19.73) the nonzero off-diagonal entries are reduced from 6 to 3 by suitably numbering the nodes.

Tinney and Walker have suggested three methods for optimal ordering.

1. Number the rows according to the number of nonzero, off-diagonal elements before elimination. Thus rows with less number of off-diagonal elements are numbered first and the rows with large number last.
2. Number the rows so that at each step of elimination the next row to be eliminated is the one having fewest nonzero terms. This method required simulation of the elimination process to take into account the changes in the nonzero connections affected at each step.
3. Number the rows so that at each step of elimination, the next row to be eliminated is the one that will introduce fewest new nonzero elements. This requires simulation of every feasible alternative at each step.

Scheme **1** is simple and fast. However, for power flow solutions, Scheme **2** has proved to be advantageous even with its additional computing time. If the number of iterations is large, Scheme **3** may prove to be advantageous.

19.8 DECOUPLED METHODS

All power systems exhibit in the steady state a strong interdependence between active powers and bus voltage angles and between reactive power and voltage magnitude. The coupling between real power and bus voltage magnitude and between reactive power and voltage phase angle are both relatively weak. This weak coupling is utilized in the development of the so-called decoupled methods. Recalling Eq. (19.54)

$$
\begin{bmatrix} \Delta P \\ \Delta Q \end{bmatrix} = \begin{bmatrix} H & N \\ M & L \end{bmatrix} \begin{bmatrix} \Delta \delta \\ |V|/\Delta|V| \end{bmatrix}
$$

by neglecting N and M submatrices as a first step, decoupling can be obtained so that

$$
|\Delta P| = |H| \cdot |\Delta \delta|
\tag{19.74}
$$

and

$$
|\Delta Q| = |L| \cdot |\Delta|V||/|V|
\tag{19.75}
$$

The decoupled method converges as reliably as the original Newton method from which it is derived. However, for very high accuracy the method requires more iterations because overall quadratic convergence is lost. The decoupled Newton method saves by a factor of four on the storage for the *J*-matrix and its triangulation. But, the overall saving is 35%−50% of storage when compared to the original Newton method. The computation per iteration is 10%−20% less than for the original Newton method.

19.9 FAST DECOUPLED METHODS

For security monitoring and outage-contingency evaluation studies, fast load flow solutions are required. A method developed for such an application is described in this section.

The elements of the submatrices H and L (Eq. 19.54) are given by

$$H_{ik} = |V_i \quad V_k \quad Y_{ik}| \sin (\delta_i - \delta_k - \theta_{ik})$$
$$= |V_i \quad V_k \quad Y_{ik}| (\sin \delta_{ik} \cos \theta_{ik} - \cos \delta_{ik} \sin \theta_{ik})$$
$$= |V_i \quad V_k| (G_{ik} \sin \delta_{ik} - B_{ik} \cos \delta_{ik})$$

where

$$\delta_i - \delta_k = \delta_{ik}$$

$$H_{ii} = - \sum_{k=1}^{n} |V_i \quad V_k \quad Y_{ik}| \sin (\delta_i - \delta_k - \theta_{ik})$$
$$= - V_i^2 Y_{ii} \sin \theta_{ii} - \sum_{\substack{k=1 \\ k \neq i}}^{n} V_i \quad V_k \quad Y_{ik} \sin (\delta_i - \delta_k - \theta_{ik})$$
$$= V_i^2 B_{ii} - Q_i$$

In a similar manner

$$L_{ik} = |V_i \quad V_k \quad Y_{ik}| \sin (\delta_i - \delta_k - \theta_{ik})$$
$$= |V_i \quad V_k| (B_{ik} \cos \theta_{ik} - G_{ik} \sin \theta_{ik})$$

$$L_{ii} = 2V_i Y_{ii} \sin \theta_{ii} + \sum_{\substack{k=1 \\ k \neq i}}^{n} V_k Y_{ik} \sin (\delta_i - \delta_k - \theta_{ik})$$

and with formulation on the right hand side

$$L_{ii} = 2|V_i \quad Y_{ii}| \sin \theta_{ii} + \sum_{\substack{k=1 \\ k \neq 1}}^{n} V_i \quad V_k \quad Y_{ik} \sin (\delta_i - \delta_k - \theta_{ik})$$
$$= |V_i^2| B_{ii} + Q_i$$

Assuming that

$$\cos \delta_{ik} \cong 1$$
$$\sin \delta_{ik} \cong 0$$
$$G_{ii} \sin \delta_{ik} \leq B_{ik}$$

and

$$Q_i \leq B_{ii} |V_i|^2$$
$$H_{ik} = -|V_i \quad V_k| B_{ik}$$

$$H_{ii} = |V_i|^2 B_{ii}$$

$$L_{ik} = |V_i V_k| B_{ik}$$

and

$$L_{ii} = |V_i|^2 B_{ii}$$

Rewriting Eqs. (19.74) and (19.75)

$$|VP| = |V||B'||V|\Delta\delta \tag{19.76}$$

$$|V| = |V||B''||V|\frac{\Delta|V|}{|V|} \tag{19.77}$$

Eq. (19.76) can be rewritten as

$$\frac{|\Delta P|}{|V|} = |V||B'|\Delta\delta$$

$$\frac{|\Delta P|}{|V|} \simeq [B'][\Delta\delta] \quad \text{setting } |V| \simeq 1 + jo \tag{19.78}$$

From Eq. (19.77)

$$\frac{|\Delta Q|}{|V|} = [B''][\Delta|V|] \tag{19.79}$$

Matrices $[B']$ and $[B'']B''$ represent constant approximations to the slopes of the tangent hyperplanes of the functions $\Delta P/|V|$ and $\Delta Q/|V|$, respectively. They are very close to the Jacobian submatrices H and L evaluated at system no-load.

Shunt reactances and off-nominal in-phase transformer taps which affect the Mvar flows are to be omitted from $[B']$ and for the same reason phase shifting elements are to be omitted from $[B'']$.

Both $[B']$ and $[B'']$ are real and "sparse" and need be triangularized only once, at the beginning of the study since they contain network admittances only.

The method converges very reliably in two to five iterations with fairly good accuracy even for large systems. A good, approximate solution is obtained after the first or second iteration. The speed per iteration is roughly five times that of the original Newton method.

19.10 LOAD FLOW SOLUTION USING *Z* BUS

19.10.1 BUS IMPEDANCE FORMATION

Any power network can be formed using the following possible methods of construction.

1. A line may be added to a reference point or bus.
2. A bus may be added to any existing bus in the system other than the reference bus through a new line.
3. A line may be added joining two existing buses in the system forming a loop.

The above three modes are illustrated in Fig. 19.7

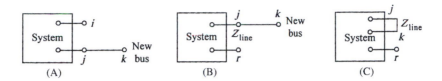

FIGURE 19.7

Building of Z bus. (A) Line added to reference bus, (B) line added to any bus other than reference line, and (C) line added joining two existing buses.

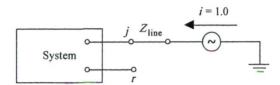

FIGURE 19.8

Addition of the line to reference line.

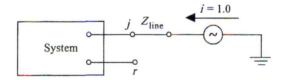

FIGURE 19.9

Addition of a radial line and new bus.

19.10.2 ADDITION OF A LINE TO THE REFERENCE BUS

If unit current is injected into bus k, no voltage will be produced at other buses of the systems.

$$Z_{ik} = Z_{ki} = 0, \quad i \neq k \tag{19.80}$$

The driving point impedance of the new bus is given by (Fig. 19.8)

$$Z_{kk} = Z_{line} \tag{19.81}$$

19.10.3 ADDITION OF A RADIAL LINE AND NEW BUS

Injection of unit current into the system through the new bus k produces voltages at all other buses of the system as shown in Fig. 19.9

These voltages would of course be same as that would be produced if the current were injected instead at bus i as shown.

Therefore

$$Z_{km} = Z_{im}$$

therefore,

$$Z_{mk} = Z_{mi}, \quad m \neq k \tag{19.82}$$

The dimension of the existing Z-Bus matrix is increased by one. The off-diagonal elements of the new row and column are the same as the elements of the row and column of bus i of the existing system.

19.10.4 ADDITION OF A LOOP CLOSING TWO EXISTING BUSES IN THE SYSTEM

Since both the buses are existing buses in the system, the dimension of the bus impedance matrix will not increase in this case. However the addition of the loop introduces a new axis which can be subsequently eliminated by Kron's reduction method.

The systems in Fig. 19.10A can be represented alternatively as in Fig. 19.10B.

The link between i and k requires a loop voltage

$$V_{\text{loop}} = 1.0 \, (Z_{ii} - Z_{2k} + Z_{kk} - Z_{ik} + Z_{\text{line}}) \tag{19.83}$$

for the circulation of unit current.

The loop impedance is

$$Z_{\text{loop}} = Z_{ii} + Z_{kk} - 2Z_{ik} + Z_{\text{line}} \tag{19.84}$$

The dimension of Z-matrix is increased due to the introduction of a new axis due to Loop 1

$$Z_{\ell\ell} = Z_{\text{loop}}$$

and

$$Z_{m-\ell} = Z_{mi} - Z_{mk}$$
$$Z_{\ell-m} = Z_{im} - Z_{km}, \quad m \neq \ell$$

The new loop axis can be eliminated now. Consider the matrix

$$\begin{bmatrix} Z_p & Z_q \\ Z_r & Z_s \end{bmatrix}$$

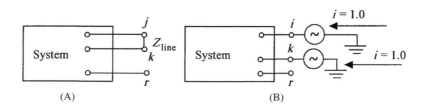

(A) (B)

FIGURE 19.10

(A) Addition of a loop and (B) equivalent representation.

It can be proved easily that

$$Z'_p = Z_p - Z_q Z_s^{-1} Z_r \tag{19.85}$$

using Eq. (19.85) all the additional elements introduced by the loop can be eliminated. The method is illustrated in Example E.19.3

19.10.5 GAUSS–SEIDEL METHOD USING *Z*-BUS FOR LOAD FLOW SOLUTION

An initial bus voltage vector is assumed as in the case of *Y*-bus method. Using these voltages, the bus currents are calculated using Eq. (19.6) or (19.7).

$$I_i = \frac{P_i - jQ_i}{V_i^*} - y_{ii} - v_i \tag{19.86}$$

where y_{ii} is the total shunt admittance at the bus i and $y_{ii} - v_i$ is the shunt current flowing from bus i to ground.

A new bus voltage estimate is obtained for an n-bus system from the following relation:

$$V_{bus} = Z_{bus} I_{bus} + V_R \tag{19.87}$$

where V_R is the $(n-1) \times 1$ dimensional reference voltage vector containing in each element the slack bus voltage. It may be noted that since the slack bus is the reference bus, the dimension of the Z_{bus} is $(n-1) \times (n-1)$.

The voltages are updated from iteration to iteration using the relation

$$V_i^{m+1} = V_S + \sum_{\substack{k=1 \\ k \neq S}}^{i-1} Z_{ik} I_k^{m+1} + \sum_{\substack{k=i \\ k \neq S}}^{n} Z_{ik} I_k^{(m)} \tag{19.88}$$

Then

$$V_k^{(m)} = \frac{P_k - jQ_k}{\left(V_k^{(m)}\right)^*} - y_{kk} V_k^{(m+1)}$$

where $i = 1, 2, \ldots, n$ and S is the slack bus.

19.11 COMPARISON OF VARIOUS METHODS FOR POWER FLOW SOLUTION

The requirements of a good power flow method are—high speed, low storage, and reliability for ill-conditioned problems. No single method meets all these requirements. It may be mentioned that for regular load flow studies NR-method in polar coordinates and for special applications fast decoupled load flow solution methods have proved to be most useful than other methods. NR method is versatile, reliable, and accurate. Fast decoupled load flow method is fast and needs the least storage.

Convergence of iterative methods depends upon the dominance of the diagonal elements of the bus admittance matrix.

Advantages of Gauss–Seidel Method:

1. The method is very simple in calculations and thus programming is easier.
2. The storage needed in the computer memory is relatively less.
3. In general the method is applicable for smaller systems.

Disadvantages of Gauss–Seidel Method:

1. The number of iterations needed is generally high and is also dependent on the acceleration factor selected.
2. For large systems, use of Gauss–Seidel method is practically prohibitive.
3. The time for convergence also increases dramatically with increase of number of buses.

Advantages of Newton–Raphson Method

1. The method is more accurate, faster, and reliable.
2. Requires less number of iterations for convergence. In fact, in three to four iterations good convergence is reached irrespective of the size of the system.
3. The number of iterations required is thus independent of the size of the system or the number of buses in the system.
4. The method is best suited for load flow solution to large size systems.
5. Decoupled and fast decoupled power flow solution can be obtained from Newton–Raphson Polar Coordinates method. Hence it also can serve as a base for security and contingency studies.

Disadvantages of Newton–Raphson Method

1. The memory needed is quite large for large size systems.
2. Calculations per iteration are also much larger than the Gauss–Seidel method.
3. Since, it is a gradient method, the method is quite involved, and hence, programming is also comparatively difficult and complicated.

WORKED EXAMPLES

E.19.1 A three-bus power system is shown in Fig E.19.1. The system parameters are given in Table E.19.1 and the load and generation data in Table E.19.2. The voltage at Bus 2 is maintained at 1.03 p.u. The maximum and minimum reactive power limits of the generation at Bus 2 are 35 and 0 Mvar, respectively. Taking Bus 1 as slack bus, obtain the load flow solution using

1. Gauss–Seidel iterative method using Y_{Bus}
2. Newton–Raphson polar coordinates method using Y_{Bus}

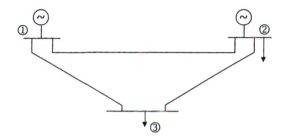

FIGURE E 19.1

A three-bus power system.

Table E.19.1 Impedance and Line Charging Admittances

Bus Code $i-k$	Impedance (p.u.) Z_{ik}	Line Charging Admittance (p.u.) y_i
1–2	0.08 + j0.24	0
1–3	0.02 + j0.06	0
2–3	0.06 + j0.018	0

Table E.19.2 Scheduled Generation, Loads, and Voltages

Bus No i	Bus Voltage V_i	Generation		Load	
		MW	Mvar	MW	Mvar
1	1.05 + j0.0	—	—	0	0
2	1.03 + j0.0	20	—	50	20
3	—	0	0	60	25

Solution:

The line admittance are obtained as

$$y_{12} = 1.25 - j3.75$$
$$y_{23} = 1.667 - j5.00$$
$$y_{13} = 5.00 - j15.00$$

The bus admittance matrix is formed using the procedure indicated in Section 2.1 as

$$Y_{Bus} = \begin{bmatrix} 6.25 & -j18.75 & -1.25 & +j3.75 & -5.0 & +j15.0 \\ -1.25 & +j3.73 & 2.9167 & -j8.75 & -j1.6667 & +j5.0 \\ -5.0 & +j15.0 & -1.6667 & +j5.0 & 6.6667 & -j20.0 \end{bmatrix}$$

1. Gauss–Seidel Iterative Method Using Y_{bus}
   The voltage at Bus 3 is assumed as $1 + j0$. The initial voltages are therefore

$$V_1^{(0)} = 1.05 + j0.0$$
$$V_2^{(0)} = 1.03 + j0.0$$
$$V_3^{(0)} = 1.00 + j0.0$$

Base MVA $= 100$

Iteration 1: It is required to calculate the reactive power Q_2 at Bus 2, which is a *P-V* or voltage-controlled bus

$$\delta_2^{(0)} = \tan^{-1}\left(\frac{e_2^a}{e_2'}\right) = 0.$$

$$e_{2(new)}' = |V_2|_{sch}\cos\delta_2 = (1.03)(1.0) = 1.03$$

$$e_{2(new)}' = V_{2\,sch}\sin\delta_2^{(0)} = (1.03)(0.0) = 0.00$$

$$Q_2^{(0)} = \left[(e_{2(new)}')^2 B_{22} + (e''_{2(new)})^2 B_{22}\right]$$
$$+ \sum_{\substack{k=1 \\ k \neq 2}}^{3}\left[(e''_{2(new)}e_k'G_{2k} + e''_k B_{2k}) - (e_{2(new)}'e''_k G_{2k} - e''_k B_{2k})\right]$$

Substituting the values

$$Q_2^{(0)} = \left[(1.03.)^2 8.75 + (0)^2 8.75\right] + 0(1.05)(-1.25) + 0.(-3.75)$$
$$-1.03\left[(0)(-1.25) - (1.05)(-3.75)\right]$$
$$+(0)\left[(1)(-1.6667) + (0)(-5.0)\right]$$
$$-1.03\left[(0)(-1.6667) - (1)(-5)\right]$$
$$= 0.07725$$

Mvar generated at Bus 2

$$= \text{Mvar injection into bus 2} + \text{Load Mvar}$$
$$= 0.07725 + 0.2 = 0.27725 \text{ p.u.}$$
$$= 27.725 \text{ Mvar}$$

This is within the limits specified.
The voltage at bus i is

$$V_i^{(m+1)} = \frac{+1}{Y_{ii}}\left[\frac{P_1 - jQ_1}{V_i^{(m)*}} - \sum_{k=1}^{i-1}y_{ik}V_k^{(m+1)} - \sum_{k=i+1}^{n1}y_{ik}V_k^{(m)}\right]$$

$$V_2^{(1)} = \frac{1}{Y_{22}}\left[\frac{P_1 - jQ_2}{V_i^{(0)*}} - Y_{21}V_1 - Y_{23}V_3^{(0)}\right]$$

$$= \frac{1}{(2.9167 - j8.75)}$$

$$\left[\frac{-0.3 - 0.07725}{1.03 - j0.0} - (-1.25 + j3.75)(1.05 + j0.0) + (-1.6667 + j5.0)(1 + j0.0)\right]$$

$$V_2^{(1)} = 1.01915 - j0.032491$$

$$= 1.0196673\angle -1.826 \text{ degrees}$$

An acceleration factor of 1.4 is used for both real and imaginary parts. The accelerated voltages is obtained using

$$v_2' = 1.03 + 1.4(1.01915 - 1.03) = 1.01481$$
$$v''_2 = 0.0 + 1.4(-0.032491 - 0.0) = -0.0454874$$

$$V_2^{(1)}(\text{accelerated}) = 1.01481 - j0.0454874$$

$$= 1.01583 \angle -2.56648 \text{ degrees}$$

The voltage at Bus 3 is given by

$$V_2^{(1)} = \frac{1}{Y_{33}}\left[\frac{P_3 - jQ_3}{V_3^{(0)*}} - Y_{31}V_1 - Y_{32}V_2^{(1)}\right]$$

$$= \frac{1}{6.6667 - j20}$$

$$\left[\left(\frac{-0.6 + j0.25}{1 - j0}\right) - (-5 + j15)(1.05 + j0) - (-1.6667 + j5)(1.01481 - j0.0454874)\right]$$
$$= 1.02093 - j0.0351381$$

The accelerated value of $V_3^{(1)}$ obtained using

$$v_3' = 1.0 + 1.4(1.02093 - 1.0) = 1.029302$$
$$v''_3 = 0 + 1.4(-0.0351384 - 0) = -0.0491933$$

$$V_3^{(1)} = 1.029302 - j0.049933$$

$$= 1.03048 \angle -2.73624 \text{ degrees}$$

The voltages at the end of the first iteration are as follows:

$$V_1 = 1.05 + j0.0$$

$$V_2^{(1)} = 1.01481 - j0.0454874$$

$$V_3^{(1)} = 1.029302 - j0.0491933$$

Check for Convergence: An accuracy of 0.001 is taken for convergence

$$[\Delta v_2']^{(0)} = [v_2']^{(1)} - [v_2']^{(0)} = 1.01481 - 1.03 = -0.0152$$

$$[\Delta v''_2]^{(0)} = [v''_2]^{(1)} - [v''_2]^{(0)} = -0.0454874 - 0.0 = -0.0454874$$

$$[\Delta v''_3]^{(0)} = [v_3']^{(1)} - [v_3']^{(0)} = 1.029302 - 1.0 = 0.029302$$

$$[\Delta v''_2]^{(0)} = [\Delta v''_2]^{(1)} - [\Delta v''_2]^{(0)} = -0.0491933 - 0.0 = -0.0491933$$

The magnitudes of all the voltage changes are greater than 0.001.

Iteration 2: The reactive power Q_2 at Bus 2 is calculated as before to give

$$\delta_2^{(1)} = \tan^{-1}\frac{[v''_2]^{(1)}}{[v'_2]^{(1)}} = \tan^{-1}\left[\frac{-0.0454874}{1.01481}\right] = -2.56648 \text{ degrees}$$

$$[v'_2]^{(1)} = |V_{2 \text{ sch}}| \cdot \cos \delta_2^{(1)} = 1.03 \cos(-2.56648 \text{ degrees}) = 1.02837$$

$$[v''_2]^{(1)} = |V_{2 \text{ sch}}| \cdot \sin \delta_2^{(1)} = 1.03 \sin(-2.56648 \text{ degrees}) = -0.046122$$

$$[V_{2 \text{ new}}]^{(1)} = 1.02897 - j0.046122$$

$$\begin{aligned}
Q_2^{(1)} = {} & (1.02897)^2(8.75) + (-0.046122)^2(8.75) \\
& + (-0.046122)[1.05(-1.25) + (0)(-3.75)] \\
& - (1.02897)[(0)(-1.25) - (1.05)(-3.75)] + \\
& - (1.02897)[(-0.0491933)(-1.6667) - (1.029302)(-5)] \\
= {} & -0.0202933
\end{aligned}$$

Mvar to be generated at Bus 2

$$= \text{Net Mvar injection into bus 2} + \text{Load Mvar}$$

$$= -0.0202933 + 0.2 = 0.1797067 \text{ p.u.} = 17.97067 \text{ Mvar}$$

This is within the specified limits. The voltages are, therefore, the same as before

$$V_1 = 1.05 + j0.0$$

$$V_2^{(1)} = 1.02897 - j0.0.46122$$

$$V_3^{(1)} = 1.029302 - j0.0491933$$

The new voltage at Bus 2 is obtained as follows:

$$\begin{aligned}
V_2^{(2)} = {} & \frac{1}{2.9167 - j8.75}\left[\frac{-0.3 + j0.0202933}{1.02827 + j0.046122}\right] \\
& -(-1.25 + j3.75)(1 - 05 + j0) \\
& -(-1.6667 + j5) \cdot (1.029302 - j0.0491933)] \\
= {} & 1.02486 - j0.0568268
\end{aligned}$$

The accelerated value of $V_2^{(2)}$ is obtained from

$$v'_2 = 1.02897 + 1.4(1.02486 - 1.02897) = 1.023216$$

$$v''_2 = -0.046122 + 1.4(-0.0568268) - (-0.046122 = -0.0611087)$$

$$v_2^{(2)\gamma} = 1.023216 - j0.0611087$$

The new voltage at Bus 3 is calculated as follows;

$$\begin{aligned}
V_3^{(2)} = {} & \frac{1}{6.6667 - j20}\left[\frac{-6.6 + j0.25}{1.029302 + j0.0491933}\right] \\
& -(-5 + j15)(1.05 + j0.0) \\
& -(-1.6667 + j5.0) \cdot (1.023216 - j0.0611)] \\
= {} & 1.0226 - j0.0368715
\end{aligned}$$

The accelerated value of $V_2^{(2)}$ obtained from

$$v_3' = 1.029302 + 1.4(1.0226 - 1.029302) = 1.02$$

$$v''_3 = (-0.0491933) + 1.4(-0.0368715) + (0.0491933) = -0.03194278$$

$$V_3^{(2)} = 1.02 - j0.03194278$$

The voltages at the end of the second iteration are as follows:

$$V_1 = 1.05 + j0.0$$

$$V_2^{(2)} = 1.023216 - j0.0611087$$

$$V_3^{(2)} = 1.02 - j0.03194278$$

The procedure is repeated till convergence is obtained at the end of the sixth iteration. The results are tabulated in Table E.19.1A.

Table E.19.1A Bus Voltage

Iteration	Bus 1	Bus 2	Bus 3
0	$1.05 + j0$	$1.03 + j0$	$1.0 + j0$
1	$1.05 + j0$	$1.01481 - j0.04548$	$1.029302 - j0.049193$
2	$1.05 + j0$	$1.023216 - j0.0611087$	$1.02 - j0.0319428$
3	$1.05 + j0$	$1.033476 - j0.0481383$	$1.027448 - j0.03508$
4	$1.05 + j0$	$1.0227564 - j0.051329$	$1.0124428 - j0.0341309$
5	$1.05 + j0$	$1.027726 - j0.0539141$	$1.0281748 - j0.0363943$
6	$1.05 + j0$	$1.029892 - j0.05062$	$1.020301 - j0.0338074$
7	$1.05 + j0$	$1.028478 - j0.0510117$	$1.02412 - j0.034802$

Line flow from Bus 1 to Bus 2

$$S_{12} = V_1(V_1^* - V_2^*)Y_{12}^* = 0.228975 + j0.017396$$

Line flow from Bus 2 to Bus 1

$$S_{21} = V_2(V_2^* - V_1^*)Y_{21}^* = -0.22518 - j0.0059178$$

Similarly the other line flows can be computed and are tabulated in Table E.19.1B. The slack bus power obtained by adding the flows in the lines terminating at the slack bus is

$$P_1 + jQ_1 = 0.228975 + j0.017396 + 0.684006 + j0.225$$

$$= (0.912981 + j0.242396)$$

Table E.19.1(B) Line Flows

Line	P	Power Flow	Q
1–2	+0.228975		0.017396
2–1	−0.225183		0.0059178
1–3	0.68396		0.224
3–1	−0.674565		−0.195845
2–3	−0.074129		0.0554
3–2	0.07461		−0.054

2. Newton−Raphson Polar Coordinates Method
   The bus admittance matrix is written in polar form as

$$Y_{\text{bus}} = \begin{bmatrix} 19.7642 \angle -71.6 \text{ degrees} & 3.95285 \angle -108.4 \text{ degrees} & 15.8114 \angle -108.4 \text{ degrees} \\ 3.95285 \angle -108.4 \text{ degrees} & 9.22331 \angle -71.6 \text{ degrees} & 5.27046 \angle -108.4 \text{ degrees} \\ 15.8114 \angle -108.4 \text{ degrees} & 5.27046 \angle -108.4 \text{ degrees} & 21.0819 \angle -71.6 \text{ degrees} \end{bmatrix}$$

Note that

$$\angle Y_{ii} = -71.6 \text{ degrees}$$

and

$$\angle Y_{ik} = -180 \text{ degrees} - 71.6 \text{ degrees} = 108.4 \text{ degrees}$$

The initial bus voltages are as follows:

$$V_1 = 1.05 \angle 0 \text{ degree}$$

$$V_2^{(0)} = 1.03 \angle 0 \text{ degree}$$

$$V_2^{(0)} = 1.0 \angle 0 \text{ degree}$$

The real and reactive powers at Bus 2 are calculated as follows:

$$P_2 = |V_2 \ V_1 \ Y_{21}| \cos(\delta_2^{(0)} - \delta_1 - \theta_{21}) + |V_2^2 \ Y_{22}| \cos(-\theta_{22}) + |V_2 \ V_3 \ Y_{23}| \cos(\delta_2^{(0)} - \delta_3^{(0)} - \theta_{23})$$

$$= (1.03)(1.05)(3.95285) \cos(108°.4) + (1.03)^2 (9.22331) \cos(-108°.4)$$

$$+ (1.03)^2 (9.22331) \cos(71°.6) + (1.03)(1.0)(5.27046) \cos(-1080°.4)$$

$$= 0.02575$$

$$Q_2 = |V_2 \ V_1 \ Y_{21}| \sin(\delta_2^{(0)} - \delta_1 - \theta_{21}) + |V_2^2 \ Y_{22}| \sin(-\theta_{22}) + |V_2 \ V_3 \ Y_{23}| \sin(\delta_2^{(0)} - \delta_3^{(0)} - \theta_{23})$$

$$= (1.03)(1.05)(3.95285) \sin(-108°.4) + (1.03)^2 (9.22331) \sin(71.6 \text{ degrees})$$

$$+ (1.03)(1.0)(5.27046) \sin(108.4 \text{ degrees})$$

$$= 0.07725$$

Generation of p.u. Mvar at Bus 2

$$= 0.2 + 0.07725$$
$$= 0.27725 = 27.725 \text{ Mvar}$$

This is within the limits specified. The real and reactive powers at Bus 3 are calculated in a similar way.

$$P_3 = \left| V_3^{(0)} \; V_1 \; Y_{31} \right| \cos(\delta_3^{(0)} - \delta_1 - \theta_{31}) + \left| V_3^{(0)} \; V_2 \; Y_{32} \right| \cos(\delta_3^{(0)} - \delta_2 - \theta_{32}) + \left| V_3^{(0)2} \; Y_{33} \right| \cos(-\theta_{33})$$

$$= (1.0)(1.05)(15.8114) \cos(-108.4 \text{ degrees}) + (1.0)(1.03)(5.27046) \cos(-108.4 \text{ degrees})$$

$$+ (1.0)^2 (21.0819) \cos(71.6 \text{ degrees})$$

$$= -0.3$$

$$Q_3 = \left| V_3^{(0)} \; V_1 \; Y_{31} \right| \sin(\delta_3^{(0)} - \delta_1 - \theta_{31}) + \left| V_3^{(0)} \; V_2 \; Y_{32} \right| \sin(\delta_3^{(0)} - \delta_2 - \theta_{32}) + \left| V_3^{(0)2} \; Y_{33} \right| \sin(-\theta_{33})$$

$$= (1.0)1.05(15.8114) \sin(-108.4 \text{ degrees}) + (1.0)(1.03)(5.27046) \sin(-108.4 \text{ degrees})$$

$$+ (1.0)^2 (21.0891) \sin(71.6 \text{ degrees})$$

$$= -0.9$$

The differences between scheduled and calculated powers are as follows:

$$\Delta P_2^{(0)} = -0.3 - 0.02575 = -0.32575$$
$$\Delta P_3^{(0)} = -0.6 - (-0.3) = -0.3$$
$$\Delta Q_2^{(0)} = -0.25 - (-0.9) = -0.65$$

It may be noted that ΔQ_2 has not been computed since Bus 2 is voltage-controlled bus. Since

$$\left| \Delta P_2^{(0)} \right|, \; \left| \Delta P_3^{(0)} \right|, \; \text{and} \; \left| \Delta Q_3^{(0)} \right|$$

are greater than the specified limit of 0.01, the next iteration is computed.

Iteration 1: Elements of the Jacobian are calculated as follows:

$$\frac{\partial P_2}{\partial \delta_2} = -\left| V_2 \; V_1 \; Y_{21} \right| \sin(\delta_2^{(0)} - \delta_1^{(0)} - \theta_{21}) + \left| V_2 \; V_3^{(0)} \; Y_{23} \right| \sin(\delta_2^{(0)} - \delta_3^{(0)} - \theta_{23})$$

$$= -(1.03)(1.05)(3.95285) \sin(108.4 \text{ degrees}) + (1.03)(1.0)(5.27046) \sin(-108.4 \text{ degrees})$$

$$= 9.2056266$$

$$\frac{\partial P_2}{\partial \delta_3} = \left| V_2 \; V_3^{(0)} \; Y_{23} \right| \sin(\delta_2^{(0)} - \delta_3^{(0)} - \theta_{23})$$

$$= (1.03)(1.0)(5.27046) \sin(-108.4 \text{ degrees}) = -5.15$$

$$\frac{\partial P_3}{\partial \delta_2} = \left| V_3^{(0)} \; V_1 \; Y_{31} \right| \sin(\delta_3^{(0)} - \delta_2^{(0)} - \theta_{32})$$

$$= (0.0)(1.03)(5.27046) \sin(-108.4 \text{ degrees})$$

$$= -5.15$$

$$\frac{\partial P_3}{\partial \delta_3} = |V_3^{(0)} \quad V_1 \quad Y_{31}| \sin(\delta_3^{(0)} - \delta_1 - \theta_{31}) + |V_3^{(0)} \quad V_2 \quad Y_{32}| \sin(\delta_3^{(0)} - \delta_2^{(0)} - \theta_{32})$$

$$= -(1.0)(1.05)(15.8114)\sin(-108.4 \text{ degrees}) - 5.15$$

$$= 20.9$$

$$\frac{\partial P_2}{\partial \delta_3} = |V_2 \quad Y_{23}| \cos(\delta_2^{(0)} - \delta_3^{(0)} - \theta_{23})$$

$$= (1.03)(5.27046)\cos(108.4 \text{ degrees})$$

$$= -1.7166724$$

$$\frac{\partial P_3}{\partial V_3} = 2|V_3 \quad Y_{33}| \cos\theta_{33} + |V_1 \quad Y_{31}| \cos(\delta_3^{(0)} - \delta_2^{(0)} - \theta_{32}) + |V_2 \quad Y_{32}| \cos(\delta_3^{(0)} - \delta_2^{(0)} - \theta_{32})$$

$$= 2(1.0)(21.0819)\cos(71.6 \text{ degrees}) + (1.05)(15.8114)\cos(-108.4 \text{ degrees})$$

$$+(1.03)(5.27046)\cos(-108.4 \text{ degrees})$$

$$= 6.366604$$

$$\frac{\partial Q_3}{\partial \delta_2} = -|V_3^{(0)} \quad V_1 \quad Y_{32}| \cos(\delta_3^{(0)} - \delta_2^{(0)} - \theta_{32})$$

$$= (1.0)(1.03)(5.27046)\cos(-108.4 \text{ degrees})$$

$$= 1.7166724$$

$$\frac{\partial Q_3}{\partial \delta_3} = |V_3^{(0)} \quad V_1 \quad Y_{32}| \cos(\delta_3^{(0)} - \delta_1 - \theta_{31}) + |V_3^{(0)} \quad V_2 \quad Y_{32}| \cos(\delta_2^{(0)} - \delta_3^{(0)} - \theta_{32})$$

$$= (1.0)(1.05)(15.8114)\cos(-108.4 \text{ degrees}) - 1.7166724$$

$$= -6.9667$$

$$\frac{\partial Q_3}{\partial V_3} = 2V_3^{(0)}Y_{33}\sin(-\theta_{33}) + |V_1 \quad Y_{31}| \sin(\delta_3^{(0)} - \delta_1 - \theta_{31}) + |V_2 \quad Y_{32}| \sin(\delta_3^{(0)} - \delta_2^{(0)} - \theta_{32})$$

$$= 2(1.0)(21.0819)\sin(71.6 \text{ degrees}) + (1.05)(15.8114)\sin(-108.4 \text{ degrees})$$

$$+(1.03)(5.27046)\sin(-108.4 \text{ degrees})$$

$$= 19.1$$

From Eq. (2.45)

$$\begin{bmatrix} -0.32575 \\ -0.3 \\ 0.65 \end{bmatrix} = \begin{bmatrix} 9.20563 & -5.15 & -1.71667 \\ -5.15 & 20.9 & 6.36660 \\ 1.71667 & -6.9967 & 19.1 \end{bmatrix} \begin{bmatrix} |\Delta\delta_2| \\ \Delta\delta_3 \\ \Delta|V_3| \end{bmatrix}$$

Following the method of triangulation and back substations

$$\begin{bmatrix} -0.35386 \\ -0.3 \\ -0.035386 \end{bmatrix} = \begin{bmatrix} 1 & -0.55944 & -0.18648 \\ -5.15 & 20.9 & 6.36660 \\ +1.71667 & -6.9667 & 19.1 \end{bmatrix} \begin{bmatrix} \Delta\delta_2 \\ \Delta\delta_3 \\ \Delta|V_3| \end{bmatrix}$$

$$\begin{bmatrix} -0.35386 \\ -0.482237 \\ +0.710746 \end{bmatrix} = \begin{bmatrix} 1 & -0.55944 & -0.18648 \\ 0 & 18.02 & 5.40623 \\ 0 & -6.006326 & 19.42012 \end{bmatrix} \begin{bmatrix} \Delta\delta_2 \\ \Delta\delta_3 \\ \Delta|V_3| \end{bmatrix}$$

Finally

$$\begin{bmatrix} -0.35386 \\ -0.0267613 \\ 0.55 \end{bmatrix} = \begin{bmatrix} 1 & -0.55944 & -0.18648 \\ 0 & 1 & 0.3 \\ 0 & 0 & 21.22202 \end{bmatrix} \begin{bmatrix} \Delta\delta_2 \\ \Delta\delta_3 \\ \Delta|V_3| \end{bmatrix}$$

Thus

$$\Delta|V_3| = \frac{(0.55)}{(21.22202)} = 0.025917$$

$$\Delta\delta_3 = -0.0267613 - (0.3)(0.025917)$$
$$= -0.0345364 \text{ rad}$$
$$= -1.98 \text{ degrees}$$

$$\Delta\delta_2 = -0.035286 - (-0.55944)(-0.034536) - (-0.18648)(0.025917)$$
$$= -0.049874 \text{ rad}$$
$$= -2.8575 \text{ degrees}$$

At the end of the first iteration, the bus voltages are as follows:

$$V_1 = 1.05 \angle 0 \text{ degree}$$

$$V_2 = 1.03 \angle 2.85757 \text{ degrees}$$

$$V_3 = 1.025917 \angle -1.9788 \text{ degrees}$$

The real and reactive powers at Bus 2 are computed as follows:

$$P_2^{(1)} = (1.03)(1.05)(3.95285)[\cos(-2.8575) - 0(-108.4 \text{ degrees})$$
$$+ (1.03)^2(1.025917)(5.27046)\cos[(-2.8575) - (-1.9788) - 108.4 \text{ degrees}$$
$$= -0.30009$$

$$Q_2^{(1)} = (1.03)(1.05)(3.95285)[\sin(-2.8575) - 0(-108.4 \text{ degrees})$$
$$+ (1.03)^2(9.22331)\sin[(-2.85757) - (-1.9788) - 108.4 \text{ degrees})]$$
$$= 0.043853$$

Generation of reactive power at Bus 2

$$= 0.2 + 0.043853 = 0.243856 \text{ p.u. Mvar}$$
$$= 24.3856 \text{ Mvar}$$

This is within the specified limits.

The real and reactive powers at Bus 3 are computed as follows:

$$P_3^{(1)} = (1.025917)(1.05)(15.8117)\cos[(-1.09788) - 0 - 108.4 \text{ degrees})]$$
$$+ (1.025917)(1.03)(5.27046)\cos[(-1.0988) - (-2.8575) - 108.4]$$
$$+ (1.025917)^2(21.0819)\cos(71.6 \text{ degrees})$$
$$= -0.60407$$

$$Q_3^{(1)} = (1.025917)(1.05)(15.8114) \sin [(-1.977) - 108.4 \text{ degrees})]$$
$$+ (1.025917)(1.03)(5.27046) \sin [(-1.9788) - (-2.8575) - 108.4 \text{ degrees})]$$
$$+ (1.025917)^2 (21.0819) \sin (71.6 \text{ degrees})$$
$$= -0.224$$

The differences between scheduled powers and calculated powers are as follows:

$$\Delta P_2^{(1)} = -0.3 - (-0.30009) = 0.00009$$
$$\Delta P_3^{(1)} = -0.6 - (-0.60407) = 0.00407$$
$$\Delta Q_3^{(1)} = -0.25 - (-0.2224) = -0.0276$$

$$-(0.3) - (0.30009)$$
$$-(0.6) - (0.60407)$$
$$-(0.25) - (0.2224)$$

Even though the first two differences are within the limits the last one, is greater than the specified limit 0.01. The next iteration is carried out in a similar manner. At the end of the second iteration, even ΔQ_3 also is found to be within the specified tolerance. The results are tabulated in Table E.19.2A and B.

Table E.19.2(A) Bus Voltages

Iteration	Bus 1	Bus 2	Bus 3
0	1.05 ∠0 degree	1.03 ∠0 degree	1. ∠0 degree
1	1.05 ∠0 degree	1.03 ∠−2.85757	1.025917∠−1.9788
2	1.05 ∠0 degree	1.03 ∠−2.8517	1.02476∠−1.947

Table E19.2(B) Line Flows

Line	P	Power Flow	Q
1−2	0.2297		0.016533
2−1	−0.22332		−0.0049313
1−3	0.68396		0.224
3−1	−0.674565		−0.0195845
2−3	−0.074126		0.0554
3−2	0.07461		−0.054

E.19.2 Obtain the load flow solution to the system given in example E.19.1 using Z-Bus. Use Gauss−Seidel method. Take accuracy for convergence as 0.0001.

Solution:

The bus impedance matrix is formed as indicated in Section 19.10. The slack bus is taken as the reference bus. In this example, as in example E.19.1 Bus 1 is chosen as the slack bus.

1. Add element 1−2. This is addition of a new bus to the reference bus.

$$Z_{bus} = \quad (2) \quad \boxed{0.05 + j0.24}$$

with (2) above the box.

2. Add element 1−3. This is also an addition of a new bus to the reference bus.

	(2)	(3)
(2)	$0.08 + j0.24$	$0.0 + j0.0$
(3)	$0.0 + j0.0$	$0.02 + j0.06$

$Z_{bus} = $

3. Add element 2−3. This is the addition of a link between two existing Buses 2 and 3.

$$Z_{2\text{-loop}} = Z_{\text{loop-2}} = Z_{22} - Z_{23} = 0.08 + j0.24$$

$$Z_{3\text{-loop}} = Z_{\text{loop-3}} = Z_{32} - Z_{33} = -(0.02 + j0.06)$$

$$Z_{\text{loop-loop}} = Z_{22} + Z_{33} - 2 Z_{23} + Z_{23,23}$$
$$= (0.08 + j0.24) + (0.02 + j0.06)(0.06 + j0.18)$$
$$= 0.16 + j0.48$$

	(2)	(3)	(ℓ)
(2)	$0.08 + j0.024$	$0 + j0$	$0.08 + j0.24$
(3)	$0.0 + j0.0$	$0.02 + j0.06$	$-(0.02 + j0.06)$
ℓ	$0.08 + j0.24$	$-(0.02 + j0.006)$	$0.16 + j0.48$

$Z_{bus} = $

The loop is now eliminated

$$Z'_{22} = Z_{22} - \frac{Z_{2-\text{loop}} Z_{\text{loop}-2}}{Z_{\text{loop}-\text{loop}}}$$
$$= (0.08 + j0.24) - \frac{(0.8 + j0.24)^2}{0.16 + j0.48}$$
$$= 0.04 + j0.12$$

$$Z'_{23} = Z'_{32} = \left[Z_{23} - \frac{Z_{2-\text{loop}} Z_{\text{loop}-3}}{Z_{\text{loop}-\text{loop}}} \right]$$
$$= (0.0 + j0.0) - \frac{(0.8 + j0.24)(-0.02 - j0.06)}{0.16 + j0.48}$$
$$= 0.01 + j0.03$$

Similarly $Z'_{33} = 0.0175 + j0.0526$

The Z-bus matrix is thus

$$Z_{Bus} = \begin{bmatrix} 0.04 + j0.12 & 0.01 + j0.03 \\ 0.01 + j0.03 & 0.017 + j0.0525 \end{bmatrix} = \begin{bmatrix} 0.1265 \angle 71.565 \text{ degrees} & 0.031623 \angle 71.565 \text{ degrees} \\ 0.031623 \angle 71.565 \text{ degrees} & 0.05534 \angle 71.565 \text{ degrees} \end{bmatrix}$$

The voltages at Buses 2 and 3 are assumed to be

$$V_2^{(0)} = 1.03 + j0.0$$

$$V_3^{(0)} = 1.0 + j0.0$$

Assuming that the reactive power injected into Bus 2 is zero,

$$Q_2 = 0.0$$

The bus currents $I_2^{(0)}$ and $I_3^{(0)}$ are computed as follows:

$$I_2^{(0)} = \frac{-0.3 + j0.0}{1.03 - j0.0} = -0.29126 - j0.0 = 0.29126 \angle 180 \text{ degrees}$$

$$I_3^{(0)} = \frac{-0.6 + j0.25}{1.0 + j0.0} = -0.6 - j0.25 = 0.65 \angle 157.38 \text{ degrees}$$

Iteration 1: The voltage at Bus 2 is computed as

$$V_2^{(1)} = V_1 + Z_{22}I_2^{(0)} + Z_{23} = I^{(0)}$$

$$= 1.05 \angle 0 \text{ degree} + (0.1265 \angle 71.565 \text{ degrees } (0.29126 \angle 180 \text{ degrees}$$

$$+ (0.031623 \angle 71.565 \text{ degrees})(0.65 \angle 157.38 \text{ degrees})$$

$$= 1.02485 - j0.05045$$

$$= 1.02609 \angle - 2.8182$$

The new bus current $I_2^{(0)}$ is now calculated as follows:

$$\Delta I_2^{(0)} = \frac{V_2^{(1)}}{Z_{22}} \left[\frac{|V_{sch}|}{|V_2^{(1)}|} - 1 \right]$$

$$= \frac{1.02609 \angle - 2.8182}{0.1265 \angle 71.565 \text{ degrees}} \times \left[\frac{1.03}{1.02609} - 1 \right] = 0.0309084 \angle - 74.3832 \text{ degrees}$$

$$Q^{(0)} = \text{Im}[V_2^{(1)} \Delta I_2^{(0)*}]$$

$$= \text{Im}[1.02609 \angle - 2.8182 \text{ degrees}](0.0309084 \angle 74.383 \text{ degrees})$$

$$= 0.03$$

$$Q_2^{(1)} = Q_2^{(0)} + \Delta Q_2^{(0)} = 0.0 + 0.03 = 0.03$$

$$I_2^{(1)} = \frac{-0.3 - j0.3}{1.02609 \angle - 2.8182 \text{ degrees}} = 0.29383 \angle 182.8918 \text{ degrees}$$

Voltage at Bus 3 is now calculated as follows:

$$V_3^{(1)} = V_1 + Z_{32}I_2^{(1)} + Z_{33}I_3^{(0)}$$

$$= 1.05 \angle 0.0 \text{ degree} + (0.031623 \angle 71.565 \text{ degree})(0.29383 \angle 182.832 \text{ degrees})$$

$$+ (0.05534 \angle 71.565 \text{ degrees})(0.65 \angle 157.38 \text{ degrees})$$

$$= (1.02389 - j0.036077) = 1.0245 \angle - 2.018 \text{ degrees}$$

$$I_3^{(1)} = \frac{0.65 \angle 157.38 \text{ degrees}}{1.0245 \angle 2.018 \text{ degrees}} = 0.634437 \angle 155.36 \text{ degrees}$$

The voltages at the end of the first iteration are as follows:

$$V_1 = 1.05 \angle 0 \text{ degree}$$

$$V_2^{(1)} = 1.02609 \angle - 2.8182 \text{ degrees}$$

$$V_3^{(1)} = 1.0245 \angle - 2.018 \text{ degrees}$$

The differences in voltages are as follows:

$$\Delta V_2^{(1)} = (1.02485 - j0.05045) - (1.03 + j0.0)$$

$$= - 0.00515 - j0.05045$$

$$\Delta V_3^{(1)} = (1.02389 - j0.0536077) - (1.0 + j0.0)$$

$$= (0.02389 - j0.036077)$$

Both the real and imaginary parts are greater than the specified limit 0.001.
Iteration 2:

$$V_2^{(2)} = V_1 + Z_{22}I_2^{(1)(1)} + Z_{23}I_3^{(1)(1)}$$

$$= 1.02 \angle 0 \text{ degree} + (0.1265 \angle 71.565 \text{ degrees})(0.29383 \angle 182.892 \text{ degrees})$$

$$+ (0.031623 \angle 71.565 \text{ degrees})(0.63447 \angle 155.36 \text{ degrees})$$

$$= 1.02634 - j0.050465$$

$$= 1.02758 \angle - 2.81495 \text{ degrees}$$

$$\Delta I_2^{(1)} = \frac{1.02758 \angle - 2.81495 \text{ degrees}}{1.1265 \angle - 71.565 \text{ degrees}} \left[\frac{1.03}{1.02758} - 1 \right]$$

$$= 0.01923 \angle - 74.38 \text{ degrees}$$

$$\Delta Q_2^{(1)} = \text{Im} \left[V_2^{(2)} \left(\Delta I_2^{(1)} \right)^* \right]$$

$$= \text{Im}(1.02758 \angle - 2.81495)(0.01913 \angle 74.38°)$$

$$= 0.0186487$$

$$Q_2^{(2)} = Q_2^{(1)} + \Delta Q_2^{(1)}$$

$$= 0.03 + 0.0186487 = 0.0486487$$

$$I_2^{(2)} = \frac{- 0.3 - j0.0486487}{1.02758 \angle 2.81495 \text{ degrees}}$$

$$= 0.295763 \angle 186.393 \text{ degrees}$$

$$V_3^{(2)} = 1.05 \angle 0 \text{ degree} + (0.31623 \angle 71.565 \text{ degrees} (0.295763 \angle 186.4 \text{ degrees}$$

$$+ 0.05534 \angle 71.565 \text{ degrees})(0.634437 \angle 155.36 \text{ degrees})$$

$$I_3^{(2)} = \frac{0.65 \angle 157.38 \text{ degrees}}{1.02466 \angle 1.9459 \text{ degrees}} = 0.6343567 \angle 155.434 \text{ degrees}$$

$$\Delta V_2^{(1)} = (1.02634 - j0.050465) - (1.02485 - j0.05041) = 0.00149 - j0.000015$$

$$\Delta V_3^{(1)} = (1.024 - j0.034793) - (1.02389 - j0.036077) = 0.00011 + j0.00128$$

As the accuracy is still not enough, another iteration is required.

Iteration 3:

$$V_2^{(3)} = 1.05 \angle 0 \text{ degree} + (0.1265 \angle 71.565 \text{ degrees})(0.295763 \angle 186.4 \text{ degrees})$$

$$+ (0.031623 \angle 71.565 \text{ degrees})(0.63487 \angle 155.434 \text{ degrees})$$

$$= 1.0285187 - j0.051262$$

$$= 1.0298 \angle - 2.853 \text{ degrees}$$

$$I_2^{(2)} = \frac{1.0298 \angle - 2.853 \text{ degree}}{0.1265 \angle 71.565 \text{ degrees}} \left[\frac{1.03}{1.0298} - 1 \right] = 0.001581 \angle 74.418 \text{ degrees}$$

$$\Delta Q_2^{(2)} = 0.00154456$$

$$Q_2^{(3)} = 0.0486487 + 0.001544 = 0.0502$$

$$I_2^{(3)} = \frac{-0.3 - j0.0502}{0.0298 \angle 2.853 \text{ degrees}} = 0.29537 \angle 186.647 \text{ degrees}$$

$$V_3^{(3)} = 1.05 \angle 0 \text{ degree} + (0.031623 \angle 71.565 \text{ degrees}) + (0.29537 \angle 186.647 \text{ degrees})$$

$$+ (0.05534 \angle 71.565 \text{ degrees})(0.634357 \angle 155.434 \text{ degrees})$$

$$= 1.024152 - j0.034817 = 1.02474 \angle - 1.9471 \text{ degrees}$$

$$I_3^{(3)} = \frac{-0.65 - \angle 157.38 \text{ degrees}}{1.02474 \angle 1.9471 \text{ degrees}} = 0.6343 \angle 155.433 \text{ degrees}$$

$$\Delta V_2^{(2)} = (1.0285187 - j0.051262) - (1.02634 - j0.050465)$$

$$= 0.0021787 - 0.000787$$

$$\Delta V_3^{(2)} = (1.024152 - j0.034817) - (1.024 - j0.034793)$$

$$= 0.000152 - j0.00002$$

Iteration 4:

$$V_2^{(4)} = 1.02996 \angle - 2.852 \text{ degrees}$$
$$\Delta I_2^{(3)} = 0.0003159 \angle - 74.417 \text{ degrees}$$
$$\Delta Q_2^{(3)} = 0.0000867$$

$$Q_2^{(4)} = 0.0505$$
$$I_2^{(4)} = 0.29537 \angle 186.7 \text{ degrees}$$
$$V_2^{(4)} = 1.02416 - j0.034816 = 1.02475 \angle - 1.947 \text{ degrees}$$
$$\Delta V_2^{(3)} = 0.000108 + j0.000016$$

$$\Delta V_3^{(3)} = 0.00058 + j0.000001$$

The final voltages are as follows:

$$V_1 = 1.05 + j0.0$$
$$V_2 = 1.02996 \angle -2.852 \text{ degrees}$$
$$V_3 = 1.02475 \angle -1.947 \text{ degrees}$$

The line flows may be calculated further if required.

E.19.3 Consider the bus system shown in Fig. E.19.3.

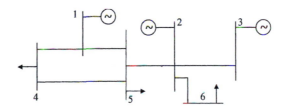

FIGURE E.19.3

A six-bus power system.

The following is the data:

Line Impedance (p.u.)	Real		Imaginary	
1.4	0.57000	E − 1	0.845	E − 1
1−5	1.33000	E − 2	3.600	E − 2
2−3	3.19999	E − 2	1.750	E − 1
2−5	1.73000	E − 2	0.560	E − 1
2−6	3.00000	E − 2	1.500	E − 1
4−5	1.94000	E − 2	0.625	E − 1

Scheduled generation and bus voltages:

Bus Code P	Assumed Bus Voltage	Generation		Load	
		MW p.u.	Mvar p.u	MW p.u.	Mvar p.u
1	1.05 + j0.0 (specified)	–	–	–	–
2	–	1.2	0.05	–	–
3	–	1.2	0.05	–	–
4	–	–	–	1.4	0.05
5	–	–	–	0.8	0.03
6	–	–	–	0.7	0.02

Taking Bus 1 as slack bus and using an accelerating factor of 1.4, perform load flow by Gauss–Seidel method. Take precision index as 0.0001.

Solution:

The bus admittance matrix is obtained as follows:

P–Q	Bus Code Real	Admittance (p.u.) Imaginary
1–1	14.516310	−32.57515
1–4	−5.486446	8.13342
1–5	−9.029870	24.44174
2–2	7.329113	−28.24106
2–3	−1.011091	5.529494
2–5	−5.035970	16.301400
2–6	−1.282051	6.410257
3–2	−1.011091	5.529404
3–3	1.011091	−5.529404
4–1	−5.486446	8.133420
4–4	10.016390	−22.727320
4–5	−4.529948	14.593900
5–1	−9.029870	24.441740
5–2	−5.035970	16.301400
5–4	−4.529948	14.593900
5–5	18.595790	−55.337050
6–2	−1.282051	6.410257
6–6	1.282051	−6.410254

All the bus voltages, $V^{(0)}$, are assumed to be $1 + j0$ except the specified voltage at Bus 1 which is kept fixed at $1.05 + j0$. The voltage equations for the first Gauss–Seidel iteration are as follows:

$$V_2^{(1)} = \frac{1}{Y_2}\left[\frac{P_2 - jQ_2}{V_2^{(0)*}} - Y_{23}V_3^{(0)} - Y_{25}V_5^{(0)} - Y_{26}V_6^{(0)}\right]$$

$$V_3^{(1)} = \frac{1}{Y_{33}} \left[\frac{P_3 - jQ_3}{V_3^{(0)*}} - Y_{32} V_2^{(1)} \right]$$

$$V_4^{(1)} = \frac{1}{Y_{44}} \left[\frac{P_4 - jQ_4}{V_4^{(0)*}} - Y_{41} V_1 - Y_{45} V_5^{(0)} \right]$$

$$V_5^{(1)} = \frac{1}{Y_{55}} \left[\frac{P_5 - jQ_5}{V_5^{(0)*}} - Y_{51} V_1 - Y_{51} V_2^{(1)} - Y_{54} V_4^{(1)} \right]$$

$$V_6^{(1)} = \frac{1}{Y_{66}} \left[\frac{P_6 - jQ_6}{V_6^{(0)*}} - Y_{62} V_2^{(1)} \right]$$

Substituting the values, the equation for solution is as follows:

$$V_2^{(1)} = \left(\frac{1}{7.329113} - j28.24100 \right) \times \left[\frac{1.2 - j0.05}{1 - j0} \right]$$
$$- (-1.011091 + j5.529404) \times (1 + j0) - (-5.03597 + j16.3014)(1 + j0)$$
$$- (1 - 282051 + j16.3014)(1 + j0)$$
$$= 1.016786 + j0.0557924$$

$$V_3^{(1)} = \left(\frac{1}{1.011091} - j5.52424 \right) \times \left[\frac{1.2 - j0.05}{1 - j0} \right]$$
$$- (-1.011091 + j5.529404) \times (1.016786 + j0.0557924)$$
$$= 1.089511 + j0.3885233$$

$$V_4^{(1)} = \left(\frac{1}{10.01639} - j22.72732 \right) \times \left[\frac{-1.4 + j0.005}{1 - j0} \right] - (-5.486446 + j8.133342) \times (1.05 + j0)$$
$$- (-4.529948 + j14.5939)(1 + j0)$$
$$= 0.992808 - j0.0658069$$

$$V_5^{(1)} = \left(\frac{1}{18.59579} - j55.33705 \right) \times \left[\frac{-0.8 + j0.03}{1 - j0} \right] - (-9.02987 + j24.44174) \times (1.05 + j0)$$
$$- (-5.03597 + j16.3014)(1.016786 + j0.0557929)$$
$$- (-4.529948 + j14.5939)(0.992808 - j0.0658069)$$
$$= 1.028669 - j0.01879179$$

$$V_6^{(1)} = \left(\frac{1}{1.282051} - j6.410257 \right) \times \left[\frac{-0.7 + j0.02}{1 - j0} \right]$$
$$- (-1.282051 - j6.410257) \times (1.016786 + j0.0557924)$$
$$= 0.989904 - j0.0669962$$

The results of these iterations is given in Table E.19.3A.

Table E.19.3A

It.No	Bus 2	Bus 3	Bus 4	Bus 5	Bus 6
0	1 + j0.0	1 + j0.0	1 + j0.0	1 + j0.0	1 + j0.0
1	1.016789 + j0.0557924	1.089511 + j0.3885233	0.992808 − j0.0658069	1.02669 − j0.01879179	0.989901 − j0.0669962
2	1.05306 + j0.1018735	1.014855 + j0.2323309	1.013552 − j0.0577213	1.042189 + j0.0177322	1.041933 + j0.0192121
3	1.043568 + j0.089733	1.054321 + j0.3276035	1.021136 − j0.0352727	1.034181 + j0.00258192	1.014571 − j0.02625271
4	1.047155 + j0.101896	1.02297 + j0.02763564	1.012207 − j0.0500558	1.035391 + j0.00526437	1.02209 + j0.00643566
5	1.040005 + j0.093791	1.03515 + j0.3050814	1.61576 − j0.0425892	0.033319 + j0.003697056	1.014416 − j0.01319787
6	1.04212 + j0.0978431	1.027151 + j0.2901358	1.013044 − j0.04646546	10.33985 + j0.004504417	1.01821 − j0.001752973
7	1.040509 + j0.0963405	1.031063 + j0.2994083	1.014418 − j0.0453101	1.033845 + j0.00430454	1.016182 − j0.00770669
8	1.041414 + j0.097518	1.028816 + j0.294465	1.013687 − j0.0456101	1.033845 + j0.004558826	1.017353 − j0.0048398
9	1.040914 + j0.097002	1.030042 + j0.2973287	1.014148 − j0.04487629	1.033711 + j0.004413647	1.016743 − j0.0060342
10	1.041203 + j0.0972818	1.02935 + j0.2973287	1.013881 − j0.04511174	1.03381 + j0.004495542	1.017089 − j0.00498989
11	1.041036 + j0.097164	1.029739 + j0.296598	1.01403 − j0.04498312	1.03374 + j0.004439559	1.016877 − j0.00558081
12	1.041127 + j0.0971998	1.029518 + j0.2960784	1.013943 − j0.04506212	1.033761 + j0.00447096	1.016997 − j0.00524855
13	1.041075 + j0.0971451	1.029642 + j0.2963715	1.019331 − j0.04501488	1.033749 + j0.004454002	1.016927 − j0.00543323
14	1.041104 + j0.0971777	1.02571 + j0.2962084	1.0013965 − j0.04504223	1.033756 + j0.004463713	1.016967 − j0.00053283

Table E.19.3B

Bus	Voltage Magnitude (p.u.)	Phase Angle (degrees)
1	1.05	0
2	1.045629	5.3326
3	1.071334	16.05058
4	1.014964	−2.543515
5	1.033765	2.473992
6	1.016981	−3.001928

In the polar form all the voltages at the end of the 14th iteration are given in Table E.19.3B.

E.19.4 For the given sample find load flow solution using N-R polar coordinates, decoupled method.

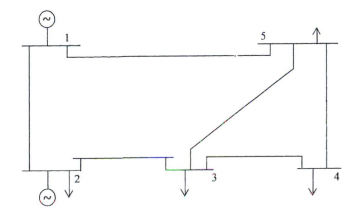

Bus Code	Line Impedance Z_{pq}	Line Charging
1−2	0.02 + j0.24	j0.02
2−3	0.04 + j0.02	j0.02
3−5	0.15 + j0.04	j0.025
3−4	0.02 + j0.06	j0.01
4−5	0.02 + j0.04	j0.01
5−1	0.08 + j0.02	j0.2

Bus Code (Slack)	Generation		Load	
	Mw	Mvar	MW	Mvar
1	0	0	0	0
2	50	25	15	10
3	0	0	45	20
4	0	0	40	15
5	0	0	50	25

	1	2	3	4	5
1	11.724 − j24.27	−10 + j20	0 + j0	0 + j0	−1.724 + j4.31
2	−10 + j20	10.962 − j24.768	−0.962 + j4.808	0 + j0	0 + j0
3	0 + j0	−0.962 + j4.808	6.783 − j21.944	−5 + j15	−0.822 + j2.192
4	0 + j0	0 + j0	−5 + j15	15 − j34.98	−10 + j20
5	−1.724 + j4.31	0 + j0	−0.82 + j2.192	−10 + j20	12.546 − j26.447

FIGURE E.19.4

Bus admittance matrix.

Solution:

The residual or mismatch vector for Iteration 1 is as follows:

$dp[2] = 0.04944$
$dp[3] = -0.041583$
$dp[4] = -0.067349$
$dp[5] = -0.047486$
$dQ[2] = -0.038605$
$dQ[3] = -0.046259$
$dQ[4] = -0.003703$
$dQ[5] = -0.058334$

The new voltage vector after Iteration 1 is as follows:

Bus no 1 E: 1.000000 F: 0.000000
Bus no 2 E: 1.984591 F: −0.008285
Bus no 3 E: 0.882096 F: −0.142226
Bus no 4 E: 0.86991 F: −0.153423
Bus no 5 E: 0.875810 F: −0.142707

The residual or mismatch vector for Iteration 2 is as follows:

$dp[2] = 0.002406$
$dp[3] = -0.001177$
$dp[4] = -0.004219$
$dp[5] = -0.000953$
$dQ[2] = -0.001087$
$dQ[3] = -0.002261$
$dQ[4] = -0.000502$
$dQ[5] = -0.002888$

The new voltage vector after Iteration 2 is as follows:

Bus no 1 E: 1.000000 F: 0.000000
Bus no 2 E: 0.984357 F: −0.008219
Bus no 3 E: 0.880951 F: −0.142953

Bus no 4 E: 0.868709 F: -0.154322
Bus no 5 E: 0.874651 F: -0.143439

The residual or mismatch vector for Iteration 3 is as follows:

$dp[2] = 0.000005$
$dp[3] = -0.000001$
$dp[4] = -0.000013$
$dp[5] = -0.000001$
$dQ[2] = -0.000002$
$dQ[3] = -0.000005$
$dQ[4] = -0.000003$
$dQ[5] = -0.000007$

The final load flow solution (for allowable error 0.0001):

Bus no 1 Slack $P = 1.089093$ $Q = 0.556063$ $E = 1.000000$ $F = 0.000000$
Bus no 2 pq $P = 0.349995$ $Q = 0.150002$ $E = 0.984357$ $F = -0.008219$
Bus no 3 pq $P = -0.449999$ $Q = -0.199995$ $E = 0.880951$ $F = -0.1429531$
Bus no 4 pq $P = -0.399987$ $Q = -0.150003$ $E = 0.868709$ $F = -0.154322$
Bus no 5 pq $P = -0.500001$ $Q = -0.249993$ $E = 0.874651$ $F = -0.143439$

Decoupled load flow solution (polar coordinate method)
The residual or mismatch vector for Iteration 0 is as follows:

$dp[2] = 0.350000$
$dp[3] = -0.450000$
$dp[4] = -0.400000$
$dp[5] = -0.500000$
$dQ[2] = -0.190000$
$dQ[3] = -0.145000$
$dQ[4] = -0.130000$
$dQ[5] = -0.195000$

The new voltage vector after Iteration 0 is as follows:

Bus no 1 E: 1.000000 F: 0.000000
Bus no 2 E: 0.997385 F: -0.014700
Bus no 3 E: 0.947017 F: -0.148655
Bus no 4 E: 0.941403 F: -0.161282
Bus no 5 E: 0.943803 F: -0.150753

The residual or mismatch vector for Iteration 1 is

$dp[2] = 0.005323$
$dp[3] = -0.008207$
$dp[4] = -0.004139$
$dp[5] = -0.019702$
$dQ[2] = -0.067713$

dQ[3] = −0.112987
dQ[4] = −0.159696
dQ[5] = −0.210557

The new voltage vector after Iteration 1 is as follows:

Bus no 1 E: 1.000000 F: 0.000000
Bus no 2 E: 0.982082 F: −0.013556
Bus no 3 E: 0.882750 F: −0.143760
Bus no 4 E: 0.870666 F: −0.154900
Bus no 5 E: 0.876161 F: −0.143484

The residual or mismatch vector for Iteration 2 is as follows:

dp[2] = 0.149314
dp[3] = −0.017905
dp[4] = −0.002305
dp[5] = −0.006964
dQ[2] = −0.009525
dQ[3] = −0.009927
dQ[4] = −0.012938
dQ[5] = 0.007721

The new voltage vector after Iteration 2 is as follows:

Bus no 1 E: 1.000000 F: 0.000000
Bus no 2 E: 0.981985 F: −0.007091
Bus no 3 E: 0.880269 F: −0.142767
Bus no 4 E: 0.868132 F: −0.154172
Bus no 5 E: 0.874339 F: −0.143109

The residual or mismatch vector for Iteration 3 is as follows:

dp[2] = 0.000138
dp[3] = 0.001304
dp[4] = 0.004522
dp[5] = −0.006315
dQ[2] = 0.066286
dQ[3] = 0.006182
dQ[4] = −0.001652
dQ[5] = −0.002233

The new voltage vector after Iteration 3 is as follows:

Bus no 1 E: 1.000000 F: 0.000000
Bus no 2 E: 0.984866 F: −0.007075
Bus no 3 E: 0.881111 F: −0.142710
Bus no 4 E: 0.868848 F: −0.154159
Bus no 5 E: 0.874862 F: −0.143429

The residual or mismatch vector for Iteration 4 is as follows:

$dp[2] = -0.031844$
$dp[3] = 0.002894$
$dp[4] = -0.000570$
$dp[5] = 0.001807$
$dQ[2] = -0.000046$
$dQ[3] = 0.000463$
$dQ[4] = 0.002409$
$dQ[5] = -0.003361$

The new voltage vector after Iteration 4 is as follows:

Bus no 1 E: 1.000000 F: 0.000000
Bus no 2 E: 0.984866 F: -0.008460
Bus no 3 E: 0.881121 F: -0.142985
Bus no 4 E: 0.868849 F: -0.1546330
Bus no 5 E: 0.874717 F: -0.143484

The residual or mismatch vector for Iteration 5 is as follows:

$dp[2] = 0.006789$
$dp[3] = -0.000528$
$dp[4] = -0.000217$
$dp[5] = -0.0000561$
$dQ[2] = -0.000059$
$dQ[3] = -0.000059$
$dQ[4] = -0.000635$
$dQ[5] = -0.000721$

The new voltage vector after Iteration 5 is as follows:

Bus no 1 E: 1.000000 F: 0.000000
Bus no 2 E: 0.984246 F: -0.008169
Bus no 3 E: 0.880907 F: -0.142947
Bus no 4 E: 0.868671 F: -0.154323
Bus no 5 E: 0.874633 F: -0.143431

The residual or mismatch vector for Iteration 6 is as follows:

$dp[2] = 0.000056$
$dp[3] = 0.000010$
$dp[4] = 0.000305$
$dp[5] = -0.000320$
$dQ[2] = 0.003032$
$dQ[3] = -0.000186$
$dQ[4] = -0.000160$
$dQ[5] = -0.000267$

The new voltage vector after Iteration 6 is as follows:

Bus no 1 E: 1.000000 F: 0.000000
Bus no 2 E: 0.984379 F: −0.008165
Bus no 3 E: 0.880954 F: −0.142941
Bus no 4 E: 0.868710 F: −0.154314
Bus no 5 E: 0.874655 F: −0.143441

The residual or mismatch vector for Iteration 7 is as follows:

$dp[2] = -0.001466$
$dp[3] = 0.000106$
$dp[4] = -0.000073$
$dp[5] = 0.000156$
$dQ[2] = 0.000033$
$dQ[3] = 0.000005$
$dQ[4] = 0.000152$
$dQ[5] = -0.000166$

The new voltage vector after iteration 7 is as follows:

Bus no 1 E: 1.000000 F: 0.000000
Bus no 2 E: 0.954381 F: −0.008230
Bus no 3 E: 0.880958 F: −0.142957
Bus no 4 E: 0.868714 F: −0.154325
Bus no 5 E: 0.874651 F: −0.143442

The residual or mismatch vector for Iteration 8 is as follows:

$dp[2] = -0.000022$
$dp[3] = 0.000001$
$dp[4] = -0.000072$
$dp[5] = -0.000074$
$dQ[2] = -0.000656$
$dQ[3] = 0.000037$
$dQ[4] = -0.000048$
$dQ[5] = -0.000074$

The new voltage vector after Iteration 8 is as follows:

Bus no 1 E: 1.000000 F: 0.000000
Bus no 2 E: 0.984352 F: −0.008231
Bus no 3 E: 0.880947 F: −0.142958
Bus no 4 E: 0.868706 F: −0.154327
Bus no 5 E: 0.874647 F: −0.143440

The residual or mismatch vector for Iteration 9 is as follows:

$dp[2] = 0.000318$
$dp[3] = -0.000022$
$dp[4] = 0.000023$
$dp[5] = -0.000041$
$dQ[2] = -0.000012$
$dQ[3] = -0.000000$
$dQ[4] = 0.000036$
$dQ[5] = -0.000038$

The new voltage vector after Iteration 9 is as follows:

Bus no 1 E: 1.000000 F: 0.000000
Bus no 2 E: 0.984352 F: -0.008217
Bus no 3 E: 0.880946 F: -0.142954
Bus no 4 E: 0.868705 F: -0.154324
Bus no 5 E: 0.874648 F: -0.143440

The residual or mismatch vector for Iteration 10 is as follows:

$dp[2] = 0.000001$
$dp[3] = -0.000001$
$dp[4] = 0.000017$
$dp[5] = -0.000017$
$dQ[2] = 0.000143$
$dQ[3] = -0.000008$
$dQ[4] = 0.000014$
$dQ[5] = -0.000020$

The new voltage vector after Iteration 10 is as follows:

Bus no 1 E: 1.000000 F: 0.000000
Bus no 2 E: 0.984658 F: -0.008216
Bus no 3 E: 0.880949 F: -0.142954
Bus no 4 E: 0.868707 F: -0.154324
Bus no 5 E: 0.874648 F: -0.143440

The residual or mismatch vector for Iteration 11 is as follows:

$dp[2] = -0.000069$
$dp[3] = 0.000005$
$dp[4] = -0.000006$
$dp[5] = 0.000011$
$dQ[2] = 0.000004$
$dQ[3] = -0.000000$
$dQ[4] = 0.000008$
$dQ[5] = -0.000009$

The final load flow solution after 11 iterations (for allowable error 0.0001) is as follows:

Bus no 1 Slack $P = 1.089043$	$Q = 0.556088$	$E = 1.000000$	$F = 0.000000$
Bus no 2 pq $P = 0.350069$	$Q = 0.150002$	$E = 0.984658$	$F = -0.008216$
Bus no 3 pq $P = -0.450005$	$Q = -0.199995$	$E = 0.880949$	$F = -0.142954$
Bus no 4 pq $P = -0.399994$	$Q = -0.150003$	$E = 0.868707$	$F = -0.154324$
Bus no 5 pq $P = -0.500011$	$Q = -0.249991$	$E = 0.874648$	$F = -0.143440$

PROBLEMS

P.19.1 Obtain a load flow solution for the system shown in Fig. P.19.1 use
1. Gauss−Seidel method
2. N-R polar coordinator method

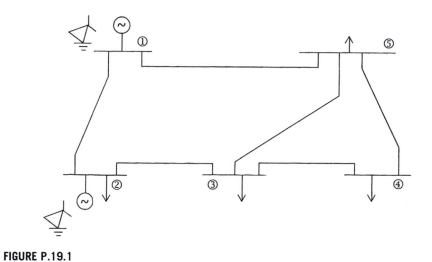

FIGURE P.19.1

Bus code $p-q$	Impedance Z_{pq}	Line charges Y pq/s
1−2	0.02 + j0.2	0.0
2−3	0.01 + j0.025	0.0
3−4	0.02 + j0.4	0.0
3−5	0.02 + 0.05	0.0
4−5	0.015 + j0.04	0.0
1−5	0.015 + j0.04	0.0

Values are given in p.u. on a base of 100 MVA.

The scheduled powers are as follows:

Bus Code (P)	Generation		Load	
	MW	Mvar	MW	Mvar
1 (slack bus)	0	0	0	0
2	80	35	25	15
3	0	0	0	0
4	0	0	45	15
5	0	0	55	20

Take voltage at Bus 1 as $1 \angle 0$ degree p.u.

P.19.2 Repeat problem P.19.1 with line charging capacitance $Y_{pq}/2 = j0.025$ for each line.

P.19.3 Obtain the decoupled and fast decoupled load flow solution for the system in P.19.1 and compare the results with the exact solution.

P.19.4 For the 51-bus system shown in Fig. P.19.4, the system data is given as follows in p.u. Perform load flow analysis for the system

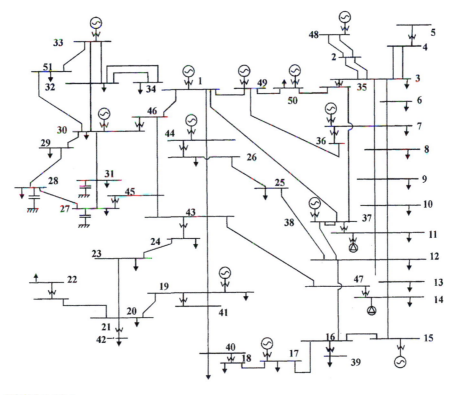

FIGURE P.19.4

51-Bus power system.

Line Data	Resistance	Reactance	Capacitance
2−3	0.0287	0.0747	0.0322
3−4	0.0028	0.0036	0.0015
3−6	0.0614	0.1400	0.0558
3−7	0.0247	0.0560	0.0397
7−8	0.0098	0.0224	0.0091
8−9	0.0190	0.0431	0.0174
9−10	0.0182	0.0413	0.0167
10−11	0.0205	0.0468	0.0190
11−12	0.0660	0.0150	0.0060
12−13	0.0455	0.0642	0.0058
13−14	0.1182	0.2360	0.0213
14−15	0.0214	0.2743	0.0267
15−16	0.1336	0.0525	0.0059
16−17	0.0580	0.3532	0.0367
17−18	0.1550	0.1532	0.0168
18−19	0.1550	0.3639	0.0350
19−20	0.1640	0.3815	0.0371
20−21	0.1136	0.3060	0.0300
20−23	0.0781	0.2000	0.0210
23−24	0.1033	0.2606	0.0282
12−25	0.0866	0.2847	0.0283
25−26	0.0159	0.0508	0.0060
26−27	0.0872	0.2870	0.0296
27−28	0.0136	0.0436	0.0045
28−29	0.0136	0.0436	0.0045
29−30	0.0125	0.0400	0.0041
30−31	0.0136	0.0436	0.0045
27−31	0.0136	0.0436	0.0045
30−32	0.0533	0.1636	0.0712
32−33	0.0311	0.1000	0.0420
32−34	0.0471	0.1511	0.0650
30−51	0.0667	0.1765	0.0734
51−33	0.0230	0.0622	0.0256
35−50	0.0240	0.1326	0.0954
35−36	0.0266	0.1418	0.1146
39−49	0.0168	0.0899	0.0726
36−38	0.0252	0.1336	0.1078
38−1	0.0200	0.1107	0.0794
38−47	0.0202	0.1076	0.0869
47−43	0.0250	0.1336	0.1078

Continued			
Line Data	**Resistance**	**Reactance**	**Capacitance**
42–43	0.0298	0.1584	0.1281
40–41	0.0254	0.1400	0.1008
41–43	0.0326	0.1807	0.1297
43–45	0.0236	0.1252	0.1011
43–44	0.0129	0.0715	0.0513
45–46	0.0054	0.0292	0.0236
44–1	0.0330	0.1818	0.1306
46–1	0.0343	0.2087	0.1686
1–49	0.0110	0.0597	0.1752
49–50	0.0071	0.0400	0.0272
37–38	0.0014	0.0077	0.0246
47–39	0.0203	0.1093	0.0879
48–2	0.0426	0.1100	0.0460
3–35	0.0000	0.0500	0.0000
7–36	0.0000	0.0450	0.0000
11–37	0.0000	0.0500	0.0000
14–47	0.0000	0.0900	0.0000
16–39	0.0000	0.0900	0.0000
18–40	0.0000	0.0400	0.0000
20–42	0.0000	0.0800	0.0000
24–43	0.0000	0.0900	0.0000
27–45	0.0000	0.0900	0.0000
26–44	0.0000	0.0500	0.0000
30–46	0.0000	0.0450	0.0000
1–34	0.0000	0.0630	0.0000
21–2	0.0000	0.2500	0.0000
4–5	0.0000	0.2085	0.0000
19–41	0.0000	0.0800	0.0000

P.19.5 The data for a 13-machine, 71-bus, 94-line system is given. Obtain the load flow solution.

Data:

No. of buses	71
No. of lines	94
Base power (MVA)	200
No. of machines	13
No. of shunt loads	23

Bus No	Generation		Load	Power
1	–	–	0.0	0.0
2	0.0	0.0	0.0	0.0
3	506.0	150.0	0.0	0.0
4	0.0	0.0	0.0	0.0
5	0.0	0.0	0.0	0.0
6	100.0	32.0	0.0	0.0
7	0.0	0.0	12.8	8.3
8	300.0	125.0	0.0	0.0
9	0.0	0.0	185.0	130.0
10	0.0	0.0	80.0	50.0
11	0.0	0.0	155.0	96.0
12	0.0	0.0	0.0	0.0
13	0.0	0.0	100.0	62.0
14	0.0	0.0	0.0	0.0
15	180.0	110.0	0.0	0.0
16	0.0	0.0	73.0	45.5
17	0.0	0.0	36.0	22.4
18	0.0	0.0	16.0	9.0
19	0.0	0.0	32.0	19.8
20	0.0	0.0	27.0	16.8
21	0.0	0.0	32.0	19.8
22	0.0	0.0	0.0	0.0
23	0.0	0.0	75.0	46.6
24	0.0	0.0	0.0	0.0
25	0.0	0.0	133.0	82.5
26	0.0	0.0	0.0	0.0
27	300.0	75.0	0.0	0.0
28	0.0	0.0	30.0	20.0
29	260.0	70.0	0.0	0.0
30	0.0	0.0	120.0	0.0
31	0.0	0.0	160.0	74.5
32	0.0	0.0	0.0	99.4
33	0.0	0.0	0.0	0.0
34	0.0	0.0	112.0	69.5
35	0.0	0.0	0.0	0.0
36	0.0	0.0	50.0	32.0
37	0.0	0.0	147.0	92.0
38	0.0	0.0	93.5	88.0
39	25.0	30.0	0.0	0.0
40	0.0	0.0	0.0	0.0
41	0.0	0.0	225.0	123.0
42	0.0	0.0	0.0	0.0

Continued				
Bus No	**Generation**		**Load**	**Power**
43	0.0	0.0	0.0	0.0
44	180.0	55.0	0.0	0.0
45	0.0	0.0	0.0	0.0
46	0.0	0.0	78.0	38.6
47	0.0	0.0	234.0	145.0
48	340.0	250.0	0.0	0.0
49	0.0	0.0	295.0	183.0
50	0.0	0.0	40.0	24.6
51	0.0	0.0	227.0	142.0
52	0.0	0.0	0.0	0.0
53	0.0	0.0	0.0	0.0
54	0.0	0.0	108.0	68.0
55	0.0	0.0	25.5	48.0
56	0.0	0.0	0.0	0.0
57	0.0	0.0	55.6	35.6
58	0.0	0.0	42.0	27.0
59	0.0	0.0	57.0	27.4
60	0.0	0.0	0.0	0.0
61	0.0	0.0	0.0	0.0
62	0.0	0.0	40.0	27.0
63	0.0	0.0	33.2	20.6
64	300.0	75.0	0.0	0.0
65	0.0	0.0	0.0	0.0
66	96.0	25.0	0.0	0.0
67	0.0	0.0	14.0	6.5
68	90.0	25.0	0.0	0.0
69	0.0	0.0	0.0	0.0
70	0.0	0.0	11.4	7.0
71	0.0	0.0	0.0	0.0

Line Data

Line No	From Bus	To Bus	Line Impedance		1/2 Y charge	Turns Ratio
1	9	8	0.0000	0.0570	0.0000	1.05
2	9	7	0.3200	0.0780	0.0090	1.00
3	9	5	0.0660	0.1600	0.0047	1.00
4	9	10	0.0520	0.1270	0.0140	1.00
5	10	11	0.0660	0.1610	0.0180	1.00
6	7	10	0.2700	0.0700	0.0070	1.00
7	12	11	0.0000	0.0530	0.0000	0.95
8	11	13	0.0600	0.1480	0.0300	1.00

(Continued)

Continued

Line No	From Bus	To Bus	Line Impedance		1/2 Y charge	Turns Ratio
9	14	13	0.0000	0.0800	0.0000	1.00
10	13	16	0.9700	0.2380	0.0270	1.00
11	17	15	0.0000	0.0920	0.0000	1.05
12	7	6	0.0000	0.2220	0.0000	1.05
13	7	4	0.0000	0.0800	0.0000	1.00
14	4	3	0.0000	0.0330	0.0000	1.05
15	4	5	0.0000	0.1600	0.0000	1.00
16	4	12	0.0160	0.0790	0.0710	1.00
17	12	14	0.0160	0.0790	0.0710	1.00
18	17	16	0.0000	0.0800	0.0000	0.95
19	2	4	0.0000	0.0620	0.0000	1.00
20	4	26	0.0190	0.0950	0.1930	0.00
21	2	1	0.0000	0.0340	0.0000	1.05
22	31	26	0.0340	0.1670	0.1500	1.00
23	26	25	0.0000	0.0800	0.0000	0.95
24	25	23	0.2400	0.5200	0.1300	1.00
25	22	23	0.0000	0.0800	0.0000	0.95
26	24	22	0.0000	0.0840	0.0000	0.95
27	22	17	0.0480	0.2500	0.0505	1.00
28	2	24	0.0100	0.1020	0.3353	1.00
29	23	21	0.0366	0.1412	0.0140	1.00
30	21	20	0.7200	0.1860	0.0050	1.00
31	20	19	0.1460	0.3740	0.0100	1.00
32	19	18	0.0590	0.1500	0.0040	1.00
33	18	16	0.0300	0.0755	0.0080	1.00
34	28	27	0.0000	0.0810	0.0000	1.05
35	30	29	0.0000	0.0610	0.0000	1.05
36	32	31	0.0000	0.0930	0.0000	0.95
37	31	30	0.0000	0.0800	0.0000	0.95
38	28	32	0.0051	0.0510	0.6706	1.00
39	3	33	0.0130	0.0640	0.0580	1.00
40	31	47	0.0110	0.0790	0.1770	1.00
41	2	32	0.0158	0.1570	0.5100	1.00
42	33	34	0.0000	0.0800	0.0000	0.95
43	35	33	0.0000	0.0840	0.0000	0.95
44	35	24	0.0062	0.0612	0.2120	1.00
45	34	36	0.0790	0.2010	0.0220	1.00
46	36	37	0.1690	0.4310	0.0110	1.00
47	37	38	0.0840	0.1880	0.0210	1.00
48	40	39	0.0000	0.3800	0.0000	1.05

Line No	From Bus	To Bus	Line Impedance		1/2 Y charge	Turns Ratio
49	40	38	0.0890	0.2170	0.0250	1.00
50	38	41	0.1090	0.1960	0.2200	1.00
51	41	51	0.2350	0.6000	0.0160	1.00
52	42	41	0.0000	0.0530	0.0000	0.95
53	45	42	0.0000	0.0840	0.0000	0.95
54	47	49	0.2100	0.1030	0.9200	1.00
55	49	48	0.0000	0.0460	0.0000	1.05
56	49	50	0.0170	0.0840	0.0760	1.00
57	49	42	0.0370	0.1950	0.0390	1.00
58	50	51	0.0000	0.0530	0.0000	0.95
59	52	50	0.0000	0.0840	0.0000	0.95
60	50	55	0.0290	0.1520	0.0300	1.00
61	50	53	0.0100	0.0520	0.0390	1.00
62	53	54	0.0000	0.0800	0.0000	0.95
63	57	54	0.0220	0.0540	0.0060	1.00
64	55	56	0.0160	0.0850	0.0170	1.00
65	56	57	0.0000	0.0800	0.0000	1.00
66	57	59	0.0280	0.0720	0.0070	1.00
67	59	58	0.0480	0.1240	0.0120	1.00
68	60	59	0.0000	0.0800	0.0000	1.00
69	53	60	0.0360	0.1840	0.3700	1.00
70	45	44	0.0000	0.1200	0.0000	1.05
71	45	46	0.0370	0.0900	0.0100	1.00
72	46	41	0.0830	0.1540	0.0170	1.00
73	46	59	0.1070	0.1970	0.0210	1.00
74	60	61	0.0160	0.0830	0.0160	1.00
75	61	62	0.0000	0.0800	0.0000	0.95
76	58	62	0.0420	0.1080	0.0020	1.00
77	62	63	0.0350	0.0890	0.0090	1.00
78	69	68	0.0000	0.2220	0.0000	1.05
79	69	61	0.0230	0.1160	0.1040	1.00
80	67	66	0.0000	0.1880	0.0000	1.05
81	65	64	0.0000	0.0630	0.0000	1.05
82	65	56	0.0280	0.1440	0.0290	1.00
83	65	61	0.0230	0.1140	0.0240	1.00
84	65	67	0.0240	0.0600	0.0950	1.00
85	67	63	0.0390	0.0990	0.0100	1.00
86	61	42	0.0230	0.2293	0.0695	1.00
87	57	67	0.0550	0.2910	0.0070	1.00
88	45	70	0.1840	0.4680	0.0120	1.00
89	70	38	0.1650	0.4220	0.0110	1.00

(Continued)

Continued

Line No	From Bus	To Bus	Line Impedance		1/2 Y charge	Turns Ratio
90	33	71	0.0570	0.2960	0.0590	1.00
91	71	37	0.0000	0.0800	0.0000	0.95
92	45	41	0.1530	0.3880	0.1000	1.00
93	35	43	0.0131	0.1306	0.4293	1.00
94	52	52	0.0164	0.1632	0.5360	1.00

Shunt Load Data

S. No	Bus No	Shunt	Load
1	2	0.00	−0.4275
2	13	0.00	0.1500
3	20	0.00	0.0800
4	24	0.00	−0.2700
5	28	0.00	−0.3375
6	31	0.00	0.2000
7	32	0.00	−0.8700
8	34	0.00	0.2250
9	35	0.00	−0.3220
10	36	0.00	0.1000
11	37	0.00	0.3500
12	38	0.00	0.2000
13	41	0.00	0.2000
14	43	0.00	−0.2170
15	46	0.00	0.1000
16	47	0.00	0.3000
17	50	0.00	0.1000
18	51	0.00	0.1750
19	52	0.00	−0.2700
20	54	0.00	0.1500
21	57	0.00	0.1000
22	59	0.00	0.0750
23	21	0.00	0.0500

QUESTIONS

Q.19.1 Explain why load flow studies are performed?

Q.19.2 Discuss the classification of buses.

Q.19.3 With a neat flowchart explain the load flow solution by Gauss−Seidel method.

Q.19.4 Explain the principle and method of solution of the load flow problem by Newton–Raphson: (1) rectangular coordinates and (2) polar coordinates methods.

Q.19.5 Compare Gauss–Seidel method and Newton–Raphson method for load flow solution.

    1. Explain

        a. Decoupled load flow

        b. Fast-decoupled load flow

    2. What are the applications of methods mentioned in 1 above.

ECONOMIC OPERATION OF POWER SYSTEMS

20

Planning operation and control of interconnected power systems present a variety of challenging problems. An important problem in this area is the economic operation of the system, which means that every step in planning, scheduling, and operation of the system, unit-wise, plant-wise, and inter connection-wise must be optimal, leading to absolute economy. In this the transmission losses too play an important role. In this chapter both thermal and hydro systems will be dealt with using suitable analytical models that result in meaningful savings.

20.1 CHARACTERISTICS OF STEAM PLANTS

In analyzing the economics of operation of thermal systems, modeling of input—output characteristics assumes significance. For this purpose a single unit comprising of boiler, turbine, and generator may be considered. The unit has to supply power to the local needs for the auxiliaries in the station. This later component may be around 2%—5%. The station auxiliaries include boiler feed pumps, condenser circulating water pumps, fans, etc. The total input to the unit could be either kcal/h in terms of heat supplied or Rs/h in terms of the cost of the fuel such as coal, gas, or fossil fuel of any other form. The net output of the unit that is supplied into the system at the generator bus will be in kilowatt or megawatt. Scheduling is the process of allocation of generation among various operating generator units. Economic scheduling is the cost effective mode of generation allocation among the various units. This can also be termed as optimal operation. The analytical solution proposed in general for optimal operation depends on the incremental cost concept (Fig. 20.1).

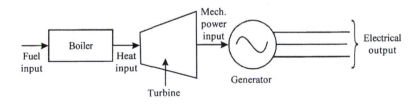

FIGURE 20.1

Boiler turbine-generator unit.

Electrical Power Systems. DOI: http://dx.doi.org/10.1016/B978-0-08-101124-9.00020-6

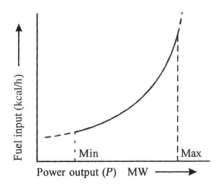

FIGURE 20.2

Thermal unit input—output characteristics.

20.2 INPUT—OUTPUT CURVES

As has been already stated the input—output characteristics for any thermal unit or units that comprise the plant can be obtained from the operating data. The input can be in kilocalories per hour and the output may be in kilowatts or preferably in megawatts. Typical characteristic is shown in Fig. 20.2 for the unit shown in Fig. 20.1.

The characteristic in practice may not be such a smooth idealized curve, and from the practical data, such an idealized curve can be interpolated.

Steam turbine generating unit characteristics may have minimum and maximum limits in operation.

They may be determined by factors such as steam cycle used, operating temperatures, and material thermal characteristics.

20.3 THE INCREMENTAL HEAT RATE CHARACTERISTICS

From the input—output characteristics, the incremental heat rate characteristic can be obtained which is the ratio of the differentials.

$$\text{Incremental fuel rate or heat rate} = \frac{d\,(\text{input})}{d\,(\text{output})} \tag{20.1}$$

By calculating the slope of the characteristic in Fig. 20.1 at every point the incremental fuel rate characteristic can be plotted as shown in Fig. 20.3. This characteristic in fact tells about the thermal efficacy of the unit under consideration that can be used for comparison with other units in performance.

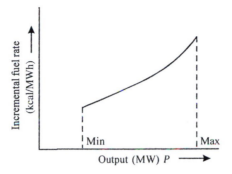

FIGURE 20.3

Incremental fuel rate characteristic for thermal unit.

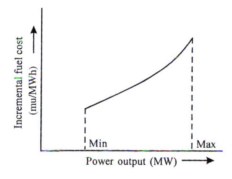

FIGURE 20.4

Incremental fuel cost characteristic for thermal unit.

20.4 **THE INCREMENTAL FUEL COST CHARACTERISTIC**

The incremental fuel rate in kcal/kWh can be multiplied by cost of the fuel in terms of mu/kcal. In any case the ordinates are in mu/MWh. or mu/MWh. The calorific value of the fuel is required in these calculations. This characteristic is shown in Fig. 20.4. Here, mu represents monetory units such as dollar, euro, and rupee.

20.5 **HEAT RATE CHARACTERISTIC**

Sometimes the unit net heat rate characteristic is also considered important. To obtain this characteristic the net heat rate in kcal/kWh is plotted against the power output (Fig. 20.5).

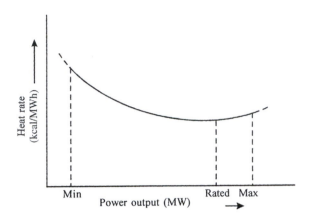

FIGURE 20.5

Heat rate characteristic.

The thermal efficiency of the unit is influenced by factors like steam condition, reheat stages, condenser pressure, and the steam cycle used. The efficiency of the units in practice is around 30%.

20.6 INCREMENTAL PRODUCTION COST CHARACTERISTICS

The production cost of the power generated actually depends on several items such as fuel cost, labor charges, cost of items such as oil, water, and other supplies needed, and also the cost of maintenance. It is well known that in thermal generation the fuel cost is by far the largest cost head and is directly related to the power generated. Even the other charges, i.e., the additional running expenses too are, more or less, related to the amount of generation. Thus it is a simple practical proposition to assume all the additional costs as a fixed percentage of the incremental fuel cost.

The sum of incremental fuel cost and other incremental running expenses is called incremental production cost. The ordinates of Fig. 20.4 will also represent incremental production cost to some other scale. An explicit mathematical relationship involving all the factors involved in power generation to the total power generated is infact a very difficult task.

In general the input—output data is fitted into a quadratic characteristic even though it is also possible to fit into any polynomial curve for ease of mathematical manipulation.

Large turbine-generator units may have several steam admission valves which are opened in a sequence to meet the increasing steam demand. The input—output characteristic for such a unit with two valves may show discontinuity as shown in Fig. 20.6.

20.7 CHARACTERISTICS OF HYDROPLANTS

The input—output characteristics for hydro units can be obtained in the same way as for thermal units on the assumption of constant water head. The input—output characteristics may be as

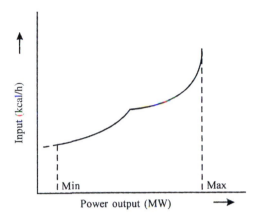

FIGURE 20.6

Input—output characteristic for multiple admission valves (for two valves).

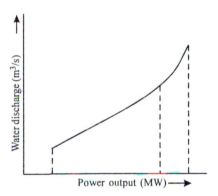

FIGURE 20.7

Input—output characteristic for hydro units.

shown in Fig. 20.7. The ordinates are water input or discharge in cubic meters per second shown against power output in megawatts. While the water requirement is nearly linear till rated load, after that the efficiency decreases and greater discharge is required to meet the increased load demand.

It may be noted that if the head varies the input—output characteristics change. It will move vertically upward, as head falls and vice versa since the hydropower generated is directly related to the head of the water level and as head falls, higher water discharge is required for the same power generation. Similarly as head rises lesser discharge is needed. The characteristic moves downward. This is shown in Fig. 20.8.

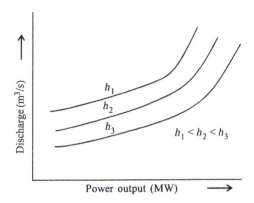

FIGURE 20.8

Effect of head on discharge for hydro units.

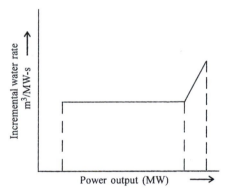

FIGURE 20.9

Incremental water rate characteristic.

20.8 INCREMENTAL WATER RATE CHARACTERISTICS

The incremental water rate curve is obtained from the water input–output characteristic in the same way as for thermal units. A typical characteristic is shown in Fig. 20.9.

As the input–output curve is linear for a greater part, the incremental water rate characteristic is a horizontal line over this region indicating constant slope, and thereafter it rises rapidly. With increase in load, more and more units will have to be brought into service.

There will be discontinuity in the characteristics in such a case, as shown in Fig. 20.10. The discontinuity generally can be neglected so that the characteristic will still have the same shape as in Fig. 20.9.

The conversion of incremental water rate into incremental production cost requires considerations of agriculture, navigation, drinking needs of water, and other riparian rights, etc., even though

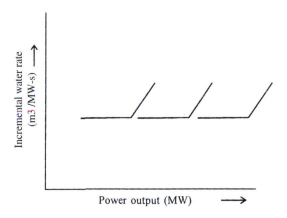

FIGURE 20.10

Multiple unit operation incremental water rate characteristic.

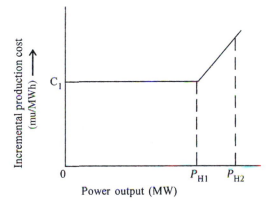

FIGURE 20.11

Incremental production cost representation.

water is available freely in nature. Further the cost of dams, canals, conduits, gates, penstocks, and other parts of hydro development are also involved.

20.9 INCREMENTAL PRODUCTION COST CHARACTERISTIC

The incremental water rate characteristic can be converted into incremental production cost characteristic by multiplying the incremental water rate characteristic by water rate or cost of water monetary units (mu) per cubic meter C_w. The incremental production cost characteristic is shown in Fig. 20.11.

The incremental production cost characteristic can be expressed analytically as follows:

$$\text{IPC} = C_1 \quad (0 \le P \le P_{H1}) \tag{20.2}$$

$$\text{IPC} = aP_H + C_1 \quad (P_{H1} \le P \le P_{H2}) \tag{20.3}$$

where a is the slope of the characteristic between P_{H1} and P_{H2}.

20.10 GENERATING COSTS AT THERMAL PLANTS

Consider any ith plant among n thermal plants supplying active power. C_i is the cost per unit of P_i in the neighborhood of P_i then the generating cost is

$$F_i = C_i P_i \tag{20.4}$$

The total cost of operating a system with n_g generating sets can be represented by

$$F = \sum_{i=1}^{N_g} C_i P_i \tag{20.5}$$

If the system operates over a time period T, then the total expenditure involved will be

$$F_T = \int_0^T \sum_{i=1}^{N_g} C_i P_i(t) \, \mathrm{d}t \tag{20.6}$$

Steam plants with partial admission nozzle governing give better performance at partial loads since the cost coefficient increases with increasing megawatt loading. However, such units cannot be shutdown, frequently, because of the complexities of steam chest. But, steam plants with throttle governing are more suitable for periodic shutdown due to their simpler steam chest. Such units are more suitable for rapid starting and loading. From minimum to maximum permissible limits of operation, the cost coefficients of the units are substantially constant.

20.11 ANALYTICAL FORM FOR INPUT–OUTPUT CHARACTERISTICS OF THERMAL UNITS

It is already pointed out that the input–output characteristic may be fitted into a polynomial of the form

$$\text{Cost } F = A + BP + CP^2 + \cdots \tag{20.7}$$

where $A, B, C, \ldots$ are constants.

If a quadratic representation is made, then denoting the cost

$$F = \frac{1}{2}aP^2 + bP + C \tag{20.8}$$

will lead to an incremental cost characteristic of the form

$$\frac{dF}{dP} = aP + b \tag{20.9}$$

a linear relationship around any operating point P.

Prices of electrical energy vary from country to country. In the United States, the prices change from state to state. The prices are also dependent on a number of factors such as generation services, transmission services, distribution services, and customer services. There could be basic hourly pricing block and index pricing by competitive retail providers. It is also a new customary to talk about location-based marginal price. They determine the two-part tariff but topically the application of linear incremental costs is valid on the generation side of the power system.

Levelized cost of electricity, where all sources are equally treated by tax code is defined an

$$\text{LCOE} = \frac{\Sigma \text{ Costs over lifetime}}{\Sigma \text{ Electrical energy produced over lifetime}}$$

The electrical energy charges in some countries are given in the following to give an idea.

France	Hydro energy	20 MWh^{-1}
	Nuclear energy	50 MWh^{-1}
Germany	Coal based	$38–80 \text{ MWh}^{-1}$
United Kingdom	Natural gas	$55–130 \text{ MWh}^{-1}$
Australia	Coal based	A 38 MWh^{-1}
United States	Conventional average	37.9 MWh^{-1}
	Conventional coal	

20.12 CONSTRAINTS IN OPERATION

The power system has to satisfy several constraints while in operation. These may be broadly divided into two types. The first of these arises out of the necessity for the system to satisfy load balance and are called equality constraints. At any bus i, P_{S_i} and Q_{S_i} correspond to the scheduled generation, P_{D_i} and Q_{D_i} are the load demands then the following equations must be satisfied at the bus i:

$$\begin{aligned} P_{S_i} - P_{D_i} - P_i = G_i = 0 \\ Q_{S_i} - Q_{D_i} - Q_i = H_i = 0 \end{aligned} \tag{20.10}$$

P_i and Q_i are given by

$$P_i = \sum_{j=1}^{n} V_i V_j Y_{ij} \cos (\delta_i - \delta_j - \theta_{ij}) \tag{20.11}$$

$$Q_i = \sum_{j=1}^{n} V_i V_j Y_{ij} \sin(\delta_i - \delta_j - \theta_{ij}), \quad i = 1, 2 \dots, n \tag{20.12}$$

with usual notation and G_i and H_i are the residuals at bus i which should become zero at the point of solution.

In addition, a number of other constraints due to physical and operational limitations of the units and components will arise in economic scheduling. These are in the form of inequality constraints.

Each generator in operation will have a minimum and maximum permissible output and the production must be constrained to ensure that

$$P_i^{\min} \le P_i \le P_i^{\max}, \quad i = 1, 2 \dots, n_g \tag{20.13}$$

where n_g is the total number of generator units.

Similarly limits may also have to be considered over the range of reactive power capabilities of the generator units requiring that

$$Q_i^{\min} \le Q_i \le Q_i^{\max}, \quad i = 1, 2 \dots, n_q \tag{20.14}$$

where n_q is the total number of reactive sources in the system.

Further, the constraint

$$P_i^2 + Q_i^2 \le (S_i^{\text{rated}})^2 \tag{20.15}$$

must be satisfied, where S is the MVA capacity of the generating unit for limiting stator heating.

Dynamic limits may also have to be considered when fast changes in generation are envisaged for picking up or for shedding down of loads. These limits put additional constraints of the form.

$$\left|\frac{dP_i(t)}{dt}\right|^{\min} \le \left|\frac{dP_i(t)}{dt}\right| \le \left|\frac{dP_i(t)}{dt}\right|^{\max} \tag{20.16}$$

The maximum and minimum operating conditions for a group of generators within a power station may be different from the respective sum of the maximum and minimum operating levels of turbines that are supplied by a single boiler. The extremes of boiler operating conditions will determine these limits. Thus groups of generators from individual boiler units may have to be subjected to additional constraints of the nature

$$P_{k_g}^{\min} \le \sum_{i \in G} P_{k_i} \le P_{k_g}^{\max}, \quad k = 1, 2, \dots, \text{GR} \tag{20.17}$$

where GR is the total number of generator groups, the outputs of which are to be separately limited.

Spare capacity is required to account for the errors in load prediction, sudden and fast changes in load demand, and the inadvertent loss of scheduled generation. Thus the total generation G available at any time must be in excess of the total anticipated load demand and system losses by an amount not less than a specified minimum spare capacity P_{SP}.

$$G \ge \sum_{i=1}^{n_g} P_i + P_{\text{SP}} \tag{20.18}$$

In a similar manner constraints may be required to be associated with groups of generators, where all plants are not equally operationally suitable for taking up additional load. If TG is the total number of groups, then

$$G_k \ge \sum_{i \in G} P_{k_i} + P_{\text{SG}} \tag{20.19}$$

where $k = 1, 2, \dots, \text{TG}$.

The summation is over the set G over which group constraints are applied.

Thermal considerations may require that the transmission lines be subjected to branch transfer constraints of the type.

$$-S_i^{max} \leq S_{b_i} \leq S_i^{max}, \quad i = 1, 2, \ldots, n_b \tag{20.20}$$

where n_b is the number of branches and S_{b_i} is the branch transfer MVA.

In addition constraints are to be imposed for bus voltage magnitudes and for phase displacements between them for maintaining voltage profile and for limiting overloading, respectively.

Thus we have

$$V_{ij}^{min} \leq V_{ij} \leq V_{ij}^{max}, \quad i = 1, 2 \ldots, n \tag{20.21}$$

$$\delta_{ij}^{min} \leq \delta_{ij} \leq \delta_{ij}^{max}, \quad i = 1, 2 \ldots, n \tag{20.22}$$

where $j = 1, 2, \ldots, m$ and $j \neq i$

where n is the total number of nodes and m is the number of nodes neighboring each node with interconnecting branches.

In case transformer tap positions are to be included for optimization, then the tap positions T_i must lie within the range available, i.e.,

$$T_i^{min} \leq T_i \leq T_i^{max} \tag{20.23}$$

Sometimes phase shifting transformers are made available in the system. If such equipment exists, then constraints of the type

$$PS_i^{min} \leq PS_i \leq PS_i^{max} \tag{20.24}$$

must be reckoned where PS_i is the phase shift obtained from the ith phase shifting transformer.

If power system security is also required to be considered in the formulation for economic operation, then power flows between certain important buses may also have to be considered for the final solution. It may be mentioned that consideration of each and every possible branch for outage will not be a feasible proposition.

20.13 PLANT SCHEDULING METHODS

At the plant level, several operating procedures were adopted in the past leading to efficient operation resulting in economy.

1. Base Loading to Capacity

   The turbo generators are successively loaded to their rated capacities in the order of their efficiencies. That is to say, that the most efficient unit will get greater share in load allocation which is a natural solution to the problem.
2. Base Loading to Most Efficient Load

   In this case the heat rate characteristics are considered and the turbo generator units are successively loaded to their most efficient loads in increasing order of their heat rates.

   In both the above methods, thermodynamic considerations assumed importance and the schedules will not differ from each other much.

3. Proportional Loading to Capacity

A third method that was considered as a thumb rule in the absence of any technical data is to load the generating units in proportion to their rated capacities as stated on the name plates.

20.14 MERIT ORDER METHOD

If the incremental cost characteristics are fairly constant over a wide range in operation, then neglecting the transmission losses and running reserve requirements, it is possible to prepare schedules for load allocation using incremental efficiencies. Merit tables based upon incremental efficiencies are prepared and each unit is loaded to its rated capacity in order of the highest incremental efficiency. Changes in fuel costs, plant cycle efficiencies, plant availabilities, etc., require the merit tables to be revised regularly to reflect these factors. Then, it is possible to look at the tables so prepared and schedule the generation to different units.

20.15 EQUAL INCREMENTAL COST METHOD: TRANSMISSION LOSSES NEGLECTED

Method of Lagrange Multipliers

It is but natural that for a given load to be allocated between several generating units, the most efficient unit identified by incremental cost of production should be the one to get priority. When this is applied repeatedly to all the units, the load allocation will become complete when all of them, which are involved in operation, are all working at the same incremental cost of production.

The above can be proved mathematically as follows:

Consider n_g generating units supplying $P_1, P_2, \ldots, P_{n_g}$ active powers to supply a total load demand P_D.

The objective function for minimization is the total input to the system in rupees per hour.

$$F(P_1, P_2, \ldots, P_{n_g}) = F(P) = \int_{i=1}^{n_g} C_i(P_i) \tag{20.25}$$

where $C_i(P_i)$ is the generation cost for the ith unit and n_g is the total number of generating units.

The equality constraint is given by

$$G(P_1, P_2, \ldots, P_{n_g}) = G(P) = P_D - \sum_{i=1}^{n_g} P_i = 0 \tag{20.26}$$

i.e., total supply = total demand neglecting losses and reserve. Using the method of Lagrange multipliers for equality constraints, the Lagrange function is defined as:

$$L(P, \lambda) = F(P) + \lambda G(P) \tag{20.27}$$

where λ is a Lagrange multiplier.

The necessary conditions are given by

$$\frac{\partial L}{\partial P_i} = 0, \text{and} \tag{20.28}$$

$$\frac{\partial L}{\partial \lambda} = 0 \tag{20.29}$$

since

$$L = \sum_{i=1}^{n_g} C_i(P_i) + \lambda \left(P_D - \sum_{i=1}^{n_g} P_i \right) \tag{20.30}$$

we obtain

$$\frac{\partial L}{\partial P_i} = \frac{\partial}{\partial P_i} C_i(P_i) + \lambda(-1) = 0 \tag{20.31}$$

and

$$\frac{\partial L}{\partial P_i} = \frac{\partial}{\partial P_i} C_i(P_i) + \lambda(-1) = 0 \tag{20.32}$$

The later equation is the load demand constraint only, whereas the former gives

$$\frac{\partial C_i(P_i)}{\partial P_i} = \frac{\partial C_2(P_2)}{\partial P_2} = \cdots = \frac{\partial C_{n_g}(P_{n_g})}{\partial P_{n_g}} = \lambda \tag{20.33}$$

Eq. (20.33) states that at the optimum all the generating stations operate at the same incremental cost for optimum economy and their incremental production cost is equal to the Lagrange multiplier λ at the optimum.

Eq. (20.32) gives

$$(P_1 + P_2 + \cdots + P_{n_g}) = P_D \tag{20.34}$$

The principle underlying the mathematical treatment is that load should be taken up always at the lowest incremental cost. It must be ensured that the generations so determined are within their capacities. Otherwise, the generation has to be kept constant at the capacity limit for that unit and eliminated from further optimum calculations.

It can be seen that in this method at the optimum the incremental cost of production is also the incremental cost of the power received. Eqs. (20.33) and (20.34) can be solved for economic scheduling analytically neglecting the effect of transmission losses.

The size of the power systems increased enormously, with long transmission lines connecting several power generating stations extending over large geographical areas transferring power to several load centers. With this development, it has become necessary to consider not only the incremental fuel costs but also incremental transmission losses incurred in these lines while power is transmitted. Initial attempts in this direction involved the development of a comprehensive formula involving the generating powers and line parameters. The most important work has come from Kirchmayer and others in 1951. Their method is based on a set of coefficients called B-coefficients. Determination of these coefficients is based on several assumptions and is mathematically quite involved requiring transformations.

However, due to the elegance of the B-coefficient formula and simplicity in application of these coefficients are widely used by a number of power companies in the past in the United States and elsewhere for economic scheduling for including the effect of transmission losses.

20.16 TRANSMISSION LOSS FORMULA—B-COEFFICIENTS

An expression for the transmission loss was derived by Kirchmayer with several assumptions made. In his derivation he used Kron's tensorial transformation. However a simplified procedure for the derivation of the B-coefficients will be presented here.

Consider a power system supplying n_1 loads. Let the load currents be i_{L1}, i_{L2}, ..., i_{Ln_1}. These loads are supplied by n_g generators. Let the generator currents be i_{g1}, i_{g2}, ..., i_{gn_g}. This is shown in Fig. 20.12.

Consider a network element K (an interconnected line in the system) carrying current i_K. Let generator 1 alone supply the entire load current I_L where

$$I_L = i_{L1} + i_{L2} + \cdots + i_{Ln_1}$$
$$= \int_{j=1}^{n_1} i_{Lj} \qquad (20.35)$$

under this condition let the current in K be i_{K1}

In a similar manner if each of the n_g generators operating alone also supply the total load current I_L while the rest of the generators are disconnected the current carried by the network element K changes from i_{K1} to i_{K2}, i_{K3} to i_{Kn_g}.

Let the ratio of i_{K1} to i_L be d_{K1}

$$d_{K1} = \frac{i_{K1}}{I_L} \qquad (20.36)$$

Also

$$d_{K2} = \frac{i_{K2}}{I_L}$$

and so on.

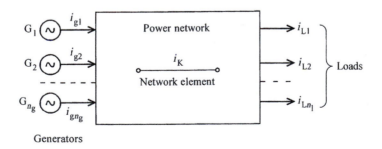

Generators

FIGURE 20.12

Power system with generator and load currents.

Now, if all the generators are connected to the power system simultaneously to supply the same load, by the principle of super position.

$$i_K = d_{K1}i_{g1} + d_{K2}i_{g2} + \cdots + d_{Kn_g}i_{gn_g} \tag{20.37}$$

Let the individual load currents remain a constant complex ratio of the total load current I_L. It is assumed that (X/R) ratio for all the line elements or branches in the network remains the same. The factors d_{Ki} will then be real and not complex.

The individual generator currents may have phase angles d_1, d_2, ..., d_{n_g} with respect to a reference axis. The generator currents can be expressed as:

$$
\begin{aligned}
i_{g1} &= \left|i_{g1}\right|\cos\delta_1 + \left|i_{g1}\right|\sin\delta_1 \\
i_{g2} &= \left|i_{g2}\right|\cos\delta_2 + \left|i_{g2}\right|\sin\delta_2 \\
&\cdots\cdots\cdots\cdots\cdots\cdots\cdots\cdots\cdots\cdots \\
i_{gng} &= \left|i_{gng}\right|\cos\delta_{ng} + \left|i_{gng}\right|\sin\delta_{ng}
\end{aligned}
\tag{20.38}
$$

For simplicity to derive the formula

Let $n_g = 3$ so that Eq. (20.37) becomes

$$i_K = d_{k1}i_{g1} + d_{k2}i_{g2} + d_{k3}i_{g3} \tag{20.39}$$

$$
\begin{aligned}
|i_K|^2 = {}&(d_{k1}|i_{g1}|\cos\delta_1 + d_{k2}|i_{g2}|\cos\delta_2 + d_{k3}|i_{g3}|\cos\delta_3) \\
&+ (d_{ki}|i_{g1}|\sin\delta_1 + d_{k2}|i_{g2}|\sin\delta_2 + d_{k3}|i_{g3}|\sin\delta_3)^2
\end{aligned}
\tag{20.40}
$$

$$
\begin{aligned}
= {}&d_{k1}^2\left|i_{g1}\right|^2 + d_{k2}^2\left|i_{g2}\right|^2 + d_{k3}^2\left|i_{g3}\right|^2 \\
&+ 2d_{k1}d_{k2}\left|i_{g1}\right|\left|i_{g2}\right|\cos\delta_1\cos\delta_2 \\
&+ 2d_{k2}d_{k3}\left|i_{g2}\right|\left|i_{g3}\right|\cos\delta_2\cos\delta_3 \\
&+ 2d_{k3}d_{k1}\left|i_{g3}\right|\left|i_{g1}\right|\cos\delta_3\cos\delta_1 \\
&+ 2d_{k1}d_{k2}\left|i_{g1}\right|\left|i_{g2}\right|\sin\delta_1\sin\delta_2 \\
&+ 2d_{k2}d_{k3}\left|i_{g2}\right|\left|i_{g3}\right|\sin\delta_2\sin\delta_3 \\
&+ 2d_{k3}d_{k1}\left|i_{g3}\right|\left|i_{g1}\right|\sin\delta_3\sin\delta_1
\end{aligned}
\tag{20.41}
$$

$$
\begin{aligned}
= {}&d_{k1}^2\left|i_{g1}\right|^2 + d_{k2}^2\left|i_{g2}\right|^2 + d_{k3}^2\left|i_{g3}\right| \\
&+ 2d_{k1}d_{k2}\left|i_{g1}\right|\left|i_{g2}\right|\cos(\delta_1 - \delta_2) \\
&+ 2d_{k2}d_{k3}\left|i_{g2}\right|\left|i_{g3}\right|\cos(\delta_2 - \delta_3) \\
&+ 2d_{k3}d_{k1}\left|i_{g3}\right|\left|i_{g1}\right|\cos(\delta_3 - \delta_1)
\end{aligned}
\tag{20.42}
$$

Eliminating currents in terms of powers supplied by the generators

$$i_{g1} = \frac{P_1}{\sqrt{3}|V_1|\cos\varphi_1}$$

$$i_{g2} = \frac{P_2}{\sqrt{3}|V_2|\cos\varphi_2}$$

and

$$i_{g3} = \frac{P_3}{\sqrt{3}|V_3|\cos\varphi_3}$$

where P_1, P_2, and P_3 are the active power supplied by the generators 1, 2, and 3 at voltages $|V_1|$, $|V_2|$, and $|V_3|$ and power factors at the generator buses being $\cos\phi_1$, $\cos\phi_2$, and $\cos\phi_3$, respectively.

$$|i_K|^2 = \frac{d_{k1}^2 P_1^2}{\left(\sqrt{3}|V_1|\cos\varphi_1\right)^2} + \frac{d_{k2}^2 P_2^2}{\left(\sqrt{3}|V_2|\cos\varphi_2\right)^2} + \frac{d_{k3}^2 P_3^2}{\left(\sqrt{3}|V_3|\cos\varphi_3\right)^2}$$

$$+ \frac{2d_{k1}d_{k2}P_1P_2\cos(\delta_1-\delta_2)}{3|V_1||V_2|\cos\varphi_1\cos\varphi_2} + \frac{2d_{k2}d_{k3}P_2P_3\cos(\delta_2-\delta_3)}{3|V_2||V_3|\cos\varphi_2\cos\varphi_3} + \frac{2d_{k3}d_{k1}P_3P_1\cos(\delta_3-\delta_1)}{3|V_3||V_1|\cos\varphi_3\cos\varphi_1}$$

The power losses in the network comprising of n_b network elements or branches P_{LOSS} is given by

$$P_L = \sum_{k=1}^{n_b} 3|i_K|^2 R_K$$

where R_K is the resistance of the element K.

$$P_L = \frac{P_1^2 \sum_{k=1}^{n_g} d_{k1}^2 R_K}{|V_1|^2(\cos\varphi_1)^2} + \frac{P_2^2 \sum_{k=1}^{n_g} d_{k2}^2 R_K}{|V_2|^2(\cos\varphi_2)^2} + \frac{P_3^2 \sum_{k=1}^{n_g} d_{k3}^2 R_K}{|V_3|^2(\cos\varphi_3)^2}$$

$$+ \frac{2P_1P_2 \sum_{k=1}^{n_g} d_{k1}d_{k2}R_K \cos(\delta_1-\delta_2)}{|V_1||V_2|\cos\varphi_1\cos\varphi_2}$$

$$+ \frac{2P_2P_3 \sum_{k=1}^{n_g} d_{k2}d_{k3}R_K \cos(\delta_2-\delta_3)}{|V_2||V_3|\cos\varphi_2\cos\varphi_3}$$ (20.43)

$$+ \frac{2P_3P_1 \sum_{k=1}^{n_g} d_{k3}d_{k1}R_K \cos(\delta_3-\delta_1)}{|V_3||V_1|\cos\varphi_3\cos\varphi_1}$$

Let

$$B_{11} = \frac{1}{|V_1|^2(\cos\varphi_1)^2} \sum d_{k1}^2 R_K$$ (20.44)

$$B_{22} = \frac{1}{|V_2|^2(\cos\varphi_2)^2} \sum d_{k2}^2 R_K$$ (20.45)

$$B_{33} = \frac{1}{|V_3|^2(\cos\varphi_3)^2} \sum d_{k3}^2 R_K$$ (20.46)

$$B_{12} = \frac{\cos(\delta_1-\delta_2) \sum_{k=1}^{n_g} d_{k1}d_{k2}R_K}{|V_1||V_2|(\cos\varphi_1)(\cos\varphi_2)}$$ (20.47)

$$B_{23} = \frac{\cos(\delta_2-\delta_3) \sum_{k=1}^{n_g} d_{k2}d_{k3}R_K}{|V_2||V_3|(\cos\varphi_2)(\cos\varphi_3)}$$ (20.48)

$$B_{31} = \frac{\cos(\delta_3-\delta_1) \sum_{k=1}^{n_g} d_{k3}d_{k1}R_K}{|V_1||V_3|(\cos\varphi_3)(\cos\varphi_1)}$$ (20.49)

$$P_{LOSS} = P_1^2 B_{11} + P_2^2 B_{22} + P_3^2 B_{33}$$
$$+ 2P_1 P_2 B_{12} + 2P_2 P_3 B_{23} + 2P_3 P_1 B_{31} + \cdots$$
$$= \sum_{m=1}^{3} \sum_{m=1}^{3} P_m B_{mn} P_n \tag{20.50}$$

In general the formula for B_{mn} coefficients can be expressed as

$$B_{mn} = \frac{\cos(\delta_m - \delta_n)}{|V_m||V_n|(\cos \varphi_m)(\cos \varphi_n)} \sum_k d_{km} d_{kn} R_K \tag{20.51}$$

In the matrix form, the loss formula is expressed for an n generator system as:

$$\begin{bmatrix} P_1 & P_2 & \cdots & P_n \end{bmatrix} \begin{bmatrix} B_{11} & B_{12} & \cdots & B_{1n} \\ B_{21} & B_{22} & \cdots & B_{2n} \\ \vdots & \vdots & \vdots & \vdots \\ B_{n1} & B_{n2} & \cdots & B_{nn} \end{bmatrix} \begin{bmatrix} P_1 \\ P_2 \\ \vdots \\ P_n \end{bmatrix} \tag{20.52}$$

The coefficients can be considered constant, if in addition to the assumptions already made we further assume that the generator voltages V_1, V_2,..., etc., remain constant in magnitude and generator bus power factors $\cos \phi_1$, $\cos \phi_2$, ... also remains constant.

20.17 ACTIVE POWER SCHEDULING

Economic Scheduling of Thermal Plants Considering Effect of Transmission Losses:
   The objective function is

$$F(P) = F(P_1, P_2, \ldots P_{n_g})$$
$$= \sum_{i=1}^{n_g} C_i(P_i) \tag{20.53}$$

which has to be minimized over a given period of time. As only active power is scheduled, the equality constraint $G(P)$ is given by

$$G(P) = P_D - P_L - \sum_{i=1}^{n_g} P_i = 0 \tag{20.54}$$

must be satisfied at every generator bus where P_i is the generation at bus i; P_D is the total load demand, and P_L is the total transmission loss in all the lines. It is desired to minimize Eq. (20.53) subject to the constraint Eq. (20.54).
   The Lagrange function L is formed as

$$L(P, \lambda) = F(P) + 1\lambda \left[P_D + P_L - \sum_{i=1}^{n_g} P_i \right] \tag{20.55}$$

Applying the necessary conditions for the minimum of L

$$\frac{\partial L}{\partial P_i}(P, \lambda) = \frac{\partial}{\partial P_i} F(P_i) + \frac{\partial}{\partial P_i} \lambda \left[P_D + P_L - \sum_{i=1}^{n_g} P_i \right] = 0 \tag{20.56}$$

i.e.,

$$\frac{\partial}{\partial P_i} C\left[P_1\ldots, P_i, \ldots P_{n_g}\right] + \frac{\partial}{\partial P_i} \lambda \left[P_D + P_L - \sum_{i=1}^{n_g} P_i \right] = 0 \tag{20.57}$$

$$\therefore \quad \frac{\partial C_i(P_i)}{\partial P_i} + \lambda \frac{\partial P_L}{\partial P_i} - \lambda = 0 \tag{20.58}$$

Further

$$\frac{\partial C_i(P_i)}{\partial P_i} = \lambda \left[1 - \frac{\partial L}{\partial P_i} \right], \quad i = 1, 2, \ldots, n_g \tag{20.59}$$

Also, it can be expressed as

$$\frac{\partial F_i}{\partial P_i} + \lambda \frac{\partial P_L}{\partial P_i} = \lambda \tag{20.60}$$

The sum of the incremental production cost of power at any plant i and the incremental transmission losses incurred due to generation P_i at bus i charged at the rate of λ must be constant for all generators and equal to λ. This constant λ is equal to the incremental cost of the received power.

In Section 20.16 the loss formula is derived as

$$P_L = \sum_i \sum_j P_i B_{ij} P_j$$
$$= B_{11} + P_1^2 + B_{22} P_2^2 + \cdots + B_{n_g} P_{n_g}^2 + 2B_{12} P_1 P_2 + 2B_{13} P_1 P_3 + \cdots + 2B_{1n_g} P_{n_g}^2 + \cdots \tag{20.61}$$

Differentiating

$$\frac{\partial P_L}{\partial P_i} = \sum_j 2B_{ij} P_j \tag{20.62}$$

Also

$$\frac{\partial C_i(P_i)}{\partial P_i} = \frac{dC_i(P_i)}{dP_i}, \quad i = 1, 2, \ldots n_g \tag{20.63}$$

$$j = 1, 2, \ldots, n_g \tag{20.64}$$

If the incremental costs are represented by a linear relationship following a quadratic input—output characteristic, then, denoting

$$\frac{dC_i(P_i)}{dP_i} = a_i P_i + b_i \tag{20.65}$$

Eq. (20.64) will become

$$a_i P_i + b_i + \lambda \sum_{j=1}^{n_g} 2B_{ij}P_j = \lambda \tag{20.66}$$

Further

$$\sum_{j=1}^{n_g} 2B_{ij}P_j = 2B_{ii}P_i + \sum_{\substack{j=1 \\ j \neq i}}^{n_g} 2B_{ij}P_j \tag{20.67}$$

Eq. (20.66) can be rewritten as

$$a_i P_i + b_i + \lambda 2B_{ii}P_i + \lambda \sum B_{ij}P_j = \lambda$$

$$a_i P_i + b_i + \lambda 2B_{ii}P_i + \lambda \sum_{\substack{i=1 \\ j \neq i}}^{n_g} B_{ij}P_j = \lambda$$

solving for P_i

$$P_i = \frac{\lambda - b_i - \lambda \sum_{\substack{i=1 \\ j \neq i}}^{n_g} 2B_{ij}P_j}{a_i + 2\lambda B_{ii}} \tag{20.68}$$

$$P_i = \frac{1 - \dfrac{b_i}{\lambda} - \sum_{\substack{i=1 \\ j \neq i}}^{n_g} 2B_{ij}P_j}{\dfrac{a_i}{\lambda} + 2B_{ii}}, \quad i = 1, 2, \ldots, n_g \tag{20.69}$$

There are n_g equations to be solved for n_g powers (P_i) for which Gauss's or Gauss–Seidel method is well suited. Knowing a_i, b_i, and B_{ij} coefficients for any assumed value of λ, P_i values may converge to a solution, giving the generator scheduled powers. It is very important that a suitable value is assumed for λ, so that a quick convergence of the equations is obtained.

20.18 PENALTY FACTOR

Consider Eq. (20.59)

$$\frac{dC_i(P_i)}{dP_i} + \lambda \frac{\partial P_L}{\partial P_i} = \lambda$$

It can be rewritten as

$$\frac{\partial C_i(P_i)}{\partial P_i} = \lambda \left[1 - \frac{\partial P_L}{\partial P_i} \right] \tag{20.70}$$

in other words

$$\frac{\partial C_i(P_i)}{\partial P_i} \left[\frac{1}{1 - \dfrac{\partial P_L}{\partial P_i}} \right] = \lambda \tag{20.71}$$

When transmission losses are included, the incremental production cost at each plant i must be multiplied by a factor $(1/(1 - (\partial P_L/\partial P_i)))$ which then will be equal to the incremental cost of power delivered. Hence the factor $(1/(1 - (\partial P_L/\partial P_i)))$ is called *penalty factor*.

Since $\partial P_L/\partial P_i$ is much less than unity, sometimes it is approximated by $1 + (\partial P_L/\partial P_i)$ so that

$$\frac{\partial C_i(P_i)}{\partial P_i}\left(1 + \frac{\partial P_L}{\partial P_i}\right) = \lambda \tag{20.72}$$

The term $1 + (\partial P_L/\partial P_i)$ is called *approximate penalty factor*.

20.19 EVALUATION OF λ FOR COMPUTATION

Consider Eq. (20.69)

$$P_i = \frac{1 - \dfrac{b_i}{\lambda} - \sum_{\substack{j=1 \\ j \neq i}}^{n_g} 2B_{ij}P_j}{\dfrac{a_i}{\lambda} + 2B_{ii}P_j}, \quad i = 1, 2, \ldots, n_g$$

Solution to P_i values depends upon the value of λ chosen. The value of λ determines a set of generations for a particular received load.

It is established that for scheduling purposes it is necessary to start the computation with two different values of λ. As the arbitrarity assigned generations are improved from iteration to iteration, a new value of λ may be computed for each new iteration using the following algorithm for a specified total received load power P_R.

$$\lambda^{(i)} = \lambda^{(i-1)} + (P_R^d - P_R^{(i-1)})\left(\frac{\lambda^{(i-1)} - \lambda^{(i-2)}}{P_R^{(i-1)} - P_R^{(i-2)}}\right) \tag{20.73}$$

where $P_R^{(i-1)}$ is the received power with $\lambda^{(i-1)}$, $P_R^{(i-2)}$ is the received power with $\lambda^{(i-2)}$, $\lambda^{(i)}$, $\lambda^{(i-1)}$, and $\lambda^{(i-2)}$ are the values of λ during iterations (i), $(i-1)$, and $(i-2)$, P_R^d is the desired total power to be received by the loads.

$$= \sum_{i=1}^{n_g} P_i - P_L$$

When two values of P_R calculated successively during the iterations converge to a single value with reasonable accuracy, the latest value of P_R calculated will also converge to P_R^d.

20.20 HYDROELECTRIC PLANT MODELS

Conventional hydroelectric plants are classified as run-of-river plants, run-of-river plants with pondage and storage type plants. In the former type water is utilized as is available in the stream, as there is no provision for storage. Where there is pondage provision, hourly fluctuations in load

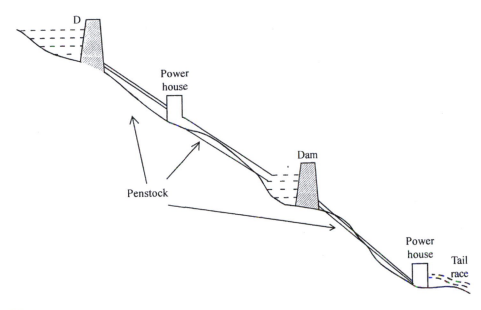

FIGURE 20.13

Cascaded hydroelectric plants.

can be met. In the later type, where storage is provided the water stored during the excess water period can be utilized during the lean season or when power demand is high. The plant may be a single development on a river or there may be several plants constructed and cascaded on the same river (Fig. 20.13). In some cases interconnection of plants on different streams is also possible.

20.21 PUMPED STORAGE PLANT

The pumped storage hydroelectric plants generally store water to supply peak load demands, so that fuel is saved at thermal plants. For this purpose, at light load periods water is pumped to the reservoir back from the tail water pond using power from the grid. The power house may either have both turbines and pumps separately or have reversible pump turbines. The operating characteristics of a pumped storage plant are shown in Fig. 20.14.

Let e_e be the energy spent to pump water to the reservoir. By releasing this water at peak load times the energy supplied to the load is e_S. The ratio (e_S/e_e) is usually of the order 60%–70%. Pumped storage plants are to be operated in such a manner that due to the peak load chipping on the load curve, the saving in fuel cost thus achieved should exceed the pumping of water charges.

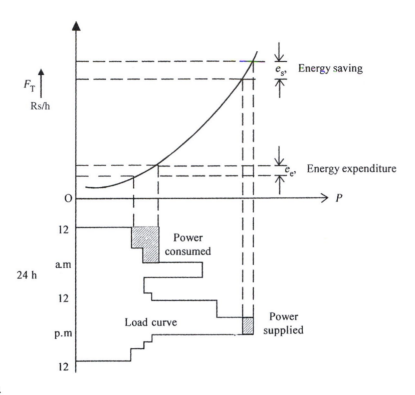

FIGURE 20.14

Dumped storage plant characteristics.

20.22 HYDROTHERMAL SCHEDULING

Most of the power systems are a mix of different modes of generating stations of which the thermal and hydrogenerating units are predominant. While in some systems hydrogeneration may be more than thermal generation in some other cases it may be the other way. The operating cost of thermal plants is high even though their capital cost is low. In case of hydroelectric plants, the running costs are very low, but the capital cost is high as construction of dams, canals, penstocks, surge tanks and other elements of development are involved in addition to the power house. The hydro plants can be started easily and can be assigned load in very short time. This is not so in case of thermal plants, as it requires several hours to bring the boiler, super heater, and turbine system ready to take the load allotment. For the reason mentioned, the hydro plants can handle fast changing loads effectively. The thermal plants in contrast are slow in response. For this reason the thermal plants are more suitable to operate as base load plants, leaving hydroplants to operate as peak load plants.

However, the exact mode of operation depends upon the type of the development, and factors such as storage and pondage, and the amount of water that is available is the most important consideration. A plant may be run-of-river, run-of-river with pondage, storage, or pumped storage type.

Whatever may be the type of plant, it is necessary to utilize the total quantity of water available in hydro development so that maximum economy is achieved. The economic scheduling in the integrated operation is however made difficult as water release policy for power is subject to a variety of constraints. There are multiple water usages which are to be satisfied. Determination of the so-called pseudofuel cost or cost for water usage for use in conjunction with incremental water rate characteristic is a formidable exercise. Nevertheless, hydrothermal economic scheduling is possible with assumptions made wherever necessary.

In systems where there is close balance between hydro and thermal generation and in systems where the hydro capacity is only a fraction of the total capacity, it is generally desired to schedule generation such that thermal generating costs are minimized.

20.23 ENERGY SCHEDULING METHOD

Consider two plants, one of hydroelectric and the other thermal. Let these both supply a common load.

Let for any time period K the maximum hydropower available.

$$P_{H_K}^{\max} \geq P_{\text{load}_K} \quad k = 1, 2, \ldots\ldots K \tag{20.74}$$

where P_{load_K} is the load power during the time period k. It is presumed that the energy available from the hydro plant is not sufficient to meet the load demand.

$$\sum_{k=1}^{K} P_{H_K} n_K \leq \sum_{k=1}^{K} P_{\text{load}_K} n_K \tag{20.75}$$

where n_K is the number of hours in period k.

Let

$$\sum_{k=1}^{K} n_K = T_{\max} \tag{20.76}$$

$$= \text{total period of time over which energy schedule is required}$$

It is proposed to utilize the hydroenergy completely in such a way that the operating cost of the thermal plant is minimized.

$$\text{Load energy to be supplied} = \sum_{k=1}^{K} P_{\text{load}_K} n_K \tag{20.77}$$

$$\text{Hydro energy to be utilized} = \sum_{k=1}^{K} P_{H_K} n_K \tag{20.78}$$

Thermal energy required

$$E_S = \sum_{k=1}^{K} P_{\text{load}_K} n_K - \sum_{k=1}^{K} P_{H_K} n_K \tag{20.79}$$

Let the thermal plant be operated for time period less than $T_{\max}$ and for number of intervals K_S. Hence

$$\sum_{k=1}^{K_S} P_{S_K} n_K = E_S \tag{20.80}$$

where P_S denotes steam power.

The problem can then be stated as

$$\text{Min } F_T = \sum_{k=1}^{K_S} F(P_{S_K}) n_K \tag{20.81}$$

subject to

$$\sum_{k=1}^{K_S} P_{S_K} n_K - E_S = 0 \tag{20.82}$$

The Lagrange function is given by

$$L = \sum_{k=1}^{K_S} F(P_E) n_K + \lambda \left(E_E - \sum_{k=1}^{K_S} P_{S_K} n_K \right) \tag{20.83}$$

$$\frac{\partial L}{\partial P_{S_K}} = \frac{dF(P_{S_K})}{dP_{S_K}} - \lambda = 0 \quad \text{for } k = 1, 2, \dots, K_S \tag{20.84}$$

i.e.,

$$\frac{dF(P_{S_K})}{dP_{S_K}} = \lambda \quad \text{for } k = 1, 2, \dots, K_S \tag{20.85}$$

Eq. (20.85) indicates that the steam plant must be run at a constant incremental production cost for the entire time period of its operation. Denoting this value as $P_{S_k}^0$, the schedule is depicted in Fig. 20.15.

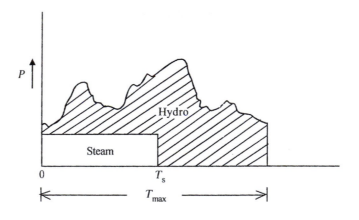

FIGURE 20.15

Hydro—thermal coordination.

Let the steam plant cost characteristic be expressed by

$$F(P_S) = aP_S^2 + bP_S + C \qquad (20.86)$$

The total cost of running the steam plant over the interval T_S

$$F_T = (aP_S^{0^2} + bP_S^0 + c)T_S \qquad (20.87)$$

Infact F_T is also given by

$$
\begin{aligned}
F_T &= \sum_{k=1}^{K_S} F(P_S^0) n_K \\
&= F(P_S^0) \sum_{k=1}^{K_S} n_K \\
&= F(P_S^0) T_S
\end{aligned}
\qquad (20.88)
$$

Further

$$\sum_{k=1}^{K_S} P_{S_K} n_K = \sum_{k=1}^{K_S} P_S^0 n_K = P_S^0 T_S = E_S \qquad (20.89)$$

Hence

$$T_S = \frac{E_S}{P_S^0} \qquad (20.90)$$

So,

$$F_T = (aP_S^{0^2} + bP_S^0 + c)\left(\frac{E_S}{P_S^0}\right) \qquad (20.91)$$

$$
\begin{aligned}
\frac{dF_T}{dP_S^0} &= \frac{d}{dP_S^0}\left[aP_S^0 E_S + bE_S + \frac{cE_S}{P_S^0}\right] \\
&= aE_S + 0 - cE_S \frac{1}{(P_S^0)^2}
\end{aligned}
\qquad (20.92)
$$

$$
\begin{aligned}
(P_S^0)^2 &= \frac{cE_S}{aE_S} = \frac{c}{a} \\
P_S^0 &= \sqrt{\left(\frac{c}{a}\right)}
\end{aligned}
\qquad (20.93)
$$

The steam unit is operated at its maximum efficiency throughout the time period T_S. This can be proved as follows:
Let f_c be the fuel cost.

$$
\begin{aligned}
F(P_S) &= aP_S^2 + bP_S + c \\
&= f_c H(P_S)
\end{aligned}
\qquad (20.94)
$$

where function H denotes the heat value.

The heat rate is then given by

$$\frac{H(P_S)}{P_S} = \frac{1}{f_c} \left[\frac{aP_S^2 + bP_S + c}{P_S} \right] \tag{20.95}$$

$$\frac{d}{dP_S} \left[\frac{H(P_S)}{P_S} \right] = \frac{d}{dP_S} \left[\frac{1}{f_c} \left(aP_S + b + \frac{c}{P_S} \right) \right]$$

$$= \left[\frac{1}{f_c} a - \frac{c}{f_c} \frac{1}{P_S^2} \right] = 0 \tag{20.96}$$

$$P_S = \sqrt{\frac{c}{a}} = P_S^0 \tag{20.97}$$

20.24 SHORT-TERM HYDROTHERMAL SCHEDULING

20.24.1 METHOD OF LAGRANGE MULTIPLIERS (LOSSES NEGLECTED)

Consider a power system which contains both hydro and steam power generating stations. Consider that the entire hydrogeneration is equal to P_H and the total steam power is P_S. With such equivalent power generating station let a load P_L be supplied as shown in Fig. 20.16.

Let the combined operation be over a period of time T. Let this time period be divided into intervals 1, 2, ..., J to suit the load curve so that

$$\sum_{j=1}^{J} n_j = T \tag{20.98}$$

The total volume of water available for discharge over this time period.

$$W = \sum_{j=1}^{J} n_j w_j \tag{20.99}$$

where w_j is the water rate for interval j. The fuel cost required to be minimized over the time period T is given as

$$F_T = \sum_{j=1}^{J} n_j F(P_{S_j}) \tag{20.100}$$

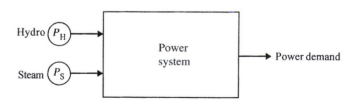

FIGURE 20.16

Hydrothermal scheduling.

For load balance, the equality constraint is

$$P_{\text{load}_j} - P_{S_j} - P_{H_j} = 0, \quad j = 1, 2, \ldots, J \tag{20.101}$$

The loads are assumed to remain constant during time intervals considered. The total value of water at the beginning and at the end of the interval T, in the reservoir are W_i and W_f, respectively.

During this period of scheduling, the head of water is assumed to remain constant.

The input−output characteristic for the equivalent hydro plant is given by

$$w = w(P_{\text{H}}) \tag{20.102}$$

The Lagrange function for minimization of Eq. (20.100) subject to the constraints (20.99) and (20.101) is

$$L = \sum_{j=1}^{J} n_j F(P_{S_j}) + \lambda_j [P_{\text{load}_j} - P_{S_j} - P_{H_j}] + \gamma \left[\sum_{j=1}^{J} n_j w_j (P_{H_j}) - W \right] \tag{20.103}$$

For any specific value of $j = k$, the necessary conditions are

$$\frac{\partial L}{\partial P_{S_K}} = 0 \quad \text{and} \quad \frac{\partial L}{\partial P_{H_K}} = 0 \tag{20.104}$$

giving

$$n_K \frac{dF_{S_k}}{dP_{S_K}} = \lambda_k$$

and

$$\gamma n_K \frac{dw_k}{dP_{H_K}} = \lambda_k \tag{20.105}$$

Solution to the above two equations gives the economic generations at steam and hydro plants over any time interval. The incremental production cost at the steam plants must be the same as incremental production cost at the hydro plants. For simplicity n_K may be taken as one unit. So that

$$\frac{dF_S}{dP_S} = \lambda$$

and

$$\gamma \cdot \frac{dw}{dP_H} = \lambda \tag{20.106}$$

20.24.2 LAGRANGE MULTIPLIERS METHOD—TRANSMISSION LOSSES CONSIDERED

If the transmission losses are considered, then the equality constraint includes P_L, the loss terms.

$$[P_{\text{load}_j} + P_{L_j} - P_{S_j} - P_{H_j}] = 0 \tag{20.107}$$

The Lagrange function changes now into

$$L = \sum_{j=1}^{J} n_j F(P_{S_j}) + \lambda_j (P_{\text{load}_j} - P_{L_j} - P_{S_j} - P_{H_j}) + \gamma \sum_{j=1}^{J} [n_j w_j (P_{H_j}) - W] \tag{20.108}$$

For any specific $j = k$, we obtain the optimality conditions as before

$$\frac{\partial L}{\partial P_{S_K}} = n_K \frac{dF(P_{S_K})}{dP_{S_K}} + \lambda_k(-1) + \lambda_k\left(\frac{-\partial P_{L_k}}{\partial P_{S_K}}\right)$$

$$\frac{\partial L}{\partial P_{H_K}} = \lambda_k(-1) + \lambda_k \frac{\partial P_{L_k}}{\partial P_{H_K}}\left[+\gamma \frac{n_K dw_k(P_{H_K})}{dP_{H_K}}\right] = 0$$

i.e.,

$$n_K \frac{dF(P_{S_K})}{dP_{H_K}} + \lambda_k \frac{\partial P_{L_k}}{\partial P_{S_K}} = \lambda_k$$

and

$$\gamma n_K \frac{dw_k}{dP_{H_K}}(P_{H_K}) + \gamma_k \frac{\partial P_{L_k}}{\partial P_{H_K}} = \lambda_k$$

since k is chosen arbitrarily, and by considering the time period $n_K = 1$.

The equations reduce to

$$\frac{dF(P_S)}{dP_S} + \lambda\frac{\partial P_L}{\partial P_S} = \lambda \tag{20.109}$$

and

$$\gamma\frac{dw(P_k)}{dP_H} + \lambda\frac{\partial P_L}{\partial P_H} = \lambda \tag{20.110}$$

It can be shown that the above equations are valid for any number of steam plants n_S and for any number of hydroplants n_H.

Hence

$$\frac{dF(P_{S_i})}{dP_{S_i}} + \lambda\frac{\partial P_L}{\partial P_{S_i}} = \lambda \quad \text{for } i - 1, 2, \ldots, n_s \tag{20.111}$$

and

$$\gamma\frac{dw(P_{H_K})}{dP_{H_K}} + \lambda\frac{\partial P_E}{\partial P_{H_K}} = \lambda \quad \text{for } k = 1, 2, \ldots, n_H \tag{20.112}$$

the above equations are called *coordination equations*, the solution to which will give the economic schedule for P_{S_i} and P_{H_K}.

20.24.3 SHORT-TERM HYDROTHERMAL SCHEDULING USING *B*-COEFFICIENTS FOR TRANSMISSION LOSSES

Let the total number of generating stations be "n." Out of these n stations, S stations are steam power stations and the remaining H stations are hydroelectric generating stations so that

$$S + H = n \tag{20.113}$$

Let the water input rate at the jth hydro plant be assumed as w_j (m²/s).

The total transmission losses in the lines of the system are given by

$$P_L = \sum_{i=1}^{s}\sum_{j=1}^{s} P_{S_i} B_{ij} P_{S_j} + \sum_{i=s+1}^{n}\sum_{j=s+1}^{n} P_{H_i} B_{ij} P_{H_j} + 2\sum_{i=1}^{s}\sum_{j=s+1}^{n} P_{S_i} B_{ij} \tag{20.114}$$

where the subscripts S and H refer to steam and hydro plants.
The total input hourly rate to all the thermal plants is

$$F_T = \sum_{i=1}^{s} F_i \ \text{Rs/h} \tag{20.115}$$

The incremental cost of production at ith steam plant

$$= \frac{dF_1}{dP_{S_i}} \ \text{Rs/MWh} \tag{20.116}$$

The incremental transmission losses at steam and hydro plants are given, respectively, by

$$\frac{\partial P_L}{\partial P_{S_i}} \ \text{and} \ \frac{\partial P_L}{\partial P_{H_j}}$$

The incremental rate of water flow at jth hydro plant

$$= \frac{\partial w_j}{\partial P_{H_j}} \ \text{m}^3/\text{s} \tag{20.117}$$

The total power received by the loads,

$$P_R = P_{S_i} + \sum_{j=s+1}^{n} P_{H_j} - P_L \tag{20.118}$$

It is desired to make the total input to the system over a period T a minimum.
From calculus of variations for F_T to be a minimum over the time period T, the first variation of Eq. (20.118) must be set equal to zero.

$$\therefore \ \sum_{i=1}^{s} \delta P_{S_i} + \sum_{j=s+1}^{n} \delta P_{H_j} - \sum_{i=1}^{s} \frac{\partial P_L}{\partial P_{S_i}} \delta P_{S_i} - \sum_{j=s+1}^{n} \frac{\partial P_L}{\partial P_{H_j}} \delta P_{H_j} = 0 \tag{20.119}$$

The quantity of water available over the period T is assumed constant.
i.e.,

$$\int_{o}^{T} w_j dt = W \quad \text{for} \quad j = s+1, \ldots, n \tag{20.120}$$

Then, it is further required to make

$$\left[\int_{0}^{T} F_T dt + \sum_{j=s+1}^{n} \gamma_j \int_{0}^{T} w_j \, dt \right] \tag{20.121}$$

a minimum; where γ_j are constant multipliers.

The first variation of the quantity in equation must vanish. Hence

$$\delta\left[\int_0^T F_T dt + \sum_{j=s+1}^n \gamma_j \int_0^T w_j \, dt\right] = 0 \tag{20.122}$$

i.e.,

$$\sum_{i=1}^s \frac{\partial F_T}{\partial P_{S_i}} \partial P_{S_i} + \sum_{j=s+1}^n r_j \frac{\partial w_j}{\partial P_{H_j}} \partial P_{H_j} = 0 \tag{20.123}$$

where ∂P_{S_i} and ∂P_{H_j} are the variations in steam power generation and hydropower generation, respectively.

From Eq. (20.119)

$$\sum_{i=1}^s \left(1 - \frac{\partial P_L}{\partial P_{S_i}}\right) \partial P_{S_i} + \sum_{j=s+1}^n \left(1 - \frac{\partial P_L}{\partial P_{H_j}}\right) \partial P_{H_j} = 0 \tag{20.124}$$

For a small variation of δP_{H_m} at the mth hydro plant, Eq. (20.124) can be split into the following form:

$$\left[1 - \frac{\partial P_L}{\partial P_{H_m}}\right] \partial P_{H_m} = - \sum_{i=1}^s \left[1 - \frac{\partial P_L}{\partial P_{S_i}}\right] \partial P_{S_i} - \sum_{\substack{j=s+1 \\ j \neq m}}^n \left(1 - \frac{\partial P_L}{\partial P_{H_j}}\right) \partial P_{H_j} \tag{20.125}$$

Again, Eq. (20.123) can be rewritten as

$$\sum_{i=1}^s \frac{\partial F_T}{\partial P_{S_i}} \partial P_{S_i} + \gamma_m \frac{\partial w_m}{\partial P_{H_m}} \partial P_{H_m} + \sum_{\substack{j=s+1 \\ j \neq m}}^n \gamma_j \frac{\partial w_j}{\partial P_{H_j}} \partial P_{H_j} = 0 \tag{20.126}$$

Hence

$$\gamma_m \frac{\partial w_m}{\partial P_{H_m}} \partial P_{H_m} = - \sum_{i=1}^\alpha \frac{\partial F_T}{\partial P_{S_i}} \partial P_{S_i} - \gamma_j \sum_{\substack{j=s+1 \\ j \neq m}}^n \frac{\partial w_j}{\partial P_{H_j}} \partial P_{H_j} \tag{20.127}$$

Multiplying Eq. (20.127) on both sides by $(1 - (\partial P_L/\partial P_{H_m}))$

$$\gamma_m \frac{\partial w_m}{\partial P_{H_m}} \partial P_{H_m} \left(1 - \frac{\partial P_L}{\partial P_{H_m}}\right) = - \sum_{i=1}^s \frac{\partial F_T}{\partial P_{S_i}} \partial P_{S_i} \left(1 - \frac{\partial P_L}{\partial P_{H_m}}\right)$$
$$- \gamma_j \sum_{\substack{i=s+1 \\ j \neq m}}^n \frac{\partial w_i}{\partial P_{H_j}} \partial P_{H_j} \left(1 - \frac{\partial P_L}{\partial P_m}\right) \tag{20.128}$$

In Eq. (20.128) if $(1 - (\partial P_L/\partial P_{H_m})) \partial P_{H_m}$ is replaced by the quantity on the right hand side of Eq. (20.125), we obtain

$$\gamma_m \frac{\partial w_m}{\partial P_{H_m}} \left[- \sum_{i=1}^s \left(1 - \frac{\partial P_L}{\partial P_{S_i}}\right) \delta P_{S_i} - \sum_{\substack{j=s+1 \\ j \neq m}}^n \left(1 - \frac{\partial P_L}{\partial P_{H_j}}\right) \partial P_{H_j} \right] \tag{20.129}$$

$$= -\sum_{i=1}^{s} \frac{\partial F_T}{\partial P_{S_i}} \delta P_{S_i} \left(1 - \frac{\partial P_L}{\partial P_{H_m}} \right) - \gamma_j \sum_{\substack{j=s+1 \\ j \neq m}}^{n} \frac{\partial w_j}{\partial P_{H_j}} \delta P_H C \left(1 - \frac{\partial P_L}{\partial P_{H_j}} \right) \tag{20.130}$$

Rearranging the term

$$\sum_{i=1}^{s} \frac{\partial F_T}{\partial P_{S_i}} \left(1 - \frac{\partial P_L}{\partial P_{H_m}} \right) \delta P_{S_i} - \sum_{i=1}^{n} \gamma_m \frac{\partial w_m}{\partial P_{H_m}} \left(1 - \frac{\partial P_L}{\partial P_{H_m}} \right) \partial P_{S_i}$$
$$+ \sum_{\substack{r=s+1 \\ j \neq m}}^{n} r_j \frac{\partial w_j}{\partial P_{H_j}} \left(1 - \frac{\partial P_L}{\partial P_{H_j}} \right) \delta P_{H_j} - \sum_{\substack{j=s+1 \\ j \neq m}}^{n} \gamma_m \frac{\partial w_m}{\partial P_{H_m}} \left(1 - \frac{\partial P_L}{\partial P_{H_j}} \right) \delta P_{H_j} \tag{20.131}$$

Eq. (20.131) can be put as

$$\sum_{i=1}^{s} \frac{\partial F_T}{\partial P_{S_i}} \left(1 - \frac{\partial P_L}{\partial P_{H_m}} \right) - \gamma_m \frac{\partial w_m}{\partial H_m} \left(1 - \frac{\partial P_L}{\partial P_{S_i}} \right) \delta P_{S_i}$$
$$+ \sum_{\substack{j=s+1 \\ j \neq m}}^{n} \gamma_j \gamma_m \frac{\partial w_j}{\partial P_{H_j}} \left(1 - \frac{\partial P_L}{\partial P_{H_m}} \right) - \lambda_m \frac{\partial w_m}{\partial P_{H_m}} \left(1 - \frac{\partial P_L}{\partial P_{H_m}} \right) \delta P_{H_j} = 0 \tag{20.132}$$

Note that ∂P_{S_i} and ∂P_{H_j} are the variations in powers at steam and hydro plants, respectively, and hence are finite. Then, to make Eq. (20.128) satisfied, each of the coefficients of the variations must be zero.

$$\therefore \frac{\partial F_T}{\partial P_{S_i}} \left(1 - \frac{\partial P_L}{\partial P_{H_j}} \right) - \gamma_m \frac{\partial w_m}{\partial P_{H_m}} \left(1 - \frac{\partial P_L}{\partial P_{S_i}} \right) = 0 \tag{20.133}$$

and

$$\gamma_j \frac{\partial w_j}{\partial P_{Hj}} \left(1 - \frac{\partial P_L}{\partial P_{Hm}} \right) - \gamma_m \frac{\partial w_m}{\partial P_{Hm}} \left(1 - \frac{\partial P_L}{\partial P_{Hm}} \right) = 0 \tag{20.134}$$

In other words

$$\frac{\dfrac{\partial F_T}{\partial P_{S_i}}}{\left(1 - \dfrac{\partial P_L}{\partial P_{S_i}} \right)} = \gamma_m \frac{\dfrac{\partial W_m}{\partial P_{H_m}}}{\left(1 - \dfrac{\partial P_L}{\partial P_{H_j}} \right)} = \gamma_j \frac{\dfrac{\partial w_i}{\partial P_{H_j}}}{\left(1 - \dfrac{\partial P_L}{\partial P_{H_j}} \right)} = \text{Constant} \tag{20.135}$$

The partial derivatives in Eq. (20.135) are also the total derivatives. It may also be recognized that each of the term in Eq. (20.135) is also the incremental cost of the received power in Rs/MW. If denotes their incremental cost, then $dF_T/dP_{S_i} = dF_i/dP_{S_i}$ is the cost of generation at each plant i and it depends on the generation at that plant only

$$\left. \begin{aligned} \frac{dF_i}{dP_{Si}} \left(\frac{1}{1 - \dfrac{\partial P_L}{\partial P_{S_i}}} \right) &= \lambda \\[2em] r_j \frac{\partial w_j}{\partial P_{H_j}} \left(\frac{1}{1 - \dfrac{\partial P_L}{\partial P_{H_j}}} \right) &= \lambda \end{aligned} \right\} \tag{20.136}$$

Also

$$\frac{\partial F_i}{\partial P_{S_i}} + \lambda \frac{\partial P_L}{\partial P_{S_i}} = \lambda \text{ and } \gamma_j \frac{dw_i}{dP_{H_j}} + \lambda \frac{\partial P_L}{\partial P_{H_j}} = \lambda \tag{20.137}$$

The above equations are called coordination equations for hydrothermal combined operation. Solution to these equations results in minimizing the input costs to supply the given load.

Let

$$\frac{dF_i}{dP_{S_i}} = a_i P_{S_i} + b_i \tag{20.138}$$

and

$$\frac{dW_j}{dP_{H_j}} = c_j P_{H_j} + d_j \tag{20.139}$$

where a and c are the slopes of the incremental cost curves and b and d are the intercepts.

Further

$$P_L = \Sigma_k P_k B_{K_l} P_l \tag{20.140}$$

Partially differentiating P_L in Eq. (20.140) and substituting in Eq. (20.137)

$$a_i P_{S_i} + b_i + 2\lambda \left(\sum_{K=1}^{s} B_{k_i} P_{S_K} + \sum_{j=s+1}^{n} B_{ij} P_{H_j} \right) = \lambda \tag{20.141}$$

$$\gamma_j c_j P_{H_j} + r_j d_j + 2\lambda \left(\sum_{k=s+1}^{n} B_{j_k} P_{H_K} + \sum_{l=1}^{n} B_{j_l} P_{S_l} \right) = \lambda \tag{20.142}$$

Rewriting the third term on the left side of Eqs. (20.141) and (20.142), as

$$\sum_{K=1}^{s} B_{k_i} P_{S_K} = B_{iz'} P_{S_i} + \sum_{\substack{k=1 \\ k \neq z'}}^{s} B_{k_i} P_{S_K} \tag{20.143}$$

and

$$\sum_{K=s+1}^{n} B_{jk} P_{H_K} = B_{ij} P_{H_j} + \sum_{\substack{K=s+1 \\ K \neq j}}^{n} B_{ki} P_{S_K} \tag{20.144}$$

combining the terms containing P_{S_i} and P_{H_j} for solution

$$P_{S_i} = \frac{1 - \frac{b_i}{\lambda} - \sum_{\substack{k=1 \\ k \neq z'}}^{s} 2B_{ki}P_{S_K} - \sum_{l=s+1}^{n} 2B_{il}P_{H_l}}{\frac{a_i}{\lambda} + 2B_{iz'}} \qquad (20.145)$$

and

$$P_{H_j} = \frac{1 - \frac{dj}{\lambda} - \sum_{k=s+1}^{n} 2B_{jk}P_{H_K} - \sum_{l=1}^{s} 2B_{jl}P_{sl}}{\gamma_j \frac{c_j}{\lambda} + 2B_{jj}} \qquad (20.146)$$

These two equations (20.145) and (20.146) can be solved iteratively till convergence is obtained assuming suitable values for λ and γ_j.

WORKED EXAMPLES

E.20.1 Consider a power system with two generating stations. The incremental production cost characteristics for the two stations are

$$\frac{\partial F_1}{\partial P_1} = (27.5 + 0.15P_1) \text{ mu/MWh}$$

$$\frac{\partial F_2}{\partial P_2} = (19.5 + 0.26P_2) \text{mu/MWh}$$

Given that the minimum and maximum powers are 10 and 100 MW at each plant schedule the generation at each plant to supply a system load given by the load curve shown in Fig. E.20.1.

Solution:

Case (i): $P_R = P_1 + P_2$; $P_R = 50$ MW; $P_1 = P_R - P_2 = 50 - P_2$

$$\frac{\partial F_1}{\partial P_1} = \frac{\partial F_2}{\partial P_2} = \lambda$$

$$27.5 + 0.15P_1 = 19.5 + 0.26P_2$$

$$27.5 + 0.15(50 - P_2) = 19.5 + 0.26P_2$$

$$27.5 - 19.5 + 7.50 = 0.26P_2 + 0.15P_2$$

$$15.5 = 0.41P_2; \quad P_2 = 37.8$$

$$P_1 = 50 - 37.8 = 12.2 \text{ MW}$$

Case (ii): $P_1 + P_2 = 100$ MW

$$P_1 = 100 - P_2$$

$$27.5 + 0.15(100 - P_2) = 19.5 + 0.26 P_2$$

$$27.5 + 15 - 19.5 = 0.26P_2 + 0.15P_2$$

$$23 = 0.41P_2$$

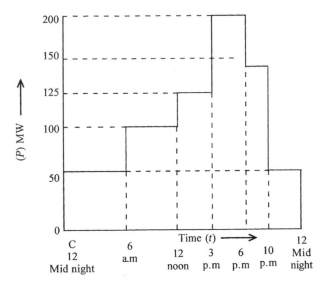

FIGURE E.20.1

Load curve.

$$P_2 = 23/0.41 = 56.01 \text{ MW}$$
$$P_1 = 100 - 56.1 = 43.90 \text{ MW}$$

Case (iii): $P_1 + P_2 = 125$ MW

$$27.5 + (0.15)(125 - P_2) = 19.5 + 0.26P_2$$
$$27.5 + 18.75 - 19.5 = 0.15P_2 + 0.26P_2 = 0.41P_2$$
$$26.75 = 0.41P_2$$

therefore and

$$P_2 = 65.24 \text{ MW}$$
$$P_1 = 59.76 \text{ MW}$$

Case (iv): $P_1 + P_2 = 150$ MW

$$27.5 + 0.15(150 - P_1) = 19.5 + 0.26P_2$$
$$27.5 + 22.5 - 19.5 = 0.26P_2 + 0.15P_2$$
$$30.5 = 0.41P_2; \quad P_2 = 74.39$$
$$P_1 = 75.61$$

Case (v): $P_1 + P_2 = 200$ MW

$$27.5 + (0.15)(200 - P_2) = 19.5 + 0.26P_2$$
$$27.5 + 30 - 19.5 = 0.41P_2$$
$$38 = 0.41P_2; \quad P_2 = 92.68; \quad P_1 = 107.32$$

The value of λ can be computed now. The results are tabulated.

Case	P_1	P_2	P_D	λ
Case (i)	12.2	37.8	50	29.33
Case (ii)	43.9	56.1	100	34.08
Case (iii)	59.76	65.24	125	36.55
Case (iv)	75.61	74.39	150	38.84
Case (v)	107.32	92.68	200	43.598

E.20.2 The fuel input characteristics for two thermal plants are given by

$$F_1 = (8P_1 + 0.024 + 80)10^6 \text{ kcal/h}$$
$$F_2 = (6P_2 + 0.004 + 120)10^6 \text{ kcal/h}$$

where P_1 and P_2 are in megawatts
  i. plot the input−output characteristic for each plant
 ii. plot the heat rate characteristic for each plant
iii. assuming the cost of fuel as mu/100/ton, calculate the incremental production cost characteristic in mu/MWh at each plant. Plot the same against power produced in MW.
Solution:
  i. Input−output characteristics
     Consider the power in steps of 10 MW
     Given that

$$F_1 = (8P_1 + 0.024 + 80)10^6 \text{ kcal/h}$$

and

$$F_2 = (6P_1 + 0.004 + 120)10^6 \text{ kcal/h}$$

Substituting P_1 and P_2 in steps of 10 MW the values of F_1 and F_2 are obtained and tabulated.

Table E.20.2A Input−Output Calculations			
P_1 (MW)	$F_1 \times 10^6$ (kcal/h)	P_2 (MW)	$F_2 \times 10^6$ (kcal/h)
10	162.4	10	184
20	250	20	256
30	342	30	336
40	438	40	424
50	540	50	520
60	646	60	624
70	758	70	736
80	874	80	856
90	994	90	984
100	1220	100	1120

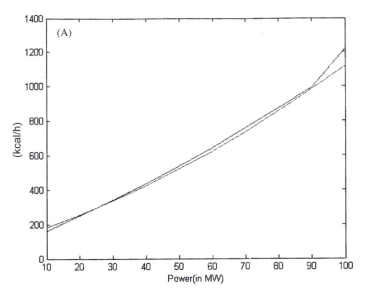

FIGURE E.20.2A

Input—output curves.

the input—output characteristics are plotted as shown in Fig. E.20.2A.

ii. Heat rate characteristic

Let $P_1 = 10$ MW at Plant 1

Heat rate at this P_1 is $\dfrac{162.4}{10} \times 10^6 \dfrac{\text{kcal/h}}{\text{MW}}$

At Plant 2 for $P_2 = 10$ MW

Heat rate $= \dfrac{184}{10} \times 10^6 = 18.4 \times 10^5 \dfrac{\text{kcal/h}}{\text{MW}}$

In a similar manner the heat rate is computed for powers in steps of 10 MW till 100 MW at both the plants. The results are tabulated and curves are plotted.

Table E.20.2B Heat Rate Calculations		
Output (MW)	Heat rate × 10⁶ at Plant 1 (kcal/MWh)	Heat rate × 10⁶ at Plant 2 (kcal/MWh)
10	16.25	18.4
20	12.48	12.8
30	11.39	11.2
40	10.96	10.6
50	10.80	10.4
60	10.77	10.4
70	10.82	10.5
80	10.92	10.7
90	11.05	10.93
100	11.2	11.2

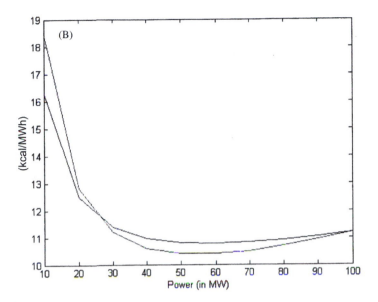

FIGURE E.20.2B

Heat rate curves.

iii. Calorific values of fuel at plant $1 = 4000$ kcal/h

$$\text{Cost of fuel} = \text{mu } 100 \text{ per ton} = \text{mu}/100/1000 \text{ per kg} \tag{1}$$

$$= \frac{100}{1000 \text{ kg}} \times \frac{1}{4000} \frac{\text{kg}}{\text{kcal}} = \frac{100}{4 \times 10^6} \frac{\text{mu}}{\text{kcal}} \tag{2}$$

$$\frac{\mathrm{d}F_1}{\mathrm{d}P_1} = (8 + 0.048P_1)10^6 \text{ kcal/MWh}$$

Incremental fuel cost

$$= (8 + 0.048\, P_1)10^6 \times \frac{100}{4 \times 10^6} \text{ mu/MWh} \tag{3}$$

$$= (2 + 0.012\, P_1) \times 100 = (200 + 1.2\, P_1) \text{ mu/MWh} \tag{4}$$

At 100 MW generation, the cost is

$$(200 + 12) = 212 \text{ mu/MWh}$$

The incremental production cost is obtained by adding the maintenance costs of 10%

Hence the incremental production cost at Plant 1

$$= (200 + 1.2\, P_1)\, 1.1 = (220 + 1.32\, P_1)$$

At $P_1 = 10$ MW; $\quad \text{IPC}_1 = 220 + 13.2 = 233.2$ mu/MWh

Similarly at Plant 2

$$\frac{\mathrm{d}F_1}{\mathrm{d}P_1} = (8 + 0.048P_1)10^6 \text{kcal/MWh}$$

Table E.20.2C Incremental Production Costs

Power (P)	IPC at Plant 1 (mu/MWh)	IPC at Plant 2 (mu/MWh)
10	233.2	149.6
20	246.4	167.2
30	259.6	184.8
40	272.8	202.4
50	286.0	220.0
60	299.2	237.2
70	312.4	255.2
80	325.6	272.4
90	338.8	290.4
100	352.0	308

incremental fuel cost

$$= (6 + 0.08\,P_2) \times 10^6 \times \frac{100}{5 \times 10^6} = (120 + 1.6\,P_2)\ \text{mu/MWh}$$

Considering maintenance costs at 10%, the incremental production cost is $(120 + 1.6P_2) \times 1.1 = (132.0 + 1.76P_2)$ mu/MWh.

At $P_2 = 10$ MW, the incremental production cost
$= (120 + 1.6 \times 10) \times 1.1 = (136 \times 1.1) = 149.6$ mu/MWh.

Likewise, the incremental production costs are calculated at all power levels and tabulated. Finally the values are plotted as shown in Table E.20.2C.

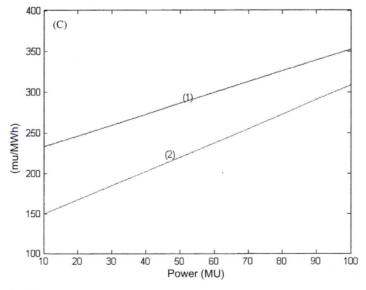

FIGURE E.20.2C

Incremental production cost curves.

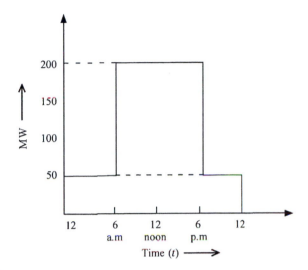

FIGURE E.20.3

Load curve.

E.20.1 For the plants in Example E.20.2 obtain the economic generation schedule. The load curve is given in Fig. E.20.3.

Solution:

For economic schedule

$$\frac{dC_1(P_1)}{dP_1} = \frac{dC_2(P_2)}{dP_2} = \lambda$$

$$(220 + 1.32P_1) = (132 + 1.762P_2)$$

Case (i) $P_1 + P_2 = 50$ MW

$$P_1 = 50 - P_2$$

$$220 + 1.32(50 - P_2) = 132 + 1.76\,P_2$$

$$220 + 66 - 132 = 1.76P_2 + 1.32P_2 = 3.082P_2$$

$$154 = 3.082\,P_2$$

$$P_2 = 49.967554 \approx 50 \text{ MW}$$

$$\therefore P_1 = 0$$

If a minimum generation limit is imposed then $P_1 = 10$ MW and $P_2 = 40$ MW

Case (ii) $P = 200$ MW

$$200 + 132(200 - P_2) = 132 + 1.76P_2$$

$$220 - 132 + 264 = 1.76\,P_2 + 1.32\,P_2 = 3.08\,P_2$$

$$352 = 3.08P_2$$

$$P_2 = 114.2857 \approx 114.3 \text{ MW}$$

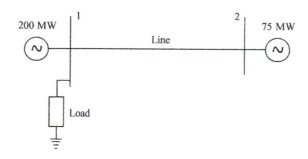

FIGURE E.20.4

$$P_1 = 200 - 114.3 = 85.7 \text{ MW}$$

If a maximum generation limit of 100 MW is imposed, then

$$P_2 = 100 \text{ MW and hence } P_1 = 100 \text{ MW}$$

E.20.2 Given a two-bus system as shown in Fig. E.20.4

It is observed that when a power of 75 MW is imported to Bus 1, the loss amounted to 5 MW. Find the generation needed from each plant and also the power received by the load, if the system λ is given by 20 mu/MWh. The incremental fuel cost at the two plants are given by

$$\frac{dC_1(P_1)}{dP_1} = 0.03P_1 + 15 \text{ mu/MWh}$$

$$\frac{dC_2(P_2)}{dP_2} = 0.05P_2 + 18 \text{ mu/MWh}$$

Solution:

The load is at Bus 1. Hence, P_1 will not have any effect on the line losses

Therefore $B_{11} = B_{12} = B_{21} = 0$

$$P_L = B_{22}P_2^2$$

$$5 = B_{22}\, 75^2$$

$$B_{22} = \frac{5}{5625} = 8.9 \times 10^{-4}$$

At Station 1

$$\frac{C_1(P_1)}{dP_1} + \lambda\frac{\partial P_L}{dP_1} = \lambda$$

$$\frac{\partial P_L}{\partial P_1} = 0.03P_1 + 15 + \lambda.0 = \lambda \tag{i}$$

At Station 2

$$\frac{\partial P_L}{\partial P_2} = 2B_{22}P_2$$

$$\frac{dC_2(P_2)}{dP_2} + \lambda\frac{\partial P_L}{\partial P_2} = \lambda$$

$$0.05\,P_2 + 18 + \lambda(2(8.9) \times 10^{-4})P_2 = \lambda \qquad\qquad (ii)$$

At $\lambda = 20$ from Eq. (i)

$$0.03P_1 + 15 = 20$$

$$0.03P_1 = 20 - 15 = 5$$

$$P_1 = 5/0.03 = 166.67 \text{ MW}$$

Again, from Eq. (ii)

$$0.05P_2 + 18 + 20(2 \times 8.9 \times 10^{-4})P_2 = 20$$

$$0.05P_2 + 0.0356P_2 = 20 - 18 = 2$$

$$0.0856P_2 = 2; \quad P_2 = 23.3644 \text{ MW}$$

$$\text{Total transmission loss} = B_{22}\,P_2^2$$

$$= 8.9 \times 10^{-4} \times 23.3644^2 = 0.4858 \text{ MW}$$

The load demand at Bus 1 is

$$166.67 + 23.3644 \times 0.4858 = 178.02 \text{ MW}$$

E.20.3 A power system with two generating stations supplied a total load of 300 MW. Neglecting transmission losses the economic schedule for the plant generation is 175 and 125 MW. Find the saving in the production cost in mu/h due to this economic schedule as compared to equal distribution of the same load between the two units.

The incremental cost characteristics are

$$\frac{dC_1(P_1)}{dP_1} = 30 + 0.3P_1$$

and

$$\frac{dC_2(P_2)}{dP_2} = 32.5 + 0.4P_2$$

Solution:

The cost of generation at Plant 1

$$C_1 = \int \frac{dC_1(P_1)}{dP_1} \cdot dP_1 = \int (0.3P_1 + 30)dP_1$$

$$= (0.3P_1^2 + 30P_1 + x) \text{ Rs/h}$$

$$C_2 = \int \frac{dC_2(P_2)}{dP_2} \cdot dP_2 = \int (32.5 + 0.4P_2)dP_2$$

$$= (0.4P_2^2 + 32.5P_2 + y)\text{mu/h}$$

x and y are constants of integration which need not be evaluated.
For equal distribution of generation

$$P_1 = 150 \text{ MW and } P_2 = 150 \text{ MW}$$

The increase in cost at Plant 1 by generating 175 MW instead of 150 MW is

$$= 0.3(175^2 - 150^2) + 30(175 - 150)$$

$$= (30, 625 - 22, 500)0.3$$

$$= 0.3 \times 8125 + 750 = 3187.5 \text{ mu/h}$$

The reduction in cost at Plant 2 by generating 125 MW only instead of 150 MW is

$$[0.4(150)^2 + 32.5 \times (150 + y)] - [0.4(125)^2 + 32.5(125) + y]$$

$$= 0.4(-15625 + 22500) + 32.5(-125 + 150)$$

$$= 0.4 \times 6875 + 812.5 = 2750 + 812.5 = 3562.5$$

Savings $= 3562.5 - 3187.5 = 372 \text{ mu/h}$
Annual savings $= 375 \times 24 \times 365 = 32, 85, 000 \text{ mu}$

E.20.4 Consider two steam power plants operating with incremental production costs

$$\frac{dC_1(P_1)}{dP_1} = (0.08P_1 + 16) \text{ mu/MW-h}$$

and

$$\frac{dC_2(P_2)}{dP_2} = (0.08P_2 + 12) \text{ mu/MW-h}$$

Given the loss coefficients

$$B_{11} = 0.001 \text{ per MW}$$

$$B_{12} = B_{21} = -0.0005 \text{ per MW}$$

$$B_{22} = 0.0024 \text{ per MW}$$

Find the economic schedule of generation for $\lambda = 20$ mu/MWh
Solution:

$$P_1 = \frac{1 - \dfrac{b_1}{\lambda} - 2B_{21}P_2}{\dfrac{a_1}{\lambda} + 2B_{11}} = \frac{1 - \dfrac{16}{20} - 2(-0.0005)P_2}{\dfrac{0.08}{20} + 2(0.001)} = \frac{0.2 + 0.001P_2}{0.006}$$

$$P_2 = \frac{1 - \dfrac{b_2}{\lambda} - 2B_{12}P_1}{\dfrac{a_2}{\lambda} + 2B_{22}} = \frac{1 - \dfrac{12}{20} - 2(-0.0005)P_1}{\dfrac{0.08}{20} + 2(0.0024)} = \frac{0.4 + 0.001P_2}{0.0088}$$

starting with $P_2 = 0$, $P_1 = 33.3$
then the value of P_2 is computed as 49.2.
Continuing the iterations the values obtained are tabulated.

Iteration	P_1	P_2
1	33.3	49.2
2	41.5	50.2
3	41.7	50.2
4	41.7	50.2

The converged values are $P_1 = 41.7$ MW; and $P_2 = 50.2$ MW.
With these generations, the total transmission losses are

$$P_L = \sum_{i=1}^{2}\sum_{j=1}^{2} P_i B_{ij} P_j$$

$$B_{11}P_1^2 + 2B_{12}P_1P_2 + B_{22}P_2^2$$

$$= (0.001) \times (41.7)^2 + 2(-0.0005)(41.7)(50.2)$$

$$+ (0.0024)(50.2)^2 = 5.7 \text{ MW}$$

$$\text{Total generation} = P_T = \sum_{i=1}^{2} P_2 = 41.7 + 50.2 = 919 \text{ MW}$$

Total power received

$$P_R = P_T - P_L = 91.9 - 5.7 = 86.2 \text{ MW}$$

E.20.5 The input−output data for a particular hydroplant is given in Fig. E.20.7A. Find the incremental water rate characteristic and convert it into an equivalent incremental production cost characteristic taking the cost of water as 10^{-3} Rs/m^3.
Solution:
The incremental water rate is obtained by finding the slope of the given input−output curve at different points. A straight line segment approximation of incremental water rate is obtained as in Fig. E.20.7B.

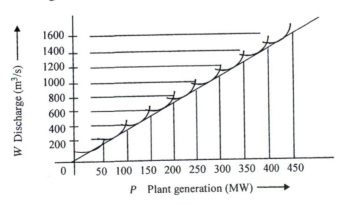

FIGURE E.20.7A

Input−output curve.

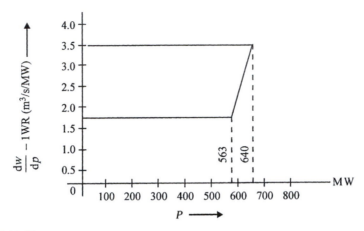

FIGURE E.20.7B

Incremental water rate.

Incremental water rate

$$= 1.7 \text{ m}^3/\text{s/MW for } P = 0 - 563 \text{ MW}$$

$$= (0.0233P - 11.4) \text{ m}^3/\text{s/MW}$$

for

$$P = 563 - 640 \text{ MW}$$

Incremental production cost

$$= \text{Incremental water rate} \times \text{Cost of water}$$

$$= 1.7 \times 3600 \times 10^{-3} \text{ mu/MW-h}$$

$$= 6.2 \text{ mu/MW-h for } P = 0 - 563 \text{ MW}$$

Incremental production cost for the range

$$P = 563 - 640 \text{ MW is given by}$$
$$(0.0876P - 41.2) \text{ mu/MW-h}$$

E.20.6 Consider the following six-bus system with three generating plants and three load buses. G_1 and G_2 are steam power plants and G_3 is hydroelectric plant.

Generating Station	Type	No. of Units	Capacity of Each Unit (MW)	Total Capacity (MW)
G_1	Thermal	3	12.5	37.5
G_2	Thermal	2	60.0	120.0
G_3	Hydro	2	60.0	120.0
Total capacity				277.5

Loads:

$$L_1 \ 160 \ \text{MW}$$

$$L_2 \ 35 \ \text{MW}$$

$$L_3 \ 72.5 \ \text{MW}$$

Total load = 267.5 MW

Base voltage = 220 kV

Base MVA = 100

The B-coefficient matrix is given as

$$
\begin{array}{c}
\begin{array}{ccc} \quad\quad G_1 \quad\quad G_2 \quad\quad G_3 \end{array} \\
B = \begin{array}{c} G_1 \\ G_2 \\ G_3 \end{array}
\begin{bmatrix}
0.0210 & 0.0034 & 0.0181 \\
0.0034 & 0.02497 & 0.003 \\
0.0181 & 0.003 & 0.050
\end{bmatrix}
\end{array}
$$

Characteristics of steam and hydro plants
G₁-thermal station:
Calorific value of coal = 5125 kcal/kg
Cost of coal = mu 45 per ton
Incremental fuel cost at near no-load = mu 21/MWh
Incremental fuel cost at 40 MW = mu 23/MWh
Slope of incremental cost curve = 0.05
    Intercept on *y*-axis = 21

$$\frac{dF_1}{dP_1} = 0.05P_{S1} + 21$$

G₂-thermal station:
Calorific value of coal = 4600 kcal/kg
Cost of coal = mu 48 per ton
Incremental fuel cost at near no-load = mu 25/MWh
Incremental fuel cost at 40 MW = mu 31.6/MWh
Slope of incremental cost curve = 0.094
Intercept on *y*-axis = 21

$$\frac{dF_2}{dP_2} = 0.094P_{S2} + 21$$

G₃—Hydroelectric plant:
    Assuming $Y_3 = 1/100$ (mu/100 m³)of water at 104-MW output, the incremental plant cost = mu 19.5/MWh, and at 120-MW output the incremental plant cost = mu 36 per MWh. Slope of the incremental cost curve

$$\gamma = \frac{16.5}{-16} = 1.03$$

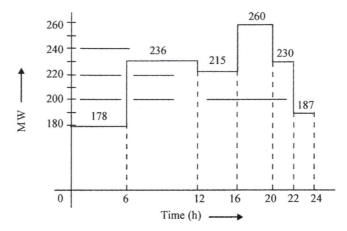

FIGURE E.20.8

Load curve.

Intercept on y-axis $= -88$

$$\gamma_3 = \frac{dW_3}{dP_{H3}} = 19.5; 0 \le P_{H3} \le 104 \text{ MW}$$

$$= 1.03 P_{H3} - 88;$$

$$104 \le P_{H3} \le 120 \text{ MW}$$

The system daily load curve is given in Fig. E.20.8. Obtain the economic schedules.
Solution:
The scheduling equations are given from Eqs. (6.102) and (6.103) as

$$P_1 = \frac{1 - \dfrac{21}{\lambda} - 0.000068 P_2 - 0.000362 P_3}{\dfrac{0.05}{\lambda} + 0.00042}$$

$$P_2 = \frac{1 - \dfrac{21}{\lambda} - 0.000068 P_1 - 0.000060 P_3}{\dfrac{0.094}{\lambda} + 0.0004994}$$

$$P_3 = \frac{1 + \dfrac{88}{\lambda} - 0.000362 P_1 - 0.000060 P_2}{\dfrac{1.03}{\lambda} + 0.0010}$$

$$P_1 + P_2 + P_3 - (0.00021 P_1^2 + 0.0002497 P_2^2 + 0.0005 P_3^2)$$
$$+ (0.000068 P_1 P_2 + 0.000362 P_1 P_3 + 0.00006 P_2 P_3) = P_R$$

in order to schedule for a specified received load, the values of l are computed from equation as

$$\lambda^{(i)} = \lambda^{(i-1)} + (P_R^{(d)} - P_R^{(i-1)})\left[\frac{\lambda^{(i-1)} - \lambda^{(i-2)}}{P_R^{(i-1)} - P_R^{(i-2)}}\right]$$

where the superscript i indicates the iteration being started; P_R is the received load and $P_R^{(d)}$ is the desired received load

To start the computation, two initial values for λ are assumed, namely

$$\lambda_1 = 22.0$$

$$\lambda_2 = 30.0$$

Economic schedules are then computed for different water rates over a wide range

$$\gamma_3 \frac{dw_3}{dP_3} = 0.515P_3 - 44 \text{ at } \gamma_3 = 0.5 \text{ mu}/100 \text{ m}$$

$$= 0.618P_3 - 52.8 \text{ at } \gamma_3 = 0.6 \text{ mu}/100 \text{ m}$$

$$= 0.742P_3 - 63.5 \text{ at } \gamma_3 = 0.7 \text{ paise}/100 \text{ m}$$

$$= 0.824P_3 - 70.4 \text{ at } \gamma_3 = 0.8 \text{ mu}/100 \text{ m}$$

$$= 0.927P_3 - 79.2 \text{ at } \gamma_3 = 0.9 \text{ mu}/100 \text{ m}$$

$$= 1.03P_3 - 88 \text{ at } \gamma_3 = 1.0 \text{ mu}/100 \text{ m}$$

$$= 1.24P_3 - 105.5 \text{ at } \gamma_3 = 1.2 \text{ mu}/100 \text{ m}$$

the results of the economic schedule are shown in (A)–(G) of Table E.20.8 corresponding to different water rates.

Table E.20.8

$P_R^{(d)}$ (MW)	P_1 (MW)	P_2 (MW)	P_3 (MW)	P_{total} (MW)	P_{loss} (MW)	P_R (MW)	λ
(A) $\gamma = 0.5$ mu/100 m³							
178.000	34.2786	28.1089	125.870	188.168	10.2033	177.965	24.2133
187.000	37.5000	33.4225	126.731	197.674	10.6672	187.006	24.8071
215.000	37.5000	58.5480	131.085	227.133	12.1321	215.001	27.5978
230.000	37.5000	72.1076	133.447	243.055	13.0705	229.984	29.1365
236.000	37.5000	77.5726	134.404	249.477	13.4780	235.999	29.7608
260.000	37.5000	99.5099	138.274	275.284	15.2841	259.999	32.3101
(B) $\gamma = 0.6$ mu/100 m³							
178.000	37.5000	30.4325	119.685	187.617	9.60962	178.007	24.4699
187.000	37.5000	38.6243	120.872	196.997	9.9923	187.004	25.3683
215.000	37.5000	64.2988	124.632	226.431	11.4309	215.000	28.2338
230.000	37.5000	78.1710	126.688	242.359	12.3596	229.999	29.8147
236.000	37.5000	83.7402	127.517	248.758	12.7620	235.996	30.4564
260.000	37.5000	106.168	130.888	274.556	14.5569	259.999	33.0778

(*Continued*)

Table E.20.8 *Continued*

$P_R^{(d)}$ (MW)	P_1 (MW)	P_2 (MW)	P_3 (MW)	P_{total} (MW)	P_{loss} (MW)	P_R (MW)	λ
(C) $\gamma = 0.7$ mu/100 m³							
178.000	37.5000	34.7763	114.802	187.080	9.07387	172.006	24.9369
187.000	37.5000	43.1259	115.827	196.453	9.44968	187.003	25.8570
215.000	37.5000	69.3002	119.072	225.872	10.8717	215.000	28.7397
230.000	37.5000	83.4432	120.848	241.791	11.7948	229.997	30.4096
236.000	37.5000	89.1077	121.564	248.171	12.1942	235.977	31.0668
260.000	37.5000	112.001	124.484	273.985	13.9878	259.998	33.7546
(D) $\gamma = 0.8$ mu/100 m³							
178.000	37.5000	37.1910	112.105	186.796	8.79129	178.005	25.1972
187.000	37.5000	45.6237	113.044	196.167	9.16186	187.003	26.1275
215.000	37.5000	72.0465	116.019	225.565	10.5818	214.984	29.0978
230.000	37.5000	86.4328	117.651	241.494	11.5045	229.989	30.7381
236.000	37.5000	92.0947	118.312	247.907	11.9066	236.000	31.4043
260.000	37.5000	115.205	120.994	273.699	13.7017	259.997	34.1278
(E) $\gamma = 0.9$ mu/100 m³							
178.000	37.5000	39.5740	109.453	186.527	8.52298	178.004	25.4547
187.000	37.5000	38.0938	110.303	195.897	8.89451	187.002	26.3966
215.000	37.5000	74.8023	112.999	225.302	10.3088	214.993	29.4050
230.000	37.5000	89.2306	114.447	241.208	11.2304	229.977	31.0671
236.000	37.5000	95.0556	115.078	247.634	11.6339	236.000	31.7422
260.000	37.5000	118.416	117.513	273.429	13.4335	259.996	34.5033
(F) $\gamma = 1.0$ mu/100 m³							
178.000	37.5000	41.5314	107.232	186.314	8.31038	178.004	25.666
187.000	35.3000	50.1235	109.682	195.682	8.68070	187.002	26.2182
215.000	37.5000	77.0672	110.525	225.048	10.0942	214.998	29.6583
230.000	35.3000	91.6397	111.878	241.018	11.0182	230.000	31.3386
236.000	37.5000	97.4943	112.426	247.421	11.4210	236.000	32.0212
260.000	37.5000	121.065	114.658	273.221	13.2262	259.995	34.8137
(G) $\gamma = 1.2$ mu/100 m³							
178.000	37.5000	44.9039	103.559	185.963	7.96039	178.003	26.0326
187.000	37.5000	53.6119	104.220	195.332	8.33065	187.001	27.0000
215.000	37.5000	80.9275	106.321	224.749	9.74867	215.000	30.0921
230.000	37.5000	95.7020	107.476	240.678	10.6780	230.000	31.8021
236.000	37.5000	101.639	107.913	247.083	11.0835	235.999	32.4970
260.000	37.5000	125.548	109.849	272.897	12.9033	259.993	35.3412

E.20.7 A system consists of two generators with the following characteristics

$$F_1 = (7P_1 + 0.03p_1^2 + 70)\,10^6$$
$$F_2 = (5P_2 + 0.05p_2^2 + 100)\,10^6$$

where F and P are fuel input in kcal/h and unit output in MW, respectively. The daily load cycle is given as follows:

Time	Load
12 midnight to 6 am	50 MW
6 am to 6 pm	150 MW
6 pm to 12 midnight	50 MW

Give the economic schedule for the three periods of the day:
Solution:
Let the cost of the fuel be the same at both the plants.

$$\frac{dF_1}{dP_1} = (7 + 0.06P_1)10^6 \text{ kcal/MWh}$$

If C is cost of fuel in mu/kcal.
The incremental production cost at Plant 1

$$= \frac{dC_1}{dP_1} = (7 + 0.06P_1)10^6 \times C/\text{mu/MWh}$$

similarly

$$= \frac{dC_2}{dP_2} = (5 + 0.1P_2)10^6 \times C/\text{mu/MWh}$$

For economy

$$\frac{dC_1}{dP_1} = \frac{dC_2}{dP_2}$$

and

$$P_1 + P_2 = 50 \text{ MW} \tag{i}$$

$$(7 + 0.06\,P_1)10^6 \times C = (5 + 0.1\,P_2)10^6 \times C \tag{ii}$$

solving (i) and (ii) for P_1 and P_2

$$P_1 = 18.75 \text{ MW}$$
$$P_2 = 31.25 \text{ MW}$$

For $P_1 + P_2 = 150$ MW
We obtain $(7 + 0.06P_1) = (5 + 0.1P_2)$

$$7 + 0.06P_1 = 5 + 0.1(150 - P_1)$$
$$0.06P_1 + 0.1P_1 = 5 - 7 + 15$$
$$0.16P_1 = 13$$

$$P_1 = 81.25 \text{ MW}$$
$$P_2 = 68.75 \text{ MW}$$

E.20.8 The incremental production cost data of two plants are $dF_1/dP_1 = 2 + P_1$ and $dF_2/dP_2 = 1.5 + P_2$ where P_1 and P_2 are expressed in per unit on a 100-MVA base. Assume that both the units are in operation and that the maximum loading of each unit is 100 MW and the minimum loading of each unit is 10 MW. The loss coefficients on a 100-MVA base are given by

$$B = \begin{bmatrix} 0.10 & -0.05 \\ -0.05 & 0.2 \end{bmatrix}$$

for $\lambda = 2.5$ solve the coordination equations, by the iterative method.
Solution:

$$\text{IPC}_i = a_i P_i + b_i$$

$$P_1 = \frac{1 - \dfrac{b_1}{\lambda} - 2B_{21}P_2}{\dfrac{a_1}{\lambda} + 2B_{11}} = \frac{1 - \dfrac{2}{2.5} - 2 \times (-0.05) \times P_2}{\dfrac{1}{2.5} + 2 \times (0.10)}$$

$$= \frac{1 - 0.8 + 0.1P_2}{0.4 + 0.2} = \frac{0.2}{0.6} = 0.333 \text{ at } P_2 = 0$$

$$P_2 = \frac{1 - \dfrac{b_2}{\lambda} - 2B_{12}P_2}{\dfrac{a_2}{\lambda} + 2B_{22}} = \frac{1 - \dfrac{1.5}{2.5} - 2 \times (-0.05) \times P_2}{\dfrac{1}{2.5} + 2 \times (0.02)} = \frac{1 - 0.6 + 0.1P_1}{0.4 + 0.4}$$

$$\text{At } P_1 = 0.333$$

$$P_2 = \frac{0.4 + 0.0333}{0.8} = \frac{0.4333}{0.8} = 0.541625$$

performing iterations further

$$P_1 = \frac{0.2 + 0.1(0.5416)}{0.6} = \frac{0.2 + 0.5416}{0.6} = \frac{0.25416}{0.6} = 0.4236$$

$$P_2 = \frac{0.4 + 0.1(0.4236)}{0.8} = \frac{0.4 + 0.04236}{0.8} = 0.55295$$

$$P_3 = \frac{0.2 + 0.055295}{0.6} = 0.42549 \quad P_2 = \frac{0.4 + 0.042549}{0.8} = 0.5531865$$

The results are:

Iteration	P_1	P_2
1	0.333	0.541625
2	0.4236	0.55295
3	0.42546	0.55318

At the end of three iterations

$$P_1 = 0.4255$$
$$P_2 = 0.5532$$

If one more iteration is carried out

$$P_1 = \frac{0.2 + 0.05532}{0.6} = 0.4255$$

$$P_2 = \frac{0.4 + 0.04255}{0.5} = 0.5532$$

No change in the values.

E.20.9 The following incremental costs pertain to a two-plant system.

$$\frac{\mathrm{d}F_1}{\mathrm{d}P_1} = 0.03P_1 + 14 \text{ mu/MWh}$$

$$\frac{\mathrm{d}F_2}{\mathrm{d}P_2} = 0.04P_2 + 10 \text{ mu/MWh}$$

The loss coefficients are $B_{11} = 0.001$ per MW; $B_{12} = B_{22} = 0$. If λ for the system is mu 30/MWh compute the required generation at the two plants and the loss in the system.
Solution:
For economy

$$P_1 = \frac{1 - \dfrac{b_1}{\lambda} - 2B_{21}P_2}{\dfrac{a_1}{\lambda} + 2B_{11}} = \frac{1 - \dfrac{14}{30} - 0}{\dfrac{0.03}{30} + 2 \times 0.001} = 177.78 \text{ MW}$$

$$P_2 = \frac{1 - \dfrac{b_2}{\lambda} - 2B_{12}P_1}{\dfrac{a_2}{\lambda} + 2B_{22}} = \frac{1 - \dfrac{10}{30}}{\dfrac{0.04}{30}} = 500 \text{ MW}$$

The transmission losses are

$$P_L = \sum_{i=1}^{2}\sum_{j=1}^{2} P_i B_{ij} P_j$$
$$= B_{11}P_1^2 + 2B_{12}P_1P_2B_{22}P_2^2$$
$$= B_{11} = 0.001 \times (177.78)^2 = 31.6 \text{ MW}$$

E.20.10 Consider a steam station with two units the input−output characteristics being specified by

$$F_1 = 80 + 8P_1 + 0.024$$
$$F_2 = 120 + 6P_2 + 0.04$$

In scheduling a load of 100 MW by equal incremental cost method, the incremental cost of unit 1 is specified wrongly by 10% more than the true value while that of unit 2 is specified by 6% less than the true value

Find

i. The change in generation schedules.

ii. The change in the total cost of generation.

Solution:

The incremental production cost at Plant 1 is $8 + 0.048P_1$

It is specified 10% more.

That is $(8 + 0.048P_1)1.1 = 8.8 + 0.0528P_1$.

The incremental production cost at Plant 2 is $6 + 0.08P_2$

It is specified 6% less

That is as $(6 + 0.08P_2)0.94 = 5.64 + 0.0752P_2$

Schedule With Correct Incremental Production Costs:

$$8 + 0.048P_1 = 6 + 0.08P_2$$

$$P_1 + P_2 = 100 \text{ MW}$$

Solving $P_1 = 46.875$ MW

$$P_2 = 53.125 \text{ MW}$$

Cost of production at Plant1

$$C_1 = 80 + 8 \times 46.875 + 0.024 \times 46.875^2 = 507.73438$$

Case of production at Plant 2

$$C_2 = 120 + 6 \times 53.125 + 0.04 \times 53.125^2 = 551.64063$$

Total cost $C = C_1 + C_2 = 1059.375$

Schedule With Incorrect Specification:

$$8.8 + 0.0528P_1 = 5.64 + 0.752P_2$$

$$P_1 + P_2 = 100$$

Solving

$$P_1 = 34.0625$$

$$P_2 = 65.9375$$

Cost of production at Plant 2

$$C_1^1 = 80 + 8 \times 34.0625 + 0.024 \times 34.0625^2 = 380.34609$$

Cost of production at Plant 1

$$C_2^1 = 120 + 6 \times 65.9375 + 6.04 \times 65.9375^2 = 569.53528$$

Total cost $C^1 = C_1^1 + C_2^1 = 949.88218$

Change in total cost of generation

$$= 1029.375 - 949.8821 = 109.49 \text{ mu}$$

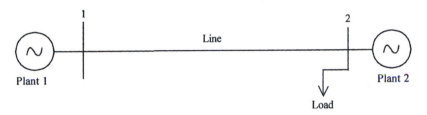

FIGURE E.20.13

E.20.11 A two-bus system shown in Fig. E.20.13 supplies a load at Bus 2. If 50 MW is transmitted from Plant 1 to load at Bus 2 over the line, the loss is 2.5 MW. The incremental production costs at both the plants are given with usual notation by

$$\frac{dC_1}{dP_1} = 0.03P_1 + 15$$

and

$$\frac{dC_2}{dP_2} = 0.05P_2 + 20$$

The value of λ is 23 mu/MWh

Determine the generation schedule for economy with losses coordinated. What will be the generation schedule if the losses are not coordinated but considered. What will be the savings by coordination of losses into economic schedule?

Solution:

The load is at Bus 2 only. There is transmission loss only from the power supplied by Plant 1. No power is transmitted over the line from Plant 2.

Hence

$$2.5 \text{ MW} = B_{11}P_1^2 + B_{22}.0 + 0a$$
$$= B_{11}50^2$$

$$B_{11} = \frac{2.5}{50 \times 50} = 0.001(\text{MW})^{-1}$$

Case (i) Losser coordinated

$$\frac{dC_1}{dP_1} + \lambda\frac{dP_L}{dP_1} = \lambda$$

$$\frac{dC_2}{dP_2} + \lambda\frac{dP_L}{dP_2} = \lambda$$

$$0.03P_1 + 15 + 2 \times 0.001P_1 = 23$$

$$P_1 = 258.06 \text{ MW}$$

$$0.05 P_2 + 20 + 0 = 23$$

$$P_2 = 3/0.05 = 60 \text{ MW}$$

Total generation $= P_1 + P_2 = 318.06$ MW

Total load demand $= P_1 + P_2 - P_{loss}$

$$= 318.06 - 0.001 \times 258.06^2 = 60.595 \text{ MW}$$

$$P_D = 318.06 - 66.595 = 251.465 \text{ MW}$$

Case (ii) Economic schedule for $P_D = 251.465$ MW with losses not coordinated.

$$0.03P_1 + 15 = 0.05P_2 + 20 \qquad\qquad\qquad \text{(i)}$$

$$P_1 + P_2 = B_{11}P_1^2 + 251.465 \qquad\qquad\qquad \text{(ii)}$$

$$0.05\, P_2 = 0.03\, P_1 + 15 - 20$$

$$P_2 = \frac{0.03P_1 - 5}{0.05} = 0.6P_1 - 100$$

Substituting this P_2 value in Eq. (ii)

$$P_1 + 0.6P_1 - 100 = 0.001P_1^2 + 251.465$$

i.e.,

$$0.001\, P_1^2 - 1.6\, P_1 + 351.465 = 0$$

Solving $P_1 = 262.88$ (valid answer)
Then, $P_2 = 157.728 - 100 = 57.728$ MW
Total generation $P_1 + P_2 = 320.608$ MW
The savings at Plant 1 due to loss coordination

$$\int_{258.06}^{262.88} (0.03P_1 + 15)dP_1 = 0.03P_1^2 + 15P_1 \Big|_{258.06}^{262.88} = [69,105.9 - 66,594.9] = 2511.0 \text{ mu/h}$$

The extra expenditure at Plant 2 due to increased production

$$\int_{60}^{57.728} (0.05P_2 + 20)dP_2 = 0.05P_2^2 + 20P_2 \Big|_{60}^{57.728} = [180 + 1154.56] = 1334.56 \text{ mu/h}$$

Net savings $= 2511 - 1334.56 = 1176.44$ mu/h

E.20.12 Consider the two-bus system shown in Fig. E.20.14. The incremental production costs at the two generating stations are given by

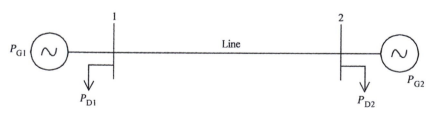

FIGURE E.20.14

$$\frac{dC_1}{dP_1} = 0.005P_1 + 5$$

and

$$\frac{dC_2}{dP_2} = 0.004P_2 + 7$$

The B-coefficients are given in matrix form as

$$B = \begin{bmatrix} 0.0002 & -0.00005 \\ -0.00005 & 0.0003 \end{bmatrix}$$

Determine the penalty factors at both the buses and also the approximate penalty factors. Given $\lambda = 8$

Solution:

$$B_{11} = 0.0002; \quad B_{22} = 0.0003; \quad B_{12} = B_{21} = -0.00005$$

$$P_L = P_1^2 B_{11} + P_2^2 B_{22} + 2P_1 P_2 B_{12}$$

$$\frac{\partial P_L}{\partial P_1} = 2P_1 B_{11} + 2P_2 B_{22} = 2 \times 0.0002 P_1 + 2 \times 0.00005 P_2$$

$$\frac{\partial P_L}{\partial P_2} = 2P_2 B_{22} + 2P_1 B_{21} = 2 \times 0.0003 P_2 + 2 \times 0.00005 P_1$$

The coordination equations are

$$(0.005P_1 + 5) + 8(0.0004P_1 + 0.0001P_2) = 8$$

$$(0.004P_2 + 6) + 8(0.0006P_2 + 0.0001P_1) = 8$$

Simplifying

$$0.005P_1 + 0.0032P_1 + 0.0008P_2 = 3$$

$$0.004P_1 + 0.0048P_1 + 0.0008P_2 = 2$$

i.e.,

$$0.0082P_1 + 0.0008P_2 = 3$$

$$0.0008P_1 + 0.0088P_2 = 2$$

Solving for P_1 and P_2

$$P_1 = 346.75 \text{ and } P_2 = 195.75$$

$$\left(1 - \frac{\partial P_L}{\partial P_1}\right) = 1 - 2 \times 346.75 \times 0.0002 + 0.0001 \times 195.75 = 0.8417$$

$$\left(1 - \frac{\partial P_L}{\partial P_2}\right) = 1 - 2 \times 0.0003 \times 195.75 + 0.0001 \times 346.75 = 0.847875$$

Penalty factor at Plant 1 $= \dfrac{1}{0.847875} = 1.17942$

Penalty factor at Plant 2 $= \dfrac{1}{0.8417} = 1.18807$

Approximate penalty factors

At Plant 1 the approximate penalty factor

$$= \left(1 + \frac{\partial P_L}{\partial P_1}\right) = 1 + 0.1387 + 0.019575 = 1.158275$$

At Plant 2 the approximate penalty factor

$$= \left(1 + \frac{\partial P_L}{\partial P_2}\right) = 1 + 0.11745 + 0.034675 = 1.152125$$

E.20.13 Given the network in the figure shown along with the currents flowing in the lines and the impedances of the lines in per unit on a 100-MVA base. Compute the *B*-coefficients for the network when the voltage at bus 1 is 1.0 ∠0 degree p.u.

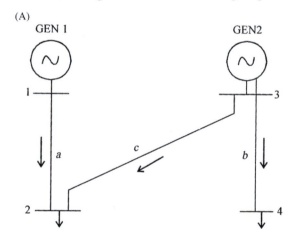

FIGURE E.20.15(A)

Current in line $a = 1.5 - j0.3$ p.u.
Current in line $b = 2 - j0.3$ p.u.
Current in line $c = 1 - j0.2$ p.u.
Load current at Bus 2 $= 2.5 - j0.5$ p.u.
Impedance of line $a = 0.01 + j0.06$ p.u.
Impedance of line $b = 0.01 + j0.05$ p.u.
Impedance of line $c = 0.01 + j0.04$ p.u.
Solution:

(B)

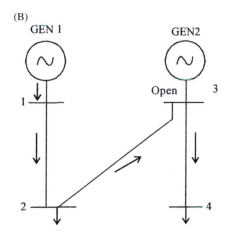

FIGURE E.20.15(B)

The current distribution factors

$$d_{a1} = 1$$

$$d_{b1} = \frac{2 - j0.2}{4.5 - j0.7} = 0.44135$$

$$d_{c1} = \frac{2 - j0.2}{4.5 - j0.7} = 0.44135$$

(C)

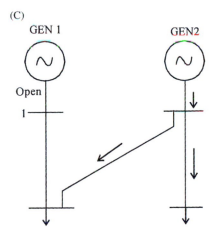

FIGURE E.20.15(C)

$$d_{a2} = 0$$

$$d_{b2} = \frac{2 - j0.2}{4.5 - j0.7} = 0.44135$$

$$d_{c2} = \frac{2.5 - j0.5}{4.5 - j0.7} = 0.559825$$

Voltages

Voltage at Bus 1

$$V_1 = 1.0 \angle 0 \text{ degree}$$

Voltage at Bus 2

$$V_2 = (1 + j0) + (0.01 + j0.06)(1.5 - j0.3)$$
$$= 1.033 + j0.087 = 1.03366571 \angle 4.814 \text{ degrees}$$

Voltage at Bus 3

$$V_3 = (1.033 + j0.087) + (1 - j0.2)(0.01 + j0.04)$$
$$= 1.051 + j0.0125 = 1.0584 \angle 6.78257 \text{ degrees}$$

Voltages at Bus 4

$$V_4 = (1.051 + j0.125) - (0.01 + j0.05)(2 - j0.2)$$
$$= 1.03 + j0.027 = 1.03035 \angle 1.50 \text{ degrees}$$

$$\delta_1 = \tan^{-1}\left(\frac{-0.3}{1.5}\right) = -11.3 \text{ degrees}$$

$$\delta_2 = \tan^{-1}\left(\frac{-0.4}{3}\right) = -7.595 \text{ degrees}$$

$$\cos(d_1 - d_2) = [-11.3 \text{ degrees} - (-7.595 \text{ degrees})] = \cos(-3.705 \text{ degrees}) = 0.9979$$

Power factor at Plant 1 $(\cos \phi_1) = \cos(0 \text{ degree} + 11.3 \text{ degrees}) = 0.9806$

Power factor at Plant 2 $(\cos \phi_2) = \cos(6.78257 + 7.595) = 0.96868$

$$B_{11} = \frac{1}{V_1^2 \cos \varphi_1^2}\left[d_{a1}^2 R_a + d_{b1}^2 R_b + d_{c1}^2 R_c\right]$$

$$= \frac{1}{1^2 \times 0.9806^2} = \left[1 \times 0.01 + 0.44135^2 \times 0.01 + 0.44135^2 \times 0.01\right]$$

$$= 0.01445$$

$$B_{22} = \frac{1}{V_2^2 \cos^2 \varphi_2}\left[d_{a2}^2 R_a + d_{b2}^2 R_b + d_{c2}^2 R_c\right]$$

$$= \frac{1}{(1.036571)^2 \times (0.96868)^2}$$

$$= [0.44135^2 \times 0.01 + 0.559825^2 \times 0.01 + 0]$$

$$= 0.00464292$$

$$B_{12} = B_{21} = \frac{\cos(\delta_1 - \delta_2)}{|V_1||V_2|\cos \varphi_1 \cos \varphi_2}$$

$$[d_{a1}d_{a2}R_a + d_{b1}d_{b2}R_b + d_{c1}d_{c2}R_c]$$

$$\frac{0.9979[0 + 0.44135 \times 0.44135 \times 0.01 + 0.44135 \times 0.559825 \times 0.01]}{(1 \times 1.036657)(0.96868 \times 0.9806)}$$

$$= 0.0044778857$$

on a 100-MVA base the loss coefficients are

$$B_{11} = 0.01445 \times 10^{-2} \; (\text{MW})^{-1}$$

$$B_{22} = 0.0046429 \times 10^{-2} \; (\text{MW})^{-1}$$

$$B_{12} = B_{21} = 0.00447788 \times 10^{-7} \; (\text{MW})^{-1}$$

PROBLEMS

P.20.1 a. A power system consists of two, 120-MW units whose input cost data are represented by the equations:

$$C_1 = 0.04P_1^2 = + 22P_1 + 800 \text{ mu/h}$$

$$C_2 = 0.04 \, P_2^2 = + 22 \, P_2 + 1000 \text{ mu/h}$$

If the total received power $P_R = 200$ MW. Determine the load sharing between units for most economic operation.

  b. Discuss the general problem of economic operation of large interconnected areas.

P.20.2 a. Derive an expression for the hourly loss in economy due to error in the representation of input data.

  b. The incremental fuel costs for two plants are given by

$$\frac{dc_1}{dP_1} = 0.1P_1 + 20 \text{ Rs/MWh}; \quad \frac{dc_2}{dP_2} = 0.15P_2 + 22.5 \text{ Rs/MWh}$$

The system is operating at the optimum condition with $p_1 = p_2 = 100$ MW and $\partial p_1 / \partial p_2 = 0.2$.

Find the penalty factor at Plant 1 and the incremental cost of received power.

P.20.3 a. The incremental fuel costs for the two plants are given by

$$\frac{dc_1}{dP_1} = 0.2P_1 + 45; \quad \frac{dc_2}{dP_2} = 0.25P_2 + 34$$

where C is in Rs/h and P is in MW. If both units operate at all times and maximum and minimum loads on each are 100 and 20 MW, respectively, determine the economic load schedule of the plants for the loads of 80 and 180 MW. Neglect the line losses.

  b. write short notes on physical interpretation of coordination equation.

P.20.4 a. Derive the conditions to be satisfied for economic operation of a loss less power system.

  b. 150, 220, and 220 MW are the ratings of three units located in a thermal power station. Their respective incremental costs are given by the following equations:

$$\frac{dc_1}{dP_1} = mu(0.11P_1 + 12); \quad \frac{dc_2}{dP_2} = mu(0.095P_2 + 14); \quad \frac{dc_3}{dP_3} = mu(0.1P_3 + 13)$$

where P_1, P_2, and P_3 are the loads in MW. Determine the economical load allocation between the three units, when the total load on the station is (1) 350 MW and (2) 500 MW

P.20.5 The equations of the input costs of three power plants operating in conjunction and supply power to a system network are obtained as follows:

$$C_1 = 0.06 + 15P_1 + 150 \text{ mu/h}$$
$$C_2 = 0.08 + 13P_2 + 180 \text{ mu/h}$$
$$C_3 = 0.10 + 10P_3 + 200 \text{ mu/h}$$

The incremental loss—rates of the network with respect to Plants 1, 2, and 3 are 0.06, 0.09, and 1.0 per MW of generation, respectively. Determine the most economical share of a total load of 120 MW which each of the plants would take up for minimum input cost of received power is mu/MWh.

P.20.6 a. The incremental costs for two generating plants are

$$IC_1 = 0.1 \, P_1 + 20 \text{ mu/MWh}$$
$$IC_2 = 0.1 \, P_2 + 15 \text{ mu/MWh}$$

where P_1 and P_2 are in MW. The loss coefficients (B_{mn}) expressed in MW^{-1} unit are $B_{11} = 0.001, B_{22} = 0.0024, B_{12} = N_{21} = -0.0005$. Compute the economical generation scheduling corresponding to the Lagrangian multiplier 1.25 mu/MWh and the corresponding system load that can be met with. If the total load is 150 MW, taking 5% change in the value of l, what should be the value of l in the next iteration?

  b. What are the assumptions made in deriving the loss coefficients?

P.20.7 The fuel inputs to two plants are given by

$$F_1 = 0.015 + 16P_1 + 50; \quad F_2 = 0.02 + 12P_2 + 30$$

The loss coefficients of the system are given by $B_{11} = 0.005$, $B_{12} = -0.0012$, and $B_{22} = 0.002$. The load to be met is 200 MW. Determine the economic operating schedule and the corresponding cost of generation if the transmission line losses are coordinated.

P.20.8 A constant load of 300 MW is supplied by two 200-MW generators 1 and 2 for which the respective incremental fuel costs are

$$\frac{dC_1}{dP_{G_1}} = 0.10P_{G_1} + 20 \quad \frac{dC_2}{dP_{G2}} = 0.10P_{G_2} + 20$$

with powers P_G in MW and costs C is mu/h, determine

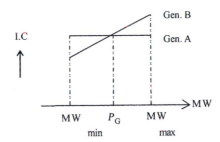

FIGURE P.20.10

a. The most economical division of load between the generators and

b. The saving in mu/day thereby obtained compared to equal load sharing between the machines.

P.20.9 A system consists of two plants connected by a tie line and a load is located at Plant 2. When 100 MW are transmitted from Plant 1, a loss of 10 MW takes place on the tie line. Determine the generation schedule at both the plants and the power received by the load when l for the system is mu25/MWh and the incremental fuel costs are given by the equations:

$$\frac{dF_1}{dP_1} = 0.03P_1 + 17 \text{ mu/MWh}; \qquad \frac{dF_2}{dP_2} = 0.06P_2 + 19 \text{ mu/MWh}$$

P.20.10 The incremental fuel cost curves of generators A and B are shown in Fig. P.20.10. How would a load (1) more than $2P_G$, (2) less than $2P_G$, and (3) equal to $2P_G$ be shared between A and B if both generators are running.

P.20.11 a. Explain heat rate curve and cost curve. Bring out the differences between them

b. Determine the economic operating point of three units supplying a load of 800 MW. The incremental fuel costs of the three units are:

$$\frac{dF_1}{dP_1} = 6.48 + 0.00256P_1 \text{ mu/MWh}; \qquad \frac{dF_2}{dP_2} = 7.85 + 0.00388P_2 \text{ mu/MWh}$$

$$\frac{dF_3}{dP_3} = 7.97 + 0.00964P_3 \text{ mu/MWh}$$

QUESTIONS

Q.20.1 Explain the following terms with reference to thermal plants

    a. heat rate curve

    b. incremental fuel rate curve

    c. incremental production cost curve.

Q.20.2 Explain the following terms with reference to hydro plants

    a. input−output curve

    b. incremental water rate curve

    c. incremental production cost curve.

Q.20.3 How is generation scheduled among various generators when transmission losses are neglected in a thermal system? Explain.

Q.20.4 Derive the transmission loss formula for a system consisting of n-generating plants supplying several loads interconnected through a transmission network.

Q.20.5 Derive expressions for economic distribution of load between generating units considering the effect of transmission losses.

Q.20.6 What are coordination equation? Give their physical significance.

Q.20.7 What is a penalty factor in economic scheduling. Explain its significance.

Q.20.8 Explain how the incremental production cost of a thermal power station can be determined.

Q.20.9 Discuss the general problem of economic operation of large interconnected areas.

Q.20.10 State what is meant by base-load and peak-load stations. Discuss the combined operation of hydroelectric and steam power stations.

Q.20.11 Derive an expression for the hourly loss in economy due to error in the representation of input data.

Q.20.12 What are the assumptions made in deriving transmission loss coefficients? Enumerate them.

Q.20.13 Explain hydrothermal economic load scheduling. Derive the necessary equations.

LOAD FREQUENCY CONTROL

21

In a power system the load demand is continuously changing. In accordance with it the power input has also to vary. If the input—output balance is not maintained, a change in frequency will occur. The control of frequency is achieved primarily through speed governor mechanism aided by supplementary means for precise control. At the outset, the speed governor mechanism and its operation will be presented.

Governor: The power system is basically dependent upon the synchronous generator and its satisfactory performance. The important control loops in the system are:

1. Frequency control
2. Automatic voltage control

In this chapter the frequency control will be discussed. Frequency control is achieved through generator control mechanism. The governing systems for thermal and hydrogenerating plants are different in nature since the inertia of water that flows into the turbine presents additional constraints which are not present with steam flow in a thermal plant. However, the basic principle is still the same; i.e., the speed of the shaft is sensed and compared with a reference, and the feedback signal is utilized to increase or decrease the power generated by controlling the inlet valve to turbine of steam or water as the case may be.

21.1 SPEED GOVERNING MECHANISM

The speed governing mechanism includes the following parts:

Speed Governor: It is an error sensing device in load frequency control. It includes all the elements that are directly responsive to speed and influences other elements of the system to initiate action.

Governor-Controlled Valves: They control the input to the turbine and are actuated by the speed control mechanism.

Speed Control Mechanism: It includes all equipment such as levers and linkages, servomotors, amplifying devices, and relays that are placed between the speed governor and the governor-controlled valves.

Speed Changer: It enables the speed governor system to adjust the speed of the turbo generator unit while in operation.

Electrical Power Systems. DOI: http://dx.doi.org/10.1016/B978-0-08-101124-9.00021-8

651

21.2 SPEED GOVERNOR

A simple schematic representation of the governor is shown in Fig. 21.1.

The pilot valve v operates to increase or decrease the opening of the steam inlet valve V. Let X_B and X_C be the changes in the position of the pilot valve v and control valve V responding to a change in governor position X_A due to load.

When the pilot valve is closed $X_B = 0$ and $X_C \cong 0$, (i.e.,) the control valve is not completely closed, as the unit has to supply its no-load losses. Let ω_0 be the no-load angular speed of the turbine. As load is applied, the speed falls and through the linkages the governor operates to move the piston P downward along with points A and B. The pilot valve v admits oil under D and lifts it up so that the input is increased and speed raised. If the link BC is removed, then the pilot valve comes to rest only when the speed returns to its original value. An "isochronous" characteristic will be obtained with such an arrangement where speed is restored to its preload-disturbance value. This is shown in Figs. 21.2 and 21.3.

With the link BC, the steady state is reached at a speed slightly lower than the no-load speed giving a drooping characteristic for the governor system. A finite value of the steady-state speed regulation is obtained with this arrangement. For a given speed changer position, the per unit steady-state speed regulation is defined by

$$\text{Steady-state speed regulation} = \frac{N_0 - N}{N_r}$$

where N_0 is the speed at no-load, N_r is the rated speed, N is the speed at rated load.

The isochronous and drooping characteristics are shown in Fig. 21.3.

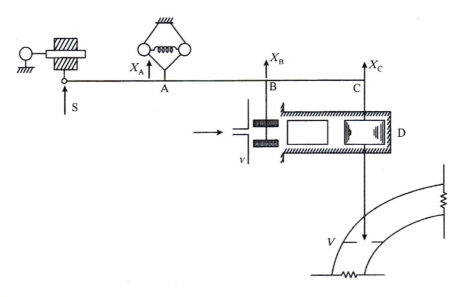

FIGURE 21.1

Speed governor.

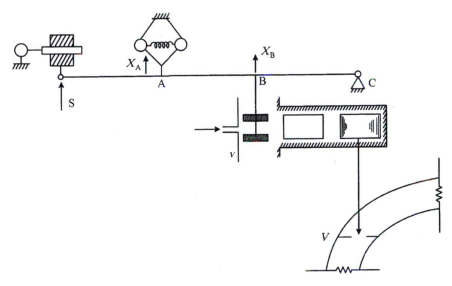

FIGURE 21.2

Isochronous governor.

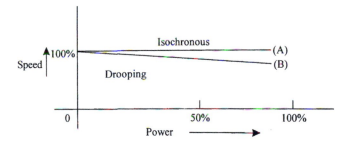

FIGURE 21.3

Governor characteristics.

21.3 STEADY-STATE SPEED REGULATION

The opening of the control valve is directly related to the power input to the unit so that

$$\frac{P}{P_r} = \frac{x_C}{C} \tag{21.1}$$

where C is the opening of the control valve corresponding to rated output P_r and x_C is the opening while the turbine delivers an output of P. The displacement at A, x_A is proportional to the change in speed so that

$$x_A = K(N_0 - N)$$
$$= K \, \Delta N \tag{21.2}$$

Defining the lever ratios

$$\frac{AB}{BC} = l_1 \tag{21.3}$$

$$\frac{BC}{AC} = l_2 \tag{21.4}$$

and

$$\frac{AB}{AC} = l_3 \tag{21.5}$$

$$\frac{x_A}{x_C} = \frac{AB}{BC} = l_1 \tag{21.6}$$

so that $x_A = l_1 \, x_C = l_1 \cdot C$

$$\Delta N = \frac{x_A}{k} = \frac{l_1}{k} \cdot \frac{P}{P_r} \cdot C = \frac{Cl_1}{k} \cdot \frac{P}{P_r} \tag{21.7}$$

The steady-state speed regulation is

$$R = \frac{N_0 - N}{N_r} = \frac{Cl_1}{k} \cdot \frac{P}{P_r N_r} = \left(\frac{Cl_1}{k} \cdot \frac{1}{P_r N_r} \right) P \tag{21.8}$$

thus, the steady-state speed regulation is directly proportional to the output power.
R has the dimension of Hz/MW.

21.4 ADJUSTMENT OF GOVERNOR CHARACTERISTICS

For a given governor, R can be changed by varying the lever ratio l_1 and the no-load speed N_0. The droop of the governor characteristics can be changed from curve B to B' as in Fig. 21.4 by changing the lever ratio l_1. Generally in order to coordinate a generating unit with the rest of the system, the lever ratio l_1 is adjusted. This is done occasionally and when the machine is cold.

When the machine is in operation, the no-load speed N_0 can be adjusted by varying an external force acting on the governor thrust sleeve at S so that the characteristic is shifted parallel to itself as shown by curves C and D. This is achieved by the speed changer (see Fig. 21.4).

21.5 TRANSFER FUNCTION OF SPEED CONTROL MECHANISM

From Eq. (21.2)

$$x_A = K(N_0 - N)$$
$$= K|(\omega_0 - \omega) = k|\Delta\omega \tag{21.9}$$

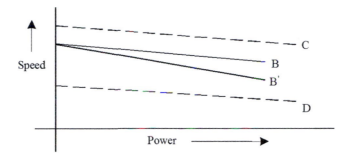

FIGURE 21.4

Action of speed changer.

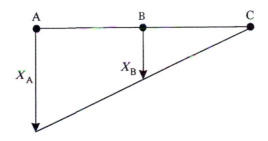

FIGURE 21.5

Deriving transfer function for speed control mechanism step (1).

If the point C is fixed, then (Fig. 21.5)

$$\frac{x_B}{x_A} = \frac{BC}{AC} = l_2 \tag{21.10}$$

$$x_B = l_2 x_A \tag{21.11}$$

If C were to move with A fixed, then from Fig. 21.6.

$$\frac{x_B}{x_C} = \frac{AB}{AC} = l_3$$

$$x_B = l_3 \cdot x_C$$

As the point C will be shifted upward by x_C due to the load increment, the actual movement of B would be

$$\therefore x_B = l_2 x_A - l_3 x_C \tag{21.12}$$

The opening of the pilot valve, x_B determines the rate of oil flow into the cylinder. The movement of the control valve, x_C is proportional to the total oil admitted under the piston P.

Therefore

$$x_C = k; \quad \int x_B \quad \text{where } k \text{ is a constant} \tag{21.13}$$

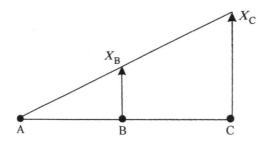

FIGURE 21.6

Deriving transfer function for speed control mechanism step (2).

Taking Laplace transform on both sides

$$X_C(S) = \frac{k}{s} X_B(S) \tag{21.14}$$

Taking Laplace transform of Eq. (21.12)

$$X_B(S) = l_2 X_A(S) - l_3 X_C(S) \tag{21.15}$$

substituting for $X_B(S)$ in (21.14)

$$\frac{s X_C(S)}{k} = l_2 X_A(S) - l_3 X_C(S) \tag{21.16}$$

$$X_C(S)\left[\frac{s}{k} + l_3\right] = l_2 X_A(S) \tag{21.17}$$

$$X_C(S) = \frac{\left(\frac{l_2}{l_3}\right) X_A(S)}{\left(\frac{s}{kl_3} + 1\right)} \tag{21.18}$$

$$= \frac{k_S}{1 + T_S s} X_A(S) \tag{21.19}$$

where $1/kl_3 = T_S$ and $l_2/l_3 = K_S$

T_S is the time constant of the speed control mechanism and is of the order of several milliseconds. The transfer function of the speed control mechanism is

$$G_{SC}(S) = \frac{X_C(S)}{X_A(S)} = \frac{K_S}{1 + sT_S} \tag{21.21}$$

21.6 TRANSFER FUNCTION OF A POWER SYSTEM

Under steady-state operating conditions, the synchronous angular frequency is

$$\omega_s = 2\pi f_S$$

where f_S is the synchronous frequency, i.e., the rated frequency.

$\Delta\delta =$ Change in angular position of the generator rotor corresponding to an increased load demand ΔP_D on the generator.

$$\delta = \delta_s + \Delta\delta \text{ (radians)} \tag{21.22}$$

$$\omega = \frac{d}{dt}(\delta) = \frac{d}{dt}(\delta_s + \Delta\delta) = \omega_s + \frac{d}{dt}\Delta\delta \tag{21.23}$$

$$\omega = \omega_s + \Delta\omega \text{ (radians/s)} \tag{21.24}$$

Let

$$f = f_S + \Delta f \tag{21.25}$$

where

$$\Delta f = \frac{1}{2\pi}\frac{d}{dt}\Delta\delta \text{ (Hz)} \tag{21.26}$$

The kinetic energy stored in the machine is

$$W = \frac{1}{2}I\omega^2 = \frac{1}{2}I(2\pi f)^2 \text{ (MJ)} \tag{21.27}$$

The kinetic energy at synchronous speed is

$$W_S = \frac{1}{2}I(2\pi f_S)^2 \text{ (MJ)} \tag{21.28}$$

Then

$$W = W_S\left(\frac{f}{f_S}\right)^2 \text{ (MJ)} \tag{21.29}$$

$$W = W_S\left(\frac{f_S + \Delta f}{f_S}\right)^2 \tag{21.30}$$

$$W \cong W_S\left(\frac{1 + 2\Delta f}{f_S}\right) \text{ for small } \frac{\Delta f}{f_S} \tag{21.31}$$

The rate of change of kinetic energy is the increase in power

$$\frac{d}{dt}(W) = \frac{2W_S}{f_S}\frac{d}{dt}(\Delta f) \tag{21.32}$$

Defining per-unit inertia constant $H = W_S/P_r$

$$\frac{d}{dt}(W) = \frac{2H}{f_S}\frac{d}{dt}(\Delta f) \text{ p.u.} \quad \text{taking } P_r = 1.0 \text{ p.u.} \tag{21.33}$$

Further, all types of composite loads experience a change in power consumption with frequency. Defining the load damping factor

$$D = \frac{\partial P_D}{\partial f} \text{ (p.u. MW/Hz)} \tag{21.34}$$

where P_D is the load demand in p.u.; the change in load demand in this case is then D. Δf p.u. MW

For a small step change in load demand by ΔP_D, the power balance equation can be written as follows:

$$\Delta P_G - \Delta P_D = D\Delta f + \frac{2H}{f_S}\frac{d}{dt}(\Delta f) \tag{21.35}$$

Taking Laplace transform of Eq. (21.35)

$$[\Delta P_G(S) - \Delta P_D(S)] = D\Delta F(S) + \frac{2H}{f_S}\cdot s\cdot \Delta F(S) \tag{21.36}$$

$$\Delta F(S) = [\Delta P_G(S) - \Delta P_D(S)]\cdot \frac{1}{\left[D + \dfrac{2H}{f_S}\right]} \tag{21.37}$$

$$\Delta F(S) = [\Delta P_G(S) - \Delta P_D(S)]\frac{\dfrac{1}{D}}{\left[1 + s\cdot \dfrac{2H}{f_S D}\right]} \tag{21.38}$$

$$\therefore \quad \Delta F(S) = [\Delta P_G(S) - \Delta P_D(S)]\left(\frac{K_P}{1 + sT_P}\right) \tag{21.39}$$

where

$$K_P = \frac{1}{D} \ (\text{Hz/p.u. MW}) \tag{21.40}$$

and

$$T_P = \frac{2H}{f_S D} \tag{21.41}$$

The transfer function relating the frequency change to the change in input–output power may be then written as

$$G_P(S) = \frac{K_P}{1 + sT_P} \tag{21.42}$$

Fig. 21.7 shows the block schematic of Eq. (21.39).

The complete block schematic of speed governing system is shown in Fig. 21.8.

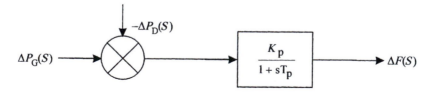

FIGURE 21.7

Block schematic for speed generator system.

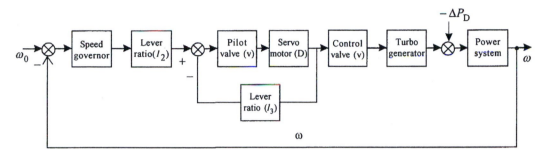

FIGURE 21.8

The block diagram of speed governing system.

21.7 TRANSFER FUNCTION OF SPEED GOVERNOR

Let m be the mass of the flying masses, c_f be the friction control, and c_s be the spring control of the governor system.

The equation of motion of the governor is

$$m\frac{d^2x_A}{dt^2} + c_f\frac{dx_A}{dt} + c_s x_A = k_s\Delta\omega \tag{21.43}$$

Taking the Laplace transform on both sides of Eq. (21.43).

$$(ms^2 + c_f s + c_s)x_A(S) = k_s\Delta w(S)$$

Rearranging

$$\frac{x_A(S)}{\Delta\omega(S)} = \frac{k_s}{ms^2 + c_f s + c_s} = \frac{k_s}{s^2 + 2\xi\nu s + \nu^2} \tag{21.44}$$

$$= G_g(S)$$

where $G_g(S)$ is the transfer function of the speed governor.

$$\nu = \sqrt{\frac{c_s}{m}} = \text{natural frequency of oscillation of flying masses with no damping}$$

$$\xi = \frac{c_f}{2\sqrt{k_s m}} = \text{damping constant} \tag{21.45}$$

$$X_A(S) = G(S)\cdot\Delta w(S)$$

21.8 MODEL FOR A STEAM VESSEL

Steam inlet piping or steam chest can be considered as a steam vessel with steam input and steam output (Fig. 21.9).

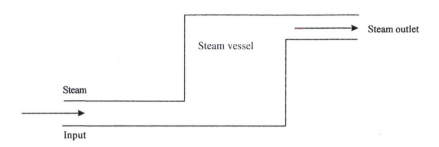

FIGURE 21.9

Steam vessel.

For changes in steam valve position, the dynamics can be represented by

$$\frac{dW}{dt} = Q_{in} - Q_{out}$$

where W is weight of steam in kg and Q_{in} and Q_{out} are the rate of flow of steam in kg/s at inlet and outlet of the steam vessel, respectively. Assuming that the flow rate is proportional to the pressure.

$$Q_{out} = \frac{P}{P_0} Q_0 \tag{21.55}$$

where P_0 and Q_0 are the steady-state pressure and rate of flow of steam in the vessel.

$$\frac{d}{dt} \cdot \frac{Q_{out}}{dt} = \frac{Q_0}{P_0} \cdot \frac{dp}{dt} \tag{21.56}$$

Assuming constant temperature in the vessel

$$\frac{dW}{dt} = \frac{dW}{dp} \cdot \frac{dp}{dt} = \frac{d}{dp}\left(W \cdot \frac{dp}{dt}\right) \tag{21.57}$$

$$= \frac{d}{dp}\left(\frac{V}{v}\right)\frac{dp}{dt}$$

where V is the volume of the vessel in m^3 and v is the specific volume of steam (m^3/kg) in the vessel.

$$Q_{in} - Q_{out} = \frac{d\omega}{dt}$$

$$= V \cdot \frac{d}{dp}\left(\frac{1}{v}\right)\frac{dp}{dt}$$

$$= V \frac{d}{dp}\left(\frac{1}{v}\right) \cdot \frac{dQ_{out}}{dt} \cdot \frac{P_0}{Q_0} \tag{21.58}$$

$$= V \frac{d}{dp}\left(\frac{1}{v}\right)\frac{P_0}{Q_0} \cdot \frac{d}{dt} Q_{out}$$

Let

$$T = \frac{P_0}{Q_0} \cdot V \frac{d}{dp} \cdot \frac{1}{v} \tag{21.59}$$

$$Q_{in} - Q_{out} = T\frac{dQ_{out}}{dt} \tag{21.60}$$

Taking Laplace transforms of Eq. (21.60)

$$Q_{in}(S) - Q_{out}(S) = TsQ_{out}(S)$$

$$\frac{Q_{out}(S)}{Q_{in}(S)} = \frac{1}{1 + sT} \tag{21.61}$$

The transfer function of the steam vessel is

$$G_{Sv}(S) = \frac{1}{1 + sT} \tag{21.62}$$

21.9 STEAM TURBINE MODEL

In normal steady state the turbine power P_T keeps balance with the electromechanical air-gap power P_G, resulting in zero acceleration and a constant speed or frequency. Changes in P_T or P_G, i.e., ΔP_T or ΔP_G or both will upset the above balance. If the difference $\Delta P_T - \Delta P_G$ is positive, the turbine generator unit will accelerate, and if negative it will decelerate.

The Turbine Generator power increment ΔP_T depends entirely upon the valve power increment ΔP_v and the response characteristics of reheat and nonreheat type of turbines are different. However, it is possible to express the turbine-generator dynamics in terms of turbine transfer function. The turbine-generator transfer function is given by

$$G_{TG}(S) = \frac{\Delta P_{TG}(S)}{\Delta P_v(S)}$$

For a simple nonreheat type turbine, the model is given by a single time constant

$$G_{TG}(S) = \frac{K_{TG}}{1 + sT_{TG}} \tag{21.63}$$

where K_{TG} is the gain constant and T_{TG} is the time constant of turbine-generator unit.

21.10 SINGLE CONTROL AREA

In the previous sections models for turbine-generator, power system and speed governing systems are obtained. In practice, rarely a single generator feeds a large area. Several generators connected in parallel, located also, at different places will supply the power needs of a geographical area. Quite normally, all these generators may have the same response characteristics for changes in load demand.

In such a case it is possible to define a control area, grouping all the generators in the area together and treating them as a single equivalent generator. For small load changes all these generators swing in unison. Putting together, the various models derived can be conceived as a single control area or simply an area as shown in Fig. 21.10.

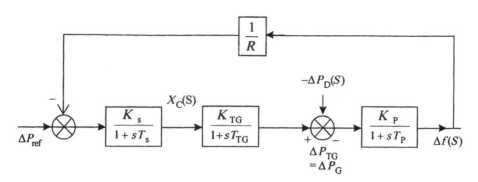

FIGURE 21.10

Block diagram of a single area system.

From Eq. (21.21)

$$X_C(S) = \frac{K_S}{1 + sT_S} X_A(S)$$

From Eq. (21.42)

$$G_p(S) = \frac{K_P}{1 + sT_P}$$

and from Eq. (21.39)

$$\Delta F(S) = [\Delta P_G(S) - \Delta P_D(S)]\left[\frac{K_P}{1 + sT_P}\right]$$

Note that $\Delta P_G(S) = \Delta P_{TG}(S)$ as generator power change which is also the turbine power change and also the turbine generator unit power change.

Also, from Eq. (21.63) $G_{TG} = K_{TG}/1 + sT_{TG}$. We obtain the block diagram of Fig. 21.10 from Eqs. (21.8), (21.21), (21.39), and (21.63).

21.11 THE BASICS OF LOAD FREQUENCY CONTROL

The following basic requirements are to be fulfilled for successful operation of the system:

1. The generation must be adequate to meet all the load demand.
2. The system frequency must be maintained within narrow and rigid limits.
3. The system voltage profile must be maintained within reasonable limits.
4. In case of interconnected operation the tie-line power flows must be maintained at the specified values.

When real power balance between generation and demand is achieved, the frequency specification is automatically satisfied. Similarly, with a balance between reactive power generation and demand, voltage profile is also maintained within the prescribed limits. Under steady-state conditions, the total real power generation in the system equals the total MW demand plus real power losses. Any difference is immediately indicated by a change in speed or frequency. Generators are fitted with speed governors which will have varying characteristics: different sensitivities, dead bands, response times, and droops. They adjust the input to match the demand within their limits. Any change in local demand within permissible limits is absorbed by generators in the system in a random fashion.

An independent aim of the *automatic generation control* is to reschedule the generation changes to preselected machines in the system after the governors have accommodated the load change in a random manner. Thus additional or supplementary regulation devices are needed along with governors for proper regulation.

The control of generation in this manner is termed *load frequency control*. For interconnected operation, the last of the four requirements mentioned earlier is fulfilled by deriving an error signal from the deviations in the specified tie-line power flows to the neighboring utilities and adding this signal to the control signal of the load-frequency control system. This last requirement will be discussed in detail later in the chapter.

Should the generation be not adequate to balance the load demand, it is imperative that one of the following alternatives be considered for keeping the system in operating condition:

1. Starting fast peaking units
2. Load shedding for unimportant loads
3. Generation rescheduling

It is apparent from the above that since the voltage specifications are not stringent, load frequency control is by far the most important in power system control. The block schematic for such a control is shown in Fig. 21.11.

To understand the mechanism of frequency control, consider a small−step increase in load. The initial distribution of the load increment is determined by the system impedance and the

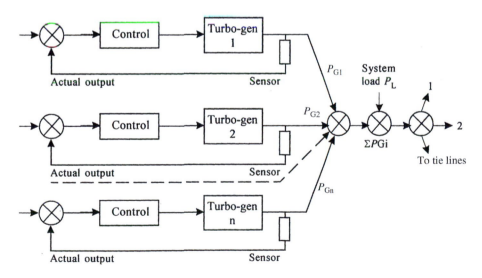

FIGURE 21.11

Block diagram for load frequency control.

instantaneous relative generator rotor positions. The energy required to supply the load increment is drawn from the kinetic energy of the rotating machines. As a result, the system frequency drops. The distribution of load during this period among the various machines is determined by the inertias of the rotors of the generators partaking in the process. This problem is studied in the stability analysis of the system.

After the speed or frequency fall due to reduction in stored energy in the rotors has taken place, the drop is sensed by the governors and they divide the load increment between the machines as determined by the droops of the respective governor characteristics.

Subsequently, secondary control restores the system frequency to its normal value by readjusting the governor characteristics.

21.12 FLAT FREQUENCY CONTROL

In this method a master machine is charged with the task of maintaining the system frequency. All the remaining machines carry constant loads. The capacity of the machine which controls the frequency should be between 5% and 10% of the entire generating capacity of the system for effective frequency control. In small, independent systems, old and inefficient machines may be assigned to frequency regulation leaving new and efficient machines to supply the load demand.

Modern power systems are so large that it is impossible to design a single central control system that would handle the overall control job. It is extremely useful to take into account the weak links in the system and then apply control through decomposition. The demarcation of load frequency control and Mvar voltage control characteristics is one such decomposition. Geographical and functional decomposition are successfully applied to power systems and this leads to the concept of area control. A modern power system can be divided into several areas for load frequency control. Each control area fulfils the following:

1. The area is a geographically contiguous portion of a large interconnected area, which adjusts its own generation to accommodate load changes within its precincts.
2. Under normal conditions of operation, it exchanges bulk power with neighboring areas.
3. Under abnormal conditions of operation, it may deviate from predetermined schedules and provide assistance to any neighboring control area in the system.
4. It is expected, in addition, to partake with the other areas in the system in a suitable manner in the system frequency regulation.

The rotors of all generators in a control area swing together for load changes. Thus a coherent group of generators within a geographical region may constitute a *control area* which is connected to other similar areas by weak tie lines as shown in Fig. 21.12.

21.13 REAL POWER BALANCE FOR LOAD CHANGES

It has been already mentioned that when an imbalance occurs between real power generation and load demand, frequency deviation takes place. Consider a single area operating in steady state and let it undergo a step change ΔP_D in load demand. Assuming that the generation changes by an amount

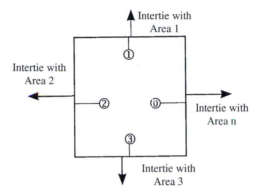

FIGURE 21.12

Single control area interties.

ΔP_G following the load change due to governor action, the power imbalance is $\Delta P_G - \Delta P_D$. There are three possible modes in which this power can be absorbed. For positive power imbalance:

1. the area kinetic energy increases (i.e., the kinetic energy of all the generator rotors in the area increases),
2. the load demand increases (due to load characteristics), and
3. the power outflow from the area to other interconnected areas increases, if such interconnections exist.

21.14 TRANSFER FUNCTION OF A SINGLE AREA SYSTEM

Under steady-state operating conditions, transfer functions for a single control area can be derived exactly in the same way as for the system in Section 21.6.

$$\omega_S = 2\pi f_S$$

where ω_s and f_S are synchronous angular speed and rated frequency, respectively.

If $\Delta\delta$ is the change in angular position of the equivalent generator representing the area corresponding to an increased load ΔP_D in the area.

Then, $\delta = \delta_S + \Delta\delta$ (radians)

$$\omega = \frac{d}{dt}(\delta) = \frac{d}{dt}(\delta_s + \Delta\delta) = \omega_S + \frac{d\Delta\delta}{dt} \qquad (21.64)$$
$$= \omega_s + \Delta\omega$$

Let

$$f = f_S + \Delta f$$

where

$$\Delta f = \frac{1}{2\pi}\frac{d}{dt}\Delta\delta \ (\text{Hz}) \qquad (21.65)$$

The area kinetic energy

$$W = \frac{1}{2}I\omega^2 = \frac{1}{2}I(2\pi f)^2 \text{ (MJ)}$$

where I is the moment of inertia of the area.

Also, the kinetic energy at synchronous speed ω_s is

$$W_S = \frac{1}{2}I(2\pi f)^2 \text{ (MJ)}$$

Thus

$$W = W_S\left(\frac{f}{f_S}\right)^2 \text{ (MJ)}$$

$$= W_S\left(\frac{f_S + \Delta f}{f_S}\right)^2$$

$$W \approx W_S\left(1 + 2\frac{\Delta f}{f_S}\right) \quad \text{for small } \frac{\Delta f}{f_S} \tag{21.66}$$

The rate of change of kinetic energy is the increase in area power which is

$$\frac{d}{dt}(W) = \frac{2W_S}{f_S}\frac{d}{dt}(\Delta f)$$

Defining per-unit inertia constant $M = W_S/P_r$

$$\frac{d}{dt}(W) = \frac{2M}{f_S}\frac{d}{dt}(\Delta f) \text{ (p.u.)}$$

All types of composite loads experience a change in power consumption with frequency. Defining the load damping factor D.

$$D = \frac{\partial P_D}{\partial f} \text{ (p.u. MW/Hz)} \tag{21.67}$$

where P_D is the load demand in p.u.; the increase in load demand in this case is then $D \cdot \Delta f$ (p.u. MW). For a small-step change in load demand, Δf (p.u. MW). For a small-step change in load demand, ΔP_D the power balance equation takes the form:

$$\Delta P_G - \Delta P_D = D \cdot \Delta f + \frac{2H}{f_S}\frac{d}{dt}(\Delta f) \tag{21.68}$$

Taking the Laplace transform of Eq. (21.68)

$$\Delta F(S) = [\Delta P_G(S) - \Delta P_D(S)]\frac{1}{\left[D + \dfrac{2Hs}{f_S}\right]}$$

or

$$= [\Delta P_G(S) - \Delta P_D(S)]\frac{\dfrac{1}{D}}{\left[1 + \dfrac{2H}{f_S D}.s\right]}$$

$$= [\Delta P_G(S) - \Delta P_D(S)] = D\Delta F(S) + \frac{2Hs}{f_S}\Delta F(S) \tag{21.69}$$

$$= [\Delta P_G(S) - \Delta P_D(S)]\frac{K_P}{[1 + sT_P]}$$

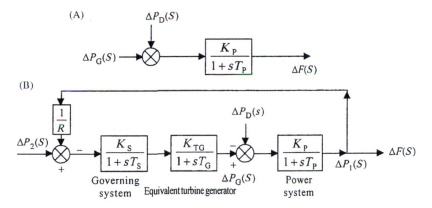

FIGURE 21.13

(A and B) Block diagram of a single area system.

where $K_P = 1/D$ (Hz/p.u. MW)

$$T_P = \frac{2H}{f_S D}$$

The transfer function relating the frequency change to the change in input/output powers may be designed by $G_P(S)$ so that

$$G_P(S) = \frac{K_P}{1 + sT_P} \tag{21.70}$$

The block schematic for a single area system is shown in Fig. 21.13A. An entire control area may be represented as in Fig. 21.13B.

21.15 ANALYSIS OF SINGLE AREA SYSTEM

Consider the single area system shown in Fig. 21.13. For a step load change in the system, the following equations can be written

$$[\Delta P_G(S) - \Delta P_G(S)]G_P(S) = \Delta F(S)$$

$$\frac{1}{RG_{ST}(S)\Delta F(S)} = \Delta P_G(S)$$

where

$$G_{ST} = \left(\frac{K_{TG}}{1 + sT_{TG}} \times \frac{K_S}{1 + sT_S} \right)$$

Solving for $\Delta F(S)$

$$\Delta F(S) = -\frac{G_P(S)}{1 + \left(\frac{1}{R}\right) G_P(S) G_{ST}(T_S)} \Delta P_D(S) \qquad (21.71)$$

For a step load change, $\Delta P_D(S) = \Delta P_D / s$
Substituting the value of $\Delta P_D(S)$

$$\Delta F(S) = -\frac{G_P(S)}{1 + \left(\frac{1}{R}\right) G_P(S) G_{ST}(T)} \cdot \frac{\Delta P_D(S)}{s}$$

Applying the final value theorem, the steady-state error Δf_{SS} would be

$$\Delta f_{SS} = \underset{s \to 0}{Lim}[s \Delta F(S)] = \frac{-\Delta P_D}{D + \dfrac{K_{TG} K_S}{R}}$$

The product $K_{TG} K_S$ can be made unity by properly selecting the units for the input–output quantities to the combined block $G_{ST}(S)$ so that

$$\Delta f_{SS} = \frac{-\Delta P_D}{D + \dfrac{1}{R}}$$

The quantity $(D + (1/R))$ is defined as *area frequency response characteristic* or *area frequency regulation characteristic* (AFRC) and denoted by β. It may be noticed that as R becomes smaller Δf_{SS} approaches zero.

21.15.1 REFERENCE POWER SETTING

In Fig. 21.13 reference power is indicated at the input summer to the system. This determines the starting point of the governor characteristic. Under static conditions setting, $K_S K_{TG} = 1$. We obtain

$$\Delta P_{ref}^0 - \frac{1}{R} \Delta f^0 = \Delta P_{TG}^0$$

Changing the reference power setting will also change the turbine generator output in a proportional manner. For instance if the machine is connected to an infinite bus in which case $\Delta f = 0$ we have the direct relationship.

$$\Delta P_{ref} = \Delta P_{TG}$$

For a fixed setting of the speed changer, the steady-state increase in power output from the turbine generator ΔP_{TG} is directly proportional to the frequency drop. As in this case $\Delta P_{ref} = 0$

$$\Delta P_{TG} = -\frac{1}{R} \Delta f$$

R has the units Hertz per MW or Hertz per-unit MW as the case may be.
If $D = 0$, then AFRC $= \beta = 1/R =$ the area speed regulation coefficient.
The time constants T_s and T_T are much smaller compared to T_P and as such, an approximate solution to the response of the system could be obtained by neglecting them. Then, where $A = K_P + R$ and $B = T_P R$, partial fractions for the expression in the bracket can be obtained as follows:

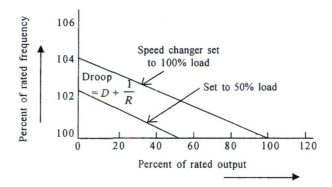

FIGURE 21.14

Speed changer setting.

$$(S) = -\frac{G_P(S)}{1 + \left(\dfrac{1}{R}\right)G_P(S)}\frac{\Delta P_D}{s}$$

$$= \frac{-\dfrac{K_P}{1 + sT_P}}{1 + \left(\dfrac{1}{R}\right)\dfrac{K_P}{1 + sT_P}}\frac{\Delta P_D}{s}$$

$$= -\Delta P_D R K_P\left[\left(\frac{1}{s}\right)\left(\frac{1}{A + Bs}\right)\right]$$

$$\frac{1}{s}\cdot\frac{1}{A + Bs} = \frac{C}{s} + \frac{D}{A + Bs}$$

At

$$s = 0; \quad C = \frac{1}{A}$$

And at

$$s = -\frac{A}{B}; \quad D = -\frac{B}{A}$$

So that

$$\Delta F(S) = -\Delta P_D R K_P\left[\frac{1}{s}\frac{1}{(K_P + R)} - \frac{T_P R}{(K_P + R)} - \frac{1}{(K_P + R) + sT_P R}\right] = -\frac{\Delta P_D R K_P}{R + K_P}\left[\frac{1}{s} - \frac{1}{\dfrac{K_P + R}{RT_P} + s}\right]$$

$$\Delta f(t) = \mathcal{L}^{-1}[\Delta F(S)]$$

$$= \mathcal{L}^{-1}\left[\frac{-\Delta P_D}{\beta}\left(\frac{1}{s} - \frac{1}{s + \dfrac{K_P\beta}{T_P}}\right)\right]$$

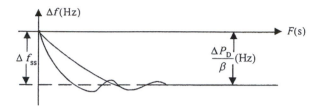

FIGURE 21.15

Uncontrolled single area system-response to step load change.

$$f(t) = \frac{-\Delta P_\mathrm{D}}{\beta}\left[1 - \exp\left(\beta\frac{K_\mathrm{P}t}{T_\mathrm{P}}\right)\right] \tag{21.72a}$$

The response is shown in Fig. 21.15.

21.16 DYNAMIC RESPONSE OF LOAD FREQUENCY CONTROL LOOP: UNCONTROLLED CASE

It has been shown that the load frequency control system possess inherently steady-state error for a step input. Applying the usual procedure, the dynamic response of the control loop can be evaluated so that the initial response also can be seen for any overshoot.

For this purpose considering the relatively larger time constant of the power system, the governor action can be neglected, treating it as an instantaneous action. Further the turbine-generator dynamics also may be neglected at the first instant to derive a simple expression for the time response.

It has been proved that (Eq. 21.71)

$$\Delta F(S) = -\frac{G_\mathrm{P}}{1 + \dfrac{1}{R}G_\mathrm{S}G_\mathrm{TG}G_\mathrm{P}}\Delta P_\mathrm{D}(S)$$

For a step load change of magnitude k

$$\Delta P_\mathrm{D}(S) = \frac{-k}{s}$$

Neglecting the governor action and turbine dynamics

$$\Delta F(S) = -\frac{G_\mathrm{P}}{1 + \dfrac{1}{R}G_\mathrm{P}}\frac{k}{s}$$

$$= -\left(\frac{K_\mathrm{P}}{1 + sT_\mathrm{P}}\right)\frac{1}{\left(1 + \dfrac{1}{R}\dfrac{K_\mathrm{P}}{1 + sT_\mathrm{P}}\right)}\frac{k}{s}$$

Applying partial fractions

$$\Delta F(S) = - \frac{K_P k}{R + K_P} \left[\frac{1}{s} + \frac{1}{s + \left(\frac{1}{T_P} + \frac{K_P}{R T_P} \right)} \right] \tag{21.72b}$$

For specific values of K_P, K, T_P, and R, $\Delta f(t)$ can be obtained by taking inverse Laplace transform of $\Delta F(S)$.

21.17 CONTROL STRATEGY

The uncontrolled system is thus subject to a steady-state error for step load changes. To reduce this error, consider first the introduction of a negative feedback signal from the frequency deviation $\Delta F(S)$ as shown in Fig. 21.16.
Then

$$\Delta P_{\text{ref}} = \Delta P_C = - K_1 \Delta F(S) \tag{21.73}$$

For this case, the steady-state frequency error has been proved in Eq. (21.64) as

$$\Delta f_{SS} = \frac{-\Delta P_D}{D + \frac{1}{R}} = \frac{-\Delta P_D}{\beta} \text{ or } \frac{-\Delta P_D}{D + K_1} \tag{21.74}$$

If K_1 is made very large, then only Δf_{SS} reduces to zero. In other words, this means that R should be made equal to zero, which is not desirable. Proportional control is not suitable for reducing the steady-state error to zero.

It is a well-known fact in control theory that integral control will improve the steady-state performance.

Let

$$\Delta P_C(t) = - K_2 \int \Delta f(t) \, dt \tag{21.75}$$

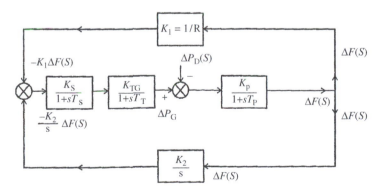

FIGURE 21.16

Proportional and integral control.

So that

$$\Delta P_C(S) = -\frac{K_2}{s}\Delta F(S) \tag{21.76}$$

The negative sign signifies that the control signal ΔP_C has to be increased for reduction in frequency. The equations for the system in Fig. 21.16 are:

$$\frac{K_P}{1 + sT_P}(\Delta P_G(S) - \Delta P_D(S)) = \Delta F(S) \tag{21.77}$$

Also, if $K_S K_{TG}$ is adjusted to be unity

$$\left(\frac{K_2}{S} + \frac{1}{R}\right)\Delta F(S)\left[\frac{1}{1 + sT_S}\right]\left(\frac{1}{1 + sT_T}\right) = \Delta P_G(S) \tag{21.78}$$

Solving both the equations above for $\Delta F(S)$,

$$\Delta F(S) = \frac{\Delta P_D K_P(1 + sT_s)(1 + sT_{TG})}{s(1 + sT_P)(1 + sT_S)(1 + sT_T) + K_P\left(K_2 + \dfrac{s}{R}\right)}$$

where $\Delta P_D(S) = \Delta P_D/s$ (step load change) and steady-state error is obtained as

$$\underset{s\to 0}{\text{Lt}}\, s\Delta F(S) = \underset{s\to 0}{\text{Lt}}\, s.\frac{\Delta P_D K_P(1 + sT_s)(1 + sT_{TG})}{s(1 + sT_P)(1 + sT_S)(1 + sT_T) + K_P\left(K_2 + \dfrac{s}{R}\right)} = 0 \tag{21.79}$$

By making use of integral control strategy, the steady-state error can be eliminated.

For the reasons explained before, T_T and T_S cam be neglected for an approximate analysis for the response. For a step load disturbance

$$F(S) = \frac{s\left(\dfrac{K_P}{T_P}\right)}{s^2 + s\left(1 + \dfrac{K_P}{R}\right)\dfrac{1}{T_P} + \dfrac{K_2 K_P}{T_P}} \times \frac{-DP_D}{s} \tag{21.80}$$

The response $\Delta f(t)$ depends on the nature of the denominator expression, i.e., the characteristic equation.

$$s^2 + 2\delta\omega_n s + \omega_n^2$$

where

$$\omega_n = \sqrt{\frac{K_2 K_P}{T_P}} \quad \text{is the natural frequency}$$

and

$$\delta = \frac{1 + \dfrac{K_P}{R}}{2T_P}\sqrt{\frac{T_P}{K_2 K_P}} = \text{damping ratio}$$

The gain for critical damping can be obtained by setting δ equal to unity.

$$\left(\left(\frac{1 + \dfrac{K_P}{R}}{2T_P}\right)^2 \left(\frac{T_P}{K_2 T_P}\right) = 1\right)$$

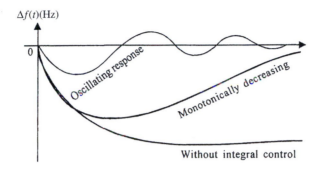

FIGURE 21.17

Response to integral control with different gain settings.

or

$$K_2 = \left(\frac{1 + \dfrac{K_P}{R}}{2T_P} \right)^2 \left(\frac{T_P}{K_P} \right) \qquad (21.81)$$

For values of K_2 greater than that given by Eq. (21.81), the response is oscillatory due to under-damping. For values of K_2 less than the critical value, the response is monotonically decreasing without oscillations or as in Fig. 21.17.

The selection of the gain for the controller should be such that the following specifications are satisfied.

1. The control loop must be stable.
2. The frequency error should return to zero following a step load change. The deviation in the transient state must also be minimized.
3. The integral of the frequency error should not exceed a certain value (say 150 Hz or 3 s)

21.18 PROPORTIONAL–INTEGRAL–DERIVATIVE CONTROLLERS

From the above analysis, it is clear that proportional, integral, and derivative control strategy can be applied for load frequency control. While proportional control is inherent in the feedback through the governor mechanism itself, derivative control when intro-duced improves transient performance and ensures better margin of stability for the system.

$$\Delta P_C(S) = -K_1 \Delta F(S) - \frac{K_2}{s} \Delta F(S) - K_3 s \Delta F(S)$$

where K_3 is the gain of derivative controller.

The response to such a control is

$$F(S) = \frac{\Delta P_D(S) \cdot \dfrac{K_P}{(T_P + K_P K_3)s}}{s^2 + \dfrac{1 + K_1 K_P + \dfrac{K_P}{R}}{T_P + K_P K_3} s + K_2 \dfrac{K_P}{(T_P + K_P K_3)}}$$

By proper choice of K_1, K_2, and K_3 all the specifications for the system performance can be satisfied.

21.19 INTERCONNECTED OPERATION

Power systems are interconnected for economy and continuity of power supply. For the interconnected operation incremental efficiencies, fuel costs, water availability, generation limits, tie-line capacities, spinning reserve allocation, and area commitments are important considerations in preparing load dispatch schedules.

In this chapter the power control of interconnected system is presented.

21.20 TWO-AREA SYSTEM—TIE-LINE POWER MODEL

Consider two interconnected areas as shown in Fig. 21.18 operating at the same frequency f^0 while a power flow from Area 1 to Area 2 let $|V_1|$ and $|V_2|$ be the voltage magnitudes at δ_1^0 and δ_2^0 be the voltage phase angles at the two ends of the tie-line while P_{12}^0 flows from Area 1 to Area 2 then

$$P_{12}^0 = \frac{|V_1| \, |V_2|}{X} \sin (\delta_1^0 - \delta_2^0) \tag{21.82}$$

where X is the reactance of the line.

If the angles change by $\Delta \delta_1$ and $\Delta \delta_2$ due to load changes in Areas 1 and 2, respectively, then the tie-line power changes by

$$\Delta P_{12} = \frac{|V_1^0| |V_2^0|}{X} \cos (\delta_1^0 - \delta_2^0)(\Delta P_1 - \Delta P_2) \tag{21.83}$$

Defining

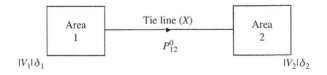

FIGURE 21.18

Two-area system.

$$\frac{\Delta P_{12}}{\Delta \delta_1 - \Delta \delta_2} = \frac{\Delta P_{12}}{\Delta \delta} \text{ (MW/radian)} \tag{21.84}$$

as a synchronizing coefficient of the tie-line or "stiffness coefficient" of the line, denoted by T^0

$$\Delta P_{12} = T^0(\Delta \delta_1^0 - \Delta \delta_2^0) \tag{21.85}$$

consider

$$\Delta \omega = \frac{d}{dt} \Delta \delta$$

$$\text{i.e.,} \quad 2\pi \Delta f = \frac{d}{dt} \Delta \delta$$

$$\Delta f = \frac{1}{2\pi} \frac{d}{dt} \Delta \delta \text{ (Hz)}$$

In other words

$$\Delta \delta = 2\pi \int \Delta f \, dt \text{ (radians)} \tag{21.86}$$

Hence

$$\Delta \delta_1 = 2\pi \int \Delta f_1 \, dt \text{ (radians)}$$

and

$$\Delta \delta_2 = 2\pi \int \Delta f_2 \, dt \text{ (radians)}$$

From Eq. (21.85)

$$\Delta P_{12} = 2\pi T^0 \left(\int \Delta f_1 \, dt - \int \Delta f_2 \, dt \right) \text{ (MW)} \tag{21.87}$$

Taking Laplace transform on both sides

$$\Delta P_{12}(S) = \frac{2\pi T^0}{s} [\Delta F_1(S) - \Delta F_2(S)] \tag{21.88}$$

Block schematic of Eq. (21.88) is shown in Fig. 21.19.

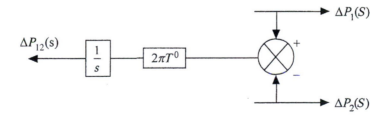

FIGURE 21.19

Block diagram for tie-line power.

If the two areas are rated at P_{r1} and P_{r2}
then if $P_{r1}/P_{r2} = a_{12}$

$$\Delta P_{21}(S) = \frac{2\pi T_{21}^0}{s}[\Delta F_2(S) - \Delta F_1(S)]$$

$$\frac{2\pi T_{21}^0 \cdot a_{12}}{s}[\Delta F_1(S) - \Delta F_2(S)]$$

21.21 BLOCK DIAGRAM FOR A TWO-AREA SYSTEM

The block diagram for a two-area system can now be developed. Each area can be represented by a block diagram as in the case of a single area system, but with suffixes 1 and 2. The block diagram for the tie-line power deviation (Fig. 21.19) can be used to interconnect both the areas as shown in Fig. 21.19.

21.22 ANALYSIS OF TWO-AREA SYSTEM

Steady-State Response:
    Consider the speed changer positions as fixed so that

$$\Delta P_{ref,1} = \Delta P_{ref,2} = 0$$

let the loads in both the areas change by $\Delta P_{D1} = k_1$ and $\Delta P_{D1} = k_2$, step changes in the two area, respectively. Let Δf^0 be the final or steady-state frequency deviation due to load changes. Similarly let ΔP_{12}^0 be the change in the tie-line power flow.
    From Fig. 21.20

$$\Delta P_{TG,1}^0 = -\frac{1}{R_1}\Delta f_0 \tag{21.89}$$

$$\Delta P_{TG,2}^0 = -\frac{1}{R_2}\Delta f_0 \tag{21.90}$$

where $\Delta P_{TG,1}^0$ and $\Delta P_{TG,2}^0$ are the steady-state changes in turbine-generator outputs.
    Also, from the same Fig. 21.20, at the summing point of load and tie-line powers, we get

$$\left(-\frac{1}{R_1}\Delta f^0 - k_1 - \Delta P_{12}^0\right)K_{P1} = \Delta f^0 \tag{21.91}$$

i.e.,

$$\left(-\frac{1}{R_1}\Delta f^0 - k_1\right) = \frac{\Delta f^0}{K_{P1}} + \Delta P_{12}^0$$

since

$$\frac{1}{K_{P1}} = D_1$$

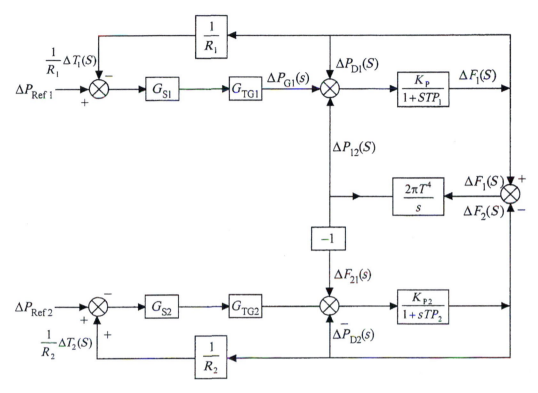

FIGURE 21.20

Block diagram for a two-area system.

$$\left(-\frac{1}{R_1}\Delta f^0 - k_1\right) = D_1\Delta f^0 + \Delta P_{12}^0 \tag{21.92}$$

$$\text{similarly for Area } 2\left(-\frac{1}{R_2}\Delta f^0 - k_2\right) = D_2\Delta f^0 + \Delta P_{12}^0 \tag{21.93}$$

solving Eqs. (21.92) and (21.93)
  we obtain

$$\Delta f^0 = -\left(\frac{k_1 + k_2}{\beta_1 + \beta_2}\right) \text{(Hz)} \tag{21.94}$$

and

$$\Delta P_{12}^0 = -\Delta P_{21}^0 = \frac{\beta_1 k_2 - \beta_2 k_1}{\beta_1 + \beta_2} \tag{21.95}$$

where

$$\beta_1 = D_1 + \frac{1}{R_1} \tag{21.96}$$

and

$$\beta_2 = D_2 + \frac{1}{R_2} \qquad (21.97)$$

which are defined as the area frequency response characteristics.

Let both the areas be identical.

Then

$$D_1 = D_2 = D$$

$$R_1 = R_2 = R$$

$$\beta_1 = \beta_2 = \beta$$

The deviations become

$$\Delta f^0 = \frac{k_1 + k_2}{2\beta} \text{ (Hz)} \qquad (21.98)$$

and

$$\Delta P^0_{12} = -\Delta P^0_{21} = \frac{k_2 - k_1}{2} \text{ (p.u. MW)} \qquad (21.99)$$

If the area ratings are different and $P_{r1}/P_{r2} = a_{12}$, then

$$\Delta f^0 = -\frac{k_2 + a_{12}k_1}{\beta_1 + a_{12}\beta_1}$$

and

$$\Delta P^0_{12} = -\Delta P^0_{21} = \frac{\beta_1 k_2 - \beta_2 k_1}{\beta_2 + a_{12}\beta_1}$$

If a load disturbance occurs in only one of the areas, it is clear that with $k_1 = 0$ or $k_2 = 0$, the frequency derivation Δf^0 is only half of the steady-state error that would have occurred had there been no interconnection. Thus, with several systems interconnected, the steady-state frequency error would be reduced.

Also from the tie-line power derivation, it can be observed that half of the load change in either area will be supplied by the other area, which demonstrates the importance of emergency assistance in interconnected or pool operation.

21.23 DYNAMIC RESPONSE

From the block diagram of Fig. 21.20, the following equation can be written:

$$\left\{ -\frac{1}{R_1}\Delta F_1(S)\left[\frac{K_{S1}}{1 + sT_{S1}}\right]\left[\frac{K_{TG1}}{1 + sT_{TG2}}\right] - \Delta P_{D1}(S) - \Delta P_{12}(S) \right\} \frac{K_{P1}}{1 + sT_{P1}} = \Delta F_1(S) \qquad (21.100)$$

$$\left\{ -\frac{1}{R_2}\Delta F_2(S)\left[\frac{K_{S2}}{1 + sT_{S2}}\right]\left[\frac{K_{TG2}}{1 + sT_{TG2}}\right] - \Delta P_{D2}(S) - \Delta P_{21}(S) \right\} \frac{K_{P2}}{1 + sT_{P2}} = \Delta F_2(S) \qquad (21.101)$$

$$\Delta P_{12}(S) = \frac{2\pi T^0}{s}[\Delta F_1(S) - \Delta F_2(S)] \tag{21.102}$$

$$\Delta P_{21}(S) = -\Delta P_{21}(S)$$

There are four equations with four variables, ΔF_1, ΔF_2, ΔP_{12}, and ΔP_{21} to be determined for given ΔP_{D1} and ΔP_{D2}. The dynamic response can be obtained; even though it is a little bit involved.

For simplicity assume that the two areas are equal. Neglect the governor and turbine dynamics, which means that the dynamics of the system under study is much slower than the fast acting turbine-governor system in a relative sense. Also assume that the load does not change with frequency ($D_1 = D_2 = D = 0$).

We obtain under these assumptions the following relations
since $D = 0$

$$G_{P1} = \frac{K_{P1}}{1 + sT_{P1}} = G_{P2} = \frac{K_{P2}}{1 + sT_{P2}} = \frac{\frac{1}{D}}{1 + s\left(\frac{2H}{f^0 D}\right)}$$

$$G_P = \frac{1}{D + \left(\frac{2Hs}{f^0}\right)} = \frac{f^0}{2Hs}$$

Eqs. (21.100) and (21.101) simplify to

$$\left[-\frac{1}{R}\Delta F_1(S) - \Delta P_{D1}(S) - \Delta P_{12}(S)\right]\frac{f^0}{2sH} = \Delta F_1(S) \tag{21.102a}$$

$$\left[-\frac{1}{R}\Delta F_2(S) - \Delta P_{D2}(S) - \Delta P_{21}(S)\right]\frac{f^0}{2sH} = \Delta F_2(S) \tag{21.102b}$$

subtracting Eq. (21.102b) from (21.102a)

$$\left[-\frac{1}{R}\Delta F_1(S) + \frac{1}{R}\Delta F_2(S) - \Delta P_{D1}(S) + \Delta P_{D2}(S) - \Delta P_{12}(S) + \Delta P_{21}(S)\right]\frac{f^0}{2sH} = \Delta F_1(S) - \Delta F_2(S)$$

$$\left[\frac{1}{R}\Delta F_2(S) + \Delta F_1(S) + \Delta P_{D2}(S) - \Delta P_{D1}(S) - 2\Delta P_{12}(S)\right]\frac{f^0}{2sH} = \frac{\Delta P_{12}(S)}{2\pi T^0} \tag{21.103}$$

since

$$(\Delta F_1(S) - \Delta F_2(S))\frac{2\pi T^0}{s} = \Delta P_{12}(S) = -\Delta P_{21}(S)$$

Therefore

$$\left[\frac{1}{R}\left(\frac{-\Delta P_{12}(S)s}{2\pi T^0}\right) + (\Delta P_{D2}(S) - \Delta P_{D1}(S)) - \Delta P_{12}(S)\right]\frac{f^0}{2sH} = \Delta P_{12}(S)\frac{s}{2\pi T^0}$$

$$\Delta P_{12}(S)\left[\frac{sf^0}{2\pi RT_0 2sH} + \frac{2f^0}{2sH} + \frac{s}{2\pi T^0}\right] = \left[\frac{\Delta P_{D2}(S) - \Delta P_{D1}(S)}{2sH}\right]f^0$$

$$\Delta P_{12}(S) = \frac{[\Delta P_{D2}(S) - \Delta P_{D1}(S)]\frac{f^0}{2sH}}{\frac{s}{2\pi T^0} + \frac{2f^0}{2\pi T^0} + \frac{sf^0}{2\pi RT^0 2sH}}$$

$$= \frac{\left[[\Delta P_{D2}(S) - \Delta P_{D1}(S)]\frac{\pi f^0 T^0}{SH}\right]}{s + \frac{2f^0 \pi T^0}{sH} + \frac{f^0}{2RH}}$$

$$= \frac{\pi f^0 T^0}{H} \frac{[\Delta P_{D2}(S) - \Delta P_{D1}(S)]}{s^2 + \left(\frac{f^0}{2RH}\right)s + \left(\frac{2f^0 \pi T^0}{H}\right)} \qquad (21.104)$$

The denominator is of the form

$$(s^2 + 2Ks + \omega^2) = (s+K)^2 + (\omega^2 - K^2)$$

where

$$K = \frac{f^0}{4RH} \quad \text{and} \quad \omega = \sqrt{\frac{2\pi f^0 T^0}{H}}$$

setting

$$\sqrt{\omega^2 - K^2} \text{ as } \omega_0$$

$$\omega_0 = \sqrt{\frac{2\pi T^0 f^0}{H} - \left(\frac{f^0}{4RH}\right)^n} \qquad (21.105)$$

Note that both K and ω^2 are positive. From the roots of the characteristic equation, we notice that the system is stable and damped. The frequency of the damped oscillations is given by ω_0.

Since H and f^0 are constant, the frequency of oscillations depends upon the regulation parameter R. Low R gives high K and high damping and vice versa. If $R \to \alpha$; $K \to 0$ is the condition for no-governor action and there will be undamped oscillations.

We thus conclude from the preceding analysis that the two-area system, just as in the case of a single area system in the uncontrolled mode, has a steady-state error but to a lesser extent and the tie-line power deviation and frequency deviation exhibit oscillations that are damped out later.

21.24 TIE-LINE BIAS CONTROL—IMPLEMENTATION

From the preceding discussion, it is clear that in interconnected operation each area in normal steady state must control the changes in such a fashion that it absorbs its own load change.

The Tie-Line Bias Control Strategy Presented Can Be Summarized: All interconnected areas must contribute their share to frequency control in addition to taking care of their own net interchange.

The control error for each area can be now defined as a linear combination of frequency and tie-line power errors.

$$\text{Area control error for Area 1; } ACE_1 = \Delta P_{12} + B_1 \Delta f_1 \tag{21.106}$$

$$\text{and area control error for Area 2; } ACE_2 = \Delta P_{21} + B_2 \Delta f_2 \tag{21.107}$$

Hence in the block diagram of Fig. 21.21, the ΔP_{ref} commands can be defined as follows:

$$\Delta P_{\text{ref},1} = -K'_I \int (\Delta P_{12} + B_1 \Delta f_1) \, dt \tag{21.108}$$

and

$$\Delta P_{\text{ref},2} = -K''_I \int (\Delta P_{21} + B_2 \Delta f_2) \, dt \tag{21.109}$$

where B_1 and B_2 are the frequency bias parameters for Areas 1 and 2 while K'_I and K''_I are the integral controller gain constants.

Note that under steady-state conditions when ΔP_{12} and Δf_1, Δf_2 become zero, from Eq. (21.108) and (21.109) $\Delta P_{12} + B_1 \Delta f_1 = 0$ and $\Delta P_{21} + B_2 \Delta f_2 = 0$ convey that the controller action in the final stage is independent of B_1 and B_2.

In fact, even one of the Bs, either B_1 or B_2 can be zero. It is suggested that only one B can be selected as equal to the area frequency response characteristic $\beta = D + (1/R)$ to give satisfactory performance.

The complete tie-line bias control of a two-area system is shown in Fig. 21.20.

If area ratings are different (-1), block is replaced by $(-a_{12}) = P_{r1}/P_{r2}$

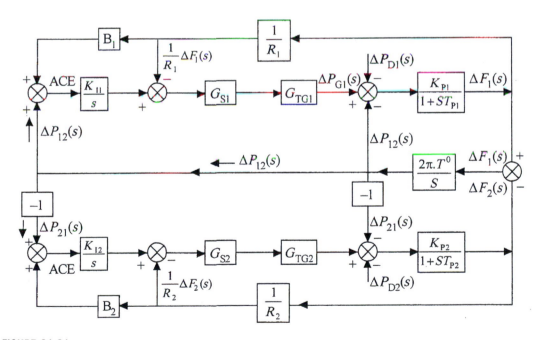

FIGURE 21.21

Tie-line bias control of a two-area system.

The selection of integral controller gains K'_I and K''_I must be such that too large a value should not lead to chasing of minor deviations of no-consequence.

21.25 THE EFFECT OF BIAS FACTOR ON SYSTEM REGULATION

Let a and b be the ratios of generation of areas A and B in terms of the total generation, i.e.,

$$a + b = 1 \text{ (p.u.)}$$

Let P_{sc} be the total system generating capacity per cycle and correspondingly, let P_{AC} and P_{BC} be the generating capacity per cycle of A and B. B_A is the tie-line bias setting at area A expressed in generation capacity per cycle and B_S is the imposed regulating characteristic for the whole system with a tie-line bias controller in A only, with no controller in B, the ratio of bias setting at A to the generation characteristic at A, r, is given by

$$r = \frac{B_A}{P_{AC}}$$

$$P_{SC} = P_{AC}a + P_{BC}(1 - a) = P_{AC} \cdot a + P_{BC} \cdot b \tag{21.110}$$

$$B_S = B_A a + P_{BC}(1 - a) = B_A \cdot a + P_{BC} \cdot b \tag{21.111}$$

The change in natural frequency Δf_n following the load change as dictated by the governor characteristic is

$$\Delta f_n = \frac{\Delta P_D}{P_{SC}}$$

If Δf_b is the system frequency deviation due to the controller, then

$$\Delta f_S = \frac{\Delta P_D}{B_S}$$

from Eqs. (21.110) and (21.111)

$$\Delta f_n = \frac{\Delta P_D}{P_{AC}a + P_{BC}(1 - a)} \tag{21.112}$$

and

$$\Delta f_S = \frac{\Delta P_D}{B_A a + P_{BC}(1 - a)} \tag{21.113}$$

The improvement change in frequency deviation, as a percentage of the initial disturbance is

$$\Delta f = \frac{\Delta f_S - \Delta f_n}{\Delta f_n} \times 100 \tag{21.114}$$

$$= \frac{\Delta P_D \left[\dfrac{1}{B_A \cdot a + P_{BC}(1 - a)} - \dfrac{1}{P_{AC} \cdot a + P_{BC}(1 - a)} \right]}{\dfrac{\Delta P_D}{P_{AC} \cdot a + P_{BC}(1 - a)}}$$

$$= \frac{(P_{AC} - B_A)a}{(B_A - P_{nc})a + P_{BC}} \times 100 \tag{21.115}$$

Since, $r = B_A/P_{AC}$ and assuming $P_{AC} = P_{BC}$.

$$\Delta f = -\frac{a(r-1)}{1 + a(r-1)} \times 100\% \tag{21.116}$$

or

$$\Delta f = \frac{(b-1)(r-1)}{r - b(r-1)} \times 100\% \tag{21.117}$$

The percentage change in frequency is independent of ΔP_D, and depends only on a, b, and r. When $r = 1$; $\Delta f = 0$, therefore

$$\Delta f_n = \Delta f_S$$

The initial ungoverned frequency deviation is equal to the subsequent governed frequency deviation.

21.26 SCOPE FOR SUPPLEMENTARY CONTROL

During operation, the system, frequency, tie-line flows, and other derived quantities will be continuously swinging with periods generally of a few seconds. Considering the time lags in measurements, the prime mover governor characteristics, etc., supplementary controllers cannot respond to rapid swings of frequency and tie-line power. Neither can they effectively regulate these changes nor is it expected of them. It is desired that supplementary controllers should regulate the frequency and tie-line deviations about a base that corresponds to the prescheduled values. In a reactive sense, the supplementary control is a steady-state control. A high-speed dynamic analysis of the system and area characteristics are required only in connection with improving system stability.

WORKED EXAMPLES

E.21.1 A 100-MW generator has a regulation parameter R of 5%. By how much will the turbine power increase if the frequency drops by 0.1 Hz with the reference unchanged.

Solution:
Actual change in frequency = 5% of 50 Hz = $0.05 \times 50 = 2.5$ Hz

$$R = 2.5 \text{ Hz}/100 \text{ MW} = 0.025 \text{ Hz/MW}$$

If $\Delta f = -0.1$ Hz, the increase in turbine power

$$\Delta P = -\frac{1}{2} = \Delta f = -\frac{1}{0.025} \times (-0.1) = 4 \text{ MW}$$

The turbine power increase = 4 MW.

E.21.2 A 100-MW generator with $R = 0.02$ Hz/MW has its frequency fallen by 0.1 Hz. By how much the reference power setting be changed if the turbine power remains unchanged.

Solution:
The signal to increase the generation is blocked. Thus at the input summing point the reference power setting must be changed. Such that

$$\Delta P_{ref} - \frac{1}{R} \Delta f = 0$$

i.e.,

$$\Delta P_{ref} - \frac{1}{R} \Delta f = \frac{1}{0.025} \times 0.1 = 4 \text{ MW}$$

E.21.3 Two generators with ratings 100 and 300 MW operate at 50 Hz frequency. The system load increases by 100 MW when both the generators are operating at about half of their capacity. The frequency then falls to 49.5 Hz.
If the generators are to share the increased load in proportion to their ratings, what should be the individual regulations? What should be regulations if expressed in per-unit Hertz/per-unit MW?

Solution:
$$\Delta P = - \frac{1}{R} \Delta f$$

$$\Delta P_1 = - \frac{1}{R_1} \Delta f$$

$$\Delta P_2 = - \frac{1}{R_2} \Delta f$$

$$\Delta f = 0.5 \text{ Hz}$$

Power is shared proportional to their ratings

$$\Delta P_1 = 100 \times \frac{100}{400} = 25 \text{ MW}$$

$$\Delta P_2 = 100 \times \frac{300}{400} = 75 \text{ MW}$$

Hence

$$R_1 = - \frac{0.5}{75} = - 0.00667$$

If regulation is expressed in per unit, then with $f = 50$ Hz

$$R_1 = - \frac{0.02}{50} \times \frac{100}{1} = 0.04$$

$$R_2 = - \frac{0.00667}{50} \times \frac{300}{1} = 0.04 \text{ p.u. Hz/p.u. MW}$$

Both have the same value, even though based on their individual ratings, they have different regulation.

E.21.4 Determine the primary load frequency control loop parameters for a control area having the following data:

Total rated area capacity $P_r = 1000$ MW
Normal operating load $= 500$ MW
Inertia constant $H = 4.0$ s
Regulation $R = 2.5$ Hz/p.u. MW

Solution:
Load damping

$$D = \frac{\partial P}{\partial f} = \frac{500 \text{ MW}}{50 \text{ Hz}} = 10 \text{ MW/Hz}$$

(Here, the load damping is assumed linear and percentage change is assumed to be the same.)

In per unit

$$D = \frac{10}{1000} = 0.01 \text{ p.u. MW/Hz}$$

$$T_P = \frac{2H}{f_S D} = \frac{2 \times 4.0}{50 \times 0.01} = 16 \text{ s}$$

$$K_P = \frac{1}{D} = \frac{1}{0.01} = 100 \text{ Hz/p.u. MW}$$

E.21.5 Determine the area frequency response characteristic and the static frequency error for a system with the following data, when 1% load change occurs?

$$D = 0.01 \text{ p.u. MW/Hz}$$
$$R = 2.5 \text{ Hz/p.u. MW}$$
$$T_P = 16 \text{ s}$$
$$K_P = 100 \text{ Hz/p.u. MW}$$

Solution:

$$\text{Area frequency response characteristic } \beta = D + \frac{1}{R}$$

$$= 0.01 + \frac{1}{2.5} = 0.41 \text{ MW/Hz}$$

$$\Delta f = -\frac{M}{\beta} = -\frac{1}{100 \times 0.41} = 0.02439 \text{ Hz}$$

E.21.6 In Example E.21.5, the governor is blocked so that it does not change the generation. In that case what would be the steady-state frequency error?

Solution:

When the governor is not acting, the feedback loop is not existing. In such a case R is infinite.

$$\beta = D + \frac{1}{R} = D = 0.01 \text{ p.u. MW/Hz}$$

Hence

$$\Delta f = -\frac{M}{\beta} = -\frac{0.01}{0.01} = -1 \text{ Hz}$$

Frequency falls by 1 Hz, i.e., $f = 50 - 1 = 49$ Hz.

It may be noted that with the generator acting the frequency from E.21.5 is

$$50 - 0.02439 = 49.9756 \text{ Hz}$$

The importance of feedback through governor mechanism can be understood from the above.

E.21.7 A 100-MVA synchronous generator operates initially at 3000 rpm, 50 Hz. A 25-MW load is suddenly applied to the machine and the steam valve to the turbine opens only after 0.5 s due to the time lag in the generator action. Calculate the frequency to which the generated voltage drops before the steam flow commences to increase to meet the new load. The value of the stored energy for the machine is 5 kW-s/KVA of generator energy. Also calculate the value of H constant for the generator.

Solution:

Stored energy = 5 kW/kVA

i.e., = 500 MW-s/100 MVA

Load increase = 25 MW

Energy required to supply this load for 0.5 s = 25 MW-s

Frequency at 500 MW-s stored energy = 50 Hz

Frequency fall = Δf

$$\frac{\Delta f}{f} = \frac{\Delta f}{50} = \frac{25 \times 0.5}{500} \text{MW-}s$$

$$\Delta f = \frac{50 \times 25 \times 0.5}{500} = 1.25 \text{ Hz}$$

Frequency falls to $50 - 1.25 = 48.75$ Hz

$$H \text{ constant} = \frac{\text{Stored kinetic energy at rated frequency}}{\text{Machine rating}} = \frac{5 \text{ MW-s}}{\text{MVA}} \times \frac{100 \text{ MVA}}{100 \text{ MVA}} = 5 \text{ s}$$

E.21.8 Given the following parameters, obtain the frequency error. Plot it when a step load disturbance of (1) 1% and (2) 2% occur in the system.

$$T_P = 22 \text{ s}$$

$$R = 2.5$$

$$K_P = 100$$

Solution:

$$\text{The expression for } \Delta F(s) = \frac{G_P}{1 + \frac{1}{R}G_S G_{TG} G_P} \Delta P_D(S)$$

Neglecting the turbine dynamics and governor action $(G_S \, G_{TG} \approx 1.0)$

$$\Delta F(S) = -\frac{K_P}{1 + ST_P} \cdot \frac{1}{1 + \frac{1}{R}\frac{K_P}{1 + ST_P}} \cdot \frac{k}{s}$$

$$k_1 = 0.01 \text{ and } k_2 = 0.02$$

simulating the transfer functions in MATLAB the response is obtained and shown in Fig. E.21.8.

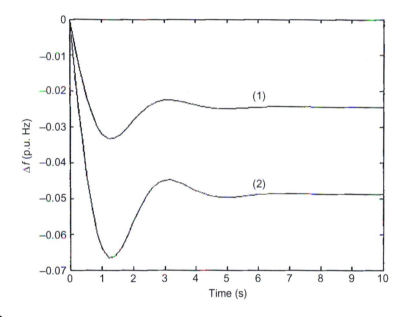

FIGURE E.21.8

Step load frequency error characteristic without supplementary control (1) $k = 0.01$; (2) $k = 0.02$ (governor action and turbine dynamics are neglected).

E.21.9 Show the effect of governor action and turbine dynamics, if they are not to be neglected in Example E.21.8 given $T_S = 100$ ms and $T_{TG} = 0.5$ s.

Solution:

For this, the exact frequency error is used

$$\Delta F(S) = -\frac{G_P}{1 + \frac{1}{R}G_S G_{TG} G_P} \cdot \Delta P_D(S)$$

$$\Delta P_{\mathrm{D}}(S) = 0.01 \text{ p.u.}; \quad G_{\mathrm{S}} = \frac{K_{\mathrm{S}}}{1 + ST_{\mathrm{S}}}; \quad G_{\mathrm{TG}} = \frac{K_{\mathrm{TG}}}{1 + ST_{\mathrm{TG}}}$$

$$K_{\mathrm{S}} K_{\mathrm{TG}} \approx 1$$

The response is shown in Fig. E.21.9.

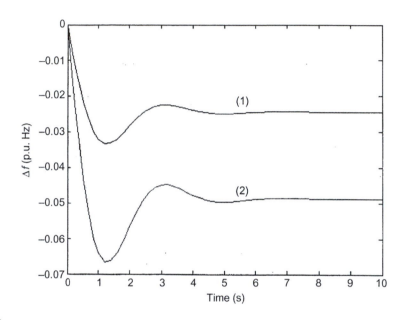

FIGURE E.21.9

Step load frequency error characteristic without supplementary control (1) $k = 0.01$; (2) $k = 0.02$; $T_{\mathrm{S}} = 100$ ms, and $T_{\mathrm{TG}} = 0.5$ s.

It can be seen that greater the step load change, larger the error and the governor action and turbine dynamics does not cause any change in the response in the steady state, except for a transient deviation at the beginning of the distribution.

E.21.10 An isolated power station has the following parameters:
Turbine time constant $T_{\mathrm{t}} = 0.5$ s
Governor time constant $T_{\mathrm{S}} = 0.2$ s
Governor inertia constant $H = 4$ s
Governor speed regulation $R = R$ per unit
The load varies by 0.8% for a 1% change in frequency, i.e., $D = 0.8$. The governor speed regulation is set to $R = 0.05$ p.u. The turbine rated output is 250 MW at a nominal frequency of 50 Hz. A sudden load change of 50 MW ($\Delta P_{\mathrm{L}} = 0.2$ p.u.) occurs.

1. Construct the SIMULINK block diagram and obtain the frequency deviation response.
2. Set integral gain to 7 and obtain the frequency deviation response and compare both the responses.

Solution:

To create a SIMULINK block diagram presentation, select new (model) file from FILE menu. This provides an untitled blank window for designing and simulating a dynamic system. Copy different blocks from the SIMULINK libraries or other previously opened windows into the new window by depressing the mouse button and dragging.

1. Open the continuous library and drag the transfer function block to the window. Double click on transfer function to open the dialog box. Enter the numerator and denominator values (the coefficients in the descending powers of s, if any power of s is missing, enter zero) of the transfer function.
2. Open the math library and drag the sum block in to the window. Open the sum dialog box and enter $+-$ under list of signs.
3. From the math library drag the Gain block into the model file right click on the gain block and click on the Flip option to rotate the gain block by 180 degrees.
4. Open the source library and drag the step input block to window. Double click on it to open its dialog box and set up the step time (step duration), initial, and final values (which will be same) to represent the step input.
5. Open the sink library and drag scope to window to observe the response.

By using the left mouse button, connect all the blocks.
Before starting simulation, set the simulation parameters. Pull down the simulation dialog box and select parameters. Set the start time, stop time and for a more accurate integration, set the maximum step size.
In this example the parameters for all the blocks for the system in the figure are initialized. Open m-file and enter the parameter values. The following m-file has to be run prior to the simulation (of model file).
Open new m-file and enter the parameter values as shown below.

$$T_g = 0.2;$$
$$T_t = 0.5;$$
$$H = 5;$$
$$D = 0.6;$$
$$R = 0.05;$$

%for integral control

$$K_i = 7;$$

Save the m-file under parameters and run the file. SIMULINK block diagram and results for LFC.
By using above procedure construct the simulink block diagram for the load frequency control of isolated power system as shown in Fig. E.21.10A.

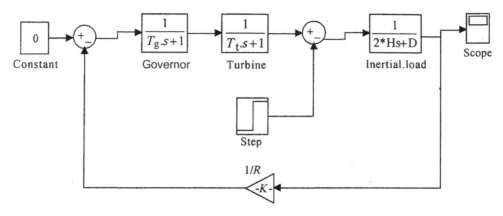

FIGURE E 21.10(A)

Block diagram model of load frequency control (isolated power system).

Pull down the file menu and use save as to save the model under AFC. Start the simulation. Double click on the scope, click on the auto scale, the result is displayed as shown in Fig. E.21.10B.

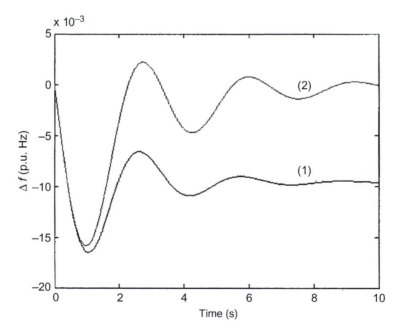

FIGURE E.21.10 (B)

Response of LFC. (1) Uncontrolled response. (2) Integral control response.

Add integral controller block for above system and save the file under LFC1.
Start the simulation. Double click on the scope, click on the autoscale, the result is displayed as shown in Fig. E.21.10B.

E.21.11 A single control area system with the following data experiences a sudden load change of 3%.

$$K_P = 100$$

$$T_P = 25$$

$$R = 2 \text{ Hz/p.u. MW}$$

with integral control using a gain of 10 obtain the frequency plot with time and show that the frequency deviation is reduced to zero.

Solution:
Using MATLAB, the solution is obtained and shown in Fig. E.21.11.

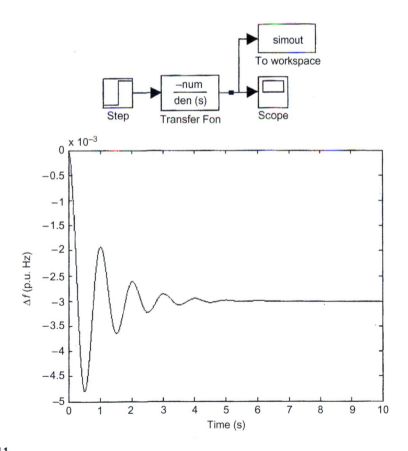

FIGURE E.21.11

$$K_i = 10 \quad K_P = 100 \quad T_P = 25 \quad R = 2 \quad \text{num} = K_P/T_P$$

$$d_1 = (1/T_P) + (K_P/(R * T_P)) \quad d_2 = K_i * K_P/T_P \quad \text{den} = [1 \; d_1 \; d_2]$$

E.21.12 Two control areas have the following characteristics

Area 1:

$$R_1 = 0.011 \text{ p.u.}$$

$$D_1 = 0.85 \text{ p.u.}$$

$$\text{Base MVA} = 1000$$

Area 2:

$$R_2 = 0.018 \text{ p.u.}$$

$$D_2 = 0.95 \text{ p.u.}$$

$$\text{Base MVA} = 1000$$

A load change of 200 MW occurs in Area 1. Determine the new steady-state frequency.

Solution:

$$\Delta f^0 = \frac{k_1 + k_2}{\beta_1 + \beta_2} \text{ (Hz)}$$

$$k_1 = 0.2 \text{ p.u.;} \quad k_2 = 0$$

$$\beta_1 = D_1 + \frac{1}{R_1} = 0.85 + \frac{1}{0.011} = 90.9091 + 0.85 = 91.7591$$

$$\beta_2 = D_2 + \frac{1}{R_2} = 0.95 + \frac{1}{0.018} = 55.55 + 0.95 = 56.405$$

$$\Delta f^0 = \frac{k_1}{(\beta_1 + \beta_2)} = \frac{0.2}{91.7591 + 56.405} = \frac{0.2}{148.16} = 0.0013683 \text{ p.u. Hz}$$

$$f = 50 - (0.0013683 \times 50) = 49.93 \text{ Hz}$$

E.21.13 In Example E.7.1, determine the tie-line power flow deviation.

Solution:

$$\Delta P_{\text{tie-line}} = \frac{\beta_1 k_2 - k_1 \beta_2}{\beta_1 + \beta_2} \text{ (p.u.MW)}$$

$$k_2 = 0$$

$$\Delta P_{\text{tie-line}} = \frac{-k_1 \beta_2}{\beta_1 + \beta_2}$$

$$= (\Delta f^0)\beta_2 = -0.0013683 \times 56.405$$

$$= -0.0771789 \text{ p.u. MW}$$

$$= -77.1789 \text{ MW}$$

E.21.14 In Example E.7.1 if the load disturbance occurs simultaneously also in Area 2 by 100 MW determine the frequency and tie-line power changes.

Solution:

$$K_2 = 0.1 \text{ p.u.}$$

$$\Delta f^0 = \frac{k_1 + k_2}{\beta_1 + \beta} = \frac{0.2 + 0.1}{91.7591 + 56.405} = 0.0020248 \text{ p.u. Hz}$$

$$\Delta f^0 = 0.10124 \text{ Hz}$$

$$f = 50 - 0.10124 = 49.89876 \cong 49.9 \text{ Hz}$$

$$\Delta P_{\text{tie-line}} = \frac{\beta_1 k_2 - k_1 \beta_2}{\beta_1 + \beta_2}$$

$$\frac{91.7591 \times 0.1 - 0.2 \times 56.405}{148.16} = \frac{9.17591 - 11.2810}{148.16}$$

$$= \frac{-2.10509}{148.16} - 0.014208 \text{ p.u. MW}$$

$$= -14.2082 \text{ MW}$$

E.21.15 Two interconnected areas A and B have capacities 2000 and 750 MW, respectively. The speed regulation coefficients are 0.1 p.u. for both the areas on their own area ratings. The damping torque coefficients are 1.0 p.u. also on their own base.
Find the steady-state change in system frequency when a load increment of 50 MW occurs in area A. Find also the tie-line power deviation. System frequency is 50 Hz.

Solution:

$$a_{12} = \frac{P_{r1}}{P_{r2}} = \frac{2000}{750} = 2.67$$

$$D = 1.0 \text{ p.u. MW/p.u.Hz} = \frac{1}{50} \text{p.u. MW/Hz}$$

$$R = 0.1 \text{ p.u.} = 0.1 \times 50 = \text{Hz/p.u. MW}$$

$$\beta_1 = \beta_2 = \beta = D + \frac{1}{R} = \frac{1}{50} + \frac{1}{5} = 0.02 + 0.02 = 0.22$$

$$\Delta f^0 = \frac{k_2 + a_{12}k_1}{\beta_2 + a_{12}\beta_1}$$

$$k_1 = \frac{50 \text{ Mw}}{2000 \text{ p.u.MW}} = \frac{1}{40} \text{ p.u.Mw}; k_2 = 0$$

$$\Delta f^0 = \frac{2.67 \times \frac{1}{40}}{0.22 + 2.67 \times 0.22} = \frac{0.06675}{0.8074} = 0.08267 \text{ Hz}$$

$$\Delta P_{12}^0 = -\frac{\beta_2 K_1}{\beta_2 + a_{12}\beta_1} = \frac{-0.22 \times \frac{1}{40}}{0.22 + 2.67 \times 0.22} = -\frac{0.0055}{0.8074} = -0.00681 \text{ p.u. MW}$$

$$= -0.00681 \times 2000 = 13.62 \text{ MW}$$

E.21.16 In Problem E.7.4, if the tie-line connection is lost while it is carrying 50-MW load, what will be frequency in system A.

Solution:
The system A will be now an isolated area.
The frequency drop

$$\Delta f_0 = -\frac{\Delta P_{DA}}{\beta} = \frac{-\frac{1}{40}}{0.22} = -0.113636 \text{ Hz}$$

E.21.17 For a two identical area system, the following data is given. Determine the frequency of oscillations when a step load disturbance occurs.
Speed regulation coefficient $= R = 4$ Hz/p.u. MW
Damping coefficient $D = 0.03$ p.u. MW/Hz
System frequency $= 50$ Hz
The tie-line has a capacity of 0.1 p.u.
The power angle is 30 degrees just before the occurrence of the load disturbance.

Solution:

$$K = \frac{f_0}{4RH} = \frac{50}{4 \times 4 \times 5} = 0.625$$

$$T^0 = 0.1 \cos 30 \text{ degrees} = 0.0866$$

$$\omega^2 = \frac{2\pi T^0 f^0}{H} = \frac{2\pi \times 0.0866 \times 50}{5} = 5.443$$

$$\omega = 2.333; \quad K^2 = 0.3906$$

$$\omega_0 = \sqrt{5.443 - 0.3906} = \sqrt{5.052375} = 2.2477489 \text{ radian/s}$$

However, if the damping coefficient is not to be neglected

$$K = \frac{f_0}{4H}\left[D + \frac{1}{R}\right] = 0.625 + \frac{50}{4 \times 5} \times 0.03 = 0.625 + 0.075 = 0.7; \quad k^2 = 0.49$$

In this case

$$\omega = \sqrt{5.443 - 0.49} = \sqrt{0.4953} = 2.2255 \text{ radian/s}$$

E.21.18 A power system consists of two areas interconnected by a tie line which has a capacity of 500 MW and is operating at a power angle of 35 degrees. If each area has a capacity of 5000 MW and the speed regulation coefficients for both the areas are also the same and are equal to $R = 2$ Hz/p.u. MW, determine the frequency of oscillation of the power for step change in load. The inertia constants are also the same for both the areas and are equal to $H = 5$ s.

Solution:

$$T^0 = \frac{500}{5000} \cos 35 \text{ degrees} = 0.819 \times 0.1 = 0.0819$$

$$\omega_0 = \sqrt{\left[\frac{2\pi \times 50 \times 0.0819}{5}\right] - \left(\frac{50}{4 \times 2 \times 5}\right)^2}$$

$$= \sqrt{5.148 - 1.5625} = \sqrt{3.5855} = 1.8935 \text{ radian/s}$$

E.21.19 In the above problem, if a step load change of 85 MW occurs in one of the areas determine the tie-line power deviation.

Solution:

The two areas are equal, the load change will be shared equally by both the areas. A power of $85/2 = 42.5$ MW will flow from the other area into the area where a load change occurs.

E.21.20 Two power stations A and B operate in parallel. They are interconnected by a short transmission line. The station capacities are 100 and 200 MW, respectively. The generators at A and B have speed regulation 3% and 2%, respectively. Calculate the output of each station and the load on the interconnector if,

1. the load on each station is 125 MW,

2. the loads on respective bus bars are 60 MW and 190 MW, and

3. The load is 150 MW at the station A bus-bar only.

Solution:

Case (a): $P_1 + P_2 = 125 + 125 = 250$ MW

$$\text{Regulation} = \frac{N_0 - N}{N_r} = \frac{f_0 - f}{f_w}$$

$$\frac{P_1}{100} = \frac{1 - f}{0.03}$$

$$\frac{P_2}{200} = \frac{1 - f}{0.02}$$

$$0.0003P_1 = (1 - f) = 0.0001P_2$$

i.e.,

$$3P_1 = P_2$$

$$P_1 + P_2 = 250$$

Solving

$$P_1 = 62.5 \text{ MW and } P_2 = 187.5 \text{ MW}$$

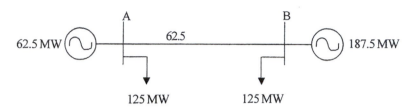

Case (b): the loads on each bus-bar is determined by the speed regulation characteristics.

$$3P_1 = P_2$$
$$P_1 + P_2 = 60 + 190 = 250 \text{ MW}$$

Solving

$$P_1 = 62.5 \text{ MW and } P_2 = 187.5 \text{ MW}$$

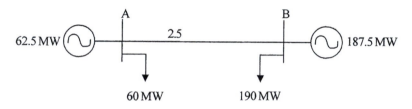

Case (c): There is load of 150 MW at A only

$$3P_1 = P_2$$
$$P_1 + P_2 = 150$$

Solving

$$P_1 = 37.5$$
$$P_2 = 112.5 \text{ MW}$$

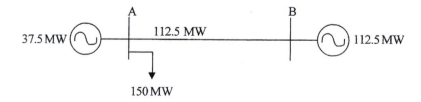

E.21.21 The turbines in a power station A have a uniform speed regulation of 2.5% from

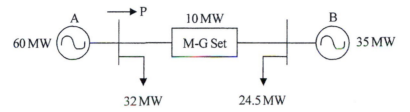

FIGURE E.21.21

no-load to full load. The rated capacity of the generators connected to the bus-bars total 60 MW and the frequency is 50 Hz. Station B has total generating capacity of 35 MW at the bus-bars and has a speed regulation of 3.0% connected through an induction motor generation set rated 10 MW, which has a full load slip of 3.0%. There are loads of 32 and 24.5 MW connected to station A and B, respectively. Find the load on the interconnector cable at this operating condition.

Solution:
Let P (MW) flow from A to B
Total load on A $= (32 + P)$ MW
Total load on B $= (24.5 - P)$
Percentage drop in speed at A $= 2.5\dfrac{(32 + P)}{60}$

Percentage drop in speed at B $= 3.5\dfrac{(24.5 - P)}{35}$

Percentage drop in speed of induction motor generator set $= \dfrac{3P}{100}$

Percentage drop in speed at B $-$ percentage drop in speed at A $=$ percentage drop in speed of induction motor generator set

$$\frac{3.5(24.5 - P)}{35} - \frac{2.5(32 + P)}{60} = \frac{3P}{10}$$

$$\frac{(24.5 - P)}{10} - \frac{3P}{10} = \frac{32 + P}{24}$$

solving $$106P = 268$$

$$P = \frac{268}{106} = 2.5283 \text{ MW}$$

P flows from A to B
Total load as $$A = 32 + 2.5283 = 34.5283 \text{ MW}$$
Total load as $$B = 24.5 - 2.5283 = 21.9717 \text{ MW}$$

PROBLEMS

P.21.1 Two generators rated 200 and 400 MW are operating in parallel. The droop characteristics of their governors are 4% and 5%, respectively, from no-load to full load. Assuming that the generators are operating at 50 Hz at no load, how would a load of 600 MW be shared between them? What will be the system frequency at this load? Assume free governor operation. Repeat the problem if both the governor have a droop of 4%.

P.21.2 A 100-MVA asynchronous generator operates on full load at a frequency of 50 Hz. The load is suddenly reduced to 50 MW. Due to time lag in the governor system, the steam valve beings to close after 0.4 s. Determine the change in frequency that occurs in this time. Given $H = 5$ KW-s/KVA of generator capacity.

P.21.3 Two generators rated 200 and 400 MW are operating in parallel. The droop characteristics of their governors are 4% and 5%, respectively, from no-load to full load. The speed changes are so set that the generators operate at 50 Hz sharing the load of 600 MW in the ratio of their ratings. If the load reduces to 400 MW, how will it be shared among the generators and what will the system frequency? Assume free governor operation. The speed changers of the governors are reset so that the load of 400 MW is shared among the generators at 50 Hz in the ratio of their ratings. What are the no-load frequencies of the generators?

P.21.4 In the single area system shown below determine
   1. The steady-state frequency error with $\Delta P_C = 0$
   2. Critical gain K of the integral control of $\Delta P_C = -\int K \Delta f$

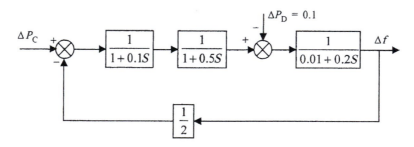

P.21.5 A 500-MW generator is operating at a load of 20 MW. A load change of 1% causes the frequency to change by 1%. If the system frequency is 50 Hz, determine the value of load damping factor in per unit.

P.21.6 Two interconnected control areas have the following characteristics
   Area 1:

$$R_1 = 0.01 \text{ p.u.}$$

$$D_1 = 0.08 \text{ p.u.}$$

$$\text{Base MVA} = 500$$

Area 2

$$R_3 = 0.02 \text{ p.u.}$$

$$D_2 = 1.0 \text{ p.u.}$$

$$\text{Base MVA} = 1000$$

A load change of 100 MW occurs in Area 1. Determine the new steady-state frequency and tie-line power deviation.

P.21.7 In P.21.6 if a load change of 200 MW occurs in Area 2, determine the frequency derivation and tie-line power deviation.

P.21.8 In P.21.6 if a load change of 150 MW in Area 1 and a load change of 250 MW in Area 2 occur simultaneously, determine the frequency and tie-line power deviation.

P.21.9 Given $R_1 = 2.4 \text{ Hz}/p.\text{u. MW}$

$$R_2 = 2.6 \text{ Hz}/p.\text{u. MW}$$

$$T_{T1} = 0.28 \text{ s}$$

$$T_{T2} = 0.31 \text{ s}$$

$$T_{S1} = 0.075 \text{ s}$$

$$T_{S2} = 0.08 \text{ s}$$

$$K_{P1} = 100 \text{ Hz}/p.\text{u. MW}$$

$$K_{P1} = 120 \text{ Hz}/p.\text{u. MW}$$

$$T_{12} = 0.07 \text{ p.u. MW}/\text{radian}$$

$$T_{21} = 0.07 \text{ p.u. MW}/\text{radian}$$

Solve the state equations for a two-area system and plot the frequency, generation, and tie-line power deviations for a step load disturbance.

P.21.10 Repeat problem P.21.9 using MATLAB compare the result.

P.21.11 Repeat problem P.21.9 with $P_{D1} = 0.01$ p.u. MW in Area 1 and $P_{D2} = 0.01$ p.u. MW in Area 2.

P.21.12 Repeat problem P.21.7 with $P_{D2} = 0.03$ p.u. MW. What are your conclusions from the study of the results of problems P.21.11 and P.21.12.

QUESTIONS

Q.21.1 Explain the necessity of maintaining a constant frequency in power system operation.

Q.21.2 With a neat diagram, explain briefly different parts of a turbine speed governing system.

Q.21.3 Derive the model of a speed governing system and represent it by a block diagram.

Q.21.4 With a block diagram explain the load frequency control for a single area system.

Q.21.5 Derive the model of a speed governing system and represent it by a block diagram.

Q.21.6 With first-order approximation explain the dynamic response of an isolated area for load frequency control.

Q.21.7 Discuss the importance of combined load frequency control and economic dispatch control with a neat block diagram.

Q.21.8 Discuss in detail the importance of load frequency problem.

Q.21.9 Distinguish between load frequency control and economic dispatch control.

Q.21.10 A synchronous generator supplies power to a synchronous motor via a transmission network. Find equivalent inertia constant of a machine connected to an infinite bus.

Q.21.11 Explain how the tie-line power deviation can be incorporated in two-area system block diagram?

Q.21.12 What is a tie-line ? Explain.

Q.21.13 Derive the relation between steady-state frequency error and tie-line deviation for step load disturbances in both areas.

Q.21.14 What is area frequency response characteristic? Explain it in the context of two-area system.

Q.21.15 What are the advantages of interconnected operation of power systems? Explain.

Q.21.16 Explain how state variable representation can be obtained for a two-area system. Write down the equations.

Q.21.17 Sketch and explain the block schematic of a two-area system.

Q.21.18 What are the features of the dynamic response of a two-area system for step load disturbances?

Q.21.19 Explain tie-line bias control applied to a two-area system.

Q.21.20 What are the considerations in selecting the frequency bias parameter?

SYNCHRONOUS MACHINE

The synchronous machine is the most important component in a power system. In fact the very existence of the power system depends on it.

A synchronous machine consists of two major components, viz., the stator and the rotor which are in relative motion.

While the field carries the direct current, the generated emf is alternating. The detailed analysis of a synchronous machine is very complicated due to the presence of harmonics on account of variation in air-gap permeance. The important qualities are the various voltages generated, the corresponding flux linkages, and the currents. The various reactances invariably play a vital role in the performance of the machine. It is an established fact that the machine analysis is better achieved through the two-axis theory. At first, a brief derivation of the two-axis voltages will be given based on the transformer and speed voltages produced. Later, a detailed analysis is given.

Much of the work carried out by early pioneers of synchronous machine analysis like Doherty and Nickle, Park, Blondel, and others will be too difficult at this stage to understand. Hence a much simplified analysis is presented that will be sufficient to obtain steady-state and transient-state phasor diagrams.

22.1 THE TWO-AXIS MODEL OF SYNCHRONOUS MACHINE

In the following the two-axis model of a three-phase armature winding of a synchronous machine will be obtained. We can replace the three-phase winding by a two-phase winding with the same exciting effect at any point around the machine air gap between the stator and the rotor. Consider the time variation of the mmf of the three-phase windings R, Y, and B.

$$\overline{M}_R = M_m \sin \omega t \cdot \cos \theta \tag{22.1}$$

$$\overline{M}_Y = M_m \sin\left(\omega t - \frac{2\pi}{3}\right) \cdot \cos\left(\theta - \frac{2\pi}{3}\right) \tag{22.2}$$

$$\overline{M}_B = M_m \sin\left(\omega t - \frac{4\pi}{3}\right) \cdot \cos\left(\theta - \frac{4\pi}{3}\right) \tag{22.3}$$

where the time variation of mmf resulting from the currents are represented by sine functions. The time zero is chosen such that the R-phase current is zero and increasing. The space distributions of the mmfs are represented by cosine functions; q is a space angle the space origin being the axis of R-winding. The stator windings are symmetrical and carry balanced three-phase currents (Figs. 22.1 and 22.2).

Electrical Power Systems. DOI: http://dx.doi.org/10.1016/B978-0-08-101124-9.00022-X

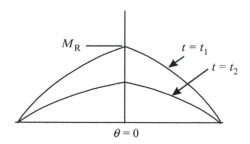

FIGURE 22.1

Rotating mmf waves.

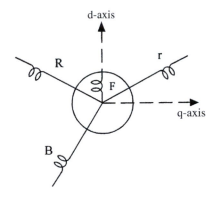

FIGURE 22.2

Symmetrical three-phase windings.

$$\overline{M}_R + \overline{M}_Y + \overline{M}_B = \frac{3}{2}M_m \sin(\omega t - \theta) \tag{22.4}$$

Consider an equivalent two-axis windings as shown in Fig. 22.3 displaced mutually in space by 90 degrees.

The mmfs are

$$M_d = \frac{3}{2}M_m \sin \omega t \cdot \cos \theta \tag{22.5}$$

$$M_q = \frac{3}{2}M_m \sin\left(\omega t - \frac{\pi}{2}\right) \cdot \cos\left(\theta - \frac{\pi}{2}\right) \tag{22.6}$$

$$\begin{aligned}\overline{M}_d + \overline{M}_q &= \frac{3}{2}\overline{M}_m \sin(\omega t - \theta)\\ &= \overline{M}_R + \overline{M}_Y + \overline{M}_B\end{aligned} \tag{22.7}$$

Thus the two windings d and q will give the same resulting mmf at any point around the air gap as the three-phase windings.

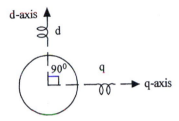

FIGURE 22.3

Two-axis representation.

22.2 DERIVATION OF PARK'S TWO-AXIS MODEL

Consider the schematic of the two-axis model of the synchronous machine shown in Fig. 22.4. For simplicity, only one damper coil is considered on each axis; d and q represent the fictitious armature windings along d- and q-axes.

D and Q represent the short-circuited damper coils along the d- and q-axes. F is the field winding on the main polar axis or d-axis.

Let V is the Voltage in volts, I is the current in amperes, λ is the flux linkages in webers with suffix f, d, and q for field, d- and q-axes, ω is the speed in radian/s $= d\theta/dt$, R_a is the armature resistance, same in both d- and q-axes, and R_f, R_d, and R_q refer to resistance of field, d-axis, and q-axis damper windings, respectively.

The generator-induced voltages in d- and q-axis windings are as follows:

$$V_d = -R_a I_d - \lambda_q \frac{d\theta}{dt} + \frac{d\lambda_d}{dt} \tag{22.8}$$

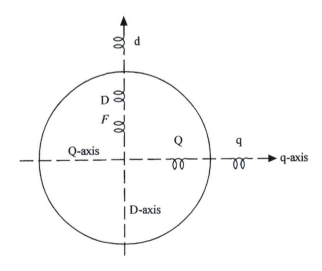

FIGURE 22.4

Two-axis model for synchronous machine.

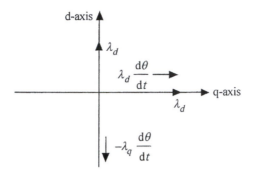

FIGURE 22.5

Induced voltages along d- and q-axes.

$$V_q = -R_a I_q + \lambda_d \frac{d\theta}{dt} + \frac{d\lambda_q}{dt} \qquad (22.9)$$

(look at Fig. 22.5 for explanation)

$$V_f = R_f I_f + \frac{d\lambda_f}{dt} \qquad (22.10)$$
$$= \text{DC supply voltage}$$

For the damper coils which are short-circuited

$$0 = R_d I_d + \frac{d\lambda_d}{dt} \qquad (22.11)$$

$$0 = R_q I_q + \frac{d\lambda_q}{dt} \qquad (22.12)$$

In the following, a detailed analysis of the machine performance is given.

22.3 SYNCHRONOUS MACHINE ANALYSIS

In the analysis it is always assumed that the rotor magnetic paths and all of its electrical circuits are symmetrical with respect to pole and interpole axes. The field winding has its axis in line with the pole axis. In reality all the damper bars are connected together to form a closed mesh. For analysis the current paths in these bars may be assumed to be symmetrical with respect to both the pole and interpole axis. Fig. 22.6 shows the arrangement.

All mutual-inductances between stator and rotor circuits are periodic functions of rotor angular positions. Also, the mutual inductances between any two stator phases are also periodic functions of rotor angular position. For ease of analysis, saturation is neglected.

It is assumed that the stator windings are sinusodially distributed. Further stator slots do not cause any change in rotor inductances with rotor angle.

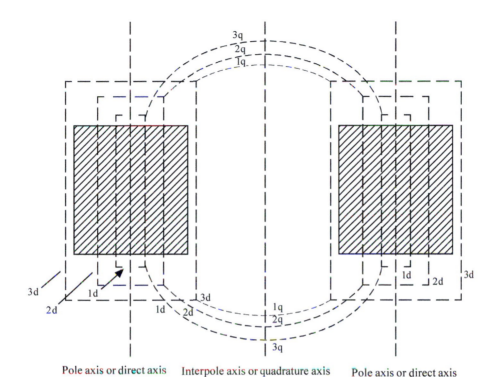

Pole axis or direct axis   Interpole axis or quadrature axis   Pole axis or direct axis

FIGURE 22.6

Arrangement of rotor circuits.

22.3.1 **VOLTAGE RELATIONS—STATOR OR ARMATURE**

$$
\left.
\begin{aligned}
e_R &= \frac{d}{dt}\Psi_R - ri_R \\
e_y &= \frac{d}{dt}\Psi_y - ri_y \\
e_B &= \frac{d}{dt}\Psi_B - ri_B
\end{aligned}
\right\}
\tag{22.13}
$$

where e_R, e_y, e_B are terminal voltages of phases R,Y,B; Ψ_R, Ψ_y, Ψ_B are flux linkages in phases R,Y,B; i_R, i_y, i_B are currents in phases R,Y,B; "r" is the resistance of each armature winding; and R,Y,B are the three phases of the stator.

Field or Rotor

$$
e_{fd} = \frac{d}{dt}\psi_{fd} + r_{fd}i_{fd}
\tag{22.14}
$$

where the suffix fd indicates the field quantities.

Direct-Axis Damper Windings

Since the windings are short circuited

$$0 = \frac{\mathrm{d}}{\mathrm{d}t}\psi_{1d} + r_{11d}i_{1d} + r_{12d}i_{2d} + \cdots$$

$$0 = \frac{\mathrm{d}}{\mathrm{d}t}\psi_{2d} + r_{21d}i_{1d} + r_{22d}i_{2d} + \cdots \tag{22.15}$$

The subscripts $12d$ and $21d$ denote the mutual effects between circuits $1d$ and $2d$. They are both resistance and inductance coupled.

Quadrature-Axis Damper Windings

Here also, as the windings are short circuited

$$0 = \frac{\mathrm{d}}{\mathrm{d}t}\psi_{1q} + r_{11q}i_{1q} + r_{12q}i_{2q} + \cdots$$

$$0 = \frac{\mathrm{d}}{\mathrm{d}t}\psi_{2q} + r_{21q}i_{1q} + r_{22q}i_{2q} + \cdots \tag{22.16}$$

22.3.2 FLUX LINKAGE RELATIONS

Armature

$$\psi_{\mathrm{R}} = -x_{\mathrm{RR}}i_{\mathrm{R}} - x_{\mathrm{Ry}}i_{\mathrm{y}} - x_{\mathrm{RB}}i_{\mathrm{B}} + x_{\mathrm{Rfd}}i_{\mathrm{fd}}$$

$$+x_{\mathrm{R1}d}i_{1d} + x_{\mathrm{R2}d}i_{2d} + \cdots + x_{\mathrm{R1}q}i_{1q} + x_{\mathrm{R2}q}i_{2q} + \cdots$$

$$\psi_{\mathrm{y}} = -x_{\mathrm{yR}}i_{\mathrm{R}} - x_{\mathrm{yy}}i_{\mathrm{y}} - x_{\mathrm{yB}}i_{\mathrm{B}} + x_{\mathrm{yfd}}i_{\mathrm{fd}}$$

$$+x_{\mathrm{y1}d}i_{1d} - x_{\mathrm{y2}d}i_{2d} + \cdots + x_{\mathrm{y}iq} + x_{\mathrm{y2}q}i_{2q} + \cdots$$

$$\psi_{\mathrm{B}} = -x_{\mathrm{BR}}i_{\mathrm{R}} - x_{\mathrm{By}}i_{\mathrm{y}} - x_{\mathrm{BB}}i_{\mathrm{B}} + x_{\mathrm{Bfd}}i_{\mathrm{fd}}$$

$$+x_{\mathrm{B1}d}i_{1d} + x_{\mathrm{B2}d}i_{2d} + \cdots + x_{\mathrm{B1}q}i_{1q} + x_{\mathrm{B2}q}i_{2q} + \cdots \tag{22.17}$$

Field

$$\psi_{\mathrm{fd}} = -x_{\mathrm{fRd}}i_{\mathrm{R}} - x_{\mathrm{fyd}}i_{\mathrm{y}} - x_{\mathrm{fBd}}i_{\mathrm{B}} + x_{\mathrm{ffd}}i_{\mathrm{fd}}$$

$$+x_{\mathrm{f1}d}i_{1d} + x_{\mathrm{f2}d}i_{2d} + \cdots + x_{\mathrm{f1}q}i_{1q} + x_{\mathrm{f2}q}i_{2q} + \cdots \tag{22.18}$$

Direct-Axis Damper Winding

$$\psi_{1d} = -x_{1\mathrm{Rd}}i_{\mathrm{R}} - x_{1\mathrm{yd}}i_{\mathrm{y}} - x_{1\mathrm{Bd}}i_{\mathrm{B}} + x_{1\mathrm{fd}}i_{\mathrm{fd}}$$

$$+x_{11d}i_{1d} + x_{12d}i_{2d} + \cdots + x_{id1q}i_{1q} + x_{ld2q}i_{2q} + \cdots \tag{22.19}$$

Quadrature-Axis Damper Winding

$$\psi_{1q} = -x_{1Rq}i_R - x_{1Yq}i_Y - x_{1Bq}i_B + x_{1q\ \text{fd}}i_{\text{fd}}$$

$$+x_{1q1d}i_{1d} + x_{1q2d}i_{2d} + \cdots + x_{11q}i_{1q} + x_{12q}i_{2q} + \cdots \tag{22.20}$$

22.3.3 INDUCTANCE RELATIONS

Self-Inductance of the Armature Windings

The self-inductance of any armature winding varies periodically from a maximum value when the pole axis is in line with the phase axis to a minimum value when the quadrature pole axis is in line with the phase axis. Because of the symmetry of the rotor, sinusoidal distribution of winding with inductance having a period of 180 electrical degrees. The expressions for self-inductances are (Fig. 22.7):

$$\left.\begin{array}{l} x_{RR} = x_{RR0} + x_{RR2}\cos 2q_R \\ x_{yy} = x_{yy0} + x_{yy2}\cos 2q_y \\ x_{BB} = x_{BB0} + x_{BB1}\cos 2q_B \end{array}\right\} \tag{22.21}$$

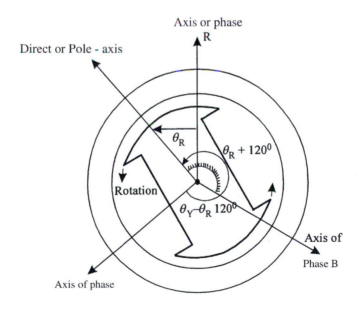

FIGURE 22.7

The rotor and stator axis.

Mutual-Inductances of the Armature Windings
These inductances can be written as

$$X_{Ry} = X_{yR} = -[x_m + x_s \cos 2 (\theta + 30 \text{ degrees})]$$
$$X_{yB} = X_{By} = -[x_m + x_s \cos 2 (\theta - 90 \text{ degrees})] \qquad (22.22)$$
$$X_{BR} = X_{RB} = -[x_m + x_s \cos (\theta + 150 \text{ degrees})]$$

The above expressions are obtained from the following considerations:

A component of mutual flux of armature phases does not link the rotor and is therefore independent of angle

$$f_d = k_d \cos \theta_R$$

$$f_q = -k_q \sin \theta_R$$

The linkage with phase "y" due to these components is proportional to

$$f_d \cos \theta_y - f_d \sin \theta_y = k_d \cos \theta_R \cos \theta_y + k_q \cos \theta_R \cos \theta_y$$

$$= k_d \cos \theta_R \cos (\theta - 120 \text{ degrees}) + k_d \sin \theta_R \sin (\theta - 120 \text{ degrees})$$

$$= \frac{k_d + k_q}{4} + \frac{k_d + k_q}{2} \cos 2(\theta - 60 \text{ degrees})$$

$$= \frac{1}{2}A + B \cos 2(\theta - 60 \text{ degrees})$$

$$= \left[\frac{1}{2}A + B \cos 2 (\theta + 30 \text{ degrees})\right]$$

Hence

$$x_{RY} = -[x_y + x_s \cos (2)(q + 30 \text{ degrees})] \text{ and so on}$$

(Note that $x_y = x_m$, the mutual reactance)

The rotor self-inductances are defined as x_{ffd}, x_{11d}, x_{22d},... and are assumed constants.

The rotor mutual-inductances: Because of the rotor symmetry, there is no mutual coupling between rotor d- and q-axes.

$$x_{f1q} = x_{f2q} = x_{id1q} = x_{1d2q} = x_{1qfd} = 0$$

$x_{f1d} = x_{1fd}$, etc., for the rotor field and d-axis damper winding.

Mutual-Inductances Between Stator and Rotor Flux

The fundamental components of the air-gap flux will link with the sinusodially distributed stator flux. The flux linkage is a maximum when the two coil axes are in line.

$$\left.\begin{array}{l} x_{Rfd} = x_{fRd} = x_{Rfd}\cos \theta \\ x_{yfd} = x_{fyd} = x_{yfd}\cos(\theta - 120 \text{ degrees}) \\ x_{Bfd} = x_{fBd} = x_{Bfd}\cos(\theta + 120 \text{ degrees}) \end{array}\right\} \qquad (22.23)$$

$$\left.\begin{array}{l} x_{R1d} = x_{1Rd} = x_{R1d}\cos \theta \\ x_{y1d} = x_{1yd} = x_{y1d}\cos(\theta - 120 \text{ degrees}) \\ x_{B1d} = x_{1Bd} = x_{B1d}\cos(\theta + 120 \text{ degrees}) \end{array}\right\} \qquad (22.24)$$

$$\left.\begin{array}{l} x_{R1q} = x_{1Rq} = -x_{R1q}\sin\theta \\ x_{y1q} = x_{1yq} = -x_{R1q}\sin(\theta - 120 \text{ degrees}) \\ x_{B1q} = x_{1Bq} = -x_{B1q}\sin(\theta + 120 \text{ degrees}) \end{array}\right\} \qquad (22.25)$$

22.3.4 FLUX LINKAGE EQUATIONS

Using the mutual-inductance relations (22.23) the flux linkage equations are rewritten as follows:

Field

$$\psi_{fd} = -x_{Rfd}[i_R \cos\theta + i_y \cos(\theta - 120 \text{ degrees}) + i_B \cos(\theta + 120 \text{ degrees}) + x_{ffd}i_{fd} + x_{f1d}i_{fd} + x_{f2d}i_{fd} + \cdots]$$

(22.26)

Direct-Axis Damper Winding

$$\psi_{1d} = -x_{R1d}[i_R \cos\theta + i_y \cos(\theta - 120 \text{ degrees}) + i_B \cos(\theta + 120 \text{ degrees}) + x_{1fd}i_{fd} + x_{11d}i_{fd} + x_{12d}i_{fd} + \cdots$$

(22.27)

Quadrature-Axis Damper Winding

$$\psi_{1q} = -x_{R1q}[i_R \sin\theta + i_y \sin(\theta - 120 \text{ degrees}) + i_B \sin(\theta + 120 \text{ degrees}) + x_{11q}i_{1q} + x_{12q}i_{2q} + \cdots$$

(22.28)

22.4 THE TRANSFORMATIONS

Now, it is possible to consider a transformation of phase quantities into "d" and "q" axes components. However, since the transformed variables for the three-phase system must also be three, we introduce the zero-sequence components for consistent transformation of variables. Thus it is defined that

$$
\left.
\begin{aligned}
i_d &= \frac{2}{3}\left[i_R \cos\theta + i_y \cos(\theta - 120 \text{ degrees}) + i_B \cos(\theta + 120 \text{ degrees})\right] \\
i_q &= -\frac{2}{3}\left[i_R \sin\theta + i_y \sin(\theta - 120 \text{ degrees}) + i_B \sin(\theta + 120 \text{ degrees})\right] \\
i_o &= \frac{1}{3}(i_R + i_y + i_B)
\end{aligned}
\right\}
$$

(22.29)

substituting Eq. (22.7) into Eqs. (22.14)−(22.16)

$$
\left.
\begin{aligned}
\psi_{fd} &= -\frac{3}{2}x_{Rfd}i_d + x_{ffd}i_{fd} + x_{f1d}i_d + \cdots \\
\psi_{1d} &= -\frac{3}{2}x_{R1d}i_d + x_{1fd}i_{fd} + x_{11d}i_d + x_{12d}i_d + \cdots \\
\psi_{1q} &= -\frac{3}{2}x_{R1q}i_q + x_{11q}i_{1q} + x_{12q}i_{2q} + \cdots
\end{aligned}
\right\}
$$

(22.30)

substituting Eqs. (22.21)−(22.25) into (22.17)

$$
\begin{aligned}
\psi_R &= -x_{RR0}i_R + x_{Ry0}(i_y + i_B) - x_{RR2}i_R \cos 2\theta \\
&\quad + x_{RR2}i_R \cos 2(\theta + 30 \text{ degrees}) + x_{RR2}i_B \cos 2(\theta + 150 \text{ degrees}) \\
&\quad + (x_{Rfd}i_{fd} + x_{R1d}i_{1d} + x_{R2d}i_{2d} + \cdots) \cos\theta - (x_{R1q}i_{1q} + x_{R2q}i_{2q} + \cdots)\sin\theta
\end{aligned}
$$

(22.31)

$$\psi_y = -x_{RR0}i_y + x_{Ry0}(i_B + i_R) + x_{RR2}i_R \cos 2(\theta + 30 \text{ degrees})$$
$$-x_{RR2}i_y \cos 2(\theta - 120 \text{ degrees}) + x_{RR2}i_B \cos 2(\theta - 90 \text{ degrees})$$
$$+(x_{Rfd}i_{fd} + x_{R1d}i_{1d} + x_{R2d}i_{2d} + \cdots) \cos(\theta - 120 \text{ degrees})$$
$$-(x_{R1q}i_{1q} + x_{R2q}i_{2q} + \cdots)\sin(\theta - 120 \text{ degrees}) \tag{22.32}$$

$$\psi_B = -x_{RR0}i_B + x_{Ry0}(i_R + i_Y) + x_{RR2}i_R \cos 2(\theta + 158 \text{ degrees})$$
$$-x_{RR2}i_y\cos 2(\theta - 90 \text{ degrees}) - x_{RR2}i_B\cos 2(\theta + 120 \text{ degrees})$$
$$+(x_{Rfd}i_{fd} + x_{R1d}i_{1d} + x_{R2d}i_{2d} + \cdots) \cos (\theta + 120 \text{ degrees})$$
$$-(x_{R1q}i_{1q} + x_{R2q}i_{2q} + \cdots) \sin (\theta + 120 \text{ degrees}) \tag{22.33}$$

Eliminating the armature phase currents i_R, i_y, and i_B by i_d, i_q and i_0, we can obtain ψ_R, ψ_Y, and ψ_B. Further, defining ψ_d, ψ_q, and ψ_0 by a similar transformation that is used for currents.

$$\psi_d = \frac{2}{3}\left[\psi_R \cos \theta + \psi_y \cos(\theta - 120 \text{ degrees}) + \psi_B\cos(\theta + 120 \text{ degrees})\right]$$
$$\psi_q = -\frac{2}{3}\left[\psi_R \sin \theta + \psi_y \sin(\theta - 120 \text{ degrees}) + \psi_B \sin(\theta + 120 \text{ degrees})\right] \tag{22.34}$$
$$\psi_0 = \frac{1}{3}\left[\psi_R + \psi_y + \psi_B\right]$$

Substituting Eqs. (22.19)–(22.21) into (22.34)

$$\psi_d = -\left(x_{RR0} + x_{Ry0} + \frac{3}{2}x_{RR2}\right)i_d + x_{Rfd}i_{fd} + x_{R1d}i_{1d} + x_{R2d}i_{2d} + \cdots \tag{22.35}$$

$$\psi_q = -\left(x_{RR0} + x_{RY0} + \frac{3}{2}x_{RR2}\right)i_q + x_{R1q}i_{1q} + x_{R2q}i_{2q} + \cdots \tag{22.36}$$

$$\psi_0 = -(x_{RR0} - 2x_{RY0})i_0 \tag{22.37}$$

Thus, for direct and quadrature axes we can define

$$x_d = \left(x_{RR0} + x_{Ry0} + \frac{3}{2}x_{RR2}\right) \tag{22.38}$$

$$x_q = \left(x_{RR0} + x_{Ry0} + \frac{3}{2}x_{RR2}\right) \tag{22.39}$$

Further

$$x_0 = x_{RR0} - 2x_{Ry0} \tag{22.40}$$

22.5 STATOR VOLTAGE EQUATIONS

Now, it is desired to eliminate i_R, i_y, and i_B from Eq. (22.13).

Just as in the case of currents and flux linkages, voltage transformation also will be now defined:

$$e_d = \frac{2}{3}\left[e_R \cos\theta + e_y \cos(\theta - 120 \text{ degrees}) + e_B \cos(\theta + 120 \text{ degrees})\right]$$

$$e_q = -\frac{2}{3}\left[e_R \sin\theta + e_y \sin(\theta - 120 \text{ degrees}) + e_B \sin(\theta + 120 \text{ degrees})\right] \tag{22.41}$$

$$e_0 = \frac{1}{3}\left[e_R + e_y + e_B\right]$$

substituting Eq. (22.13) into (22.40) and using (22.29)

$$e_d = \frac{2}{3}\left[\cos\theta\frac{d\psi_R}{dt} + \cos(\theta - 120 \text{ degrees})\frac{d\psi_y}{dt} + \cos(\theta + 120 \text{ degrees})\frac{d\psi_B}{dt}\right] - ri_d$$

$$e_q = -\frac{2}{3}\left[\sin\theta\frac{d\psi_R}{dt} + \sin(\theta - 120 \text{ degrees})\frac{d\psi_y}{dt} + \sin(\theta + 120 \text{ degrees})\frac{d\psi_B}{dt}\right] - ri_q \tag{22.42}$$

$$e_0 = \frac{d}{dt}\psi_0 - ri_0$$

differentiating ψ_d from (22.34)

$$\frac{d\psi_d}{dt} = \frac{2}{3}\left[\cos\theta\frac{d\psi_R}{dt} + \cos(\theta - 120 \text{ degrees})\frac{d\psi_y}{dt} + \cos(\theta + 120 \text{ degrees})\frac{d\psi_B}{dt}\right]$$

$$-\frac{2}{3}\left[\psi_R \sin\theta\frac{d\vartheta}{dt} + \psi_y\sin(\theta - 120 \text{ degrees})\frac{d\vartheta}{dt} + \psi_B\sin(\theta + 120 \text{ degrees})\frac{d\vartheta}{dt}\right] \tag{22.43}$$

Substituting ψ_d from (22.34) further

$$\frac{d\psi_d}{dt} = \frac{2}{3}\left[\cos\theta\frac{d\psi_R}{dt} + \cos(\theta - 120 \text{ degrees})\frac{d\psi_y}{dt} + \cos(\theta + 120 \text{ degrees})\frac{d\psi_B}{dt} + \psi_q\frac{d\vartheta}{dt}\right] \tag{22.44}$$

In a similar way

$$\frac{d}{dt}\psi_q = -\frac{2}{3}\left[\sin\theta\frac{d\psi_R}{dt} + \sin(\theta - 120 \text{ degrees})\frac{d\psi_y}{dt} + \sin(\theta + 120 \text{ degrees})\frac{d\psi_B}{dt}\right] \tag{22.45}$$

Eq. (22.42) reduces to

$$e_0 = \frac{d}{dt}\psi_0 - ri_0$$

$$e_q = \dot\psi_q = \psi_d\dot\theta - ri_q \tag{22.46}$$

$$e_0 = \dot\psi_0 - ri_0$$

Eq. (22.32) is the same as Eqs. (22.8) and (22.9).

22.6 STEADY-STATE EQUATION

In the steady state, the flux linkages are

$$\left.\begin{array}{l} \psi_d = -x_d i_d + x_{rfd}I_{fd} \\ \psi_{fd} = -x_{rfd}i_d + x_{ffd}I_{fd} \\ \psi_q = -x_q i_q \end{array}\right\} \tag{22.47}$$

the voltage relation for the field is then

$$e_{fd} = \dot{\psi}_{fd} + R_{fd}I_{fd}$$

In the steady state from Eq. (22.46) at $w = 1.0$ p.u.

$$e_d = -\psi_d - ri_d(\dot{\psi}_d = 0) \tag{22.48}$$

$$e_q = -\psi_q - ri_q(\dot{\psi}_q = 0) \tag{22.49}$$

$$\psi_d = x_{Rfd}I_{fd} - x_d i_d = \frac{x_{Rfd}e_{fd}}{R_{fd}} - x_d i_d \tag{22.50}$$

$$= E - x_d\, i_d$$

where E is the field excitation measured in terms of terminal voltage that it produces on open circuit at a normal speed.

$$e_d = x_q\, i_q - r\, i_d \tag{22.51}$$

$$e_q = E - x_d\, i_d - r\, i_q \tag{22.52}$$

so that, neglecting resistances

$$i_d = \frac{E - e \cos \delta}{x_d} \tag{22.53}$$

and

$$i_q = \frac{e \operatorname{Sin} \delta}{x_q} \tag{22.54}$$

where the open-circuit voltage E is ahead of the corresponding bus voltages by an angle δ.

22.7 STEADY-STATE VECTOR DIAGRAM

$$e_d = x_q\, i_q - r_{id} \tag{22.55}$$

$$e_q = E - x_d\, i_d - r\, i_q \tag{22.56}$$

An voltage E_q is defined as

$$E_q = E - (x_d - x_q)i_d \tag{22.57}$$

Then Eq. (22.56) can be put (Fig. 22.8)

$$e_q = E_q - x_d\, i_q - r\, i_d \tag{22.58}$$

Defining

$$\left.\begin{aligned}\bar{e} &= e_d + jl_q \\ \bar{i} &= i_d + ji_q\end{aligned}\right\} \tag{22.59}$$

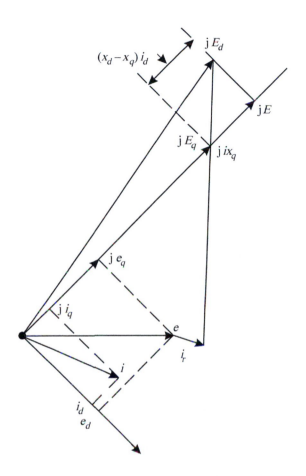

FIGURE 22.8

Synchronous generator in steady-state phasor diagram.

Then

$$\bar{e} = jE_q - (r + jx_q)\bar{i} \tag{22.60}$$

It may be noted that is also the terminal voltage of the machine v_t.
E is the voltage behind excitation.

22.8 REACTANCES

Whenever a three-phase short circuit occurs at the terminals of an alternator, the current in the armature circuit increases suddenly to a large value and since the resistance of the circuit is small compared to its reactance, the current is highly lagging and the p.f. is approximately zero. Due to this sudden switching, there are two components of currents:

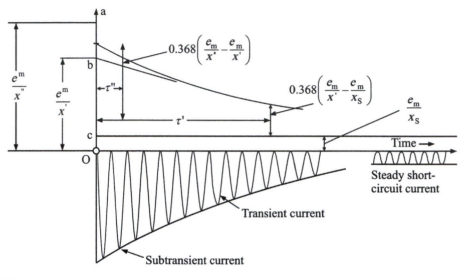

FIGURE 22.9

Analysis of symmetrical short-circuit current. Oa, subtransient current; Ob, transient current; Oc, steady-state current.

(1) AC component, (2) DC component (decaying).

The oscillogram of current variation as a function of time is shown in Fig. 22.9 for a three-phase fault at the terminals of an alternator.

The rotor rotates at zero speed with respect to the field due to AC component of current in the stator, whereas it rotates at a synchronous speed with respect to the field due to the DC component of current in the stator conductors. The rotor winding acts as the secondary of a transformer for which the primary is the stator winding. Similarly in case of the rotor that has damper winding fixed on its poles, the whole system will work as a three-winding transformer in which stator is the primary and the rotor field winding and damper windings form the secondary of a transformer. It is to be noted that the transformer action is there with respect to the DC component of current only. The AC component of current being highly lagging tries to demagnetize, i.e., reduce the flux in air gap.

This reduction of flux from the instant of short circuit to the steady-state operation cannot take place instantaneously because of the large amount of energy stored by the inductance of the corresponding system. So this change in flux is slow and depends upon the time constant of the system.

To balance the suddenly increased demagnetizing mmf of the armature current, the exciting current, i.e., the field winding current must increase in the same direction of flow as before the fault.

This happens due to the transformer action. At the same time, the current in the damper and the eddy currents in the adjacent metal parts increase in obedience to Lenz's law, this assists the rotor field winding to sustain the flux in the air gap. At the instant of the short circuit, there is mutual coupling between the stator winding, rotor winding, and the damper winding and the equivalent circuit is represented in Fig. 22.10A−C.

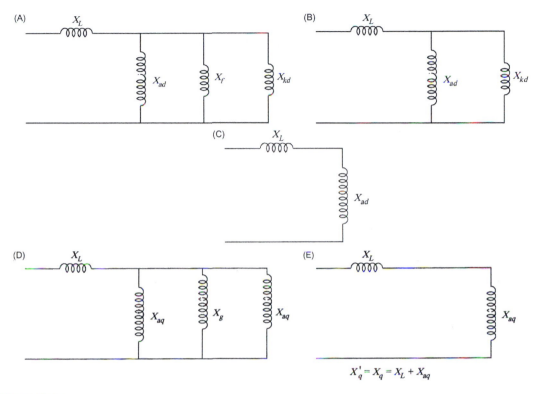

$$X'_q = X_q = X_L + X_{aq}$$

FIGURE 22.10

Synchronous equivalent circuit—d-axis: subtransient state (A), transient state (B), and steady state (C), and q-axis: subtransient state (D), transient state (E).

Since the equivalent resistance of the damper winding when referred to the stator is more as compared to the rotor winding, the time constant of damper winding t'' is smaller than the rotor field winding. Therefore the effect of damper winding and the eddy current in the pole faces disappears after the first few cycles. Accordingly, the equivalent circuit after first few cycles reduces to the one shown in Fig. 22.10B. After a few more cycles depending upon the time constant of the field winding t', the effect of the DC. component will die down and steady-state condition will prevail for which the equivalent circuit is shown in Fig. 22.10C. The inductance increases from the initial state to the final steady state.

i.e., synchronous reactance > transient reactance > subtransient reactance.

In the subtransient state

$$X_d^{||} = X_L + \cfrac{1}{\cfrac{1}{X_{ad}} + \cfrac{1}{X_f} + \cfrac{1}{X_{kd}}} \tag{22.61}$$

In the transient state

$$X_d^| = X_L + \cfrac{1}{\cfrac{1}{X_{ad}} + \cfrac{1}{X_f}} \tag{22.62}$$

In the steady state $X_d = X_L + X_{ad}$

In a similar way, if X_g represents a fictitious coil g to represent the effect of flux linkages along the q-axis, then

$$X_q^{||} = X_L + \cfrac{1}{\cfrac{1}{X_{aq}} + \cfrac{1}{X_g} + \cfrac{1}{X_{kq}}} \tag{22.63}$$

This subtransient reactance $X_q^{||}$ is very small. In a very short time circuit changes to (Fig. 22.10E) showing that $X_q = X_q^|$

$$X_q^| = X_L + \cfrac{1}{\cfrac{1}{X_{aq}}} = X_q = X_L + X_{aq} \tag{22.64}$$

In the steady state

$$X_q = X_L + X_{aq} \tag{22.65}$$

The following table shows the typical values of constants for different types of synchronous machines.

	Turbogenerator		Synchronous	Waterwheel generator	
	Two-Pole	**Four-Pole**		**With Dampers**	**Without Dampers**
Synchronous	2.0	1.45	1.25	0.9	0.9
x_{sd}	1.4–2.5	1.35–1.65	0.75–1.8	0.5–1.5	0.5–1.5
Transient	0.19	0.27	0.21	0.23	0.23
$x_d^|$	0.11–0.25	0.24–0.31	0.12–0.27	0.14–0.32	0.14–0.32
Subtransient	0.13	0.19	0.13	0.16	0.18
$X_d^{||}$	0.08–0.18	0.15–0.23	0.09–0.15	0.1–0.27	0.16–0.3
Negative-sequence	0.16	0.28	0.12	0.16	0.23
x_2	0.09–0.23	0.24–0.31		0.1–0.27	0.12–0.37
Zero-sequence	0.08	0.28	0.03	0.08	0.08
x_0	0.02–0.15	0.22–0.31		0.06–0.1	0.06–0.1
Inertia constant	4.7	4.8	1.2	3.4	3.4
H	2.6–6	4.3–5.7	0.7–1.8	2–5	2–5
Open-circuit time constant	9.5	5.5			
t_{ds}	3.5–16				

Note : *Average values are given in each block.*

22.9 EQUIVALENT CIRCUITS AND PHASOR DIAGRAMS

Model for Transient Stability

For study periods of the order of 1 s or less, the synchronous machine can be represented by a voltage behind a transient reactance that is constant in magnitude but changes its angular position. This is shown in Fig. 22.11.

The voltage relation is expressed by

$$E = V_t + r_a I_t + jx I_t \tag{22.66}$$

where I_t is the terminal current and V_t is the machine terminal voltage.

The phasor diagram is shown in Fig. 22.12.

This model can be extended to include the effect of saliency and the effect of field flux linkages considering the d and q-axis quantities.

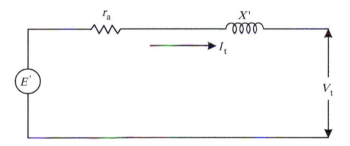

FIGURE 22.11

Synchronous machine equivalent circuit for transient stability study.

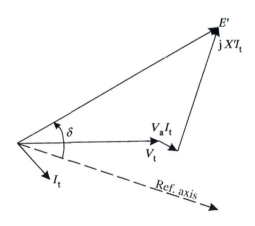

FIGURE 22.12

Phasor diagram for transient stability.

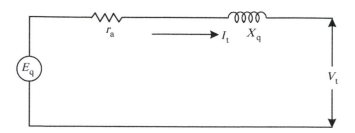

FIGURE 22.13

Synchronous machine q-axis equivalent circuit.

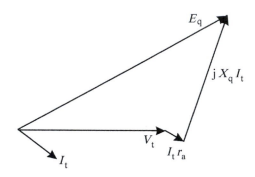

FIGURE 22.14

Synchronous machine q-axis phasor diagram.

The direct axis is chosen as the center line passing through the machine pole and the quadrature axis 90 degrees (clc) behind this axis. The position of the q-axis can be determined by calculating a fictitious voltage located on the q-axis called E_q (voltage behind q-axis reactance). The equivalent circuit and the phasor diagram are shown in Figs. 22.13 and 22.14.

It may be noted that the sinusoidal flux produced by the field current acts along the main or d-axis. The voltage induced by the field current lags behind this flux by 90 degrees and is therefore on the q-axis. This voltage can be obtained by adding the voltage drop across r_a to v_t and the voltage drops representing the demagnetizing effects along the d- and q-axes.

22.10 TRANSIENT STATE PHASOR DIAGRAM

$$\left.\begin{array}{l} \psi_d = -x_d i_d + x_{Rfd}\, i_{fd} \\ \psi_{fd} = -x_{Rfd} i_d + x_{ffd}\, i_{fd} \end{array}\right\} \tag{22.67}$$

Eliminating the field current i_{fd}

$$i_{fd} = \frac{\psi_d + x_d i_d}{x_{Rfd}} \tag{22.68}$$

Again from Eq. (22.47)

$$\psi_{fd} = \frac{x_{ffd}}{x_{Rfd}}\left[\psi_d + \left(x_d - \frac{x_{Rfd}^2}{x_{ffd}}\right)i_d\right] \tag{22.69}$$

the quantity $\left(x_d - \dfrac{x_{Rfd}^2}{x_{ffd}}\right)$ is a short-circuit reactance of the armature direct-axis circuit and is defined.

Direct-Axes Transient Reactance

$$x_d^! = x_d - \frac{x_{Rfd}^2}{x_{ffd}} \tag{22.70}$$

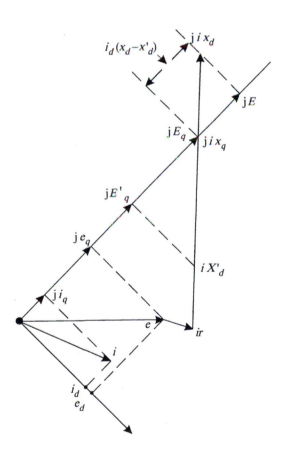

FIGURE 22.15

Transient state phasor diagram of synchronous machine.

The quantity $\left(\dfrac{X_{Rfd}}{x_{ffd}}\right)\psi_d = E_d^|$ is the voltage behind transient reactance $x_d^|$ (22.71)

Hence, Eq. (22.71) becomes

$$E_q^| = \psi_d + x_d^| \, i_d \tag{22.72}$$

Since

$$e_q = Y_d - r \, i_q$$

$$e_q = E_q^| - x_d^| \, i_d - r i_q \tag{22.73}$$

Again from Eq. (22.72)

$$E_q^| = \frac{x_{Rfd}}{x_{ffd}}\psi_{fd} = -\frac{x_{Rfd}^2}{x_{ffd}}i_d + E \tag{22.74}$$

$$= (x_d - x_d^|)i_d \tag{22.75}$$

The transient state phasor diagram is given in Fig. 22.15.
Also

$$E_q^| = E_q - \bar{I}(x_q - x_d^|) \tag{22.76}$$

which can be verified from the phasor diagram.

22.11 POWER RELATIONS

Now, expressions for the real and reactive powers developed by the synchronous generator will be derived.

The phasor diagram for a salient-pole synchronous machine neglecting the resistance is given in Fig. 22.16. The complex power generated is given by

$$S = P + jQ \tag{22.76}$$

$$= |e| \, |i| \cos \phi + j \, |e| \, |i| \sin \phi \tag{22.77}$$

From the figure

$$e \cos \delta = jE - i_d \, x_d \tag{22.78}$$

$$e \sin \delta = i_q \, x_q \tag{22.79}$$

$$i_d = i \sin \beta \tag{22.80}$$

$$i_q = i \cos \beta \tag{22.81}$$

where

$$\beta - \delta = \phi$$

$$\cos \phi = \cos(\beta - \delta) = \cos \beta \cos \delta + \sin \beta \sin \delta$$

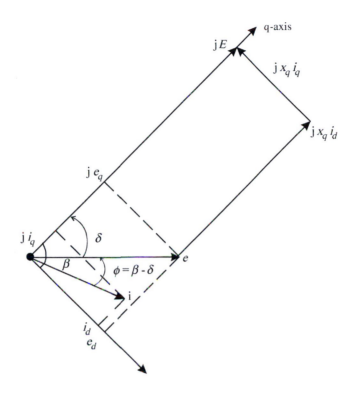

FIGURE 22.16

Salient-pole synchronous machine—resistance neglected.

$$\sin \phi = \sin(\beta - \delta) = \sin \beta \cos \delta - \cos \beta \sin \delta$$

Multiplying the above equation by i

$$i \cos \phi = i \cos \beta \cos \delta + i \sin \beta \sin \delta \qquad (22.82)$$

and, also

$$i \sin \phi = i \sin \beta \cos \delta - i \cos \beta \sin \delta \qquad (22.83)$$

utilizing Eqs. (22.80) and (22.81)

$$i \cos \phi = i_q \cos \delta + i_d \sin \delta \qquad (22.84)$$

$$i \sin \phi = i_d \cos \delta + i_q \sin \delta \qquad (22.85)$$

From Eq. (22.77)

$$S = e(i_q \cos \delta + i_d \sin \delta) + j \, e \, (i_d \cos \delta - i_q \sin \delta) \qquad (22.86)$$

From Eqs. (22.78) and (22.79)

$$i_d = \frac{E - e \cos \delta}{x_d} \qquad (22.87)$$

and

$$i_q = \frac{e \sin \delta}{x_d} \tag{22.88}$$

$$S = e \left[\frac{e \sin \delta}{x_q} \cos \delta + \frac{E - e \cos \delta}{x_d \cdot} \sin \delta \right] + je \left[\frac{E - e \cos \delta}{x_d} \cos \delta - \frac{e \sin \delta}{x_q} \sin \delta \right] \tag{22.89}$$

The real power

$$P = \frac{Ee}{x_d} \sin \delta + e^2 \left[\frac{\sin \delta \cos \delta}{x_q} - \frac{\cos \delta \sin \delta}{x_d} \right]$$

$$P = \frac{Ee}{x_d} \sin \delta + \frac{e^2}{2} \left[\frac{1}{x_q} - \frac{1}{x_d} \right] \sin 2\delta \tag{22.90}$$

The reactive power, in a similar way, is obtained as

$$Q = \frac{Ee}{x_d} \cos \delta - e^2 \left(\frac{\cos^2 \delta}{x_d} + \frac{\sin^2 \delta}{x_q} \right) \tag{22.91}$$

the second term in Eq. (22.90) is called the saliency torque as it is due to the difference in the reluctance of pole region and interpole region in the air gap.

In case of cylindrical rotor construction $x_d = x_q$ and hence Eqs. (22.90) and (22.91) will reduce to

$$P = \frac{Ee}{x_s} \sin \delta \tag{22.92}$$

and

$$Q = \frac{Ee}{x_s} \cos \delta - \frac{e^2}{x_s} \tag{22.93}$$

where x_s is the synchronous reactance.

22.12 SYNCHRONOUS MACHINE CONNECTED THROUGH AN EXTERNAL REACTANCE

Consider a synchronous generator connected to a bus through an external reactance x_1 (Fig. 22.17); the machine may have a synchronous reactance of x_s or d- and q-axes reactances x_d and x_q. The voltage phasor diagram of Fig. 22.16 will simply change into Fig. 22.18.

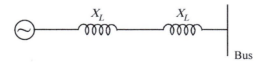

FIGURE 22.17

Synchronous machine with external reactance.

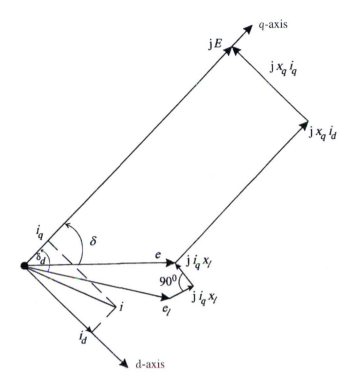

FIGURE 22.18

Effect of external reactance x_1.

Thus it is possible to include the effect of an external reactance by considering the values in Fig. 22.19 in the following way:

$$\delta \rightarrow \delta_1$$

$$x_d \rightarrow x_d + x_1$$

$$x_q \rightarrow x_q + x_1$$

22.13 THE SWING EQUATION

The interconnection between electrical and mechanical side of the synchronous machine is provided by the dynamic equation for the acceleration or deceleration of the combined-prime mover (turbine)—synchronous machine rotor. This is usually called swing equation.

The net torque acting on the rotor of a synchronous machine

$$T = \frac{WR^2}{g}\alpha \tag{22.94}$$

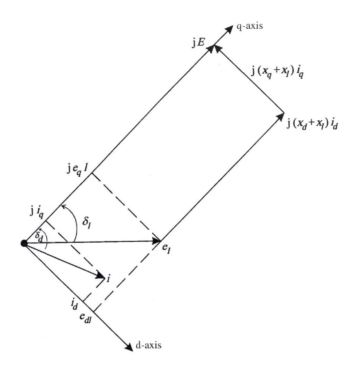

FIGURE 22.19

Equivalent phasor diagram with external reactance x_l.

where T is the algebraic sum of all torques in kg-m; a is the mechanical angular acceleration; and WR^2 is the moment of inertia in kg-m^2.

$$\text{Electrical angle } \vartheta_e = \vartheta_m \frac{P}{Z} \tag{22.95}$$

where ϑ_m is mechanical angle and P is the number of poles.

The frequency

$$f = \frac{PN}{120} \tag{22.96}$$

where N is the rpm.

$$f = \frac{P}{2}\left(\frac{\text{rpm}}{60}\right)$$
$$\frac{60 f}{\text{rpm}} = \frac{P}{2} \tag{22.97}$$
$$\vartheta_e = \left(\frac{60f}{\text{rpm}}\right)\vartheta_m$$

The electrical angular position d in radians of the rotor with respect to a synchronously rotating reference axis is

$$\delta_e = \vartheta_e - \omega_0 t \tag{22.98}$$

where ω_0 is the rated synchronous speed in radian/s and t is the time in seconds (Note : $\delta + \omega_0 t = \vartheta_e$).

The angular acceleration taking the second derivative of Eq. (22.98) is given by

$$\frac{d^2\delta}{dt^2} = \frac{d^2\vartheta_e}{dt^2}$$

From Eq. (22.97) differentiating twice

$$\frac{d^2\vartheta_e}{dt^2} = \left(\frac{60f}{\text{rpm}}\right)\frac{d^2\vartheta_m}{dt^2}$$

$$\therefore \frac{d^2\vartheta_m}{dt^2} = \alpha = \left(\frac{\text{rpm}}{60f}\right)\frac{d^2\vartheta_e}{dt^2}$$

From Eq. (22.94)

$$T = \frac{WR^2}{g}\left(\frac{\text{rpm}}{60f}\right)\frac{d^2\vartheta_e}{dt^2} = \frac{WR^2}{g}\left(\frac{\text{rpm}}{60f}\right)\frac{d^2\delta}{dt^2} \tag{22.99}$$

Let the base torque be defined as

$$T_{\text{Base}} = \frac{\text{Base KVA}}{2\pi\left(\frac{\text{rpm}}{60}\right)} \tag{22.100}$$

torque in per unit

$$\begin{aligned}T \text{ p.u.} &= \frac{T}{T_{\text{Base}}} = \frac{WR^2}{g}\left(\frac{\text{rpm}}{60f}\right)\frac{d^2\delta}{dt^2} \cdot \frac{2\pi\left(\frac{\text{rpm}}{60}\right)}{\text{Base KVA}} \\ &= \frac{WR^2}{g}\left(\frac{\text{rpm}}{60}\right)^2 \frac{2\pi}{f} \cdot \frac{1}{\text{Base KVA}}\frac{d^2\delta}{dt^2}\end{aligned} \tag{22.101}$$

$$\text{Kinetic energy KE} = \frac{1}{2}\frac{WR^2}{g}\omega_0^2 \tag{22.102}$$

where

$$\omega_0 = 2\pi\frac{\text{rpm}}{60}$$

Defining

$$H = \frac{\text{Kinetic energy at rated speed}}{\text{Base KVA}}$$

$$= \underbrace{\frac{1}{2}\frac{WR^2}{g}\left(2\pi\tfrac{\text{rpm}}{60}\right)^2}_{\text{KE at rated speed}}\frac{1}{\text{Base KVA}}$$

$$T = \frac{H}{\pi f}\cdot\frac{d^2\delta}{dt^2} \tag{22.103}$$

The torque acting on the rotor of a generator includes the mechanical input torque from the prime mover, torque due to rotational losses (i.e., friction, windage, and core loss), electrical output torque and damping torques due to prime mover generator, and power system.

The electrical and mechanical torques acting on the rotor of a motor are of opposite sign and are a result of the electrical input and mechanical load. We may neglect the damping and rotational losses, so that the accelerating torque.

$$T_a = T_m - T_e$$

where T_e is the air-gap electrical torque and T_m the mechanical shaft torque.

$$\frac{H}{\pi f}\frac{d^2\delta}{dt^2} = T_m - T_e \tag{22.104}$$

i.e.,

$$\frac{d^2\delta}{dt^2} = \frac{\pi f}{H}(T_m - T_e) \tag{22.105}$$

Torque in per unit is equal to power in per unit if speed deviations are neglected. Then

$$\frac{d^2\delta}{dt^2} = \frac{\pi f}{H}(P_m - P_e) \tag{22.106}$$

Eqs. (22.105) and (22.106) are called swing equations.

It may be noted, that, since $\delta = \vartheta - \omega_0 t$

$$\frac{d\delta}{dt} = \frac{d\vartheta}{dt} - \omega_0$$

Since the rated synchronous speed in radian/s is $2\pi f$

$$\frac{d\vartheta}{dt} = \frac{d\delta}{dt} + \omega_0$$

we may put the equation in another way.

$$\text{Kinetic energy} = \frac{1}{2}I\omega^2 \text{ J}$$

The moment of inertia I may be expressed in $\text{Js}^2/(\text{rad})^2$ since ω is in radian/s. The stored energy of an electrical machine is more usually expressed in MJ and angles in degrees. Angular momentum M is thus described by MJs/electrical degree

$$M = I \cdot \omega$$

where ω is the synchronous speed of the machine and M is called inertia constant. In practice ω is not synchronous speed while the machine swings and hence M is not strictly a constant.

The quantity H defined earlier as inertia constant has the units MJ.

$$H = \frac{\text{Stored energy in MJ}}{\text{Machine rating in MVA }(G)} \tag{22.107}$$

$$\text{but stored energy} = \frac{1}{2}I\omega^2 = \frac{1}{2}M\omega$$

In electrical degrees $\omega = 360\,f(=2\pi f)$ $\tag{22.108}$

$$GH = \frac{1}{2}M(360\,f) = \frac{1}{2}M2\pi f = M\pi f$$

$$M = \frac{GH}{\pi f} \quad \text{(MJs/electrical degree)} \tag{22.109}$$

In the per-unit systems

$$M = \frac{H}{\pi f} \tag{22.110}$$

so that

$$\frac{d^2\delta}{dt^2} = \frac{\pi f}{H}(P_m - P_e) \tag{22.111}$$

which may be written also as

$$M\frac{d^2\delta}{dt^2} = P_m - P_e \tag{22.112}$$

This is another form of swing equation.
Further

$$P_e = \frac{EV}{X}\sin\delta$$

So that

$$M\frac{d^2\delta}{dt^2} = P_m - \frac{EV}{X}\sin\delta \tag{22.113}$$

with usual notation.

WORKED EXAMPLES

E.22.1 A salient-pole synchronous generator is operated with $E = 1.1$ p.u. and $e = 1.0$ p.u. (line). Given that $x_d = 0.8$ p.u. and $x_q = 0.6$ p.u. for each phase. The synchronous reactance is estimated at 0.7 p.u. per phase, What is the power developed. If the saliency is neglected, what will be the power generated ?

Solution:

$$P = \frac{Ev}{x_d} \sin \delta + \frac{e^2}{2} \left(\frac{1}{x_q} - \frac{1}{x_d} \right) \sin 2\delta$$

$$= \frac{(1.1)(1.0)}{0.8} \sin \delta + \frac{1}{2} \left(\frac{1}{0.65} - \frac{1}{0.8} \right) \sin 2\delta$$

$$= 1.375 \sin \delta + (0.7692 - 0.625) \sin 2\delta$$

$$= 1.375 \sin \delta + 0.1442 \sin 2\delta$$

δ	P_0
0	0
10 degrees	0.2880879
20 degrees	0.5629663
30 degrees	0.8123772
40 degrees	1.0258305
50 degrees	1.195259

Taking x_s, neglecting saliency

$$P = \frac{E_e}{x_s} \sin \delta = \frac{(1.1)(1.0)}{(0.7)} \sin \delta = 1.5714286 \sin \delta$$

δ	P
0	0
10 degrees	0.272875
20 degrees	0.53746
30 degrees	0.7857143
40 degrees	1.0100829
50 degrees	1.2037143

The two powers do not differ much and hence for all practical purposes saliency can be neglected in power calculations without causing much error.

E.22.2 Consider a synchronous generator with x_d 0.8 p.u./phase and $x_q = 0.6$ p.u./phase. The machine is connected to a bus through a reactor of impedance j0.1 p.u. on generator rating. The generator is excited to a voltage of $|E| = 1.2$ p.u. and delivers a load of 10 MW to the bus at 1 p.u. voltage. The generator is rated at 15 MW with a line voltage of 13.6 kV. *Determine δ, δ_l, I, and e*

Solution:

$$P = \frac{E_e}{x_d + x_l} \sin \delta_l + \frac{e^2}{2} \left(\frac{1}{x_q + x_l} - \frac{1}{x_d + x_l} \right) \sin 2\delta_l$$

In per-unit system

$$P = \frac{10 \text{ MW}}{15 \text{ MW}} = 0.0667 \text{ p.u.}$$

$$0.667 = \frac{1.2 \times 1.0}{0.8 + 0.1} \sin \delta_l + \frac{1.0^2}{2} \left(\frac{1}{0.6 + 0.1} - \frac{1}{0.8 + 0.1} \right) \sin 2\delta_l a$$

$$= \frac{1.2}{0.9} \sin \delta_l + \frac{1}{2}(1.4285714 - 1.111111)\sin 2\delta_l$$

$$\delta_l = 24.4 \text{ degrees}$$

$$I_q = \frac{1.0(\sin 24.4 \text{ degrees})}{0.6 + 0.1} = \frac{0.4131}{0.7} = 0.590$$

$$I_d = \frac{1.2 - 1.0 \cos 24.4 \text{ degrees}}{0.8 + 0.1} = \frac{1.2 - 0.9107}{0.9} = \frac{0.2893}{0.9} = 0.3214$$

$$I = \sqrt{(0.3214)^2 + (0.59)^2} \tan^{-1} \left(\frac{0.3214}{0.59} \right)$$

$$= 0.67186 \angle \tan^{-1} 0.5447457$$

$$= 0.67186 \angle -28.58 \text{ degrees}$$

The terminal voltage $v_t = v + j$

$$= 1.0 \angle -24.4 \text{ degrees} + j0.1 \, X \, 0.67186 \angle -28.58$$

$$= (0.9106836 - j0.4131) + 0.03214 + j0.0589993$$

$$= 0.9428236 - j0.3541$$

$$= 1.0071264 \angle -20.58 \text{ degrees}$$

QUESTIONS

Q.22.1 Explain the steady-state modeling of a synchronous machine.

Q.22.2 State and derive the swing equation.

Q.22.3 Explain the steady-state and transient modes of generator modeling.

Q.22.4 Clearly explain how a synchronous generator is modeled for steady-state analysis. Draw the phasor diagram and obtain the power angle equation for a nonsalient pole synchronous generator connected to an infinite bus. Sketch the power angle curve.

Q.22.5 What is the advantage of transformation of phase quantities into d−q−0 components?

Q.22.6 What is the significance of inertia constant?

Q.22.7 In swing equation what is the reactance used for computation of electrical power.

Q.22.8 Sketch the oscillogram of synchronous machine currents on short circuit.

Q.22.9 From the oscillogram of Q.22.8 explain the following quantities.

  (i) subtransient reactance
  (ii) transient reactance
  (iii) steady-state reactance

Q.22.10 Represent the reactances in Q.22.9 by equivalent circuits for both direct and quadrate axes.

PROBLEMS

P.22.1 A Synchronous generator is rated at 50 Hz 22 kv and 100 MVA. Based upon its own ratings the reactances of the machine are $x_d = 1.00$ p.u. and $x_q = 0.65$ p.u.

The machine is delivering 50 MW at 1.0 p.u. terminal voltage with an excitation of 1.4 p.u. Find δ and the reactive power generated.

P.22.2 A synchronous generator has $x_d = 0.95$ p.u. and $= 0.3$ p.u. on its own ratings. The excitation is maintained at 1.45 p.u. The generator delivers 0.7 p.u. MW to a bus at 1.0 p.u voltage. If the generator voltage falls by 50% suddenly determine an expression for the power generated subsequently.

VOLTAGE AND REACTIVE POWER CONTROL

23

Industrial and domestic loads, both, require real and reactive power. Hence generators have to produce both real and reactive power. Reactive power is required to excite various types of electrical equipment as well as transmission network. The reactive power requirement of consumers arises mostly from the lagging vars needed to supply magnetizing current to transformers and induction motors.

23.1 IMPEDANCE AND REACTIVE POWER

In the transmission network requirement is the difference between that absorbed in the series inductance (I^2X) and that produced in the shunt capacitance (V^2B). There is a level of loading at which the leading vars of charging current balance the lagging vars of the inductive lines called the natural or surge impedance loading (SIL) of the system. This natural load for a transmission line is given approximately as (B/X)p.u. In the case of cables the shunt capacitance is higher and series inductance is lower. As a consequence of which the natural loading is higher. However, this limit is generally above the thermal limit of the cable. The natural impedance loading for overhead lines is shown in Fig. 23.1 as a function of system voltage on a logarithmic scale. The conductors are

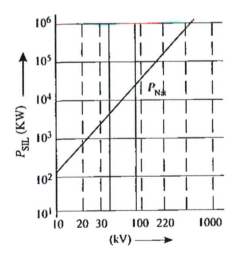

FIGURE 23.1

Natural impedance loadings as a function of system voltage.

Electrical Power Systems. DOI: http://dx.doi.org/10.1016/B978-0-08-101124-9.00023-1

731

bundled two together for 220 kV and four for 380 kV. Table 23.1 shows the reactive power requirements of overhead lines, cables, and transformers. The reactive power compensation of the transmission system depends on the load and its power factor. When the line is operated at no-load, the full charging power occurs and would result in considerable increase in voltage unless some compensating device is used. With full compensation at no-load, the line may be operated at any partial load between no-load and full-load with the voltage not exceeding the permissible limits.

The loss in a line is given by

$$P_L = K \frac{P^2 \ell}{V^2 \cos^2 \varphi}$$

where P is the power transmitted at voltage V over a line of length at a power factor $\cos \phi$. Higher voltages are selected for transmission to keep the losses in an economically justifiable relationship to the power P. In view of the inverse square relationship reduction of reactive power becomes an essential factor for obtaining efficient operation of the high-voltage lines. The reactive power requirement of overhead lines are shown in Fig. 23.2A and B.

In order to supply quality service to customers reliably and economically, voltage and (or) var control play (s) a leading role. Such a control has to be exercised at all over the power system, i.e., right from the generating point to the consumer terminals.

Rapid changes in voltage (flicker) can result due to some industrial loads such as arc furnaces, arc welders, and wood chippers. Fig. 23.3 shows the real and reactive power demand following the daily load cycle supplying a composite system load, which may create relatively large variations in voltage if control is not exercised. Also, there may be cyclic and noncyclic loads that create voltage disturbance at both transmission and distribution levels.

In addition, events such as planned line switching, unplanned line trips, planned and unforeseen generator trips, and equipment failure may produce voltage and var variations. Unless proper voltage support is given at strategic locations in the system, the aforesaid events may result in loss of stability and possible loss of service to a large number of consumers.

Table 23.1 Var Requirements of Overhead Lines, Cables, and Power Transformers

Equipment	MVAr Requirements No-Load	+Inductive, − Capacitive Full Load
Overhead lines		
400 kV, 4×0.4 in^2, 100 mi	− 105.0	+1251.0
275 kV, 2×0.4 in^2, 50 mi	− 22.0	+174.0
132 kV, 1×0.175 in^2, 20 mi	− 1.6	+10.0
Undertaken cables		
400 kV, 2×3.0 in^2, 10 mi	− 540	− 487.0
275 kV, 1×3.0 in^2, 10 mi	− 170	− 155.0
132 kV, 1×0.5 in^2, 5 mi	− 14.5	− 13.6
Transformers		
400/275 kV, 750 MVA	+0.5	+90.0
275/132 kV, 240 MVA	+0.3	+48.0
132/33 kV, 90 mVA	+ 20.5	+0.2

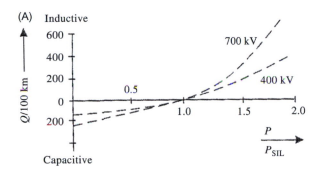

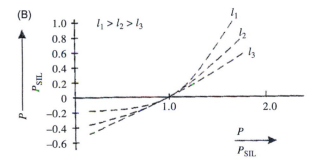

FIGURE 23.2

Reactive power requirement of overhead lines (four-conductor bundle) (A) Variation with line voltage and (B) variation with line length.

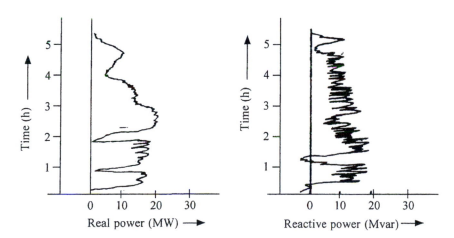

FIGURE 23.3

Arc furnace consumption.

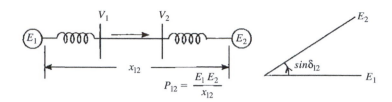

FIGURE 23.4

Power transfer with no intermediate voltage support.

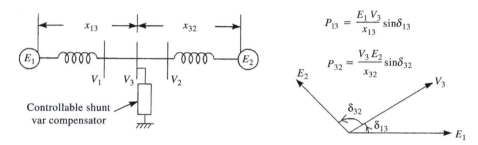

FIGURE 23.5

Power transfer with intermediate var compensator.

For long-distance transmission of power, the use of high-voltage direct current (HVDC) transmission has proved economical in certain cases. The var demand of DC terminals varies usually from 0% to 60% of the MW rating of the DC lines as power transfer is varied over its full range. When a fault takes place on the nearby AC system, the var demand of the DC link may reach a high value and unless compensated may produce large AC voltage variations.

Continuous control can be achieved by means of var compensators installed at line ends and/or in the intermediate substations.

The use of shunt-connected controllable var compensation to improve the power transfer capability and stability is an acknowledged fact. From Fig. 23.4, it can be seen that the theoretical maximum power transfer takes place at a power angle of $\delta_{12} = 90$.

With an intermediate, controllable, shunt var compensator, the angle could be increased, in principle to 180 degrees across the line (Fig. 23.5).

23.2 SYSTEM VOLTAGE AND REACTIVE POWER

Consider the system shown in Fig. 23.6. The voltage at Bus 2 is related to the voltage at Bus 1 by the relation.

$$V_2 = V_1 - IZ \tag{23.1}$$

Also,

$$V_1 \cdot I* = P + jQ \tag{23.2}$$

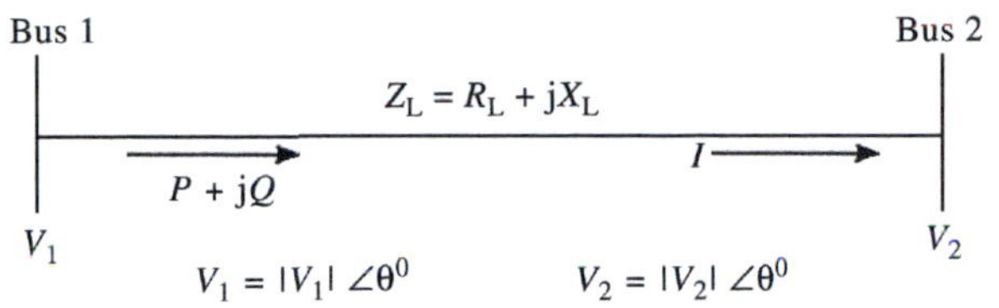

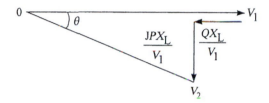

FIGURE 23.6

Power flow across a short line.

FIGURE 23.7

Phasor representation of Eq. (23.3).

So that

$$I = \frac{P - jQ}{V_1^*} = \frac{P - jQ}{V_1} \tag{23.3}$$

since V_1 is the reference phasor.

From Eqs. (23.1) and (23.2)

$$V_2 = V_1 - \left(\frac{P - jQ}{V_1}\right) Z$$

Neglecting the line resistance

$$V_2 = V_1 - j\frac{P}{V_1} X_L - \frac{Q}{V_1} X_L \tag{23.4}$$

Eq. (23.4) is illustrated by the phasor diagram in Fig. 23.7. It can be inferred from Eq. (23.3) or Fig. 23.7 that the voltage level is influenced largely by the reactive power drop QX_L/V_1, since the quadrature component PX_L/V_1 does not materially affect the voltage profile (both the drops are only small fractions of the bus voltage magnitudes).

23.3 REACTIVE POWER GENERATION BY SYNCHRONOUS MACHINES

Synchronous generators are able to produce both lagging and leading vars. Overexcitation of a generator field produces vars, whereas underexcited field causes vars to be absorbed. At lagging power factors, the limit on var generation is imposed by either rotor heating (due to maximum excitation current limit) or by stator heating (thermal MVA loading limit of the stator) consideration.

Generators are invariably fitted with automatic voltage regulators which maintain the thermal voltage at its normal value by adjustment of excitation. The operating charts for salient- and nonsalient-pole synchronous machines are shown in Fig. 23.8.

23.4 EFFECT OF EXCITATION CONTROL

Consider a synchronous machine with terminal voltage V_t. The direct-axis rotor angle with respect to a synchronously revolving axis is δ. The voltage due to excitation acting along the quadrature axis is E_q and E'_q is the transient voltage along this axis. If a load change occurs and the field current, I_f, is not changed, then the various quantities mentioned change with P, the real power as shown in Fig. 23.9A.

In case the field current I_f is changed such that the transient flux linkages along the q-axis E'_q proportional to Ψ_f, the field flux linkages is maintained constant, the power transfer could be

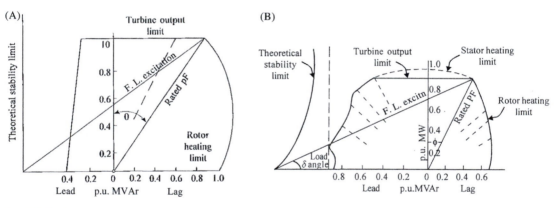

FIGURE 23.8

(A) Operating chart for nonsalient-pole synchronous machine. (B) Operating chart for salient-pole machine.

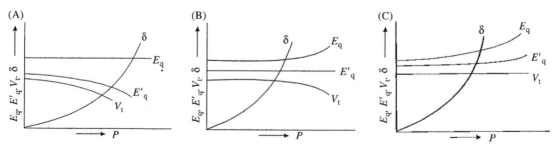

FIGURE 23.9

Effect of excitation control—E_q constant (A), E'_q constant (B), and V_t constant (C).

increased by 30%−60% greater than in case (A) and the quantities are plotted for this case in Fig. 23.9B.

If the field current I_f is changed along with P simultaneously so that V_t is maintained constant, then it is possible to increase power delivery by 50%−80% more than case (A). This is shown in Fig. 23.9C.

It can be concluded that excitation control has a great role to play in power system operation.

23.5 **VOLTAGE REGULATION AND POWER TRANSFER**

If the fall of terminal voltage is assumed to be linear, then the graphs of machine terminal voltage with load P can be represented by the relation.

$$E = E_0(1 - KP)$$

where E_0 is the no-load terminal voltage and K is a coefficient of regulation (Fig. 23.10). Since

$$P = \frac{EV}{X}\sin\delta$$
$$= \frac{E_0 V}{X}\sin\delta - \frac{E_0 V}{X}KP\sin\delta$$

(23.5)

where δ is the angle between E and V. Substituting P_m for $E_0 V/X$ and solving for P

$$P = \frac{P_m \sin\delta}{1 + P_m K \sin\delta}$$

which is a maximum at $\delta = 90$ degrees having the value

$$P_{\max} = \frac{P_m}{1 + Kp_m}$$

Under ideal conditions, where there is perfect control of excitation, K is zero.

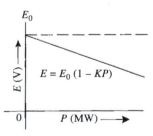

FIGURE 23.10

Variation regulation.

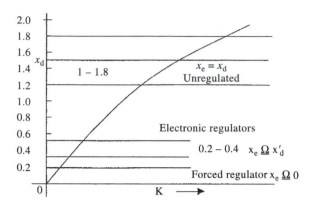

FIGURE 23.11

Effect of regulators.

If the machine is represented by a certain voltage E behind a certain reactance X_e, then Fig. 23.11 illustrates the effect of various types of regulators on its performance.

It can be proved for an ideal, voltage-actuated automatic regulator that the maximum value of the gain is

$$K_{v_{max}} = \frac{x_d - x_d'}{x_d'} \tag{23.6}$$

where x_d and x_d' are direct-axis synchronous and transient reactances of the machine, respectively.

There is also a minimum value for the gain to ensure stability at large load angles. For better performance, forced regulation using the derivatives of current and/or voltage is recommended.

The effect of different types of regulators is shown in Fig. 23.11.

23.6 EXCITER AND VOLTAGE REGULATOR

The function of an exciter is to increase the excitation current for voltage drop and decrease the same for voltage rise. The voltage change is defined as

$$\Delta V \underline{\Delta} (V_t - V_{ref})$$

where V_t is the terminal voltage and V_{ref} is the reference voltage.

Exciter Ceiling Voltage: It is defined as the maximum voltage that may be attained by an exciter with specified conditions of load.

Exciter Response: It is the rate of increase or decrease of the exciter voltage when a change in this voltage is demanded. As an example consider the response curve shown in Fig. 23.12.

$$\text{Response} = \frac{100 \text{ V}}{0.4 \text{ s}} = 250 \text{ V}/s$$

Exciter Buildup: The exciter buildup depends upon the field resistance and the changing of its value by cutting or adding. The greatest possible control effort is the complete shorting of the field rheostat when maximum current value is reached in the field circuit. This can be done by closing the contact to C shown in Fig. 23.13.

When the exciter is operated at a rated speed at no-load, the record of voltage as function of time with a step change that drives the exciter to its ceiling voltage is called the exciter buildup curve. Such a response curve is shown in Fig. 23.14.

Line ac represents the excitation system voltage response (Table 23.2).

$$\text{Response ratio} = \frac{Cd}{0a(0.5)}\text{p.u. } V/s \tag{23.7}$$

In general the present day practice is to use 125 V excitation up to 10 MVA units and 250 V systems up to 100 MVA units. Units generating power beyond 100 MVA have excitation system voltages variedly. Some use 350 and 375 V system while some go up to 500-V excitation system.

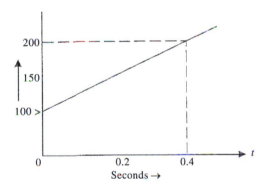

FIGURE 23.12

Excitation response.

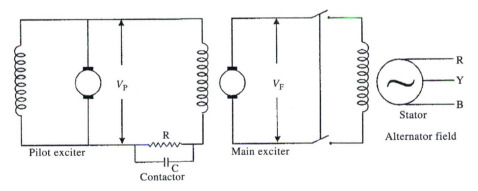

FIGURE 23.13

Excitation control.

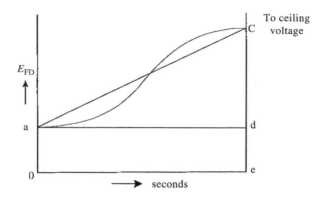

FIGURE 23.14

Exciter buildup curve.

Table 23.2 Typical Ceiling Voltages		
Response Ratio	**Conventional Exciter**	**SCR Exciter**
0.5	1.25–1.35	1.2
1.0	1.4–1.5	1.2–1.25
1.5	1.55–1.65	1.3–1.4
2.0	1.7–1.8	1.45–1.55
4.0	. . .	2.0–2.1

23.7 BLOCK SCHEMATIC OF EXCITATION CONTROL

A typical excitation control system is shown in Fig. 23.15. The terminal voltage of the alternator is sampled, rectified, and compared with a reference voltage, the difference is amplified and fed back to the exciter field winding to change the excitation current. An excitation system using an ampli-dyne is shown in Fig. 23.16.

23.8 STATIC EXCITATION SYSTEM

In the static excitation system the generator field is fed from a thyristor network shown in Fig. 23.17. It is just sufficient to adjust the thyristor firing angle to vary the excitation level. A major advantage of such a system is that when required the field voltage can be varied through a full range of positive to negative values very rapidly with the ultimate benefit of generator voltage regulation during transient disturbances. The thyristor network consists of either three-phase fully controlled or semi-controlled bridge rectifiers. Field suppression resistor dissipates energy in the field circuit while the field breaker ensures field isolation during generator faults.

A compact AC static excitation scheme is shown in Fig. 23.18.

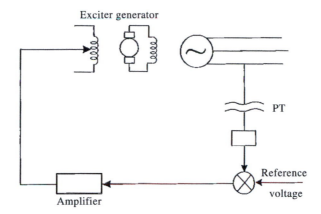

FIGURE 23.15

Excitation control.

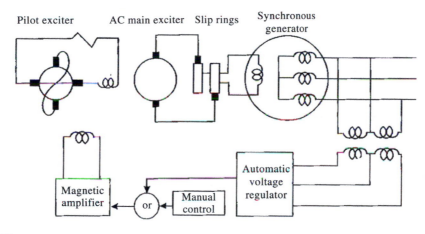

FIGURE 23.16

DC type excitation systems.

23.9 BRUSHLESS EXCITATION SYSTEM

In the brushless excitation system of Fig. 23.19, an alternator with rotating armature and stationary field is employed as the main exciter. Direct voltage for the generator excitation is obtained by rectification through a rotating, semiconductor diode network which is mounted on the generator shaft itself.

Thus, the excited armature, the diode network and the generator field are rigidly connected in series. The advantage of this method of excitation is that the moving contacts such as slip rings and brushes are completely eliminated thus offering smooth and maintenance-free operation. A permanent magnet generator serves as the power source for the exciter field. The output of the permanent

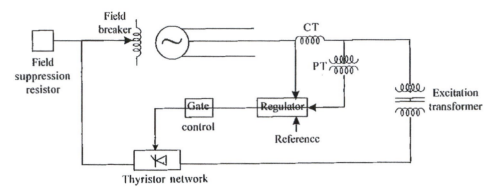

FIGURE 23.17

Static excitation scheme.

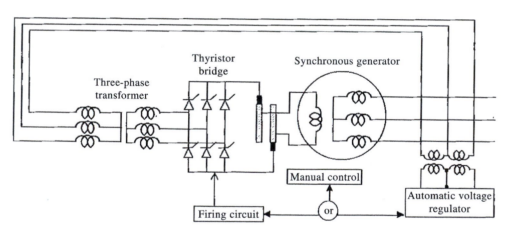

FIGURE 23.18

AC static-type excitation systems.

magnet generator is rectified with a thyristor network and is applied to the exciter field. The voltage regulator measures the output or terminal voltage, compares it with a set reference, and utilizes the error signal, if any, to control the gate pulses of the thyristor network.

A detailed brushless excitation scheme is shown in Fig. 23.20.

23.10 AUTOMATIC VOLTAGE REGULATORS FOR ALTERNATORS

For an isolated generator feeding a load, the automatic voltage regulator (AVR) functions to maintain the bus-bar voltage constant. However, on dynamic, interconnected systems the AVR has the following objectives:

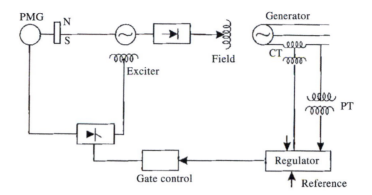

FIGURE 23.19

Brushless excitation system.

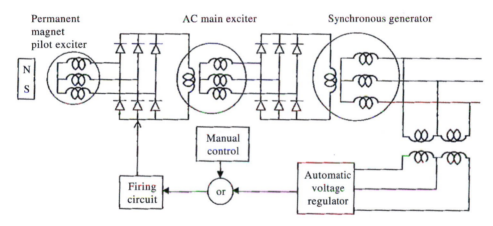

FIGURE 23.20

AC alternator-type excitation system.

1. To keep the system voltage constant so that the connected equipment operates satisfactorily.
2. To obtain a suitable distribution of reactive load between machines working in parallel.
3. To improve stability.

Under normal working conditions, it is not difficult to maintain generator terminal voltage at any specified level. However, with sudden disturbances like load fluctuations, the AVR is required to reduce the magnitude and duration of voltage variations as effectively as possible.

The magnitude of the voltage dip is primarily determined by the transient reactance of the machine while the duration of the dip depends on the time constant of the generator and rapidity of regulation. Static regulators are more useful in this context as they have lesser time delays.

The AVR which senses the terminal voltage and adjusts the excitation to maintain a constant terminal voltage also maintains the reactive output at the required level since the latter depends on the effective voltage difference between generator terminals and its point of connection to the main system. Thus, while AVR is allowed to maintain the voltage on the low-voltage side, a change in reactive output to suit a change in system conditions is obtained by tap changing a generator transformer. With several generators synchronized on a single, low-impedance bus section, one generator with the AVR may be permitted to maintain the bus-bar voltage while the excitation of the rest are used to distribute properly the vars among the generators.

23.11 ANALYSIS OF GENERATOR VOLTAGE CONTROL

Consider the excitation system shown in Fig. 23.21; the block diagram representation is given in Fig. 23.22.

From the block diagram

$$|V|_{\text{ref}} - |V| = \Delta e$$

The amplifier is generally assumed to be instantaneous in action so that Δe_A, the output from the amplifier is $K_A \Delta e$ where K_A is the amplifier gain. The input to the exciter field will

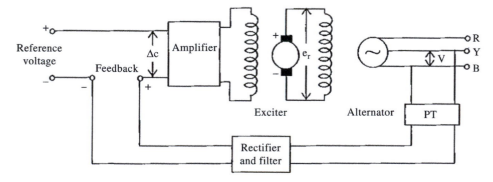

FIGURE 23.21

Excitation system.

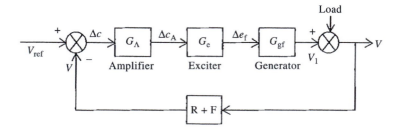

FIGURE 23.22

Block schematic of an excitation system.

be absorbed by the resistance and inductance of the exciter field R_e and L_e, respectively, so that

$$\Delta e_A = R_e \Delta i_e + L_e \frac{d}{dt}(\Delta i_e)$$

where Δi_e is the change in exciter field current. If 1 A change in field current produces K_i volt change in the output, then

$$\Delta e_f = K_i \Delta i_e$$

The transfer function of the exciter can be obtained as follows:

$$\frac{\Delta e_f}{\Delta e_A} = \frac{K_i \Delta i_e}{\Delta i_e R_e + L_e \frac{d}{dt}(\Delta i_e)}$$

In the frequency domain, taking Laplace transform

$$\frac{\Delta E_f(S)}{\Delta E_A(S)} = \frac{K_i}{R_e + L_e s} = \frac{\frac{K_i}{R_e}}{1 + s\frac{L_e}{R_e}} = \frac{K_e}{1 + sT_e} \tag{23.8}$$

Hence $G_e = K_e/(1 + sT_e)$ is the transfer function of the exciter with the time constant $T_e = (L_e/R_e)$ and gain $K_e = (K_i/R_e)$.

The value of the amplifier time constant may be of the order of 0.02–0.1 s while T_e may be 0.5–1.0 s for conventional machines.

At a later stage it will be shown that the exciter block is more conveniently represented by $1/(K_E + sT_E)$ in accordance with the IEEE recommendation.

The input voltage signal Δe_f to the generator field, when applied to the circuit results in the following Kirchhoff's voltage equation.

$$\Delta e_f = R_f \Delta i_f + L_{ff} \frac{d}{dt} \Delta i_f$$

where R_f and L_{ff} are the alternator field resistance and self–inductance, respectively, and Δi_f is the change in the field current. Taking Laplace transform.

$$\Delta E_f(S) = (R_f + L_{ff}s)\Delta I_f(S)$$

If the output voltage changes by $\Delta|V|$ then
$\Delta I_f(S) = \sqrt{2}/\omega L_{fa} \Delta|V|$ where L_{fa} is the mutual-inductance between the field and stator phase winding. Hence the transfer function for the generator block will be

$$\frac{\Delta V(S)}{\Delta E_f(S)} = \frac{\Delta|V|}{R_f + L_{ff}S} \frac{\sqrt{2}}{\omega L_{fa}} = \frac{\Delta|V|}{R_f + L_{ff}S} \cdot \left(\frac{\omega L_{fa}}{\sqrt{2}\,\Delta|V|}\right) = \frac{K_{gf}}{1 + sT_{gf}} \tag{23.9}$$

where T_{gf} is the direct-axis open-circuit time constant also denoted more commonly by $T_{d0} = L_{ff}/R_f$ seconds, and $k_{gf} = \omega L_{fa}/\sqrt{2}R_f$

The voltage regulator loop can be represented by the block diagram shown in Fig. 23.23.

The three cascaded transfer function blocks G_A, G_e, and G_{gf} can be combined into a single block so that the feedback control loop can be further simplified as in Fig. 23.24.

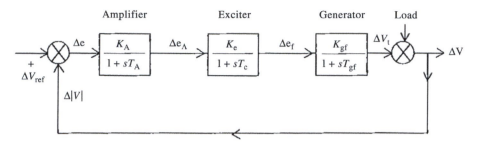

FIGURE 23.23

Voltage regulator block diagram.

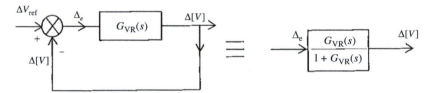

FIGURE 23.24

Simplified block diagram.

23.12 STEADY-STATE PERFORMANCE EVALUATION

The control loop must regulate the output voltage V_t so that the error is made equal to zero. It is also imperative that the response must be reasonably fast, yet not cause any instability problem.

The steady-state voltage error Δe_{ss} is given by

$$\Delta e_{ss} = \Delta |V|^0_{ref} - \Delta V_{ss}$$

$$= \Delta |V|^0_{ref} - \frac{G(0)}{1 + G(0)} \Delta |V|^0_{ref}$$

where $G(0)$ is the value of $G(s)$ as $s \to 0$, i.e., the steady-state value

$$= \frac{1 + G(0) - G(0)}{1 + G(0)} \Delta |V|^0_{ref}$$

$$= \frac{1}{1 + G(0)} \Delta |V|^0_{ref}$$

$$G(0) = \frac{K_A}{(1+0)} \frac{K_e}{(1+0)} \frac{K_{gf}}{(1+0)} = K$$

$$\Delta e_{ss} = \frac{1}{1 + K} \tag{23.10}$$

Larger the overall gain of the forward block gain K smaller is the steady-state error. But too large a gain K can cause instability.

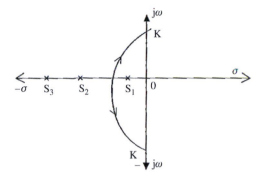

FIGURE 23.25

Root locus plot.

23.13 DYNAMIC RESPONSE OF VOLTAGE REGULATION CONTROL

Consider

$$\Delta V(t) = L^{-1}\left[\Delta V_{\text{ref}}(S)\frac{G(S)}{1 + G(S)}\right]$$

The response depends upon the roots of the characteristic eqn. $1 + G(S) = 0$.

As there are three time constants, we may write the three roots as S_1, S_2, and S_3. A typical root locus plot is shown in Fig. 23.25.

From the plot, it can be observed that at gain higher than K the control loop becomes unstable.

23.14 STABILITY COMPENSATION FOR VOLTAGE CONTROL

Since at higher gains the voltage control loop tends to oscillate and even may become unstable, it requires phase lead compensation or derivative control. As there are three time constants contributing to phase lag, a phase lead compensator with transfer function.

$$G_C(S) = 1 + sT_C$$

will alter the open-loop transfer function to

$$G_{\text{VR}}(S) = \frac{K(1 + sT_C)}{(1 + sT_A)(1 + sT_e)(1 + sT_{\text{gf}})} \tag{23.11}$$

It will not affect the steady-state accuracy as K is kept constant.

23.15 STABILIZING TRANSFORMER

Consider the excitation system with a stabilizing transformer as shown in Fig. 23.26.

(A)

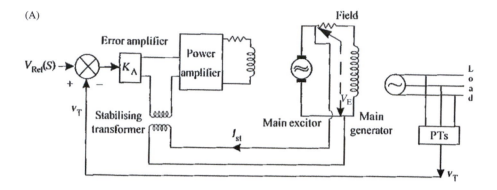

(B)

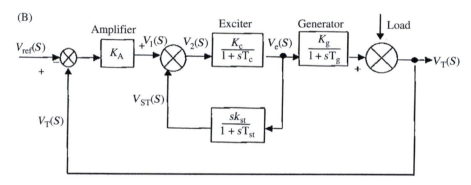

FIGURE 23.26

Excitation control with stabilizing transformer: circuit (A) and block diagram (B).

K_A = Error amplifier gain
K_{st}, T_{st} = Stabilizing transformer gain and time constants.

A stabilizing transformer is introduced in the exciter circuit as shown in Fig. 23.26A and B. The input voltage V_e causes a current I_{st} to flow through the resistance and inductance of the feedback loop.

$$V_e = R_f I_{st}(S) + L_f \frac{dI_{st}(S)}{dt}$$

Taking Laplace transform

$$V_e(S) = R_f I_{st}(S) + L_f s I_{st}(S)$$
$$= (R_f + sL_f) I_{st}(S)$$

If the mutual-inductance is M for the stabilizing transformer, then

$$V_{st}(t) = M \frac{d}{dt} i_{st}(t)$$

Taking Laplace transform

Hence

$$\frac{V_{st}(S)}{V_e(S)} = \frac{sM}{R_f + L_f s} = \frac{\dfrac{sM}{R_f}}{1 + \dfrac{L_f}{R_f}s}$$

Letting

$$\frac{L_f}{R_f} = T_{st} \quad \text{and} \quad \frac{M}{R_f} = K_{st}$$

$$\frac{V_{st}(S)}{V_e(S)} = \frac{sK_{st}}{1 + T_{st}s} \tag{23.12}$$

It may be noted that since the secondary of the stabilizing transformer is connected to the input terminals of an amplifier, it draws zero current. Thus the stabilizing transformer improves the response of the excitation system giving derivative control.

23.16 VOLTAGE REGULATORS

The heart of the excitation system lies in the voltage regulator. It is a device that serves the output voltage change (sometimes current or both) and provides corrective action to take place.

The early voltage regulators were all of electromechanical type. Electronic voltage regulators came in 1930s. They provide smoother and faster voltage control than the electromechanical type. Another development resulted in the use of rotating amplifiers. Magnetic amplifiers were also used for voltage regulators.

Later versions of voltage controllers used solid-state active circuits. Whatever may be the hardware used, there was consistent effort to provide faster response, without dead band, backlash, and absolutely reliable. Excitation systems which include voltage regulators also are classified into continuously acting type and noncontinuously acting type.

All the systems have a potential transformer and rectifier to sample the output AC voltage and provide DC signal to the control loop.

Another component is a comparator that compares the DC signal with reference value and provides an error signal.

An amplifier amplifies the error signal and brings it to the operational level.

The exciter output is manipulated by the amplified error signal to suitably modify the output voltage of the generator.

As a simple proportional feedback results in a steady-state error, a rate feedback or derivative control is proposed by Kron in 1954 and this was discussed in Section 23.15.

23.16.1 COMPUTER REPRESENTATION OF EXCITATION SYSTEM

For synchronous machines, several models were developed, and linear representations of the machine using state-space methods needed suitable models for excitation system for computer simulation to perform dynamic studies. IEEE has suggested certain standard models for different types of excitation system.

1. *Type 1 System:* It is a continuously acting regulator and exciter system. It will be explained in detail later.

2. *Type 1S system:* This is a special case of continuously acting system. In this system excitation is obtained through rectification of the terminal voltage. While in type 1 system regulator voltage has maximum and minimum limits, in this case, the maximum regulator voltage is made proportional to V_t, the terminal voltage of the generator.

3. *Type 2 System:* In this system rotating rectifiers are used and is brushless. Alternator–exciter and diode rectifiers are rotating on the same shaft and hence has no slip rings. Being brushless, the excitation voltage (E_{FD}) is not available for feedback. In this there are two damping loops giving good performance.

4. *Type 3 System:* This type of model is evolved for systems where both current and voltage signals are used for feedback.

5. *Type 4 System:* All the previously described systems are continuously acting type with high gain and fast action. Several of the earlier systems are noncontinuous acting type. In such cases the presence of dead zone results in open-loop operation. They are also slow due to friction and inertia. Rheostat systems are examples for such an operation. In Type 4 systems now there are two speeds of operation depending upon the magnitude of the voltage error. Type 4 system hence is nonlinear and as such is not represented in state variable form. Westinghouse and General electric have developed such excitation systems.

23.17 IEEE TYPE 1 EXCITATION SYSTEM

Consider Fig. 23.27. Let V_f be the voltage across the field winding and i_f the current through it λ_f are the flux linkages in the field winding.

$$v_f = i_f r_f + \frac{d}{dt} \lambda_f(i_f) \tag{23.13}$$

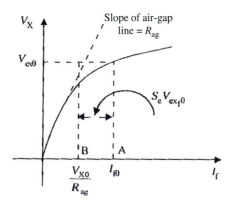

FIGURE 23.27

No-load magnetization curve.

consider the no-load magnetization curve shown in Fig. 23.27 for the exciter voltage buildup.

S_e is the saturation coefficient $(A - B)/B$

The exciter voltage

$$v_{ex}^0 = f(i_f, i_{ex}^0) \tag{23.14}$$

$$i_f = \frac{v_{ex}^0}{R_{ag}} + S_e v_{ex}^0 \tag{23.15}$$

where R_{ag} is the resistance given by the air-gap line.

The saturation may be represented by

$$S_e = A_{ex}^0 e^{B_{ex}^0 V_{ex}^0} \tag{23.16}$$

an exponential relation where, A and B are to be determined.

$$v_f = \frac{r_f v_{ex}^0}{R_{ag}} + S_e v_{ex}^0 r_f + \frac{d}{dt} \lambda_f(v_{ex}^0) \tag{23.17}$$

from Eq. (23.15)

$$= \left[\frac{r_f}{R_{ag}} + S_e(i_f) r_f \right] v_{ex}^0 + \frac{d}{dt} \lambda_f(v_{ex}^0) \tag{23.18}$$

$$= \frac{r_f}{R_{ag}} \left[1 + R_{ag} S_e(i_f) \right] v_{ex}^0 + \frac{d}{dv_{ex}^0} \lambda_f(v_{ex}^0) \cdot \frac{dv_{ex}^0}{dt}$$

In per-unit system (see Fig. 23.28)

$$v_f(\text{p.u.}) = \frac{r_f}{R_{ag}} \left[1 + R_{ag} S_e(i_f) \right] v_{ex}^0 \text{ p.u.} + \frac{d\lambda_f(v_{ex})}{dv_{ex}} \frac{dv_{ex}(p.u.)}{dt}$$

$$= \frac{r_f}{R_{ag}} [1 + S_e(\text{p.u.}) \, v_{ex}(\text{p.u.})] v_{ex}(\text{p.u.}) + \tau_E \frac{d}{dt} v_{ex}(p.u.)$$

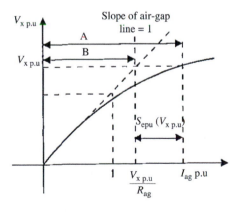

FIGURE 23.28

Magnetization curve in per unit.

Let

$$S_e(\text{p.u.}) = \frac{i_f(\text{p.u.}) - \dfrac{v_{ex}(\text{p.u.})}{R_{ag}}}{\dfrac{v_{ex}(\text{p.u.})}{R_{ag}}} = \frac{A - B}{B} \tag{23.19}$$

$$\tau_E = \frac{d\lambda_f(v_{ex})}{dv_{ex}} = \frac{d\lambda_f(v_{ex}(\text{p.u.}))}{dv_{ex}(\text{p.u.})}$$

$$K_E = \frac{r_f}{R_{ag}}$$

$$S_E = \frac{r_f}{R_{ag}} S_e(\text{p.u.}) \, (V_{ex} \text{ p.u.})$$

Then the transfer function for the exciter block can be represented by Fig. 23.29. IEEE Type 1 excitation system representation is shown in Fig. 23.30.

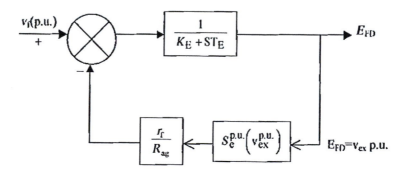

FIGURE 23.29

Block diagram for Eq. 23.18 in per unit.

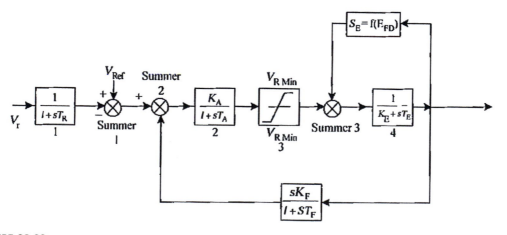

FIGURE 23.30

IEEE Type I excitation system.

23.18 POWER SYSTEM STABILIZER

It can be proved that an voltage regulator in the forward path of the exciter generator systems will introduce a damping torque and under heavy loading conditions this damping torque becomes negative. This is a situation where dynamic instability may cause concern. Further, it has been also pointed out that in the forward path of excitation control loop, the three time constants introduce a large phase lag at low frequencies just above the natural frequency of the excitation systems. To overcome this effect and improve damping, compensating networks are introduced to produce torque in phase with the speed. Such a network is called *"power system stabilizer"* (PSS)

Fig. 23.31 shows the block schematic of PSS at the appropriate point in the control loop.

PSS is generally shown as a feedback element from the shaft speed and has a transfer function of the form

$$G_{PSS}(S) = \frac{K_0 \tau_0 s}{1 + \tau_0 s} \left[\frac{[1 + T_1 s][1 + T_3 s]}{[1 + T_2 s][1 + T_4 s]} \right] \tag{23.20}$$

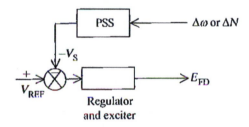

FIGURE 23.31

Connecting power system stabilizer.

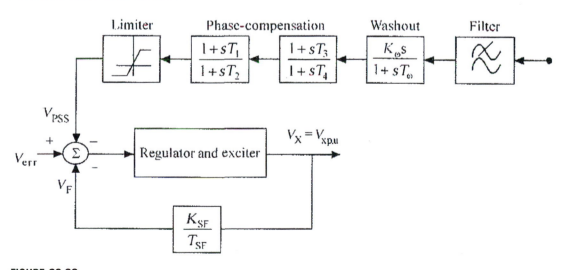

FIGURE 23.32

Power system stabilizer.

A washout circuit for reset action is used to eliminate steady effect after a time lag T_0 (4—30 s). This control circuit ensures that there is no permanent offset in the terminal voltage due to a prolonged error in frequency. The later may occur due to prolonged overload or islanding situation. A lead compensation pair is used at two stages with center frequencies $(1/2\pi)\sqrt{T_1 T_2}$ and $(1/2\pi)\sqrt{T_3 T_4}$. This part of the stabilizer circuit will improve the phase lag.

A filter section is generally added as shown in Fig. 23.32 so that undesirable frequency components are suppressed and thus eliminate the possibility of undesirable interaction. Limits are placed so that the output signal of the PSS is prevented from driving the excitation into heavy saturation.

The output signal of the PSS is fed as a supplementary input signal V_{PSS} for the regulation of the excitation system. Design of the PSS circuit is beyond the scope of this book.

23.19 REACTIVE POWER GENERATION BY TURBO GENERATOR

The reactive power generation capability of a steam turbo generator is dependent on the type of construction.

Unit Construction: In this type, where there is no steam interconnection, the generator can absorb and produce reactive power while supplying the rated real power to the load. A reduction of active power load on the generator increases both its reactive absorption and production capability. However, the active power scheduling is quite often based on an optimal criterion as discussed in Chapter 19, Load Flow Analysis, overlooking which may add to the costs. The level to which real power loading may be reduced is mostly determined by considerations of boiler performance. Minimum permissible loadings usually fall in the range of 50%—70% of rating.

Steam Range Construction: In the steam range construction having steam interconnection between generators, the minimum loading constraint imposed by the boiler is of less significance since steam from one boiler can be used to supply several generators.

As an example, if four sets are supplied by a single boiler, then with three sets drawing steam just for gland ceiling, maintaining vacuum and cooling while the fourth set at the partial load, a wide range of reactive generation or absorption capability can be obtained. However, this arrangement is not in use now.

23.20 SYNCHRONOUS COMPENSATORS

These are rotating machines which are connected to the system at appropriate places only for the purpose of controlling vars. They are synchronous motors operating on load with adjustable excitation. They are normally fitted with voltage regulators which operate when the voltage deviates by a certain predetermined percentage of system voltage.

These machines are also capable of giving dynamic reactive power compensation, i.e., they make available the required reactive power within a short time. The prime consideration for installing synchronous compensator is its flexibility of operation under all-load conditions. This aspect justifies its use in certain systems even though the initial cost is high. Another important

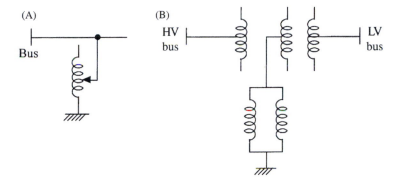

FIGURE 23.33

Shunt reactors. (A) With taps and (B) in parallel.

consideration for using the synchronous compensator is the occurrence of sudden voltage dips due to sudden short circuits or overloads.

Under such conditions, the synchronous compensator performs better than other apparatus. With additional excitation windings through which excitation can be forced momentarily to counteract the transient effects, the synchronous compensator can even improve stability. The synchronous compensator may be supplied from the tertiary winding of the transformer as shown in Fig. 23.33B, so that additional cost on transformer is not incurred. Nowadays static devices replaced rotating compensators.

23.21 REACTORS

Inductive reactors absorb reactive power and may be used in circuits, series or shunt connected. While series-connected reactors are used to limit fault currents, shunt reactors are used for var control. Reactors installed at line ends and intermediate substations can compensate up to 70% of charging power while the remaining 30% power at no-load can be provided by the underexcited operation of the generator.

With increase in load, generator excitation may be increased with reactors gradually cut-out. Fig. 23.33 shows some typical shunt reactor arrangements.

23.22 CAPACITORS

Capacitors produce vars and may be connected in series or shunt in the system. Series capacitors compensate the line reactance in long overhead lines and thus improve the stability limit. However, they give rise to additional problems like high-voltage transients, subsynchronous resonance, etc. Shunt capacitors are used for reactive compensation. Simplicity and low cost are the chief considerations for using shunt capacitor.

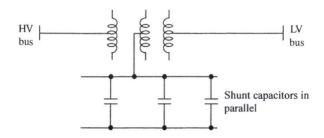

FIGURE 23.34

Shunt capacitors.

Table 23.3 Comparison of Shunt Capacitor and Synchronous Condensers

Criteria	Shunt Capacitors	Synchronous Condenser
Initial cost	1.0	1.0–1.2
Losses	Less than 1.5%	1.5%–4%
Variation of reactive components of current	In steps only with possible surges of voltage and current	Continuous
Overload	Not possible	Possible with reduced reacting
Installation	Simple	Relatively difficult
Increase of rating	Possible by adding more units	Not possible unless another unit is installed
Effect of harmonic voltage	Possibility of resonance	No effect

Further, for expanding systems additions can be made. Fig. 23.34 shows the connection of shunt capacitors through the tertiary of a transformer (Table 23.3).

23.23 TAP-CHANGING TRANSFORMERS

Tap-changing transformers with variable transformation ratio can cause substantial change in the flow of vars. The tap-changing transformers when used in radial lines maintain voltage at their secondary terminals or at load terminals within limits. When used in tie lines, the tap-changer can regulate vars substantially. In case of weaker tie lines, active power may also change to some extent. Lastly, when used in networks or loops, circulating vars can be controlled by tap-changing transformers.

Consider a tie line with sending end and receiving end, both having tap-changing transformer connected to them.

Let V_s and V_r be the voltages at either end at nominal conditions.

Let $t_s V_s$ be the actual voltage at the sending end and $t_r V_r$ be the actual voltage at the receiving end.

$$V_s - V_r = V_1 - V_2 \cong (IR \cos \phi_r + IX \sin \phi_r)$$

$$\Delta V \cong \frac{PR + QX}{V_r}$$

Hence

$$(t_s V_s - t_r V_r) \cong \frac{PR + QX}{V_r}$$

Hence

$$t_s V_s \cong t_r V_r + \frac{PR + QX}{V_r}$$

To ensure that the same overall voltage prevails all over the line, minimum range of taps is used so that $t_s\, t_r = 1$

Substituting this condition and solving for t_s

$$t_s = \frac{1}{V_s}\left(\frac{V_r}{t_s} + \frac{PR + QX}{\dfrac{V_r}{t_s}} \right)$$

$$t_s^2 = \frac{V_r}{t_s} + \frac{PR + QX}{V_r V_s} t_s^2$$

$$t_s^2\left[1 - \left(\frac{PR + QX}{V_r V_s} \right) \right] = \frac{V_r}{V_s} \qquad\qquad (23.21)$$

$$t_s = \frac{\dfrac{V_r}{V_s}}{\sqrt{\left[1 - \dfrac{PR + QX}{V_r V_s} \right]}}$$

23.24 TAP-STAGGERING METHOD

At low loads the reactive losses in transformers are also low. The surplus generated by the system at such low loads can be absorbed by increased reactive power losses in transformers using tap-staggering. Consider the pair of transformers shown in Fig. 23.35 connected between HV and LV buses.

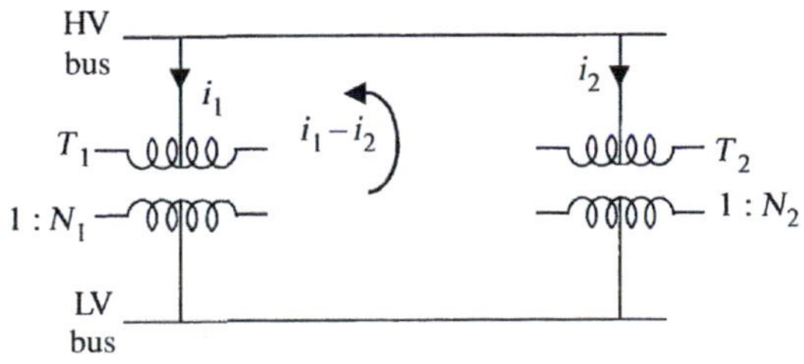

FIGURE 23.35

Tap staggering.

If the taps on the transformers are staggered, a quadrature current circulates around the transformers and carries additional reactive (I^2X) losses. With more number of transformers operating, the losses could be increased by this method.

23.25 VOLTAGE REGULATION AND SHORT-CIRCUIT CAPACITY

Consider the short-line voltage regulation

$$\Delta V = \frac{IR}{V_r}\cos\varphi_r + \frac{IX}{V_r}\sin\varphi_r$$

where V_r is the receiving end voltage and ϕ_r the angle between V_r and I_r, I_r being the current in the line. This voltage drop can be obtained alternatively from the short-circuit capacity of the bus. Since $R + jX$ is the line impedance, the voltage drop is also equal to

$$\Delta V = (R+jX)\cdot I_r$$

$$= (R+jX)\frac{P-jQ}{V} - \left(\frac{PR+QX}{V}\right) + j\left(\frac{PX-QR}{V}\right)$$

when a three-phase fault occurs at the load bus, the short-circuit volt amperes into the bus.

$$S_{SC} = V\cdot I_{SC} = V\cdot\frac{V}{Z_{SC}} = \frac{V^2}{(R+jX)}$$

$$R+jX = \frac{V^2}{S_{SC}}$$

$$R = \frac{V^2}{S_{SC}}\cos\phi_{SC}$$

and

$$X = \frac{V^2}{S_{SC}}\sin\phi_{SC}$$

$$\Delta V = \left[P\frac{V^2}{S_{SC}}\cos\varphi_{SC} + Q\frac{V^2}{S_{SC}}\sin\varphi_{SC}\right]\frac{1}{V} + j\left[P\frac{V^2}{S_{SC}}si\,n\varphi_{SC} - Q\frac{V^2}{S_{SC}}\cos\varphi_{SC}\right]\frac{1}{V}$$

It can be shown that the magnitude of voltage regulation given by $IR\cos\phi + IX\sin\phi$ or $(PR+QX)/V$ as the imaginary component gives only phase shift. Hence, the real part of equation for ΔV above gives the change in voltage magnitude.

Hence

$$\Delta V = \left[\frac{P\cos\varphi_{SC}}{S_{SC}} + \frac{P\sin\varphi_{SC}}{S_{SC}}\right]V$$

For small changes in P and Q

$$\frac{\Delta V}{V} = \frac{1}{S_{SC}}(\Delta P\cos\varphi_{SC} + \Delta Q\cos\varphi_{SC})$$

Once again, remembering that the real power does not affect the voltage magnitude much

$$\frac{\Delta V}{V} \cong \frac{1}{S_{SC}} \Delta Q \sin \varphi_{SC}$$

Under short-circuit conditions the current is almost lagging by 90 degrees to the voltage. Hence

$$\sin \phi_{SC} \cong 1$$

Thus

$$\frac{\Delta V}{V} \cong \frac{\Delta Q}{S_{SC}} \tag{23.22}$$

Per-unit change in voltage magnitude is equal to the rate of change in reactive power to the short-circuit capacity of the bus.

It may be noted in this context that for a system with $X \gg R$, the in-phase voltage drop would be approximately equal to $(I \sin \phi_r)X$. An in-phase voltage compensator or booster would control the reactive power flow in the system. The quadrature voltage drop would be approximately equal to $(I \sin \phi_r)$. The quadrature voltage compensator would control the active power flow in the system.

The ratio of X/R is very high for EHV and UHV lines.

23.26 LOADING CAPABILITY OF A LINE

There are four kinds of limitations identified for the loading of a transmission line.

1. Thermal limit
2. Dielectric limit
3. Stability limit
4. Natural loading limit.

Of all the above limits, natural loading limit is the lowest limit and thermal limit is the highest.

1. *Thermal Limit:* This limit depends upon
  • atmospheric condition including ambient temperature and wind conditions,
  • condition of the conductor, and
  • ground clearance.
  Normal loading at the design and planning level is determined by loss evaluation. Any increase in line capacity with thermal consideration should also consider transformers connected at either end also.
2. *Dielectric Limit:* Line design on the basis of insulation requirement is always on the conservative side. Line voltage can always be increased by 10% for increased power transfer. This requires proper consideration of transients and arresters in operation.
3. *Stability Limit:* There are several stability considerations for increased transmission capability. They are:
  • Transient stability
  • Dynamic stability
  • Steady-state stability

- Frequency collapse
- Voltage collapse
- Subsynchronous resonance

23.27 COMPENSATION IN POWER SYSTEMS

Electrical power demand is growing at a great rate day by day and the generation is, in general, not able to cope up with the demand. Several ways of increasing the power generation are investigated including many nonconventional modes. Again, transmission of increased power over the existing lines is considered to meet the increasing demand as laying of new lines and acquisition of right of way are too expensive in developed areas. This necessitated the implementation of compensation in power systems. For many years, series and shunt compensators are in use. Electrical energy cannot be stored in bulk quantities. There must always exist a balance between generation and demand.

23.28 LOAD COMPENSATION

Utilization of reactive power to improve voltage profile and power factor is termed as *load compensation*. Load compensation thus refers to improving power quality. Reactive power can be injected by installing various sources such as reactors and capacitors in shunt or in series with the system at appropriate places. Another aspect to be considered is that as far as possible negative-sequence currents are to be avoided so that power loss is reduced. This is achieved by operating the system in a balanced condition.

Line Compensation: SIL of a line is the power delivered by a line to a purely resistive load equal to its surge impedance.

$$\text{SIL} = \frac{V_L^2}{\sqrt{\frac{L}{C}}} MW = \frac{V_L^2}{R_C}$$

Power transmitted by a line is usually expressed in terms of this power. When surge impedance is the load impedance, the voltage profile of the line is flat. This is not practicable. However, it is possible to modify the characteristics of the line by using such elements as capacitors, inductors, or synchronous machines. These elements can be connected in series or in parallel. Shunt compensation is similar to load compensation. Shunt-connected capacitors improve the power factor, reduce the reactive power, and thus increase the real power transmission. However, this method has limitations and the size of the condenser banks increase to a prohibitive level. Further they are to be switched off under light load conditions as otherwise the voltage rise may not be acceptable. Series compensations using capacitors cancel a part of the line reactance and increase the maximum power transfer. Such a compensation will reduce the power angle for a given power transfer and enables increased loading. Excessive series compensation using capacitors may result in resonance. Subsynchronous resonance at frequencies less than 50 Hz occur with series compensation if care is not taken. This is discussed later in the chapter.

Line compensation results in

1. minimization of Ferranti effect,
2. elimination of the need for under excited operation of generators, and
3. enhanced power transfer capability.

From the receiving-end power circle diagram, we have

$$P_R = \frac{|V_S||V_R|}{|B|} \cos(\beta - \delta) - \frac{|A||V_R^2|}{|B|} \cos(\beta - \alpha) \tag{23.23}$$

with usual notation.

P_R is a maximum when $\delta = \beta$

$$P_{R,max} = \frac{|V_S||V_R|}{|B|} - \frac{|A||V_R^2|}{|B|} \cos(\beta - \Delta) \tag{23.24}$$

Similarly

$$P_{S,max} = \frac{|V_S||V_R|}{|B|} + \frac{D|V_S^2|}{|B|} \cos(\beta - \Delta) \tag{23.25}$$

For a tie line with series impedance

$$B = Z = R + jX = \sqrt{R^2 + X^2}$$

$$D = 1.0$$

$$\alpha = 0$$

$$\Delta = 0$$

and

$$\beta = \tan^{-1} \frac{X}{R}$$

$$\therefore P_{R,max} = \frac{V_S V_R}{\sqrt{R^2 + X^2}} - \frac{V_R^2 R}{R^2 + X^2}$$

and

$$P_{S,max} = \frac{V_S V_R}{\sqrt{R^2 + X^2}} + \frac{V_S^2 R}{R^2 + X^2}$$

differentiating $P_{R,max}$ with respective to X

$$\frac{dP_{R,max}}{dX} = -V_S V_R \left[\frac{(R^2 + X^2)}{2}\right]^{-3/2} 2X + \frac{V_R^2 R}{(R^2 + X^2)^2} 2X = 0$$

$$\therefore \left(-\frac{V_S}{V_R}\right) + \frac{2R}{\sqrt{(R^2 + X^2)}} = 0$$

$$\frac{4R^2}{(R^2 + X^2)} = \left(\frac{V_S}{V_R}\right)^2; \quad X = R\sqrt{4\left(\frac{V_R}{V_S}\right)^2 - 1}$$

when

$$V_S = V_R; \quad X = \sqrt{3}R \tag{23.26}$$

In overhead lines X/R ratio is much greater than this optimum value needed to maintain the same voltage at both the ends. For 132-kV lines X/R ratio is from 2.5 to 3.5; for 275-kV lines it is about 8 and for 400 kV lines it is as high as 16. Hence to maintain voltage profile X/R ratio has to be decreased. This can be effectively done by series capacitor compensator so that line inductance is compensated.

Nevertheless, series capacitors are to be protected against overcurrents and overvoltages during fault conditions.

23.29 STATIC COMPENSATORS

In recent years reactive compensation of charging power is made feasible with the application of three-phase, thyristor, power controller circuits with automatic control functions.

The term static var compensator (SVC) is applied to a number of static var compensation devices for use in shunt reactive control.

These devices consist of shunt-connected, static reactive element (linear or nonlinear reactors and capacitors) configured into a var compensating system. Some possible configurations are shown in Fig. 23.36. They comprise parallel combinations of ideal inductance and capacitance elements which are either fixed or variable as indicated in Fig. 23.36 (A) (B) and (C).

Even though the capacitors and reactors in Fig. 23.36 are shown connected to the low-voltage side of a step-down transformer, the capacitor banks may be distributed between high- and low-voltage buses.

The capacitor bank often includes, in part, harmonic filters which prevent harmonic currents from flowing in the transformer and the high-voltage system. Filters for the fifth and seventh harmonics are generally provided. Fig. 23.37 shows one type of SVC. The thyristor-controlled reactor (TCR) is operated on the low-voltage bus.

In another form of the compensator illustrated in Fig. 23.38, the reactor compensator is connected to the secondary of a transformer.

With this transformer, the reactive power can be adjusted to anywhere between 10% to the rated value. With a capacitor bank provided with steps, a full control range from capacitive to inductive

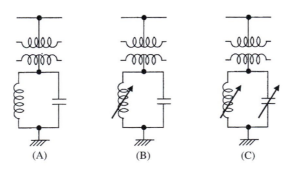

(A)   (B)   (C)

FIGURE 23.36

Idealized var compensators.

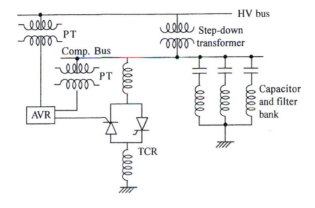

FIGURE 23.37

Thyristor-controlled reactor compensator system.

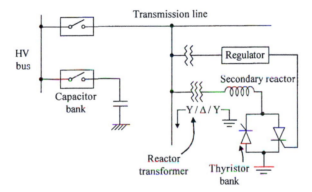

FIGURE 23.38

Reactor—transformer compensator.

power can be obtained. The reactor's transformer is directly connected to the line, so that no circuit breaker is needed.

The primary winding is star connected with neutral grounded, suitable to the thyristor network. The secondary reactor is normally nonexistent, as it is more economical to design the reactor transformer with 200% leakage impedance between primary and secondary windings. The delta-connected tertiary winding will effectively compensate the triplen harmonics.

The capacitor bank is normally subdivided and connected to the substation bus bar via one circuit breaker per subbank. The regulator generates firing pulses for the thyristor network in such a way that the reactive power required to meet the control objective at the primary side of the compensator is obtained. The reactor transformer has a practically linear characteristic from no-load to full-load condition. Thus even under sustained overvoltages, hardly any harmonic content is generated due to saturation. The transformer core has nonferromagnetic gaps to give the required linearity.

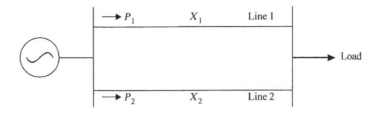

FIGURE 23.39

Transfer of power.

The following requirements are to be borne in mind while designing a compensator.

1. Reaction should be possible, fast or slow, whenever demanded. No switching of a capacitor should take place at that time to avoid additional transients in the system. Commutation from capacitor to reactor and vice versa should be fast.
2. No switching of the capacitors at the high-voltage bus bar, so that no higher frequency transients are produced at the EHV level.
3. Elimination of higher harmonics on the secondary side and blocking them from entering the system.

In a three-phase system the thyristor-controlled inductors are normally delta connected as shown in Fig. 23.39 to compensate unbalanced loads and the capacitors may be star or delta connected.

23.30 FLEXIBLE ALTERNATING CURRENT TRANSMISSION SYSTEM CONTROLLERS

In Chapter 8, High-Voltage Direct Current Transmission, SVCs are discussed. Subsequent developments lead to the emergence of Flexible AC Transmission System (FACTS) controller. Hingorani and Gyugi are pioneers in this area. FACTS controllers is a fast developing area with potential applications.

The basic principle of the flexibility in the transmission of power can be understood easily from Fig. 23.39.

The power flow from the generator to the load through the parallel lines is dependent upon the impedances or reactances when resistances are neglected. If $X_1 = X_2$, the transfer of power is 50% on each of the two lines. If $X_2 = 2X_1$, then the power transfer will be in the ratio of 2:1. Thus the higher impedance line carries less load and the lower impedance line may even get overloaded. If one of the lines, say Line 1 is a HVDC line then power flow is electrically controlled and the above problem is eliminated. Further, with HVDC the stability problem also is controlled because of the speed of control. However, HVDC is expensive in a relative sense.

A FACTS controller can overcome the problem discussed earlier by controlling the impedance, or phase angle. A FACTS controller can control the power flow in any manner that is desired. It is possible to inject desired voltage in series with the line.

Types of FACTS Controllers
Generally, FACTS controllers are divided into four categories. They are briefly discussed.

23.30.1 SERIES CONTROLLERS

They are either variable reactor or capacitor or a power electronic–based variable source so as to serve the need. The drop of voltage across the variable impedance can be considered as series injected voltage (Fig. 23.40).

23.30.2 SHUNT CONTROLLER

The shunt controller, also, may be either a variable source or a variable impedance or a combination connected in shunt as shown and they all inject current into the line (Fig. 23.41).

23.30.3 SERIES–SERIES CONTROLLERS

Two different types of configurations are feasible. A combination of separate series controllers operated in a coordinated manner in the network or otherwise, it could be a unified controller wherein series controllers provide independent series reactive power control as well provide real power transfer through a power link. This is shown in Fig. 23.42.

In FACTS terminology the term unified is used to indicate that the DC terminals of all controller converters are connected together for real power transfer.

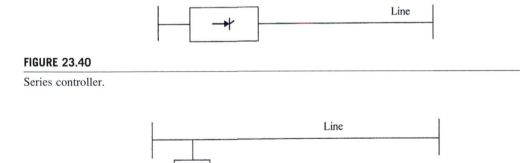

FIGURE 23.40

Series controller.

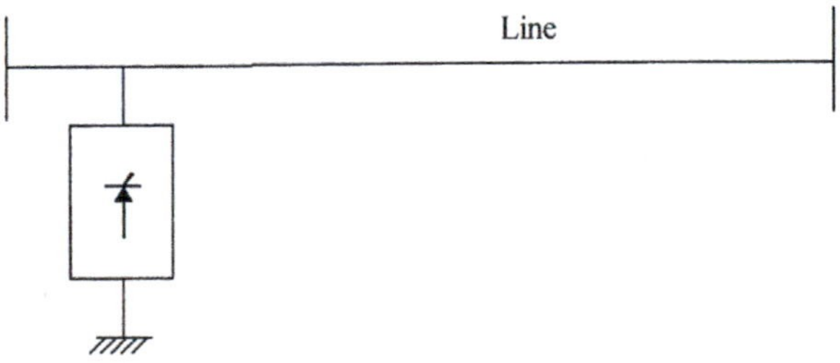

FIGURE 23.41

Shunt controller.

23.30.4 SERIES—SHUNT CONTROLLERS

In the same way as series—series, this could be also a combination of separate series and shunt controller—controlled in a coordinated manner or a unified power controller with series and shunt elements. The basic scheme is shown in Fig. 23.43.

23.30.5 POWER FLOW CONTROL

There are two types of converters—voltage sourced and current sourced, using gate turn-off devices. The voltage-sourced converter consists of a gate turn-off device paralleled with a reverse diode. A DC capacitor is used as a voltage source. The current-sourced converter has a gate turn-off device in series with a diode. It has a DC reactor as a current source.

Voltage-sourced converters are preferred in FACTS controllers. Sequential switching of the device will give AC voltage from the DC source. They are called thus static synchronous compensator. It is operated as a shunt-connected SVC whose output voltage can be controlled independent of AC system voltage. Such a device is also called STATCOM. The structure is shown in Fig. 23.44.

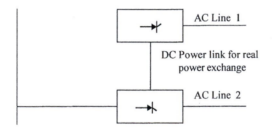

FIGURE 23.42

Series—series controllers.

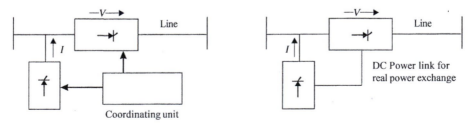

FIGURE 23.43

Series—shunt controllers.

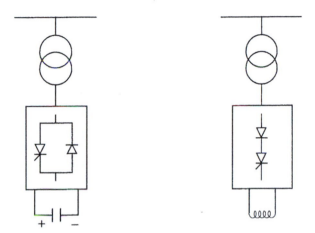

FIGURE 23.44

Static synchronous compensator.

23.30.6 **STATIC VAR COMPENSATOR**

It is a device that supplies capacitive or inductive current so as to maintain voltage at any bus. This has been already discussed in Chapter 8, High-Voltage Direct Current Transmission. The device is shunt connected.

In a similar manner a TCR, shunt connected has the effective reactance varied in a continuous manner by conduction control of the thyristor. TCR naturally belongs to the class of SVC.

If the reactance is varied in a stepped manner instead of continuous mode, the device becomes thyristor-switched reactor. Several shunt-connected reactors may be switched in or out.

Likewise, thyristor-switched capacitor is operated such that shunt capacitor units are switched in and out depending upon the need. It should be noted that in this case the thyristors are operated without firing angle control. They operate with full or zero conduction.

A thyristor-controlled resistor (TCBR) with shunt connection can be used to brake a generator during accelerating period when a fault occurs. They are called then thyristor-controlled braking resistors.

In the same way as STATCOM a static synchronous series compensator (SSSC) operates with a voltage-sourced converter or current-sourced converter giving output voltage in series with the line. Thyristor-controlled series reactor, thyristor-controlled series capacitor, thyristor-controlled switched series reactor are similar devices in series mode of operation.

23.30.7 **UNIFIED POWER FLOW CONTROLLER**

This is a combination of series and shunt-connected static compensators (STATCOM and SSSC). They are coupled through a common DC link. The controller permits bidirectional flow of active power between both the series output terminals of the SSSC and the shunt output terminals of the STATCOM. Without any external electric energy source, they provide simultaneously real and reactive series line compensation.

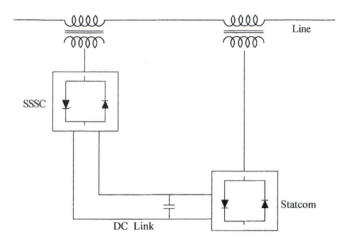

FIGURE 23.45

United power flow controller.

UPFC is also capable of providing independently controllable shunt reactive compensation. This is schematically shown in Fig. 23.45.

There are several more devices such as thyristor-controlled phase shifting transformer, interphase power controller, thyristor-controlled voltage limiter, thyristor-controlled voltage regulator, etc.

23.30.8 ADVANTAGES DUE TO FLEXIBLE ALTERNATING CURRENT TRANSMISSION SYSTEMS DEVICES

The following are some of the advantages that can be obtained by using FACTS controller.

1. Power in lines is controlled in any desired manner.
2. Line capacity is increased, practically up to thermal limits.
3. By raising transient stability limit, system security is enhanced.
4. Upgradation of lines is easier.
5. Reduced reactive power flow, thereby permitting greater active power flow.
6. Reduced cost of energy received due to enhanced line capacity (Fig. 23.45).

23.31 EFFECT OF SHUNT COMPENSATION

Consider the single line diagram of a generating station supplying load connected at the other end of a transmission line of reactance X_L with negligible resistance. Consider a reactive power source connected to the system at the midpoint of the transmission line as shown. The compensator supports the line such that the sending-end voltage, the receiving-end voltage, and the voltage at the midpoint of line are all equal in magnitude ($V_R = V_S = V$). This means that no current enters the compensator branch so that $I_S = I_R = I$. Let the phase angle between V_S and V_R be δ. Then the power transferred from the sending end to the receiving end is

$$P = \frac{V_R V_S}{X} \sin \delta = \frac{V_R V_S}{X} 2 \sin \frac{\delta}{2} \cos \frac{\delta}{2} = \frac{2V^2}{X} \sin \frac{\delta}{2} \cos \frac{\delta}{2}$$

This is a maximum when

$$\delta = \frac{\pi}{2}; \ P_{\max} = \frac{V^2}{X} \qquad \text{(i)}$$

From the phasor diagram

$$ab = \frac{IX}{4} = bc$$

and

$$I = BV \ \ ac = \frac{IX}{2} = cf$$

$$I_S \cdot \frac{X}{4} = \frac{IX}{4} = V \sin \frac{\delta}{4}$$

Hence

$$I = \frac{4V}{X} \sin \frac{\delta}{4} \qquad \text{(ii)}$$

Power at sending end, midpoint, and receiving end will be the same since the line is lossless and the compensator is ideal.

Real power transferred

$$P = \frac{V^2}{X_L} \sin \delta$$

Also

$$P = \frac{V_S \cdot V}{\frac{X_L}{2}} \sin \frac{\delta}{2} = \frac{2V^2}{X_L} \sin \frac{\delta}{2} \qquad \text{(iii)}$$

It is a maximum at $\delta/2 = \pi/2$ or $\delta = \pi$ and

$$P_{\max,C} = \frac{2V^2}{X_L} \qquad \text{(iv)}$$

The reactive power with compensator is

$$\begin{aligned} Q_C &= \left(V \sin \frac{\delta}{4} \right) \frac{4V}{X_L} \left(\sin \frac{\delta}{4} \right) = \frac{4V^2}{X_L} \sin^2 \frac{\delta}{4} \\ &= \frac{2V^2}{X_L} \left(1 - \cos \frac{\delta}{2} \right) \end{aligned} \qquad \text{(v)}$$

This is a maximum when $\cos \delta/2 = 0$ or $\delta/2 = 2\pi$.

It can be observed from the *curves* in Fig. 23.46C that midpoint shunt compensation significantly increases the power transmission capacity. However, the reactive power to be supplied by the compensator increases very rapidly.

(A)

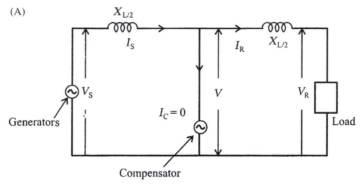

(B)

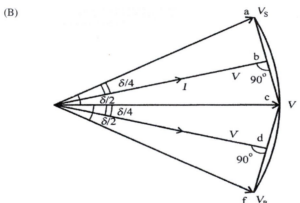

(C)

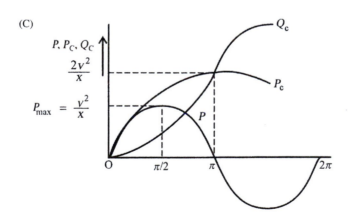

FIGURE 23.46

Shunt compensation: circuit (A), phasor diagram (B), and power transfer capability (C).

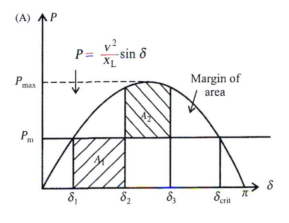

(A)

$$P = \frac{v^2}{x_L}\sin \delta$$

Margin of area

P_{max}

A_2

P_m

A_1

δ_1 δ_2 δ_3 δ_{crit} π δ

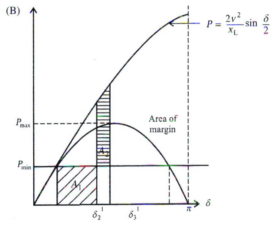

(B)

$$P = \frac{2v^2}{x_L}\sin \frac{\delta}{2}$$

Area of margin

P_{max}

A_2

P_{min}

A_1

δ_2' δ_3' π δ

FIGURE 23.47

Shunt compensation and equal area criterion: system without compensation (A) and with shunt compensation (B).

Since shunt compensators can provide effective voltage support, it can increase transmission capability for a system during transient condition, especially in postfault condition.

This can be demonstrated for a two-machine system employing equal area criteria. In Fig. 23.47A transient stability margin is indicated for an uncompensated system with usual notation.

With shunt compensator, the area of margin for stability is increased as shown in Fig. 23.47B.

WORKED EXAMPLES

E.23.1 Two substations A and B are interconnected by a line having an impedance of $(0.03 + j0.12)$ p.u. the substation voltages are $33 \angle 2$ degrees kV and $33 \angle 0$ degree kV, respectively. In phase and quadrature boosters are installed at A. Determine their output voltage ratings and MVA ratings in order to supply 5 MVA at 0.8 p.f. lagging at substation B.

Solution:

Let

$$\text{Base MVA} = 500$$

$$\text{Base KV} = 33 \text{ kV}$$

$$V_A = 1.0 \angle 2 \text{ degrees p.u. and } V_B = 1.0 \angle 0 \text{ degree p.u.}$$

Let current through the interconnector $= I$

Let load current at B $= I_B$

$$I_B = (0.8 - j0.6) = (0.8 - j0.6)\text{p.u.}$$

Without boosting voltage drop in the line

$$= ZI_B = (0.03 + j0.12)(0.8 - j0.6)$$

$$= (0.096 + j0.078) \text{ p.u.}$$

voltage drop in the line with boosters $= V_A - V_B$

$$= 1.0 \angle 2 \text{ degree} - 1.0 \angle 0 \text{ degree}$$

$$= 0.99939 + j0.0348994 - 1.0$$

$$= (-0.00061 + j0.0348997) \text{ p.u.}$$

Hence, in-phase component of boosted voltage

$$= +0.096 - (-0.00061)$$

$$= 0.09661 \text{ p.u.}$$

Quadratic component of boosted voltage

$$= 0.078 - 0.3349 = 0.0431 \text{ p.u.}$$

Rating of the in-phase booster $= 0.09661 \times 1.00$

$$= 0.09661 \text{ p.u.} = 0.09661 \times 500 = 48.3 \text{ MVA}$$

Rating of the quadrature booster $= 0.0431 \times 1.00$

$$= 0.0431 \text{ p.u.} = 0.0431 \times 500 = 21.55 \text{ MVA}$$

E.23.2 Two substations A and B operating at 11 kV three-phase are connected by two parallel lines 1 and 2. Each line has a 11/132 kV transformer and a 132/11 kV substation. Each line has an equivalent impedance of $Z_1 = 0.2 + j0.4$ and $Z_2 = 0.2 + j0.6$ ohms per phase which includes both the transformers and the line, referred to the 11-kV side.

1. If the bus bar A is at 11 kV and is sending 30 MW at 0.8 p.f leading, find the individual currents into each transformer and the powers at the station A.

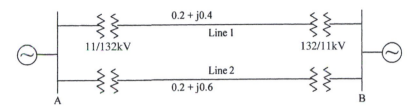

FIGURE E.23.2

2. If the transformer at A in Line 2 is fitted with tappings on the 11-kV side, what percentage tapping would be required to make each line carry equal reactive powers? What would be the power sent by Line 2 in this case?

Solution:
Let base voltage be 11 kV
Let

$$\text{Base MVA} = \frac{30.0}{0.8} = 37.5 \text{ MVA}$$

Total current flowing through both the lines

$$= \frac{37.5}{37.5}(0.8 + j0.6) - (0.8 + j0.6)\text{p.u.}$$

$$= 100 \angle 36.87 \text{ degrees (since 11 kV = 1 p.u.)}$$

$$\dot{Z}_1 \dot{I}_1 = \frac{\dot{Z}_2}{\dot{Z}_1 + \dot{Z}_2} \dot{I}_{\text{Total}}$$

Base impedance

$$= \frac{11 \text{ kV} \times 11 \text{ kV}}{37.5} = 3.227 \text{ ohms}$$

$$\dot{Z}_1 = \frac{0.2 + j0.40}{3.227} = (0.0627)0.124 \text{ dpu}$$

$$= 0.1386 \angle 63°.43$$

$$\angle 71.56 \text{ degrees p.u.}$$

Hence

$$\dot{Z}_2 = \frac{0.2 + j0.60}{3.227} = 0.0627 + j0.186 = 0.1961 \angle 71.56 \text{ degrees p.u.}$$

$$\dot{Z}_1 + \dot{Z}_2 = \frac{0.5 + j1.0}{3227} = 0.124 + j0.310 \text{ p.u.}$$

$$= 0.334 \angle 68.20 \text{ degrees p.u.}$$

$$\therefore \quad I_1 = \frac{0.1961 \times 1.0}{0.334} \angle (71°.56 + 36°.87 - 68°.2) = 0.584 \angle 40°.23 \text{ p.u.}$$

$$I_2 = \frac{0.1386 \times 1.0}{0.334} \angle (68°.43 + 36°.87 - 68°.2) = 0.415 \angle 32°.1 \text{ p.u.}$$

Base current

$$= \frac{37.5 \times 10^3}{\sqrt{3} \times 11} = 1968 \text{ A}$$

$$I_1 = 1155 \angle 40.23 \text{ degrees } (1968 \times 0.587 \angle 40°.23)$$
$$\therefore \quad I_2 = 517 \angle 32.10 \text{ degrees } (1968 \times 0.415 \angle 32°.10)$$

The per-unit MVA from Line 1

$$S_1 = P_1 + jQ_1 = 1.00 \angle 0 \text{ degree} \times 0.587 \angle -40.23 \text{ degrees} = (0.448 - j0.379)\text{p.u.}$$

The per-unit MVA from Line 2

$$S_2 = P_2 + jQ_2 = 1.00 \angle 0 \text{ degree} \times 0.415 \angle -32.1 \text{ degree} = (0.352 - j0.2205)\text{p.u.}$$

$$P_1 = 0.448 \times 37.5 = 16.8 \text{ MW}$$

$$P_2 = 0.352 \times 37.5 = 13.2 \text{ MW}$$

$$Q_1 = 0.379 \times 37.5 = 14.21 \text{ MVAr (leading)}$$

$$Q_2 = 0.2205 \times 37.5 = 8.27 \text{ MVAr (leading)}$$

3. The tap changer at station A on the sending-end transformer in Line 2 will alter the reactive power loading.
   Total load

$$= (P_1 + P_2 + jQ_1 + jQ_2) = (30 - j22.5) \text{ MVA}$$

Equal reactive powers means

$$Q_1 = Q_2 = -\frac{22.5}{2} = -11.25 \text{ MVAR}$$

$$P_1 = 16.8 \text{ MW}$$

and

$$P_2 = 13.2 \text{ MW}$$

The change in voltage ΔV is given by the approximate expression

$$\Delta V = ZI = \frac{PR + QX}{V}$$

$$= tV - V = \frac{PR + QX}{V}$$

where t is the tap-changer setting ratio

$$(t-1)V = \frac{PR + QX}{V}$$

$$(t-1) = \frac{PR + QX}{V^2}$$

$$t = 1 + \frac{PR + QX}{V^2}$$

$$P = P_2 = 0.352 \text{ p.u. } (= 13.2 \text{ MW}/37.5)$$

$$Q_2 = -0.30 \text{ p.u. } (= 11.25 \text{ MVAr}/37.5)$$

$$X_2 = 0.186 \text{ p.u.}$$

$$R_2 = 0.062 \text{ p.u.}$$

$$t = 1 + \frac{0.352 \times 0.062 + (-0.3)0.186}{1.00 \times 1.00} = 1 - 0.034 = 0.966$$

$$\text{Percentage tap change} = 96.6 - 100 = -3.4$$

Power sent through Line 2

$$S_2 = P_2 + jQ_2 = (0.352 - j0.30) \text{ p.u.}$$

$$S_1 = S - S_2 = (0.8 - j0.6) - (0.352 - j0.3)$$

$$= (0.448 - j0.3) \text{ p.u.}$$

$$S_1 = P_1 + jQ_1 = 16.8 - j11.25 \text{ MVA}$$

$$S_2 = P_2 + jQ_2 = 13.2 - j11.25 \text{ MVA}$$

$$P_1 = 16.8 \text{ MW}$$

$$P_2 = 13.2 \text{ MW}$$

$$Q_1 = -11.25 \text{ MVAr}$$

$$Q_2 = -11.25 \text{ MVAr}$$

E.23.3 A load of $(15 + j10)$ MVA is supplied with power from a generating station from a line at 110-kV three-phase 50 Hz. The line is of 100 km length. The line is represented by π model with the parameters

$$R = 26.4 \text{ ohms}$$

$$X = 33.9 \text{ ohms}$$

$$B = 219 \times 10^{-6}$$

Voltage at the generator end is in 116 kV. Determine the power supplied by the generating station.

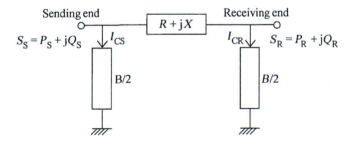

FIGURE E.23.3

$$S_R = P_R + jQ_R = \frac{15+j10}{3}\text{MVA/Phase}$$

$$V_R = \frac{110}{\sqrt{3}} = 63.5\,\text{kV}$$

Reactive power through $B/2$ at the receiving end

$$Q_{CR} = 63.5 \times 63.5 \times \frac{219 \times 10^{-6}}{2} \times 10^6\,\text{VAR}$$

$$= \frac{883,062.75}{2} = 441,531.37\,\text{VAR} = 0.44153\,\text{MVAR}$$

Receiving end power

$$S = P_R + Q_R = \frac{15+j10}{3} - j0.44153$$

$$= (5 + j2.8918)\text{MVA}$$

Real power loss in the line $= I^2 R$

$$I = \frac{S_R}{V_R} = \frac{(5+j2.8918)10^6}{(63.5)^2 \times 10^3}$$

$$= \frac{\sqrt{5^2 + 2.8918^2} \times 10^6}{63.5 \times 10^3} = 90.96$$

$$= 0.2184313\,\text{MW}$$

$I^2 X =$ Reactive power loss $= \frac{5^2 + 2.89^2 \times 10^3}{63.5^2} \times 33.9 = (90.96)^2 \times 33.9\,\text{VA}$

Reactive power in the sending end capacitance $B/2$

$$Q_{CS} = V^2 X_{CS}$$

$$V_S = \frac{116}{\sqrt{3}} = 66.9746$$

$$Q_{CS} = 66.9746^2 \times \frac{219 \times 10^{-6}}{2} \times \frac{10^6}{10^6} = 0.491172.87\,\text{MVAR}$$

Power at the sending end

$$S_S = S + I^2 R + I^2 X + Q_{CR} + Q_{CS}$$

$$= (5 + j2.89) + 0.218 + j0.279 - j0.04415$$

$$= (5.218 + j1.9574)\,\text{MVA}$$

power consumed from the station

$$3 \times S_S = 3[5.218 + j1.9574]\,\text{MVA} = (15.54 + j5.8722)\,\text{MVA}$$

E.23.4 A short line having an impedance of $(2 + j3)$ ohm interconnects two power stations A and B both operating at 11 kV; equal in magnitude and phase. To transfer 25 MW at 0.8 power factor lagging from A to B determine the voltage boost required at Plant A.

Solution:

$$I = \frac{25 \times 10^3}{\sqrt{3} \times 11 \times 0.8} = 1640.2476 \angle 36.87 \text{ degrees}$$

$$= 1640.2976(0.8 - j0.6) = (1312 - j984.1486) \text{ A}$$

voltage drop in the short line $= (1312.2 - j984.1486)(2 + j3)$

$$= 2624.4 - j1968.297 + j3936.6 + 2952 - 446$$

$$= 5576.8458 + j1968.3$$

This voltage drop must be compensated by the booster so that the voltages are maintain the same at both the ends

$$\text{Voltage boost} = (5576.85 + j1968.3) \text{ V}$$

E.23.5 A load of $(66 + j60)$ MVA at the receiving end is being transmitted via a single-circuit 220-kV line having a resistance of 21 ohm and reactance of 34 ohm. The sending end voltage is maintained at 220 kV. The operating conditions of power consumers require that at this load voltage drop across the line should not exceed 5%. To reduce voltage drop, standard single-phase 660-V, 40-kVAR capacitors are to be switched in series in each phase of the line. Determine the required number of capacitors, and rated voltage neglects the losses.

Solution:
Three-phase load $= (66 + j60)$ MVA

$$\text{Per phase load} = \frac{66 + j60}{3} = (22 + j20) \text{ MVA}$$

$$\text{The impedance } Z = R + jX = (21 + j34) \text{ohm}$$

$$\text{Phase voltage } V_p = \frac{220}{\sqrt{3}} = 127.02 \text{ kV}$$

Without series capacitor

$$\Delta V = \frac{P_R R + Q_R X}{V_R} = \frac{22 \times 21 + 20 \times 34}{127.02} = \frac{462 + 680}{127.02} = 8.9907 \text{ kV}$$

Permitted drop in voltage $= 5\%$ of 127.02

$$= \frac{5}{100} \times 127.02 = 6.351 \text{ kV}$$

Let X_C be the capacitive reactance to bring down the drop from 8.9907 to 6.351 kV.

$$6.351 = \frac{P_R R + (X - X_C)Q_R}{V_R} = \frac{22 \times 21 + 20(34 - X_C)}{127.02}$$

Solving for X_C

$$X_C = 16.65 \text{ ohm}$$

Given that 40-kVAR capacitor at 0.66 kV are to be switched.

$$I_C = \frac{40 \text{ kVAR}}{0.66 \text{ kV}} = 60.606 \text{ A}$$

$$V_R I_R = (22 + j20) \text{ MVA}$$

$$I_R = \frac{\sqrt{22^2 + 20^2} \times 10^6}{127.02 \times 10^3} = \frac{\sqrt{884} \times 10^3}{127.02} = 234.07 \text{ A}$$

capacitors per phase in parallel

$$n = \frac{234.07}{60.606} = 3.862$$

The number of capacitors must be either 4 or more than 4.

$$X_C = \frac{V_C}{I_C} = \frac{660 \text{ V}}{60.606 \text{ A}} = 10.89 \text{ ohm}$$

Each capacitor has 10.89 ohm reactance.
Four in parallel makes $X_{C \text{ parallel}} = \frac{10.89}{4}$ ohm
Since total X_C has to be 16.65 ohm, the series units n_s will be

$$\frac{10.89}{4} \times n_s = 16.65$$

Hence

$$n_s = \frac{16.65 \times 4}{10.89} = 6.1157$$

As n_s must be a whole number $n_s = 7$
Installed capacity of the capacitor bank

$$Q_C = \text{Rated voltage of the capacitor bank}$$
$$= 0.66 \text{ kV} \times 7 = 4.62 \text{ kV}$$

E.23.6 Find the capacity of a static VAR compensator to be installed at a bus with $\pm 5\%$ voltage fluctuation. The short circuit capacity $= 7000$ MVA.

Solution:

$$\Delta Q = \text{capacity of the compensator}$$
$$S_{SC} = \text{Short-circuit capacity}$$
$$\Delta V = \text{Voltage fluctuation}$$

$$\Delta V = \frac{\Delta Q}{S_{SC}}$$

$$\Delta Q = \Delta V S_{SC} = \pm 0.05 \times 7000 = \pm 350 \text{ MVAr}$$

Capacity of the static VAR compensator $= \pm 350 \text{MVAr}$

E.23.7 A 400-kV line is fed through a 132/400-kV transformer from a constant 132 kV supply. At the load end of the line, another transformer of nominal ratio 400/132 kV is used to reduce the voltage. The total impedance of the line and transformers at 400 kV is $(50 + j100)$ ohm. Both transformers are equipped with tap-changing so arranged that the product of the off-nominal setting is unity. If the load on the systems is 250 MW at 0.8 p.f. lagging. Calculate the settings of the tap changer required to maintain the voltage of the load bus at 132 kV.

Solution:

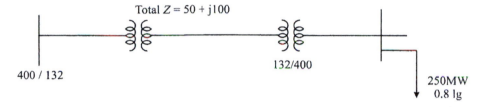

Total $Z = 50 + j100$

400 / 132

132/400

250MW
0.8 lg

FIGURE E.23.7

$$\text{Power per phase } (P) = \frac{250}{3} = 83.33 \text{ MW}$$

$$\text{Reactive power per phase} = \frac{250}{0.8} \times \frac{0.6}{3} = 62.5 \text{ MVAr}$$

$$V_R = V_S = \frac{132}{\sqrt{3}} = 76.21247 \text{ kV} = 76.21247 \times 10^3 \text{ V}$$

we have

$$t_S^2 = \frac{\dfrac{V_R}{V_S}}{\left[1 - \left(\dfrac{PR + QX}{V_R V_S}\right)\right]}$$

$$V_R = V_S = \frac{132}{\sqrt{3}} \text{ kV}$$

when the secondary is maintained at 132 kV the primary is kept at 400 kV. Since total impedance is referred to 400 kV side, we obtain

$$t_S^2 = \left[\frac{V_R}{V_S} \frac{1}{1 - \dfrac{PR + QS}{V_R V_S}}\right]$$

$$= \frac{1}{1 - \dfrac{83.33 \times 10^6 \times 50 + 62.5 \times 10^6 \times 100}{\dfrac{400}{\sqrt{3}} \times \dfrac{400}{\sqrt{3}} \times 10^6}}$$

$$= \cfrac{1}{1 - \cfrac{4166.5 + 6250}{53,333.333}} = \cfrac{1}{1 - \cfrac{10416.5}{53,333.333}} = \frac{1}{1 - 0.1953093} = \frac{1}{0.8046906} = 1.2427136$$

$$t_s = \sqrt{1.2427136} = 1.1147706$$

$$t_s = \frac{1}{1.1147706} = 0.8970455$$

PROBLEMS

P.23.1 Find the regulation and efficiency of an 80-km, 3-phase, 50-c/s transmission line delivering 24,000 kVA at a power factor of 0.8 lagging and 66 kV to a balanced load. The conductors are of copper, each having a resistance of 0.12 ohm/km, 1.5 c, outside diameter, spaced equilaterally 2.5 m between centers. Neglect leakance and use the nominal-π method.

P.23.2 A three-phase, overhead line has resistance and reactance of 6 and 20 ohm, respectively, per phase. The sending end voltage is 66 kV while the receiving end voltage is maintained at 66 kV by a synchronous phase-modifier. Determine the kVAr of the modifier when the load at the receiving end is 75 MW at a power factor of 0.8 lagging; also the maximum load that can be transmitted.

P.23.3 A long line from a hydroelectric station operating at 132 kV feeds a 66-kV system through a transformer. The load taken from the 66-kV windings of the transformer is 50 MVA at a power factor of 0.8 lagging. A tertiary winding on the transformer feeds a synchronous condenser at 11 kV. If the power factor at the receiving end of the line is to be unity, calculate the rating of each of the three windings. Neglect losses.

P.23.4 Find the rating of synchronous compensator connected to the tertiary winding of a 132-kV star-connected, 33-kV star connected, 11-kV delta-connected three-winding transformer to supply a load of 66 MW at 0.8 power factor lagging at 33 kV across the secondary. Equivalent primary and tertiary winding reactances are 32 and 0.16 ohm, respectively, while the secondary winding reactance is negligible. Assume that the primary side voltage is essentially constant at 132 kV and maximum of nominal setting between transformer primary and secondary is 1:1.

P.23.5 A single-circuit, three phase, 220-kV line runs at no-load. Voltage at the receiving end of the line is 210 kV. Find the sending end voltage, if the line has a resistance of 20.5 ohms, reactance of 81.3 ohms, and the total susceptance as 5.45×10^{-4} mho. The transmission line is to be represented by π-model.

P.23.6 A power system is operating at 1000 MW, 132 kV, 50 Hz, with 0.8 p.f. lagging in parallel with another line at 750 MW, 132 kV, 50 Hz with 0.707 p.f. lagging. Both are interconnected at the station and when the compensating device is on, the overall power factor is improved to 0.9 lagging. Suggest suitable capacitors, shunt, and series. Individual loads and combined load are to improve power factor to 0.9 in all cases.

QUESTIONS

Q.23.1 Discuss in detail about the generation and absorption of reactive power in power system components.

Q.23.2 Explain reason for variations of voltages in power systems and explain any one method to improve voltage profile.

Q.23.3 Explain clearly what do you mean by compensation of line and discuss briefly different methods of compensation.

Q.23.4 What is a static compensator? Explain with diagrams, the working principle of various types of static compensators.

Q.23.5 Explain with diagrams, the operation of a fixed capacitor and thyristor-controlled reactor.

Q.23.6 Discuss in detail about the generation and absorption of reactive power in power system components.

Q.23.7 Discuss in detail about the generation and absorption of reactive power system components.

Q.23.8 Describe the effect of connecting series capacitors in the transmission system.

Q.23.9 .

  1. What does one mean by load compensation?
  2. With neat diagrams discuss shunt and series compensation.
  3. What are the specifications of load compensator?

Q.23.10 Discuss the advantages and disadvantages of different types of compensating equipment for transmission systems.

Q.23.11 Discuss in detail the voltage stability problem in power systems.

RENEWABLE ENERGY SOURCES 24

Energy that can be realized from natural resources such as sunlight, wind, rain, tides, waves and geothermal is called renewable energy. All naturally occurring energy resources which are repetitive and consistent in availability may be called as renewable energy resources. They are supplied and sustained by the local environment. By far sun is the main source of energy. Wind power is also a derivative of solar power. At present solar power and wind power constitute the major components of renewable power generated and utilized more successfully. Countries all around the world are increasingly depending on these modes of energy for future power need in a big way. The reasons for this are depletion of fossil fuels, environmental considerations, and economy. In this context it is necessary to remember the importance of sustainable development.

Regeneration of the consumed energy from nature is necessary to protect the interests of the future generations' consumption of energy must correspond to natural replenishment. So, energy security is to be considered with all relevant constraints. Climate change is one of the issues that was responsible for the search for renewable energy sources. The average surface temperature of earth is raising and if this is allowed disastrous consequences will follow. The greenhouse gas emissions (carbon dioxide, methane, nitrous oxide, hydrofluorocarbons, and silicon hexafluoride) are the causes for the abnormal temperature rise. Another important issue is the continuous increase in the price of oil (at 10% per annum) for the last 15 years. This is further aided by an increase in the consumption of oil than it is produced. Fossil fuels are limited and are getting fast depleted. Hence the need for sustainable, secure, and competitive alternate energy resources other than the conventional sources arose. A lightly flexible electric power system supported by renewable energy sources is very much desired. International Electrochemical Commission standards contain a special section for renewable energy standards.

24.1 SOLAR POWER

Solar power is the electricity produced from sunlight. On a cloudless sky the power of sunlight at noon is about 1 kW/m^2. It is estimated that 86,000 TW of power reaches the Earth from the Sun. There are two main processes through which solar power is produced at present. They are *concentrated solar power* (CSP) technology and *solar photovoltaic* method. Both are in use. The International Energy Agency projected that by 2050 the share of the total solar power will increase to 27% of the total power with photovoltaic generation sharing 16% and the rest by CSP. A 1-kW installation may generate about 3–5 kWh of energy in a day. There are many factors that influence solar power received from the sun such as latitude of the place, season, and daily insolation. In case of CSP generation the incident solar rays are concentrated and the heat produced thereby is

Electrical Power Systems. DOI: http://dx.doi.org/10.1016/B978-0-08-101124-9.00024-3

utilized to generate power. Photovoltaic method is dependent on direct conversion of sunlight into electricity. Some materials like silicon absorb photons of light and release electrons to contribute to the flow of current. The photon in a light ray possesses energy proportional to the frequency of light. If an electron of an element absorbs the energy of one photon and acquires more energy than the work function (electron binding energy) of the material, then it is released from the element. This is called photoelectric effect. This principle is utilized in solar photovoltaic generation of electric power. CSP systems use solar thermal energy to raise steam to produce power. PV systems use solar panels on rooftops or ground-mounted solar farm.

24.1.1 CONCENTRATED SOLAR POWER GENERATION

In this type of generation of power solar rays are concentrated to be incident on insulated steam tubes to heat a working thermal fluid. Absorbers are located at the focal line of the reflecting mirrors. They run parallel to and above the reflectors.

Tracking the sun's rays to be incident more or less normal onto the reflecting mirrors is an important task in this case.

There are different types of generating systems with this principle in operation. The important methods are: (1) parabolic trough reflector systems, (2) compact linear Fresnel reflector (CLFR) systems.

Solar energy generator system (SEGS) which uses parabolic reflectors is explained in the following. Parabolic reflectors rotate the collector and receiver together through the day on a horizontal axis fixed in North–South direction. Fig. 24.1 shows the schematic arrangement. Solar rays fall on

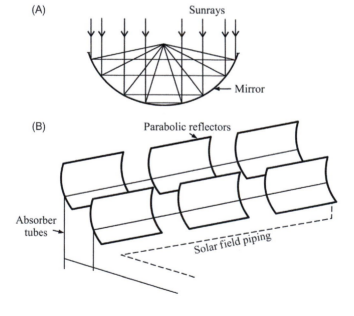

FIGURE 24.1

Concentrated solar power generation. (A) Parabolic reflectors and (B) CSP system. *CSP*, concentrated solar power.

parabolic mirror reflectors made of glass. They reflect back as much as 94% of the radiation to the focal line shown in figure along a tube that carries synthetic oil, which is used as a heat exchange medium. A temperature of the order of 400°C is reached. The reflected light is 70−80 times more intense than the ordinary light. The synthetic oil transfers its heat to water. The water produces steam and drives a turbine. If the primary working fluid is water, then management of pressure becomes a problem. Use of synthetic oil eliminates the pressure problem. The mirrors are made to track the sun all along the day using an appropriate tracking mechanism. Wind damages the mirrors. They require replacement of the damaged reflecting mirrors every year. In recent years thin film nanotechnology is used to reduce the cost of the parabolic reflectors.

24.1.2 LINEAR FRESNEL REFLECTOR SYSTEMS

Flat reflecting segments of mirrors are so arranged to match the curvature of a corresponding focusing mirror in either two or three dimensions. The former arrangement is referred to as linear Fresnel mirror. With such mirrors there are several advantages such as better focusing, easier fabrication, lesser damage due to wind and easy cleaning when compared to parabolic mirrors. The first Fresnel reflector was developed by Giovanni Francia in 1961. These reflectors use long thin, segments of mirrors to focus sunlight onto a fixed absorber. The absorbers are located at the focal point. The concentrated solar energy from the reflectors supply solar energy 30 times intense than the normal radiation. The energy is then transferred to the working fluid of the system. The basic design of the absorber for a CLFR is an inverted air cavity with a glass cover that encloses steam tubes (Fig. 24.2). CLFRs use multiple absorbers and the system was developed in 1993 at the University of Sydney and patented in 1995. The size and curvature of heliostats (mirrors) is variable as well as the heights at which they are placed. The reflectors are arranged in a North−South

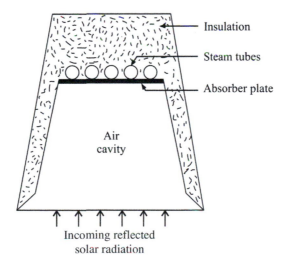

FIGURE 24.2

CLFR system. *CLFR*, Compact linear Fresnel reflector.

orientation. Changing the angles of the incident rays is a major challenge. The optimal performance of the CLFR depends on the heat transfer between the absorber and the thermal fluid. The absorber surface should also absorb solar radiation uniformly. A computer-controlled tracking system is used which maintains proper incident angle for the sun's rays to fall on the reflectors. The thickness of insulation above plate, the size of the aperture of the absorber, the shape and depth of the cavity are the important parameters in the design. Utilizing Fresnel lens effect reduces volume of reflector material, which in turn reduces the cost of the reflector.

24.1.3 PHOTOVOLTAIC GENERATION OF POWER

Photovoltaic power is one of the fastest growing energy technologies. The installed capacity increased from 200 MW in 1990 to more than 80,000 MW by 2012. Until the year AD 2000, photovoltaic power was limited to standalone systems. Now commercial PV modules are available which provide trouble-free service for an average life period of 20 years.

The first PV module was built by Bell laboratories in 1954. The space program accelerated the development of PV modules during 1960s and 1970s. Now reliable PV modules are available and developments in this field are still taking place. The basic principle of a solar cell is described in the following. Light is converted into electricity at the atomic level. A specially treated semiconductor (Silicon) wafer is provided with two contacts on either side of it as shown in Fig. 24.3A.

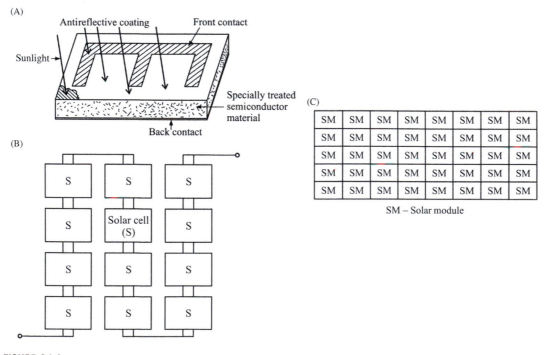

FIGURE 24.3

(A) Basic structure of a solar cell, (B) solar module, and (C) solar array or panel.

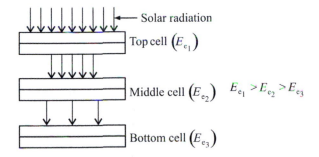

FIGURE 24.4

Multijunction solar cell.

Light rays incident on the semiconductor material remove the electrons and current flows into the external circuit through the two contacts provided on the two sides of the cell. Direct light rays at 90 degrees to the panel give more energy than through an angled PV panel. A PV module is constructed by electrically connecting a large number of such solar cells mounting them on a supporting structure (Fig. 24.3B). Each PV module is designed for a specific voltage (12 V). The current produced is then dependent on the amount of solar light reaching the module. In practice multiple modules are connected to become an array as shown in Fig. 24.3C. It is clear that larger the area of an array more will be the electricity produced.

Most of the common PV arrays use single-junction devices. In a single-cell module, photons whose energy is greater than the work function of the material can free the electrons. Thus only a part of the light is utilized. Multijunction (cascaded or tandem) cells are built with two or more different cells with materials of different work functions so that more of the incident light utilized for conversion thereby increasing the efficiency of the module. The top cell captures high-energy photons and the rest of the photons are allowed to be absorbed by lower energy band gap cells as shown in Fig. 24.4. A multijunction cell is thus built into a stack of single-junction cell in descending order of energy band gaps (E_e).

The modern PV cell materials include gallium arsenide, amorphous silicon, copper indium, and diselenide. There is a continuous research in developing more efficient materials. The semiconductor wafers are available in thicknesses of 100, 130, 150, 200, and 300 mm. The present cost of the PV cells is about $3.29 per Watt in the United States while in Europe, it is priced at about €1.24 per Watt. India has a plan to supply solar power at the same rate as grid power by 2020. The solar power project financing is supported by United Nations Environment program in addition to national institutions.

PV modules are used in general on rooftops. Solar trees are conceived in a tree-like structure. This has the double advantage of providing shade in the day and light in the night. There is no end to the use of solar energy.

24.1.4 TRACKING SYSTEMS

The solar tracking system tilts a solar panel throughout the day. They aim the panel to face directly the sun's rays. If the sky is cloudy, the tracker aims at the brighter part of the sky or cloudless area of the sky. Maximum power point tracking is a technique wherein the position of the panel and the power output are digitally tracked continuously so that optimal power output is obtained.

24.1.5 BATTERIES

Lead-acid batteries, Nickel—Cadmium batteries, Lithium-ion batteries are commonly used in solar plants. Lead-acid batteries have shorter life span. In recent years rechargeable batteries of Sodium Sulfur and Vanadium Redox batteries are being used in distributed PV systems. Zinc Air Incorporation has developed Zinc Iron Redox energy bank and are available in the markets.

Floatovoltaics: The PV systems can be made to float on the surface of lakes, irrigation canals, water reservoirs and even quarry lakes, and tail ponds. Such floatovoltaic systems are practiced in France, India, Japan, Korea, United Kingdom, and United States. They not only produce solar power but also by covering the water surface reduce the evaporation losses.

Hybrid systems are being developed to combine solar power systems with other modes of generation of power such as gas turbines, thermoelectric, and even photovoltaic with CSP.

24.1.6 SOLAR POWER INSTALLATIONS

The world's largest (CSP) power plant is the Ivanpah plant located in Mojave Desert of California with a capacity of 392 MW. It was commissioned in 2014 at a cost of $2.2 billion. It uses high-temperature steam to run steam turbines. Thousands of mirrors called heliostats are used to reflect solar light onto a receiver filled with water. The water receiver is located on the top of a tower. This plant generates about 1080 GWh annually. Each mirror is of 75.6 ft^2 in size. It has 173,000 such heliostats.

The other important solar power installations are the SEGS also located in the same desert with 354 MW capacity. Solnova and Andasol plants each with a capacity of 150 MW each in Spain also fall in this category.

The 579-MW Solar Star power station is the world's largest PV-based plant. It is located near Rosamond in California commissioned recently (June 2015). The plant has 1.7 million solar panels mounted on single-axis trackers stretching over an area of 13 km^2. The 550-MW Topaz Farms at San Luis Obispo county in California and the 290-MW plant at Aqua Caliente, Yuma county in Arizona are some other important PV-based power generating stations. The Charanka solar power plant in the Indian state of Gujarat has a rating of 270 MW (peak DC) as on 2012. The Charanka solar part has 17 projects which when completed will reach a capacity of 605 MW overtaking Ivanpah plant capacity to become the largest plant in the world. It spreads over an area of 20 km^2.

24.1.7 OPERATION WITH GRID SYSTEM

The primary problem with solar power connected to the distribution system is significant voltage changes that forces on line tap changers and line voltage regulators to act continuously. This causes faster deterioration of the voltage control mechanism. Conventional and noncoordinated reactive power control equipment operate at the "control limits" which is called "run away condition." An optimal reactive power controller may be used to overcome this situation. There are several var control options, which are suggested for practical application that satisfy all the constraints. One such method includes maximum power point tracking of the photovoltaic arrays using the allowed limits for voltage of the feeder, in the formulation of the controller algorithm.

24.2 WIND POWER

Wind power is a derivative of solar power. Wind is caused by difference in pressure due to the heat from the sun. Solar radiation heats both the land and water masses on the Earth. The heating however is not uniform and also not at the same rate. Dissimilar type of surfaces on earth containing sand, clay, stone, water, vegetative, and other types absorb, retain, and reflect sun's heat differently and at different rates. The surface of the earth becomes warmer during the day time and gets cooler in the nights. As a consequence of this the air above the earth surface gets heated up and cooled down at varying rates. The pressure differential thus produced causes wind to blow. The velocity of the blowing wind depends upon the pressure difference of the wind. Air when moving as a wind possesses kinetic energy. This kinetic energy of the blowing wind can be converted into other forms of energy provided it is properly harnessed. Kinetic energy is given by

$$KE = \frac{1}{2}mv^2$$

where m is the mass and v is the velocity.

$$= \frac{1}{2}Avt\rho v^2 = \frac{1}{2}At\rho v^3$$

where A is the area of the surface through which wind is passing, ρ is the density, and t is the time. Thus the power in the wind P is

$$P = \frac{E}{t} = \frac{1}{2}A\rho v^3$$

The power in the wind is proportional to cube of the velocity.

Advantages of Wind Power

1. Wind power is free.
2. Wind power is renewable and cannot be exhausted.
3. Conversion of wind power into electrical power does not create greenhouse gases and environmental pollution.
4. Wind turbines can be installed even at remote locations where other form of power generation is not feasible.
5. Wind turbines cost less and require less space than solar panels (PV) per kilowatt generated.

24.2.1 WIND TURBINE

A wind turbine is a device that converts kinetic energy from the wind into mechanical energy which is further converted into electrical power in a wind power plant.

24.2.2 TYPES OF WIND TURBINES

Wind turbines are built with their axis of rotation either vertical or horizontal with respect to ground. Each wind turbine has its own advantages and disadvantages.

24.2.3 HORIZONTAL AXIS WIND TURBINE

The blades of upwind turbines are mounted and pointed upwind so that the turbulence effect caused by the tower is reduced. For this reason the blades are designed to be stiff, so that they are not bent by strong winds.

Downwind turbines have their blades less stiff so that they can bend but not toward the tower. Most of the wind turbines are designed for upwind operation.

Horizontal axis wind turbine (HAWT) are more stable as the blades rotate close to the center of gravity. Also, since the blades rotate normal to the wind speed and extract more power from the wind, their efficiency is higher. Long towers further help to increase the efficiency. Control of blade pitch angle prevents failure during storms. Against these merits, HAWTs need more control such as yaw control, pitch control, and brake system for high-speed winds.

A typical HAWT structure is shown in Figs. 24.5 and 24.6. A vertical tower made of conical steel tube or prestressed concrete is used to support the turbine. A strong foundation is given to the tower to withstand the uneven pressure exerted by the turbine mounted on the top. Along the length of the tower, an access ladder is provided for service purposes (not shown in the figure). The generator, gear box, drive train, and brake assembly are housed inside a cover called nacelle. The brake assembly contains either a mechanical or an electrical breaking system. On the rotor hub, an anemometer is located. Offshore wind turbines are provided with platforms on which service personnel can be deployed from helicopters. Gearbox is used when the wind turbine size is large. It controls the speed and reduces the stress and vibration.

24.2.4 VERTICAL AXIS WIND TURBINES

These turbines have their rotor shafts mounted vertically. The blades rotate around a vertical shaft. It follows the principle of "air foil." The first model is called Darrieus turbine. It has C-shaped blades. This type of turbine has no starting capability (Fig. 24.7). An improved version of it is Giromil vertical axis wind turbine shown in Fig. 24.8. This has two or three vertically straight blades. Helical blades are used to minimize the pulsating torque generated. Savonius vertical axis wind turbine is a slow rotating high-torque generating machine. It uses the drag principle. It cannot rotate faster than the wind speed. Nevertheless, vertical axis wind turbine (VAWT) needs less control, easy to transport, and construct compared to HAWT. It has also low-speed startup capability. But the blades cause a drag force, and the machine is less efficient.

24.2.5 ROTOR BLADES

In the design of the blades aerodynamic considerations are applied. Most of the wind turbines have either two or three turbine blades. The aerodynamic efficiency increases with the number of blades. The aerodynamic efficiency increases with the number of blades. But lower the number of blades

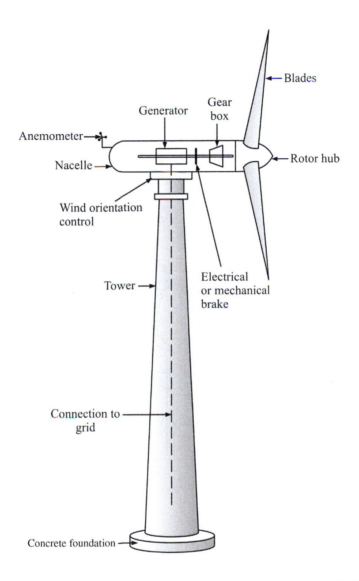

FIGURE 24.5

Horizontal axis wind turbine.

higher the value of the speed. Since the power developed is dependent highly on the velocity of the turbine, it has to be seen in the design that speed is not compromised. If the blades are increased from one to two, the aerodynamic efficiency increases by 6%. When increased from two to three, the corresponding increase is only 3%. Thus the returns are diminishing. The dynamic loading of the rotor of the turbine system transferred onto the drive train and further to the tower is important. System reliability is influenced by the blade. Lower the number of blades less will be the material and manufacturing costs and higher will be the speed of the turbine system and power output, Each

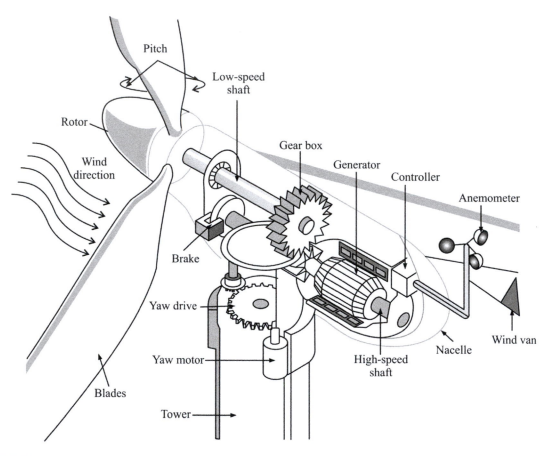

FIGURE 24.6

Components of wind turbine in the housing hub.

blade experiences a cyclic load and a better balance is obtained when three blades are used for the turbine.

Blade Materials: Blades are nowadays manufactured using fiber-reinforced polymers, composites of polymers, and fibers. Glass fiber-reinforced plastics and carbon fiber–reinforced plastics are also used.

24.2.6 GENERATORS

Earlier induction generators were used for wind power generation. They generate alternating current when the generator is rotated faster than the synchronous speed. But induction generator requires externally supplied armature current. It cannot start on its own excitation. Capacitor banks were used to supply the magnetizing current or VARs required. However, present day wind

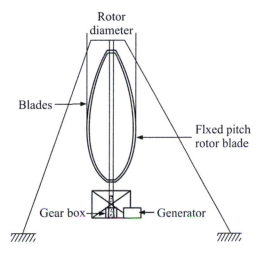

FIGURE 24.7

Vertical axis wind turbine Darocieus type.

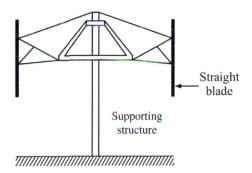

FIGURE 24.8

Vertical axis wind turbine (Giromil type).

turbines use variable speed generators combined with partial scale power converters called doubly fed induction generators (DFIG), variable speed wind turbine generators with full-scale power converters and synchronous generators. For units above 1 MW, DFIGs are used. With these motors only a percentage of power generated (20%−30%) will be passed through the converters compared with 100% power for synchronous generator−based wind turbines. Hence DFIG turbine costs less.

Wind power suppliers to the grid are required to specify the power factor, frequency, and the dynamic characteristics of the wind turbine generator during fault conditions. The wind turbine generators are expected to provide low-voltage ride through capability when connected to the grid. Induction generators cannot support the system voltage during faults.

Alta wind energy center in California is the largest in the United States with a capacity of 1320 MW rating. By 2030 in the United States it is desired to increase the production of wind energy to 20% of the total demand. The European Union has a target growth of 10% addition for every year with an ultimate goal of 8% to be wind-based generation for entire EU. China is the largest producer of wind energy with an installed capacity of 41.8 GW and has plans to increase its share to 11.8% of the demand. The installed capacity in India by the end of 2014 is 22.465 GW. The largest wind power generating plant in India is located at Muppandal in Tamilnadu state with an installed capacity of 1500 MW.

24.3 GEOTHERMAL ENERGY

Geothermal energy is an inexhaustible renewable energy source. It is environmentally friendly. There are four different possibilities as of today, through which this energy can be tapped. Some of them are already in use.

24.3.1 MODES OF GEOTHERMAL POWER EXTRACTION

Fig. 24.9 shows the possibilities of power extraction deep from inside the Earth's crust.

1. Surface geothermal energy available in layers of Earth's crust between 150 and 600 m. This energy is mainly used for heating and cooling of residential homes.
2. Geothermal energy at much deeper depths of about 2000 m and again used for residential heating.
3. Hydrothermal systems using aquifers (hot waters) in deep water−bearing layers between 2500 and 4000 m inside the Earth's crust utilized for both electricity and heating (e.g., Larderello in Italy).
4. Petrothermal systems using a mechanical process of rock drilling. In this method two holes are drilled into the hard granite rocks 5000 m and beyond into the earth's surface and water at high pressure is forced into the holes splitting the rocks along their naturally occurring cracks and returns to the surface through the adjacent hole after exchanging the heat with hot rocks. With an edge length of 1 km ($1 m^3$) at 200°C this process yields as much as 10 MW of electric power for 20 years.

A radioactive decay of Uranium, Thorium, and other fission products is active inside the earth and constantly generate new heat. Even if some of the heat generated is lost to atmosphere through radiation, this loss is marginal and earth generates heat within it constantly. It is estimated that to cool by 300°C earth took over 3 billion years.

The first three modes of producing geothermal energy based on the depth are in general for heating and cooling purposes. But it is the last method that has the potential to give continuous energy to mankind. An experimental plant is set up at Soultz Sous Forets in France to investigate the feasibility of the power generation by this process. Another experimental plant is located in Germany at Landau Pfalz. This mode of generation has the potential to supply the entire world, all its energy needs without getting depleted ever. Earth's crust is about 20−65 km in thickness and it is less under the oceans (5−6 km). The average gradient of temperature is about 2.5−4°C for every

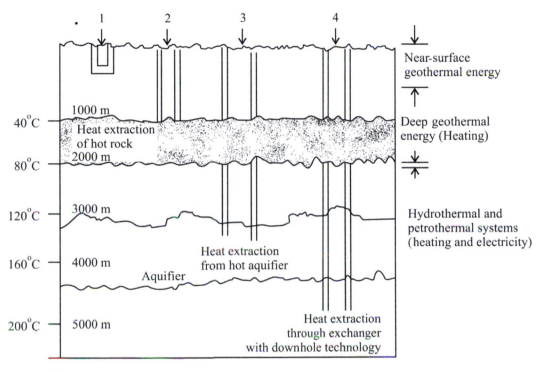

FIGURE 24.9

Geothermal energy.

100 m of depth. The thermal efficiency is about 7%−10%. Geothermal power investigations reached as of now only a depth of 3 km.

24.3.2 TYPES OF PRODUCTION PROCESSES

There are now three production processes developed for harnessing geothermal power. They are dry steam power plants, flash steam stations, and binary cycle type.

In *a dry steam power plant* a geothermal reservoir produces steam. The steam is passed through a turbine generator to produce electricity. The hot water coming out at much lower temperature is sent into the ground by subsurface injection as shown in Fig. 24.10.

In *flash type power stations* a steam separator is used to separate out water and steam and the left over water is bypassed and injected back into the well while steam is passed into the turbine generator (Fig. 24.11). Here high-pressure water is pulled up into low-pressure tanks and the resulting flashed steam is used to generate power. The water flows up through wells by its own pressure and in the process loses some of its pressure and steam is formed.

Binary power plants utilize a combination of flash and dry modes of generation. They are the most recent plants and can utilize water at low temperatures (57°C). The efficiency is higher (10%−13%). As on 2014 the geothermal power plant at Geysers located 72 miles north of San

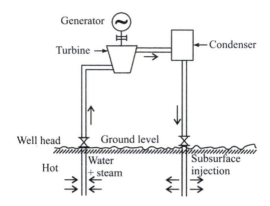

FIGURE 24.10

Dry steam station.

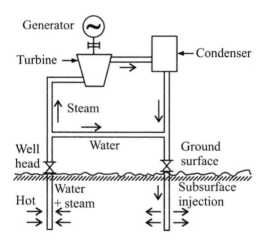

FIGURE 24.11

Flash steam plant.

Francisco, California in the United States is the largest in the world. There are also geothermal power plants in New Zealand and Russia.

24.4 BIOMASS

Biomass is a name used in general to denote materials derived from plants and animals. Biofuels are the man's oldest source for heating. In fact, 10%−15% of the energy demand is met by biofuels in the world. It is the most important source of renewable energy. In the industrialized countries 9%−13% of the energy needs are met through biomass technologies. In the developing countries the usage is as high as 20%–33%. Unfortunately a large part of this source is utilized for cooking and space heating by the poor and also in an inefficient manner. The first-generation biofuels obtained from sugar, starch, and vegetable oils are ethanol, biodiesel, butanol, biofuel gasoline, green diesel, bioethers biogas, syngas, biogas, etc. Advanced or second-generation biofuels are made from lignocelluloses or woody crops, and agricultural waste. The process of production is called cellulosic technology. The major applications of biomass fuels for electricity generation are listed by EUREL:

- co-firing of biomass with coal and coal-fired heating plants,
- biomass-fueled heating combined with small-scale electricity production,
- gasification in a combined cycle for electricity generation, and
- electricity production from waste biomass.

Biomass plants reduce atmospheric release of methane, a potent greenhouse gas. Biomass generally results in sulfur and methane emission. Biomass contributes to air pollution through airborne particles of carbon and high content of nitrous oxides. To utilize biomass safely innovative methods are required to be developed. In a small scale, power is generated from biomass in several countries. Bagasse cogeneration is another related mode of power generation.

24.5 TIDAL POWER

The kinetic energy possessed by the waves in the sea is tapped in several ways to run turbines and generate power. There are two practical ways that this energy is obtained as of now.

Tidal Stream Generator: In this method the kinetic energy in the waves is directly converted into mechanical power in turbine and then into electrical energy from generators coupled to the turbine. The turbine generators can be made part of existing bridge structures or can be arranged at ocean straits and inlets. The turbines may be vertical, horizontal, or open. They are placed at the bottom of the water to realize maximum benefit.

Barrage Generators: In this type of generation the potential head that is naturally available between the high tide and low tide is exploited for generation of power. The barrage is a dam constructed along the width of the estuary. The difference of head in water level (potential energy) is utilized to run the turbines.

There are other modes suggested yet to be made practical such as dynamic tidal power generation and tidal lagoons.

It is claimed that tides are more predictable than wind power and solar power. Tidal power is mainly an outcome of the relative motion of the Earth-Moon system.

24.6 DISTRIBUTED GENERATION

Energy generated or stored by a variety of small grid–connected devices is called distributed energy and the devices are referred to as *distributed energy resources* (DERs). Through appropriate interface, the devices are integrated into the grid for mutual benefit. In this manner energy from a large number of sources can be collected. DER improve the supply security and are environmentally friendly in general. Modem localized small-scale grids are referred as *microgrids*. In the United States microgrid capacity is projected to be 1.8 GW by 2018. The European Union of (27+) countries plans to increase the renewable energy generation by 40% over the next decade.

24.7 MICROGRID MANAGEMENT

Microgrid management is identified with several functions such as: demand side management to synchronize local demand to local generation, energy storage management to store maximum amount of electricity for plug-in electric vehicles, local batteries, or illumination, supply side management (SSM) to utilize any balance of energy and to assist the grid under emergency or blackout conditions. Microgrid technology has to play in future a vital role in managing renewable energy resources and their end use which includes grid balancing, and vehicle-to-grid, and grid-to-vehicle management. Emergency situations can be significantly reduced.

24.8 ONLINE VOLTAGE STABILITY MONITORING AND CONTROL

The existing automatic voltage control systems can be augmented by additional closed-loop control systems. It is required to predict all the important or credible contingencies that are likely to breach system security. Adopting sensitivity-based linear control analysis, a robust control solution can determined. By redefining the voltage profile for each contingency, voltage collapse can be avoided.

24.9 DEMAND RESPONSE

Giving importance to demand response provides an opportunity for consumers to play an important role in the operation of the grid. It enables reduction or shifting of the power consumption during peak load periods. This happens as a consequence of time-based rates and similar financial incentives. Demand response programs are used by electric system planners and operators for resource options and for balancing supply and demand. This leads to lower cost of production and lower prices in both wholesale markets and retail. Time-based rates include time of use pricing, critical peak pricing, variable peak pricing, real-time pricing, and critical peak rebates. In case of direct load control air conditioning, water heating systems and pumping motors to be rescheduled in operation in exchange for financial incentives.

24.10 ADVANTAGES OF RENEWABLE ENERGY SOURCE INTEGRATION WITH GRID

When renewable energy sources such as solar, wind, and other forms, distributed generation, energy storage and their related technologies, and demand responses are integrated with transmission and distribution systems, a number of benefits can be realized. By using advanced system design, planning, and operation, we reap a number of benefits.

1. Utilization of clean power instead of conventional fuel-based power generation carbon emission is reduced.
2. Better utilization of assets results as cost of production is reduced and peak demand is also reduced.
3. The system reliability, resiliency, and security are increased.
4. Oil consumption can be reduced by plug-in electric vehicle operation.

24.11 GRID INTEGRATION CONSIDERATIONS

When renewable energy sources are connected to the grid, several important issues arise. They are:

1. cost, reliability, and efficiency of the interface required for proper integration;
2. congestion on the grid and the nature of the grid (strong or weak);
3. the varying nature of the renewable energy generation; and
4. voltage stability problem that may arise out of the need for bidirectional flow of power.

To overcome the problems, enhancing the transmission line capability for active grid control may not be economical. Energy storage including pumped storage hydropower to cater to the fluctuations in demand supply balance is one feasible way. For stability, fast acting storage is necessary. Load management and demand response are also important aspects for consideration.

The average number of minutes of power outages in Germany reduced very much due to inclusion of renewable energy sources in the grid. The grid was found to be more stable when records for a period of 7 years from 2006–13 were examined. The reliability index SAIDI was reduced from 15.91 in 2012 to 15.32 in 2013. From 2006 onward Germany is using SAIDI to rate the grid stability. The average value of SAIDI over the period of 2006–12 was 16.91. There was an addition of 35 GW of solar power and 35 GW of wind power by 2013. Despite overall improvements, there were more interruptions on low-voltage levels and this was attributed to distributed renewable generation. There is an opinion that aggressive addition of renewable energy sources could cause grid instability. Correct prediction, planning, operation and management with right technology, and proper regulatory and economic incentives, it is felt that renewable energy sources will contribute substantially to the improvement of the grid.

24.12 BATTERY STORAGE SYSTEM

Grid stability with battery system is reported from the studies of California Independent System Operator. Advanced nanophosphate Li-ion technology–based batteries are used successfully. The enhanced dynamic characteristics of these batteries enabled the grid to regulate the frequency and meet all emergency grid challenges.

Consider Fig. 24.12 where in the hourly schedule, the 5-min schedule and the actual load curve for a network over a short period are shown separately. The energy to be provided for regulation is

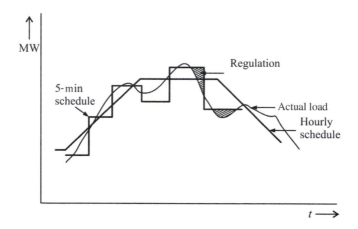

FIGURE 24.12

Energy regulation with batteries.

also shown for an interval. Fast supply or absorption of power by the battery is determined by its dynamic characteristic. The AGC takes control of the frequency fluctuation around the normal system frequency. Even though the battery system may have limited storage capacity, since power flow is bidirectional over a period of time, it is enough to regulate the frequency continuously. The state of California in the United States has 20% renewable energy with targeted goal of 33%. The generation contains a mix of solar, geothermal, wind, and biomass. It is reported through fault and outage simulation studies; operation with storage batteries is more safe.

Small, mini, and micro hydro plants are also considered under renewable energy sources and their operation with the grid improves its stability further.

Large-scale renewable power generation presents the problem of predicting the fluctuations in generation both day-to-day and minute-to-minute extremely difficult. There has to be a dependable base generation such as nuclear power in an integrated power system. With the research going on in altogether new algorithm development to analyze the contingencies that include weather changes and implications, fast acting energy storage system technologies and hardware embedded control are going to play a vital role in the utilization of renewable energy sources and their integration with the grid.

RESTRUCTURING OF ELECTRICAL POWER SYSTEMS

The vertically integrated electric power systems were separated out into independently operating generation, transmission, and distribution entities, each entity being given specific responsibilities for improved economy, efficiency, maximum asset utilization, and customer satisfaction. The increased contribution of power from nonutility generators and the need for increasing the share of generation from renewable energy resources to limit the greenhouse gas emission paved the way for reforms in electric power industry. The United States Energy Policy Act (1992) provided incentives for the development of competitive electric power market. Federal Electricity Regulatory authority in the United States issued several orders introducing open transmission access and other reforms. Restructuring of power systems took place all over the world in both the developed and developing countries. In Sweden Svenska Kraftnaet and in Norway Statt Net were created to operate transmission systems independently. In the United Kingdom the Central Electricity Generating Board was broken and National Grid Corporation was created. In all deregulated markets the emergence of Independent System Operator (ISO) is an important development. Two important market systems emerged from deregulation. The pool system is followed in the United Kingdom, Australia, Latin American countries, and in some US markets. In this system both market settlement and transmission system management including security, reliability, fair and equitable transmission tariffs, and other system services are handed over to the ISO. On the other hand the second market system termed open access system is dominated by bilateral contracts between the generators and the customers. This system is followed in the Nordic countries (Norway, Sweden, Denmark, and Finland). In this system the ISO has a limited role and is responsible for system security and reliability. The transmission authority plays the role of ISO in several utilities.

The automatic generation control (AGC) and voltage control were initially strengthened by supervisory control and data acquisition (SCADA). While the earlier SCADA utilized LAN and shared information in real time, the later version utilized open architecture. It is used for communication between master station and others' Internet protocol (WAN). Later developments necessitated the automation of the distribution system also. Distribution system automation provided minute to minute efficient management, maximum plant utilization, high-quality power, and enabled expansion easier due to the modular structure adopted. Automation functions include automatic bus sectionalization and service restoration, Volt-VAR control, substation transformer load balancing, data acquisition and processing, and other management processes. In the deregulated environment regulation of frequency and tie-line powers, voltage and reactive power control, system stability considerations, maintenance of generation and transmission reserves, and other related services are designated as ancillary services. The North American Electric Reliability Council defined ancillary service in the context of interconnected operation as a service necessary to effect a transfer of electricity between purchasing and selling entities and in which a transmitter must provide an open access transmission tariff.

Computer control, real-time operation, and growth of complexity in electric energy transactions due to multiple players paved the way for the emergence of smart grid which depends on smart meters, intelligent appliances, advanced controls, and two-way communication system.

25.1 SMART GRID

The AGC and the reactive power control used originally with power system operation were integrated later into SCADA in the vertically integrated systems of the past. With the recent restructuring of power systems that has taken place in the last few decades wherein the monolithic system was separated into generation system, transmission system, and distribution system, new operating strategies were introduced. The increase in the proportion of the contribution of generated power from renewable energy sources and distributed generation necessitated open access to transmission systems. Independent system operator emerged as an important entity. Distribution automation has also begun. Instrument transformers with protective relays are used to sense the power system voltage and current values system wise. Intelligent electronic devices (IEDs) with microprocessors located at remote locations called remote terminal units are used to monitor. Digital fault recorders are used, which are also IEDs to record all power system disturbances. Programmable logic controllers are used to perform all logical control operations. All protective devices and load tap changers are controlled by IEDs. Reclosers are also managed by IEDs. The use of computer-based remote control system and automation in utility electricity delivery resulted in the development of smart grid system, which incorporates a two-way digital communication system and computer processing between the consumers and the power suppliers. Smart grid means real-time operation of electrical power transmission and distribution systems which cater to the needs of their customers with security, reliability, and flexibility that results in economy. Sensors are used extensively to gather relevant information. Automation facilitates adjustments and controlling of millions of devices connected to the utility throughout the network. A smart grid is capable of providing power from multiple and widely distributed sources. Smart grid technology enabled integration of renewable energy sources such as photovoltaic cells, concentrated solar power panels, wind energy, geothermal, cogeneration, and others with the conventional power systems. Again, the power availability from the renewable energy sources varies widely with time. Sunny days and windy days may not match the requirements of peak demand that follows a general pattern superimposed by fluctuations. The availability of solar and wind power, which constitutes the major part of renewable energy of now, is unpredictable. Therefore a smart grid is expected to store energy when available and use it later when demand requires. The devices that store energy include batteries, flywheels, super capacitors, super conducting magnetic energy storage systems (SMES), etc. Plug-in hybrid electric vehicles are also considered for storage purpose. Increased reliability demands highly sophisticated adaptive generation and distribution control algorithms. Siemens, one of the technology suppliers in this area, holds the opinion that smart grid should involve improved energy delivery and informed consumption with reduced environmental impact. A smart grid may be considered as a secure integration of two infrastructures—namely electrical power system and information. It is an enabling engine that is intelligent, autonomous, quality focused, motivating, integrating, efficient, flexible, and accommodating. In view of the various complexities that exist inherently in a power system with reference to its enlarged field of operation systems' approach is more appropriate in dealing with smart grid. The United States has passed in December 2007 "Energy Independence and Security Act (EISA)." Office of Electricity Delivery and Energy Reliability (OE) of the United States of America has created a vision for its future energy needs. In *International Energy Outlook* the United States predicted that the energy consumption globally would increase by 50% from 2005 to 2030. The IEEE smart grid research group classified the necessary areas for advanced research and development in control systems, communications, power, computing, consumer socialization, cybersecurity, and vehicular needs. The IEEE in its Grid Vision 2050 has provided an insight into the various aspects concerning the smart grid projects which are summarized in the following.

25.2 SMART GRID COMPONENTS

1. *Intelligent Appliances*: Appliances are interfaced with instrumentation that decides when to consume power based on preset customer preferences. This will help the generators reduce their peak load requirements. It also reduces or postpones the need for new generating capacity. When renewable energy sources are used the greenhouse effect will also be reduced. Smart grid−enabled devices include water heaters and devices such as thermostats.
2. *Smart Power Meters*: Digital meters with two-way communication facility enable automatic billing data collection, notice outages and arrange for dispatch of repair crews to reach the affected spots in the system faster. Smart meters record usage in real time and utilize wires, fibers, Wi Fi, cellular and power line carrier communication methods. PLC with a spectrum of 1−30 MHz and a data rate of 2-3 Mbps is however noisy and harsh with a range of 1−3 km only. In contrast GSM in the spectral range of 900−1800 MHz and data rate up to 14.4 kbps has a coverage of 1−10 km. Communication protocol is given by IEC 61850 for secure substation metering.
3. *Smart Substations*: Substation equipment is connected to monitoring and controlling circuits to collect operational data that includes power factor, breaker, transformer, and battery status, etc., for operational security and emergency control.
4. *Smart Distribution*: It is aided by automation and tools for analysis which have the capability of detecting and even predicting line or cable failures based on real-time data pertaining to weather, outage history, etc., so that the system is self-healing, self-balancing, and self-optimizing. The distribution system may contain superconducting cables. (High-temperature super conductor cables (HTS) enable cables to carry 3−5 times the load carried by AC cables and up to 10 times of DC cables).
5. *Smart Generation*: The power generating units are enabled to optimize energy production and automatically maintain voltage, frequency, and power factor at various specified points in the grid as per the prescribed standards.
6. *Universal Access*: The grid is made accessible to all the players including all renewable energy sources, storage devices, and plug-in electrical vehicles.

25.3 SMART GRID BENEFITS

The following are the benefits that can be harnessed by developing a smart grid.

1. *Self-Healing*: Since a smart grid detects and responds to all power disturbance events and enables fast recovery in case of occurrence, it minimizes downtime and thus reduces financial losses and customer inconvenience. Locating and connecting abundant sensors at all strategic points between the consumers and the suppliers enables dispatching of millions of data points back to the control centers and contributes to self-healing. Complete distribution automation is desired for beneficial smart grid operation.

2. *Customer Participation*: When connected, all types of consumers whether industrial, commercial, or residential are allowed to have a visible or transparent real-time pricing and are given the option to choose the volume of consumption and price that best suits their requirement. There exists consumer participation with demand response.

3. *Robust and Resilient Grid*: It has a built-in security environment and thus resists both physical and cyberattacks.

4. *Quality Power*: It is possible to supply power free of voltage sags, spikes disturbances, and interruptions to the consumers. All modern engineering industrial needs can be met by smart grid. Smart grid provides enhanced power quality.

5. *Flexible Generations and Storage*: Smart grid is capable of getting connected into the network by all types of distributed generation and storage in a "plug and play" mode. It provides a wide variety of generation options.

6. *Market Supportive*: Smart grid by virtue of its various pro-consumer advantages encourages energy market both wholesale and retail, and thus induces greater investments and innovations. Open access market facilitates the integrated operation.

7. *Asset Optimization*: In view of the greater flexibility and "plug-in and play" provision, adjustments to varying power demands are easily met than otherwise and thus require less new infrastructure and helps economy.

8. *Efficient Operation*: Since expenditure is reduced due to flexibility, smart grid operation is more efficient.

25.4 SMART GRID TECHNOLOGIES

1. *Integrated Two-Way Communication System*: This is the most important aspect in developing a smart grid. The communication system includes the tasks of data acquisition, protection, and control and also provides for the user intelligent electronic appliances to interact with the integrated system. Unified communication includes audio, video and data power line communication, broadband over power line, and distribution line carrier systems which operate on high-, medium-, and low-voltage lines respectively. Wireless communication system may involve, worldwide Interoperability for Microwave, (Wi Max), Wi-Fi, ZigBee, cellular, and so on.

2. *Sensors and Measurements*: Sensors predict peak load issues and utilize automatic switching to divert or reduce power at predetermined strategic locations on the network so that the chance of

overload is eliminated or reduced and the occurrence of supply disconnection overcomes. Advanced metering infrastructure (AMI) increases the range for time-based rate programs. These time-based rate programs permit smart customers to change their demand patterns and reduce their own power bills as well as reduce the peak demand on the system. Various measurements, which are the outputs of the sensors, are utilized to assess the health and integrity of the grid continuously. Automatic meter reading and prevention of energy theft are the outcomes of this technology.

3. *Advanced Components*: The successes in fundamental research and development in areas of power electronic and microelectronic devices, superconducting materials, and chemistry will determine the success and affordability of smart grid to a greater extent. The advanced components and devices will have to enable the complex functions required to be performed by a smart grid reliably.

4. *Advanced Controls*: Continuous analysis, diagnosis, and prediction of the grid conditions and initiation and completion of appropriate corrective actions through devices and algorithms form the most essential aspect of the smart grid. By such actions power disturbances can be eliminated, mitigated, and prevented with power quality assured. For this to happen, advanced control strategies are to be considered. Distribution Automation, Energy Management, Geographic Information System, Demand Side Management System, Wide Area Measurement System (WAMS), and many other constitute components of advanced control systems.

5. *Simple Display for Decision Making*: The complex information about the power system in its operation must be displayed to the operators in such a manner that they can understand easily through appropriate interfaces.

25.5 PHASOR MEASUREMENT UNITS

A phasor is a three-phase voltage or current quantity and could be also a phase angle. Each phase requires three separate connections, one for each phase. The phase measurement units (PMUs) are distributed all over the network. Phasor data concentrators (PDCs) are used to collect the information. At a central control center, SCADA is provided. For the first time the Bonneville Power Administration utilized such a system called "WAMS." The entire network needs rapid data transfer within the frequency of sampling of the phasor. Geopositional Satellite timestamping is used to provide a time accuracy of synchronization better than 1 μs. The clocks used are required to be accurate to the extent of ± 500 ns so that the desired 1 μs standard is reached. Each device performs synchrophasor measurement. Thus the PMUs send information in several synchronous reports per second (25—50 samples per second) depending on the frequency of the system and the application. The PDC correlates data and controls and monitors the PMUs. The central control station is provided with the system-wide data from all the generators and substations every few seconds. The dynamic system condition is assessed through the high-resolution data ability. Time-synchronized stamped data obtained from voltage, current and phase angles plays the most vital function. High-bandwidth communication is used in WAMS. The data provides the basis to calculate percent damping for inter area and local area oscillations, sensitivities such as $\Delta V/\Delta P$, $\Delta \delta P$; angular difference between bus voltages and transmission loadability.

The realization of smart grid is not without challenges. It is necessary to bring several aspects such as AMI, meter data acquisition system, data concentrator units, meter data management system (MDMS), and many others are to be integrated on to a single platform. The technologies mentioned are all cost intensive and require huge financial resources. Further, individual operators to overcome legal and other obstacles need government support so that appropriate legislative measures are taken. The R&D support to develop, create, and supply compatible equipment for successful operation is another challenge. The speed at which the R&D can support is another significant factor in hastening the implementation of smart grids. There has to be a consistent policy formulation and setting up of regulations in this direction from the various players and mutual cooperation to realize smart grids.

Cybersecurity is another important consideration. Secure telecommunication, end-point protection, identity management, and secure event management are some of the features that arise. Hacking may affect thousands of customers.

There are several smart grid projects being pursued all over the world. Southern California Edison, Pacific gas and Electric, and American Electrical power are some of the projects in the United States. There are also ventures going on in Germany, Spain, India, China, and other countries.

The European Union (presently consisting of about 27 + countries) is also engaged in reforming the electrical energy industries of individual member countries and formed European Energy Union' striving for secure, affordable, and climate friendly energy for all its member states. Their

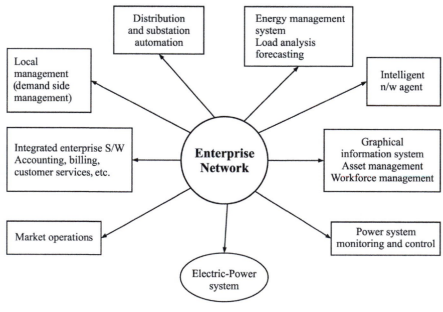

FIGURE 25.1

Electrical power grid of future.

strategy is directed toward security of supply, competitiveness, and sustainability. India is fast progressing in developing smart grids to be integrated later into a giant smart network.

European Union Energy Targets: By 2020 EU desires to reduce greenhouse gases, increase the share of renewable energy component, and increase energy efficiency by 20% compared to the level of 1990. The corresponding figures contemplated for the year 2030 stand at 40%, 27%, and 27%−30%, respectively. By 2050 EU envisages to cut the greenhouse gases by 80%−95% of the 1990 levels. The tracking record is satisfactory and EU is hopeful of reaching the targets. The plan foresees 15% of the total energy produced to be available in the interconnected grid, so that other members of EU can import wherever and whenever necessary.

Siemens has developed and is marketing Distribution management Systems with intelligent substation integrated according to international standards (IEC 61850). Distribution feeder automation allows, as pointed out earlier, utilities to create a comprehensive solution to smart grid operation. The European Standards Organization (ESO) is standardizing data communication interfaces, smart grid information security, dependability and functional safety, smart metering and a host of other aspects. There is an International Smart Grid Action network where many nations are participating and sharing their expertise. Functional block diagram for future power system network (Fig. 25.1).

25.6 OPEN SMART GRID PROTOCOL

It is a family of specifications from "European Telecommunications Standards Institution" (ETSI) used in conjunction with ISO/IEC 14908 control networking standards, for smart grid application. Open smart grid protocol (OSGP) is so optimized as to provide reliable and efficient delivery of command and control information for smart meters, load control modules, solar panels, and all other smart grid devices.

IEC 61968 helps in integration and harmonization in addition. For enterprise data sharing IEC 61968 and IEC 61970 are helpful. ANSIC12 is useful for revenue metering. Cloud computing is necessary for smart grid. It is a general term used to describe a class of network-based computing that takes place over the Internet. It denotes a collection of integrated and networked software, hardware, and Internet infrastructure called a "platform."

Further Reading

1. Kraus JD, Carver KR. *Electromagnetics*. New York: McGraw Hill Book Co; 1984.
2. Boast WB. *Vector fields: a vector foundation of electric and magnetic fields*. NY: Harper; 1964.
3. Stevenson W. *Elements of power system analysis*. New York: McGraw Hill; 1982.
4. Starr AT. *Generation transmission and utilization of electric power*. 2nd ed. London: Sir Isaac Pitman and Sons Ltd; 1946.
5. Cotton H, Barber H. *Transmission and distribution of electrical energy*. London: Hodder Arnold; 1970.
6. Begamudre RD. *Extra high voltage AC transmission engineering*. New Delhi: New Age International; 2006.
7. Kimbark EW. *Direct current transmission*, vol. 1. New York: Wiley Interscience; 1971.
8. Hingorani NG. High voltage direct current transmission. *IEEE Spectrum* 1996;**33**:63−72.
9. Peterson HA. *Transients in power systems*. New York: Wiley & Sons Dover Pub; 1966.
10. Badri Ram, Viswakarma DN. *Power system protection and switchgear*. New Delhi: Tata McGraw Hill; 2001.
11. Russell Mason C. *The art and science of protective relaying*. General electric; 2009.
12. Van Warrington. *Protective relays, theory and practice*, vols. I & II. London: Chapman and Hall; 1962.
13. Wagner CF, Evans RD. *Symmetrical components*. New York: McGraw Hill book Co; 1961.
14. Clarke E. *Circuit analysis of alternating current power systems*, vol. I. New York: McGraw Hill; 1961.
15. Kimbark EW. *Elements of stability calculations. Power system stability*, vol. I. London: Chapman and Hall; 1948.
16. Kimbark EW. *Power circuit breakers and protective relays. Power system stability*, vol. II. London: Chapman and Hall; 1950.
17. Kimbark EW. *Synchronous machines. Power system stability*, vol. III. London: Chapman and Hall; 1956.
18. Concordia C. *Synchronous machines*. New York: John Wiley and sons; 1958.
19. Elgard OI. *Electric energy systems theory: an introduction*. New York: McGraw Hill; 1971.
20. Venikov VA. *Transient phenomena in electrical power systems. International series of Monographs on electronics and instrumentation*. London: Pergamon Press: Elsevier; 1964. Earlier edition.
21. Westinghouse Electric Corporation. *Electrical transmission and distribution reference book*; 1964.
22. Kirchmayer LK. *Economic operation of power system*. New York: John Wiley and Sons; 1958.
23. Kirchmayer LK. *Economic control of interconnected systems*. New York: John Wiley and Sons; 1959.
24. Miller TJ. *Reactive control in electric systems*. New York: Wiley; 1982.
25. Say MG. *Performance and design of alternating current machines*. London: Sir Isaac Pitman and Sons; 1955.
26. Zaborszky L, Rittenhouse JW. *Electric power transmission*. New York: The Ronald Press Co; 1954.
27. VDE Verlag GmBH−Blindleistung, (German) Berlin 1963.
28. Hau E. *Wind turbines: fundamentals, technologies, application, economics*. New York: Springer; 2013.
29. Clark W, Gellings PE. *The smart grid, first Indian reprint*. Lilburn: The Fairmont Press Inc; 2015.

Periodicals, Conference Proceedings, and Other References
30. Hart CE, Tinney WF. Power flow solution by Newton's method. In: *IEEE transactions on power apparatus and systems, PAS-86*; 1967. p. 1449−60.
31. Stott B. Decoupled Newton load flow. In: *IEEE transactions on power apparatus and systems, PAS-91*; 1972. p. 1955−59.

32. Stott B, Alsac O. Fast decoupled load flow. In: *IEEE transactions on power apparatus and systems, PAS-93*; 1974.
33. Tinney WF, Walker JW. Direct solution of sparse network equations by optimally ordered triangular factorization. In: *Proc. IEEE*, vol. 55; 1967.
34. Brown HE. Z matrix algorithms in load flow programs. In: *IEEE transactions on power apparatus and systems, PAS-87*; 1968.
35. Murty PSR. *Optimal load flow studies in inter connected networks (German)* [Doctoral thesis]. Berlin: Technical University; 1972.
36. Bauman R. Power flow solution with optimal reactive power flow. *Arch Elektrotech* 1963;**48**.
37. El Abiad AH, Jaimes FJ. A method for optimal scheduling of power and voltage magnitude. In: *IEEE transactions on power apparatus and systems, PAS-88*; 1969.
38. CLFR power plant technology Applied and plasma Physics, School of Physics, University of Sidney; June 2002.
39. Taylor CE, et al. Concentrating solar Power in 2001. *An IEA−solar paces summary of present status and future prospects, solar power, trends in photovoltaic applications. Survey report of selected IEA countries between 1992 and 2009, IEA −PVPS, paces task1,Electrical power system.*
40. Murty PSR. Load modeling for power flow solution. *J Inst Eng* Dec. 1977;**58**(Pt EI3).
41. *Geothermal basics.* Geothermal Association.
42. *Electrical Power Generation from Geothermal sources.* Electropaedia.
43. *2050 Energy Strategy. European commission report.*

Index